Contents

1 Functions   1

2 Mathematics of Finance   98

▌3▐ Systems of Equations and Matrices   148

▌4▐ Linear Programming   244

*These two sections provide *alternate* methods for solving the *same* types of problems. The choice of which to cover is left to the instructor.

▌5▐ Probability   350

6 Statistics 426

7 Markov Chains 476

8 Game Theory 522

Available on the Web   Go to math.college.hmco.com and follow the links to this text.

 Logic

Preface

This book is an introduction to finite mathematics and its applications for students in the management, social, behavioral, and biomedical sciences, as well as other applied fields and general liberal arts studies. Chapter 1 provides a brief review of algebra and functions, and Chapter 2 applies exponential and logarithmic functions to the mathematics of finance. Chapters 3 and 4 explore matrices and linear programming problems, while Chapters 5 and 6 develop probability and the foundations of statistics. Chapter 7 is an extended application of matrices to the study of Markov chains, and Chapter 8 is an introduction to game theory.

Our presentation endeavors to exhibit some of the many fascinating and useful applications of these areas of mathematics in business, the sciences, and everyday life. But beyond its utility, there is beauty to mathematics and we hope to convey some of its elegance and simplicity.

CHANGES IN THE SECOND EDITION

First, what has *not* changed is the basic character of the book: simple, clear, and correct explanations of mathematics, alternating with relevant and engaging examples.

We have tightened the exposition in every section and simplified most examples to better emphasize the essential ideas.

We have updated the exercises that use real-world data and added new exercises in many sections.

We have added new **Spreadsheet Explorations** (using Excel).

We have moved most of the Application Previews to the Web (keeping only one per chapter), along with all of the Projects and Essays. They can be found at **math.college.hmco.com**. Also on the Web is the chapter from the first edition on Logic, which has been replaced by a new chapter on Game Theory.

We have removed the scientific calculator symbol (but kept the graphing calculator symbol) since calculator use is now universal among today's college students. To improve readability, we have added spaces around in-line equations.

FEATURES

Realistic Applications   Since courses using this book are by their basic nature very applied, every chapter contains an unusually large

vii

number of applications, many appearing in no other textbook, both in the development of the material and in the exercises. We explore learning curves in airplane production (pages 26–27 and 30–31), analyze long term trends in higher education (page 62), find the age of the Dead Sea Scrolls (page 86), model data flow in a computer network (pages 189–190), manage an investment portfolio (pages 318–320 and 334–335), analyze home run records (pages 418–419), and make projections of farm size distributions (pages 509–510). These and many other applications convincingly show that mathematics is more than just the manipulation of abstract symbols and is deeply connected to everyday life.

Graphing Calculators (Optional)　Calculators with capabilities previously available only on computers have changed the way mathematics can be taught and used. Reading this book does not require a graphing calculator, but having one will enable you to do many problems more easily and at the same time deepen your understanding by allowing you to concentrate on concepts. Throughout the book are Graphing Calculator Explorations and Exercises, which

> *explore new topics* (such as the graphs of $y = \sqrt[n]{x}$ on page 23 and then the "rule of .6" in Exercises 81–84 on page 29),

> *carry out otherwise "messy" calculations* (such as the population growth comparison on page 71 and then Exercises 27–28 on page 75), and

> *show the advantages and limitations of technology* (such as the differences between $\ln x^2$ and $2 \ln x$ on page 84).

While any graphing calculator (or a computer) may be used, the displays shown in the text are from the Texas Instruments TI-83 and TI-89. A discussion of the essentials of graphing calculators follows this preface. For those not using a graphing calculator, the Graphing Calculator Explorations have been carefully planned so that *most can also be read simply for enrichment* (as with the histograms and box-and-whisker plots on pages 434–435 and 451–452).

Graphing Calculator Programs (Optional)　For certain topics, the computational effort may be reduced and the understanding enhanced by using the optional graphing calculator programs we created for use with this book. These programs are free and easy to obtain (see "How to Obtain Graphing Calculator Programs" later in this Preface). The topics covered are: amortization tables (pages 136–137), matrix row operations (pages 183–184), pivot operations (page 283), normal approximations of binomial distributions (pages 466–467), steady-state distributions for regular Markov chains (pages 498–499), and simplex method solutions of games (page 551).

Spreadsheets (Optional) Conceptual understanding may also be gained from numerical explorations using computer spreadsheet programs. We have included spreadsheet examples of: the relation between revenues, costs, and profits (page 42), amortization tables (pages 132–133), matrix arithmetic (page 203), matrix inverses (pages 213–214), solving a transportation problem (pages 306–307), probabilities in coin tossing (page 413), graphs of statistical data (page 432), the duration of a Markov chain in a particular state (pages 486–487), and analyzing a Sherlock Holmes story (page 539). Ancillary materials for Microsoft Excel are also available (see "Supplements for the Student" later in this Preface).

Enhanced Readability An elegant four-color design increases the visual appeal and readability of this book. Color is frequently used for pedagogical purposes, such as in the examples of the Gauss–Jordan method (pages 188–189), matrix arithmetic (pages 201 and 202), and linear programming (pages 280, 282, and 315). For the sake of continuity, references to earlier material are minimized by restating results whenever they are used. Where references are necessary, explicit page numbers are given.

Application Previews Each chapter begins with an Application Preview that presents an interesting application of the mathematics developed in that chapter. Each is self-contained (although some exercises may later refer to it) and serves to motivate interest in the coming material.

Practice Problems Learning mathematics requires your active participation—"mathematics is not a spectator sport." Throughout the readings are short pencil and paper Practice Problems designed to consolidate your understanding of one topic before moving ahead to another, such as finding the slope of a line (page 13) or constructing a dual linear programming problem (page 297). Complete solutions to all Practice Problems are provided at the end of each section.

Annotations Notes to the right of many mathematical formulas and manipulations state the results in words, assisting the important skill of "reading mathematics," as well as providing explanations and justifications for the steps in calculations and the interpretations of the results (for instance, see page 50). Notes and diagrams to the left of new concepts and results provide visual clarification (see the Gauss–Jordan method on page 182 and the pivot operation on page 281).

Extensive Exercises   Anyone who ever learned any mathematics did so by solving many many problems, and the exercises are the most essential part of the learning process. The exercise sets (see, for instance, pages 397–400) are graded from routine drills to significant applications and some conclude with **Explorations and Excursions** that extend and augment the material presented in the text. Exercises marked with the symbol [calculator icon] require a graphing calculator. Answers to odd-numbered exercises and answers to all Chapter Summary exercises are given at the end of the book (full solutions are given in the Student Solutions Manual).

Levels of Reinforcement   Because there are many new ideas and techniques in this book, learning checks are provided at several different levels. As noted above, **Practice Problems** encourage mastery of new skills directly after they are demonstrated. **Section Summaries** briefly state both essential formulas and key concepts (see pages 415–416). **Chapter Summaries** review the major developments of the chapter and are keyed to particular chapter review exercises (see pages 420–425). **Hints and Suggestions** at the end of each chapter summary unify the chapter, give specific reminders of essential facts or "tricks" that sometimes are otherwise overlooked or forgotten, and list a selection of the review exercises for a **Practice Test** of the chapter material (see page 422).

Explorations and Excursions   At the end of some exercise sets are optional problems of a more advanced nature that carry the development of certain topics beyond the level of the text, such as: round-off errors in systems of equations (page 165), Klee-Minty linear programming problems (page 294), the Binomial Theorem and Pascal's Triangle (pages 373–374), and the mean of grouped data (pages 448–449).

Accuracy and Proofs   All of the answers and other mathematics have been checked carefully by several mathematicians. The statements of definitions and theorems are mathematically accurate. Since the treatment is applied rather than theoretical, intuitive and geometric justifications have often been preferred to formal proofs. When proofs are given, however, they are correct and "honest."

Philosophy   We wrote this book with several principles in mind. One is that to learn something, it is best to begin doing it as soon as possible: the preliminary material is brief, so you can begin doing useful mathematics immediately. An early start allows more time during the course for interesting applications and necessary review. Another principle is that the mathematics should be done together with the applications. Consequently every section contains applications (there are no "pure math" sections).

Prerequisites   The only prerequisite for this course is some knowledge of algebra, functions, and graphing, and these are reviewed in Chapter 1. Other review material has been placed in relevant locations throughout the book.

Resources on the Web   Additional materials available on the Web at **math.college.hmco.com** include:

- A chapter on **Logic** consisting of five sections: *Statements and Connectives, Truth Tables, Implications, Valid Arguments,* and *Quantifiers and Euler Diagrams.* This is a modified version of Chapter 8 from the first edition.

- Suggestions for **Projects and Essays,** open-ended topics that ask students (individually or in groups) to research a relevant person or an idea, to compare several different mathematical ideas, or to relate a concept to their lives (such as AIDS and false positive tests). These are in keeping with the recommendations from the Mathematical Association of America and the National Council of Teachers of Mathematics. Other more challenging projects can be found in the highly recommended *MAA Notes Numbers 27–30,* available from the Mathematical Association of America in Washington, DC.

- An expanded collection of **Application Previews,** short essays that were used in the first edition to introduce each section. Topics include: *Exponential Functions and the World's Worst Currency, Coincidences, Airline Safety,* and *Pain, Suffering, and Statistics.*

HOW TO OBTAIN GRAPHING CALCULATOR PROGRAMS

The optional graphing calculator programs used in the text have been written for a variety of Texas Instruments graphing calculators (including the *TI-82, TI-83, TI-83 Plus, TI-85, TI-86, TI-89* and *TI-92*), and may be obtained for free in any of the following ways:

- If you know someone who already has the programs on a Texas Instruments graphing calculator of the same model as yours, you can easily transfer the programs from their calculator to yours using the black cable that came with the calculator and the LINK button. (Even if you use one of the following methods to get a "first" copy of the programs, you can use this method to transfer the programs to others in your class.)

- If you have access to the Internet, you may download the programs (together with descriptive materials) from the **www.hmco.com** Houghton Mifflin website for this book. Then use the TI-GRAPH LINK™ or TI-CONNECT™ software and cable (available for purchase in most stores that sell graphing calculators) to transfer the programs from your computer to your calculator.

- You may send a $3\frac{1}{2}$-inch or zip disk, carefully packed, to the authors at the following address, specifying the title of your textbook, the type of your computer (Windows or Macintosh), and your Texas Instruments calculator (*TI-82, TI-83, TI-83 Plus, TI-85, TI-86, TI-89,* or *TI-92*). We will return a disk containing the appropriate programs and a packet of descriptive information. After you load the disk and programs into your computer, you will need the TI-GRAPH LINK™ or TI-CONNECT™ software and cable (available for purchase in most stores that sell graphing calculators) to transfer the programs to your calculator.

- You may send your calculator (*TI-82, TI-83, TI-83 Plus, TI-85, TI-86, TI-89,* or *TI-92*), carefully packed, to the authors at the following address, and we will return it loaded with the appropriate programs, along with a packet of descriptive information.

Authors' Address: Dr. G. C. Berresford and Dr. A. M. Rockett
Department of Mathematics
C. W. Post Campus of Long Island University
720 Northern Boulevard
Brookville, New York 11548–1300

e-mail: gberres@liu.edu *or* rockett@liu.edu

HOW TO OBTAIN EXCEL SPREADSHEET EXPLORATIONS

The MicroSoft Excel spreadsheets used in the Spreadsheet Explorations may be obtained for free in either of the following ways:

- You may download the spreadsheet files from the Houghton Mifflin website at **math.college.hmco.com** onto a computer.

- You may send a $3\frac{1}{2}$-inch or zip disk to the authors at the above address, specifying both the title of your textbook and the type of your computer (IBM-compatible or Macintosh), and we will return it to you with the appropriate Excel files.

SUPPLEMENTS FOR THE INSTRUCTOR

Instructor's Resource Manual (with Solutions and Printed Test Bank) Contains complete solutions to all problems in the text and a printed Test Bank available in computerized form on the CD. Two pre-made Chapter Tests for each chapter are also included.

HM ClassPrep with HM Testing CD-ROM combines all the resources of ClassPrep plus the algorithmically generated tests and gradebook features of HM Testing. This CD-ROM includes a variety of supplements: an electronic Instructor's Solutions Manual, Digital Figures and Tables from the book, chapter tests, the resources available on the student CD, and end of chapter Projects and Essays.

- Create and print quizzes and tests quickly and easily
- Deliver tests and record results via a LAN or the Internet
- Record and tabulate student test grades in an electronic gradebook

Eduspace® is a text-specific online learning environment that combines algorithmic tutorials with homework capabilities and classroom management functions.

- Electronically grade and record student results
- Manage a lecture-based or distance learning course online

Companion Website for Instructors offers additional teaching resources such as Digital Lesson Slides. Visit **math.college.hmco.com/ instructors** and choose this textbook from the list provided.

SUPPLEMENTS FOR THE STUDENT

HM mathSpace® Tutorial CD-ROM is a new tutorial CD that allows students to practice skills and review concepts. The CD contains algorithmically generated exercises and step-by-step solutions for student practice, along with an Excel and Graphing Calculator Guide, and end of chapter Projects and Essays. Also included is an algebra tutorial review that provides a skill refresher.

SMARTHINKING™ live online tutoring allows students to communicate in real-time with e-structors who will help them understand difficult concepts and guide them through the problem solving process.

Eduspace® is a text-specific online learning environment that combines algorithmic tutorials with homework capabilities.

- Practice skills with additional exercises and quizzes
- Get tutorial help outside class
- Complete homework assignments online if assigned by the instructor

Excel Guide for Finite Mathematics and Applied Calculus by Revathi Narasimhan, Kean University, provides Excel information, including step-by-step examples and sample exercises.

Companion Website for Students offers additional learning resources such as the Graphing Calculator Programs that accompany this text. Visit the text-specific website at **math.college.hmco.com/students** and choose this textbook from the list provided.

Houghton Mifflin Mathematics Instructional Videos and DVDs provide a lecture series featuring Dana Mosely in which he provides careful explanations of key concepts, examples, exercises, and applications in a lecture-based format.

Student Solutions Manual contains complete solutions to all odd-numbered problems in the text and solutions to *all* Chapter Review questions.

ACKNOWLEDGMENTS

We are indebted to numerous people for useful suggestions, conversations, and correspondence during the writing of this book, but most of all to our many "Math 5" students at the C. W. Post campus of Long Island University over the last several years for their questions, comments and enthusiasm for learning mathematics.

We have had the good fortune to have had wonderfully supportive and expert editors at Houghton Mifflin: Lauren Schultz (Sponsoring Editor), Jennifer King (Associate Editor), Kasey McGarrigle (Editorial Associate), and Tracy Patruno (Senior Project Editor). They made the most difficult tasks easy, and helped beyond words. Helen Medley (accuracy reviewer) and Deana Richmond (accuracy reviewer and solutions manual author) saved us from many mistakes and confusions. We also express our gratitude to the many others at Houghton Mifflin who made many important contributions too numerous to mention. Thanks also to Janet Nuciforo and others at Nesbitt Graphics for their attention to the finer details of production.

The following reviewers have contributed greatly to the progression from first draft to finished book:
Yossef Balas, *University of Southern Mississippi;*
Rajanikant Bhatt, *Texas A&M University;*
Wesley T. Black, *Illinois Valley Community College;*
Richard Blecksmith, *Northern Illinois University;*
William L. Blubaugh, *University of Northern Colorado;*

Larry Bouldin, *Roane State Community College, Tennessee;*
Bob Bradshaw, *Ohlone College, Fremont, California;*
Stephen Brick, *University of South Alabama;*
Dean Brown, *Youngstown State University, Ohio;*
Roxanne Byrne, *University of Colorado at Denver;*
Jeremy Case, *Taylor University, Indiana;*
Candy Giovanni, *Michigan State University;*
Tom Greenwood, *Bakersfield College, California;*
W. Thomas Kiley, *George Mason University, Virginia;*
M. Maheswaran, *University of Wisconsin, Marathon County;*
Gregory McColm, *University of South Florida;*
Diana McCoy, *Florida International University;*
Ho Kuen Ng, *San Jose State University, California;*
Richard Pellerin, *Northern Virginia Community College;*
Armando I. Perez, *Laredo Community College, Texas;*
Thomas Riedel, *University of Louisville, Kentucky;*
Stephen B. Rodi, *Austin Community College, Texas;*
Ben Rushing, Jr., *Northwestern State University of Louisiana;*
Patty Schovanec, *Texas Tech University;*
Kimberly A. Smith, *University of Tennessee;*
Jennifer Stevens Fowler, *University of Tennessee, Knoxville*

DEDICATION

We dedicate this book to our wives, Barbara and Kathryn, and our children, Lee, Chris, Justin, and Joshua, for their understanding and patience, without which this book would not exist.

COMMENTS WELCOMED

With the knowledge that any book can always be improved, we welcome corrections, constructive criticisms and suggestions from every reader.

A User's Guide to Features

▼ Applications Preview

Found on every chapter opening page, Application Previews serve to motivate the chapter. They offer unique "mathematics in your world" applications or interesting historical notes. Pages with further information on the topics, and often related exercises, are referenced.

2 Mathematics of Finance

2.1 Simple Interest
2.2 Compound Interest
2.3 Annuities
2.4 Amortization

Application Preview

Musical Instruments as Investments*

No two guitars sound alike—good ones, anyway. Over the last 30 years, prices for high-quality acoustic instruments have soared. In 1993, a John D'Angelico New Yorker 18-inch-wide cutaway archtop ("jazz") guitar sold for $150,000, setting the record for the price of an acoustic fretted instrument not previously owned by a deceased superstar. Many pre-World War II instruments are valued not only for their craftsmanship but also for their distinctive sound. A mint condition Martin D-45 made in 1932–1942 that might have sold for $3500 in 1971 could be sold for $225,000 in 2003, while a Gibson 1932–1939 Mastertone Granada 5-string (replacement neck) banjo selling for just $1200 in 1971 would fetch $28,000 in 2003.

How can these price increases be compared to other investments such as stocks, bonds, or bank savings accounts? The dots on the graph show the typical selling prices over the 30-year period from 1971 to 2001 for a signed 1922 Gibson F-5 mandolin, together with a curve showing the increasing value of an investment earning 11% compound interest.

Selling Prices for a Gibson Mandolin

The dots closely match the curve, showing that this mandolin has increased in value by about 11% annually, better than many bond and stock investments. The values of fine musical instruments and other investments can be found by the methods of this section.

*Based on material originally provided by Stanley M. Jay and Larry Wexer of mandolin Bros. Ltd. (Staten Island, New York), dealers in fine mandolins, guitars, and banjos.

99

7.3 ABSORBING MARKOV CHAINS **503**

Solution

a. This *is* an absorbing Markov chain since it has an absorbing state, S_3, which can be reached from S_1 in 2 steps and from S_2 in 1 step.

b. This is *not* an absorbing Markov chain. Although S_3 is an absorbing state, it cannot be reached from either S_1 or S_2 (the nonabsorbing states).

EXAMPLE 2 **WETLANDS POLLUTION**

The Southern Electric Company has filed an application with the Environmental Protection Agency to store heavy-metal waste (mostly cadmium and mercury) in a clay-lined storage pool at its manufacturing plant in Mayfield. An independent engineering consultant has certified that only 1% of the waste will leach out of the pool into the ground water each year. But a hydrologist working with the local chapter of People for a Cleaner Planet has pointed out that each year 2% of the contaminants in the ground water reach the Marlin Memorial Wetlands Conservation Area, where they remain indefinitely. Represent this situation by a transition diagram and matrix. Show that the wetlands are an absorbing state for the pollution.

Solution

The waste can be in one of three locations: at the plant (state P), in the ground water (state G), or in the wetlands (state W). The given information is shown in the following state-transition diagram:

Since these changes occur each year, we have a Markov chain with transition matrix T. W is an absorbing state because once the contamination reaches the Wetlands, it does not leave, and the other states P and G are nonabsorbing. The state-transition diagram shows that it is possible for some of the waste to move from the plant storage pool to the groundwater and on to the wetlands, so this *is* an absorbing Markov chain.

We could also use the transition matrix to show that from each nonabsorbing state it is possible to reach the absorbing state. Squaring the transition matrix T gives the matrix on the left, and since the last

◄ Real World Icon

This globe icon marks examples in which mathematics is connected to every-day life.

▼ Practice Problems

Students can check their understanding of a topic as they read the text or do homework by working out a Practice Problem. *Complete solutions* are found at the end of each section, just after the Section Summary.

▼ Graphing Calculator Explorations

To allow for optional use of the graphing calculator, the Explorations are boxed. Exercises and examples that are designed to be done with a graphing calculator are marked with an icon.

Spreadsheet Explorations ▶

Boxed for optional use, these spreadsheets enhance students' understanding of the material using Excel, an alternative for those who prefer spreadsheet technology.

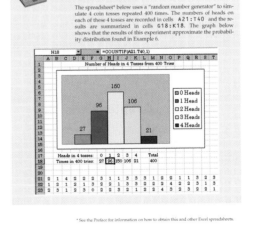

▼ Section Summary

Found at the end of every section, these summaries briefly state the main ideas of the section, providing a study tool or reminder for students. They are followed by complete solutions for the Practice Problems within that section.

▼ Applied Exercises

The applications in the exercise sets are labeled with a general title as well as a more specific title. The instructor can easily assign homework that is relevant to the students in that class.

5.5 Section Summary

For pairwise disjoint sets $S_1, S_2, \ldots, S_n$ whose union is the entire sample space $S_1 \cup S_2 \cup \cdots \cup S_n = S$, Bayes' formula reverses the order in conditional probability:

$$P(S_1 \text{ given } A) = \frac{P(S_1) \cdot P(A \text{ given } S_1)}{P(S_1) \cdot P(A \text{ given } S_1) + \cdots + P(S_n) \cdot P(A \text{ given } S_n)}$$

In practice, we carry out this calculation using a tree diagram, where the numerator is the product along one branch and the denominator is the sum of several branches, using just the definition of conditional probability.

➤ Solutions to Practice Problems

1. P (you are rich given that you win the lottery) should be very high since you certainly will be rich if you win the lottery.

   P (you win the lottery given that you are rich) should be very small since your chances of winning the lottery are very small regardless of how much money you have.

2. Using the notation from the solution to Example 2, we have

$$P(I \text{ given } V) = \frac{P(I \cap V)}{P(V)} = \frac{0.225}{0.315 + .024 + 0.225} = \frac{0.225}{0.780} \approx 0.288$$

   The probability that a voter in the last election was an Independent is about 29%.

5.5 Exercises

1. **SOCIAL SCIENCE: Voting** In a town, 60% of the citizens are Republicans and 40% are Democrats. In the last election 55% of the Republicans voted and 65% of the Democrats voted. If a voter is randomly selected, what is the probability that the person is a Republican?

2. **BIOMEDICAL: Colorblindness** An estimated 8% of men and 0.5% of women are colorblind. If a colorblind person is selected at random, what is the probability that the person is a man? (Assume that men and women occur in equal numbers.)

3. **BIOMEDICAL: Medical Testing** A new test is developed to test for a certain disease, giving "positive" or "negative" results to indicate that the person does or does not have the disease. For a person who actually *has* the disease, the test will give a positive result with probability 0.95 and a negative result with probability 0.05 (a so-called "false negative"). For a person who does *not* have the disease, the test will give a positive result with probability 0.05 (a "false positive") and a negative result with probability 0.95. Furthermore, only

one person in 1000 actually has this disease. If a randomly selected person is given the test and tests positive, what is the probability that the person actually has the disease?

4. **BIOMEDICAL: Home Pregnancy Testing** Suppose that 400 pregnant women take a home pregnancy test, and 397 of them test "positive" and the other 3 test "negative." Suppose also that 200 nonpregnant women take the test, and 184 of them test "negative" and the remaining 16 test "positive." What is the probability that a woman who tests positive is actually pregnant?

5. **MANAGEMENT SCIENCE: Manufacturing Defects** A computer chip factory has three machines, A, B, and C, for producing the memory chips. Machine A produces 50% of the factory's chips, machine B produces 30%, and machine C produces 20%. It is known that 3% of the chips produced by machine A are defective, as are 2% of chips produced by machine B and 1% of the chips from machine C. If a randomly selected chip from the factory's output is found to be defective, what is the probability that it was produced by machine B?

6. **MANAGEMENT SCIENCE: Manufacturing Defects** For the information in Exercise 5, if a randomly selected chip is found *not* to be defective, what is the probability that it came from machine B?

7. **SOCIAL SCIENCE: Airline Hijacking** The "hijacker profile" developed by the Federal Aviation Administration fits 90% of hijackers and only 0.05% of legitimate passengers. Based on historical data, assume that only 30 of 300 million passengers are hijackers. What is the probability that a person who fits the profile is actually a hijacker?

8. **SOCIAL SCIENCE: Polygraph Tests** Although seldom admissible in courts, polygraphs (sometimes called "lie detectors") are used by many businesses and

pose that you are trying to find a single thief in your company of 100 employees. What is the probability that a person who fails a polygraph test is *not* the thief?

9. **BIOMEDICAL: Random Drug Testing** In 1986 the Reagan administration issued an executive order allowing agency heads to subject all employees to urine tests for drugs. Suppose that the test is 95% accurate both in identifying drug users and in clearing nonusers. Suppose also that 1% of employees use drugs. What is the probability that a person who tests positive is *not* a drug user?

10. **GENERAL: Grades** A probability class consists of 5 math majors, 3 science majors, and 2 whose major is "undecided." If the probability of earning an A is 90% for math majors, 85% for science majors, and 70% for "undecideds," what is the probability that a randomly selected A student is a math major?

11. **GENERAL: Grades** For the information in Exercise 10, what is the probability that a randomly selected A student is "undecided"?

12. **ATHLETICS: Olympic Athletes** Suppose that 10% of olympic athletes use steroids. A blood test is 95% correct in identifying drug users and has a 2% rate of false positives (that is, of *incorrectly* indicating steroid use). What is the probability that a champion cycler who tests positive uses steroids?

13. **BIOMEDICAL: Mammograms** Many doctors recommend that women in their forties have annual mammograms to detect breast cancer. Approximately 2% of women in their forties will develop breast cancer during that decade, and the early mammograms (no longer used) for that age group had a 30% rate of false positives and a 25% rate of false negatives. What is the probability that a woman who tested positive on this test had breast cancer?

14. **SOCIAL SCIENCE: Unemployment** A city

an impressive first three months with sales of $300,000, $480,000, and $600,000. If he can keep improving this much every month, how much will he sell in his fifth month?

28. **BUSINESS: Commodity Futures** A soybean speculator has been nervously watching the December delivery price per bushel drop from $6.59 on Monday to $6.41 on Tuesday and then to $6.29 on Wednesday. If this trend continues, what will the price be on Friday?

29. **BEHAVIORAL SCIENCE: Price–Demand** The market research division of a large candy manufacturer has test-marketed the new candy YummieCrunchies in five different markets at five different prices per 8-ounce bag to determine the relation between the selling price and the sales volume. The results of its study are presented in the table and are normalized to give the weekly sales per 20,000 consumers. How many weekly sales per 20,000 consumers can the manufacturer expect when it begins national distribution next week with an introductory price of 79¢ per 8-ounce bag?

Selling Price	Sales Volume
70¢	2050
55¢	2675
75¢	1975
90¢	1250
85¢	1425

30. **MANAGEMENT SCIENCE: Advertising** The accounts manager at television station WXXB claims that every dollar spent on commercials by local retailers generates $23 of sales. As proof, she shows prospective advertisers the following table of advertising budgets and gross receipts for five area companies that aired commercials last month. Is her claim justified?

Dollars Spent on WXXB Commercials	Gross Receipts
$8000	$245,000
$5000	$215,000
$2000	$125,000
$9000	$305,000
$6000	$230,000

Chapter Summary with Hints and Suggestions

Reading the text and doing the exercises in this chapter have helped you to master the following skills, which are listed by section (in case you need to review them) and are keyed to particular Review Exercises. Answers for all Review Exercises are given at the back of the book, and full solutions can be found in the Student Solutions Manual.

3.1 Systems of Two Linear Equations in Two Variables
- Represent a pair of statements as a system of two linear equations in two variables. *(Review Exercises 1–2.)*
$$\begin{cases} ax + by = h \\ cx + dy = k \end{cases}$$
- Solve a system of two linear equations in two variables by graphing. *(Review Exercises 3–4.)*

Unique   Inconsistent   Dependent

End of Chapter Material ➤

To help students study, each chapter ends with a **Chapter Summary with Hints and Suggestions** and **Review Exercises**. The last bullet of the Hints and Suggestions lists the Review Exercises that a student could use to self-test. Both even and odd answers are supplied in the back of the book.

Integrating Excel

If you would like to use Excel or another spreadsheet software when working the exercises in this text, refer to the chart below. It lists exercises from many sections that you might find instructive to do with spreadsheet technology. Please note that none of these exercises are dependent on Excel. If you would like help using Excel, please consider the *Excel Guide for Finite Mathematics and Applied Calculus*, 2nd Edition, which is available from Houghton Mifflin. Additionally, the *Getting Started with Excel* chapter of the guide is available on the CD and website.

Section	Suggested Exercises
1.1	79 and 80
1.2	95–98
1.3	77–85
1.4	71–78
1.5	27–44
1.6	29–36
2.1	1–41
2.2	1–44
2.3	1–30
2.4	1–38
3.4	21–30 and 41–50
3.5	1–8 and 37–50
3.6	9–30
4.3	31–40
4.4	9–34
4.5	9–34
4.6	9–34
5.2	1–44
5.3	31–40
5.4	26 and 33
5.5	17
5.6	21–50
6.1	15–25
6.2	1–26
6.3	1–31
6.4	1–40
7.1	13–16 and 23–25
7.2	9–30
7.3	5–8, 13–40, 42, 44, 46, and 47
8.3	1–18

Graphing Calculator Terminology

The Graphing Calculator Explorations and Exercises have been kept as generic as possible for use with any of the popular graphing calculators. We assume that you or your teacher are familiar with the sequence of button pushes necessary to accomplish various basic operations on your calculator. Certain standard calculator terms are capitalized in this book and are described below. Your calculator may use slightly different terminology.

The viewing window or graphing **WINDOW** is the part of the Cartesian plane shown in the display screen of your graphing calculator. **XMIN** and **XMAX** are the smallest and largest x-values shown, and **YMIN** and **YMAX** are the smallest and largest y-values shown. These can be set by using the **WINDOW** or **RANGE** command and are changed automatically by using any of the **ZOOM** operations. **XSCALE** and **YSCALE** define the distance between tick marks on the x- and y-axes.

XSCALE and YSCALE are each set at 1, so the tick marks are 1 unit apart. The unit distances in the x- and y-directions on the screen may differ.

Viewing Window [−10, 10] by [−10, 10]

The viewing window is always [XMIN, XMAX] by [YMIN, YMAX]. We will set XSCALE and YSCALE so that there are a reasonable number of tick marks (generally 2 to 20) on each axis. Either or both of the x- and y-axes may not be visible when the viewing window does not include the origin.

Pixel, an abbreviation for *pic*ture *el*ement, refers to a tiny rectangle on the screen that can be darkened to represent a dot on a graph. Pixels are arranged in a rectangular array on the screen. In the above window, the axes and tick marks are formed by darkened pixels. The size of the screen and the number of pixels vary with different calculators.

TRACE allows you to move a flashing pixel (or *cursor*) along a curve in the viewing window, with the x- and y-coordinates shown at the bottom of the screen.

Useful Hint: To make the *x*-values in TRACE take simple values like 0.1, 0.2, and 0.3, choose XMIN and XMAX to be multiples of one less than the number of pixels across the screen. For example, on the *TI-83* and *TI-83 Plus*, which have 95 pixels across the screen, using an *x*-window like $[-9.4, 9.4]$ or $[-4.7, 4.7]$ or $[-940, 940]$ will TRACE with simpler *x*-values than the standard windows used in this book.

ZOOM IN allows you to magnify any part of the viewing window to see finer detail around a chosen point. **ZOOM OUT** does the opposite, like stepping back to see a larger portion of a picture but with less detail. These and any of several other **ZOOM** commands change the viewing window.

VALUE or **EVALUATE** finds the value of a previously entered expression at a specified *x*-value.

SOLVE or **ROOT** or **ZERO** finds the *x*-value that solves $f(x) = 0$, or equivalently, the *x*-intercepts of a curve $y = f(x)$. When applied to a difference $f(x) - g(x)$, it finds the *x*-value where the two curves meet (as also done by the **INTERSECT** command).

MAX and **MIN** find the highest and lowest points of a (previously entered) curve between specified *x*-values.

In **CONNECTED MODE** your calculator will darken pixels to connect calculated points on a graph in an attempt to show it as a continuous or "unbroken" curve. However, this may lead to "false lines" in a graph that should have breaks or "jumps." Such false lines can be eliminated by using **DOT MODE**.

The **TABLE** command on some calculators lists in table form the values of a function, just as you have probably done when graphing a curve. The *x*-values may be chosen by you or by the calculator.

The **Order of Operations** used by most calculators evaluates operations in the following order: first powers and roots, then operations like **LN** and **LOG,** then multiplications and divisions, and finally additions and subtractions, always working from left to right within each level. For example, 5^2x means $(5^2)x$, *not* $5^{(2x)}$. Also, $1/x + 1$ means $(1/x) + 1$, *not* $1/(x + 1)$. See your calculator's instruction manual for further information. *Be careful:* Some calculators evaluate $1/2x$ as $(1/2)x$ and some as $1/(2x)$. When in doubt, use parentheses to clarify the expression.

Much more information can be found in the manual for your graphing calculator and other features will be discussed later when they are needed.

1 Functions

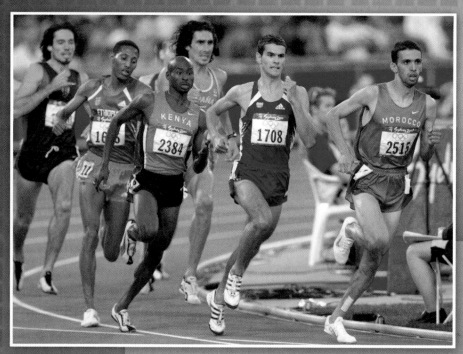

Moroccan runner
Hicham El Guerrouj,
current world
record holder for
the mile run,
bested the record
set 6 years earlier
by 1.26 seconds.

Application Preview

World Record Mile Runs

The dots on the graph below show the world record times for the mile run from 1865 to the 1999 world record of 3 minutes 43.13 seconds, set by the Moroccan runner Hicham El Guerrouj. These points fall roughly along a line, called the **regression line.** The regression line is easily found using a graphing calculator, based on a method called **least squares,** which is explained in Section 6 of Chapter 3. Notice that the times do not level off as you might expect, but continue to decrease.

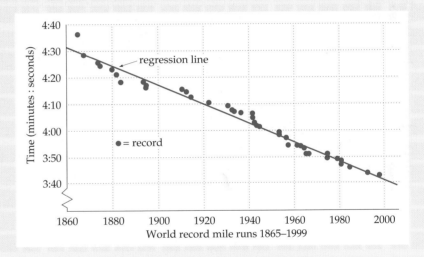

World record mile runs 1865–1999

History of the Record for the Mile Run

Time	Year	Athlete	Time	Year	Athlete
4:36.5	1865	Richard Webster	4:12.6	1915	Norman Taber
4:29.0	1868	William Chinnery	4:10.4	1923	Paavo Nurmi
4:28.8	1868	Walter Gibbs	4:09.2	1931	Jules Ladoumegue
4:26.0	1874	Walter Slade	4:07.6	1933	Jack Lovelock
4:24.5	1875	Walter Slade	4:06.8	1934	Glenn Cunningham
4:23.2	1880	Walter George	4:06.4	1937	Sydney Wooderson
4:21.4	1882	Walter George	4:06.2	1942	Gunder Hägg
4:18.4	1884	Walter George	4:06.2	1942	Arne Andersson
4:18.2	1894	Fred Bacon	4:04.6	1942	Gunder Hägg
4:17.0	1895	Fred Bacon	4:02.6	1943	Arne Andersson
4:15.6	1895	Thomas Conneff	4:01.6	1944	Arne Andersson
4:15.4	1911	John Paul Jones	4:01.4	1945	Gunder Hägg
4:14.4	1913	John Paul Jones	3:59.4	1954	Roger Bannister

World Record Mile Runs
(continued)

Time	Year	Athlete	Time	Year	Athlete
3:58.0	1954	John Landy	3:49.4	1975	John Walker
3:57.2	1957	Derek Ibbotson	3:49.0	1979	Sebastian Coe
3:54.5	1958	Herb Elliott	3:48.8	1980	Steve Ovett
3:54.4	1962	Peter Snell	3:48.53	1981	Sebastian Coe
3:54.1	1964	Peter Snell	3:48.40	1981	Steve Ovett
3:53.6	1965	Michel Jazy	3:47.33	1981	Sebastian Coe
3:51.3	1966	Jim Ryun	3:46.31	1985	Steve Cram
3:51.1	1967	Jim Ryun	3:44.39	1993	Noureddine Morceli
3:51.0	1975	Filbert Bayi	3:43.13	1999	Hicham El Guerrouj

Source: USA Track & Field

The equation of the regression line shown in the graph is $y = -0.356x + 257.44$, where x represents years after 1900 and y is in seconds. The regression line can be used to predict the world mile record in future years. Notice that the most recent world record would have been predicted quite accurately by this line, since the rightmost dot falls almost exactly on the line. Linear trends, however, must not be extended too far. The downward slope of this line means that it will eventually "predict" mile runs in a fraction of a second, or even in *negative* time (see page 16). *Moral:* In the real world, linear trends do not continue indefinitely. This and other topics in "linear" mathematics will be developed in Section 1.1.

1.1 REAL NUMBERS, INEQUALITIES, AND LINES

Introduction

In this section we will study *linear* relationships between two variable quantities—that is, relationships that can be represented by *lines*. In later sections we will study *nonlinear* relationships, which can be represented by *curves*.

When reading this book, it will be helpful (but not necessary) to have a graphing calculator. The **Graphing Calculator Explorations** show how to use a graphing calculator to explore concepts more deeply or to analyze applications in more detail. The parts of the book that require graphing calculators are marked by the symbol ▦ . If you do not have a *graphing* calculator, you will need a *scientific* or *business* calculator with keys like $\boxed{y^x}$ and $\boxed{\ln x}$ for powers and logarithms. For those with access to a computer, the **Spreadsheet Explorations** show how spreadsheets can be used for some of the operations of a graphing calculator.

Real Numbers and Inequalities

In this book the word "number" means *real* number, a number that can be represented by a point on the number line (also called the *real line*).

The *order* of the real numbers is expressed by inequalities, with $a < b$ meaning "a is to the *left* of b," and $a > b$ meaning "a is to the *right* of b."

Inequalities	
Inequality	*In Words*
$a < b$	a is less than (smaller than) b
$a \leq b$	a is less than or equal to b
$a > b$	a is greater than (larger than) b
$a \geq b$	a is greater than or equal to b

The inequalities $a < b$ and $a > b$ are called "strict" inequalities, and $a \leq b$ and $a \geq b$ are called "nonstrict" inequalities.

EXAMPLE 1

INEQUALITIES BETWEEN NUMBERS

a. $3 \leq 5$ **b.** $6 > -2$ **c.** $-10 < -5$

-10 is less than (smaller than) -5

Throughout this book are many **Practice Problems**—short questions designed to check your understanding of a topic before moving on to new material. Full solutions are given at the end of the section. Solve the following Practice Problem and then check your answer.

Practice Problem 1   Which number is smaller: $\dfrac{1}{100}$ or $-1,000,000$?   ➤ Solution on page 13

Multiplying or dividing both sides of an inequality by a negative number reverses the direction of the inequality:

$$-3 < 2 \qquad \text{but} \qquad 3 > -2 \qquad \text{Multiplying by } -1$$

A *double* inequality, such as $a < x < b$, means that *both* the inequalities $a < x$ and $x < b$ hold. The inequality $a < x < b$ can be interpreted graphically as "x is between a and b."

$$a < x < b$$

Sets and Intervals

Braces { } are read "the set of all" and a vertical bar | is read "such that."

EXAMPLE **2**

INTERPRETING SETS

The set of all

a. $\{x \mid x > 3\}$ means "the set of all x such that x is greater than 3."

Such that

b. $\{x \mid -2 < x < 5\}$ means "the set of all x such that x is between -2 and 5."

Practice Problem **2**

a. Write in set notation "the set of all x such that x is greater than or equal to -7."

b. Express in words: $\{x \mid x < -1\}$. ➤ Solutions on page 13

The set $\{x \mid 2 \leq x \leq 5\}$ can be expressed in *interval* notation by enclosing the endpoints 2 and 5 in square brackets: [2, 5]. The *square* brackets indicate that the endpoints are *in*cluded. The set $\{x \mid 2 < x < 5\}$ can be written (2, 5). The *parentheses* indicate that the endpoints 2 and 5 are *ex*cluded. An interval is *closed* if it includes both endpoints, and *open* if it includes neither endpoint. The four types of intervals are shown below: a *solid* dot • on the graph indicates that the point is *in*cluded in the interval; a *hollow* dot ○ indicates that the point is *ex*cluded.

Finite Intervals

Interval Notation	Set Notation	Graph	Type
[a, b]	$\{x \mid a \leq x \leq b\}$		Closed
(a, b)	$\{x \mid a < x < b\}$		Open

Finite Intervals (continued)

Interval Notation	Set Notation	Graph	Type
$[a, b)$	$\{x \mid a \le x < b\}$		Half-open
			or
$(a, b]$	$\{x \mid a < x \le b\}$		Half-closed

An interval may extend infinitely far to the right (indicated by the symbol ∞ for "infinity") or infinitely far to the left (indicated by $-\infty$ for "negative infinity"). Note that ∞ and $-\infty$ are not numbers, but are merely symbols to indicate that the interval extends endlessly in one direction or the other. The infinite intervals in the next box are said to be *closed* or *open* depending on whether they *include* or *exclude* their single endpoint.

Infinite Intervals

Interval Notation	Set Notation	Graph	Type
$[a, \infty)$	$\{x \mid x \ge a\}$		Closed
(a, ∞)	$\{x \mid x > a\}$		Open
$(-\infty, a]$	$\{x \mid x \le a\}$		Closed
$(-\infty, a)$	$\{x \mid x < a\}$		Open

EXAMPLE 3

GRAPHING SETS AND INTERVALS

Interval Notation	Set Notation	Graph
$(-2, 5]$	$\{x \mid -2 < x \le 5\}$	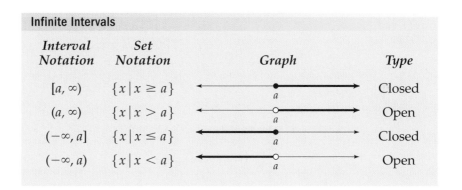
$[3, \infty)$	$\{x \mid x \ge 3\}$	

We use *parentheses* rather than square brackets with ∞ and $-\infty$ since they are not actual numbers.

The interval $(-\infty, \infty)$ extends infinitely far in *both* directions (meaning the entire real line) and is also denoted by $\mathbb{R}$ (the set of all real numbers).

$$\mathbb{R} = (-\infty, \infty)$$

Cartesian Plane

Two real lines or *axes*, one horizontal and one vertical, intersecting at their zero points, define the *Cartesian plane.** The axes divide the plane into four *quadrants*, I through IV, as shown below.

Any point in the Cartesian plane can be specified uniquely by an ordered pair of numbers (x, y); x, called the *abscissa* or *x-coordinate*, is the number on the horizontal axis corresponding to the point; y, called the *ordinate* or *y-coordinate*, is the number on the vertical axis corresponding to the point.

The Cartesian plane

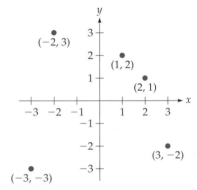

The Cartesian plane with several points. Order matters: $(1, 2)$ is not the same as $(2, 1)$

Lines and Slopes

The symbol Δ (read "delta," the Greek letter D) means "the change in." For any two points (x_1, y_1) and (x_2, y_2) we define

$$\Delta x = x_2 - x_1 \qquad \text{The change in } x \text{ is the difference in the } x\text{-coordinates}$$
$$\Delta y = y_2 - y_1 \qquad \text{The change in } y \text{ is the difference in the } y\text{-coordinates}$$

Any two distinct points determine a line. A nonvertical line has a *slope* that measures the steepness of the line, defined as *the change in y divided by the change in x* for any two points on the line.

---

* So named because it was originated by the French philosopher and mathematician René Descartes (1596–1650). Following the custom of the day, Descartes signed his scholarly papers with his Latin name Cartesius, hence "Cartesian" plane.

Slope of Line Through (x_1, y_1) and (x_2, y_2)

$$m = \frac{\Delta y}{\Delta x} = \frac{y_2 - y_1}{x_2 - x_1}$$

Slope is the change in y over the change in x $(x_2 \neq x_1)$

The changes Δy and Δx are often called, respectively, the "rise" and the "run," with the understanding that a negative "rise" means a "fall." Slope is then "rise over run."

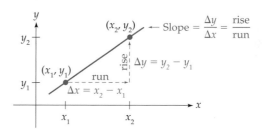

EXAMPLE 4

FINDING SLOPES AND GRAPHING LINES

Find the slope of the line through each pair of points, and graph the line.

a. $(1, 3), (2, 5)$ **b.** $(2, 4), (3, 1)$

c. $(-1, 3), (2, 3)$ **d.** $(2, -1), (2, 3)$

Solution

We use the slope formula $m = \dfrac{y_2 - y_1}{x_2 - x_1}$ for each pair $(x_1, y_1), (x_2, y_2)$.

a. For $(1, 3)$ and $(2, 5)$ the slope is
$$\frac{5 - 3}{2 - 1} = \frac{2}{1} = 2.$$

b. For $(2, 4)$ and $(3, 1)$ the slope is
$$\frac{1 - 4}{3 - 2} = \frac{-3}{1} = -3.$$

c. For $(-1, 3)$ and $(2, 3)$ the slope is $\dfrac{3 - 3}{2 - (-1)} = \dfrac{0}{3} = 0$.

d. For $(2, -1)$ and $(2, 3)$ the slope is *undefined:* $\dfrac{3 - (-1)}{2 - 2} = \dfrac{4}{0}$.

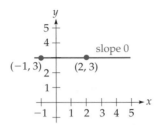

If $\Delta x = 1$, as in Examples 4a and 4b, then the slope is just the "rise," giving

$$\text{Slope} = \left(\begin{matrix}\text{Amount that the line rises}\\ \text{when } x \text{ increases by 1}\end{matrix}\right)$$

Practice Problem 3 A company president is considering four different business strategies, called S_1, S_2, S_3, and S_4, each with different projected future profits. The graph on the right shows the annual projected profit for the first few years for each of the strategies. Which strategy will yield:

a. The highest projected profit in year 1?

b. The highest projected profit in the long run? ➤ **Solutions on page 13**

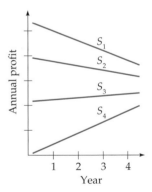

Equations of Lines

The point where a nonvertical line crosses the *y*-axis is called the *y-intercept* of the line. The *y*-intercept can be given either as the *y*-coordinate b or as the point $(0, b)$. Such a line can be expressed very simply in terms of its slope and *y*-intercept, representing the points by variable coordinates (or "variables") x and y.

Slope-Intercept Form of a Line

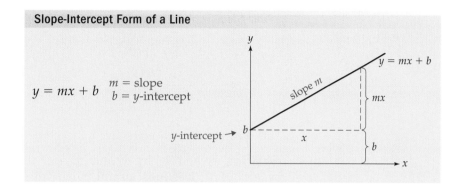

$$y = mx + b \quad \begin{array}{l} m = \text{slope} \\ b = y\text{-intercept} \end{array}$$

EXAMPLE 5

USING THE SLOPE-INTERCEPT FORM

Find an equation of the line with slope -2 and y-intercept 4, and graph it.

Solution

$$y = -2x + 4 \qquad \begin{array}{l} y = mx + b \text{ with} \\ m = -2 \text{ and } b = 4 \end{array}$$

We graph the line by first plotting the y-intercept $(0, 4)$. Using the slope $m = -2$, we plot another point 1 unit to the right and 2 units *down* from the y-intercept. We then draw the line through these two points, as shown on the left.

Graphing Calculator Exploration

$y_1 = x$, $y_2 = 2x$, and $y_3 = 3x$ on $[-10, 10]$ by $[-10, 10]$

a. Use a graphing calculator to graph the lines $y_1 = x$, $y_2 = 2x$, and $y_3 = 3x$ simultaneously on the window $[-10, 10]$ by $[-10, 10]$. How do the graphs change as the coefficient of x increases from 1 to 2 to 3?

b. Predict what the graph of $y = 0.5x$ would look like. What about $y = -2x$? Check your predictions by graphing them.

c. Describe the graph of the line $y = mx$ for any number m.

> **Point-Slope Form of a Line**
>
> $$y - y_1 = m(x - x_1)$$
>
> (x_1, y_1) = point on the line
> m = slope

This form comes directly from the slope formula $m = \dfrac{y_2 - y_1}{x_2 - x_1}$ by replacing x_2 and y_2 by x and y, and then multiplying each side by $(x - x_1)$. It is most useful when you know the slope of the line and a point on it.

EXAMPLE 6

USING THE POINT-SLOPE FORM

Find an equation of the line through $(6, -2)$ with slope $-\frac{1}{2}$.

Solution

$$y - (-2) = -\frac{1}{2}(x - 6)$$

$y - y_1 = m(x - x_1)$ with
$y_1 = -2$, $m = -\frac{1}{2}$, and $x_1 = 6$

$$y + 2 = -\frac{1}{2}x + 3$$

Eliminating parentheses

$$y = -\frac{1}{2}x + 1$$

Subtracting 2 from each side

Alternatively, we could have found this equation using $y = mx + b$, replacing m by the given slope $-\frac{1}{2}$, and then substituting the given $x = 6$ and $y = -2$ to evaluate b.

EXAMPLE 7

FINDING AN EQUATION FOR A LINE THROUGH TWO POINTS

Find an equation for the line through the points $(4, 1)$ and $(7, -2)$.

Solution

The slope is not given, so we calculate it from the two points.

$$m = \frac{-2 - 1}{7 - 4} = \frac{-3}{3} = -1 \qquad m = \frac{y_2 - y_1}{x_2 - x_1} \text{ with } (4, 1) \text{ and } (7, -2)$$

Then we use the point-slope formula with this slope and either of the two points.

$$y - 1 = -1(x - 4)$$

$y - y_1 = m(x - x_1)$ with slope -1 and point $(4, 1)$

$$y - 1 = -x + 4$$

Eliminating parentheses

$$y = -x + 5$$

Adding 1 to each side

Practice Problem **4**

Find the slope-intercept form of the line through the points $(2, 1)$ and $(4, 7)$.

➤ **Solution on page 13**

Vertical and horizontal lines have particularly simple equations: a variable equaling a constant.

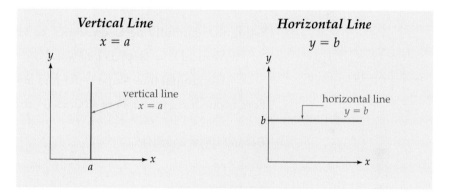

Vertical Line
$$x = a$$

Horizontal Line
$$y = b$$

EXAMPLE 8

GRAPHING VERTICAL AND HORIZONTAL LINES

Graph the lines $x = 2$ and $y = 6$.

Solution

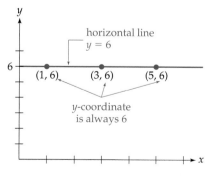

EXAMPLE 9

FINDING EQUATIONS OF VERTICAL AND HORIZONTAL LINES

a. Find an equation for the *vertical* line through the point (3, 2).

b. Find an equation for the *horizontal* line through the point (3, 2).

Solution

a. Vertical line $\qquad x = 3$ $\qquad\qquad$ $x = a$, with a being the x-coordinate from (3, 2)

b. Horizontal line $\quad y = 2$ $\qquad\qquad$ $y = b$, with b being the y-coordinate from (3, 2)

Practice Problem 5 Find an equation for the vertical line through the point (−2, 10).

➤ **Solution on next page**

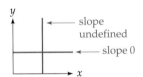

Distinguish carefully between slopes of vertical and horizontal lines:

Vertical line: slope is *undefined*.

Horizontal line: slope *is* defined and is *zero*.

There is one form that covers *all* lines, vertical and nonvertical.

General Linear Equation

$$ax + by = c$$

For constants a, b, c, with a and b not both zero

Any equation that can be written in this form is called a *linear equation*, and the variables are said to *depend linearly* on each other.

EXAMPLE 10

FINDING THE SLOPE AND THE y-INTERCEPT FROM A LINEAR EQUATION

Find the slope and y-intercept of the line $2x + 3y = 12$.

Solution We write the line in slope-intercept form. Solving for y:

$$3y = -2x + 12$$

Subtracting $2x$ from both sides of $2x + 3y = 12$

$$y = -\frac{2}{3}x + 4$$

Dividing each side by 3 gives the slope-intercept form $y = mx + b$

Therefore, the slope is $-\frac{2}{3}$ and the y-intercept is (0, 4).

Practice Problem 6   Find the slope and y-intercept of the line $x - \dfrac{y}{3} = 2$.

➤ **Solution on next page**

1.1 Section Summary

An *interval* is a set of real numbers corresponding to a section of the real line. The interval is *closed* if it contains all of its endpoints, and *open* if it contains none of its endpoints.

The nonvertical line through two points (x_1, y_1) and (x_2, y_2) has slope

$$m = \frac{\Delta y}{\Delta x} = \frac{y_2 - y_1}{x_2 - x_1} \qquad x_1 \neq x_2$$

There are five equations or "forms" for lines:

$y = mx + b$	Slope-intercept form m = slope, b = y-intercept
$y - y_1 = m(x - x_1)$	Point-slope form (x_1, y_1) = point, m = slope
$x = a$	Vertical line (slope undefined) a = x-intercept
$y = b$	Horizontal line (slope zero) b = y-intercept
$ax + by = c$	General linear equation

➤ **Solutions to Practice Problems**

1. $-1{,}000{,}000$ [the negative sign makes it less than (to the left of) the positive number $\frac{1}{100}$]

2. **a.** $\{x \mid x \geq -7\}$
   b. The set of all x such that x is less than -1

3. **a.** S_1
   b. S_4

4. $m = \dfrac{7 - 1}{4 - 2} = \dfrac{6}{2} = 3$ From points (2, 1) and (4, 7)

   $y - 1 = 3(x - 2)$ Using the point-slope form with $(x_1, y_1) = (2, 1)$
   $y - 1 = 3x - 6$
   $\qquad y = 3x - 5$

5. $x = -2$

6. $x - \dfrac{y}{3} = 2$

$\qquad -\dfrac{y}{3} = -x + 2 \qquad$ Subtracting x from each side

$\qquad y = 3x - 6 \qquad$ Multiplying each side by -3

Slope is $\ m = 3\ $ and y-intercept is $(0, -6)$.

1.1 Exercises

Write each interval in set notation and graph it on the real line.

1. $[0, 6)$ **2.** $(-3, 5]$ **3.** $(-\infty, 2]$ **4.** $[7, \infty)$

5. Given the equation $\ y = 5x - 12,\ $ how will y change if x:
 a. Increases by 3 units?
 b. Decreases by 2 units?

6. Given the equation $\ y = -2x + 7,\ $ how will y change if x:
 a. Increases by 5 units?
 b. Decreases by 4 units?

Find the slope (if it is defined) of the line determined by each pair of points.

7. $(2, 3)$ and $(4, -1)$ **8.** $(3, -1)$ and $(5, 7)$

9. $(-4, 0)$ and $(2, 2)$ **10.** $(-1, 4)$ and $(5, 1)$

11. $(0, -1)$ and $(4, -1)$ **12.** $\left(-2, \frac{1}{2}\right)$ and $\left(5, \frac{1}{2}\right)$

13. $(2, -1)$ and $(2, 5)$ **14.** $(6, -4)$ and $(6, -3)$

For each equation, find the slope m and y-intercept $(0, b)$ (when they exist) and draw the graph.

15. $y = 3x - 4$ **16.** $y = 2x$

17. $y = -\frac{1}{2}x$ **18.** $y = -\frac{1}{3}x + 2$

19. $y = 4$ **20.** $y = -3$

21. $x = 4$ **22.** $x = -3$

23. $2x - 3y = 12$ **24.** $3x + 2y = 18$

25. $x + y = 0$ **26.** $x = 2y + 4$

27. $x - y = 0$ **28.** $y = \frac{2}{3}(x - 3)$

29. $y = \dfrac{x + 2}{3}$ **30.** $\dfrac{x}{2} + \dfrac{y}{3} = 1$

31. $\dfrac{2x}{3} - y = 1$ **32.** $\dfrac{x + 1}{2} + \dfrac{y + 1}{2} = 1$

Use a graphing calculator to graph each line. [*Note:* Your graph will depend on the viewing window you choose. Begin with a "standard" window such as $[-10, 10]$ by $[-10, 10]$, and choose a larger window (or "zoom out") if the line does not appear.]

33. $y = 2x - 8$ **34.** $y = 3x - 6$

35. $y = 7 - 3x$ **36.** $y = 5 - 2x$

37. $y = 50 - x$ **38.** $y = x - 40$

Write an equation of the line satisfying the following conditions. If possible, write your answer in the form $\ y = mx + b$.

39. Slope -2.25 and y-intercept 3

40. Slope $\frac{2}{3}$ and y-intercept -8

41. Slope 5 and passing through the point $(-1, -2)$

42. Slope -1 and passing through the point $(4, 3)$

43. Horizontal and passing through the point $(1.5, -4)$

44. Horizontal and passing through the point $\left(\frac{1}{2}, \frac{3}{4}\right)$

45. Vertical and passing through the point $(1.5, -4)$

46. Vertical and passing through the point $\left(\frac{1}{2}, \frac{3}{4}\right)$

47. Passing through the points $(5, 3)$ and $(7, -1)$

48. Passing through the points $(3, -1)$ and $(6, 0)$

49. Passing through the points $(1, -1)$ and $(5, -1)$

50. Passing through the points $(2, 0)$ and $(2, -4)$

Write an equation of the form $y = mx + b$ for each line in the following graphs. [*Hint:* Either find the slope and y-intercept or use any two points on the line.]

51.

52.

53.

54.

Write equations for the lines determining the four sides of each figure.

55.

56.

57. Show that $y - y_1 = m(x - x_1)$ simplifies to $y = mx + b$ if the point (x_1, y_1) is the y-intercept $(0, b)$.

58. Show that the linear equation $\dfrac{x}{a} + \dfrac{y}{b} = 1$ has x-intercept $(a, 0)$ and y-intercept $(0, b)$. (The

x-intercept is the point where the line crosses the x-axis.)

59. Find the x-intercept $(a, 0)$ where the line $y = mx + b$ crosses the x-axis. Under what condition on m will a single x-intercept exist?

60. **i.** Show that the general linear equation $ax + by = c$ with $b \neq 0$ can be written as $y = -\dfrac{a}{b}x + \dfrac{c}{b}$, which is the equation of a line in slope-intercept form.
  ii. Show that the general linear equation $ax + by = c$ with $b = 0$ but $a \neq 0$ can be written as $x = \dfrac{c}{a}$, which is the equation of a vertical line.

  [*Note:* Since these steps are reversible, parts (i) and (ii) together show that the general linear equation $ax + by = c$ (for a and b not both zero) includes vertical and nonvertical lines.]

61. a. Graph the lines $y_1 = -x$, $y_2 = -2x$, and $y_3 = -3x$ on the window $[-5, 5]$ by $[-5, 5]$ (using the *negation* key $(-)$, not the *subtraction* key $-$). Observe how the coefficient of x changes the slope of the line.
  b. Predict what the line $y = -9x$ would look like, and then check your prediction by graphing it.

62. a. Graph the lines $y_1 = x + 2$, $y_2 = x + 1$, $y_3 = x$, $y_4 = x - 1$, and $y_5 = x - 2$ on the window $[-5, 5]$ by $[-5, 5]$. Observe how the constant changes the position of the line.
  b. Predict what the lines $y = x + 4$ and $y = x - 4$ would look like, and then check your prediction by graphing them.

APPLIED EXERCISES

63. BUSINESS: Energy Usage A utility considers demand for electricity "low" if it is below 8 mkW (million kilowatts), "average" if it is at least 8 mkW but below 20 mkW, "high" if it is at least 20 mkW but below 40 mkW, and "critical" if it is 40 mkW or more. Express these demand levels in interval notation. [*Hint:* The interval for "low" is $[0, 8)$.]

64. GENERAL: Grades If a grade of 90 through 100 is an A, at least 80 but less than 90 is a B, at least 70 but less than 80 a C, at least 60 but less than 70 a D, and below 60 an F, write these grade levels in interval form (ignoring rounding). [*Hint:* F would be [0, 60).]

65. ATHLETICS: Mile Run Read the Application Preview on pages 1–2.

a. Use the regression line $y = -0.356x + 257.44$ to predict the world record in the year 2010. [*Hint:* If x represents years after 1900, what value of x corresponds to the year 2010? The resulting y will be in seconds, and should be converted to minutes and seconds.]

b. According to this formula, when will the record be 3 minutes 30 seconds? [*Hint:* Set the formula equal to 210 seconds and solve. What year corresponds to this x-value?]

66. ATHLETICS: Mile Run Read the Application Preview on pages 1–2. Evaluate the regression line $y = -0.356x + 257.44$ at $x = 720$ and at $x = 724$ (corresponding to the years 2620 and 2624). Does the formula give reasonable times for the mile record in these years? [*Moral:* Linear trends may not continue indefinitely.]

67. BUSINESS: Corporate Profit A company's profit increased linearly from $6 million at the end of year 1 to $14 million at the end of year 3.

a. Use the two (year, profit) data points (1, 6) and (3, 14) to find the linear relationship $y = mx + b$ between $x = $ year and $y = $ profit.

b. Find the company's profit at the end of 2 years.

c. Predict the company's profit at the end of 5 years.

68. ECONOMICS: Per Capita Personal Income In the short run, per capita personal income (PCPI) in the United States grows approximately linearly. In 1990 PCPI was 19.1, and in 2000 it had grown to 29.7 (both in thousands of dollars). (*Source:* Bureau of Economic Analysis)

a. Use the two given (year, PCPI) data points (0, 19.1) and (10, 29.7) to find the linear relationship $y = mx + b$ between $x = $ years since 1990 and $y = $ PCPI.

b. Use your linear relationship to predict PCPI in 2010.

69. GENERAL: Temperature On the Fahrenheit temperature scale, water freezes at 32° and boils at 212°. On the Celsius (centigrade) scale, water freezes at 0° and boils at 100°.

a. Use the two (Celsius, Fahrenheit) data points (0, 32) and (100, 212) to find the linear relationship $y = mx + b$ between $x = $ Celsius temperature and $y = $ Fahrenheit temperature.

b. Find the Fahrenheit temperature that corresponds to 20° Celsius.

70. ECOLOGY: Waste Disposal The amount of municipal solid waste generated per person per day in the United States has increased approximately linearly, from 2.7 pounds in 1960 to 4.6 pounds in 2000.

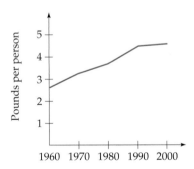

Municipal Solid Waste Generated Per Capita Per Day
(*Source*: U.S. Environmental Protection Agency)

a. Use the two (year, pound) data points (0, 2.7) and (40, 4.6) to find the linear relationship $y = mx + b$ between $x = $ years since 1960 and $y = $ per capita waste.

b. Use your formula to predict the amount in the year 2010.

71–72. BUSINESS: Straight-Line Depreciation Straight-line depreciation is a method for estimating the value of an asset (such as a piece of machinery) as it loses value ("depreciates") through use. Given the original *price* of an asset, its *useful lifetime,* and its *scrap value* (its value at the end of its useful lifetime), the value of the asset after t years is given by the formula

$$\text{Value} = (\text{price}) - \left(\frac{(\text{price}) - (\text{scrap value})}{(\text{useful lifetime})}\right) \cdot t$$

$$\text{for} \quad 0 \le t \le (\text{useful lifetime})$$

71. a. A farmer buys a harvester for \$50,000 and estimates its useful life to be 20 years, after which its scrap value will be \$6000. Use the formula above to find a formula for the value V of the harvester after t years, for $0 \le t \le 20$.

b. Use your formula to find the value of the harvester after 5 years.

c. Graph the function found in part (a) on a graphing calculator on the window [0, 20] by [0, 50,000]. [*Hint:* Use x instead of t.]

72. a. A newspaper buys a printing press for \$800,000 and estimates its useful life to be 20 years, after which its scrap value will be \$60,000. Use the formula above Exercise 71 to find a formula for the value V of the press after t years, for $0 \le t \le 20$.

b. Use your formula to find the value of the press after 10 years.

c. Graph the function found in part (a) on a graphing calculator on the window [0, 20] by [0, 800,000]. [*Hint:* Use x instead of t.]

73–74. ENVIRONMENTAL SCIENCE: Beverton-Holt Recruitment Curve Some organisms exhibit a density-dependent mortality from one generation to the next. Let $R > 1$ be the net reproductive rate (that is, the number of surviving offspring per parent), let $x > 0$ be the density of parents and y be the density of surviving offspring. The *Beverton-Holt recruitment curve* is

$$y = \frac{Rx}{1 + \left(\dfrac{R - 1}{K}\right)x}$$

where $K > 0$ is the *carrying capacity* of the organism's environment. Notice that if $x = K$, then $y = K$.

73. Show that if $x < K$, then $x < y < K$. Explain what this means about the population size over successive generations if the initial population is smaller than the carrying capacity of the environment.

74. Show that if $x > K$, then $K < y < x$. Explain what this means about the population size over successive generations if the initial population is larger than the carrying capacity of the environment.

75. SOCIAL SCIENCE: Age at First Marriage Americans are marrying later and later. Based on data from the U.S. Bureau of the Census for the years 1980 to 2000, the median age at first marriage for men is $y_1 = 24.8 + 0.13x$, and for women it is $y_2 = 22.1 + 0.17x$, where x is the number of years since 1980.

a. Graph these lines on the window [0, 30] by [20, 30].

b. Use these lines to predict the median marriage ages for men and women in the year 2010. [*Hint:* Which x-value corresponds to 2010? Then use TRACE, EVALUATE, or TABLE.]

c. Predict the median marriage ages for men and women in the year 2020.

76. SOCIAL SCIENCE: Equal Pay for Equal Work Women's pay has often lagged behind men's, although Title VII of the Civil Rights Act requires equal pay for equal work. Based on data from 1980–2000, women's annual earnings as a percent of men's can be approximated by the formula $y = 0.97x + 59.5$, where x is the number of years since 1980. (For example, $x = 10$ gives $y = 69.2$, so in 1990 women's wages were about 69.2% of men's wages.)

a. Graph this line on the window [0, 30] by [0, 100].

b. Use this line to predict the percentage in the year 2005. [*Hint:* Which x-value corresponds to 2005? Then use TRACE, EVALUATE, or TABLE.]

c. Predict the percentage in the year 2010. (*Source:* U.S. Department of Labor—Women's Bureau)

77. GENERAL: Smoking and Education
According to a recent study,* the probability that a smoker will quit smoking increases with the smoker's educational level. The probability (expressed as a percent) that a smoker with x years of education will quit is approximately $y = 0.831x^2 - 18.1x + 137.3$ (for $10 \le x \le 16$).

  a. Graph this curve on the window [10, 16] by [0, 100].
  b. Find the probability that a high school graduate smoker $(x = 12)$ will quit.
  c. Find the probability that a college graduate smoker $(x = 16)$ will quit.

78. ENVIRONMENTAL SCIENCE: Wind Energy The use of wind power is growing rapidly after a slow start, especially in Europe, where it is seen as an efficient and renewable source of energy. Global wind power generating capacity for the years 1980 to 2000 is given approximately by $y = 30.5x^2 - 112x + 250$ megawatts, where x is the number of years after 1980. (One megawatt would supply the electrical needs of approximately 100 homes).

  a. Graph this curve on the window [0, 30] by [0, 2500].
  b. Use this curve to predict the global wind power generating capacity in the year 2005. [*Hint:* Which x-value corresponds to 2005? Then use TRACE, EVALUATE, or TABLE.]
  c. Predict the global wind power generating capacity in the year 2010.
  (*Source:* Worldwatch Institute)

* William Sander, "Schooling and Quitting Smoking," *The Review of Economics and Statistics* LXXVII(1):191–199, February 1995.

79–80. GENERAL: Life Expectancy The following tables give the life expectancy (years of life expected) for a newborn child born in the indicated year (Exercise 79 is for males, Exercise 80 for females). For each exercise:

  a. Enter the data into a graphing calculator and make a plot of the resulting points, with Years Since 1960 on the x-axis and Life Expectancy on the y-axis.
  b. Use the graphing calculator to find the linear regression line for these points. Enter the resulting function as y_1, which then estimates life expectancy based on the year of birth. Graph the points together with the regression line.
  c. Use your line y_1 to estimate the life expectancy of a child born in the year 2025. (This might be your child or grandchild.) [*Hint:* What x-value corresponds to 2025?]

79.

Birth Year	(years since 1960)	Life Expectancy (male)
1960	0	66.6
1970	10	67.1
1980	20	70.0
1990	30	71.8
2000	40	74.4

80.

Birth Year	(years since 1960)	Life Expectancy (female)
1960	0	73.1
1970	10	74.7
1980	20	77.5
1990	30	78.8
2000	40	79.6

1.2 EXPONENTS

Introduction

Not all variables are related linearly. In this section we will discuss exponents, which will enable us to express many nonlinear relationships.

Positive Integer Exponents

Numbers may be expressed with exponents, as in $2^3 = 2 \cdot 2 \cdot 2 = 8$. More generally, for any positive integer n, x^n means the product of n x's.

$$x^n = \overbrace{x \cdot x \cdots x}^{n}$$

The number being raised to the power is called the *base* and the power is the *exponent:*

        Exponent or power

$$x^n$$

        Base

    There are several *properties of exponents* for simplifying expressions. The first three are known, respectively, as the addition, subtraction, and multiplication properties of exponents.

Properties of Exponents

$x^m \cdot x^n = x^{m+n}$	To *multiply* powers of the same base, *add* the exponents
$\dfrac{x^m}{x^n} = x^{m-n}$	To *divide* powers of the same base, *subtract* the exponents (top exponent minus bottom exponent)
$(x^m)^n = x^{m \cdot n}$	To raise a power to a power, *multiply* the powers
$(xy)^n = x^n \cdot y^n$	To raise a product to a power, raise *each factor* to the power
$\left(\dfrac{x}{y}\right)^n = \dfrac{x^n}{y^n}$	To raise a fraction to a power, raise the numerator *and* denominator to the power

---

EXAMPLE **1**

SIMPLIFYING EXPONENTS

a. $x^2 \cdot x^3 = x^5$ (2 + 3)

                             Since $x^2 \cdot x^3 = \underbrace{\overbrace{x \cdot x}^{2} \cdot \overbrace{x \cdot x \cdot x}^{3}}_{5}$

b. $\dfrac{x^5}{x^3} = x^2$ (5 − 3)

                             Since $\dfrac{x^5}{x^3} = \dfrac{\overbrace{x \cdot x \cdot \cancel{x} \cdot \cancel{x} \cdot \cancel{x}}^{2}}{\cancel{x} \cdot \cancel{x} \cdot \cancel{x}}$

c. $(x^3)^2 = x^6$ ⟵ $2 \cdot 3$ Since $(x^3)^2 = x^3 \cdot x^3$

d. $(2w)^3 = 2^3 w^3 = 8w^3$ Since $(2w)^3 = 2w \cdot 2w \cdot 2w = 2^3 w^3$

e. $\left(\dfrac{x}{5}\right)^3 = \dfrac{x^3}{5^3} = \dfrac{x^3}{125}$ Since $\left(\dfrac{x}{5}\right)^3 = \dfrac{x}{5} \cdot \dfrac{x}{5} \cdot \dfrac{x}{5}$

f. $\dfrac{[(x^2)^3]^4}{x^5 \cdot x^7 \cdot x} = \dfrac{x^{24}}{x^{13}} = x^{11}$ Combining all the rules

($2 \cdot 3 \cdot 4$ ⟶ $24 - 13$, $5 + 7 + 1$)

Practice Problem 1   Simplify: **a.** $\dfrac{x^5 \cdot x}{x^2}$ **b.** $[(x^3)^2]^2$ ➤ **Solutions on page 28**

Remember: For exponents in the form $x^2 \cdot x^3 = x^5$,   *add* exponents.
For exponents in the form $(x^2)^3 = x^6$,   *multiply* exponents.

Graphing Calculator Exploration

a. Use a graphing calculator to graph $y_1 = x$, $y_2 = x^2$, $y_3 = x^3$, and $y_4 = x^4$ on the viewing window $[0, 2]$ by $[0, 2]$. Use TRACE to identify which curve goes with which power.

b. Which curve is highest for values of x between 0 and 1? Which is lowest?

c. Which curve is highest for values of x greater than 1? Which is lowest?

d. Predict what the curve $y = x^5$ would look like. Check your prediction by graphing it.

e. Predict which of these curves will be positive when x is negative. Check your prediction by changing the viewing window to $[-2, 2]$ by $[-2, 2]$.

Zero and Negative Exponents

For any number x other than zero, we define

$$x^0 = 1 \qquad \text{x to the power 0 is 1}$$

$$x^{-1} = \frac{1}{x} \qquad \text{x to the power -1 is one over x}$$

$$x^{-2} = \frac{1}{x^2} \qquad \text{x to the power -2 is one over x squared}$$

$$x^{-n} = \frac{1}{x^n} \qquad \text{x to a negative power is one over x to the positive power}$$

The definitions of x^0 and x^{-n} are motivated by the following calculations.

$$1 = \frac{x^2}{x^2} = x^{2-2} = x^0 \qquad \text{The subtraction property of exponents leads to $x^0 = 1$}$$

$$\frac{1}{x^n} = \frac{x^0}{x^n} = x^{0-n} = x^{-n} \qquad \text{$x^0 = 1$ and the subtraction property of exponents lead to $x^{-n} = \frac{1}{x^n}$}$$

EXAMPLE 2

SIMPLIFYING ZERO AND NEGATIVE EXPONENTS

a. $5^0 = 1$

b. $7^{-1} = \frac{1}{7}$

c. $3^{-2} = \frac{1}{3^2} = \frac{1}{9}$

d. $(-2)^{-3} = \frac{1}{(-2)^3} = \frac{1}{-8} = -\frac{1}{8}$

e. 0^0 and 0^{-3} are undefined.

Practice Problem 2 Evaluate: **a.** 2^0 **b.** 2^{-4} ➤ Solutions on page 28

A fraction to a negative power means *division* by the fraction, so we "invert and multiply."

$$\left(\frac{x}{y}\right)^{-1} = \frac{1}{\dfrac{x}{y}} = 1 \cdot \frac{y}{x} = \frac{y}{x}$$

Reciprocal of the original fraction

Therefore, for $x \neq 0$ and $y \neq 0$,

$$\left(\frac{x}{y}\right)^{-1} = \frac{y}{x}$$
A fraction to the power -1 is the reciprocal of the fraction

$$\left(\frac{x}{y}\right)^{-n} = \left(\frac{y}{x}\right)^{n}$$
A fraction to the negative power is the reciprocal of the fraction to the positive power

EXAMPLE 3

SIMPLIFYING FRACTIONS TO NEGATIVE EXPONENTS

a. $\left(\dfrac{3}{2}\right)^{-1} = \dfrac{2}{3}$

b. $\left(\dfrac{1}{2}\right)^{-3} = \left(\dfrac{2}{1}\right)^{3} = \dfrac{2^3}{1^3} = 8$

$\uparrow$

Reciprocal of $\dfrac{3}{2}$

Practice Problem 3 Simplify: $\left(\dfrac{2}{3}\right)^{-2}$

➤ Solution on page 28

Roots and Fractional Exponents

We may take the square root of any *nonnegative* number, and the cube root of *any* number.

EXAMPLE 4

EVALUATING ROOTS

a. $\sqrt{9} = 3$

b. $\sqrt{-9}$ is undefined.
Square roots of negative numbers are not defined

c. $\sqrt[3]{8} = 2$

d. $\sqrt[3]{-8} = -2$
Cube roots of negative numbers *are* defined

e. $\sqrt[3]{\dfrac{27}{8}} = \dfrac{\sqrt[3]{27}}{\sqrt[3]{8}} = \dfrac{3}{2}$

There are *two* square roots of 9, namely 3 and -3, but the radical sign $\sqrt{\ }$ means just the *positive* one (the "principal" square root).

$\sqrt[n]{a}$ means the principal nth root of a.
Principal means the positive root if there are two

In general, we may take *odd* roots of *any* number, but *even* roots only if the number is positive or zero.

EXAMPLE 5

EVALUATING ROOTS OF POSITIVE AND NEGATIVE NUMBERS

Odd roots of negative
numbers *are* defined

a. $\sqrt[4]{81} = 3$ **b.** $\sqrt[5]{-32} = -2$ Since $(-2)^5 = -32$

Graphing Calculator Exploration

a. Use a graphing calculator to graph $y_1 = x$, $y_2 = \sqrt{x}$, $y_3 = \sqrt[3]{x}$, and $y_4 = \sqrt[4]{x}$ simultaneously on the viewing window [0, 3] by [0, 2]. Use TRACE to identify which curve goes with which root.

b. Which curve is highest for values of x between 0 and 1? Which is lowest?

c. Which curve is highest for values of x greater than 1? Which is lowest?

d. Predict what the curve $y = \sqrt[7]{x}$ would look like. Check your prediction by graphing it.

e. Which of these roots are defined for *negative* values of x? Check your answer by changing the window to [−3, 3] by [−2, 2] and using TRACE where x is negative.

Fractional Exponents

Fractional exponents are defined as follows:

$$x^{\frac{1}{2}} = \sqrt{x}$$ Power $\frac{1}{2}$ means the principal square root

$$x^{\frac{1}{3}} = \sqrt[3]{x}$$ Power $\frac{1}{3}$ means the cube root

$$x^{\frac{1}{n}} = \sqrt[n]{x}$$ Power $\frac{1}{n}$ means the principal nth root (for a positive integer n)

The definition of $x^{\frac{1}{2}}$ is motivated by the multiplication property of exponents:

$$\left(x^{\frac{1}{2}}\right)^2 = x^{\frac{1}{2}\cdot 2} = x^1 = x$$

Taking square roots of each side of $\left(x^{\frac{1}{2}}\right)^2 = x$ gives

$$x^{\frac{1}{2}} = \sqrt{x}$$

x to the half power means the square root of x

EXAMPLE 6

EVALUATING FRACTIONAL EXPONENTS

a. $9^{\frac{1}{2}} = \sqrt{9} = 3$ **b.** $125^{\frac{1}{3}} = \sqrt[3]{125} = 5$

c. $81^{\frac{1}{4}} = \sqrt[4]{81} = 3$ **d.** $(-32)^{\frac{1}{5}} = \sqrt[5]{-32} = -2$

e. $\left(-\dfrac{27}{8}\right)^{\frac{1}{3}} = \sqrt[3]{-\dfrac{27}{8}} = -\dfrac{\sqrt[3]{27}}{\sqrt[3]{8}} = -\dfrac{3}{2}$

Practice Problem 4   Evaluate: **a.** $(-27)^{\frac{1}{3}}$ **b.** $\left(\dfrac{16}{81}\right)^{\frac{1}{4}}$ ► **Solutions on page 28**

To define $x^{\frac{m}{n}}$ for positive integers m and n, the exponent $\frac{m}{n}$ must be fully reduced (for example, $\frac{4}{6}$ must be reduced to $\frac{2}{3}$). Then

$$x^{\frac{m}{n}} = \left(x^{\frac{1}{n}}\right)^m = (x^m)^{\frac{1}{n}}$$

Since in both cases the exponents multiply to $\frac{m}{n}$

Therefore we define:

Fractional Exponents

$$x^{\frac{m}{n}} = \left(\sqrt[n]{x}\right)^m = \sqrt[n]{x^m}$$

$x^{m/n}$ means the mth power of the nth root, or equivalently, the nth root of the mth power

Both expressions, $\left(\sqrt[n]{x}\right)^m$ and $\sqrt[n]{x^m}$, will give the same answer. In either case the numerator determines the power and the denominator determines the root.

Power

$$x^{\frac{m}{n}}$$

Root

Power over root

EXAMPLE **7**

EVALUATING FRACTIONAL EXPONENTS

a. $8^{2/3} = \sqrt[3]{8^2} = \sqrt[3]{64} = 4$ } same   First the power, then the root

b. $8^{2/3} = \left(\sqrt[3]{8}\right)^2 = (2)^2 = 4$ First the root, then the power

c. $25^{3/2} = \left(\sqrt{25}\right)^3 = (5)^3 = 125$

d. $\left(\dfrac{-27}{8}\right)^{2/3} = \left(\sqrt[3]{\dfrac{-27}{8}}\right)^2 = \left(\dfrac{-3}{2}\right)^2 = \dfrac{9}{4}$

Practice Problem **5**   Evaluate: **a.** $16^{3/2}$ **b.** $(-8)^{2/3}$ ➤ **Solutions on page 28**

Graphing Calculator Exploration

a. Use a graphing calculator to evaluate $25^{3/2}$. [On some calculators, press $25\wedge(3 \div 2)$.] Your answer should agree with Example 7c above.

b. Evaluate $(-8)^{2/3}$. Use the $\boxed{(-)}$ key for negation, and parentheses around the exponent. Your answer should be 4. If you get an "error," try evaluating the expression as $((-8)^{1/3})^2$ or $((-8)^2)^{1/3}$. Whichever way works, remember it for evaluating negative numbers to fractional powers in the future.

EXAMPLE **8**

EVALUATING NEGATIVE FRACTIONAL EXPONENTS

a. $8^{-2/3} = \dfrac{1}{8^{2/3}} = \dfrac{1}{(\sqrt[3]{8})^2} = \dfrac{1}{2^2} = \dfrac{1}{4}$
A negative exponent means the reciprocal of the number to the positive exponent, which is then evaluated as before

b. $\left(\dfrac{9}{4}\right)^{-3/2} = \left(\dfrac{4}{9}\right)^{3/2} = \left(\sqrt{\dfrac{4}{9}}\right)^3 = \left(\dfrac{2}{3}\right)^3 = \dfrac{8}{27}$

Interpreting the power 3/2
Reciprocal to the positive exponent
Negative exponent

Practice Problem 6   Evaluate: **a.** $25^{-3/2}$ **b.** $\left(\dfrac{1}{4}\right)^{-1/2}$ **c.** $5^{1.3}$ [*Hint:* Use a calculator.]

➤ **Solutions on page 28**

Avoiding Pitfalls in Simplifying

The square root of a product is equal to the product of the square roots:

$$\sqrt{a \cdot b} = \sqrt{a} \cdot \sqrt{b}$$

However, the corresponding statement for *sums* is *not* true:

$$\sqrt{a + b} \quad \text{is } not \text{ equal to} \quad \sqrt{a} + \sqrt{b}$$

For example,

$$\underbrace{\sqrt{9 + 16}}_{\sqrt{25}} \neq \underbrace{\sqrt{9}}_{3} + \underbrace{\sqrt{16}}_{4}$$

The two sides are not equal: one is 5 and the other is 7

Therefore, do not "simplify" $\sqrt{x^2 + 9}$ into $x + 3$. The expression $\sqrt{x^2 + 9}$ *cannot be simplified.* Similarly,

$$(x + y)^2 \quad \text{is } not \text{ equal to} \quad x^2 + y^2$$

The expression $(x + y)^2$ means $(x + y)$ times itself:

$$(x + y)^2 = (x + y)(x + y) = x^2 + xy + yx + y^2 = x^2 + 2xy + y^2$$

$$(x + y)^2 = x^2 + 2xy + y^2$$

$(x + y)^2$ is the first number squared plus twice the product of the numbers plus the second number squared

Learning Curves in Airplane Production

Time / Repetitions

It is a truism that the more you practice a task, the faster you can do it. Successive repetitions generally take less time, following a "learning curve" like that on the left. Learning curves are used in industrial production. For example, it took 150,000 work-hours to build the first Boeing 707 airliner, while later planes $(n = 2, 3, \ldots, 300)$ took less time.*

$$\left(\begin{array}{c}\text{Time to build}\\ \text{plane number } n\end{array}\right) = 150n^{-0.322} \quad \text{thousand work-hours}$$

* A work-hour is the amount of work that a person can do in 1 hour. For further information on learning curves in industrial production, see J. M. Dutton et al., "The History of Progress Functions as a Managerial Technology," *Business History Review* **58**:204–233.

The time for the 10th Boeing 707 is found by substituting $n = 10$:

$$\left(\begin{array}{c}\text{Time to build}\\\text{plane 10}\end{array}\right) = 150(10)^{-0.322} \qquad \begin{array}{l}150n^{-0.322}\\\text{with}\quad n = 10\end{array}$$

$$\approx 71.46 \text{ thousand work-hours} \qquad \text{Using a calculator}$$

This shows that building the 10th Boeing 707 took about 71,460 work-hours, which is less than half of the 150,000 work-hours needed for the first. For the 100th 707:

$$\left(\begin{array}{c}\text{Time to build}\\\text{plane 100}\end{array}\right) = 150(100)^{-0.322} \qquad \begin{array}{l}150n^{-0.322}\\\text{with}\quad n = 100\end{array}$$

$$\approx 34.05 \text{ thousand work-hours}$$

or about 34,050 work-hours, which is less than half the time needed to build the 10th. Such learning curves are used for determining the cost of a contract to build several planes.

Notice that the learning curve graphed on the previous page decreases less steeply as the number of repetitions increases. This means that while construction time continues to decrease, it does so more slowly for later planes. This behavior, called "diminishing returns," is typical of learning curves.

1.2 Section Summary

We defined zero, negative, and fractional exponents as follows:

$$x^0 = 1 \qquad\qquad \text{for } x \neq 0$$

$$x^{-n} = \frac{1}{x^n} \qquad\qquad \text{for } x \neq 0$$

$$x^{\frac{m}{n}} = \left(\sqrt[n]{x}\right)^m = \sqrt[n]{x^m} \qquad m > 0, \ n > 0, \ \frac{m}{n} \text{ fully reduced}$$

With these definitions, the following properties of exponents hold for *all* exponents, whether integral or fractional, positive or negative.

$$x^m \cdot x^n = x^{m+n} \qquad (x^m)^n = x^{m \cdot n} \qquad \left(\frac{x}{y}\right)^n = \frac{x^n}{y^n}$$

$$\frac{x^m}{x^n} = x^{m-n} \qquad (xy)^n = x^n \cdot y^n$$

➤ **Solutions to Practice Problems**

1. **a.** $\dfrac{x^5 \cdot x}{x^2} = \dfrac{x^6}{x^2} = x^4$

   b. $[(x^3)^2]^2 = x^{3 \cdot 2 \cdot 2} = x^{12}$

2. **a.** $2^0 = 1$

   b. $2^{-4} = \dfrac{1}{2^4} = \dfrac{1}{16}$

3. $\left(\dfrac{2}{3}\right)^{-2} = \left(\dfrac{3}{2}\right)^2 = \dfrac{9}{4}$

4. **a.** $(-27)^{1/3} = \sqrt[3]{-27} = -3$

   b. $\left(\dfrac{16}{81}\right)^{1/4} = \sqrt[4]{\dfrac{16}{81}} = \dfrac{\sqrt[4]{16}}{\sqrt[4]{81}} = \dfrac{2}{3}$

5. **a.** $16^{3/2} = \left(\sqrt{16}\right)^3 = 4^3 = 64$

   b. $(-8)^{2/3} = \left(\sqrt[3]{-8}\right)^2 = (-2)^2 = 4$

6. **a.** $25^{-3/2} = \dfrac{1}{25^{3/2}} = \dfrac{1}{(\sqrt{25})^3} = \dfrac{1}{5^3} = \dfrac{1}{125}$

   b. $\left(\dfrac{1}{4}\right)^{-1/2} = \left(\dfrac{4}{1}\right)^{1/2} = \sqrt{4} = 2$

   c. $5^{1.3} \approx 8.103$

1.2 Exercises

Evaluate each expression *without* using a calculator.

1. $(2^2 \cdot 2)^2$ **2.** $(5^2 \cdot 4)^2$ **3.** 2^{-4}

4. 3^{-3} **5.** $\left(\dfrac{1}{2}\right)^{-3}$ **6.** $\left(\dfrac{1}{3}\right)^{-2}$

7. $\left(\dfrac{5}{8}\right)^{-1}$ **8.** $\left(\dfrac{3}{4}\right)^{-1}$ **9.** $4^{-2} \cdot 2^{-1}$

10. $3^{-2} \cdot 9^{-1}$ **11.** $\left(\dfrac{3}{2}\right)^{-3}$ **12.** $\left(\dfrac{2}{3}\right)^{-3}$

13. $\left(\dfrac{1}{3}\right)^{-2} - \left(\dfrac{1}{2}\right)^{-3}$ **14.** $\left(\dfrac{1}{3}\right)^{-2} - \left(\dfrac{1}{2}\right)^{-2}$

15. $\left[\left(\dfrac{2}{3}\right)^{-2}\right]^{-1}$ **16.** $\left[\left(\dfrac{2}{5}\right)^{-2}\right]^{-1}$

17. $25^{1/2}$ **18.** $36^{1/2}$ **19.** $25^{3/2}$

20. $16^{3/2}$ **21.** $16^{3/4}$ **22.** $27^{2/3}$

23. $(-8)^{2/3}$ **24.** $(-27)^{2/3}$ **25.** $(-8)^{5/3}$

26. $(-27)^{5/3}$ **27.** $\left(\dfrac{25}{36}\right)^{3/2}$ **28.** $\left(\dfrac{16}{25}\right)^{3/2}$

29. $\left(\dfrac{27}{125}\right)^{2/3}$ **30.** $\left(\dfrac{125}{8}\right)^{2/3}$ **31.** $\left(\dfrac{1}{32}\right)^{2/5}$

32. $\left(\dfrac{1}{32}\right)^{3/5}$ **33.** $4^{-1/2}$ **34.** $9^{-1/2}$

35. $4^{-3/2}$ **36.** $9^{-3/2}$ **37.** $8^{-2/3}$

38. $16^{-3/4}$ **39.** $(-8)^{-1/3}$ **40.** $(-27)^{-1/3}$

41. $(-8)^{-2/3}$ **42.** $(-27)^{-2/3}$ **43.** $\left(\dfrac{25}{16}\right)^{-1/2}$

44. $\left(\dfrac{16}{9}\right)^{-1/2}$ **45.** $\left(\dfrac{25}{16}\right)^{-3/2}$ **46.** $\left(\dfrac{16}{9}\right)^{-3/2}$

47. $\left(-\dfrac{1}{27}\right)^{-5/3}$ **48.** $\left(-\dfrac{1}{8}\right)^{-5/3}$

Use a calculator to evaluate each expression. Round answers to two decimal places.

49. $7^{0.39}$ **50.** $5^{0.47}$ **51.** $8^{2.7}$ **52.** $5^{3.9}$

Use a graphing calculator to evaluate each expression.

53. $(-8)^{7/3}$ **54.** $(-8)^{5/3}$ **55.** $\left[\left(\dfrac{5}{2}\right)^{-1}\right]^{-2}$

56. $\left[\left(\dfrac{3}{2}\right)^{-2}\right]^{-1}$ **57.** $[(4)^{-1}]^{0.5}$ **58.** $[(0.25)^{-1}]^{0.5}$

59. $(0.4^{-7})^{-1/7}$ **60.** $[(0.5^{-1})^{-2}]^{-3}$

61. $[(0.1)^{0.1}]^{0.1}$ **62.** $\left(1 + \dfrac{1}{1000}\right)^{1000}$

63. $\left(1 - \dfrac{1}{1000}\right)^{-1000}$ **64.** $(1 + 10^{-6})^{10^6}$

Simplify.

65. $(x^3 \cdot x^2)^2$ **66.** $(x^4 \cdot x^3)^2$ **67.** $[z^2(z \cdot z^2)^2 z]^3$

68. $[z(z^3 \cdot z)^2 z^2]^2$ **69.** $[(x^2)^2]^2$ **70.** $[(x^3)^3]^3$

71. $\dfrac{(ww^2)^3}{w^3 w}$ **72.** $\dfrac{(ww^3)^2}{w^3 w^2}$ **73.** $\dfrac{(5xy^4)^2}{25x^3 y^3}$

74. $\dfrac{(4x^3 y)^2}{8x^2 y^3}$ **75.** $\dfrac{(9xy^3 z)^2}{3(xyz)^2}$ **76.** $\dfrac{(5x^2 y^3 z)^2}{5(xyz)^2}$

77. $\dfrac{(2u^2 vw^3)^2}{4(uw^2)^2}$ **78.** $\dfrac{(u^3 vw^2)^2}{9(u^2 w)^2}$

APPLIED EXERCISES

79–80. ALLOMETRY: Dinosaurs The study of size and shape is called "allometry," and many allometric relationships involve exponents that are fractions or decimals. For example, the body measurements of most four-legged animals, from mice to elephants, obey (approximately) the following power law:

$$\left(\begin{array}{c} \text{Average body} \\ \text{thickness} \end{array} \right) = 0.4 \,(\text{hip-to-shoulder length})^{3/2}$$

where body thickness is measured vertically and all measurements are in feet. Assuming that this same relationship held for dinosaurs, find the average body thickness of the following dinosaurs, whose hip-to-shoulder length can be measured from their skeletons:

79. Diplodocus, whose hip-to-shoulder length was 16 feet.

80. Triceratops, whose hip-to-shoulder length was 14 feet.

81–82. BUSINESS: The Rule of .6 Many chemical and refining companies use "the rule of point six" to estimate the cost of new equipment. According to this rule, if a piece of equipment (such as a storage tank) originally cost C dollars, then the cost of similar equipment that is x times as large will be approximately $x^{0.6} C$ dollars. For example, if the original equipment cost C dollars, then new equipment with twice the capacity of the old equipment $(x = 2)$ would cost $2^{0.6} C = 1.516C$ dollars—that is, about 1.5 times as much. Therefore, to increase capacity by 100% costs only about 50% more.*

* Although the rule of .6 is only a rough "rule of thumb," it can be somewhat justified on the basis that the equipment of such industries consists mainly of containers, and the cost of a container depends on its surface area (square units), which increases more slowly than its capacity (cubic units).

81. Use the rule of .6 to find how costs change if a company wants to quadruple $(x = 4)$ its capacity.

82. Use the rule of .6 to find how costs change if a company wants to triple $(x = 3)$ its capacity.

83–84. BUSINESS: The Rule of .6 (*continuation*) Use a graphing calculator to graph $y = x^{0.6}$, expressing y, the cost multiple for larger equipment, in terms of x, the size multiple. Use the viewing window [0, 5] by [0, 3].

83. By how much can a company multiply its capacity for twice the money? That is, find the value of x that satisfies $x^{0.6} = 2$. [*Hint:* Either use TRACE or find where $y_1 = x^{0.6}$ INTERSECTs $y_2 = 2$.]

84. Does the curve rise more steeply or less steeply as x increases? What does this mean about how rapidly cost increases as equipment size increases?

85–86. ALLOMETRY: Heart Rate It is well known that the hearts of smaller animals beat faster than the hearts of larger animals. The actual relationship is approximately

$$(\text{Heart rate}) = 250(\text{weight})^{-1/4}$$

where the heart rate is in beats per minute and the weight is in pounds. Use this relationship to estimate the heart rate of:

85. A 16-pound dog.

86. A 625-pound grizzly bear.

87–88. ALLOMETRY: Heart Rate (*continuation*) Use a graphing calculator to graph $y = 250x^{-0.25}$, which expresses y, heartbeats per minute, in terms of x, the animal's weight. Use the viewing window [0, 200] by [0, 150].

87. Notice that the curve decreases less steeply for larger values of x. Explain what this means about how rapidly heart rate decreases as body weight increases.

88. Evaluate this formula at your own weight, x, to find your predicted heart rate. Then take your pulse and see if the numbers (roughly) agree.

89–90. BUSINESS AND PSYCHOLOGY: Learning Curves in Airplane Production Recall (pages 26–27) that the learning curve for the production of Boeing 707 airplanes is $150n^{-0.322}$ (thousand work-hours). Find how many work-hours it took to build:

89. The 50th Boeing 707.

90. The 250th Boeing 707.

91. GENERAL: Richter Scale The Richter scale (developed by Charles Richter in 1935) is widely used to measure the strength of earthquakes. Every increase of 1 on the Richter scale corresponds to a 10-fold increase in ground motion. Therefore, an increase on the Richter scale from A to B means that ground motion increases by a factor of 10^{B-A} (for $B > A$). Find the increase in ground motion between the following earthquakes:

  a. The 1994 Northridge, California, earthquake, measuring 6.8 on the Richter scale, and the 1906 San Francisco earthquake, measuring 8.3. (The San Francisco earthquake resulted in 500 deaths and a 3-day fire that destroyed 4 square miles of San Francisco.)
  b. The 1995 Kobe (Japan) earthquake, measuring 7.2 on the Richter scale, and the 1933 Miyagi earthquake, measuring 8.1. (The Miyagi earthquake caused a 90-foot-high tsunami, or "tidal wave," that killed 3064 people. The death toll from the Kobe earthquake was more than 5000.)

92. GENERAL: Richter Scale (*continuation*) Every increase of 1 on the Richter scale corresponds to an approximately *30-fold* increase in *energy released*. Therefore, an increase on the Richter scale from A to B means that the energy released increases by a factor of 30^{B-A} (for $B > A$).

  a. Find the increase in *energy released* between the earthquakes in Exercise 91a.

  b. Find the increase in *energy released* between the earthquakes in Exercise 91b.

93–94. GENERAL: Waterfalls Water falling from a waterfall that is x feet high will hit the ground with speed $\frac{60}{11}x^{0.5}$ miles per hour (neglecting air resistance).

93. Find the speed of the water at the bottom of the highest waterfall in the world, Angel Falls in Venezuela (3281 feet high).

94. Find the speed of the water at the bottom of the highest waterfall in the United States, Ribbon Falls in Yosemite, California (1650 feet high).

95–96. ENVIRONMENTAL SCIENCE: Biodiversity It is well known that larger land areas can support larger numbers of species. According to one study,[*] multiplying the land area by a factor of x multiplies the number of species by a factor of $x^{0.239}$. Use a graphing calculator to graph $y = x^{0.239}$. Use the viewing window [0, 100] by [0, 4].

95. Find the multiple x for the land area that leads to *double* the number of species. That is, find the value of x such that $x^{0.239} = 2$. [*Hint:* Either use TRACE or find where $y_1 = x^{0.239}$ INTERSECTs $y_2 = 2$.]

96. Find the multiple x for the land area that leads to triple the number of species. That is, find the value of x such that $x^{0.239} = 3$. [*Hint:* Either use TRACE or find where $y_1 = x^{0.239}$ INTERSECTs $y_2 = 3$.]

97. BUSINESS: Learning Curves A manufacturer of supercomputers finds that the number of work-hours required to build the 1st, the 10th, the 20th, and the 30th supercomputers are as follows:

Supercomputer Number	Work-Hours Required
1	3200
10	1900
20	1400
30	1300

* Rober H. MacArthur and Edward O. Wilson, *The Theory of Island Biogeography* (Princeton University Press, 1967).

a. Enter these numbers into a graphing calcu-
lator and make a plot of the resulting points
(Supercomputer Number on the x-axis and
Work-Hours Required on the y-axis).

b. Have the calculator find the power regression formula for these data, fitting a curve
of the form $y = ax^b$ to the points. Enter
the results in y_1. Plot the points together
with the regression curve. Observe that the
curve fits the points rather well.

c. Evaluate y_1 at $x = 50$ to predict the number of work-hours required to build the
50th supercomputer.

98. GENERAL: Paper Stacking Suppose that
you take an ordinary piece of paper (about $\frac{1}{250}$

of an inch thick), cut it in half, stack the two
halves, cut them in two and stack the pieces,
and repeat this cutting and stacking operation
many times. Each operation doubles the
height of the stack, so that after a total of
25 such operations, the stack would be
$2^{25} \cdot \frac{1}{250} \cdot \frac{1}{12} \cdot \frac{1}{5280}$ miles high.

a. Evaluate this height.

b. Use a graphing calculator to make a TABLE
showing the height of the stack when the
number of operations is 15 or more. [*Hint:*
Use $y_1 = 2^x / (250 \cdot 12 \cdot 5280)$ for values of
x beginning at 15.] For what number of operations is the stack over 1 mile high? over
10 miles high? over 100 miles high?

1.3 FUNCTIONS

Introduction

In the previous section we saw that the time required to build a Boeing
707 airliner will vary, depending on the number that have already been
built. Mathematical relationships such as this, in which one number depends on another, are called *functions,* and are central to the study of
mathematics. In this section we define and give some applications of
functions.

Functions

A *function* is a rule or procedure for finding, from a given number, a new
number.* If the function is denoted by f and the given number by x, then
the resulting number is written $f(x)$ (read "f of x") and is called *the value of
the function f at x.* The set of numbers x for which a function f is defined is
called the *domain* of f, and the set of all resulting function values $f(x)$ is
called the *range* of f. For any x in the domain, $f(x)$ must be a *single* number.

Function

A *function* f is a rule that assigns to each number x in a set a number $f(x)$. The set of all allowable values of x is called the *domain,*
and the set of all values $f(x)$ for x in the domain is called the *range.*

---

* In this chapter the word "function" will mean *function of one variable.* In Chapter 4 we
will discuss functions of more than one variable.

For example, recording the temperature at a given location through-out a particular day would define a *temperature* function:

$$f(x) = \begin{pmatrix} \text{Temperature at} \\ \text{time } x \text{ hours} \end{pmatrix}$$

Domain would be [0, 24]

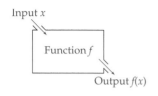

Input x

Function f

Output $f(x)$

A function f may be thought of as a numerical procedure or "machine" that takes an "input" number x and produces an "output" number $f(x)$, as shown on the left. The permissible input numbers form the do-main, and the resulting output numbers form the range.

We will be mostly concerned with functions that are defined by *formulas* for calculating $f(x)$ from x. If the domain of such a function is not stated, then it is always taken to be the *largest* set of numbers for which the function is defined, called the *natural domain* of the function. To *graph* a function f, we plot all points (x, y) such that x is in the do-main and $y = f(x)$. We call x the *independent variable* and y the *de-pendent variable*, since y *depends on* (is calculated from) x. The domain and range can be illustrated graphically.

The domain of a function $y = f(x)$ is the set of all allowable x-values, and the range is the set of all corresponding y-values.

Practice Problem 1 Find the domain and range of the function graphed below.

➤ Solution on page 44

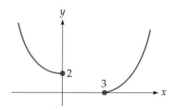

EXAMPLE **1**

FINDING THE DOMAIN AND RANGE

For the function $f(x) = \dfrac{1}{x-1}$, find:

a. $f(5)$ **b.** the domain   **c.** the range

Solution

a. $f(5) = \dfrac{1}{5-1} = \dfrac{1}{4}$

$f(x) = \dfrac{1}{x-1}$ with $x = 5$

$f(x) = \dfrac{1}{x-1}$ is defined for all x except $x = 1$.

b. Domain $= \{x \mid x \neq 1\}$

c. The graph of the function (from a graphing calculator) is shown on the right. From it, and realizing that the curve continues upward and downward (as may be verified by zooming out), it is clear that *every y value is taken except for* $y = 0$ (since the curve does not touch the x-axis). Therefore:

$f(x) = \dfrac{1}{x-1}$ on $[-4, 6]$ by $[-5, 5]$

$$\text{Range} = \{y \mid y \neq 0\}$$

May also be written $\{z \mid z \neq 0\}$ or with any other letter

---

The range could also be found by solving $y = \dfrac{1}{x-1}$ for x, giving $x = \dfrac{1}{y} + 1$, which again shows that y can take any value except 0.

EXAMPLE **2**

FINDING THE DOMAIN AND RANGE

For $f(x) = 2x^2 + 4x - 5$, determine:

a. $f(-3)$ **b.** the domain   **c.** the range

Solution

a. $f(-3) = 2(-3)^2 + 4(-3) - 5$
$\quad\quad\quad = 18 - 12 - 5 = 1$

$f(x) = 2x^2 + 4x - 5$ with each x replaced by -3

b. Domain $= \mathbb{R}$

$2x^2 + 4x - 5$ is defined for *all* real numbers

c. From the graph of $f(x) = 2x^2 + 4x - 5$ on the right, the lowest y value is -7 (as can be found from TRACE or MINIMUM), and all higher y values are taken (since the curve is a parabola opening upward). Therefore:

$$\text{Range} = \{y \mid y \geq -7\}$$

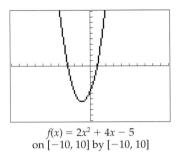

$f(x) = 2x^2 + 4x - 5$
on $[-10, 10]$ by $[-10, 10]$

Any letters may be used for defining a function or describing the domain and the range. For example, since the range $\{y \mid y \geq -7\}$ is a set of numbers, it could also be written $\{w \mid w \geq -7\}$ or, in interval notation without *any* variables, as $[-7, \infty)$.

Practice Problem 2   For $g(z) = \sqrt{z - 2}$, determine:

a. $g(27)$ **b.** the domain   **c.** the range   ➤ Solutions on page 44

For each x in the domain of a function there must be a *single* number $y = f(x)$, so the graph of a function cannot have two points (x, y) with the same x value but different y values. This leads to the following *graphical* test for functions.

Vertical Line Test for Functions

A curve in the Cartesian plane is the graph of a *function* if and only if no vertical line intersects the curve at more than one point.

EXAMPLE 3

USING THE VERTICAL LINE TEST

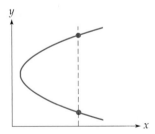

This is *not* the graph of a function of x because there is a vertical line (shown dashed) that intersects the curve twice.

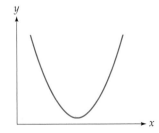

This *is* the graph of a function of x because no vertical line intersects the curve more than once.

A graph that has two or more points (x, y) with the same x-value but different y-values, such as the one on the left in Example 3, defines a *relation* rather than a function. We will be concerned exclusively with *functions,* and so we will use the terms "function," "graph," and "curve" interchangeably.

Functions can be classified into several types.

Linear Function

A *linear function* is a function that can be expressed in the form

$$f(x) = mx + b$$

with constants m and b. Its graph is a line with slope m and y-intercept b.

EXAMPLE 4

FINDING A COMPANY'S COST FUNCTION

An electronics company manufactures pocket calculators at a cost of $9 each, and the company's fixed costs (such as rent) amount to $400 per day. Find a function $C(x)$ that gives the total cost of producing x pocket calculators in a day.

Solution

Each calculator costs $9 to produce, so x calculators will cost $9x$ dollars, to which we must add the fixed costs of $400.

$$
\underbrace{C(x)}_{\substack{\text{Total} \\ \text{cost}}} = \underbrace{9x}_{\substack{\nearrow \quad \nwarrow \\ \text{Unit \quad Number} \\ \text{cost \quad of units}}} + \underbrace{400}_{\substack{\text{Fixed} \\ \text{cost}}}
$$

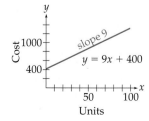

The graph of $C(x) = 9x + 400$ is a line with slope 9 and y-intercept 400, as shown on the left. Notice that the *slope* is the same as the *rate of change* of the cost (increasing at the rate of $9 per additional calculator), which is also the company's *marginal cost* (the cost of producing one more calculator is $9). For a linear cost function, the *slope,* the *rate of change,* and the *marginal cost* are always the same.

Practice Problem 3

A trucking company will deliver furniture for a charge of $25 plus 5% of the purchase price of the furniture. Find a function $D(x)$ that gives the delivery charge for a piece of furniture that costs x dollars.

➤ Solution on page 44

A mathematical description of a real-world situation is called a *mathematical model*. For example, the cost function $C(x) = 9x + 400$ from the previous example is a mathematical model for the cost of manufacturing calculators. In this model, x, the number of calculators, should take only whole-number values $(0, 1, 2, 3, \ldots)$, and the graph should consist of discrete dots rather than a continuous curve. Instead, we will find it easier to let x take *continuous* values, and round up or down as necessary at the end.

Quadratic Function

A *quadratic function* is a function that can be expressed in the form

$$f(x) = ax^2 + bx + c$$

with constants ("coefficients") $a \neq 0$, b, and c. Its graph is called a *parabola*.

The condition $a \neq 0$ keeps the function from becoming $f(x) = bx + c$, which would be linear. Many familiar curves are parabolas.

The center of gravity of a diver describes a parabola.

A stream of water from a hose takes the shape of a parabola.

The parabola $f(x) = ax^2 + bx + c$ opens *upward* if the constant a is *positive* and opens *downward* if the constant a is *negative*. The *vertex* of a parabola is its "central" point. The vertex is the *lowest* point on the parabola if it opens *up*, and the *highest* point if it opens *down*.

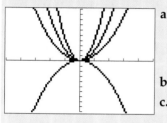

Graphing Calculator Exploration

a. Graph the parabolas $y_1 = x^2$, $y_2 = 2x^2$, and $y_3 = 4x^2$ on the graphing window $[-5, 5]$ by $[-10, 10]$. Use TRACE to identify which curve goes with which formula. How does the shape of the parabola change when the coefficient of x^2 increases?

b. Graph $y_4 = -x^2$. What did the negative sign do to the parabola?

c. Predict the shape of the parabolas $y_5 = -2x^2$ and $y_6 = \frac{1}{3}x^2$. Then check your predictions by graphing the functions.

The x-coordinate of the vertex of a parabola may be found by the following formula, which will be derived in Exercises 88–90 on page 49.

Vertex Formula for a Parabola

The vertex of the parabola $f(x) = ax^2 + bx + c$ has x-coordinate

$$x = \frac{-b}{2a}$$

For example, the vertex of the parabola $y = 2x^2 + 4x - 5$ has x-coordinate

$$x = \frac{-4}{2 \cdot 2} = \frac{-4}{4} = -1 \qquad \begin{array}{l} x = \dfrac{-b}{2a} \text{ with} \\ a = 2 \text{ and } b = 4 \end{array}$$

as we can see from the graph on page 34. The y-coordinate of the vertex comes from evaluating $y = 2x^2 + 4x - 5$ at this x-coordinate.

EXAMPLE **5** **GRAPHING A QUADRATIC FUNCTION**

Graph the quadratic function $f(x) = 2x^2 - 40x + 104$.

Solution

Graphing using a graphing calculator is largely a matter of finding an appropriate viewing window, as the following three unsatisfactory windows show.

on $[-10, 10]$ by $[-10, 10]$ on $[0, 20]$ by $[-10, 10]$ on $[0, 20]$ by $[-105, -95]$

To find an appropriate viewing window, we use the vertex formula:

$$x = \frac{-b}{2a} = \frac{-(-40)}{2 \cdot 2} = \frac{40}{4} = 10 \qquad \begin{array}{l} x\text{-coordinate of the vertex, from} \\ x = \dfrac{-b}{2a} \text{ with } a = 2 \text{ and } b = -40 \end{array}$$

We move a few units, say 5, to either side of $x = 10$, making the x-window $[5, 15]$. Using the calculator to EVALUATE the given function at $x = 10$ (or evaluating by hand) gives $y(10) = -96$. Since the parabola opens upward (the coefficient of x^2 is positive), the curve rises up from its vertex, so we select a y-interval from -96 upward, say $[-96, -70]$. Graphing the function on the window $[5, 15]$ by $[-96, -70]$ gives the following result. (Some other graphing windows are just as good.)

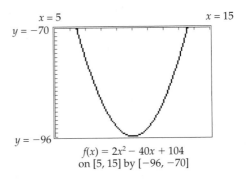

$f(x) = 2x^2 - 40x + 104$
on $[5, 15]$ by $[-96, -70]$

Solving Quadratic Equations

A value of x that solves an equation $f(x) = 0$ is called a *root* of the equation, or a *zero* of the function, or an *x-intercept* of the graph of $y = f(x)$. The roots of a quadratic equation can often be found by factoring.

EXAMPLE 6

SOLVING A QUADRATIC EQUATION BY FACTORING

Solve $2x^2 - 4x = 6$.

Solution

$$2x^2 - 4x - 6 = 0 \qquad \text{Subtracting 6 from each side to get zero on the right}$$

$$2(x^2 - 2x - 3) = 0 \qquad \text{Factoring out a 2}$$

$$2\underbrace{(x - 3)} \cdot \underbrace{(x + 1)} = 0 \qquad \text{Factoring } x^2 - 2x - 3$$

$$\begin{matrix} \text{Equals 0} & \text{Equals 0} \\ \text{at } x = 3 & \text{at } x = -1 \end{matrix} \qquad \text{Finding } x\text{-values that make each factor zero}$$

$$x = 3, \quad x = -1 \qquad \text{Solutions}$$

Graphing Calculator Exploration

```
Zero
X=-1        Y=0
```

Find the solutions to the equation in Example 6 by graphing the function $f(x) = 2x^2 - 4x - 6$ and using ZERO or TRACE to find where the curve crosses the x-axis. Your answers should agree with those found in Example 6.

Practice Problem 4

Solve by factoring or graphing: $9x - 3x^2 = -30$

➤ **Solution on page 44**

Quadratic equations can often be solved by the "Quadratic Formula." A derivation of this formula is given on pages 43–44.

Quadratic Formula

The solutions to $ax^2 + bx + c = 0$ are

$$x = \frac{-b \pm \sqrt{b^2 - 4ac}}{2a}$$

The "plus or minus" $\pm$ sign means calculate *both* ways, first using the + sign and then using the $-$ sign

---

In a business, it is often important to find a company's *break-even points*, the numbers of units of production where the company's costs are equal to its revenue.

EXAMPLE 7

FINDING BREAK-EVEN POINTS

A company that installs automobile compact disc (CD) players finds that if it installs x CD players per day, then its costs will be $C(x) = 120x + 4800$ and its revenue will be $R(x) = -2x^2 + 400x$ (both in dollars). Find the company's break-even points.

Solution

We set the cost and revenue functions equal to each other and solve.

$$120x + 4800 = -2x^2 + 400x \qquad \text{Setting } C(x) = R(x)$$

$$2x^2 - 280x + 4800 = 0 \qquad \text{Combining all terms on one side}$$

$$x = \frac{280 \pm \sqrt{(-280)^2 - 4 \cdot 2 \cdot 4800}}{2 \cdot 2} \qquad \text{Quadratic Formula with } a = 2, \; b = -280, \text{ and } c = 4800$$

$$= \frac{280 \pm \sqrt{40{,}000}}{4} = \frac{280 \pm 200}{4}$$

Working out the formula on a calculator

$$= \frac{480}{4} \text{ or } \frac{80}{4} = 120 \text{ or } 20$$

The company will break even when it installs either 20 or 120 CD players (as shown in the graph on the left).

Although it is important for a company to know where its break-even points are, most companies want to do better than break even—they want to maximize their profit. Profit is defined as *revenue minus cost* (since profit is what is left over after subtracting expenses from income).

Profit

$$\text{Profit} = \text{Revenue} - \text{Cost}$$

EXAMPLE **8**

MAXIMIZING PROFIT

For the CD installer whose daily revenue and cost functions were given in Example 7, find the number of units that maximizes profit, and the maximum profit.

Solution

The profit function is the revenue function minus the cost function.

$$P(x) = \underbrace{-2x^2 + 400x}_{R(x)} - \underbrace{(120x + 4800)}_{C(x)} \qquad \begin{aligned} P(x) &= R(x) - C(x) \ \text{ with} \\ R(x) &= -2x^2 + 400x \ \text{ and} \\ C(x) &= 120x + 4800 \end{aligned}$$

$$= -2x^2 + 280x - 4800 \qquad \text{Simplifying}$$

Since this function represents a parabola opening downward (because of the -2), it is maximized at its vertex, which is found using the vertex formula.

$$x = \frac{-280}{2(-2)} = \frac{-280}{-4} = 70 \qquad \begin{aligned} x &= \frac{-b}{2a} \ \text{ with} \\ a &= -2 \ \text{ and } \ b = 280 \end{aligned}$$

Thus, profit is maximized when 70 CD players are installed. For the maximum profit, we substitute $x = 70$ into the profit function:

$$P(70) = -2(70)^2 + 280\cdot70 - 4800 \qquad \begin{aligned} P(x) &= -2x^2 + 280x - 4800 \\ \text{with } \ x &= 70 \end{aligned}$$

$$= 5000 \qquad \text{Multiplying and combining}$$

Therefore, the company will maximize its profit when it installs 70 CD players per day. Its maximum profit will be $5000 per day.

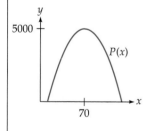

Why doesn't a company make more profit the more it sells? Because to increase its sales it must lower its prices, which eventually leads to lower profits. The relationship among the cost, revenue, and profit functions can be seen graphically as follows.

Spreadsheet Exploration

The following spreadsheet* shows the graphs of the functions $R(x)$, $C(x)$, and $P(x)$ from Examples 7 and 8. The values for x were made by entering **0** in **A3** and then copying the formula **=A3+10** for **A4** into cells **A5** through **A18**. The values for $R(x)$ were found by entering the formula **=-2*A3*A3+400*A3** in **B3** and then copying it into cells **B4** through **B18**. Then $C(x)$ was similarly found by starting with **C3** being **=120*A3+4800**. $P(x)$, the difference $R(x) - C(x)$, was found with **D3** being **=B3-C3**. The chart was made by plotting the values for the three columns corresponding to the revenue, cost, and profit.

	A	B	C	D	E	F	G	H	I	J
1	Number of units	Revenue	Cost	Profit						
2	x	$R(x)$	$C(x)$	$P(x)$						
3	0	0	4800	-4800						
4	10	3800	6000	-2200						
5	20	7200	7200	0						
6	30	10200	8400	1800						
7	40	12800	9600	3200						
8	50	15000	10800	4200						
9	60	16800	12000	4800						
10	70	18200	13200	5000						
11	80	19200	14400	4800						
12	90	19800	15600	4200						
13	100	20000	16800	3200						
14	110	19800	18000	1800						
15	120	19200	19200	0						
16	130	18200	20400	-2200						
17	140	16800	21600	-4800						
18	150	15000	22800	-7800						

Notice that the break-even points from Example 7 ($x = 20$ and $x = 120$) correspond to a profit of zero, and that the maximum profit (at $x = 70$) occurs halfway between the two break-even points.

Not all quadratic equations have (real) solutions.

* See the Preface for how to obtain this and other Excel spreadsheets.

EXAMPLE **9**

USING THE QUADRATIC FORMULA

Solve $\frac{1}{2}x^2 - 3x + 5 = 0$.

Solution

The Quadratic Formula with $a = \frac{1}{2}$, $b = -3$, and $c = 5$ gives

$$x = \frac{3 \pm \sqrt{9 - 4(\frac{1}{2})(5)}}{2(\frac{1}{2})} = \frac{3 \pm \sqrt{9 - 10}}{1} = 3 \pm \sqrt{-1} \quad \text{Undefined}$$

Therefore, the equation $\frac{1}{2}x^2 - 3x + 5 = 0$ *has no real solutions* (because of the undefined $\sqrt{-1}$). The geometrical reason that there are no solutions can be seen in the graph on the left: The curve never reaches the x-axis, so the function never equals zero.

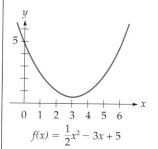

$f(x) = \frac{1}{2}x^2 - 3x + 5$

The quantity $b^2 - 4ac$, whose square root appears in the Quadratic Formula, is called the *discriminant*. If the discriminant is *positive* (as in Example 7), the equation $ax^2 + bx + c = 0$ has *two* solutions (since the square root is added and subtracted). If the discriminant is *zero*, there is only *one* root (since adding and subtracting zero gives the same answer). If the discriminant is *negative* (as in Example 9), then the equation has *no* real roots. Therefore, the discriminant being positive, zero, or negative corresponds to the parabola meeting the x-axis at 2, 1, or 0 points, as shown below.

two real roots

one real root

no real roots

Derivation of the Quadratic Formula

$$ax^2 + bx + c = 0 \qquad \text{The quadratic set equal to zero}$$
$$ax^2 + bx = -c \qquad \text{Subtracting } c$$
$$4a^2x^2 + 4abx = -4ac \qquad \text{Multiplying by } 4a$$

$$4a^2x^2 + 4abx + b^2 = b^2 - 4ac \qquad \text{Adding } b^2$$

$$(2ax + b)^2 = b^2 - 4ac \qquad \text{Since } 4a^2x^2 + 4abx + b^2 = (2ax + b)^2$$

$$2ax + b = \pm \sqrt{b^2 - 4ac} \qquad \text{Taking square roots}$$

$$2ax = -b \pm \sqrt{b^2 - 4ac} \qquad \text{Subtracting } b$$

$$x = \frac{-b \pm \sqrt{b^2 - 4ac}}{2a} \qquad \text{Dividing by } 2a \text{ gives the Quadratic Formula}$$

1.3 Section Summary

In this section we defined and gave examples of *functions*, and saw how to find their domains and ranges. The most important characteristic of a function f is that for any given "input" number x in the domain, there is exactly one "output" number $f(x)$. This requirement is stated geometrically in the *vertical line test*, that no vertical line can intersect the graph of a function at more than one point. We then defined *linear functions* (whose graphs are lines) and *quadratic functions* (whose graphs are parabolas). We solved quadratic equations by factoring, graphing, and using the Quadratic Formula. We maximized and minimized quadratic functions using the vertex formula.

➤ **Solutions to Practice Problems**

1. Domain: $\{x \mid x \leq 0 \text{ or } x \geq 3\}$; Range: $\{y \mid y \geq 0\}$

2. a. $g(27) = \sqrt{27 - 2} = \sqrt{25} = 5$

    b. Domain: $\{z \mid z \geq 2\}$

    c. Range: $\{y \mid y \geq 0\}$

$y_1 = \sqrt{x - 2}$ on $[-1, 10]$ by $[-1, 10]$

3. $D(x) = 25 + 0.05x$

4. $9x - 3x^2 = -30$
$$-3x^2 + 9x + 30 = 0$$
$$-3(x^2 - 3x - 10) = 0$$
$$-3(x - 5)(x + 2) = 0$$
$$x = 5, \ x = -2 \qquad \text{or from:}$$

1.3 Exercises

Determine whether each graph defines a function of x.

1.

2.

3.

4.

5.

6.

7.

8.

Find the domain and range of each function from its graph.

9.

10.

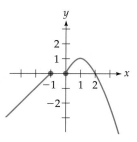

11–22. For each function:

a. Evaluate the given expression.

b. Find the domain of the function.

c. Find the range.
    [*Hint:* Use a graphing calculator.]

11. $f(x) = \sqrt{x - 1}$; find $f(10)$

12. $f(x) = \sqrt{x - 4}$; find $f(40)$

13. $h(z) = \dfrac{1}{z + 4}$; find $h(-5)$

14. $h(z) = \dfrac{1}{z + 7}$; find $h(-8)$

15. $h(x) = x^{1/4}$; find $h(81)$

16. $h(x) = x^{1/6}$; find $h(64)$

17. $h(x) = x^{2/3}$; find $f(-8)$

[*Hint for Exercises 17 and 18:* You may need to enter $x^{m/n}$ as $(x^m)^{1/n}$ or as $(x^{1/n})^m$, as discussed on page 25.]

18. $f(x) = x^{4/5}$; find $f(-32)$

19. $f(x) = \sqrt{4 - x^2}$; find $f(0)$

20. $f(x) = \dfrac{1}{\sqrt{x}}$; find $f(4)$

21. $f(x) = \sqrt{-x}$; find $f(-25)$

22. $f(x) = -\sqrt{-x}$; find $f(-100)$

23–30. Graph each function "by hand." [*Note:* Even if you have a graphing calculator, it is important to be able to sketch simple curves by finding a few important points.]

23. $f(x) = 3x - 2$ **24.** $f(x) = 2x - 3$

25. $f(x) = -x + 1$ **26.** $f(x) = -3x + 5$

27. $f(x) = 2x^2 + 4x - 16$

28. $f(x) = 3x^2 - 6x - 9$

29. $f(x) = -3x^2 + 6x + 9$

30. $f(x) = -2x^2 + 4x + 16$

 31–34. For each quadratic function:
a. Find the vertex using the vertex formula.
b. Graph the function on an appropriate viewing window. (Answers may differ.)

31. $f(x) = x^2 - 40x + 500$

32. $f(x) = x^2 + 40x + 500$

33. $f(x) = -x^2 - 80x - 1800$

34. $f(x) = -x^2 + 80x - 1800$

35–52. Solve each equation by factoring or the Quadratic Formula, as appropriate.

35. $x^2 - 6x - 7 = 0$ 36. $x^2 - x - 20 = 0$

37. $x^2 + 2x = 15$ 38. $x^2 - 3x = 54$

39. $2x^2 + 40 = 18x$ 40. $3x^2 + 18 = 15x$

41. $5x^2 - 50x = 0$ 42. $3x^2 - 36x = 0$

43. $2x^2 - 50 = 0$ 44. $3x^2 - 27 = 0$

45. $4x^2 + 24x + 40 = 4$ 46. $3x^2 - 6x + 9 = 6$

47. $-4x^2 + 12x = 8$ 48. $-3x^2 + 6x = -24$

49. $2x^2 - 12x + 20 = 0$ 50. $2x^2 - 8x + 10 = 0$

51. $3x^2 + 12 = 0$ 52. $5x^2 + 20 = 0$

53–62. Solve each equation using a graphing calculator. [*Hint:* Begin with the viewing window $[-10, 10]$ by $[-10, 10]$ or another of your choice (see Useful Hint in Graphing Calculator Terminology following the Preface) and use ZERO, SOLVE, or TRACE and ZOOM IN.] (In Exercises 61 and 62, round answers to two decimal places.)

53. $x^2 - x - 20 = 0$ 54. $x^2 + 2x - 15 = 0$

55. $2x^2 + 40 = 18x$ 56. $3x^2 + 18 = 15x$

57. $4x^2 + 24x + 45 = 9$ 58. $3x^2 - 6x + 5 = 2$

59. $3x^2 + 7x + 12 = 0$ 60. $5x^2 + 14x + 20 = 0$

61. $2x^2 + 3x - 6 = 0$ 62. $3x^2 + 5x - 7 = 0$

 63. Use your graphing calculator to graph the following four equations simultaneously on the viewing window $[-10, 10]$ by $[-10, 10]$:

$$y_1 = 2x + 6$$
$$y_2 = 2x + 2$$
$$y_3 = 2x - 2$$
$$y_4 = 2x - 6$$

a. What do the lines have in common and how do they differ?
b. Write the equation of another line with the same slope that lies 2 units below the lowest line. Then check your answer by graphing it with the others.

64. Use your graphing calculator to graph the following four equations simultaneously on the viewing window $[-10, 10]$ by $[-10, 10]$:

$$y_1 = 3x + 4$$
$$y_2 = 1x + 4$$
$$y_3 = -1x + 4 \quad \text{(Use } (-) \text{ to get } -1x.)$$
$$y_4 = -3x + 4$$

a. What do the lines have in common and how do they differ?
b. Write the equation of a line through this y-intercept with slope $\frac{1}{2}$. Then check your answer by graphing it with the others.

APPLIED EXERCISES

65. **BUSINESS: Cost Functions** A lumberyard will deliver wood for $4 per board foot plus a delivery charge of $20. Find a function $C(x)$ for the cost of having x board feet of lumber delivered.

66. **BUSINESS: Cost Functions** A company manufactures bicycles at a cost of $55 each. If the company's fixed costs are $900, express the company's costs as a linear function of x, the number of bicycles produced.

67. BUSINESS: Salary An employee's weekly salary is $500 plus $15 per hour of overtime. Find a function $P(x)$ giving his pay for a week in which he worked x hours of overtime.

68. BUSINESS: Salary A sales clerk's weekly salary is $300 plus 2% of her total week's sales. Find a function $P(x)$ for her pay for a week in which she sold x dollars of merchandise.

69. GENERAL: Water Pressure At a depth of d feet underwater, the water pressure is $p(d) = 0.45d + 15$ pounds per square inch. Find the pressure at:

a. The bottom of a 6-foot-deep swimming pool.
b. The maximum ocean depth of 35,000 feet.

70. GENERAL: Boiling Point At higher altitudes, water boils at lower temperatures. This is why at high altitudes foods must be boiled for longer times—the lower boiling point imparts less heat to the food. At an altitude of h thousand feet above sea level, water boils at a temperature of $B(h) = -1.8h + 212$ degrees Fahrenheit. Find the altitude at which water boils at 98.6 degrees Fahrenheit. (Your answer will show that at a high enough altitude, water boils at normal body temperature. This is why airplane cabins must be pressurized—at high enough altitudes one's blood would boil.)

71–72. GENERAL: Stopping Distance According to data from the National Transportation Safety Board, a car traveling at speed v miles per hour should be able to come to a full stop in a distance of

$$D(v) = 0.055v^2 + 1.1v \quad \text{feet}$$

Find the stopping distance required for a car traveling at:

71. 40 mph.   **72.** 60 mph.

73. BIOMEDICAL: Cell Growth The number of cells in a culture after t days is given by $N(t) = 200 + 50t^2$. Find the size of the culture after:

a. 2 days.   **b.** 10 days.

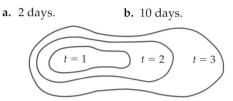

74. GENERAL: Juggling If you toss a ball h feet straight up, it will return to your hand after $T(h) = 0.5\sqrt{h}$ seconds. This leads to the *juggler's dilemma*: Juggling more balls means tossing them higher. However, the square root in the above formula means that tossing them twice as high does not gain twice as much time, but only $\sqrt{2} \approx 1.4$ times as much time. Because of this, there is a limit to the number of balls that a person can juggle, which seems to be about ten. Use this formula to find:

a. How long will a ball spend in the air if it is tossed to a height of 4 feet? 8 feet?
b. How high must it be tossed to spend 2 seconds in the air? 3 seconds in the air?

75. GENERAL: Impact Velocity If a marble is dropped from a height of x feet, it will hit the ground with velocity $v(x) = \frac{60}{11}\sqrt{x}$ miles per hour (neglecting air resistance). Use this formula to find the velocity with which a marble will strike the ground if it is dropped from the top of the tallest building in the United States, the 1454-foot Sears Tower in Chicago.

76. GENERAL: Tsunamis The speed of a tsunami (popularly known as a tidal wave, although it has nothing whatever to do with tides) depends on the depth of the water through which it is traveling. At a depth of d feet, the speed of a tsunami will be $s(d) = 3.86\sqrt{d}$ miles per hour. Find the speed of a tsunami in the Pacific basin where the average depth is 15,000 feet.

 77–78. GENERAL: Impact Time of a Projectile If an object is thrown upward so that its height (in feet) above the ground t seconds after it is thrown is given by the function $h(t)$, find when the object hits the ground. That is, find the positive value of t

such that $h(t) = 0$. Give the answer correct to two decimal places. [*Hint:* Enter the function in terms of x rather than t. Use the ZERO operation, or TRACE and ZOOM IN, or similar operations.]

77. $h(t) = -16t^2 + 45t + 5$

78. $h(t) = -16t^2 + 40t + 4$

79. BUSINESS: Break-Even Points and Maximum Profit A company that produces tracking devices for computer disk drives finds that if it produces x devices per week, its costs will be $C(x) = 180x + 16{,}000$ and its revenue will be $R(x) = -2x^2 + 660x$ (both in dollars).

  a. Find the company's break-even points.
  b. Find the number of devices that will maximize profit, and the maximum profit.

80. BUSINESS: Break-Even Points and Maximum Profit City and Country Cycles finds that if it sells x racing bicycles per month, its costs will be $C(x) = 420x + 72{,}000$ and its revenue will be $R(x) = -3x^2 + 1800x$ (both in dollars).

  a. Find the store's break-even points.
  b. Find the number of bicycles that will maximize profit, and the maximum profit.

81. BUSINESS: Break-Even Points and Maximum Profit A sporting goods store finds that if it sells x exercise machines per day, its costs will be $C(x) = 100x + 3200$ and its revenue will be $R(x) = -2x^2 + 300x$ (both in dollars).

  a. Find the store's break-even points.
  b. Find the number of sales that will maximize profit, and the maximum profit.

82. BUSINESS: Break-Even Points and Maximum Profit A company that installs car alarm systems finds that if it installs x systems per week, its costs will be $C(x) = 210x + 72{,}000$ and its revenue will be $R(x) = -3x^2 + 1230x$ (both in dollars).

  a. Find the company's break-even points.
  b. Find the number of installations that will maximize profit, and the maximum profit.

83. BIOMEDICAL: Muscle Contraction The fundamental equation of muscle contraction is of the form $(w + a)(v + b) = c$, where w is

the weight placed on the muscle, v is the velocity of contraction of the muscle, and a, b, and c are constants that depend upon the muscle and the units of measurement. Solve this equation for v as a function of w, a, b, and c.

84. GENERAL: Longevity According to insurance data, when a person reaches age 65, the probability of living for another x decades is approximated by the function $f(x) = -0.077x^2 - 0.057x + 1$ (for $0 \leq x \leq 3$). Find the probability that such a person will live for another:

  a. one decade
  b. two decades
  c. three decades

85. BUSINESS: Sales The following table gives a company's annual sales (in millions of units) at the ends of its first through fourth years.

Year	Sales] (millions)
1	3.8
2	3.6
3	3.7
4	4.0

  a. Enter the numbers from the table into a graphing calculator and make a plot of the resulting points (Year on the x-axis and Sales on the y-axis).
  b. Have your calculator find the quadratic (parabolic) regression formula for these data. Then enter the result in y_1, which gives a formula for sales each year. Plot the points together with the regression line.
  c. Predict the sales at the end of year 5 by evaluating $y_1(5)$.

Explorations and Excursions

The following problems extend and augment the material presented in the text.

The Vertex of a Parabola

86. a. Sketch $y = x^2 + 1$ and explain why 1 is the smallest possible value for y. [*Hint:* Can x^2 ever be negative?]
  b. Sketch $y = -x^2 + 1$ and explain why 1 is the largest possible value for y.

87. a. Sketch $y = (x - 1)^2 + 2$ and explain why 2 is the smallest possible value for y. [*Hint:* Can $(x - 1)^2$ ever be negative?]
   b. Sketch $y = (x + 2)^2 - 1$ and explain why -1 is the smallest possible value for y.

88. Verify that

$$a\left(x + \frac{b}{2a}\right)^2 + c - \frac{b^2}{4a} = ax^2 + bx + c$$

by "multiplying out" the expression on the left and collecting "like terms."

89. a. For $a > 0$, show that the smallest possible value of $f(x) = a\left(x + \frac{b}{2a}\right)^2 + c - \frac{b^2}{4a}$ is obtained when $x = \dfrac{-b}{2a}$.

   [*Hint:* Can $a\left(x + \dfrac{b}{2a}\right)^2$ ever be negative? When is it zero?]

   b. For $a < 0$, show that the largest possible value of $f(x) = a\left(x + \dfrac{b}{2a}\right)^2 + c - \dfrac{b^2}{4a}$ is obtained when $x = \dfrac{-b}{2a}$. [*Hint:* Can $a\left(x + \dfrac{b}{2a}\right)^2$ ever be positive? When is it zero?]

90. Vertex Formula Explain why Exercises 88 and 89 show that the vertex of the parabola $f(x) = ax^2 + bx + c$ occurs at $x = \dfrac{-b}{2a}$.
   [*Hint:* Recall that the vertex is the lowest point on the parabola if it opens up and is the highest point if the parabola opens down.]

FUNCTIONS, CONTINUED

Introduction

In this section we will define other useful types of functions and an important operation, the *composition* of two functions.

Polynomial Functions

A *polynomial function* (or simply a *polynomial*) is a function that can be written in the form

$$f(x) = a_n x^n + a_{n-1} x^{n-1} + \cdots + a_2 x^2 + a_1 x + a_0$$

where n is a nonnegative integer and $a_0, a_1, \ldots, a_n$ are (real) numbers, called *coefficients*. The *domain* of a polynomial is $\mathbb{R}$, the set of all (real) numbers. The *degree* of a polynomial is the highest power of the variable. The following are polynomials.

$f(x) = 2x^8 - 3x^7 + 4x^5 - 5$ A polynomial of degree 8 (since the highest power of x is 8)

$f(x) = -4x^2 - \frac{1}{3}x + 19$ A polynomial of degree 2 (a quadratic function)

$f(x) = x - 1$ A polynomial of degree 1 (a linear function)

$f(x) = 6$ A polynomial of degree 0 (a constant function)

Units produced

A cost function may increase at different rates at different production levels.

Polynomials are used to model many situations in which change occurs at different rates. For example, the polynomial in the graph on the left might represent the total cost of manufacturing x units of a product. At first, costs rise quite steeply as a result of high start-up expenses, then they rise more slowly as the economies of mass production come into play, and finally they rise more steeply as new production facilities need to be built.

Polynomial equations can often be solved by factoring (just as with quadratic equations).

EXAMPLE 1

SOLVING A POLYNOMIAL EQUATION BY FACTORING

Solve $3x^4 - 6x^3 = 24x^2$

Solution

$$3x^4 - 6x^3 - 24x^2 = 0$$ Rewritten with all the terms on the left side

$$3x^2(x^2 - 2x - 8) = 0$$ Factoring out $3x^2$

$$3x^2 \ (x - 4) \ (x + 2) = 0$$ Factoring further

Equals   Equals   Equals   Finding the zeros
zero at   zero at   zero at   of each factor
$x = 0$ $x = 4$ $x = -2$

$$x = 0, \quad x = 4, \quad x = -2$$ Solutions

As in this example, if a positive power of x can be factored out of a polynomial, then $x = 0$ is one of the roots.

Practice Problem 1 Solve $2x^3 - 4x^2 = 48x$ ➤ Solution on page 59

Rational Functions

The word "ratio" means fraction or quotient, and the quotient of two polynomials is called a *rational function*. The following are rational functions.

$$f(x) = \frac{4x^3 + 3x^2}{x^2 - 2x + 1} \qquad g(x) = \frac{1}{x^2 + 1}$$

A rational function is a polynomial over a polynomial

The domain of a rational function is the set of all numbers for which the denominator is not zero. For example, the domain of the function on the bottom left of the previous page is $\{x \mid x \neq 1\}$ (since $x = 1$ makes the denominator zero), and the domain of the function on the right is $\mathbb{R}$ (since $x^2 + 1$ is never zero).

Practice Problem 2   What is the domain of the rational function $f(x) = \dfrac{18}{(x + 2)(x - 4)}$?

➤ Solution on page 59

Simplifying a rational function by canceling a common factor from the numerator and the denominator can change the domain of the function, so that the "simplified" and "original" versions may not be equal (since they have different domains). For example, the rational function on the left below is not defined at $x = 1$, but the simplified version on the right *is* defined at $x = 1$, so that the two functions are technically not equal.

$$\underbrace{\frac{x^2 - 1}{x - 1}}_{\substack{\text{Not defined at } x = 1, \\ \text{so the domain is } \{x \mid x \neq 1\}}} = \frac{(x + 1)(x - 1)}{x - 1} \neq \underbrace{x + 1}_{\substack{\textit{Is} \text{ defined at } x = 1, \\ \text{so the domain is } \mathbb{R}}}$$

However, the functions *are* equal at every x-value *except* $x = 1$, and the graphs (shown below) are the same except that the rational function omits the point at $x = 1$.

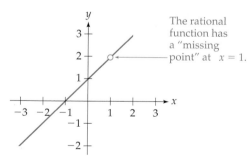

The rational function has a "missing point" at $x = 1$.

Graph of $y = \dfrac{x^2 - 1}{x - 1}$

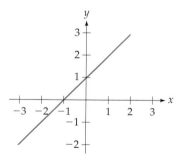

Graph of $y = x + 1$

Piecewise Linear Functions

The rule for calculating the values of a function may be given in several parts. If each part is linear, the function is called a *piecewise linear function*, and its graph consists of "pieces" of straight lines.

EXAMPLE 2

GRAPHING A PIECEWISE LINEAR FUNCTION

$$\text{Graph } f(x) = \begin{cases} 5 - 2x & \text{if } x \geq 2 \\ x + 3 & \text{if } x < 2 \end{cases}$$

This notation means: Use the top formula for $x \geq 2$ and the bottom formula for $x < 2$

Solution

We graph one "piece" at a time.

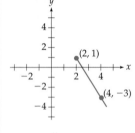

Step 1: To graph the first part, $f(x) = 5 - 2x$ if $x \geq 2$, we use the "endpoint" $x = 2$ and also $x = 4$ (or any other x-value satisfying $x \geq 2$). The points are (2, 1) and (4, −3), with the y-coordinates calculated from $f(x) = 5 - 2x$. Draw the line through these two points, but only for $x \geq 2$ (from $x = 2$ to the *right*), as shown on the left.

Step 2: For the second part, $f(x) = x + 3$ if $x < 2$, the restriction $x < 2$ means that the line ends just *before* $x = 2$. We mark this "missing point" (2, 5) by an "open circle" ○ to indicate that it is *not* included in the graph (the y-coordinate comes from $f(x) = x + 3$). For a second point, choose $x = 0$ (or any other $x < 2$), giving (0, 3). Draw the line through these two points, but only for $x < 2$ (to the *left* of $x = 2$), completing the graph of the function.

An important piecewise linear function is the *absolute value* function.

EXAMPLE 3

THE ABSOLUTE VALUE FUNCTION

The absolute value function is $f(x) = |x|$ defined as

$$f(x) = \begin{cases} x & \text{if } x \geq 0 \\ -x & \text{if } x < 0 \end{cases}$$

The second line, for *negative x*, attaches a *second* negative sign to make the result *positive*

For example, when applied to either 3 or -3, the function gives *positive* 3:

$$f(3) = 3$$

Using the top formula (since $3 \geq 0$)

$$f(-3) = -(-3) = 3$$

Using the bottom formula (since $-3 < 0$)

To graph the absolute value function, we may proceed as in Example 2, or simply observe that for $x \geq 0$, the function gives $y = x$ (a half-line from the origin with slope 1), and for $x < 0$, it gives $y = -x$ (a half-line on the other side of the origin with slope -1), as shown in the following graph.

Absolute Value Function

$$f(x) = |x| = \begin{cases} x & \text{if } x \geq 0 \\ -x & \text{if } x < 0 \end{cases}$$

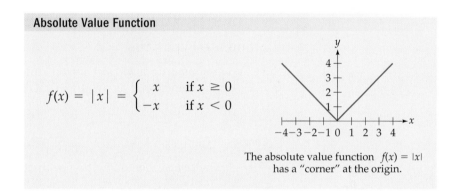

The absolute value function $f(x) = |x|$ has a "corner" at the origin.

Examples 2 and 3 show that the "pieces" of a piecewise linear function may or may not be connected.

EXAMPLE 4

GRAPHING AN INCOME TAX FUNCTION

Federal income taxes are "progressive," meaning that they take a higher percentage of higher incomes. For example, the 2001 federal income tax for a single taxpayer whose taxable income was no more than $136,750 was determined by a three-part rule: 15% of income up to $27,050, plus 27.5% of any amount over $27,050 up to $65,550, plus 30.5% of any amount over $65,550 up to $136,750. For an income of x dollars, the tax $f(x)$ may be expressed as follows:

$$f(x) = \begin{cases} 0.15x & \text{if } 0 \leq x \leq 27{,}050 \\ 4057.50 + 0.275(x - 27{,}050) & \text{if } 27{,}050 < x \leq 65{,}550 \\ 14{,}645 + 0.305(x - 65{,}550) & \text{if } 65{,}550 < x \leq 136{,}750 \end{cases}$$

Graphing this by the same technique as before leads to the following graph. The slopes 0.15, 0.275, and 0.305 are called the *marginal tax rates*.

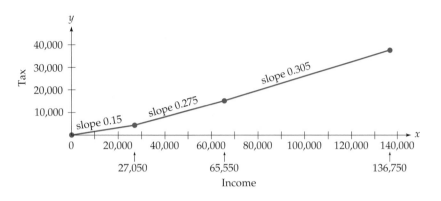

Composite Functions

Just as we substitute a *number* into a function, we may substitute a *function* into a function. For two functions f and g, evaluating f at $g(x)$ gives $f(g(x))$, called the *composition of f with g evaluated at x*.*

Composite Functions

The *composition* of f with g evaluated at x is $f(g(x))$.

The *domain* of $f(g(x))$ is the set of all numbers x in the domain of g such that $g(x)$ is in the domain of f. If we think of the functions f and g as "numerical machines," then the composition $f(g(x))$ may be thought of as a *combined* machine in which the output of g is connected to the input of f.

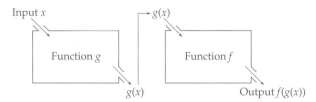

A "machine" for generating the composition of f with g.
A number x is fed into the function g, and the output
$g(x)$ is then fed into the function f, resulting in $f(g(x))$.

---

* The composition $f(g(x))$ may also be written $(f \circ g)(x)$, although we will not use this notation.

EXAMPLE **5**

FINDING A COMPOSITE FUNCTION

If $f(x) = x^7$ and $g(x) = x^3 - 2x$, find the composition $f(g(x))$.

Solution

$$f(g(x)) \quad = \quad \underbrace{[g(x)]^7}_{\substack{f(x) = x^7 \text{ with } x \\ \text{replaced by } g(x)}} \quad = \quad \underbrace{(x^3 - 2x)^7}_{\substack{\text{Using} \\ g(x) = x^3 - 2x}}$$

EXAMPLE **6**

FINDING BOTH COMPOSITE FUNCTIONS

If $f(x) = \dfrac{x + 8}{x - 1}$ and $g(x) = \sqrt{x}$, find $f(g(x))$ and $g(f(x))$.

Solution

$$f(g(x)) = \frac{g(x) + 8}{g(x) - 1} = \frac{\sqrt{x} + 8}{\sqrt{x} - 1} \qquad \begin{array}{l} f(x) = \dfrac{x + 8}{x - 1} \text{ with } x \\ \text{replaced by } g(x) = \sqrt{x} \end{array}$$

$$g(f(x)) = \sqrt{f(x)} = \sqrt{\frac{x + 8}{x - 1}} \qquad \begin{array}{l} g(x) = \sqrt{x} \text{ with } x \\ \text{replaced by } f(x) = \dfrac{x + 8}{x - 1} \end{array}$$

The order of composition is important: $f(g(x))$ is *not* the same as $g(f(x))$. To show this, we evaluate the above $f(g(x))$ and $g(f(x))$ at $x = 4$:

$$f(g(4)) = \frac{\sqrt{4} + 8}{\sqrt{4} - 1} = \frac{2 + 8}{2 - 1} = \frac{10}{1} = 10 \qquad \begin{array}{l} f(g(x)) = \dfrac{\sqrt{x} + 8}{\sqrt{x} - 1} \\ \text{at } x = 4 \end{array}$$

Different answers

$$g(f(4)) = \sqrt{\frac{4 + 8}{4 - 1}} = \sqrt{\frac{12}{3}} = \sqrt{4} = 2 \qquad \begin{array}{l} g(f(x)) = \sqrt{\dfrac{x + 8}{x - 1}} \\ \text{at } x = 4 \end{array}$$

Practice Problem 3   If $f(x) = x^2 + 1$ and $g(x) = \sqrt[3]{x}$, find: **a.** $f(g(x))$,   **b.** $g(f(x))$.

➤ Solutions on page 59

EXAMPLE **7** **PREDICTING WATER USAGE**

A planning commission estimates that if a city's population is p thousand people, its daily water usage will be $W(p) = 30p^{1.2}$ thousand gallons. The commission further predicts that the population in t years will be $p(t) = 60 + 2t$ thousand people. Express the water usage W as a function of t, the number of years from now, and find the water usage 10 years from now.

Solution

Water usage W as a function of t is the *composition* of $W(p)$ with $p(t)$:

$$W(p(t)) = 30[p(t)]^{1.2} = 30(60 + 2t)^{1.2} \qquad \begin{array}{l} W = 30p^{1.2} \text{ with } p \\ \text{replaced by } p(t) = 60 + 2t \end{array}$$

To find water usage in 10 years, we evaluate $W(p(t))$ at $t = 10$:

$$\begin{aligned} W(p(10)) &= 30(60 + 2 \cdot 10)^{1.2} \qquad & 30(60 + 2t)^{1.2} \text{ with } t = 10 \\ &= 30(80)^{1.2} \approx 5765 \qquad & \text{Using a calculator} \end{aligned}$$

Thousand gallons

Therefore, in 10 years the city will need about 5,765,000 gallons of water per day.

Shifts of Graphs

Sometimes the composition of two functions is just a horizontal or vertical shift of an original graph. This occurs when one of the functions is simply the addition or subtraction of a constant. The following diagrams show the graph of $y = x^2$ together with various shifts and the functions that generate them.

Vertical shifts

Horizontal shifts

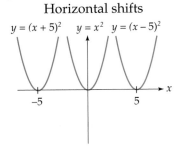

In general, adding to or subtracting from the *x-value* means a *horizontal* shift, while adding to or subtracting from the *function* means a *vertical* shift. These same ideas hold for *any* function: given the graph of

$y = f(x)$, adding or subtracting a positive number a to the function $f(x)$ or to the variable x shifts the graph as follows:

Shifts of Graphs

Function	Shift	
$y = f(x) + a$	shifted *up* by a units	} Vertical
$y = f(x) - a$	shifted *down* by a units	} shifts
$y = f(x + a)$	shifted *left* by a units	} Horizontal
$y = f(x - a)$	shifted *right* by a units	} shifts

Of course, a graph can be shifted both horizontally *and* vertically:

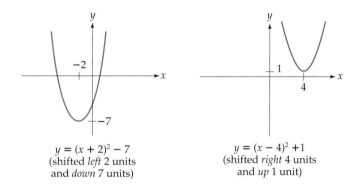

$y = (x + 2)^2 - 7$
(shifted *left* 2 units
and *down* 7 units)

$y = (x - 4)^2 + 1$
(shifted *right* 4 units
and *up* 1 unit)

Such double shifts can be applied to *any* function $y = f(x)$: the graph of $y = f(x + a) + b$ is shifted *left* a units and *up* b units (with the understanding that a *negative* a or b means that the direction is reversed).

Be careful: Remember that adding a *positive* number to x means a *left* shift.

Graphing Calculator Exploration

The absolute value function $y = |x|$ may be graphed on some graphing calculator as $y_1 = \text{ABS}(x)$.

a. Graph $y_1 = \text{ABS}(x - 2) - 6$ and observe that the absolute value function is shifted *right* 2 units and *down* 6 units. (The graph shown is drawn using ZOOM ZSquare.)

b. Predict the shift of $y_1 = \text{ABS}(x + 4) + 2$ and then verify your prediction by graphing the function on your calculator.

1.4 Section Summary

We have introduced a variety of functions: polynomials (which include linear and quadratic functions), rational functions, and piecewise linear functions. You should be able to identify these basic types of functions from their algebraic forms. We also added constants to perform horizontal and vertical *shifts* of graphs of functions, and combined functions by using the "output" of one as the "input" of the other, resulting in *composite* functions.

A Gallery of Functions

POLYNOMIALS

Linear function
$f(x) = mx + b$

Quadratic functions
$f(x) = ax^2 + bx + c$

$f(x) = ax^4 + bx^3 + cx^2 + dx + e$

RATIONAL FUNCTIONS

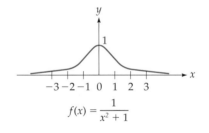

$$f(x) = \frac{1}{x - a}$$

$$f(x) = \frac{1}{x^2 + 1}$$

PIECEWISE LINEAR FUNCTIONS

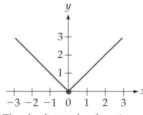

The absolute value function
$f(x) = |x|$

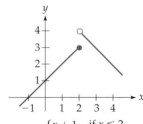

$$f(x) = \begin{cases} x + 1 & \text{if } x \le 2 \\ 6 - x & \text{if } x > 2 \end{cases}$$

➤ **Solutions to Practice Problems**

1. $2x^3 - 4x^2 - 48x = 0$

   $2x(x^2 - 2x - 24) = 0$

   $2x(x + 4)(x - 6) = 0$

   $x = 0, \quad x = -4, \quad x = 6$

2. $\{x \mid x \neq -2, x \neq 4\}$

3. **a.** $f(g(x)) = [g(x)]^2 + 1 = \left(\sqrt[3]{x}\right)^2 + 1$ or $x^{2/3} + 1$

   b. $g(f(x)) = \sqrt[3]{f(x)} = \sqrt[3]{x^2 + 1}$ or $(x^2 + 1)^{1/3}$

1.4   Exercises

Find the domain and range of each function graphed below.

1.

2.

For each function in Exercises 3–10:

  a. Evaluate the given expression.

  b. Find the domain of the function.

  c. Find the range. [*Hint:* Use a graphing calculator. You may have to ignore some false lines on the graph. Graphing in "dot mode" will also eliminate false lines.]

3. $f(x) = \dfrac{1}{x + 4}$;   find $f(-3)$

4. $f(x) = \dfrac{1}{(x - 1)^2}$;   find $f(-1)$

5. $f(x) = \dfrac{x^2}{x - 1}$;   find $f(-1)$

6. $f(x) = \dfrac{x^2}{x + 2}$;   find $f(2)$

7. $f(x) = \dfrac{12}{x(x + 4)}$;   find $f(2)$

8. $f(x) = \dfrac{16}{x(x - 4)}$;   find $f(-4)$

9. $g(x) = |x + 2|$;   find $g(-5)$

10. $g(x) = |x| + 2$;   find $g(-5)$

Solve each equation by factoring.

11. $x^5 + 2x^4 - 3x^3 = 0$

12. $x^6 - x^5 - 6x^4 = 0$

13. $5x^3 - 20x = 0$

14. $2x^5 - 50x^3 = 0$

15. $2x^3 + 18x = 12x^2$

16. $3x^4 + 12x^2 = 12x^3$

17. $6x^5 = 30x^4$

18. $5x^4 = 20x^3$

19. $3x^{5/2} - 6x^{3/2} = 9x^{1/2}$ [*Hint:* First factor out

20. $2x^{7/2} + 8x^{5/2} = 24x^{3/2}$ a fractional power.]

Solve each equation using a graphing calculator. (For Exercises 31 and 32, round answers to two decimal places.)

21. $x^3 - 2x^2 - 8x = 0$

22. $x^3 + 2x^2 - 8x = 0$

23. $x^5 - 2x^4 - 3x^3 = 0$

24. $x^6 + x^5 - 6x^4 = 0$

25. $2x^3 = 12x^2 - 18x$

26. $3x^4 + 12x^3 + 12x^2 = 0$

27. $6x^5 + 30x^4 = 0$

28. $5x^4 + 20x^3 = 0$

29. $2x^{5/2} + 4x^{3/2} = 6x^{1/2}$

30. $3x^{7/2} - 12x^{5/2} = 36x^{3/2}$

31. $x^5 - x^4 - 5x^3 = 0$ **32.** $x^6 + 2x^5 - 5x^4 = 0$

Graph each function.

33. $f(x) = |x - 3| - 3$ [*Hint:* Think of shifted graphs.]

34. $f(x) = |x + 2| - 2$

35. $f(x) = \begin{cases} 2x - 7 & \text{if } x \geq 4 \\ 2 - x & \text{if } x < 4 \end{cases}$

36. $f(x) = \begin{cases} 2 - x & \text{if } x \geq 3 \\ 2x - 4 & \text{if } x < 3 \end{cases}$

37. $f(x) = \begin{cases} 8 - 2x & \text{if } x \geq 2 \\ x + 2 & \text{if } x < 2 \end{cases}$

38. $f(x) = \begin{cases} 2x - 4 & \text{if } x > 3 \\ 5 - x & \text{if } x \leq 3 \end{cases}$

Identify each function as a polynomial, a rational function, a piecewise linear function, or none of these. (Do not graph the functions; just identify their types.)

39. $f(x) = x^5$

40. $f(x) = 3|x|$

41. $f(x) = 1 - |x|$

42. $f(x) = x^4$

43. $f(x) = x + 2$

44. $f(x) = \begin{cases} 3x - 1 & \text{if } x \geq 2 \\ 1 - x & \text{if } x < 2 \end{cases}$

45. $f(x) = \dfrac{1}{x + 2}$

46. $f(x) = x^2 + 9$

47. $f(x) = \begin{cases} x - 2 & \text{if } x < 3 \\ 7 - 4x & \text{if } x \geq 3 \end{cases}$

48. $f(x) = \dfrac{x}{x^2 + 9}$

49. $f(x) = 3x^2 - 2x$

50. $f(x) = x^3 - x^{2/3}$

51. $f(x) = x^2 + x^{1/2}$

52. $f(x) = 5$

53. For the functions $y_1 = 100x$, $y_2 = 10x^2$, $y_3 = x^3$:

  a. Predict which curve will be the highest for large values of x.

  b. Predict which curve will be the lowest for large values of x.

  c. Check your predictions by graphing the functions on the window $[0, 12]$ by $[0, 2000]$.

  d. From your graph, where do all these curves meet?

54. Graph the parabola $y_1 = 1 - x^2$ and the semicircle $y_2 = \sqrt{1 - x^2}$ on the graphing window $[-1, 1]$ by $[0, 1]$. Use ZOOM ZSQUARE to make the semicircle look more like a semicircle. Use TRACE to determine which is the "inside" curve (the parabola or the semicircle) and which is the "outside" curve. These graphs show that when you graph a parabola, you should draw the curve near the vertex to be slightly more "pointed" than a circular curve.

55. For any x, the function INT(x) is defined as the greatest integer less than or equal to x. For example, INT$(3.7) = 3$ and INT$(-4.2) = -5$.

  a. Use a graphing calculator to graph the function $y_1 = \text{INT}(x)$. (You may need to graph it in DOT mode to eliminate false connecting lines.)

  b. From your graph, what are the domain and range of this function?

56. **a.** Use a graphing calculator to graph the function $y_1 = 2\,\text{INT}(x)$. (See the previous exercise for a definition of INT(x).)

  b. From your graph, what are the domain and range of this function?

For each pair $f(x)$ and $g(x)$, find **a.** $f(g(x))$, **b.** $g(f(x))$.

57. $f(x) = x^5$; $g(x) = 7x - 1$

58. $f(x) = x^8$; $g(x) = 2x + 5$

59. $f(x) = \dfrac{1}{x}$; $g(x) = x^2 + 1$

60. $f(x) = \sqrt{x}$; $g(x) = x^3 - 1$

61. $f(x) = x^3 - x^2$; $g(x) = \sqrt{x} - 1$

62. $f(x) = x - \sqrt{x}$; $g(x) = x^2 + 1$

63. $f(x) = \dfrac{x^3 - 1}{x^3 + 1}$; $g(x) = x^2 - x$

64. $f(x) = \dfrac{x^4 + 1}{x^4 - 1}; \quad g(x) = x^3 + x$

65. a. Find the composition $f(g(x))$ of the two linear functions $f(x) = ax + b$ and $g(x) = cx + d$ (for constants a, b, c, and d).
  b. Is the composition of two linear functions always a linear function?

66. a. Is the composition of two quadratic functions always a quadratic function? [*Hint:* Find the composition of $f(x) = x^2$ and $g(x) = x^2$.]
  b. Is the composition of two polynomials always a polynomial?

67. Find, rounding to five decimal places:
  a. $\left(1 + \dfrac{1}{100}\right)^{100}$

  b. $\left(1 + \dfrac{1}{10{,}000}\right)^{10{,}000}$

  c. $\left(1 + \dfrac{1}{1{,}000{,}000}\right)^{1{,}000{,}000}$

  d. Do the resulting numbers seem to be approaching a limiting value? Estimate the limiting value to five decimal places. The number that you have approximated is denoted e. You will use it in the next section.

68. Use the TABLE feature of your graphing calculator to evaluate $\left(1 + \dfrac{1}{x}\right)^x$ for values of x like 100, 10,000, and 1,000,000 and higher. Do the resulting numbers seem to be approaching a limiting value? Estimate the limiting value to five decimal places. The number that you have approximated is denoted e. You will use it in the next section.

69. How will the graph of $y = (x + 3)^3 + 6$ differ from the graph of $y = x^3$? Check by graphing both functions together.

70. How will the graph of $y = -(x - 4)^2 + 8$ differ from the graph of $y = -x^2$? Check by graphing both functions together.

APPLIED EXERCISES

71. ECONOMICS: Income Tax The function below expresses an income tax that is 10% for incomes up to $5000, and otherwise is $500 plus 30% of income in excess of $5000.

$$f(x) = \begin{cases} 0.10x & \text{if } 0 \le x \le 5000 \\ 500 + 0.30(x - 5000) & \text{if } x > 5000 \end{cases}$$

  a. Calculate the tax on an income of $3000.
  b. Calculate the tax on an income of $5000.
  c. Calculate the tax on an income of $10,000.
  d. Graph the function.

72. ECONOMICS: Income Tax The following function expresses an income tax that is 15% for incomes up to $6000, and otherwise is $900 plus 40% of income in excess of $6000.

$$f(x) = \begin{cases} 0.15x & \text{if } 0 \le x \le 6000 \\ 900 + 0.40(x - 6000) & \text{if } x > 6000 \end{cases}$$

  a. Calculate the tax on an income of $3000.
  b. Calculate the tax on an income of $6000.
  c. Calculate the tax on an income of $10,000.
  d. Graph the function.

73–74. GENERAL: Dog Years The usual estimate that each human-year corresponds to 7 dog-years is not very accurate for young dogs, since they quickly reach adulthood. Exercises 73 and 74 give more accurate formulas for converting human-years x into dog-years. For each conversion formula:

  a. Find the number of dog-years corresponding to the following amounts of human time: 8 months, 1 year and 4 months, 4 years, 10 years.
  b. Graph the function.

73. $\quad f(x) = \begin{cases} 10.5x & \text{if } 0 \le x \le 2 \\ 21 + 4(x - 2) & \text{if } x > 2 \end{cases}$

(This function expresses dog-years as $10\frac{1}{2}$ dog-years per human-year for the first 2 years and then 4 dog-years per human-year for each year thereafter.)

74. $f(x) = \begin{cases} 15x & \text{if } 0 \leq x \leq 1 \\ 15 + 9(x - 1) & \text{if } 1 < x \leq 2 \\ 24 + 4(x - 2) & \text{if } x > 2 \end{cases}$

(This function expresses dog-years as 15 dog-years per human-year for the first year, 9 dog-years per human-year for the second year, and then 4 dog-years per human-year for each year thereafter.)

75. BUSINESS: Insurance Reserves An insurance company keeps reserves (money to pay claims) of $R(v) = 2v^{0.3}$, where v is the value of all of its policies, and the value of its policies is predicted to be $v(t) = 60 + 3t$, where t is the number of years from now. (Both R and v are in millions of dollars.) Express the reserves R as a function of t, and evaluate the function at $t = 10$.

76. BUSINESS: Research Expenditures An electronics company's research budget is $R(p) = 3p^{0.25}$, where p is the company's profit, and the profit is predicted to be $p(t) = 55 + 4t$, where t is the number of years from now. (Both R and p are in millions of dollars.) Express the research expenditure R as a function of t, and evaluate the function at $t = 5$.

77. GENERAL: Oil Imports and Fuel Efficiency The average fuel efficiency of all cars in America, called the *fleet mpg*, is 21.6 mpg. The amount of crude oil that would be saved if the fleet mpg were increased to a value of x is

$$S(x) = 3208 - \frac{69{,}300}{x} \quad \begin{array}{l} \text{million barrels} \\ \text{annually} \end{array}$$

a. Graph the function $S(x)$ on the viewing window [21.6, 40] by [0, 2000].
b. For what fleet mpg x will the savings reach 720 million barrels, which is the amount annually imported from the Middle East OPEC nations? [*Hint:* Either

ZOOM IN around the point at $y = 720$ or, better, use INTERSECT with $y_2 = 720$.] (*Note:* New cars average 29 mpg, but fleet mpg always lags. Ten manufacturers have already built and tested prototype cars getting from 67 to 138 mpg.)
(*Source:* Federal Highway Administration and the Rocky Mountain Institute.)

78. GENERAL: Long-Term Trends in Higher Education The following table gives the percentage of adults in the United States who were college graduates, for the years 1940 to 2000.

Years Since 1940	Percentage of College Graduates
1940 0	4
1960 20	9
1980 40	18
2000 60	26

a. Enter these data into a graphing calculator and make a plot of the resulting points (Years since 1940 on the x-axis and Percentage on the y-axis).
b. Have your calculator find the cubic (third-order polynomial) regression formula for these data. Then enter the result as y_1, which gives a formula for the percentage for each year. Plot the points together with the regression curve.
c. Use the regression formula to predict the percentage of college graduates in the year 2010 $(x = 70)$.
d. Use the regression formula to predict the percentage of college graduates in the year 2060 $(x = 120)$. The answer shows that polynomial predictions, if extended too far, give nonsensical results.

1.5 **EXPONENTIAL FUNCTIONS**

Introduction

We now introduce exponential functions, showing how they are used to model the processes of growth and decay.

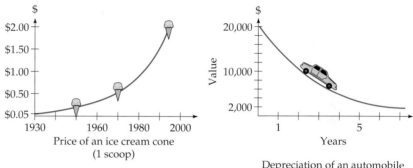

Price of an ice cream cone
(1 scoop)

Depreciation of an automobile

We will also define the very important mathematical constant e.

Exponential Functions

A function that has a variable in an exponent, such as $f(x) = 2^x$, is called an *exponential function*.

$$f(x) = 2^x$$

Exponent

Base

The table below shows some values of the exponential function $f(x) = 2^x$, and its graph (based on these points) is shown on the right.

x	$y = 2^x$
-3	$2^{-3} = \frac{1}{8}$
-2	$2^{-2} = \frac{1}{4}$
-1	$2^{-1} = \frac{1}{2}$
0	$2^0 = 1$
1	$2^1 = 2$
2	$2^2 = 4$
3	$2^3 = 8$

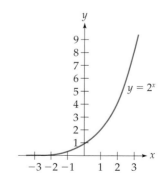

Domain of $y = 2^x$ is $\mathbb{R} = (-\infty, \infty)$ and range is $(0, \infty)$.

Clearly, the graph of the exponential function 2^x is quite different from the parabola x^2.

The exponential function $f(x) = \left(\frac{1}{2}\right)^x$ has base $\frac{1}{2}$. The following table shows some of its values, and its graph is shown to the right of the table. Notice that it is the mirror image of the curve $y = 2^x$.

x	$y = \left(\frac{1}{2}\right)^x$
-3	$\left(\frac{1}{2}\right)^{-3} = 8$
-2	$\left(\frac{1}{2}\right)^{-2} = 4$
-1	$\left(\frac{1}{2}\right)^{-1} = 2$
0	$\left(\frac{1}{2}\right)^{0} = 1$
1	$\left(\frac{1}{2}\right)^{1} = \frac{1}{2}$
2	$\left(\frac{1}{2}\right)^{2} = \frac{1}{4}$
3	$\left(\frac{1}{2}\right)^{3} = \frac{1}{8}$

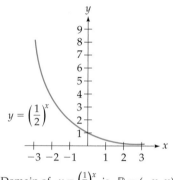

Domain of $y = \left(\frac{1}{2}\right)^x$ is $\mathbb{R} = (-\infty, \infty)$ and range is $(0, \infty)$.

We can define an exponential function $f(x) = a^x$ for any positive base a. We always take the base to be positive, so for the rest of this section *the letter a will stand for a positive constant.*

Exponential functions with bases $a > 1$ are used to model *growth*, as in populations or savings accounts, and exponential functions with bases $a < 1$ are used to model *decay*, as in depreciation. (For base $a = 1$, the graph is a horizontal line, since $1^x = 1$ for all x.)

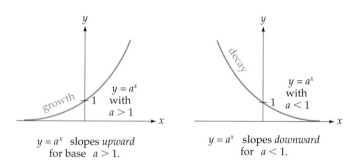

$y = a^x$ slopes *upward* for base $a > 1$.

$y = a^x$ slopes *downward* for $a < 1$.

Compound Interest

Money invested at compound interest grows exponentially. (The word "compound" means that the interest is added to the account, earning more interest.) Banks always state *annual* interest rates, but the compounding may be done more frequently. For example, if a bank offers 8% compounded quarterly, then each quarter you get 2% (one quarter of the annual 8%), so that 2% of your money is added to the account each quarter. If you begin with P dollars (the "principal"), at the end of the first quarter you would have P dollars plus 2% of P dollars:

$$\begin{pmatrix} \text{Value after} \\ \text{1 quarter} \end{pmatrix} = P + 0.02P = P \cdot (1 + 0.02)$$

Notice that increasing a quantity by 2% is the same as multiplying it by $(1 + 0.02)$. Since a year has 4 quarters, t years will have $4t$ quarters. Therefore, to find the value of your account after t years, we simply multiply the principal by $(1 + 0.02)$ a total of $4t$ times, obtaining:

$$\begin{pmatrix} \text{Value after} \\ t \text{ years} \end{pmatrix} = P \cdot \overbrace{(1 + 0.02) \cdot (1 + 0.02) \cdots (1 + 0.02)}^{4t \text{ times}}$$

$$= P \cdot (1 + 0.02)^{4t}$$

The 8%, which gave the $\frac{0.08}{4} = 0.02$ quarterly rate, can be replaced by any interest rate r (written in decimal form), and the 4 can be replaced by any number m of compounding periods per year, leading to the following general formula.

Compound Interest

For P dollars invested at interest rate r compounded m times a year,

$$\begin{pmatrix} \text{Value after} \\ t \text{ years} \end{pmatrix} = P \cdot \left(1 + \frac{r}{m}\right)^{mt}$$

r = annual rate
m = periods per year
t = number of years

For example, for monthly compounding we would use $m = 12$ and for daily compounding $m = 365$ (the number of days in the year).

EXAMPLE 1

FINDING A VALUE UNDER COMPOUND INTEREST

Find the value of $4000 invested for 2 years at 12% compounded quarterly.

Solution

$$4000 \cdot \left(1 + \underbrace{\frac{0.12}{4}}_{0.03}\right)^{4 \cdot 2} = 4000(1 + 0.03)^8$$

$P \cdot \left(1 + \frac{r}{m}\right)^{mt}$
with $P = 4000$,
$r = 0.12$, $m = 4$,
and $t = 2$

$$= 4000 \cdot 1.03^8 \approx 5067.08$$

Using a calculator

The value after two years will be $5067.08.

We may interpret the formula $P(1 + 0.03)^8$ intuitively as follows: multiplying the principal by $(1 + 0.03)$ means that you keep the original amount (the "1") plus some interest (the 0.03), and the exponent 8 means that this is done a total of 8 times.

Practice Problem 1

Find the value of $2000 invested for 3 years at 24% compounded monthly.

➤ Solution on page 73

Much more will be said about compound interest in Section 2.2 of Chapter 2.

Graphing Calculator Exploration

In the long run, which is more important—principal or interest rate? The graph shows the value of $1000 at 5% interest, together with the value of a mere $200 at the higher rate of 10% (both compounded annually). The fact that the (initially) lower curve eventually surpasses the higher one illustrates a general fact: the higher interest rate will eventually prevail, regardless of the size of the initial investment.

Find how soon the $200 at 10% will surpass the $1000 at 5% by graphing $y_1 = 1000(1 + 0.05)^x$ and $y_2 = 200(1 + 0.10)^x$ on the window $[0, 40]$ by $[-2000, 8000]$ and using INTERSECT. What if the higher rate is only 9%? (You will have to extend your window.)

The graph labels: $200 at 10%, $1000 at 5%

Depreciation by a Fixed Percentage

Depreciation by a fixed percentage means that a piece of equipment loses a fixed percentage (say 30%) of its value each year. Losing a percentage of value is like compound interest but with a *negative* interest rate. Therefore, we use the compound interest formula with $m = 1$ (since depreciation is annual) and with r being *negative*.

EXAMPLE 2 **DEPRECIATING AN ASSET**

A car worth $15,000 depreciates in value by 40% each year. How much is it worth after 3 years?

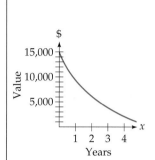

Solution

The car loses 40% of its value each year, which is equivalent to an annual interest rate of *negative* 40%. The compound interest formula gives

$$15{,}000(1 - 0.40)^3 = 15{,}000(0.60)^3 = \$3240$$

$P(1 + r/m)^{mt}$ with
$P = 15{,}000$, $r = -0.40$,
$m = 1$, and $t = 3$

Using a calculator

The exponential function $f(x) = 15{,}000(0.60)^x$, giving the value of the car after x years of depreciation, is graphed on the left.

Practice Problem 2

A printing press, originally worth $50,000, loses 20% of its value each year. What is its value after 4 years? ➤ **Solution on page 73**

The above graph shows that depreciation by a fixed percentage is quite different from "straight-line" depreciation. Under straight-line depreciation the same *dollar* value is lost each year, while under fixed-percentage depreciation the same *percentage* of value is lost each year, resulting in larger dollar losses in the early years and smaller dollar losses in later years. Depreciation by a fixed percentage (also called the "declining balance" method) is one type of "accelerated" depreciation. The method of depreciation that one uses depends on how one chooses to estimate value, and in practice is often determined by the tax laws.

The Number *e*

Imagine that a bank offers 100% interest, and that you deposit $1 for 1 year. Let us see how the value changes under different types of compounding.

For *annual* compounding, your $1 would in a year grow to $2 (the original dollar plus a dollar interest).

For *quarterly* compounding, we use the compound interest formula with $P = 1$, $r = 1$ (for 100%), $m = 4$, and $t = 1$:

$$1\left(1 + \frac{1}{4}\right)^{1 \cdot 4} = (1 + 0.25)^4 = (1.25)^4 \approx 2.44 \qquad P(1 + r/m)^{mt}$$

or 2 dollars and 44 cents, an improvement of 44 cents over annual compounding.

For *daily* compounding, the value after a year would be

$$\left(1 + \frac{1}{365}\right)^{365} \approx 2.71$$

$$m = 365 \text{ periods}$$
$$\frac{r}{m} = \frac{100\%}{365} = \frac{1}{365}$$

an increase of 27 cents over quarterly compounding. Clearly, if the interest rate, the principal, and the amount of time stay the same, the value increases as the compounding is done more frequently.

In general, if the compounding is done m times a year, the value of the dollar after a year will be

$$\left(\begin{array}{c}\text{Value of \$1 after 1 year at 100\%}\\ \text{interest compounded } m \text{ times a year}\end{array}\right) = \left(1 + \frac{1}{m}\right)^{m}$$

The following table shows the value of $\left(1 + \dfrac{1}{m}\right)^{m}$ for various values of m.

Value of \$1 at 100% Interest Compounded _m_ Times a Year for 1 Year			
m	$\left(1 + \dfrac{1}{m}\right)^{m}$	*Answer (rounded)*	
1	$\left(1 + \dfrac{1}{1}\right)^{1}$	$= 2.00000$	Annual compounding
4	$\left(1 + \dfrac{1}{4}\right)^{4}$	≈ 2.44141	Quarterly compounding
365	$\left(1 + \dfrac{1}{365}\right)^{365}$	≈ 2.71457	Daily compounding
10,000	$\left(1 + \dfrac{1}{10,000}\right)^{10,000}$	≈ 2.71815	
100,000	$\left(1 + \dfrac{1}{100,000}\right)^{100,000}$	≈ 2.71827	
1,000,000	$\left(1 + \dfrac{1}{1,000,000}\right)^{1,000,000}$	≈ 2.71828	
10,000,000	$\left(1 + \dfrac{1}{10,000,000}\right)^{10,000,000}$	≈ 2.71828	

Answers agree to five decimal places

Notice that as m increases, the values in the right-hand column seem to be settling down to a definite value, approximately 2.71828. That is, as m approaches infinity, the limit of $\left(1 + \dfrac{1}{m}\right)^m$ is approximately 2.71828. This particular number is very important in mathematics, and is given the name e (just as 3.14159... is given the name π). In the following definition, we use the letter n to state the definition in its traditional form.

The Constant e

$$e = \lim_{n \to \infty}\left(1 + \frac{1}{n}\right)^n = 2.71828\ldots$$

$n \to \infty$ is read "n approaches infinity." The dots mean that the decimal expansion goes on forever

---

The same e appears in probability and statistics in the formula for the "bell-shaped" or "normal" curve (see page 461). Its value has been calculated to many thousands of decimal places, and its value to 15 decimal places is $e \approx 2.718281828459045$.

Continuous Compounding of Interest

This kind of compound interest, the limit as the compounding frequency approaches infinity, is called *continuous* compounding. We have shown that $1 at 100% interest compounded continuously for 1 year would be worth precisely e dollars (about $2.72). The formula for continuous compound interest at other rates is as follows (a justification for it is given at the end of this section).

Continuous Compounding

For P dollars invested at interest rate r compounded continuously,

$$\left(\begin{matrix}\text{Value after} \\ t \text{ years}\end{matrix}\right) = Pe^{rt}$$

EXAMPLE 3   **FINDING VALUE WITH CONTINUOUS COMPOUNDING**

Find the value of $1000 at 8% interest compounded continuously for 20 years.

Solution

We use the formula Pe^{rt} with $P = 1000$, $r = 0.08$, and $t = 20$.

$$Pe^{rt} = \underbrace{1000}_{P} \cdot e^{\overset{\uparrow}{(0.08)}\overset{\uparrow}{(20)}} = 1000 \cdot \underbrace{e^{1.6}}_{4.95303} \approx \$4953.03$$

$e^{1.6}$ is usually found using the [2nd] and [In] keys

Intuitive Meaning of Continuous Compounding

Under quarterly compounding, your money is, in a sense, earning interest throughout the quarter, but the interest is not added to your account until the end of the quarter. Under continuous compounding, the interest is added to your account *as it is earned*, with no delay. The extra earnings in continuous compounding come from this "instant crediting" of interest, since then your interest starts earning more interest immediately.

The Function $y = e^x$

The number e gives us a new exponential function $f(x) = e^x$. This function is used extensively in business, economics, and all areas of science. The table below shows the values of e^x for various values of x. These values lead to the graph of $f(x) = e^x$ shown on the right.

x	$y = e^x$
-3	$e^{-3} \approx 0.05$
-2	$e^{-2} \approx 0.14$
-1	$e^{-1} \approx 0.37$
0	$e^0 = 1$
1	$e^1 \approx 2.72$
2	$e^2 \approx 7.39$
3	$e^3 \approx 20.09$

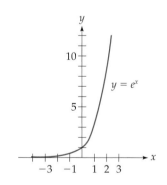

Domain of $y = e^x$ is $\mathbb{R} = (-\infty, \infty)$ and range is $(0, \infty)$.

Notice that e^x is never zero, and is positive for all values of x, even when x is negative. We restate this important observation as follows:

e to any power is positive.

The following graph shows the function $f(x) = e^{kx}$ for various values of the constant k. For positive values of k the curve rises, and for negative values of k the curve falls (as you move to the right). For higher values of k the curve rises more steeply. Each curve crosses the y-axis at 1.

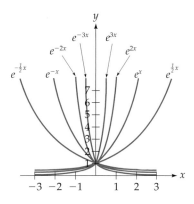

$f(x) = e^{kx}$ for various values of k.

Graphing Calculator Exploration

The most populous states are California and Texas, with New York third and Florida fourth but gaining. According to data from the Census Bureau, x years after 2000 the population of New York will be $19e^{0.0053x}$ and the population of Florida will be $15.9e^{0.0211x}$ (both in millions).

a. Graph these two functions on the window $[0, 20]$ by $[0, 25]$. [Use the 2nd and ln keys for entering e to powers.]

b. Use INTERSECT to find the x-value where the curves intersect.

c. From your answer to part (b), in which year is Florida projected to overtake New York as the third largest state? [*Hint:* x is years after 2000.]

Exponential Growth

All exponential growth, whether continuous or discrete, has one common characteristic: the amount of growth is proportional to the size. For example, the interest that a bank account earns is proportional to the size of the account, and the growth of a population is proportional to the size of the population. This is in contrast, for example, to a person's height, which does not increase exponentially. That is, exponential growth occurs in those situations where a quantity grows *in proportion to its size.*

Justification of the Formula for Continuous Compounding

The compound interest formula Pe^{rt} is derived as follows: P dollars invested for t years at interest rate r compounded m times a year yields

$$P\left(1 + \frac{r}{m}\right)^{mt}$$

From the formula
on page 65

Define $n = m/r$, so that $m = rn$. Replacing m by rn and letting m (and therefore n) approach ∞, this becomes

$$P\left(1 + \frac{r}{rn}\right)^{rnt} = P\left[\underbrace{\left(1 + \frac{1}{n}\right)^{n}}\right]^{rt} \rightarrow Pe^{rt}$$

Letting $n \rightarrow \infty$

Approaches
e as $n \rightarrow \infty$

The limit shown on the right is the continuous compounding formula Pe^{rt}.

1.5 Section Summary

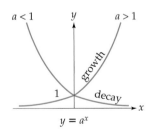

$a < 1$ y $a > 1$

growth

1 decay

x

$y = a^x$

Exponential functions have exponents that involve variables. The exponential functions $f(x) = a^x$ slope *upward* or *downward* depending upon whether the base a (which must be positive) satisfies $a > 1$ or $a < 1$.

The formula $P(1 + r/m)^{mt}$ gives the value after t years of an investment of P dollars that increases at annual rate r compounded m times a year. This same formula, with a *negative* growth rate r and $m = 1$, governs depreciation by a fixed percentage.

By considering compound interest that is compounded more and more frequently, we defined a new constant e:

$$e = \lim_{n \to \infty}\left(1 + \frac{1}{n}\right)^{n} \approx 2.71828$$

Exponential functions with base e are used extensively in modeling many types of growth, such as the growth of populations and interest that is compounded continuously. For interest compounded *continuously*, the formula is Pe^{rt}.

> ➤ **Solutions to Practice Problems**

1. $2000\left(1 + \dfrac{0.24}{12}\right)^{12\cdot3} = 2000(1 + 0.02)^{36} = 2000(1.02)^{36} \approx \$4079.77$

2. $50{,}000(1 - 0.20)^4 = 50{,}000(0.8)^4 = 50{,}000(0.4096) = \$20{,}480$

1.5 Exercises

Use a calculator to evaluate, rounding to three decimal places.

1. a. e^2 **b.** e^{-2} **c.** $e^{1/2}$

2. a. e^3 **b.** e^{-3} **c.** $e^{1/3}$

Express as a power of e.

3. a. $e^5 e^{-2}$ **b.** $\dfrac{e^5}{e^3}$ **c.** $\dfrac{e^5 e^{-1}}{e^{-2} e}$

4. a. $e^{-3} e^5$ **b.** $\dfrac{e^4}{e}$ **c.** $\dfrac{e^2 e}{e^{-2} e^{-1}}$

Graph each function. If you are using a graphing calculator, make a hand-drawn sketch from the screen.

5. $y = 3^x$ **6.** $y = 5^x$

7. $y = \left(\frac{1}{3}\right)^x$ **8.** $y = \left(\frac{1}{5}\right)^x$

Calculate each value of e^x using a calculator.

9. $e^{1.74}$ **10.** $e^{-0.09}$

11. e^x **Versus** x^n Which curve is eventually higher, x to a power or e^x?

a. Graph x^2 and e^x on the window $[0, 5]$ by $[0, 20]$. Which curve is higher?

b. Graph x^3 and e^x on the window $[0, 6]$ by $[0, 200]$. Which curve is higher for large values of x?

c. Graph x^4 and e^x on the window $[0, 10]$ by $[0, 10{,}000]$. Which curve is higher for large values of x?

d. Graph x^5 and e^x on the window $[0, 15]$ by $[0, 1{,}000{,}000]$. Which curve is higher for large values of x?

e. Do you think that e^x will exceed x^6 for large values of x? Based on these observations, can you make a conjecture about e^x and *any* power of x?

12. Linear Versus Exponential Growth

a. Graph $y_1 = x$ and $y_2 = e^{0.01x}$ in the window $[0, 10]$ by $[0, 10]$. Which curve is higher for x near 10?

b. Then graph the same curves on the window $[0, 1000]$ by $[0, 1000]$. Which curve is higher for x near 1000?

A function such as y_1 represents *linear* growth, and y_2 represents *exponential* growth, and the result here is true in general: exponential growth always beats linear growth (eventually, no matter what the constants).*

---

* The realization that populations grow exponentially while food supplies grow only linearly caused the great nineteenth-century essayist Thomas Carlyle to dub economics the "dismal science." He was commenting not on how interesting economics is, but on the grim conclusions that follow from populations outstripping their food supplies.

APPLIED EXERCISES

13. BUSINESS: Interest Find the value of $1000 deposited in a bank at 10% interest for 8 years compounded:

a. annually. **b.** quarterly. **c.** continuously.

14. BUSINESS: Interest Find the value of $1000 deposited in a bank at 12% interest for 8 years compounded:

a. annually. **b.** quarterly. **c.** continuously.

15. GENERAL: Interest A loan shark lends you $100 at 2% compound interest per week (this is a *weekly*, not an annual rate).

a. How much will you owe after 3 years?

b. In "street" language, the profit on such a loan is known as the "vigorish" or the "vig." Find the shark's vig.

16. GENERAL: Compound Interest In 1626, Peter Minuit purchased Manhattan Island from the native Americans for $24 worth of trinkets and beads. Find what the $24 would be worth in the year 2000 if it had been deposited in a bank paying 5% interest compounded quarterly.

17. PERSONAL FINANCE: Millionaires Your rich uncle, wanting to make you a millionaire, deposited $5810 at the time of your birth in a trust fund paying 8% compounded quarterly. Will he succeed by the time you retire at age 65?

18. PERSONAL FINANCE: College Funding To pay for a college education, you deposit $12,000 in a tax-free education trust fund when your child is born. If the interest rate is 8% compounded daily, when your child is 18 will the fund have grown to the estimated $50,000 cost of a college education?

19. PERSONAL FINANCE: Comparing Interest Rates Which is better, 10% compounded quarterly or 9.9% compounded continuously? [*Hint:* Which will yield more for a deposit of $1 for one year?]

20. PERSONAL FINANCE: Comparing Interest Rates Which is better, 8% compounded quarterly or 7.9% compounded continuously? [*Hint:* Which will yield more for a deposit of $1 for one year?]

21. BUSINESS: Zero-Coupon Bonds A bond trader prices a zero-coupon bond* at $560 and the amount grows at interest rate 5.8% compounded continuously. Will it reach its "par" value of $1000 in ten years? (Continuous compounding is frequently used in bond trading.)

22. BUSINESS: Zero-Coupon Bonds A bond trader prices a zero-coupon bond* at $625 and the amount grows at interest rate 4.7% compounded continuously. Will it reach its "par" value of $1000 in ten years? (Continuous compounding is frequently used in bond trading.)

23. BUSINESS: Options Trading The Black–Scholes** formula for pricing options involves continuous compounding. If an option is now worth $2000, and its value grows at interest rate 5.5% compounded continuously, what will be its value in 8 years?

24. BUSINESS: Options Trading The Black–Scholes** formula for pricing options involves continuous compounding. If an option is now worth $10,000, and its value grows at interest rate 6.1% compounded continuously, what will be its value in 5 years?

25. GENERAL: Automobile Depreciation A $20,000 automobile depreciates by 35% each year. Find its value after:

a. 4 years. **b.** 6 months.

26. GENERAL: Vehicle Depreciation A $25,000 pickup truck depreciates in value by 20% each year. Find its value after:

a. 3 years. **b.** 6 months.

---

* A *zero-coupon bond* is a bond that makes no payments (coupons) until it matures, at which time it pays its "face value" of $1000. (You buy it for much less than $1000.)

** An "option" is an offer to buy or sell an asset at some time in the future. The Black–Scholes formula provides an accurate way to price options and led to the award of the Nobel Memorial Prize in 1997 to Robert Merton and Myron Scholes. (Fischer Black was deceased and therefore ineligible for the Nobel Prize.)

27. GENERAL: Population According to the United Nations Fund for Population Activities, the population of the world x years after the year 2000 will be $5.89e^{0.0175x}$ billion people (for $0 \leq x \leq 20$). Use this formula to predict the world population in the year 2010.

28. GENERAL: Population The most populous country is China, with a (2000) population of 1.26 billion, expected to grow to $1.26 \cdot 1.008^x$ billion x years later. The 2000 population of India was 1.02 billion, expected to grow to $1.02 \cdot 1.017^x$ billion x years later. According to these predictions, which population will be larger in the year 2025?

29. GENERAL: Nuclear Meltdown According to the Nuclear Regulatory Commission, the probability of a "severe core meltdown accident" at a nuclear reactor in the United States within the next n years is $1 - (0.9997)^{100n}$. Find the probability of a meltdown:

a. within 25 years.
b. within 40 years.

(The 1986 core meltdown in the Chernobyl reactor in the Soviet Union spread radiation over much of Eastern Europe, leading to an undetermined number of fatalities.)

30. GENERAL: Mosquitoes Female mosquitoes (*Culex pipiens*) feed on blood (only the females drink blood) and then lay several hundred eggs. In this way each mosquito can, on the average, breed another 300 mosquitoes in about 9 days. Find the number of great-grandchildren mosquitoes that will be descended from one female mosquito, assuming that all eggs hatch and mature.

31. ENVIRONMENTAL SCIENCE: Light According to the Bouguer–Lambert Law, the proportion of light that penetrates ordinary seawater to a depth of x feet is $e^{-0.44x}$. Find the proportion of light that penetrates to a depth of:

a. 3 feet. **b.** 10 feet.

32–33. BIOMEDICAL: Drug Dosage If a dosage d of a drug is administered to a patient, the amount of the drug remaining in the tissues t hours later

will be $f(t) = de^{-kt}$, where k (the "absorption constant") depends on the drug.*

32. For the immunosuppressant cyclosporine, the absorption constant is $k = 0.012$. For a dose of $d = 400$ milligrams, use the preceding formula to find the amount of cyclosporine remaining in the tissues after:

a. 24 hours. **b.** 48 hours.

33. For the cardioregulator digoxin, the absorption constant is $k = 0.018$. For a dose of $d = 2$ milligrams, use the preceding formula to find the amount remaining in the tissues after:

a. 24 hours. **b.** 48 hours.

34. BIOMEDICAL: Bacterial Growth A colony of bacteria in a petri dish doubles in size every hour. At noon the petri dish is just covered with bacteria. At what time was the petri dish:

a. 50% covered. [*Hint:* No calculation needed.]
b. 25% covered?

35. BUSINESS: Advertising A company finds that x days after the conclusion of an advertising campaign the daily sales of a new product are $S(x) = 100 + 800e^{-0.2x}$. Find the daily sales 10 days after the end of the advertising campaign.

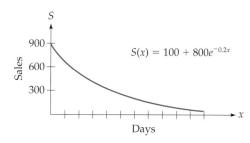

36. BUSINESS: Quality Control A company finds that the proportion of its light bulbs that will burn continuously for longer than t weeks is $e^{-0.01t}$. Find the proportion of bulbs that burn for longer than 10 weeks.

---

* For further details, see T. R. Harrison, ed., *Principles of Internal Medicine*, 15th ed. (New York: McGraw-Hill, 2001).

37. GENERAL: Temperature A covered mug of coffee originally at 200 degrees Fahrenheit, if left for t hours in a room whose temperature is 70 degrees, will cool to a temperature of $70 + 130e^{-1.8t}$ degrees. Find the temperature of the coffee after:

a. 15 minutes.   **b.** half an hour.

38. BEHAVIORAL SCIENCE: Learning In certain experiments the percentage of items that are remembered after t time units is

$$p(t) = 100\frac{1 + e}{1 + e^{t+1}}$$

Such curves are called "forgetting" curves.

Find the percentage remembered after:

a. 0 time units.   **b.** 2 time units.

39. BIOMEDICAL: Epidemics The Reed–Frost model for the spread of an epidemic predicts that the number I of newly infected people is $I = S(1 - e^{-rx})$, where S is the number of susceptible people, r is the effective contact rate, and x is the number of infectious people. Suppose that a school reports an outbreak of measles with $x = 10$ cases, and that the effective contact rate is $r = 0.01$. If the number of susceptibles is $S = 400$, use the Reed–Frost model to estimate how many students will be newly infected during this stage of the epidemic.

40. SOCIAL SCIENCE: Election Cost The cost of winning a seat in the House of Representatives in recent years has been approximately $805e^{0.0625x}$ thousand dollars, where x is the number of years since 2000. Estimate the cost of winning a House seat in the year 2010. (*Source:* Center for Responsive Politics)

41. ATHLETICS: Olympic Games When the Olympic Games were held near Mexico City in the summer of 1968, many athletes were concerned that the high elevation would affect their performance. Air pressure decreases exponentially by 0.4% for each 100 feet of altitude, so changing the altitude by x feet means multiplying the air pressure by $(1 - 0.004)^{x/100}$. By what percentage did the air pressure decrease in moving from Tokyo (the site of the 1964 Summer Olympics, at altitude 30 feet) to Mexico City (altitude 7347 feet)?

42. BIOMEDICAL: Radioactive Contamination The core meltdown and explosions at the nuclear reactor in Chernobyl in 1986 released large amounts of strontium 90, which decays exponentially at the rate of 2.5% per year so that after x years the proportion remaining will be $(1 - 0.025)^x$. Areas downwind of the reactor will be uninhabitable for 100 years. What percent of the original strontium 90 contamination will still be present after:

a. 50 years?   **b.** 100 years?

43. GENERAL: Population As stated earlier, the most populous state is California, with Texas second but gaining. According to the Census Bureau, x years after 2000 the population of California will be $34e^{0.013x}$ and the population of Texas will be $21e^{0.021x}$ (all in millions).

a. Graph these two functions on a calculator on the window [0, 100] by [0, 100].

b. In which year is Texas projected to overtake California as the most populous state? [*Hint:* Use INTERSECT.]

44. GENERAL: St. Louis Arch The Gateway Arch in St. Louis is built around a mathematical curve called a "catenary." The height of this catenary above the ground at a point x feet from the center line is

$$y = 688 - 31.5(e^{0.01033x} + e^{-0.01033x})$$

a. Graph this curve on a calculator on the window $[-400, 400]$ by $[0, 700]$.

b. Find the height of the Gateway Arch at its highest point, using the fact that the top of the arch is 5 feet higher than the top of the central catenary.

1.6   LOGARITHMIC FUNCTIONS

Introduction

In this section we will introduce *logarithmic* functions, concentrating on *common* (base 10) and *natural* (base *e*) logarithms. We will then use common and natural logarithms to solve problems about depreciation and carbon-14 dating. Common logarithms will be particularly useful in Chapter 2.

Common Logarithms

The word "logarithm" (abbreviated "log") means *power* or *exponent*. The number being raised to the power is called the *base* and is written as a subscript. For example, the expression

$$\log_{10} 1000 \qquad \text{Read: log (base 10) of 1000}$$

$$\underset{\text{base}}{\longmapsto}$$

means the *exponent* to which we have to raise 10 to get 1000. Since $10^3 = 1000,$ the exponent is 3, so the *logarithm* is 3.

$$\log_{10} 1000 = 3 \qquad \text{Since } 10^3 = 1000$$

Logarithms with base 10 are called *common logarithms.* For common logarithms we often omit the subscript, with base 10 understood.

Common Logarithms

$$\log x = y \qquad \text{is equivalent to} \qquad 10^y = x \qquad \begin{array}{l}\log x \text{ means } \log_{10} x \\ \text{(base 10)}\end{array}$$

Since $10^y = x$ is positive for every value of *y*, $\log x$ is defined only for *positive* values of *x*. The common logarithm of a number can often be found by expressing the number as a power of 10 and then taking the exponent.

EXAMPLE 1

FINDING A COMMON LOGARITHM

Evaluate $\log 100.$

Solution

$$\log 100 = y \quad \text{is equivalent to} \quad 10^y = 100 \qquad \begin{array}{l} y = 2 \text{ works, so} \\ 2 \text{ is the logarithm} \end{array}$$

The logarithm y is the exponent that solves

Therefore,

$$\log 100 = 2 \qquad \text{Since} \quad 10^2 = 100$$

EXAMPLE 2

FINDING A COMMON LOGARITHM

Evaluate $\log \dfrac{1}{10}$.

Solution

$$\log \frac{1}{10} = y \quad \text{is equivalent to} \quad 10^y = \frac{1}{10} \qquad \begin{array}{l} y = -1 \text{ works, so} \\ -1 \text{ is the logarithm} \end{array}$$

The logarithm y is the exponent that solves

Therefore,

$$\log \frac{1}{10} = -1 \qquad \text{Since} \quad 10^{-1} = \frac{1}{10}$$

Practice Problem 1   Evaluate $\log 10{,}000$.   ➤ Solution on page 88

Graphing Calculator Exploration

```
log(100)
                    2
log(1/10)
                   -1
log(10000)
                    4
```

On a graphing calculator, common logarithms are found using the [LOG] key. Verify that the last three common logarithms we considered can be found in this way.

Properties of Common Logarithms

Because logarithms are exponents, the properties of exponents can be restated as properties of logarithms. For positive numbers M and N and any number P:

Properties of Common Logarithms

1. $\log 1 = 0$	The log of 1 is 0 (since $10^0 = 1$)
2. $\log 10 = 1$	The log of 10 is 1 (since $10^1 = 10$)
3. $\log 10^x = x$	The log of 10 to a power is just the power (since $10^x = 10^x$)
4. $10^{\log x} = x$	10 raised to the log of a number is just the number $(x > 0)$
5. $\log (M \cdot N) = \log M + \log N$	The log of a product is the sum of the logs
6. $\log \left(\dfrac{1}{N}\right) = -\log N$	The log of 1 over a number is minus the log of the number
7. $\log \left(\dfrac{M}{N}\right) = \log M - \log N$	The log of a quotient is the difference of the logs
8. $\log (M^P) = P \cdot \log M$	The log of a number to a power is the power times the log

The first two properties are simply special cases of the third (with $x = 0$ and $x = 1$). Because logs are exponents, the third property simply says that the exponent of 10 that gives 10^x is x, which is obvious when you think about it. Because $\log x$ is the power of 10 that gives x, raising 10 to that power must give x, which is the fourth property. Justifications for properties 5–8 are given on pages 86–87. Property 8 will be particularly useful in applications, and can be summarized: *Logarithms bring down exponents.*

EXAMPLE **3**

USING THE PROPERTIES OF COMMON LOGARITHMS

a. $\log 10^7 = 7$	Property 3
b. $10^{\log 13} = 13$	Property 4
c. $\log (3 \cdot 4) = \log 3 + \log 4$	Property 5
d. $\log \left(\dfrac{1}{4}\right) = -\log 4$	Property 6

e. $\log\left(\dfrac{3}{4}\right) = \log 3 - \log 4$ Property 7

f. $\log(5^3) = 3\log 5$ Property 8: $\log(5^3) = 3\log 5$

EXAMPLE 4   **FINDING WHEN A CAR DEPRECIATES BY HALF**

A car depreciates by 20% per year. When will it be worth only half its original value?

Solution

We know from page 66 that depreciation follows the compound interest formula but with a *negative* interest rate (since it *loses* value). Since we are not told the price of the car, we let P stand for the price. The compound interest formula with negative interest rate $r = -0.20$, set equal to $0.5P$ (representing *half* the original price) gives:

$$P(1 - 0.20)^t = 0.5P$$ $P(1 + r/m)^{mt}$ with $r = -0.20$ and $m = 1$

$$(0.80)^t = 0.5$$ Canceling Ps and simplifying

$$\log(0.80)^t = \log 0.5$$ Taking log of each side

$$t \cdot \log 0.80 = \log 0.5$$ Bringing down the power (property 8)

$$t = \frac{\log 0.5}{\log 0.80}$$ Dividing by $\log 0.80$

$$t \approx 3.1$$ Using a calculator

The car will be worth half its value in about 3.1 years.

Incidentally, in another 3.1 years the car will again fall to half its value, thereby reaching *one quarter* of its original value in 6.2 years. This halving of value will continue every 3.1 years.

Graphs of Logarithmic and Exponential Functions

If a point (x, y) lies on the graph of $y = \log x$ or, equivalently, $x = 10^y$, then, reversing x and y, the point (y, x) lies on the graph of $y = 10^x$. That is, the curves $y = \log x$ and $y = 10^x$ are related by having their x- and y-coordinates *reversed*, so the curves are *mirror images* of each other across the line $y = x$.

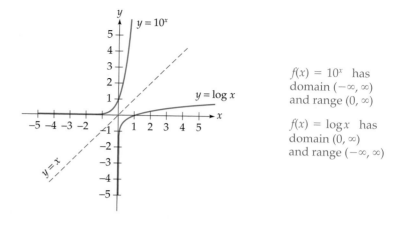

$f(x) = 10^x$ has domain $(-\infty, \infty)$ and range $(0, \infty)$

$f(x) = \log x$ has domain $(0, \infty)$ and range $(-\infty, \infty)$

This graphical relationship is equivalent to the fact that the functions $\log x$ and 10^x "undo" or "reverse" each other, as is shown by properties 3 and 4 on page 79. Such functions are called *inverse functions*:

$$\log x \quad \text{and} \quad 10^x \quad \text{are inverse functions}$$

Logarithms to Other Bases

We may calculate logarithms to bases other than 10. In fact, *any* positive number other than 1 may be used as a base for logs, using the following definition.

Base a Logarithms

$$\log_a x = y \quad \text{is equivalent to} \quad a^y = x \qquad (x > 0)$$

For example,

$$\log_2 8 = 3 \qquad \text{Since } 2^3 = 8$$

and

$$\log_9 3 = \tfrac{1}{2} \qquad \text{Since } 9^{1/2} = \sqrt{9} = 3$$

Natural Logarithms

We will use only one other base, the number e (approximately 2.718) that we defined on page 69. Logarithms to the base e are called *natural* or *Napierian* logarithms.* The natural logarithm of a positive number x is written $\ln x$ ("n" for "natural") and may be found using the $\boxed{\text{ln}}$ key on a calculator.

Natural Logarithms

$\ln x = y$ is equivalent to $e^y = x$ $\ln x$ means $\log_e x$ (base e)

Practice Problem 2   Use a calculator to find $\ln 8.34$.   Solution on page 88

The properties of natural logarithms are similar to those of common logarithms but with $\ln$ instead of $\log$ and e instead of 10. For positive numbers M and N and any number P:

Properties of Natural Logarithms

1. $\ln 1 = 0$	The natural log of 1 is 0 (since $e^0 = 1$)
2. $\ln e = 1$	The natural log of e is 1 (since $e^1 = e$)
3. $\ln e^x = x$	The natural log of e to a power is just the power (since $e^x = e^x$)
4. $e^{\ln x} = x$	e raised to the natural log of a number is just the number $(x > 0)$
5. $\ln (M \cdot N) = \ln M + \ln N$	The natural log of a product is the sum of the logs
6. $\ln \left(\dfrac{1}{N}\right) = -\ln N$	The natural log of 1 over a number is minus the log of the number
7. $\ln \left(\dfrac{M}{N}\right) = \ln M - \ln N$	The natural log of a quotient is the difference of the logs
8. $\ln (M^P) = P \cdot \ln M$	The natural log of a number to a power is the power times the log

---

As with common logs, the first two properties are special cases of the third (with $x = 0$ and $x = 1$). The third and fourth properties have interpretations analogous to their "common" counterparts (for example, the third says that the exponent of e that gives e^x is x). As before, property 8 can be summarized: *Logarithms bring down exponents.*

---

* After John Napier (1550–1617), a Scottish mathematician who, incidentally, invented the decimal point.

EXAMPLE 5

USING THE PROPERTIES OF NATURAL LOGARITHMS

a. $\ln e^7 = 7$ Property 3

b. $e^{\ln 13} = 13$ Property 4

c. $\ln (3 \cdot 4) = \ln 3 + \ln 4$ Property 5

d. $\ln \left(\dfrac{1}{4} \right) = -\ln 4$ Property 6

e. $\ln \left(\dfrac{3}{4} \right) = \ln 3 - \ln 4$ Property 7

f. $\ln (5^3) = 3 \ln 5$ Property 8: $\ln(5^3) = 3 \ln 5$

Graphing Calculator Exploration

```
ln(e^(17))
                17
e^(ln(29))
                29
```

a. Evaluate $\ln e^{17}$ to verify that the answer is 17. Change 17 to other numbers (positive, negative, or zero) in order to verify that $\ln e^x = x$ for any x.

b. Evaluate $e^{\ln 29}$ to verify that the answer is 29. Change the 29 to another positive number to verify that $e^{\ln x} = x$. What about negative numbers, or zero?

Properties 3 and 4 show that $y = \ln x$ and $y = e^x$ are *inverse functions*, so their graphs are reflections of each other in the diagonal line $y = x$.

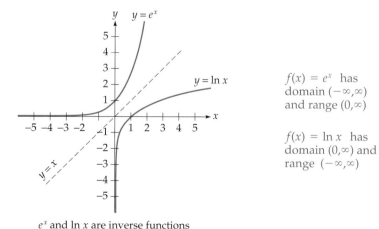

$f(x) = e^x$ has domain $(-\infty, \infty)$ and range $(0, \infty)$

$f(x) = \ln x$ has domain $(0, \infty)$ and range $(-\infty, \infty)$

e^x and $\ln x$ are inverse functions

The properties of natural logarithms are helpful for simplifying functions.

Graphing Calculator Exploration

a. Graph the function $y = \ln x^2$ on the window $[-5, 5]$ by $[-5, 5]$.

b. Change the function to $y = 2 \ln x$ and explain why the two graphs are different. (Doesn't property 8 say that the two functions should be the same?)

EXAMPLE 6   **SIMPLIFYING A FUNCTION**

$$f(x) = \ln (2x) - \ln 2$$

$$= \ln 2 + \ln x - \ln 2 \qquad \text{Since } \ln (2x) = \ln 2 + \ln x \text{ by property 5}$$

$$= \ln x \qquad \text{Canceling}$$

EXAMPLE 7   **SIMPLIFYING A FUNCTION**

$$f(x) = \ln \left(\frac{x}{e} \right) + 1$$

$$= \ln x - \ln e + 1 \qquad \text{Since } \ln (x/e) = \ln x - \ln e \text{ by property 7}$$

$$= \ln x - 1 + 1 \qquad \text{Since } \ln e = 1 \text{ by property 2}$$

$$= \ln x \qquad \text{Canceling}$$

EXAMPLE 8

SIMPLIFYING A FUNCTION

$$f(x) = \ln(x^5) - \ln(x^3) = 5\ln x - 3\ln x \qquad \text{Bringing down exponents by property 8}$$

$$= 2\ln x \qquad \text{Combining}$$

Graphing Calculator Exploration

Some advanced graphing calculators have computer algebra systems that can simplify algebraic expressions. For example, the Texas Instruments TI-89 graphing calculator simplifies logarithmic expressions, but only if the condition $x > 0$ is included so that the logarithms are defined.

Entered →

← Not simplified;
← but if we require $x > 0$,
← it is simplified!

Carbon-14 Dating

All living things absorb small amounts of radioactive carbon-14 from the atmosphere. When they die, the carbon-14 stops being absorbed and decays exponentially into ordinary carbon. Therefore, the proportion of carbon-14 still present in a fossil or other ancient remain can be used to estimate how old it is. The proportion of the original carbon-14 that will be present after t years is

$$\left(\begin{array}{l}\text{Proportion of carbon-14} \\ \text{remaining after } t \text{ years}\end{array}\right) = e^{-0.00012t}$$

EXAMPLE 9

DATING BY CARBON-14

The Dead Sea Scrolls, discovered in a cave near the Dead Sea in what was then Jordan, are among the earliest documents of Western civilization. Estimate the age of the Dead Sea Scrolls if the animal skins on which some were written contain 78% of their original carbon-14.

Solution

The proportion of carbon-14 remaining after t years is $e^{-0.00012t}$. We equate this formula to the actual proportion (expressed as a decimal).

$$e^{-0.00012t} = 0.78 \qquad \text{Equating the proportions}$$

$$\ln e^{-0.00012t} = \ln 0.78 \qquad \text{Taking natural logs}$$

$$-0.00012t = \ln 0.78 \qquad \begin{array}{l}\ln e^{-0.00012t} = -0.00012t \\ \text{by property 3}\end{array}$$

$$t = \frac{\ln 0.78}{-0.00012} \approx \frac{-0.24846}{-0.00012} \approx 2071 \qquad \begin{array}{l}\text{Solving for } t \text{ and using a} \\ \text{calculator}\end{array}$$

Therefore, the Dead Sea Scrolls are approximately 2070 years old.

Graphing Calculator Exploration

Solve Example 9 by graphing $y_1 = e^{-0.00012x}$ and $y_2 = 0.78$ on the window [0, 10,000] by [0, 1] and using INTERSECT to find where they meet.

Both here and in Example 4 on page 80 we solved for the variable in the exponent by taking logarithms, but here we used *natural* logarithms and there we used *common* logarithms. Is there a difference? Not really—you can use logarithms to *any* base to solve for exponents, and the final answer will be the same (see Exercise 37). However, when e is involved, it is usually easier to use *natural* logs.

Justification of Properties 5–8 of Logarithms

Properties 1–4 of logarithms (both common and natural) were justified earlier. Properties 5–8 of common logarithms are justified below, with the "natural" justifications obtained simply by replacing 10 by e.

The addition law of exponents, $10^x \cdot 10^y = 10^{x+y}$, can be stated in words: "The exponent of a product is the sum of the exponents." Because logs are exponents, this can be restated as "The log of a product is the sum of the logs," which is just property 5.

The subtraction law of exponents, $10^x / 10^y = 10^{x-y}$ can be stated, "The exponent of a quotient is the difference of the exponents." This translates into "The log of a quotient is the difference of the logs," which is just property 7. Property 6 is simply a special case of property 7 with $M = 1$ (and using property 1).

The law of exponents $(10^x)^y = 10^{x \cdot y}$ says that the exponent y can be "brought down" and multiplied by the x. Because logs are exponents, this says that in $\log(M^P)$, the exponent P can be brought down and multiplied by the logarithm $\log M$, giving $P \cdot \log M$, which is just property 8.

1.6 Section Summary

Logarithms are exponents: Common logs are exponents of 10, and natural logs are exponents of e. That is,

$y = \log x$ is equivalent to $x = 10^y$ $\log x$ means $\log_{10} x$ (base 10)

$y = \ln x$ is equivalent to $x = e^y$ $\ln x$ means $\log_e x$ (base e)

Each property of logarithms is equivalent to a property of exponents.

Logarithmic Property	*Exponential Property*
$\log 1 = 0$	$10^0 = 1$
$\log(M \cdot N) = \log M + \log N$	$10^x \cdot 10^y = 10^{x+y}$
$\log\left(\dfrac{1}{N}\right) = -\log N$	$\dfrac{1}{10^y} = 10^{-y}$
$\log\left(\dfrac{M}{N}\right) = \log M - \log N$	$\dfrac{10^x}{10^y} = 10^{x-y}$
$\log(M^P) = P \cdot \log M$	$(10^x)^y = 10^{x \cdot y}$

In practice, logarithms are found using the LOG and LN keys on a calculator. The properties of exponents lead to properties of logarithms, which are listed on pages 79 and 82. Property 8, that *logs bring down exponents,* is particularly useful in applications that require solving for a variable in the exponent.

▶ **Solutions to Practice Problems**

1. $\log 10,000 = 4$ (Since $10^4 = 10,000$)

2. $\ln 8.34 \approx 2.121$ (Using a calculator)

1.6 Exercises

Find each logarithm *without* using a calculator.

1. a. $\log 100,000$ **b.** $\log \frac{1}{100}$ **c.** $\log \sqrt{10}$

2. a. $\log 1000$ **b.** $\log \frac{1}{1000}$ **c.** $\log \sqrt[3]{10}$

3. a. $\ln e^5$ **b.** $\ln \frac{1}{e}$ **c.** $\ln \sqrt[3]{e}$

4. a. $\ln e^3$ **b.** $\ln \frac{1}{e^2}$ **c.** $\ln \sqrt{e}$

5. a. $\ln 1$ **b.** $\ln (\ln e^e)$ **c.** $\ln \sqrt[3]{e^2}$

6. a. $\ln e$ **b.** $\ln (\ln e)$ **c.** $\ln \sqrt{e^3}$

7. a. $\log_4 16$ **b.** $\log_4 \frac{1}{4}$ **c.** $\log_4 2$

8. a. $\log_9 81$ **b.** $\log_9 3$ **c.** $\log_9 \frac{1}{9}$

For each logarithm:

i. Find the logarithm using a calculator, rounding your answer to three decimal places.

ii. Raise the appropriate base (either 10 or e) to the power you found in part (i) and check that the result agrees with the number in the original problem. (*Note:* For the power, include *all* of the digits that your calculator showed for part (i) to minimize the error.)

9. a. $\log 22.3$ **b.** $\ln 22.3$

10. a. $\log 44.9$ **b.** $\ln 44.9$

Use the properties of natural logarithms to simplify each function.

11. $f(x) = \ln (9x) - \ln 9$ **12.** $f(x) = \ln \left(\frac{x}{2}\right) + \ln 2$

13. $f(x) = \ln (x^3) - \ln x$ **14.** $f(x) = \ln (4x) - \ln 4$

15. $f(x) = \ln \left(\frac{x}{4}\right) + \ln 4$

16. $f(x) = \ln (x^5) - 3 \ln x$

17. $f(x) = \ln (e^{5x}) - 2x - \ln 1$

18. $f(x) = \ln (e^{-2x}) + 3x + \ln 1$

19. $f(x) = 8x - e^{\ln x}$

20. $f(x) = e^{\ln x} + \ln (e^{-x})$

21. Without using a calculator, sketch the graph of $f(x) = \log (x + 1)$.

22. Without using a calculator, sketch the graph of $f(x) = \ln (x + e)$.

23. Find the domain and range and graph the function $f(x) = \ln (x^2 - 1)$.

24. Find the domain and range and graph the function $f(x) = \ln (1 - x^2)$.

APPLIED EXERCISES

25. GENERAL: Depreciation A car depreciates by 30% per year. When will it be worth only:

a. half its original value?
b. one quarter its original value?

26. GENERAL: Depreciation A truck depreciates by 25% per year. When will it be worth only:

a. half its original value?
b. one quarter its original value?

27. BUSINESS: Depreciation An industrial printing press depreciates by 15% per year. When will it be worth two thirds of its original value?

28. BUSINESS: Depreciation A supercomputer depreciates by 40% per year. When will it be worth only one tenth of its original value?

29–30. GENERAL: Carbon-14 Dating The proportion of carbon-14 still present in a sample after t years is $e^{-0.00012t}$.

29. Use the preceding formula to estimate the age of the cave paintings discovered in 1994 in the Ardèche region of France if the carbon with which they were drawn contains only 2.3% of its original carbon-14. They are the oldest known paintings in the world.

30. Use the preceding formula to estimate the age of the Shroud of Turin, believed by many to be the burial cloth of Christ, from the fact that its linen fibers contained only 92.3% of their original carbon-14.

31–32. GENERAL: Potassium-40 Dating The radioactive isotope potassium-40 is used to date very old remains. The proportion of potassium-40 that remains after t million years is $e^{-0.00054t}$. Use this function to estimate the age of the following fossils.

31. The most nearly complete skeleton of an early human ancestor ever found was discovered in Kenya in 1984. Use the above formula to estimate the age of the remains if they contained 99.91% of their original potassium-40.

32. Dating Older Women Use the above formula to estimate the age of the partial skeleton of *Australopithecus afarensis,* known as "Lucy," which was found in Ethiopia in 1977, if it had 99.82% of its original potassium-40.

33–34. BUSINESS: Depreciation Solve the following exercises on a graphing calculator by graphing a function for the depreciated value (using x for ease of entry) together with a constant function and using INTERSECT to find where they meet. You will have to choose an appropriate window.

33. A company jet that originally cost $5 million depreciates by 25% per year. When will it be worth only $500,000, at which time it can be written off as a tax loss?

34. A $400,000 fire engine depreciates by 15% per year. When will it be worth only $40,000, at which time it will be sold for scrap?

35. GENERAL: Potassium-40 Dating Estimate the age of the oldest known dinosaur, a dog-sized creature called *Herrerasaurus* found in Argentina in 1988, if volcanic material found with it contained 88.4% of its original potassium-40. (Use the potassium-40 decay function given in the directions for Exercises 31–32.)

36. GENERAL: Carbon-14 Dating In 1991 two hikers in the Italian Alps found the frozen but well-preserved body of the most ancient human ever found, dubbed "Iceman." Estimate the age of Iceman if his grass cape contained 53% of its original carbon-14. (Use the carbon-14 decay function stated in the directions for Exercises 29–30.)

Explorations and Excursions

The following problem extends and augments the material presented in the text.

Change-of-Base Formula for Logarithms

37. Let a and b be any two bases, and let x be any positive number.

a. Give a justification for each numbered equals sign.

$$\log_a x \overset{1}{=} \log_a b^{\log_b x} \overset{2}{=} (\log_b x)\cdot(\log_a b)$$

b. Show that the result can be written as

$$\log_a x = (\log_a b)\cdot(\log_b x)$$

This is the "change-of-base" formula for logarithms, enabling one to express $\log_b x$ in terms of $\log_a x$.

c. Use the change-of-base formula in the numerator and denominator of the following fraction to justify each numbered equals sign.

$$\frac{\log_a x}{\log_a y} \overset{3}{=} \frac{(\log_a b)\cdot(\log_b x)}{(\log_a b)\cdot(\log_b y)} \overset{4}{=} \frac{\log_b x}{\log_b y}$$

This equation shows that when you are finding a *ratio* of logarithms, using one base gives the same result as using any other base. Because solving for a variable in the exponent involves calculating *ratios* of logarithms, you may do so using logarithms to *any* base.

Chapter Summary with Hints and Suggestions

Reading the text and doing the exercises in this chapter have helped you to master the following skills, which are listed by section (in case you need to review them) and are keyed to particular Review Exercises. Answers for all Review Exercises are given at the back of the book, and full solutions can be found in the Student Solutions Manual.

1.1 Real Numbers, Inequalities, and Lines

- Translate an interval into set notation and graph it on the real line. (*Review Exercises 1–4.*)

$$[a, b] \quad (a, b) \quad [a, b) \quad (a, b]$$

$$(-\infty, b] \quad (-\infty, b) \quad [a, \infty) \quad (a, \infty) \quad (-\infty, \infty)$$

- Express given information in interval form. (*Review Exercises 5–6.*)

- Find an equation for a line that satisfies certain conditions. (*Review Exercises 7–12.*)

$$m = \frac{y_2 - y_1}{x_2 - x_1} \qquad y = mx + b$$

$$y - y_1 = m(x - x_1) \qquad x = a \qquad y = b$$

$$ax + by = c$$

- Find an equation of a line from its graph. (*Review Exercises 13–14.*)

- Use straight-line depreciation to find the value of an asset. (*Review Exercises 15–16.*)

- Use real-world data to find a regression line and make a prediction. (*Review Exercises 17–18.*)

1.2 Exponents

- Evaluate negative and fractional exponents without a calculator. (*Review Exercises 19–26.*)

$$x^0 = 1 \qquad x^{-n} = \frac{1}{x^n} \qquad x^{m/n} = \sqrt[n]{x^m} = \left(\sqrt[n]{x}\right)^m$$

- Evaluate an exponential expression using a calculator. (*Review Exercises 27–28.*)

1.3 Functions

- Evaluate and find the domain and range of a function. (*Review Exercises 29–32.*)

    A function f is a rule that assigns to each number x in a set (the domain) a (single) number $f(x)$. The range is the set of all resulting values $f(x)$.

- Use the vertical line test to see if a graph defines a function. (*Review Exercises 33–34.*)

- Graph a linear function. (*Review Exercises 35–36.*)

$$f(x) = mx + b$$

- Graph a quadratic function. (*Review Exercises 37–38.*)

$$f(x) = ax^2 + bx + c$$

- Solve a quadratic equation by factoring and by the Quadratic Formula. (*Review Exercises 39–42.*)

Vertex	x-intercepts
$x = \dfrac{-b}{2a}$	$x = \dfrac{-b \pm \sqrt{b^2 - 4ac}}{2a}$

- Use a graphing calculator to graph a quadratic function. (*Review Exercises 43–44.*)

- Construct a linear function from a word problem or from real-life data, and then use the function in an application. (*Review Exercises 45–48.*)

- For given cost and revenue functions, calculate the break-even points and maximum profit. (*Review Exercises 49–50.*)

1.4 Functions, Continued

- Evaluate and find the domain and range of a more complicated function. (*Review Exercises 51–54.*)

- Solve a polynomial equation by factoring. (*Review Exercises 55–58.*)

- Graph a "shifted" function. (*Review Exercises 59–60.*)

- Graph a piecewise linear function. (*Review Exercises 61–62.*)

- Given two functions, find their composition. (*Review Exercises 63–68.*)

$$f(g(x)) \qquad g(f(x))$$

- Solve an applied problem involving the composition of functions. (*Review Exercise 69.*)

- Solve a polynomial equation. (*Review Exercises 70–71.*)

- Fit a curve to real-life data and make a prediction. (*Review Exercise 72.*)

1.5 Exponential Functions

- Sketch the graph of an exponential function. (*Review Exercises 73–74.*)

- Find the value of money invested at compound interest. (*Review Exercises 75–76.*)

$$P\left(1 + \frac{r}{m}\right)^{mt}$$

- Depreciate an asset by a fixed percentage per year. (*Review Exercises 77–78.*)

$$\left(\begin{array}{l}\text{Above formula with } m = 1 \\ \text{and a } \textit{negative} \text{ value for } r\end{array}\right)$$

- Find the value of a deposit invested with continuous compounding. (*Review Exercises 79–80.*)

$$Pe^{rt}$$

1.6 Logarithmic Functions

- Evaluate common and natural logarithms *without* using a calculator. (*Review Exercises 81–82.*)

$$\log 1 = 0 \quad \log 10 = 1 \quad \log 10^x = x \quad 10^{\log x} = x$$

$$\log(M \cdot N) = \log M + \log N$$

$$\log\left(\frac{M}{N}\right) = \log M - \log N$$

$$\log\left(\frac{1}{N}\right) = -\log N \qquad \log(M^P) = P \cdot \log M$$

$$\ln 1 = 0 \quad \ln e = 1 \quad \ln e^x = x \quad e^{\ln x} = x$$

$$\ln(M \cdot N) = \ln M + \ln N$$

$$\ln\left(\frac{M}{N}\right) = \ln M - \ln N$$

$$\ln\left(\frac{1}{N}\right) = -\ln N \qquad \ln(M^P) = P \cdot \ln M$$

- Use the properties of natural logarithms to simplify a function. (*Review Exercises 83–84.*)

- Determine when an asset depreciates to a fraction of its value. (*Review Exercises 85–86.*)

- Estimate the age of a fossil. (*Review Exercises 87–88.*)

Hints and Suggestions

- (*Overview*) In reviewing this chapter, notice the difference between *geometric* objects (points, curves, etc.) and *analytic* objects (numbers, functions, etc.), and the connections between them. Descartes first made this connection: by drawing axes, he saw that points could be specified by numerical coordinates, and so *curves* could be specified by *equations* governing their coordinates. This idea connected geometry to algebra, previously distinct subjects. You should be able to express geometric objects analytically, and vice versa. For example, given a *graph* of a line, you should be able to find an *equation* for it, and given a quadratic *function*, you should be able to *graph* it.

- A graphing calculator or a computer with appropriate software can help you to *explore* a concept more fully (for example, seeing how a curve changes as a coefficient or exponent changes) and also to *solve* a problem (for example, eliminating the point-plotting aspect of graphing, or finding a regression line).

- If you don't have a graphing calculator, you should have a scientific or business calculator to carry out calculations, especially in later chapters.

- The Practice Problems help you to check your mastery of the skills presented. Complete solutions are given at the end of each section.

- The Student Solutions Manual, available separately from your bookstore, provides fully worked-out solutions to selected exercises.

- Interest rates are always *annual* (unless clearly stated otherwise) but the *compounding* may be done more frequently (in the formula, m times a year).

- When do you use the formula $P(1 + r/m)^{mt}$ and when do you use Pe^{rt}? Use Pe^{rt} if the word "continuous" occurs and use $P(1 + r/m)^{mt}$ if it does not.

- The formula Pe^{rt} has no m because there is no period to wait for the interest to be compounded—interest is added to the account *continuously*.

- For depreciation, use $P(1 + r/m)^{mt}$ with a *negative* value for r. For example, depreciating by 15% annually would mean $m = 1$ and $r = -0.15$.

- **Practice for Test:** Review Exercises, 1, 9, 11, 13, 15, 17, 19, 31, 33, 35, 37, 43, 47, 49, 55, 61, 65, 71, 75, 77, 79, 81, 83, 85, and 87.

Review Exercises for Chapter 1

Practice test exercise numbers are in green.

1.1 Real Numbers, Inequalities, and Lines

Write each interval in set notation and graph it on the real line.

1. (2, 5]   2. [−2, 0)

3. [100, ∞)   4. (−∞, 6]

5. **GENERAL: Wind Speed** The United States Coast Guard defines a "hurricane" as winds of at least 74 mph, a "storm" as winds of at least 55 mph but less than 74 mph, a "gale" as winds of at least 38 mph but less than 55 mph, and a "small craft warning" as winds of at least 21 mph but less than 38 mph. Express each of these wind conditions in interval form. [*Hint:* A small craft warning is [21, 38).]

6. State in interval form:
   a. The set of all positive numbers.
   b. The set of all negative numbers.
   c. The set of all nonnegative numbers.
   d. The set of all nonpositive numbers.

Write an equation of the line satisfying each of the following conditions. If possible, write your answer in the form $y = mx + b$.

7. Slope 2 and passing through the point $(1, -3)$

8. Slope -3 and passing through the point $(-1, 6)$

9. Vertical and passing through the point $(2, 3)$

10. Horizontal and passing through the point $(2, 3)$

11. Passing through the points $(-1, 3)$ and $(2, -3)$

12. Passing through the points $(1, -2)$ and $(3, 4)$

Write an equation of the form $y = mx + b$ for each line graphed below.

13.

14.

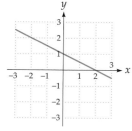

15. BUSINESS: Straight-Line Depreciation A contractor buys a backhoe for $25,000 and estimates its useful life to be 8 years, after which its scrap value will be $1000.

   a. Use straight-line depreciation to find a formula for the value V of the backhoe after t years, for $0 \le t \le 8$.

   b. Use your formula to find the value of the backhoe after 4 years.

16. BUSINESS: Straight-Line Depreciation A trucking company buys a satellite communication system for $78,000 and estimates its useful life to be 15 years, after which its scrap value will be $3000.

   a. Use straight-line depreciation to find a formula for the value V of the system after t years, for $0 \le t \le 15$.

   b. Use your formula to find the value of the system after 8 years.

17. ECOLOGY: Sulfur Dioxide Pollution Sulfur dioxide pollution has decreased significantly in the United States during the last 25 years, mostly because of antipollution devices on automobiles and on coal- and oil-fired power plants. The following table shows sulfur dioxide emissions (in millions of tons) in the United States from 1975 to 2000. To avoid large numbers, years are listed in the table as years since 1975.

	Years Since 1975	Sulfur Dioxide Emissions
1975	0	26.0
1980	5	23.5
1985	10	21.6
1990	15	19.3
1995	20	18.2
2000	25	18.1

Sources: Worldwatch Institute and U.S. Environmental Protection Agency

   a. Enter the table numbers into a graphing calculator and make a plot of the resulting points (Years Since 1975 on the x-axis and Sulfur Dioxide Emissions on the y-axis).

   b. Have your calculator find the linear regression formula for these data. Then enter the result in y_1, which gives a formula for sulfur dioxide emissions in each year. Plot the points together with the regression line. How well does the line fit the data?

   c. Use your formula to predict the sulfur dioxide pollution in the year 2010 (assuming that the past trend continues).

18. SOCIAL SCIENCE: Gap Between Rich and Poor During the last few decades, the richest 20% of the world's people have been growing richer, while the poorest 20% have been growing poorer. The probable consequences of this growing gap are not only social instability but also environmental decline, since the richest consume more wastefully while the poorest must cut down rainforests and overgraze land just to survive. The following table shows the ratio of income of the richest 20% to the poorest 20% from 1960 to 2000. To avoid large numbers, years are listed in the table as years since 1960.

	Years Since 1960	Ratio of Richest to Poorest
1960	0	30 to 1
1970	10	32 to 1
1980	20	45 to 1
1990	30	60 to 1
2000	40	72 to 1

Source: United Nations Development Programme

a. Enter the table numbers into a graphing calculator and make a plot of the resulting points [Years Since 1960 on the x-axis and the larger number in the Ratio column (the 30, 32, etc.) on the y-axis].

b. Have your calculator find the linear regression formula for these data. Then enter the result in y_1, which gives a formula for the ratio of richest to poorest in each year. Plot the points together with the regression line. How well does the line fit the data?

c. Use your formula to predict the ratio in the year 2010 (assuming that the past trend continues).

1.2 Exponents

Evaluate each expression without using a calculator.

19. $\left(\frac{1}{6}\right)^{-2}$ **20.** $\left(\frac{4}{3}\right)^{-1}$

21. $64^{1/2}$ **22.** $1000^{1/3}$

23. $81^{-3/4}$ **24.** $100^{-3/2}$

25. $\left(-\frac{8}{27}\right)^{-2/3}$ **26.** $\left(\frac{9}{16}\right)^{-3/2}$

Use a calculator to evaluate each expression. Round answers to two decimal places.

27. $3^{2.4}$ **28.** $12^{1.9}$

1.3 Functions

For each function in Exercises 29–32:

a. Evaluate the given expression.
b. Find the domain.
c. Find the range.

29. $f(x) = \sqrt{x-7}$; find $f(11)$

30. $g(t) = \dfrac{1}{t+3}$; find $g(-1)$ See instructions in previous column.

31. $h(w) = w^{-3/4}$; find $h(16)$

32. $w(z) = z^{-4/3}$; find $w(8)$

Determine whether each graph defines a function of x.

33.   **34.**

Graph each function.

35. $f(x) = 4x - 8$ **36.** $f(x) = 6 - 2x$

37. $f(x) = -2x^2 - 4x + 6$

38. $f(x) = 3x^2 - 6x$

Solve each equation by
a. factoring   b. the Quadratic Formula.

39. $3x^2 + 9x = 0$ **40.** $2x^2 - 8x - 10 = 0$

41. $3x^2 + 3x + 5 = 11$ **42.** $4x^2 - 2 = 2$

For each quadratic function in Exercises 43–44:
a. Find the vertex using the vertex formula.
b. Graph the function on an appropriate viewing window. (Answers may vary.)

43. $f(x) = x^2 - 10x - 25$

44. $f(x) = x^2 + 14x - 15$

45. BUSINESS: Car Rentals A rental company rents cars for $45 per day and $0.12 per mile. Find a function $C(x)$ for the cost of a rented car driven for x miles in a day.

46. BUSINESS: Simple Interest If money is borrowed for a short period of time, generally less than a year, the interest is often calculated as *simple* interest, according to the formula Interest $= P \cdot r \cdot t$, where P is the principal, r is the rate (expressed as a decimal), and t is the time (in years). Find a function $I(t)$ for the interest charged on a loan of $10,000 at an interest rate of 8% for t years. Simplify your answer.

47. **GENERAL: Air Temperature** The air temperature decreases by about 1 degree Fahrenheit for each 300 feet of altitude. Find a function $T(x)$ for the temperature at an altitude of x feet if the sea level temperature is 70°.

48. **ECOLOGY: Carbon Dioxide Pollution** The burning of fossil fuels (such as oil and coal) added 27 billion tons of carbon dioxide to the atmosphere during 2000, and this annual amount is growing by 0.58 billion tons per year. Find a function $C(t)$ for the amount of carbon dioxide added during the year t years after 2000, and use the formula to find how soon this annual amount will reach 30 billion tons. [*Note:* Carbon dioxide traps solar heat, increasing the earth's temperature, and may lead to flooding of lowland areas by melting the polar ice.] *Source:* U.S. Environmental Protection Agency.

49. **BUSINESS: Break-Even Points and Maximum Profit** A store that installs satellite TV receivers finds that if it installs x receivers per week, then its costs will be $C(x) = 80x + 1950$ and its revenue will be $R(x) = -2x^2 + 240x$ (both in dollars).
   a. Find the store's break-even points.
   b. Find the number of receivers the store should install to maximize profit, and the maximum profit.

50. **BUSINESS: Break-Even Points and Maximum Profit** An air conditioner outlet finds that if it sells x air conditioners per month, its costs will be $C(x) = 220x + 202,500$ and its revenue will be $R(x) = -3x^2 + 2020x$ (both in dollars).
   a. Find the outlet's break-even points.
   b. Find the number of air conditioners the outlet should sell to maximize profit, and the maximum profit.

1.4 Functions, Continued

For each function in Exercises 51–54:
   a. Evaluate the given expression.

   b. Find the domain.
   c. Find the range.

51. $f(x) = \dfrac{3}{x(x-2)}$; find $f(-1)$

52. $f(x) = \dfrac{16}{x(x+4)}$; find $f(-8)$

53. $g(x) = |x + 2| - 2$; find $g(-4)$

54. $g(x) = x - |x|$; find $g(-5)$

Solve each equation by factoring.

55. $5x^4 + 10x^3 = 15x^2$ 56. $4x^5 + 8x^4 = 32x^3$

57. $2x^{5/2} - 8x^{3/2} = 10x^{1/2}$

58. $3x^{5/2} + 3x^{3/2} = 18x^{1/2}$

Graph each function.

59. $f(x) = (x + 1)^2 - 1$ 60. $f(x) = (x - 2)^2 - 4$

61. $f(x) = \begin{cases} 3x - 7 & \text{if } x \geq 2 \\ -x - 1 & \text{if } x < 2 \end{cases}$

(If you use a graphing calculator for Exercises 61 and 62, be sure to indicate any missing points.)

62. $f(x) = \begin{cases} 6 - 2x & \text{if } x > 2 \\ 2x - 1 & \text{if } x \leq 2 \end{cases}$

For each pair of functions $f(x)$ and $g(x)$, find
   a. $f(g(x))$, b. $g(f(x))$.

63. $f(x) = x^2 + 1$; $g(x) = \dfrac{1}{x}$

64. $f(x) = \sqrt{x}$; $g(x) = 5x - 4$

65. $f(x) = \dfrac{x + 1}{x - 1}$; $g(x) = x^3$

66. $f(x) = |x|$; $g(x) = x + 2$

67. $f(x) = \sqrt[3]{x + 1}$; $g(x) = x^3 - 1$

68. $f(x) = 5$; $g(x) = 12$

69. **BUSINESS: Advertising Budget** A company's advertising budget is $A(p) = 2p^{0.15}$, where p is the company's profit, and the profit is predicted to be $p(t) = 18 + 2t$, where t is the number of years from now. (Both A and p

are in millions of dollars.) Express the advertising budget A as a function of t, and evaluate the function at $t = 4$.

70. **a.** Solve the equation $x^4 - 2x^3 - 3x^2 = 0$ by factoring.
    b. Use a graphing calculator to graph $y = x^4 - 2x^3 - 3x^2$ and find the x-intercepts of the graph. Be sure that you understand why your answers to parts (a) and (b) agree.

71. **a.** Solve the equation $x^3 + 2x^2 - 3x = 0$ by factoring.
    b. Use a graphing calculator to graph $y = x^3 + 2x^2 - 3x$ and find the x-intercepts of the graph. Be sure that you understand why your answers to parts (a) and (b) agree.

72. **BUSINESS: Revenue** The following table gives a company's annual revenue (in millions of dollars) from its overseas operations during its first 5 years.

Year	Revenue
1	2.0
2	1.8
3	1.9
4	2.1
5	2.8

a. Enter the table numbers into a graphing calculator and make a plot of the resulting points (Years on the x-axis and Revenue on the y-axis). What kind of curve do the points suggest?
b. Have your calculator fit such a curve to the data. Then enter the result in y_1, which gives a formula for the annual revenue each year. Plot the points together with the regression curve.
c. Use your formula to predict the revenue in years 6 and 7.

1.5 Exponential Functions

Graph each function.

73. $f(x) = 4^x$

74. $f(x) = \left(\frac{1}{4}\right)^x$

75. **PERSONAL FINANCE: Compound Interest** Find the value of $10,000 invested at 6% interest compounded quarterly for 5 years.

76. **PERSONAL FINANCE: Compound Interest** Find the value of $1,500 invested at 5% interest compounded daily for 10 years.

77. **PERSONAL FINANCE: Depreciation** A $21,000 car depreciates by 20% per year. Find the value after:
    a. 4 years.   **b.** 6 months.

78. **BUSINESS: Depreciation** A $70,000 ambulance depreciates by 30% per year. Find the value after:
    a. 5 years.   **b.** 6 months.

79. **BUSINESS: Continuous Compounding** A bond selling for $1000 has interest rate 4.8% compounded continuously. Find its value after ten years.

80. **BUSINESS: Continuous Compounding** A bond selling for $1,000,000 has interest rate 5.1% compounded continuously. Find its value after five years.

1.6 Logarithmic Functions

81. Find each logarithm *without* using a calculator.
    a. $\log 1000$ **b.** $\log \frac{1}{1000}$
    c. $\ln e^3$ **d.** $\ln \sqrt[4]{e}$

82. Find each logarithm *without* using a calculator.
    a. $\log \sqrt{10}$ **b.** $\log 10^8$
    c. $\ln \frac{1}{e}$ **d.** $\ln e^{3/2}$

Use the properties of natural logarithms to simplify each function.

83. $f(x) = \ln x^4 - \ln x^3 - \ln 1$

84. $f(x) = \ln e^{7x} - 5x - \ln e$

85. **PERSONAL FINANCE: Depreciation** A car depreciates by 22% per year. When will it be worth:
    a. half its original value?
    b. one quarter of its original value?

86. BUSINESS: Depreciation A corporate heli-
copter depreciates by 35% per year. When will
it be worth:

   a. half its original value?
   b. one quarter of its original value?

87–88. GENERAL: Fossils In the following
exercises, use the fact that the proportion of
potassium-40 remaining after t million years is
$e^{-0.00054t}$.

87. In 1984 in the Wind River Basin of Wyoming,
scientists discovered a fossil of a small, three-
toed horse, an ancestor of the modern horse.
Use the preceding formula to estimate the age
of this fossil if it contained 97.3% of its original
potassium-40.

88. Use the preceding formula to estimate the age
of a skull found in 1959 in Tanzania (dubbed
"Nutcracker Man" because of its huge jaw-
bone) that contained 99.9% of its original
potassium-40.

2 Mathematics of Finance

Application Preview

Musical Instruments as Investments*

No two guitars sound alike—good ones, anyway. Over the last 30 years, prices for high-quality acoustic instruments have soared. In 1993, a John D'Angelico New Yorker 18-inch-wide cutaway archtop ("jazz") guitar sold for $150,000, setting the record for the price of an acoustic fretted instrument not previously owned by a deceased super-star. Many pre-World War II instruments are valued not only for their craftsmanship but also for their distinctive sound. A mint condition Martin D-45 made in 1932–1942 that might have sold for $3500 in 1971 could be sold for $225,000 in 2003, while a Gibson 1932–1939 Master-tone Granada 5-string (replacement neck) banjo selling for just $1200 in 1971 would fetch $28,000 in 2003.

How can these price increases be compared to other investments such as stocks, bonds, or bank savings accounts? The dots on the graph show the typical selling prices over the 30-year period from 1971 to 2001 for a signed 1922 Gibson F-5 mandolin, together with a curve showing the increasing value of an investment earning 11% compound interest.

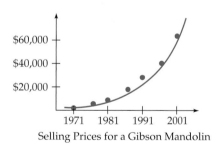

Selling Prices for a Gibson Mandolin

The dots closely match the curve, showing that this mandolin has increased in value by about 11% annually, better than many bond and stock investments. The values of fine musical instruments and other investments can be found by the methods of this section.

---

* Based on material originally provided by Stanley M. Jay and Larry Wexer of *mandolin bros. Ltd.* (Staten Island, New York), dealers in fine mandolins, guitars, and banjos.

2.1 SIMPLE INTEREST

Introduction

In the modern credit world, the old adage "time is money" has become a basic fact of economic life. When you open a savings account or take out a car loan, you directly experience the "time value of money." This chapter covers the basic financial properties of a loan between a lender and a borrower and the calculation of the interest and payments. We begin with simple interest.

Simple Interest

The *principal* of a loan is the amount of money borrowed from the lender, the *term* is the time the borrower has the money, and the *interest* is the additional money paid by the borrower for the use of the lender's money. The interest is called *simple interest* if it is calculated as a fixed percentage of the principal and is paid at the end of the term. The *interest rate* of the loan is the dollars of interest per $100 of principal per year (or "per annum"). It is usually stated as a percentage but is always written as a decimal in calculations. Simple interest is calculated as follows:

Simple Interest Formula

The interest I on a loan of P dollars at simple interest rate r for t years is

$$I = Prt$$

P = principal
r = rate
t = term

EXAMPLE 1

FINDING SIMPLE INTEREST

Find the interest on a loan of two million dollars at 7.2% simple interest for 4 months.

Solution

Writing the interest rate in decimal form and changing the term to years, we find that

$$I = 2{,}000{,}000 \cdot 0.072 \cdot \frac{4}{12} = 48{,}000$$

$I = Prt$ with
$P = 2{,}000{,}000,$
$r = 0.072,$ and $t = 4/12$

The interest is $48,000.

Be Careful! When using a calculator, *round off only your final answer.* For example, if you use $t = 0.33$ in the previous example instead of $\frac{4}{12}$ or $\frac{1}{3}$, you would get the value $2{,}000{,}000 \cdot 0.072 \cdot 0.33 = \$47{,}520$, which is wrong by $480.

Practice Problem 1

Find the interest on a loan of $50,000 at 19.8% simple interest for 3 months.

➤ Solution on page 106

If you have values for any three unknowns in the interest formula $I = Prt$, you can solve for the fourth one.

EXAMPLE 2

FINDING THE SIMPLE INTEREST RATE

What is the interest rate of a loan charging $18 simple interest on a principal of $150 after 2 years?

Solution

We solve $I = Prt$ for interest rate r:

$$r = \frac{I}{Pt} \qquad\qquad I = Prt \text{ divided by } Pt$$

Then

$$r = \frac{18}{150 \cdot 2} = \frac{18}{300} = 0.06 \qquad\qquad \begin{array}{l}\text{Substituting } I = 18,\\ P = 150, \text{ and } t = 2\end{array}$$

The interest rate is 6%.

Total Amount Due on a Loan

When the term of a loan is over, the borrower repays the principal and interest, so the total amount due is

$$\underbrace{P}_{\text{Principal}} + \underbrace{Prt}_{\text{Interest}} = \underbrace{P(1 + rt)}_{\text{Total amount}} \qquad \begin{array}{l}\text{Factoring out the}\\ \text{common term}\end{array}$$

This gives the following formula:

Total Amount Due for Simple Interest

The total amount A due at the end of a loan of P dollars at simple interest rate r for t years is

$$A = P(1 + rt)$$

Future value

The amount due may also be regarded as the *accumulated value* of an investment or the *future value* of the principal. As before, this formula may be used as it is or it may be solved for any one of the other variables, as the next few examples will show.

EXAMPLE 3

FINDING THE TOTAL AMOUNT DUE

What is the total amount due on a loan of $3000 at 6% simple interest for 4 years?

```
3000 → P
              3000
.06 → R
               .06
4 → T
                 4
P(1+RT)
              3720
```

Solution

$$A = 3000(1 + 0.06 \cdot 4)$$
$$= 3000 \cdot 1.24 = 3720$$

$A = P(1 + rt)$ with $P = 3000$, $r = 0.06$, and $t = 4$

The total due on the loan is $3720.

Present value

The $3000 loan in the above example grew to $3720 in 4 years. The $3720 is sometimes called the *future value* of the original $3000. Reversing our viewpoint, we say that the principal of $3000 is the *present value* of the later $3720. To find a formula for the present value, we solve the "total amount due" formula (on the previous page) for P by dividing by $1 + rt$:

$$P = \frac{A}{1 + rt}$$

Present value of the future amount A

EXAMPLE 4

FINDING A PRESENT VALUE

How much should be invested now at 8.6% simple interest if $10,000 is needed in 6 years?

Solution

The amount to invest now means the *present value* of $10,000 in 6 years. Using the above formula:

$$P = \frac{10,000}{1 + 0.086 \cdot 6} = \frac{10,000}{1.516} \approx 6596.31$$

$P = A/(1 + rt)$ with $A = 10,000$, $r = 0.086$, and $t = 6$

The amount required is $6596.31.

We emphasize that the *present value* is the amount that will grow to the required sum in the given time period (at the stated interest rate). For this reason, it gives the actual value *now* of an amount to be paid later.

Amounts like this, to be paid at some time in the future, occur in many situations from personal credit card balances to government bonds. Such "debts" are frequently bought and sold by banks and companies, and the present value gives the current value of such future payments. For example, at the interest rate stated in the preceding example, a payment of $10,000 to be received in 6 years is worth exactly $6596.31 now, and so should be bought or sold for this price.

Practice Problem 2

Find the present value of a "promissory note" that will pay $5000 in 4 years at 12% simple interest.   ➤ Solution on page 106

We found the present value by deriving a formula for it. We could instead have substituted the given numbers into the "total amount due" formula and then solved for P (which we may think of as standing for *principal* or *present value*). We will do the next example in this way, substituting numbers into the original formula and then solving for the remaining variable. You can use either method to solve these problems.

EXAMPLE 5

FINDING THE TERM OF A SIMPLE INTEREST LOAN

What is the term of a loan of $2000 at 4% simple interest if the amount due is $2400?

Solution

Substituting the given numbers for the appropriate variables in the "total amount due" formula (page 101) gives

$$2400 = 2000(1 + 0.04t)$$ $A = P(1 + rt)$ with $A = 2400$, $P = 2000$, and $r = 0.04$

$$1.2 = 1 + 0.04t$$ Dividing by 2000

$$0.2 = 0.04t$$ Subtracting 1

$$t = \frac{0.2}{0.04} = 5$$ Dividing by 0.04 and reversing sides

The term is 5 years.

Graphing Calculator Exploration

The formula $A = P(1 + rt)$ may be written as $A = (Pr)t + P$ so that it has the familiar $y = mx + b$ form of a straight line with slope Pr and y-intercept P (but with t and A instead of x and y).

a. Using the values $P = 2000$ and $P \cdot r = 2000 \cdot 0.04 = 80$ from the preceding example, graph the line $y_1 = 80x + 2000$ on the window $[0, 8]$ by $[0, 4000]$, so that y gives the amount of the loan for any term x.

b. Use TRACE to estimate the term x that gives $y = 2400$. How does your "graphical" answer compare to the answer for Example 5? Try ZOOMing IN to improve your estimate.

c. Add the horizontal line $y_2 = 2400$ to your graph and use INTERSECT to find the intersection point. Does the x-value of this point exactly match the answer for Example 5?

Discounted Loans and Effective Interest Rates

In a *discounted loan* the lender deducts the interest from the amount the borrower receives at the start. A discounted loan is better for the lender because getting the money early (so it can earn interest somewhere else) is always better than later. Since the borrower actually is receiving a smaller amount, we may recalculate the interest rate as a "standard" loan on this smaller amount. The resulting rate is called the *effective* simple interest rate of the loan.

EXAMPLE **6**

FINDING THE EFFECTIVE SIMPLE INTEREST RATE

Determine the effective simple interest rate on a discounted loan of $1000 at 6% simple interest for 2 years.

Solution

The interest is $\$1000 \cdot 0.06 \cdot 2 = \$120$, so as a simple interest loan, the borrower receives only $P = 1000 - 120 = \$880$ and agrees to pay back $1000 at the end of 2 years. If we write r_s for the simple interest rate of this loan, the "total amount due" formula on page 101 gives

$$1000 = 880(1 + r_s \cdot 2)$$

$A = P(1 + rt)$ with $A = 1000$, $P = 880$, and $t = 2$

Solving for r_s,

$$r_s = \frac{1}{2}\left(\frac{1000}{880} - 1\right) \approx 0.0682$$

Dividing by 880, subtracting 1, and dividing by 2

The effective simple interest rate on this discounted loan is 6.82%.

Notice that the effective rate of 6.82% is significantly greater than the stated (or "nominal") rate of 6%. In general, the nominal rate in a discounted loan may be used to conceal the true cost to the borrower (and similarly, the true benefit to the lender). This is why effective rates are so important.

Using the same method as in Example 6, we can find a formula for the effective rate. For a discounted loan of amount A at interest rate r for term t, the borrower receives $A - Art = A(1 - rt)$ dollars. Therefore, the "total amount due" formula from page 101, but with the simple interest rate replaced by the effective rate r_s, gives

$$A = \underbrace{A(1 - rt)}_{\text{Principal}}(1 + r_s t)$$

$A = P(1 + rt)$ with $A(1 - rt)$ for P

Solving for r_s:

$$\frac{1}{1 - rt} = 1 + r_s t$$

Canceling the A's and dividing by $(1 - rt)$

$$r_s t = \frac{1}{1 - rt} - 1 = \frac{1 - (1 - rt)}{1 - rt} = \frac{rt}{1 - rt}$$

Reversing sides, subtracting 1, and simplifying

$$r_s = \frac{1}{t}\frac{rt}{1 - rt} = \frac{r}{1 - rt}$$

Dividing by t and then canceling

This gives the following formula for r_s.

> **Effective Simple Interest Rate for a Discounted Loan**
>
> For a discounted loan at interest rate r for t years, the effective simple interest rate r_s is
>
> $$r_s = \frac{r}{1 - rt}$$

Notice that the effective rate r_s will be larger than r (since the denominator is less than 1) and that it depends only on the *rate* and *term* of the loan, and not on its amount. For the loan in Example 6, our formula gives the answer that we found:

$$r_s = \frac{0.06}{1 - 0.06 \cdot 2} = \frac{0.06}{0.88} \approx 0.0682 \qquad r_s = \frac{r}{1 - rt} \text{ with } r = 0.06 \text{ and } t = 2$$

Practice Problem 3   What is the effective simple interest rate of a discounted loan at 6% interest for 3 years? for 5 years?

➤ Solution below

2.1 Section Summary

The simple interest formula is

$$I = Prt$$

I = simple interest
P = principal
r = interest rate
t = term in years

We can solve for any one of the variables if the others are known. The total amount due at the end of the loan (principal plus interest) is

$$A = P(1 + rt)$$

A = total amount due

This equation may be solved for any one of the variables, keeping in mind that the amount A at the end of the loan is the *future value* of the principal P, and P is the *present value* of the future amount A.

For a *discounted* loan, the interest is subtracted from the principal at the beginning of the loan. The effective simple interest rate r_s for a discounted loan at rate r for t years is

$$r_s = \frac{r}{1 - rt}$$

➤ **Solutions to Practice Problems**

1. $I = 50{,}000 \cdot 0.198 \cdot \frac{3}{12} = 2475.$

2. $P = \frac{5000}{1 + 4 \cdot 0.12} = \frac{5000}{1.48} \approx 3378.38$

3. $r_s = \frac{0.06}{1 - 0.06 \cdot 3} \approx 0.0732.$ The effective simple interest rate is 7.32%.

For 5 years, $r_s = \frac{0.06}{1 - 0.06 \cdot 5} \approx 0.0857.$ The effective simple interest rate is 8.57%. Can you think of an intuitive reason for the effective rate to be higher if the term is longer? [*Hint:* Think of how much earlier the lender gets the money or of how much less the borrower really gets.]

2.1 **Exercises**

Find the simple interest on each loan.

1. $1500 at 7% for 10 years.

2. $2000 at 9% for 7 years.

3. $6000 at 6.5% for 8 years.

4. $4500 at 4.25% for 9 years.

5. $825 at 6.58% for 5 years 6 months.

6. $950 at 5.87% for 6 years 3 months.

7. $1280 at 4.8% for 3 months.

8. $5275 at 5.3% for 2 months.

Find the total amount due for each simple interest loan.

9. $1500 at 7% for 10 years.

10. $2000 at 9% for 7 years.

11. $6100 at 5.7% for 4 years 9 months.

12. $4500 at 6.3% for 3 years 6 months.

13. $3125 at 4.81% for 10 months.

14. $8775 at 13.11% for 7 months.

APPLIED EXERCISES

15. Interest Rate Find the interest rate on a loan charging $704 simple interest on a principal of $2750 after 4 years.

16. Interest Rate Find the interest rate on a loan charging $1127 simple interest on a principal of $4900 after 5 years.

17. Principal Find the principal of a loan at 8.4% if the simple interest after 5 years 6 months is $1155.

18. Principal Find the principal of a loan at 7.6% if the simple interest after 9 years 3 months is $2109.

19. Term Find the term of a loan of $175 at 9% if the simple interest is $63.

20. Term Find the term of a loan of $225 at 7% if the simple interest is $94.50.

21. Present Value How much should be invested now at 5.2% simple interest if $8670 is needed in 3 years?

22. Present Value How much should be invested now at 4.8% simple interest if $4530 is needed in 4 years 4 months?

ZERO COUPON BONDS A *zero coupon bond* pays only its *face value* on maturity (getting its name be-

cause it has no "coupons" for interest payments before that date). The *fair market price* is the present value of the face value at the current interest rate.

23. What is the fair market price of a $10,000 zero coupon bond due in 1 year if today's long-term simple interest rate is 5.81%?

24. What is the fair market price of a $15,000 zero coupon bond due in 1 year if today's long-term simple interest rate is 7.23%?

25. What is the fair market price of a $5000 zero coupon bond due in 2 years if today's long-term simple interest rate is 3.54%?

26. How Interest Rates Affect Present Value The present value formula $P = A/(1 + rt)$ can be viewed on your graphing calculator as a function y of the simple interest rate x by rewriting it as $y_1 = A/(1 + xt)$. Using $A = $10,000$ and $t = 1$ year, graph this expression with window [0, 1] by [0, 12,000] to see the present value of a $10,000 zero coupon bond due in 1 year as a function of the interest rate. Use TRACE or VALUE to find the present value of the bond for interest rates of 4%, 5%, 6%, 7%, 8%, and 9%. Does a 1% increase in the interest rate always cause the same decrease in the present value of the bond?

27. **Term** What should be the term for a loan of $6500 at 7.3% simple interest if the lender wants to receive $9347 when the loan is paid off?

28. **Term** What should be the term for a loan of $5400 at 5.8% simple interest if the lender wants to receive $6966 when the loan is paid off?

29. **Term** How long will it take an investment at 8% simple interest to increase by 70%?

30. **Term** How long will it take an investment at 4.2% simple interest to increase by 26.6%?

DOUBLING TIME The *doubling time* of an investment is the number of years it takes for the value to double. This is the same as the number of years for the value to increase by 100%.

31. What is the doubling time of a 5% simple interest investment?

32. Show that the doubling time of a simple interest investment is $1/r$ where r is the simple interest rate.

33. **Effective Rate** What is the effective simple interest rate of a discounted loan at 4.6% interest for 3 years 6 months?

34. **Effective Rate** What is the effective simple interest rate of a discounted loan at 7.2% interest for 2 years 10 months?

35. **Discounted Loan** The Dewey, Cheetham, and Howe Loan Corporation offers discounted loans of $1000 at 5% for 20 years. Explain why their name is appropriate.

36. **How the Term Affects the Discounted Rate** The effective rate formula $r_s = r/(1 - rt)$ can be viewed on your graphing calculator as a function of the term t by rewriting it as $y_1 = r/(1 - rx)$. Using $r = 0.05$, graph this expression with window [0, 20] by [0, 1] to see

the effective rate y of the discounted loan as a function of the term x. Use TRACE or VALUE to find the effective rate of the loan for terms of 5, 10, and 15 years. Does a 1-year increase in the term always cause the same increase in the effective rate?

37. **Bridge Loan** You have a buyer for your condominium, but the seller of your dream house wants to close now. In order to get enough money to go to the closing, you take out a 2-month bridge loan for $72,000 at 7.92% simple interest. You assume that you will close on the sale of your condominium before the 2 months are up. How much interest will you have to pay on the loan?

38. **Home Improvements** Because the contractor was short of cash when the job was finished, a plumber accepted as payment a promissory note for $2000 plus 18% simple interest due in 2 months. Needing cash himself, the plumber sold the note 1 month later for $2000 to a local loan agency. Find the simple interest rate the agency will earn on its investment.

BROKERAGE COMMISSIONS An Internet discount brokerage firm charges a commission of 10% on the first $20,000 plus 5% of the excess over $20,000 on each buy or sell transaction. Find the simple interest rate earned by each of the following investments after including the commissions paid to the brokerage firm.

39. Purchase 900 shares of American WebWide Education at $18.50 per share and sell them 4 months later at $26.75 per share.

40. Purchase 1500 shares of Well Care Deluxe at $11.90 per share and sell them 10 months later at $14.80 per share.

41. Purchase 800 shares of DucoFood Services at $28.70 per share and sell them 7 months later at $37.80 per share.

2.2 COMPOUND INTEREST

Introduction

When a simple interest loan is not paid off at the end of its term but is taken out as a new loan, the borrower owes the principal plus the interest. The interest on the new loan is called *compound interest* because it combines interest on both the original principal and on the unpaid interest. For example, if you borrow $8000 at 5% simple interest for 1 year, the formula on page 101 says that the amount due will be $8000(1 + 0.05) = $8400. If you do not pay it off, but borrow this sum for the next year, the amount due after 2 years will be $8000(1 + 0.05)(1 + 0.05) = $8000(1 + 0.05)^2 = $8820. After a third year, the amount due would become $8000(1 + 0.05)^3 = $9261. For each subsequent year, this amount would be multiplied by another (1 + 0.05), which is 1 plus the interest rate. Compounding clearly increases the amount of interest paid because of the interest on the interest.

Compound Interest

In general, the amount A due on a compound interest loan of P dollars at interest rate r per year compounded annually for t years is found by repeatedly multiplying the principal P by $(1 + r)$, once for each year:

$$A = \underbrace{P(1 + r)(1 + r) \cdots (1 + r)}_{\substack{t \text{ multiplications} \\ \text{by } (1 + r)}} = P(1 + r)^t$$

Graphing Calculator Exploration

To see that compound interest eventually surpasses simple interest (even with a higher rate and principal), compare $500 invested at 4% compounded annually with $2500 invested at 8% simple interest.

a. Graph the simple interest amount $A = P(1 + rt)$ as $y_1 = 2500(1 + .08x)$ on the window [0, 150] by [0, 35,000] so that y_1 is the amount of the investment after x years.

b. Graph the compound interest amount $A = P(1 + r)^t$ as $y_2 = 500(1 + .04)^x$ on the same window.

(continues)

c. Use TRACE or INTERSECT to find when the compound interest amount equals the simple interest amount. What happens after this intersection point? Try your graphs with the larger windows [0, 300] by [0, 70,000] and by [0, 500,000].

Notice how the formulas for the amounts due under simple and compound interest differ:

$$\text{Simple} \qquad P(1 + rt)$$
$$\text{Compound} \qquad P(1 + r)^t$$

The formula for *compound* interest has t in the *exponent*, which is why it grows so much faster (eventually).

Banks always state *annual* interest rates, but the compounding can be done more than once a year. Some standard compounding periods are:

Compounding Frequency	Periods per Year
Annual	1
Semiannual	2
Quarterly	4
Monthly	12
Daily	365

Given the same annual interest rate, more frequent compounding is better since your money starts earning interest sooner. For example, if a bank offers 8% compounded quarterly, then the interest is calculated on a *quarterly* basis: each quarter you get 2%, (one quarter of the 8%), and this is done every *quarter*, so the exponent is the number of quarters, which is *four times* the number of years. In general, if the annual interest rate is r with m compoundings per year, then the interest rate per period is $\frac{r}{m}$ and the number of compounding periods in t years is mt. This leads to the following general formula for compound interest:

Compound Interest Formula

The amount A due on a loan of P dollars at yearly interest rate r compounded m times per year for t years is

$$A = P\left(1 + \frac{r}{m}\right)^{mt}$$

r = annual rate
m = periods per year
t = term (in years)

The stated yearly interest rate r is called the *nominal rate* of the loan.

EXAMPLE 1   **FINDING THE AMOUNT DUE ON A COMPOUND INTEREST LOAN**

Find the amount due on a loan of $1500 at 4.8% compounded monthly for 2 years.

Solution

The nominal rate of 4.8% expressed as a decimal is 0.048, and monthly compounding means that $m = 12$, so we have

$$A = 1500\left(1 + \frac{0.048}{12}\right)^{12 \cdot 2}$$

$A = P(1 + r/m)^{mt}$
with $P = 1500$, $r = 0.048$,
$m = 12$, and $t = 2$

$$= 1500(1.004)^{24} \approx 1650.82$$

The amount due is $1650.82.

Practice Problem 1   Find the amount that is due on a loan of $3500 at 5.1% compounded quarterly for 3 years.   ➤ Solution on page 117

A loan is really an amount that a bank *invests* in a borrower, so any question about a loan can be rephrased as a question about an *investment*. Example 1 could have asked for the value of an investment of $1500 that grows by 4.8% compounded monthly for 2 years—the answer would still be $1650.82. In this section we will speak interchangeably of *loans* and *investments*, since they are the same but from different viewpoints, depending whether you are the borrower or the lender.

In Example 1, the amount $1650.82 may be called the *future value* of the original $1500, and conversely, the $1500 is the *present value* of the later amount $1650.82. As before, the present value gives the value *today* of a payment that will be received at some time in the future. To find a formula for the present value, we solve the compound interest formula (on the previous page) for P, using the familiar $1/x^n = x^{-n}$ rule of exponents.

Now t years later

P
$\longrightarrow A = P\left(1 + \frac{r}{m}\right)^{mt}$

Future value

Now t years later

$P = \dfrac{A}{\left(1 + \frac{r}{m}\right)^{mt}}$ A

Present value

$$P = \frac{A}{\left(1 + \dfrac{r}{m}\right)^{mt}} = A\left(1 + \frac{r}{m}\right)^{-mt}$$ Present value

This formula should be intuitively reasonable: To find *future* value you *multiply* by $(1 + \frac{r}{m})^{mt}$, so to find *present* value you should *divide* by the same quantity.

EXAMPLE 2

FINDING THE PRESENT VALUE

How much should be invested now at 8.6% compounded weekly if $10,000 is needed in 6 years?

Solution

$$P = 10{,}000\left(1 + \frac{0.086}{52}\right)^{-52 \cdot 6} \approx 5971.58$$

$P = A(1 + r/m)^{-mt}$ with
$A = 10{,}000$, $r = 0.086$,
$m = 52$, and $t = 6$

The amount required is $5971.58.

Growth Times

We solved the preceding example by using the formula for present value. Instead, we could have substituted the given numbers into the compound interest formula and *then* solved for the remaining variable. We will do the next example in this way, finding a "growth time," the time for a loan to reach a given value. We will use the rule of logarithms $\log(x^n) = n \log x$ to "bring down the power." Logarithms with *any* base may be used in such calculations.

EXAMPLE 3

FINDING THE TERM OF A COMPOUND INTEREST LOAN

What is the term of a loan of $2000 at 6% compounded monthly that will have an amount due of $2400?

Solution

We substitute the given numbers into the compound interest formula (see page 110):

$$2400 = 2000\left(1 + \frac{0.06}{12}\right)^{12 \cdot t}$$

$A = P(1 + r/m)^{mt}$ with
$A = 2400$, $P = 2000$,
$r = 0.06$, and $m = 12$

$$1.2 = 1.005^{12t}$$

Dividing by 2000 and simplifying

$$\log 1.2 = \log 1.005^{12t}$$

Taking logarithms

$$\log 1.2 = 12t \log 1.005$$

Using $\log(x^n) = n \log x$ to bring down the power

$$12t = \frac{\log 1.2}{\log 1.005} \approx 36.555$$

Dividing by $\log 1.005$, reversing sides, and calculating logarithms

—————— Months

This amount of time, $12t = 36.555$, is in *months* (t is years, but $12t$ is *months*). We round *up* to the nearest month (because a shorter term will not reach the needed amount), so the term needed is 37 months, or 3 years 1 month.

Hint: Instead of solving for t, if the compounding is monthly stop when you find $12t$ (the number of months), if quarterly stop when you find $4t$ (the number of quarters), if weekly stop when you find $52t$ (the number of weeks), and so on. Then round *up* to the next whole number of periods to ensure that the needed amount will actually be reached.

Notice that the actual amounts $2400 and $2000 did not matter in this calculation, only their *ratio* $\frac{2400}{2000} = 1.2$. Thus, a loan of $2000 grows to $2400 in the same amount of time that a loan of 20 *million* dollars would grow to 24 million dollars (at the stated interest rate) since the ratios are the same.

Therefore, in Example 3 we could have omitted the dollar amounts and simply asked for the term of a loan that would *multiply the principal by 1.2*. Furthermore, multiplying by 1.2 means *increasing by 0.2, or 20%*, so we could have asked for the term of a loan that would *increase the principal by 20%*—all three formulations are equivalent.

Be careful! Distinguish carefully between the *increase* and the *multiplier*. Increasing by 20% means *multiplying by 1.2* (the "1" keeps the original amount and the ".2" increases it by 20%). For example, to *increase* an amount by 35%, you would *multiply* it by 1.35; to increase an amount by 75%, you would multiply by 1.75; and to increase it by 100%, you would multiply by 2.

Practice Problem 2

What is the multiplier that corresponds to increasing the amount by 50%?

➤ Solution on page 117

EXAMPLE 4

FINDING THE TERM TO INCREASE THE PRINCIPAL

What is the term of a loan at 6% compounded weekly that will double the principal?

Solution

Doubling means multiplying the amount by 2. Using P for the unknown principal, we must solve

$$2P = P\left(1 + \frac{0.06}{52}\right)^{52 \cdot t}$$

P (for principal) on both sides

$$2 = (1 + 0.06/52)^{52t}$$

Canceling the P's and simplifying

$$\log 2 = 52t \log (1 + 0.06/52)$$

Taking logarithms and bringing down the power

$$52t = \frac{\log 2}{\log (1 + 0.06/52)} \approx 601.1$$

Dividing by $\log (1 + 0.06/52)$, reversing sides, and calculating logarithms

Weeks

Rounding *up*, the term needed is 602 weeks, or 11 years 30 weeks.

Practice Problem 3

What is the term of a loan at 6% compounded quarterly that increases the principal by 50%?

➤ **Solution on page 117**

Rule of 72

The *doubling time* of an investment or a loan is the number of years it takes for the value to double. Doubling times are often estimated by using the *rule of 72:*

$$\left(\begin{array}{c} \text{Doubling} \\ \text{time} \end{array} \right) \approx \frac{72}{r \times 100}$$

Divide 72 by the rate times 100

The rule of 72, however, gives only an *approximation* for the doubling time, but it is often quite accurate. (A justification of this rule is given in Exercise 50.) For example, for a loan at 6% compounded quarterly the rule of 72 gives

$$\left(\begin{array}{c} \text{Doubling} \\ \text{time} \end{array} \right) \approx \frac{72}{0.06 \times 100} = \frac{72}{6} = 12 \text{ years}$$

Rule of 72 estimate

In Example 4 we found the correct doubling time to be 11 years 30 weeks, so the rule of 72 is not far off.

Effective Rates

Suppose that you are considering two certificates of deposit (CDs), one offering 6.2% compounded semiannually and the other offering 6.15% compounded monthly. Which should you choose, the one with the higher rate or the one with more frequent compounding? More generally, how can we compare any two different interest rates with different compoundings? We do so by calculating the actual percentage increase that each will generate in a year. This number is called the *annual percentage yield* or *APY*. More formally, the *APY* is also called the *effective rate of interest*, denoted r_e, which is the simple interest rate that will return the same amount on a one-year loan. Therefore, r_e can be found by solving

$$P(1 + r_e) = P\left(1 + \frac{r}{m}\right)^m$$

Finding the "simple" rate that gives the "compound" rate for $t = 1$

Canceling the P's and subtracting the 1 on the left gives the following.

Effective Rate for a Compound Interest Loan

For a compound interest loan at interest rate r compounded m times per year, the effective rate r_e is

$$r_e = \left(1 + \frac{r}{m}\right)^m - 1 \qquad r_e = APY$$

The effective rate can be interpreted as the interest earned by \$1 for 1 year.

EXAMPLE 5

USING EFFECTIVE RATES TO COMPARE INVESTMENTS

A self-employed carpenter setting up a Keogh retirement plan must choose between a certificate of deposit (CD) with the First & Federal Bank at 6.2% compounded semiannually or a CD with the Chicago Nationswide Bank at 6.15% compounded monthly. Which is better?

Solution

The effective rate for 6.2% compounded semiannually (First & Federal) is

$$r_e = \left(1 + \frac{0.062}{2}\right)^2 - 1 \approx 0.0630 \qquad r_e \approx 6.30\%$$

For 6.15% compounded monthly (Chicago Nationswide), it is

$$r_e = \left(1 + \frac{0.0615}{12}\right)^{12} - 1 \approx 0.0633 \qquad r_e \approx 6.33\%$$

The CD at the Chicago Nationswide Bank is better.

Sometimes the interest rate is not given, and only the amounts at the beginning and end of the loan are known. How can we find the interest rate (called the *effective* rate of interest)?

The effective rate of return for an investment of P dollars that returns an amount A after t years can be found from the compound interest formula (page 110) with $m = 1$:

$$A = P(1 + r_e)^t$$

$A = P(1 + r/m)^{mt}$
with $m = 1$ and $r = r_e$

An example of this is a *zero coupon bond*, which pays its *face value* on maturity and sells now for a lower price. This kind of bond makes no payments before maturity and so has no interest "coupons."

EXAMPLE 6

EFFECTIVE RATE FOR A ZERO COUPON BOND

What is the effective rate of return of a $10,000 zero coupon bond maturing in 6 years and offered now for sale at $5500?

Solution

We solve

$$10,000 = 5500(1 + r_e)^6$$

$A = P(1 + r_e)^t$ with
$A = 10,000$, $P = 5500$,
and $t = 6$

$$\frac{10,000}{5500} = (1 + r_e)^6$$

Dividing by 5500

$$(100/55)^{1/6} = 1 + r_e$$

Raising to power $\frac{1}{6}$

$$r_e = (100/55)^{1/6} - 1 \approx 0.1048$$

Solving for r_e

The effective rate is 10.48%.

2.2 Section Summary

The *compound interest formula* is

$$A = P\left(1 + \frac{r}{m}\right)^{mt}$$

$A =$ amount due
$P =$ principal
$r =$ interest rate
$m =$ compoundings per year
$t =$ term in years

A is the *future value* of P, and P is the *present value* of A. Given values for four of the variables, we can solve for the fifth. Remember, however, to round the term *up* to the next whole number of compounding periods. The *doubling time*, the time for an amount to double $(A = 2P)$, can be approximated by the *rule of 72*: $t \approx 72/(r \times 100)$ years. Remember the difference between *increasing* by a percentage and *multiplying* by a number: Increasing by 50% means multiplying by 1.50.

The *effective rate* or *APY* of a loan is

$$r_e = \left(1 + \frac{r}{m}\right)^m - 1$$

r_e = effective rate of interest at rate r compounded m times each year

The effective rate of return for *any* investment of P dollars that returns A dollars after t years (such as a zero coupon bond) is found from the compound interest formula with $m = 1$, $A = P(1 + r_e)^t$.

> ➤ **Solutions to Practice Problems**

1. $A = 3500\left(1 + \dfrac{0.051}{4}\right)^{4 \cdot 3} \approx 4074.69.$ The amount due is \$4074.69.

2. 1.5

3. $1.5P = P\left(1 + \dfrac{0.06}{4}\right)^{4t}$ simplifies to $1.5 = (1 + 0.06/4)^{4t}$.

Taking logarithms: $\log 1.5 = \log (1 + 0.06/4)^{4t} = 4t \log (1 + 0.06/4)$

so $4t = \dfrac{\log 1.5}{\log (1 + 0.06/4)} \approx 27.2$ (quarters!). Rounding up gives 28 quarters, or 7 years.

2.2 **Exercises**

Determine the amount due on each compound interest loan.

1. \$15,000 at 8% for 10 years if the interest is compounded:

   a. annually.   **b.** quarterly.

2. \$7500 at 6.5% for 8 years 6 months if the interest is compounded:

   a. annually.   **b.** quarterly.

3. \$17,500 at 8.5% for 6 years 9 months if the interest is compounded:

   a. annually.   **b.** weekly.

4. \$25,000 at 9% for 5 years if the interest is compounded:

   a. semiannually.   **b.** monthly.

5. \$12,000 at 7.5% for 4 years 6 months if the interest is compounded:

   a. semiannually.   **b.** monthly.

6. \$16,750 at 4.5% for 6 years 6 months if the interest is compounded:

   a. semiannually.   **b.** monthly.

Calculate the present value of each compound interest loan.

7. \$25,000 after 7 years at 12% if the interest is compounded:

   a. annually.   **b.** quarterly.

8. \$9500 after 4 years 6 months at 8.4% if the interest is compounded:

   a. annually.   **b.** quarterly.

9. $19,500 after 6 years 9 months at 7.3% if the interest is compounded:

   a. annually.   **b.** weekly.

10. $16,500 after 5 years at 9% if the interest is compounded:

    a. semiannually.   **b.** monthly.

11. $11,500 after 4 years 3 months at 8.4% if the interest is compounded:

    a. semiannually.   **b.** monthly.

12. $2200 after 2 years 6 months at 7.5% if the interest is compounded:

    a. semiannually.   **b.** monthly.

Find the term of each compound interest loan.

13. 8.2% compounded quarterly to obtain $8400 from a principal of $2000.

14. 8.2% compounded annually to obtain $8400 from a principal of $2000.

15. 6.8% compounded quarterly to multiply the principal by 1.8.

16. 5.48% compounded daily to multiply the principal by 1.5.

17. 8.5% compounded monthly to increase the principal by 65%.

18. 8.5% compounded semiannually to increase the principal by 65%.

Use the "rule of 72" to estimate the doubling time (in years) for each interest rate, and then calculate it exactly.

19. 9% compounded annually.

20. 6% compounded quarterly.

21. 7.9% compounded weekly.

22. 9% compounded semiannually.

23. 6.1% compounded monthly.

24. 5.9% compounded daily.

Find the effective rate of each compound interest rate or investment.

25. 18% compounded monthly. [*Note:* This is a typical credit card interest rate, often stated as 1.5% per month.]

26. 4.3% compounded weekly.

27. 8.57% compounded semiannually.

28. 7.5% compounded quarterly.

29. A $50,000 zero coupon bond maturing in 8 years and selling now for $23,500.

30. A $10,000 zero coupon bond maturing in 12 years and selling now for $6400.

APPLIED EXERCISES

31. **Bond Funds** During the mid-1990s, the T. Rowe Price International Bond fund returned 10.43% compounded monthly. How much would a $5000 investment in this fund have been worth after 3 years?

32. **Mutual Funds** During the first half of the 1990s, the Twentieth Century Ultra aggressive growth mutual fund returned 19.83% compounded quarterly. How much would a $10,000 investment in this fund have been worth after 5 years?

33. **College Savings** How much would your parents have needed to set aside 17 years ago at 7.3% compounded weekly to give you $50,000 for college expenses today?

34. **College Savings** How much would your parents have needed to set aside 16 years ago at 6.7% compounded semiannually to give you $60,000 for college expenses today?

35. **Bond Funds** You have just received $125,000 from the estate of a long-lost rich uncle. If you invest all of your inheritance in a tax-free bond fund earning 6.9% compounded quarterly, how long do you have to wait to become a millionaire?

36. Home Buying You and your new spouse have decided to use all $6500 of your wedding present monies as a nest egg for a house down payment. Investing at 11.47% compounded monthly, how long must you wait to have enough to put 10% down on a $140,000 house?

37. Mutual Funds A $10,000 investment in the Fidelity Blue Chip Growth mutual fund in 1987 would have been worth $30,832 seven years later. What was the effective rate of this investment?

38. Rate Comparisons The First Federal Bank offers 4.7% compounded weekly passbook savings accounts while Consolidated Nationwide Savings offers 4.73% compounded quarterly. Which bank offers the better rate?

39. Rate Comparisons The Second Peoples National Bank offers a long-term certificate of deposit earning 6.43% compounded monthly. Your broker locates a $20,000 zero coupon bond rated AA by Standard & Poor's for $7965 and maturing in 14 years. Which investment will give the greater rate of return?

40. Horse Trading In March 1995, a descendant of an early Texas settler sent $100 to Sam Houston IV to make good on a $100 debt (possibly from the sale of a horse) owed for 160 years to Sam Houston, the hero of Texas independence. The check was donated to the Sam Houston Museum and the debt considered as settled. However, newspapers reported that a banker had estimated the accumulated interest on the loan to be $420 million. What yearly interest rate did the banker use in her calculations?

41. Lottery Winnings You have just won $100,000 from a lottery. If you invest all this amount in a tax-free money market fund earning 7% compounded weekly, how long do you have to wait to become a millionaire?

42. Becoming a Millionaire How much would you have to invest now at 6.5% interest compounded semiannually to have a million dollars in 40 years?

43. Rate Comparisons The People's State Bank offers 4.2% compounded quarterly, while Statewide Federal offers 4.1% compounded daily. Which bank offers the better rate?

44. Rate Comparisons The Southwestern Savings and Loan Bank offers a 10-year certificate of deposit earning 6.4% compounded monthly. Your broker offers you a $10,000 zero coupon bond costing $5325 and maturing in 10 years. Which gives the greater rate of return?

Explorations and Excursions

The following problems extend and augment the material presented in the text.

More About Compound Interest

45. Show that after two years, the (yearly) compound interest amount $A = P(1 + r)^2$ exceeds the simple interest amount $A = P(1 + 2r)$ by $P(r^2)$.

46. Show that after three years, the (yearly) compound interest amount $A = P(1 + r)^3$ exceeds the simple interest amount $A = P(1 + 3r)$ by $P(3r^2 + r^3)$.

47. Show that $P(1 + r/m)^{mt} = P(1 + r_e)^t$.

48. Show that the *effective m-compound rate* r_m for an investment compounded m times per year and returning A dollars from P dollars after t years is $r_m = m((A/P)^{1/mt} - 1)$. Then check this formula by showing that the rate r_m compounded m times for one year gives the effective rate $r_e = (A/P)^{1/t} - 1$.

More About the Rule of 72

49. Compare the "rule of 72" with the exact formula for the doubling time by graphing both as functions of the interest rate as follows. For interest compounded monthly, the number of years y for the loan to double at interest rate x is $y_1 = \log(2)/(12 \log(1 + x/12))$ while the "rule of 72" estimate is $y_2 = 72/(x \times 100)$. Graph both curves on the window [0, .5] by [0, 20] and explore them using the TRACE command. How closely do these curves match each other?

Note: Exercise 50 requires the *continuous* compounding formula from page 69.

50. a. Show that the doubling time for interest rate r compounded continuously is $\frac{1}{r} \ln 2$.

   b. How does the numerical value of $\ln 2$ explain the "rule of 72"?

2.3 ANNUITIES

Introduction

An *annuity* is a scheduled sequence of payments. Some annuities, such as retirement pensions, you receive the payments, while in others, such as car loans and home mortgages, you pay someone else. In this section we will consider an *ordinary annuity,* which is an annuity with equal payments made at the end of regular intervals, with the interest compounded at the end of each. For example, a car loan for 4 years with monthly payments of $200 at 6% compounded monthly is an ordinary annuity.

A First Example

While having a new car is nice, when the first payment of the above loan comes due at the end of the first month, you may wish that instead you were saving that $200 for yourself. If you were, then 4 years later your first $200 at 6% compounded monthly would have grown to

$$200\left(1 + \underbrace{\frac{0.06}{12}}_{1\, +\, 0.005\, =\, 1.005}\right)^{47} = 200 \cdot 1.005^{47}$$

The exponent is 47 rather than 48 because the first payment was made at the end of the first month and so earns interest for only 47 months. Similarly, your second payment would grow to $200 \cdot 1.005^{46}$, your third to $200 \cdot 1.005^{45}$, and so on down to your last payment of 200 (at the end of the last month, so earning no interest).

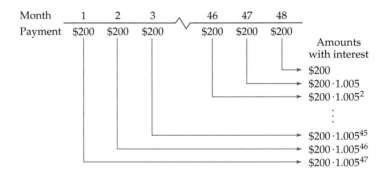

With patience and care we could evaluate and add up the numbers in the right side of this diagram, but there is an easier way. We will develop a formula for such sums and then use it to find the total amount of this annuity.

Geometric Series

The numbers in the right-hand side of the preceding diagram are found by successively multiplying by 1.005. In general, a sum of numbers each of which is a fixed multiple of the one before it is called a *geometric series*. The simplest geometric series is of the form

$$S = 1 + x + x^2 + \cdots + x^{n-1}$$

Multiplying by x gives the next number, and S is the sum of the series

If we multiply the sum by x, (which increases the exponent of each term) and subtract the original sum, we obtain

$$x \cdot S = \qquad x + x^2 + \cdots + x^{n-1} + x^n \qquad \text{x times S}$$
$$\underline{S = 1 + x + x^2 + \cdots + x^{n-1}} \qquad \text{S (lining up "like" terms)}$$

$$\underbrace{x \cdot S - S}_{S(x-1)} = -1 \qquad\qquad\qquad + x^n \qquad \text{Subtracting (most terms cancel)}$$

This last line can be written

$$S(x - 1) = x^n - 1$$

Factoring on the left, reversing the order on the right

so that

$$S = \frac{x^n - 1}{x - 1}$$

Dividing both sides by $x - 1$ (for $x \neq 1$)

Since S was defined as the sum $S = 1 + x + x^2 + \cdots + x^{n-1}$, this gives a formula for this sum:

$$1 + x + x^2 + \cdots + x^{n-1} = \frac{x^n - 1}{x - 1}$$

Multiplying both sides by a gives a formula for the sum of a geometric series:

Sum of a Geometric Series

$$a + ax + ax^2 + \cdots + ax^{n-1} = a \frac{x^n - 1}{x - 1} \qquad\qquad x \neq 1$$

For instance, the geometric series $3 + 6 + 12 + 24 + 48$, in which $a = 3$, $x = 2$, and the number of terms is $n = 5$, has the value

$$3 \cdot \frac{2^5 - 1}{2 - 1} = 93,$$

$a \frac{x^n - 1}{x - 1}$ with $a = 3$, $x = 2$, and $n = 5$

as you should check by adding up the five numbers.

Accumulated Amount Formula

For the payments at 6% compounded monthly for 4 years on page 120, we added terms of the form $200(1 + \frac{0.06}{12})^k$ with the exponent k taking values from 0 to $47 = 12 \cdot 4 - 1$. More generally, for payments of P dollars m times a year for t years at interest rate r, we would need to sum the geometric series

$$P + P\left(1 + \frac{r}{m}\right) + P\left(1 + \frac{r}{m}\right)^2 + \cdots + P\left(1 + \frac{r}{m}\right)^{mt-1}$$

Using the geometric series sum formula on the previous page, the sum of this series is

$$P\frac{\left(1 + \dfrac{r}{m}\right)^{mt} - 1}{\left(1 + \dfrac{r}{m}\right) - 1} \qquad a\frac{x^n - 1}{x - 1} \text{ with } a = P, \\ x = (1 + \tfrac{r}{m}), \text{ and } n = mt$$

Simplifying the denominator to $\frac{r}{m}$ gives the following very useful formula.

Accumulated Amount of an Annuity

For an ordinary annuity with payments of P dollars m times a year for t years at interest rate r compounded at the end of each payment period, the accumulated amount (payments plus interest) will be

$$A = P\frac{(1 + r/m)^{mt} - 1}{r/m}$$

Although we will not use it, the archaic symbol $s_{\overline{n}|i}$ (pronounced "s sub n angle i") sometimes still occurs in business textbooks and tables to denote the value of $((1 + i)^n - 1)/i$. In this notation, the above formula takes the form $A = Ps_{\overline{n}|i}$ with $n = mt$ and $i = r/m$.

EXAMPLE 1

FINDING THE ACCUMULATED AMOUNT OF AN ANNUITY

Find the accumulated amount of the annuity of monthly payments for 4 years of $200 at 6% compounded monthly (the car loan annuity from page 120).

Solution

$$A = 200\frac{(1 + 0.06/12)^{12\cdot 4} - 1}{0.06/12} \approx 10{,}819.57$$

$P\dfrac{(1 + r/m)^{mt} - 1}{r/m}$
with $P = 200$, $r = 0.06$,
$m = 12$, and $t = 4$

Therefore, regular $200 payments at the end of each month for 4 years into a savings account earning 6% compounded monthly will total $10,819.57.

The amount of this annuity is significantly more than just the sum of 48 payments of $200, which would total $9600. The difference, $10{,}819.57 - 9600 = \$1219.57$, is the total interest earned by the payments.

Practice Problem 1

What is the final balance of a retirement account earning 5% interest compounded weekly if $40 is deposited at the end of every week for 35 years? How much of this final balance is interest?

➤ Solution on page 127

Sinking Funds

A *sinking fund* is a regular savings plan designed to provide a given amount after a certain number of years. For example, suppose you decide to save up for a $18,000 car instead of taking out a loan. What amount should you save each month to have the $18,000 in 3 years? If you just put your money in a shoe box, you will need to set aside $18{,}000 \div (12 \times 3) = \$500$ each month. It would make more sense to put your money into a savings account each month and let it earn compound interest. This is the same as setting up an annuity with a given accumulated amount and asking what the regular payment should be. Solving the accumulated amount formula (on the previous page) for P by dividing by the fraction following P gives the following formula.

Sinking Fund Payment

The payment P to make m times per year for t years at interest rate r compounded at each payment to accumulate amount A is

$$P = A\frac{r/m}{(1 + r/m)^{mt} - 1}$$

This formula may also be written $P = A/s_{\overline{n}|i}$ with $n = mt$ and $i = r/m$.

EXAMPLE **2**

FINDING A SINKING FUND PAYMENT

What amount should be deposited at the end of each month for 3 years in a savings account earning 4.5% interest compounded monthly to accumulate $18,000 to buy a new car?

Solution

$$P = 18{,}000 \, \frac{0.045/12}{(1 + 0.045/12)^{12 \cdot 3} - 1} \approx 467.945$$

$P = A \dfrac{r/m}{(1 + r/m)^{mt} - 1}$
with $A = 18{,}000$,
$r = 0.045$, $m = 12$,
and $t = 3$

Rounding *up* to the next penny, $467.95 should be saved each month. (If we rounded down to $467.94, the savings account would accumulate slightly less than the amount needed.)

Notice that this amount is significantly lower than the $500 that would be needed *without* compound interest.

This example could also have been done using the accumulated amount formula (page 122) by substituting $A = 18{,}000$, $r = 0.045$, $m = 12$, and $t = 3$, and then solving for the remaining variable to find $P = \$467.95$.

Practice Problem **2**

For wage earners paid every other week instead of monthly, Example 2 becomes: "What amount should be deposited biweekly for 3 years in a savings account earning 4.5% interest compounded biweekly to accumulate $18,000 for a new car?" Find this amount.

➤ Solution on page 127

How Long Will It Take?

Continuing with our sinking fund example, suppose you could save only $200 each month. How long would it take to accumulate the $18,000?

We solve the accumulated amount formula (page 122) for the number of periods mt.

$$\frac{A}{P}\frac{r}{m} + 1 = \left(1 + \frac{r}{m}\right)^{mt}$$

$A = P\dfrac{(1 + r/m)^{mt} - 1}{r/m}$

dividing by P, multiplying by r/m, and adding 1

$$mt \log\left(1 + \frac{r}{m}\right) = \log\left(\frac{A}{P}\frac{r}{m} + 1\right)$$

Switching sides and taking logs of both to bring down the exponent

Dividing by $\log(1 + r/m)$, we have:

Number of Periods

$$mt = \frac{\log\left(\dfrac{A}{P}\dfrac{r}{m} + 1\right)}{\log\left(1 + \dfrac{r}{m}\right)} \quad \text{periods}$$

As usual, we round *up* to the next whole number of compounding periods and then express the answer as years plus any extra periods. Notice that the time t depends only on the ratio $\frac{A}{P}$ between the accumulated amount A and the regular payment P.

EXAMPLE 3

FINDING THE TERM FOR A SINKING FUND

How long will it take to accumulate $18,000 by depositing $200 at the end of each month in a savings account earning 4.5% interest compounded monthly?

Solution

$$12t = \frac{\log\left(\dfrac{18{,}000}{200}\dfrac{0.045}{12} + 1\right)}{\log\left(1 + \dfrac{0.045}{12}\right)} \approx 77.7$$

$mt = \dfrac{\log\left(\dfrac{A}{P}\dfrac{r}{m} + 1\right)}{\log\left(1 + \dfrac{r}{m}\right)}$

with $A = 18{,}000$, $P = 200$, $r = 0.045$, and $m = 12$

The time is 77.7 months, which we round *up* to 78 months. Therefore, it will take 6 years 6 months to accumulate the $18,000.

This example could also have been done using the accumulated amount formula (page 122) by substituting $A = 18{,}000$, $P = 200$, $r = 0.045$, and $m = 12$, and then solving (using logarithms) for the remaining variable to find that $12t$ is 78 months.

Practice Problem 3

How long will it take to accumulate $18,000 by depositing $250 each month in a savings account earning 4.5% interest compounded monthly?

➤ Solution on next page

Graphing Calculator Exploration

Depending on the size of the monthly deposit, the time needed to accumulate a given amount may vary greatly.

a. Using the values $r = 0.045$, $m = 12$, and $A = 18{,}000$ from Example 3, and replacing the payment amount P by x, graph $y_1 = \log((18000/x)(.045/12) + 1)/\log(1 + .045/12)$ on the window $[0, 1000]$ by $[0, 240]$. The graph shows how y, the number of months needed to accumulate $18,000, decreases as the monthly payment x increases.

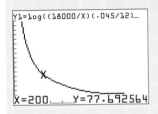

b. Use TRACE or VALUE to find y when $x = 200$ and when $x = 250$. How do your answers compare to the answers to Example 3 and Practice Problem 3?

c. Use TRACE or VALUE to find the years y needed when the monthly deposits x are $100, $150, $200, $250, and $300. Does each additional $50 reduce the time needed by the same amount? How does the graph show this?

Overview

In the previous section we found the eventual value of an investment, the initial amount necessary to reach that later amount, and the time it takes, just as we did in this section. What's the difference between the two sections? *There* we were considering investments of a *single* payment of P dollars, while *here* we are considering *multiple and regular payments* of P dollars.

Section Summary

For an *ordinary annuity* with regular payments of P dollars m times per year for t years at interest rate r compounded at the end of each payment, the accumulated amount A is

$$A = P\frac{(1 + r/m)^{mt} - 1}{r/m}$$

A = accumulated amount,
P = payment, r = interest rate,
m = times per year, t = years

To find how much of this amount is interest, subtract the total of all the payments, $P \cdot m \cdot t$, from it.

The annuity formula may be solved for the other variables. Solving for P finds the regular payment to make into a *sinking fund* to accumulate the amount A after t years:

$$P = A\frac{r/m}{(1 + r/m)^{mt} - 1}$$

Round *up*

Solving for mt finds the number of periods required for the payments into a sinking fund to accumulate to a final amount A:

$$mt = \frac{\log\left(\dfrac{A}{P}\dfrac{r}{m} + 1\right)}{\log\left(1 + \dfrac{r}{m}\right)}$$

mt = number of periods
(other variables as above)

Round *up* to find the number of periods and then express your answer as a number of years plus any additional periods.

> ### Solutions to Practice Problems

1. $A = 40\dfrac{(1 + 0.05/52)^{52 \cdot 35} - 1}{0.05/52} \approx 197{,}590.27.$ The final balance is $197,590.27. Because the deposits total $40 \cdot 52 \cdot 35 = \$72{,}800,$ the final balance contains $124,790.27 interest.

2. $P = 18{,}000\dfrac{0.045/26}{(1 + 0.045/26)^{26 \cdot 3} - 1} \approx 215.742.$ $215.75 should be deposited every other week. (Compare double this amount, $431.50, to the $467.95 monthly payment found in Example 2 on page 124.)

3. $12t = \dfrac{\log\left(\dfrac{18{,}000}{250}\dfrac{0.045}{12} + 1\right)}{\log\left(1 + \dfrac{0.045}{12}\right)} \approx 63.9$ months, which rounds up to 64 months. It will take 5 years 4 months.

2.3 Exercises

In the following ordinary annuities, the interest is compounded with each payment, and the payment is made at the end of the compounding period.

Find the accumulated amount of each annuity.

1. $1500 annually at 7% for 10 years.

2. $1750 annually at 8% for 7 years.

3. $2000 semiannually at 6% for 12 years.

4. $950 quarterly at 10.7% for 6 years.

5. $1000 monthly at 6.9% for 20 years.

6. $200 weekly at 11.3% for 5 years.

Find the required payment for each sinking fund.

7. Monthly deposits earning 5% to accumulate $5000 after 10 years.

8. Monthly deposits earning 4% to accumulate $6000 after 15 years.

9. Monthly deposits earning 11.7% to accumulate $14,000 after 5 years.

10. Quarterly deposits earning 6.4% to accumulate $50,000 after 20 years.

11. Yearly deposits earning 12.3% to accumulate $8500 after 12 years.

12. Weekly deposits earning 9.8% to accumulate $15,000 after 6 years.

Find the amount of time needed for each sinking fund to reach the given accumulated amount.

13. $1500 yearly at 8% to accumulate $100,000.

14. $2000 yearly at 9% to accumulate $125,000.

15. $800 semiannually at 6.6% to accumulate $80,000.

16. $500 quarterly at 8.2% to accumulate $19,500.

17. $235 monthly at 5.9% to accumulate $25,000.

18. $70 weekly at 10.9% to accumulate $7500.

APPLIED EXERCISES

19. **Retirement Savings** An individual retirement account, or IRA, earns tax-deferred interest and allows the owner to invest up to $3000 each year. Joe and Jill both will make IRA deposits for 30 years (from age 35 to 65) into stock mutual funds yielding 9.8%. Joe deposits $3000 once each year while Jill has $57.69 (which is 3000/52) withheld from her weekly paycheck and deposited automatically. How much will each have at age 65?

20. **Retirement Savings**
    a. Steve opens a retirement account yielding 9% and deposits $250 each month for the next 30 years (from age 35 to 65). How much will he have at age 65?
    b. Sue opens her own retirement account yielding 8% and deposits $250 each month for the 10 years from age 25 to 35. She then makes no further deposits but lets the accumulated amount earn compound interest for the next 30 years. How much will she have at age 65?
    c. Who has more money for retirement?

21. **Mutual Funds** How much must you invest each month in a mutual fund yielding 13.4% compounded monthly to become a millionaire in 10 years?

22. **Bond Funds** How much must you invest each quarter in a bond fund yielding 9.7% compounded quarterly to become a millionaire in 20 years?

23. **Lifetime Savings** The Oseola McCarty Scholarship Fund at the University of Southern Mississippi was established by a $150,000 gift from an 87-year-old woman who had dropped

out of sixth grade and worked for most of her life as a washerwoman. How much would she have had to save each week in a bank account earning 3.9% compounded weekly to have $150,000 after 75 years?

24. Lifetime Savings If each day your fairy godmother put $1 into a savings account at 8.37% compounded daily, how much would the account be worth when you were 65 years old?

25. Home Buying You and your new spouse each bring home $1500 each month after taxes and other payroll deductions. By living frugally, you intend to live on just one paycheck and save the other in a bond fund yielding 7.86% compounded monthly. How long will it take to have enough for a 20% down payment on a $165,000 condo in the city?

26. Boat Buying How long will it take to save $16,000 for a new motorboat if you deposit $375 each month into a money market fund yielding 5.73% compounded monthly?

Explorations and Excursions

The following problems extend and augment the material presented in the text.

More About Annuities

27. Internal Rate of Return of an Annuity
Suppose that you make quarterly investments of $500 into a stock fund, and after 8 years the fund is worth $27,660. How can you find the "rate of return" on your money? If x is the (annual) rate of return, then you need to solve

$$500 \frac{(1 + x/4)^{4 \cdot 8} - 1}{x/4} = 27{,}660$$

Enter the left-hand side of this equation into your calculator as y_1 and GRAPH it on the window $[0, .2]$ by $[0, 30{,}000]$. Then use TRACE (or INTERSECT with $y_2 = 27{,}660$) to find the value of x where y_1 equals 27,660. The resulting value of x is called the *internal rate of return* of your investment.

28. Stock Market According to the *Wall Street Journal*, the $16\frac{1}{2}$ years from February 1966 to August 1982 was perhaps the most brutal period for stock market investors since the Great Depression in the 1930s. Ibbotson Associates estimates that if you had invested $100 each month into the stocks that make up the S&P 500 during this period, by the market low at mid-1982 your portfolio would have been worth almost $32,600 (with all dividends reinvested). Use the method from Exercise 27 (but now with *monthly* compounding) to find the internal rate of return for such an investment.

29. Solve any three of Exercises 1–6 as follows: Enter the accumulated amount formula (page 122) as $y = P((1 + r/m)^{mt} - 1)/(r/m)$. Then for each exercise, STORE the values for P, r, m, and t, and evaluate y to find the amount due.

More About Geometric Series

30. For each choice of values for a, x, and n, write out the geometric series

$$a + ax + ax^2 + \cdots + ax^{n-2} + ax^{n-1},$$

find the sum by adding up the terms, and then evaluate the corresponding $a\frac{x^n - 1}{x - 1}$ expression to check that both have the same value.

a. $a = 2, \quad x = 3, \quad n = 5.$
b. $a = 162, \quad x = \frac{1}{3}, \quad n = 5.$
How do parts (a) and (b) differ?
c. $a = 3, \quad x = 10, \quad n = 6$

31. *Requires a calculator with series operations.* Solve Exercise 30 as follows. Your calculator can LIST the numbers making up a geometric series as a *sequence* of values by using SEQUENCE. For the values of a, x, and n in Exercise 30a, list the numbers as the SEQUENCE $2(3^n)$ for n starting at 0, ending at 4, and increasing by 1. This list of numbers can be added using SUM. In this way, check each of the SUMs in Exercise 30. Then check that the geometric SEQUENCE $200(1 + 0.06/12)^n$ on page 120 for n from 0 to 47 SUMs to $10,819.57.

32. Infinite Geometric Series *Requires a calculator with series operations.* With the values $a = 1$ and $x = \frac{1}{2}$, the geometric series formula becomes

$$1 + \tfrac{1}{2} + \left(\tfrac{1}{2}\right)^2 + \cdots + \left(\tfrac{1}{2}\right)^{n-1} = \frac{\left(\tfrac{1}{2}\right)^n - 1}{\tfrac{1}{2} - 1}$$

a. Check this formula by evaluating both sides for $n = 5$, $n = 10$, and $n = 15$. As n gets larger, do these values get closer to 2?

b. Graph the SUM y of the first x terms of this geometric SEQUENCE on the window [0, 25] by [0, 3]. Use TRACE to explore the graph. As x gets larger, do these values get closer to 2?

An *infinite geometric series* is a sum of the form $a + ax + ax^2 + ax^3 + ax^4 + \cdots$, where the $+ \cdots$ indicates that the sum continues on to include every power of x. For $a = 1$ and $x = \frac{1}{2}$, we suspect from parts (a) and (b) that

$$1 + \tfrac{1}{2} + \left(\tfrac{1}{2}\right)^2 + \left(\tfrac{1}{2}\right)^3 + \left(\tfrac{1}{2}\right)^4 + \cdots = 2$$

c. Check this arithmetic: Because $\left(\tfrac{1}{2}\right)^n$ gets closer to 0 as n gets larger, the geometric series formula $\dfrac{\left(\tfrac{1}{2}\right)^n - 1}{\tfrac{1}{2} - 1}$ gets closer to

$$\dfrac{0 - 1}{\tfrac{1}{2} - 1} \quad \text{as } n \text{ gets larger.}$$

d. If $1 + \tfrac{1}{2} + \left(\tfrac{1}{2}\right)^2 + \left(\tfrac{1}{2}\right)^3 + \left(\tfrac{1}{2}\right)^4 + \cdots$ really is 2, it should behave just like 2. Show that

$$\tfrac{1}{2}\left(1 + \tfrac{1}{2} + \left(\tfrac{1}{2}\right)^2 + \left(\tfrac{1}{2}\right)^3 + \left(\tfrac{1}{2}\right)^4 + \cdots \right) = 1$$

by multiplying out the left side and using the fact that

$$1 + \tfrac{1}{2} + \left(\tfrac{1}{2}\right)^2 + \left(\tfrac{1}{2}\right)^3 + \left(\tfrac{1}{2}\right)^4 + \cdots = 2.$$

If $|x| < 1$, then x^n gets closer to 0 as n increases and we have the following formula for an *infinite* geometric series:

$$a + ax + ax^2 + \cdots = a \cdot \frac{1}{1 - x}$$

e. Find the value of

$$1 + \tfrac{1}{3} + \left(\tfrac{1}{3}\right)^2 + \left(\tfrac{1}{3}\right)^3 + \left(\tfrac{1}{3}\right)^4 + \cdots.$$

f. Find the value of

$$9 + 9\left(\tfrac{1}{10}\right) + 9\left(\tfrac{1}{10}\right)^2 + 9\left(\tfrac{1}{10}\right)^3 + 9\left(\tfrac{1}{10}\right)^4 + \cdots.$$

Can you write this geometric series as a decimal?

2.4 AMORTIZATION

Introduction

In this section we will find the *present value of an annuity*—the total value *now* of all the future payments. Such calculations are important for buying and selling annuities and for using them to pay off debts (amortization). For example, on pages 122–123 we found the accumulated amount for monthly payments of $200 for 4 years at 6% compounded monthly. But how much would a loan company give *right now* for that promise of future payments?

Present Value of an Annuity

We know that for an ordinary annuity with payments of P dollars m times per year for t years at interest rate r compounded at each payment, the accumulated amount A is

$$A = P\frac{(1 + r/m)^{mt} - 1}{r/m} \qquad \text{From page 122}$$

On page 111 we saw that to find the present value of an amount, you simply divide by $(1 + r/m)^{mt}$. Therefore, the present value of this amount (denoted PV to distinguish it from P for payment) is

$$PV = P\frac{(1 + r/m)^{mt} - 1}{(r/m)(1 + r/m)^{mt}}$$ Dividing by $(1 + r/m)^{mt}$

$$= P\frac{1 - (1 + r/m)^{-mt}}{r/m}$$ Dividing numerator and denominator by $(1 + r/m)^{mt}$

This gives the formula:

Present Value of an Annuity

The present value PV of an ordinary annuity with payments of P dollars m times per year for t years at interest rate r compounded at each payment is

$$PV = P\frac{1 - (1 + r/m)^{-mt}}{r/m}$$

Although we will not use it, the archaic symbol $a_{n\rceil i}$ (pronounced "a sub n angle i") sometimes still occurs in business textbooks and tables to denote the value of $(1 - (1 + i)^{-n})/i$. In this notation, the above formula takes the form $PV = Pa_{n\rceil i}$ with $n = mt$ and $i = r/m$.

EXAMPLE **1**

FINDING THE PRESENT VALUE OF AN ANNUITY

What is the present value of a 6% car loan for 4 years with monthly payments of $200?

Solution

$$PV = 200\frac{1 - (1 + 0.06/12)^{-12\cdot4}}{0.06/12} \approx 8516.06$$

$$PV = P\frac{1 - (1 + r/m)^{-mt}}{r/m}$$
with $P = 200$, $r = 0.06$, $m = 12$, and $t = 4$

The present value is $8516.06.

In general, how can we interpret the present value of an annuity? It is the sum that must be deposited in a bank *now* (at the stated interest rate) to grow to exactly the amount of the annuity at the end of its term. For the loan in Example 1, in 4 years the present value will grow to $\$8516.06(1 + 0.06/12)^{12\cdot4} = \$10,819.57$, which is exactly the sum of the annuity that we found on page 123. Therefore, the present value of $\$8516.06$ is exactly the price for which the sequence of car payments should be sold, and the lender will not care which she receives.

Practice Problem 1	What is the present value of a 20-year retirement annuity paying $850 per month if the current long-term interest rate is 7.53%?

➤ Solution on page 138

Be Careful! The formulas for the *present value* and the *accumulated* or *future value* of an annuity (derived on page 122) look similar but are not the same, and should not be confused:

Present Value $\qquad\qquad P\dfrac{1 - (1 + r/m)^{-mt}}{r/m}$

Future (or Accumulated) Value $\qquad P\dfrac{(1 + r/m)^{mt} - 1}{r/m}$

As we saw on page 111, the future value formula divided by $(1 + r/m)^{mt}$ gives the present value formula.

Amortization

Suppose that you took out a loan to buy a house, and wanted to pay off the debt in a fixed number of payments, with the bank charging interest on the unpaid balance. A debt is *amortized* (or "killed off") if it is repaid by a regular sequence of payments.

Spreadsheet Exploration

The following spreadsheet* shows an "amortization table" for a debt of $100,000 to be paid off in five annual payments at 10% compounded yearly. Notice that the last payment is adjusted to *exactly* pay off the remaining debt and interest.

* See the Preface for how to obtain this and other Excel spreadsheets.

C2		▼	= =PMT(A5/(A11),A8*A11,A2)				
	A	B	C	D	E	F	G
1	Debt		Payment				
2	100000		(26,379.75)				
3							
4	Annual Rate						
5	10%		Year	Payment	Interest	Debt Reduction	Outstanding Debt
6			0.00	0.00	0.00	0.00	100,000.00
7	Term in Years		1.00	(26,379.75)	10,000.00	(16,379.75)	83,620.25
8	5		2.00	(26,379.75)	8,362.03	(18,017.72)	65,602.53
9			3.00	(26,379.75)	6,560.25	(19,819.50)	45,783.03
10	Compoundings per Year		4.00	(26,379.75)	4,578.30	(21,801.45)	23,981.58
11	1		5.00	(26,379.74)	2,398.16	(23,981.58)	0.00
12							
13			Totals:	(131,898.74)	31,898.74	(100,000.00)	

The payment amount in cell **C2** was found using the spreadsheet **PMT** function. The table shows how each payment is allocated to pay the interest and some of the debt, with the early payments paying more for interest than the later payments, which go almost entirely for debt.

How do we calculate the correct payments to amortize a debt? The payments form an annuity with present value PV equal to the debt D. To find the payment P to amortize the debt, we solve for P in the "present value of an annuity" formula on page 131 (dividing by the fraction following the P) and then replacing PV by D.

Amortization Payment

The payment P to make m times per year for t years at interest rate r compounded at each payment to amortize a debt of D dollars is

$$P = D\frac{r/m}{1 - (1 + r/m)^{-mt}}$$

This formula may also be written $P = D/a_{\overline{n}|i}$ with $n = mt$ and $i = r/m$.

EXAMPLE **2**

FINDING THE PAYMENT TO AMORTIZE A DEBT

What monthly payment will amortize a $150,000 home mortgage at 8.6% in 30 years?

Solution

$$P = 150{,}000 \frac{0.086/12}{1 - (1 + 0.086/12)^{-12 \cdot 30}} \approx 1164.02$$

$P = D\dfrac{r/m}{1 - (1 + r/m)^{-mt}}$
with $D = 150{,}000$,
$r = 0.086$, $m = 12$,
and $t = 30$

The required payment is $1164.02 each month.

Notice that the borrower will pay a total of $1164.02 \cdot 12 \cdot 30 =$ $419,047.20 during the 30 years of the loan to pay off a $150,000 debt. The difference, $269,047.20, is the interest paid to the bank.

Practice Problem 2

What monthly payment will amortize a $150,000 home mortgage at 8.6% in 25 years? What is the total amount the borrower will pay?

➤ Solution on page 138

Graphing Calculator Exploration

When amortizing a loan, a longer term means smaller payments, but the total amount the borrower pays is much larger.

a. To find the monthly payment y to amortize a debt over x years using the values $D = 150{,}000$, $r = 0.086$, and $m = 12$ from Practice Problem 2, graph the curve $y_1 = 150000(.086/12)/(1 - (1 + .086/12)^{-12x})$ on the window $[0, 50]$ by $[0, 5000]$.

b. Notice that as the term increases the payment decreases, at first rapidly but later more slowly. Use TRACE or VALUE to find the payments for mortgage terms of 15, 20, 25, and 30 years. How do your answers compare to the answers to Example 2 and Practice Problem 2?

Y1=12X(150000)(.086/12)/_

X=30   Y=419046.33

c. To find the total amount the borrower pays, multiply the payment by m and t (here replaced by x). Graph the new curve $y_1 = 12x(150000)(.086/12)/(1 - (1 + .086/12)^{-12x})$, along with the line $y_2 = 150000$ to represent the original debt, on the window [0, 50] by [0, 600,000] to find the total amount paid on a loan for x years.

d. Notice that as the term increases the total amount paid increases, well above the original debt of $150,000. Use TRACE or VALUE to find the total amount paid for mortgage terms of 15, 20, 25, and 30 years. Based on these two screens, does it make financial sense to extend the term of this loan beyond 30 years?

Unpaid Balance

How much does the borrower still owe partway through an agreed amortization schedule? Suppose you have made mortgage payments for 10 years on a 30-year loan and now you must move. How much should you pay to settle your debt? Your remaining payments form another annuity and are worth precisely the present value of this new annuity.

EXAMPLE 3

FINDING THE UNPAID BALANCE

What is the unpaid balance after 10 years of monthly payments on a 30-year mortgage of $150,000 at 8.6%?

Solution

First we must calculate the monthly payments. But for this problem, we know from Example 2 (on the previous page) that the monthly payment to amortize the original mortgage is $1164.02. The remaining payments thus form an annuity with $P = \$1164.02$, $r = 0.086$, $m = 12$, and $t = 20$ years. The amount still owed on the mortgage is the present value of this annuity, which we can calculate by the formula on page 131.

$$PV = 1164.02 \frac{1 - (1 + 0.086/12)^{-12 \cdot 20}}{0.086/12}$$

$$PV = P \frac{1 - (1 + r/m)^{-mt}}{r/m}$$
with $P = 1164.02$, $r = 0.086$, $m = 12$, and $t = 20$

$$\approx 133{,}158.27$$

The unpaid balance after 10 years of payments on this $150,000 mortgage is $133,158.27.

Notice that after the first 10 of 30 years, much less than one third of the mortgage has been paid. The early payments of any loan pay mostly interest and only a little of the principal.

As in this example, if you already know the payments, you use the same payments to calculate the unpaid balance. However, if you do *not* know the payments, you would first have to calculate them using the payment formula on page 133.

Practice Problem 3

How much is still owed on a 25-year mortgage of $150,000 at 8.6% after 13 years of monthly payments? [*Hint:* Use the result of Practice Problem 2 on page 134.] ➤ **Solution on page 139**

Equity

One often hears the term "equity" used with homes, such as a "home equity loan." *Equity* is defined as the *value of the home minus any unpaid mortgage balance,* and is a measure of the value that you (rather than the bank) have in your house. If the house with the mortgage in Example 3 now has a market value of $300,000, then the homeowner's equity after 10 years of payments is this value minus the unpaid balance: $300,000 − 133,158 = $166,842 (ignoring cents). This equity may be used to take out a new loan for a larger house.

Amortization Tables

The Spreadsheet Exploration on pages 132–133 showed an amortization table on a spreadsheet. Amortization tables can also be constructed on graphing calculators.

Graphing Calculator Exploration

The program* AMORTABL constructs an *amortization table* that shows how each payment is allocated to paying the interest and reducing the outstanding debt. To explore the table for the situation used in Examples 2 and 3, run this program by selecting it from the program menu and proceed as follows:

---

* See the Preface for information on how to obtain this and other graphing calculator programs.

```
DEBT (DOLLARS)
  150000
RATE (DECIMAL)
  .086
NUMBER OF YEARS
  30
COMPOUNDINGS/YR
  12
```

```
    VIEW TABLE
1:FROM START
2:JUMP TO YEAR
3:AT END
4:FINISHED
```

a. Enter 150000 for the debt, .086 for the interest rate, 30 for the number of years, and 12 for the number of compoundings per year (because the payments are monthly).

b. After the table of values has been calculated by finding the payment rounded to the nearest penny and then applying each payment to the rounded interest due on the debt for that period and reducing the debt by the excess, you may choose which part of the table you wish to see.

c. The amortization table contains five columns: X is the number of the payment (negative X's or X's larger than the number of payments display zeros in the other columns), Y_1 is the payment, Y_2 is the interest part of the payment, Y_3 is the remaining part used to reduce the debt, and Y_4 is the debt remaining after this reduction. Use the arrow keys to scroll right and left through these columns.

X	Y₁	Y₂	
0	0	0	
1	1164	1075	
2	1164	1074.4	
3	1164	1073.7	
4	1164	1073.1	
5	1164	1072.4	
6	1164	1071.8	
X=0			

X	Y₃	Y₄	
0	0	150000	
1	89.02	149911	
2	89.66	149821	
3	90.3	149731	
4	90.95	149640	
5	91.6	149548	
6	92.26	149456	
Y₄=150000			

X	Y₁	Y₂	
356	1164	40.8	
357	1164	32.75	
358	1164	24.64	
359	1164	16.48	
360	1160	8.25	
361	0	0	
362	419043	269043	
X=361			

X	Y₃	Y₄	
356	1123.2	4570	
357	1131.3	3438.7	
358	1139.4	2299.3	
359	1147.5	1151.8	
360	1151.8	0	
361	0	0	
362	150000	0	
Y₄=0			

The second line after the end of the amortization (line 362) contains the total payments, interest, and debt reduction. Notice that the final payment is adjusted to correct for rounding errors and the debt is reduced to zero.

X	Y₃	Y₄	
118	205.28	133573	
119	206.75	133366	
120	208.23	133158	
121	209.72	132948	
122	211.23	132737	
123	212.74	132524	
124	214.27	132310	
Y₄=133157.56			

d. To use the table to find the remaining debt after a given number of years, "jump" to that position in the table, and scroll to the Y_4 column. After 10 years of payments, the table shows \$133,157.56 remaining.

2.4 Section Summary

In this section we made extensive use of the formula for the accumulated amount of an annuity (page 122) that we derived in the previous section. The *present value* of an ordinary annuity compounded at each payment is

$$PV = P\frac{1 - (1 + r/m)^{-mt}}{r/m}$$

PV = present value,
P = payment, r = rate,
m = times per year, and
t = years

The payment to *amortize a debt* of D dollars is found by solving this formula for P.

$$P = D\frac{r/m}{1 - (1 + r/m)^{-mt}}$$

D = debt
(other variables are as before)

The amount still owed after making amortization payments for part of the term is the present value of a new annuity formed by the remaining payments and is found by the first formula.

This chapter has developed several formulas. The following diagram may help to organize some of them.

Single Payment
  { Simple Interest $A = P(1 + rt)$ Section 2.1
  { Compound Interest $A = P(1 + r/m)^{mt}$ Section 2.2

Multiple Payments (Annuity)
  { Future Value $A = P\dfrac{(1 + r/m)^{mt} - 1}{r/m}$ Section 2.3
  { Present Value $PV = P\dfrac{1 - (1 + r/m)^{-mt}}{r/m}$ Section 2.4

➤ **Solutions to Practice Problems**

1. $PV = 850\dfrac{1 - (1 + 0.0753/12)^{-12\cdot20}}{0.0753/12} \approx 105{,}272.468.$
The present value is $105,272.47.

2. $P = 150{,}000\dfrac{0.086/12}{1 - (1 + 0.086/12)^{-12\cdot25}} \approx 1217.97.$
The required payment is $1217.97 each month. The borrower will pay a total of $1217.97 \cdot 12 \cdot 25 = $365,391.

3. The remaining payments form a 12-year annuity at 8.6% with monthly payments of $1217.97 (from Practice Problem 2). The present value of this annuity is

$$PV = 1217.97 \frac{1 - (1 + 0.086/12)^{-12 \cdot 12}}{0.086/12} \approx 109{,}174.18$$

so the amount still owed is $109,174.18. Notice that after more than half the term, less than half of the $150,000 loan is paid off.

2.4 Exercises

Calculate the present value of each annuity.

1. $15,000 annually at 7% for 10 years.

2. $18,000 annually at 6% for 15 years.

3. $10,000 semiannually at 11% for 7 years.

4. $3000 quarterly at 8% for 8 years.

5. $1400 monthly at 6.9% for 30 years.

6. $25 weekly at 7.7% for 6 years.

Determine the payment to amortize each debt.

7. Monthly payments on $100,000 at 5% for 25 years.

8. Monthly payments on $125,000 at 5.4% for 20 years.

9. Monthly payments on $6000 at 6% for 5 years.

10. Annual payments on $50,000 at 8.2% for 10 years.

11. Quarterly payments on $14,500 at 12.7% for 6 years.

12. Weekly payments on $2500 at 18% for 3 years.

Find the unpaid balance on each debt (you should already know the payments from Exercises 7–12).

13. After 6 years of monthly payments on $100,000 at 5% for 25 years.

14. After 15 years of monthly payments on $125,000 at 5.4% for 20 years.

15. After 2 years 11 months of monthly payments on $6000 at 6% for 5 years.

16. After 3 years of annual payments on $50,000 at 8.2% for 10 years.

17. After 2 years 3 months of quarterly payments on $14,500 at 12.7% for 6 years.

18. After 2 years of weekly payments on $2500 at 18% for 3 years.

APPLIED EXERCISES

19. Contest Prizes The super prize in a contest is $10 million. This prize will be paid out in equal yearly payments over the next 25 years. If the prize money is guaranteed by U.S. Treasury bonds yielding 7.9% and is placed into an escrow account when the contest is announced 1 year before the first payment, how much do the contest sponsors have to deposit in the escrow account?

20. Grant Funding In September 1994, the New York City public schools system was awarded a grant of $50 million over 5 years from the Annenberg Foundation. If the grant paid equal yearly amounts for the next 5 years and was financed at 6.8%, how much did this grant cost the Foundation when it was announced?

21. **NFL Contracts** When Michael Westbrook signed a 7-year, $18 million contract with the Washington Redskins, it included a $6.5 million signing bonus that was the largest for a wide receiver in NFL history. If the $18 million was paid out in equal quarterly payments for the 7 years and the current long-term interest rate was 6.85%, how much was the contract worth when it was signed?

22. **Escrow Accounts** On September 1, 1994, Judge Sam C. Pointer, Jr., of the Federal District Court in Birmingham approved a $4.25 billion settlement against the manufacturers of silicon breast implants to be paid out over the next 30 years. If this money was to be paid out in equal semiannual amounts from an escrow account earning 8.7% interest, how much would the manufacturers have to deposit into the account (6 months before the first payment) to meet the terms of the settlement?

23. **College Savings** When I graduated from high school last spring, my dear Aunt Sallie gave me a savings account passbook for an account earning 5.61%. She told me that starting next fall, there would be enough money for me to take out $10,000 every 6 months for the next 4 years to pay for my college expenses. I thought "Wow! Thanks! That's $80,000!" but she just smiled, shook her head "No," and told me to look in the passbook. How much was in the account when she gave it to me?

24. **Car Buying** A Cadillac dealer offers the following terms on a 2-year lease of a new Eldorado: either $1995 down and $339 per month for 24 months or $9599 down and nothing more to pay for 24 months. (a) If the current interest rate were 5.2%, which option would be cheaper? (b) Since the car dealer is an expert money manager, what must the current interest rate be if both offers are equally good for the dealer?

25. **Life Insurance** Just before his first attempt at bungee jumping, John decides to buy a life insurance policy. His annual income at age 30 is $35,000, so he figures he should get enough insurance to provide his wife and new baby with that amount each year for the next 35 years. If the long-term interest rate is 6.7%,

what is the present value of John's future annual earnings? Rounding up to the next $50,000, how much life insurance should he buy?

26. **Life Insurance** A 60-year-old grandmother wants a life insurance policy that could replace her annual $55,000 earnings for the next 10 years. If the long-term interest rate is now 6.2%, how much life insurance (rounded up to the next $10,000) should she buy?

27. **Mortgages** Real estate prices are so high in Tokyo that some mortgages are written for 99 years in an attempt to keep down the monthly payments. (a) What is the monthly payment on a $500,000 mortgage at 8% for 99 years? (b) How much does this save the borrower each month compared to a 30-year mortgage at 8% for the same amount? (c) Which term results in the higher total payment?

28. **Apartment Rents** The Associates for International Research have estimated the following monthly rents for a two-bedroom apartment in major cities around the world.

Sydney	$1100	London	$2950
Cairo	$1200	Moscow	$5000
Buenos Aires	$1500	Shanghai	$6300
Paris	$2150	Tokyo	$7100
New York	$2300	Hong Kong	$7200

Assuming that the monthly rent represents the payment on the current market value of the apartment as a 30-year investment and that the long-term interest rate is 6.52%, estimate the current market value for an apartment in (a) Sydney, (b) Paris, (c) Moscow, and (d) Tokyo.

29. **Credit Cards** A MasterCard statement shows a balance of $560 at 13.9% compounded monthly. What monthly payment will pay off this debt in 1 year 8 months?

30. **Credit Cards** The MasterCard statement in Exercise 29 also states that the "minimum payment" is $15. Find, in two different ways, how long it will take to pay off this debt by making this minimum payment each month.

    a. To estimate the number of years required, STORE the values 15 for P, 0.139 for r, and

12 for m before graphing the present value formula $y = P(1 - (1 + r/m)^{-mx})/(r/m)$ on the window [0, 10] by [0, 1000]. Use TRACE or INTERSECT to estimate the number of years x that corresponds to the debt's present value $y = \$560$. Be sure to change this number of years into the number of months the payments must be made.

b. To estimate the number of years required, STORE the values 560 for D, 0.139 for r, and 12 for m before graphing the amortization payment formula

$$y = D(r/m)/(1 - (1 + r/m)^{-mx})$$

on the window [0, 10] by [0, 20]. Use TRACE or INTERSECT to estimate the number of years x that corresponds to the payment $y = \$15$. Be sure to change this number of years into the number of months the payments must be made.

c. Why are the answers from parts (a) and (b) exactly the same?

31. Credit Cards What monthly payment should you make on your Visa card to pay off a new $1575 stereo in 15 months if the interest rate is 14.8% compounded monthly?

32. Mortgages The *Town Gossip* local newspaper lost the libel suit filed by the village's former mayor. What quarterly payment will pay off the $150,000 judgment in 5 years at 11.6% compounded quarterly?

33. College Costs When Jill graduated from college, her loans with interest totaled $58,720. Because her last co-op employer invited her to stay on as a full-time employee, she was able to handle the monthly payments even though the interest rate was 9.4% and the bank expected everything to be paid back in 8 years. How much debt remained after 3 years of payments?

34. Car Buying Tom bought a new turbo Ford Thunderbird and financed it with a new-car loan of $18,000 at 13.78% for 4 years. After making the monthly payments for 7 months, he decided he couldn't study for his business courses at college and put in enough hours at his part-time job to keep up his grades, so he offered to give the car to his sister if she would take over the payments. What purchase price would she be paying for the car?

35. Automobile Financing A car dealer's newspaper ad offers "0% financing with guaranteed credit approval" and includes an example: "Finance $10,000 for 48 months—normal payment is $262.84 while 0% financing is $208.33—you save $2616." What is the "normal" finance rate? To estimate the answer, graph the payment formula

$$y = 10000(x/12)/(1 - (1 + x/12)^{-48})$$

on the window [0, .2] by [0, 500] to see the monthly payment y for a 4-year loan of $10,000 at interest rate x compounded monthly. Use TRACE or INTERSECT to find the rate x that gives a y-value near $262.84. (By the way, the small print at the bottom of the ad contains the statement "0% financing may affect selling price of car.")

36. Computer Financing A Radio Shack back-to-school sale ad offers a "complete PC system for $39 per month after no payments for 6 months." The ad explains that "if you do not pay the full amount of purchase [$1270] by the end of the deferred period," then "finance charges will accrue from the date of purchase and will be added to the purchase balance" and states that the APY is 22.55%.

a. Show that the nominal interest rate for this loan is 20.508%.

b. How much will the balance be after 6 months if you make no payments?

c. How long will it then take to pay off the debt by paying the $39 each month? To estimate the number of years required, STORE the values 39 for P, 0.20508 for r, and 12 for m before graphing the present value formula $y = P(1 - (1 + r/m)^{-mx})/(r/m)$ on the window [0, 20] by [0, 2000]. Use TRACE or INTERSECT to estimate the number of years x that corresponds to the present value y from part (b). Be sure to change this number of years into the number of months the payments must be made.

d. Using the advertised financing, what is the buyer's total cost?

37. Yearly Interest A home owner finances a new luxury car using a "home equity" loan so that she can deduct the interest each year when calculating her taxable income. If she plans to repay the $40,000 in equal monthly payments over 4 years and the interest rate is fixed at 9.35%, how much interest will she pay in the second year of the loan?

38. **Home Refinancing** A widow needs cash and decides to refinance the home she and her husband bought 20 years ago for $35,000 with a monthly payment, 30-year mortgage at 6.8% on 80% of the purchase price. The Home Sweet Home Mortgage Corporation has appraised her house at $125,000 and has agreed to lend her 70% of this value. How much cash will she have after paying off the old mortgage?

Explorations and Excursions

The following problems extend and augment the material presented in the text.

More About Amortization

39. Solve any three of Exercises 1–6 as follows: Enter the "present value of an annuity" formula as $y = P(1 - (1 + r/m)^{-mt})/(r/m)$. Then for each exercise, STORE the values for P, r, m, and t, and evaluate y to find the present value.

40. **Unpaid Balance Formula** Combine the "present value of an annuity" formula with the amortization payment formula to show that the unpaid balance after x years of payments made m times each year for t years on a debt of D dollars at interest rate r is

$$D(1 + r/m)^{mx} \frac{(1 + r/m)^{m(t-x)} - 1}{(1 + r/m)^{mt} - 1}$$

41. Solve any three of Exercises 7–12 as follows: Enter the amortization payment formula as $y = D(r/m)/(1 - (1 + r/m)^{-mt})$. Then for each exercise, STORE the values for m, D, r, and t, and evaluate y to find the required payment.

42. **Present Value of an Annuity** Show that the present value of an annuity formula may be rewritten in the form

$$PV = P \frac{1 - (1 + i)^{-n}}{i}$$

where $i = r/m$ and $n = mt$.

43. Show that the present value of an annuity is the sum of the present values of the payments. [*Hint:* Use the notation of Exercise 42 and show that the present value of the first payment is $P(1 + i)^{-1}$, that of the second is $P(1 + i)^{-2}$, and so on for the n payments. Factor out the common terms from the sum and use the geometric series formula on page 121.]

44. Use the program AMORTABL to solve Exercise 13. How does the amortization table solution differ from your previous solution to Exercise 13? How much does the final payment in the table differ from the payments made during the rest of the table?

45. Use the program AMORTABL to solve Exercise 16. Be sure to enter "1" for the "compoundings per year." How does the amortization table solution differ from your previous solution to Exercise 16? How much does the final payment in the table differ from the payments made during the rest of the table?

Chapter Summary with Hints and Suggestions

Reading the text and doing the exercises in this chapter have helped you to master the following skills, which are listed by section (in case you need to review them) and are keyed to particular Review Exercises. Answers for all Review Exercises are given at the back of the book, and full solutions can be found in the Student Solutions Manual.

2.1 Simple Interest

- Find the interest due on a simple interest loan. *(Review Exercises 1–4.)*

$$I = Prt$$

- Find the total amount due on a simple interest loan. *(Review Exercises 5–8.)*

$$A = P(1 + rt)$$

- Solve a simple interest situation for the interest, the interest rate, the principal, or the term. *(Review Exercises 9–13.)*

- Find a simple interest future or present value. *(Review Exercises 14–16.)*

$$A = P(1 + rt) \qquad P = \frac{A}{1 + rt}$$

- Find the effective simple interest rate of a discounted loan. *(Review Exercises 17–18.)*

$$r_s = \frac{r}{1 - rt}$$

- Solve an applied simple-interest problem. *(Review Exercises 19–21.)*

2.2 Compound Interest

- Determine the amount due on a compound interest loan. *(Review Exercises 22–26.)*

$$A = P(1 + r/m)^{mt}$$

- Find a future or present value. *(Review Exercises 27–31.)*

$$A = P(1 + r/m)^{mt} \qquad P = \frac{A}{(1 + r/m)^{mt}}$$

- Find the term needed for a given principal to grow to a future value. *(Review Exercises 32–36.)*

- Use the "rule of 72" to estimate a doubling time. *(Review Exercises 37–41.)*

$$\left(\begin{array}{c}\text{Doubling}\\\text{time}\end{array}\right) \approx \frac{72}{r \times 100}$$

- Find the effective rate of a loan or investment. *(Review Exercises 42–46.)*

$$r_e = (1 + r/m)^m - 1$$

2.3 Annuities

- Find the accumulated amount of an ordinary annuity. *(Review Exercises 47–51.)*

$$A = P\frac{(1 + r/m)^{mt} - 1}{r/m}$$

- Calculate the regular payment to make into a sinking fund to accumulate a given amount. *(Review Exercises 52–56.)*

$$P = A\frac{r/m}{(1 + r/m)^{mt} - 1}$$

- Find the number of periods to make payments into a sinking fund to accumulate a given amount. *(Review Exercises 57–61.)*

$$mt = \frac{\log\left((A/P)(r/m) + 1\right)}{\log(1 + r/m)}$$

- Estimate the internal rate of return of regular investments using a graphing calculator. *(Review Exercises 62–66.)*

2.4 Amortization

- Find the present value of an ordinary annuity. *(Review Exercises 67–71.)*

$$PV = P\frac{1 - (1 + r/m)^{-mt}}{r/m}$$

- Find the regular payment to amortize a debt. *(Review Exercises 72–76.)*

$$P = D\frac{r/m}{1 - (1 + r/m)^{-mt}}$$

- Find the amount still owed after making amortization payments for part of the term. *(Review Exercises 77–81.)*

Hints and Suggestions

- *(Overview)* Compound interest is repeated simple interest with the interest added to the principal in each successive period. The *future value* is the amount the principal will become after all the interest is included. Reversing the point of view, the *present value* is the principal needed now that will grow to the final amount.

The three basic amount formulas are: $P(1 + rt)$ for simple interest, $P(1 + r/m)^{mt}$ for compound interest, and $P((1 + r/m)^{mt} - 1)/(r/m)$ for an annuity (multiple payments). The other formulas all follow from these basic formulas.

- The rate is usually stated as a percent but is always used in decimal form for calculations: 6.7% in decimal form is 0.067.

- Round off only your final answer when using your calculator. Don't use the decimal 0.33 for the fraction $\frac{1}{3}$.

- Don't confuse a *percent increase* with the *multiplier. Increasing* an amount by 25% means *multiplying* it by 1.25.

- Round *up* to find the whole number of compoundings needed to reach a future value because a smaller number won't reach the stated goal.

- The "rule of 72" is only an approximation but is a helpful check that can catch "button pushing" errors when using your calculator.

- **Practice for Test:** Review Exercises 1, 5, 9, 13, 18, 24, 27, 32, 34, 40, 43, 47, 50, 54, 58, 68, 73, 78.

Review Exercises for Chapter 2

Practice test exercise numbers are in green.

Round all dollar amounts to the nearest penny, all times in years to two decimal places, and all terms to the appropriate number of compounding periods.

2.1 Simple Interest

Find the simple interest on each loan.

1. $1875 at 5.8% for 2 years.

2. $1150 at 9.2% for 6 months.

3. $8000 at $7\frac{1}{2}$% for 3 years 9 months.

4. $2385 at 11.3% for 1 year 3 months.

Find the total amount due on each simple interest loan.

5. $8900 at 5.9% for 2 years 6 months.

6. $1375 at 11.3% for 9 months.

7. $1795 at 6.38% for 3 years.

8. $3700 at 8.3% for 1 year 11 months.

Solve each problem.

9. Find the interest rate on a loan charging $272 simple interest on a principal of $2000 after 2 years.

10. Find the principal of a loan at 9.3% if the simple interest after 1 year 8 months is $279.

11. Find the term of a loan of $7600 at 9% if the simple interest is $1026.

12. What should be the term for a loan of $5500 at 8.7% simple interest if the lender wants to receive $7414 when the loan is paid off?

13. How long will it take an investment at 10% simple interest to increase by 50%?

14. How much should be invested now at 5.9% simple interest if $5208 is needed in 2 years 8 months?

15. What would be the fair market price of a $10,000 zero coupon bond due in 1 year if today's long-term simple interest rate is 6.45%?

16. What would be the fair market price of a $50,000 zero coupon bond due in 1 year if today's long-term simple interest rate is 5.78%?

17. What is the effective simple interest rate of a discounted loan at 12% interest for 6 years?

18. What is the effective simple interest rate of a discounted loan at 15% interest for 2 years?

19. **Treasury Bonds** *Forbes* magazine ranked John W. Kluge as the richest American, with a net worth of $5.2 billion, at the time of his divorce from Patricia Kluge. A newspaper reported that besides a 45-room Georgian mansion, Mrs. Kluge "... received $1 billion that invested conservatively in 30-year Treasury bonds would throw off $66 million a year." What was the interest rate when the article was written?

20. **Insurance Settlements** How large an insurance settlement does an accident victim need for a yearly income of $29,000 from long-term bond investments paying 5.8%?

21. **Furniture Buying** A furniture store offers a complete living room suite for $999 and will finance the entire price as a discounted loan at 10% for 1 year. What is the effective simple interest rate for this loan and how much will the buyer need to pay at the end of the year?

2.2 Compound Interest

Find the amount due on each compound interest loan.

22. $15,000 at 7.5% compounded quarterly for 10 years.

23. $65,000 at 8.25% compounded daily for 17 years.

24. **Mutual Funds** From 1970 to 1995, the AIM Charter Fund posted an average annual return of 15.55% including sales charges. How much would a $10,000 investment in this fund have been worth after 25 years?

25. **Mutual Funds** During the mid-1990s, the Brandywine long-term growth fund returned 19.6% compounded quarterly. How much would a $5000 investment in this fund have been worth after 4 years?

26. **National Debt** From 1835 to 1837, the United States not only was free of debt but actually had a surplus in the Treasury. On January 1, 1837, after $5 million had been set aside as a reserve fund, there remained $37,468,859. If that money had not been distributed to the 26 states but instead had been invested at 5% compounded monthly, would there have been enough by 1997 to pay off a $4.3 trillion national debt?

Find the present value of each compound interest loan.

27. $25,000 after 10 years at 9% compounded monthly.

28. $30,000 after 8 years at 6% compounded daily.

29. $175,000 after 16 years 6 months at 5.5% compounded quarterly.

30. **Certificates of Deposit** From 1968 to 1995, the average annual return on 6-month CDs was 7.79% compounded semiannually. How much would one have to have invested in 1968 in order to have $100,000 in 1995?

31. **Bond Funds** During the mid-1990s, the Northeast Investors Trust general-term bond fund returned 15.7% compounded monthly. How much would you need to invest in this fund to have $15,000 after 3 years 8 months?

Find the term of each compound interest loan.

32. 8.9% compounded quarterly to obtain $10,000 from a principal of $2000.

33. 11.5% compounded monthly to multiply the principal by 1.60.

34. How long will it take the value of an apartment building to increase by 80% if real estate prices are increasing 8% annually?

35. **Treasury Bonds** How long will you have to wait to become a millionaire if you invest all of your $850,000 lottery winnings in Treasury bonds paying 5.6% compounded semiannually?

36. **Art Appreciation** How long will it take for the value of an early Picasso pencil sketch to triple if its market value is increasing 6.5% annually?

For Exercises 37–41, use the "rule of 72" to estimate the doubling time (in years) for each interest rate and then calculate it exactly.

37. 8% compounded semiannually.

38. 6% compounded monthly.

39. 11.9% compounded weekly.

40. **Real Estate** How long will it take for the value of a house to double if real estate prices are increasing 12% each year?

41. **Index Funds** How long will it take an investment in a stock market index fund yielding 10% annually to double?

42. Find the effective rate of 13.25% compounded quarterly.

43. **Baseball Cards** A mint condition 1955 Topps "Sandy Koufax" baseball card selling for $500 in 1989 was worth $1350 in 1993. What is the effective rate of this price increase?

44. **Stocks** One of Wall Street's most cherished buys in the 1950s was Texas Instruments stock, which rose spectacularly from $72\frac{1}{8}$ to $214\frac{1}{4}$ in 18 months. What is the effective rate of this price increase?

45. **Baseball Teams** The Haas family bought the Oakland Athletics from Charles O. Finley for $12.7 million in 1980 and sold it in 1995 for $85 million. Neglecting all other expenses, what is the effective rate of return on their investment?

46. **Pianos** A properly maintained 90-year-old Steinway grand piano is now worth 13.6 times its initial purchase price. What is the effective rate of this increase?

2.3 Annuities

Find the accumulated amount of each annuity.

47. $1500 annually at 11% for 20 years.

48. $500 semiannually at 8% for 12 years.

49. **Home Buying** How much will you have for a vacation home if you save $25 each week for 16 years in a 5.5% passbook savings account?

50. **Retirement Savings** How much will an IRA stock fund earning 11.2% be worth if you deposit $180 each month for 35 years?

51. **Apartments** When André-François Raffray died at age 77 in Arles, France, he had been making $500 monthly payments to Jeanne Calment for 30 years for the right to take over her apartment when she died. Mrs. Calment, who at 120 was then the world's oldest person with the records to prove it, remarked that "In life, one sometimes makes bad deals." How much would Mr. Raffray's money be worth if instead he had deposited it into a 7.8% bond fund?

Find the required payment for each sinking fund.

52. Yearly deposits earning 7% to accumulate $16,000 after 8 years.

53. Quarterly deposits earning 4.9% to accumulate $40,000 after 25 years.

54. How much must you pay biweekly to pay off a 13.2% car loan of $19,500 in 5 years?

55. **Stock Funds** How much must you deposit each month into a stock fund earning 11.2% to accumulate $150,000 in 20 years?

56. **Home Buying** How much must you save each week in a 5.5% passbook savings account to have $26,000 for a vacation home in 16 years?

Find the number of years needed for each sinking fund to reach the given accumulated amount.

57. $1000 yearly at 10% to accumulate $100,000.

58. $250 quarterly at 8.1% to accumulate $75,000.

59. **Money Market Funds** How long must you save $275 each month in a 6.2% money market fund to accumulate $19,500 for a new car?

60. **Stock Funds** How long must you deposit $300 each month into a stock fund earning 11.2% to accumulate $150,000?

61. **Home Buying** How long must you save $25 each week in a 5.5% passbook savings account to have $26,000 for a vacation home?

Follow the method of Exercise 27 on page 129 to estimate the internal rate of return for each investment.

62. $150 quarterly to accumulate $10,000 after 10 years.

63. $225 monthly to accumulate $100,000 after 25 years.

64. $500 monthly to accumulate $1,000,000 after 30 years.

65. $1.25 weekly to accumulate $50,000 after 60 years.

66. $10 daily to accumulate $100,000 after 20 years.

2.4 Amortization

67. Find the present value of an annuity of $25,000 annually at 6% for 25 years.

68. Football Contracts Deion Sanders became the highest-paid defensive player in football when he signed a $25 million, 5-year contract with the Dallas Cowboys. If the $25 million was paid out in equal weekly payments over the 5 years and the current interest rate was 7.32%, how much was the contract worth when it was signed?

69. Grant Funding A foundation is funding a new program that will award twenty $250,000 grants each year for the next 8 years to promote art and music instruction in elementary schools. How much must be placed in a money market account paying 6.75% to fund this initiative?

70. Contest Prizes The grand prize in a lottery drawing is $14 million to be paid out in equal quarterly payments over the next 20 years. If the prize money is guaranteed by U.S. Treasury bonds yielding 7.6% and is placed into an escrow account when the lottery is announced 1 year before the first payment, how much do the lottery sponsors have to deposit in the escrow account?

71. Home Buying Before looking for their first house, the Jones family calculates that they could afford a $1250 monthly mortgage payment. How large a mortgage could they afford at 8.75% for 30 years?

72. Find the weekly payments to amortize a debt of $45,000 at 9.15% over 15 years.

73. Credit Cards A Discover card statement shows a balance of $945 at 15.7% compounded monthly. What monthly payment will pay off this debt in 2 years?

74. Credit Cards What monthly payment should you make on your Optima card to pay off $2784 of spring break vacation charges in 7 months if the interest rate is 18.9% compounded monthly?

75. Home Buying The Jones family has found a nice house in the suburbs. What is the monthly payment on their $150,000 mortgage at 8.75% for 30 years?

76. Stock Funds How much must you save each week in a stock fund yielding 10.1% to become a millionaire in 15 years?

Find the unpaid balance on each debt.

77. After 5 years of annual payments on $85,000 at 9.1% for 20 years.

78. After 3 years 8 months of monthly payments on $150,000 at 8.7% for 25 years.

79. Car Buying After 1 year 5 months of $297.92 monthly car payments on a 3-year loan at 4.6%, Karen wants to get rid of her sedan and get a red sports car. How much should she pay to settle her debt?

80. Contract Buyout The Middleville Central School Board made a mistake when they hired the new superintendent and agreed to a 6-year contract paying $125,000 per year. With current interest rates at 6.2%, how much should they offer to break the contract after the first year?

81. Home Buying After 4 years 9 months of monthly mortgage payments on their 30-year, $150,000 mortgage at 8.75%, the Jones family has to relocate to Atlanta and must sell their nice house in the suburbs. How much do they still owe on their mortgage?

3 Systems of Equations and Matrices

Application Preview

A "Taxing" Problem

For a tax law to be acceptable, it must be perceived as fair by the taxpayers. When federal and state governments both try to tax the same income, abhorrence of "double jeopardy" suggests that the taxes paid to one entity should not be further taxed by the other.

Consider a simplified situation in which the federal tax is 20% of your taxable income after the state tax has first been deducted and the state tax is 4% of your taxable income after the federal tax has first been deducted. How can you find the correct taxes? There is a widespread misconception that since neither amount can be calculated first, the only way is by a process of successive approximation: First estimate the state tax, then subtract it from your income to calculate the federal tax, then readjust the state tax estimate based on the federal tax and begin again. The problem becomes even more complicated if city taxes are included. Perhaps as a result of this confusion, state income taxes do not allow the deduction of federal taxes although federal taxes *do* allow the deduction of state taxes. Contrary to this misconception, these taxes *can* be found exactly by the methods of this chapter (see pages 159–160), and the same methods can be used to solve many other problems.

Form **W-4**	**Employee's Withholding Allowance Certificate**	OMB No. 1545-0010
Department of the Treasury Internal Revenue Service	▶ For Privacy Act and Paperwork Reduction Act Notice, see reverse.	2004

1 Type or print your first name and middle initial	Last name	2 Your social security number

Home address (number and street or rural route)

3 ☐ Single ☐ Married ☐ Married, but withhold at higher Single rate.
Note: *If married, but legally separated, or spouse is a nonresident alien, check the Single box.*

City or town, state, and ZIP code

4 If your last name differs from that on your social security card, check here and call 1-800-772-1213 for a new card . . . . . ▶ ☐

5 Total number of allowances you are claiming (from line G above or from the worksheets on page 2 if they apply) . **5**
6 Additional amount, if any, you want withheld from each paycheck . . . . **6** $
7 I claim exemption from withholding for 1995 and I certify that I meet **BOTH** of the following conditions for exemption:
 • Last year I had a right to a refund of **ALL** Federal income tax withheld because I had **NO** tax liability; **AND**

3.1 SYSTEMS OF TWO LINEAR EQUATIONS IN TWO VARIABLES

Introduction

Neither of the statements "Bob and Sue together have $100" and "Bob has $20 less than Sue" tells us how much either has. But taken together, they force the conclusion that Bob has $40 and Sue has $60. Various methods of solving such simple problems have been used since antiquity, but many become unworkable as the problems become more complicated. This chapter provides a method that solves such problems and has the pleasing property that the method for complicated problems is an easy extension of the method for the simplest.

Systems of Equations

We begin with the simplest form of these problems.

Systems of Two Linear Equations in Two Variables

A system of two linear equations in two variables is any problem expressible in the form

$$\begin{cases} ax + by = h \\ cx + dy = k \end{cases}$$

where x and y are the variables and the constants a, b, c, d, h, k are such that at least one of the coefficients a, b, c, d is not zero.

Such equations are called *linear* because $ax + by = h$ (with at least one of a and b not equal to zero) is the same as the *general linear equation* (see page 12) whose graph is a line.

EXAMPLE 1

TWO EQUATIONS IN TWO VARIABLES

Express the statements "Bob and Sue together have $100" and "Bob has $20 less than Sue" as a system of two equations in two variables.

Solution

Let x represent the amount of money that Bob has and y represent the amount that Sue has. The statement that together they have $100 may be written "$x + y = 100$" and the statement that Bob has $20 less than Sue as "$x = y - 20$". Rearranging this second equation by subtracting y from both sides, we obtain the system of equations

$$\begin{cases} x + y = 100 \\ x - y = -20 \end{cases}$$

$ax + by = h$ with
$a = 1, \quad b = 1, \quad h = 100$

$cx + dy = k$ with
$c = 1, \quad d = -1, \quad k = -20$

There are many other ways to express the two statements as a system of equations. For example, we could write the same equations but in the opposite order

$$\begin{cases} x - y = -20 \\ x + y = 100 \end{cases}$$

Switching the order
of the equations

or instead we could multiply one of the equations by 2

$$\begin{cases} 2x + 2y = 200 \\ x - y = -20 \end{cases}$$

$x + y = 100$
multiplied by 2

We could even add (or subtract) the equations and use the result to replace one of the original equations:

$$\begin{cases} x + y = 100 \\ 2x \quad\;\; = 80 \end{cases}$$

$x + y = 100$
$\underline{x - y = -20}$ Adding
$2x \quad\;\; = \quad 80$

It is an easy matter to check that each of these three systems has the same solution, $x = 40$ and $y = 60$. Notice that the second equation in the above system immediately gives $x = 40$, which is "half" of the solution. In fact, if we could write the system in the following way, we could immediately read off the *entire* solution:

$$\begin{cases} 1x + 0y = 40 \\ 0x + 1y = 60 \end{cases}$$

⟵ gives $x = 40$
⟵ gives $y = 60$

This will be our goal—to simplify a system of equations until its solution is obvious (or until it is clear that there is no solution).

Practice Problem 1

Express the statements "a jar of pennies and nickels contains 80 coins" and "the coins in the jar are worth \$1.60" as a system of two equations in two variables. ➤ Solution on page 162

A *solution* of a system of equations in two variables is a pair of values for the variables that satisfy all the equations (such as $x = 40$, $y = 60$ for Example 1). The *solution set* is the collection of all solutions. *Solving* the system of equations means finding this solution set. In this section we will solve systems of linear equations by three different methods: *graphing*, *substitution*, and *elimination*. Each has its own advantages and disadvantages.

Graphical Representations of Equations

An equation of the form $ax + by = h$ (with at least one of a and b not zero) is the equation of a line written in *general form*. We begin by graphing linear equations.

EXAMPLE 2

GRAPHING A LINEAR EQUATION IN TWO VARIABLES

Sketch the graph of each equation.

a. $2x + 3y = 12$ **b.** $2x - 3y = 0$

c. $3y = 12$ **d.** $2x = 12$

Solution

a. To graph $2x + 3y = 12$ we find the intercepts by setting each of the variables in turn equal to zero:

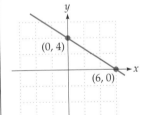

$$3y = 12$$

$$y = 4 \quad \text{for the point } (0, 4)$$

$$2x = 12$$

$$x = 6 \quad \text{for the point } (6, 0)$$

$2x + 3y = 12$ with $x = 0$

Since we began with $x = 0$

The original $2x + 3y = 12$

Since we began with $y = 0$

The line through $(0, 4)$ and $(6, 0)$ is shown on the left.

b. To graph $2x - 3y = 0$ we begin by setting $x = 0$:

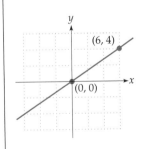

$$-3y = 0$$

$$y = 0 \quad \text{for the point } (0, 0)$$

$2x - 3y = 0$ with $x = 0$
Now divide by -3

Since we began with $x = 0$

For another point we choose any other value of x.

$$12 - 3y = 0$$

$$y = 4 \quad \text{for the point } (6, 4)$$

$2x - 3y = 0$ with $x = 6$
Now subtract 12 and
then divide by -3

Since we began with $x = 6$

The line through $(0, 0)$ and $(6, 4)$ is shown on the left.

c. To graph $3y = 12$:

$$y = 4 \qquad\qquad \text{Dividing by 3}$$

$y = 4$ is a horizontal line. Its graph is shown on the left.

d. To graph $2x = 12$:

$$x = 6 \qquad\qquad \text{Dividing by 2}$$

$x = 6$ is a vertical line. Its graph is shown on the left.

The graph of a *pair* of equations $\begin{cases} ax + by = h \\ cx + dy = k \end{cases}$ may take three different forms: two lines intersecting at just one point (a *unique* solution), two lines that don't intersect (*no* solutions), or two lines that are the *same* (*infinitely many* solutions), as shown below. For lines that don't intersect, we say that the equations are *inconsistent,* and for two lines that are the *same* we call the equations *dependent.*

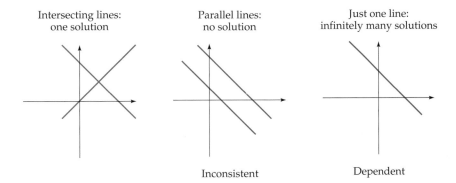

The type and solution (if any) of a system of equations can be determined from the graph if it is drawn with sufficient accuracy.

EXAMPLE **3**

SOLVING BY GRAPHING

Solve each system of equations by graphing the lines.

a. $\begin{cases} 2x + 3y = 12 \\ 2x - 3y = 0 \end{cases}$ **b.** $\begin{cases} 2x + 3y = 12 \\ 4x + 6y = 12 \end{cases}$ **c.** $\begin{cases} 2x + 3y = 12 \\ 4x + 6y = 24 \end{cases}$

Solution

The lines $2x + 3y = 12$ and $2x - 3y = 0$ were graphed in Example 2. The other lines may be graphed similarly.

a.

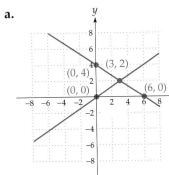

Intersect at $x = 3$, $y = 2$: one solution

b.

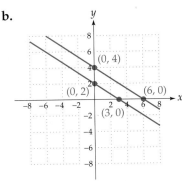

Lines are parallel: no solution (inconsistent)

c.

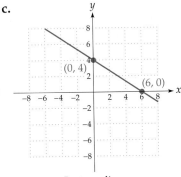

Just one line: infinitely many solutions (dependent)

In part (a), the unique solution $x = 3$, $y = 2$ is easily checked by substituting these values into the two equations:

$$2 \cdot 3 + 3 \cdot 2 = 12$$

$2x + 3y = 12$ with $x = 3$ and $y = 2$ (It checks: $12 = 12$)

$$2 \cdot 3 - 3 \cdot 2 = 0$$

$2x - 3y = 0$ with $x = 3$ and $y = 2$ (It checks: $0 = 0$)

In part (b) the lines do not intersect, so there is *no solution* (the equations are *inconsistent*).

In part (c) the lines are the *same* (the equations are *dependent*), so *every* point on the line $2x + 3y = 12$ is a solution.

For situations with infinitely many solutions, as in part (c), we may express *all* of these solutions at once as follows. We begin by solving either of the equations for one of the variables:

$$x = \frac{12 - 3y}{2} = 6 - \frac{3}{2}y$$

Solving $2x + 3y = 12$ for x

and writing the solution as

$$\begin{cases} x = 6 - \frac{3}{2}t \\ y = t \end{cases}$$

Replacing y by t

t may take any value

where t is any number (called a *parameter*). For example, taking $t = 2$ gives $x = 3$, $y = 2$ and the point (3, 2) is a solution because it satisfies both equations. All other solutions can be found by choosing other values for the parameter t.

Find the solution corresponding to the parameter value $t = -2$ and verify that the resulting values satisfy the equations in Example 3c.

➤ Solution on page 162

Graphing Calculator Exploration

You can view a nonvertical line $ax + by = h$ by entering it as $y = (h - ax)/b$. To see the system $\begin{cases} 2x + 3y = 12 \\ 2x - 3y = 0 \end{cases}$ from Example 3a:

Intersection
X=3   Y=2

a. Enter the first equation as $y_1 = (12 - 2x)/3$ and the second as $y_2 = (-2x)/(-3)$. Graph them on the window $[-3, 7]$ by $[-3, 5]$.

b. Use TRACE or EVALUATE to check the x- and y-intercepts.

c. Use TRACE or INTERSECT to find the intersection point of these two lines.

Substitution Method

The graphical method is limited by the accuracy of your sketch. Our second way of solving a system of linear equations is called the *substitution method* and involves solving one of the equations for a variable and then substituting the result into the other equation.

EXAMPLE 4

SOLVING BY THE SUBSTITUTION METHOD

Solve $\begin{cases} x + 3y = 15 \\ 2x - 5y = 8 \end{cases}$ by the substitution method.

Solution

The first equation can easily be solved for x as $x = 15 - 3y$. We then substitute $15 - 3y$ for x in the second equation.

$$2(15 - 3y) - 5y = 8 \qquad \text{Replacing } x \text{ with } 15 - 3y$$
$$30 - 6y - 5y = 8 \qquad \text{Multiplying out}$$
$$-11y = -22 \qquad \text{Collecting like terms}$$
$$y = 2 \qquad \begin{array}{l}\text{Dividing by } -11 \text{ gives}\\ \text{the } y\text{-value}\end{array}$$

Then

$$x = 15 - 3(2) = 15 - 6 = 9 \qquad x = 15 - 3y \text{ with } y = 2$$

The solution is $x = 9$, $y = 2$ as may be easily checked by substituting into the original equations.

The substitution method is particularly useful if one of the equations is easily solved for x or y.

Practice Problem 3   Solve $\begin{cases} 2x + y = 10 \\ x + 2y = 8 \end{cases}$ by the substitution method.

➤ Solution on page 162

What happens if we try to solve an *inconsistent* system, such as $\begin{cases} x + y = 10 \\ x + y = 20 \end{cases}$ (which is inconsistent since the same quantity can't equal both 10 and 20)? The first equation gives

$$x = 10 - y \qquad x + y = 10 \text{ solved for } x$$

Substituting this into the second equation

$$(10 - y) + y = 20 \qquad \begin{array}{l} x + y = 20 \text{ with } x \\ \text{replaced by } x = 10 - y \end{array}$$

which simplifies to

$$10 = 20 \qquad \text{A contradiction}$$

Interpret this result as follows: Looking for a value for y, we were led to the "impossible" equation $10 = 20$, meaning that *no* value of y works, so the original system has *no solutions* and is *inconsistent*. Whenever you find an "impossible" equation (saying that two different numbers are equal), the original system is inconsistent (has no solutions).

What happens if we try to solve a *dependent* system, such as $\begin{cases} x + y = 10 \\ 2x + 2y = 20 \end{cases}$ (which is dependent since one equation is twice the other)? The first equation gives

$$x = 10 - y \qquad\qquad x + y = 10 \text{ solved for } x$$

Substituting this into the second:

$$2(10 - y) + 2y = 20 \qquad\qquad \begin{array}{l} 2x + 2y = 20 \text{ with } x \\ \text{replaced by } x = 10 - y \end{array}$$

After multiplying out and simplifying, we obtain:

$$20 = 20 \qquad\qquad \text{True but useless}$$

Interpret this result as follows: Looking for a value for y, we were led to the "useless" equation $20 = 20$ that is *always* true, meaning that *all* values of y work, so the original system has *infinitely many solutions* and is *dependent*. Whenever you find a "useless" equation (saying that a number equals itself), the original system is dependent.

Equivalent Systems of Equations

Two systems of equations are *equivalent* if they have the same solution. On page 151 we saw that we obtained equivalent systems by applying any of the following operations: switching the order of the equations, multiplying one equation by a (nonzero) constant, or adding or sub-tracting two equations. By combining the last two operations, we may even add a *multiple* of an equation to another. For example:

1. Switch the order of the equations. $\quad\begin{cases} x + y = 100 \\ x - y = -20 \end{cases} \xrightarrow[\text{equations}]{\text{Switch}} \begin{cases} x - y = -20 \\ x + y = 100 \end{cases}$

2. Multiply or divide one of the equations by a nonzero number. $\quad\begin{cases} x + y = 100 \\ x - y = -20 \end{cases} \xrightarrow[\text{equation by } -1]{\text{Multiply second}} \begin{cases} x + y = 100 \\ -x + y = 20 \end{cases}$

3. Add (or subtract) a multiple of one equation to (or from) the other. $\quad\begin{cases} x + y = 100 \\ x - y = -20 \end{cases} \xrightarrow[\text{to the second}]{\text{Add twice the first}} \begin{cases} x + y = 100 \\ 3x + y = 180 \end{cases}$

These systems are all equivalent to each other because they all have exactly the same solutions.

Elimination Method

Our third way of solving equations, the *elimination* method, attempts to remove one variable from each and write them as an equivalent system of the form $\begin{cases} 1x + 0y = p \\ 0x + 1y = q \end{cases}$ so that the solution can be read off as $x = p, \ y = q$. If this cannot be done, the method will identify the system as inconsistent or dependent.

In these problems we may carry out subtractions in either order, *top line minus bottom* or *bottom line minus top*:

$$\begin{array}{c} 6 \\ \underline{15} \\ -9 \end{array} \leftarrow \begin{array}{l} \text{top minus bottom} \\ (6 - 15 = -9) \end{array} \qquad \text{or} \qquad \begin{array}{c} 6 \\ \underline{15} \\ 9 \end{array} \leftarrow \begin{array}{l} \text{bottom minus top} \\ (15 - 6 = 9) \end{array}$$

EXAMPLE 5

SOLVING BY THE ELIMINATION METHOD

Solve $\begin{cases} x - 3y = 3 \\ 2x + 5y = 28 \end{cases}$ by the elimination method.

Solution

First we eliminate the x from the second equation by subtracting an appropriate multiple of the first (namely, *twice* the first in order to match the $2x$ in the second equation). Beginning with the original system:

$$\begin{cases} x - 3y = 3 \\ 2x + 5y = 28 \end{cases} \quad \begin{array}{l} \times 2 \rightarrow \\ \rightarrow \end{array} \quad \begin{array}{r} 2x - 6y = 6 \\ 2x + 5y = 28 \\ \hline 11y = 22 \end{array} \leftarrow \text{bottom minus top}$$

$$\begin{cases} x - 3y = 3 \\ \quad 11y = 22 \end{cases} \qquad \text{Unchanged}$$

$$\begin{cases} x - 3y = 3 \\ \quad y = 2 \end{cases} \qquad \begin{array}{l} \text{Unchanged} \\ \text{Dividing} \quad 11y = 22 \quad \text{by 11} \end{array}$$

Next we eliminate the y from the first equation by subtracting an appropriate multiple of the second (namely, *three times* the second to match the $3y$ in the first).

$$\begin{cases} x - 3y = 3 \\ \quad y = 2 \end{cases} \quad \begin{array}{l} \rightarrow \\ \times 3 \rightarrow \end{array} \quad \begin{array}{r} x - 3y = 3 \\ 3y = 6 \\ \hline x \qquad = 9 \end{array} \leftarrow \text{top plus bottom}$$

$$\begin{cases} x = 9 \\ y = 2 \end{cases} \qquad \text{Unchanged} \qquad \text{Same as} \quad \begin{cases} 1x + 0y = 9 \\ 0x + 1y = 2 \end{cases}$$

The original system has a unique solution $x = 9, \ y = 2$.

An inconsistent system of equations always leads to the impossible equation that zero equals a nonzero number. For instance, solving Example 3b by the elimination method could be done as follows:

$$\begin{cases} 2x + 3y = 12 \\ 4x + 6y = 12 \end{cases} \xrightarrow[\text{equation by 2}]{\text{Divide second}} \begin{cases} 2x + 3y = 12 \\ 2x + 3y = 6 \end{cases} \xrightarrow[\text{from second}]{\text{Subtract first}} \begin{cases} 2x + 3y = 12 \\ 0x + 0y = -6 \end{cases}$$

Since the last equation says $0 = -6$, which is contradictory, there is no solution and the equations are inconsistent.

A dependent system of equations always results in one equation becoming $0x + 0y = 0$. For instance, solving Example 3c by the elimination method could be done as follows:

$$\begin{cases} 2x + 3y = 12 \\ 4x + 6y = 24 \end{cases} \xrightarrow[\text{equation by 2}]{\text{Divide second}} \begin{cases} 2x + 3y = 12 \\ 2x + 3y = 12 \end{cases} \xrightarrow[\text{from second}]{\text{Subtract first}} \begin{cases} 2x + 3y = 12 \\ 0x + 0y = 0 \end{cases}$$

Since the last equation says $0 = 0$, which is uninformative but not contradictory, the solutions are all the points on the line $2x + 3y = 12$ and the system is dependent.

Practice Problem 4 Solve $\begin{cases} x + y = 100 \\ x - y = -20 \end{cases}$ by the elimination method.

➤ Solution on page 162

We conclude this section with the solution of the tax problem posed in the Application Preview on page 149.

EXAMPLE 6   **A TAXING PROBLEM**

What are the federal and state taxes on an income of $34,100 if the federal tax is 20% of the income after first deducting the state tax, and the state tax is 4% of the income after first deducting the federal tax?

Solution

Let x represent the federal tax and y represent the state tax. Then

$$\begin{cases} x = 0.20(34{,}100 - y) \\ y = 0.04(34{,}100 - x) \end{cases}$$

Federal tax is 20% after first subtracting the state tax
State tax is 4% after first subtracting the federal tax

$$\begin{cases} x = 6820 - 0.2y \\ y = 1364 - 0.04x \end{cases}$$

Multiplying out

Moving all variables to the left side of the equations gives the following system. To solve it, we eliminate the x from the second equation by subtracting 0.04 times the first equation (the 0.04 is chosen to match the $0.04x$ in the second equation):

$$\begin{cases} x + 0.2y = 6820 \\ 0.04x + y = 1364 \end{cases}$$

$\times\,0.04 \rightarrow$

$$\begin{aligned} 0.04x + 0.008y &= 272.8 \\ 0.04x + y &= 1364 \\ \hline 0.992y &= 1091.2 \end{aligned}$$

$\leftarrow$ bottom minus top

$$\begin{cases} x + 0.2y = 6820 \\ 0.992y = 1091.2 \end{cases}$$

Unchanged

$$\begin{cases} x + 0.2y = 6820 \\ y = 1100 \end{cases}$$

Dividing $0.992y = 1091.2$ by 0.992

Now we eliminate the y from the first equation by subtracting 0.2 times the second (the 0.2 chosen to match the $0.2y$ in the first):

$$\begin{cases} x + 0.2y = 6820 \\ y = 1100 \end{cases}$$

$\times\,0.2 \rightarrow$

$$\begin{aligned} x + 0.2y &= 6820 \\ 0.2y &= 220 \end{aligned}$$

$$\begin{cases} x = 6600 \\ y = 1100 \end{cases}$$

Unchanged

The solution $x = 6600$, $y = 1100$ says that the federal tax is $6600 and the state tax is $1100. You should check that each of these taxes is the correct percentage of the $34,100 income after the other tax has been subtracted.

3.1 Section Summary

A *system of two linear equations in two variables* can be written in the form

$$\begin{cases} ax + by = h \\ cx + dy = k \end{cases}$$

and will be one of three types.

$\begin{cases} ax + by = h \\ cx + dy = k \end{cases}$	Graph	Number of Solutions
Has a unique solution	Lines intersect	One
Inconsistent	Parallel lines	None
Dependent	Just one line	Infinitely many

Solving by *graphing* is only as accurate as your sketch.

The *substitution method* solves one of the equations for a variable and uses this new expression to rewrite the other equation with just one variable. Solving this new equation and then finding the first variable solves the system of equations. This method is most useful when one of the coefficients of the variables (a, b, c, or d) is 1.

Equivalent systems of equations have the same solution and can be found by

1. Switching the order of the equations,
2. Multiplying or dividing one equation by a nonzero number, or
3. Replacing an equation by its sum or difference with a multiple of the other.

The *elimination method* solves the system by attempting to change it into an equivalent system of the form $\begin{cases} 1x + 0y = p \\ 0x + 1y = q \end{cases}$. If this can be done, the unique solution is $x = p$, $y = q$. If an equation of the form $0x + 0y = 1$ is obtained, there is *no solution,* and the system is *inconsistent.* If an equation of the form $0x + 0y = 0$ is obtained, there are *infinitely many solutions* and the system is *dependent.*

▶ **Solutions to Practice Problems**

1. Let x be the number of pennies in the jar and y be the number of nickels. Then the first statement may be expressed as $x + y = 80$ and the second as $x + 5y = 160$ (in cents, since each penny is worth 1¢ and each nickel is worth 5¢). The system of equations is $\begin{cases} x + y = 80 \\ x + 5y = 160 \end{cases}$

2. $\begin{cases} x = 6 - \frac{3}{2}(-2) = 9 \\ y = -2 \end{cases}$ Substituting these numbers into Example 3c:

   $$\begin{cases} 2(9) + 3(-2) = 18 - 6 = 12 \ ✔ \\ 4(9) + 6(-2) = 36 - 12 = 24 \ ✔ \end{cases}$$

   It checks!

3. Since the first equation can be solved for y as $y = 10 - 2x$, we can substitute $10 - 2x$ for y in the second equation: $x + 2(10 - 2x) = 8$. Multiplying out and collecting like terms, $x + 20 - 4x = 8$ so $-3x = -12$ giving $x = 4$. Substituting $x = 4$ into $y = 10 - 2x$ gives $y = 10 - 2(4) = 10 - 8 = 2$. The solution is $x = 4$, $y = 2$. There are several other ways of solving this problem by the substitution method, and all reach the same conclusion.

4. $\begin{cases} x + y = 100 \\ x - y = -20 \end{cases}$ $\xrightarrow[\text{to first}]{\text{Add second}}$ $\begin{cases} 2x + 0y = 80 \\ x - y = -20 \end{cases}$ $\xrightarrow[\text{by 2}]{\text{Divide first}}$

   $\begin{cases} 1x + 0y = 40 \\ x - y = -20 \end{cases}$ $\xrightarrow[\text{from second}]{\text{Subtract first}}$ $\begin{cases} 1x + 0y = 40 \\ 0x - y = -60 \end{cases}$ $\xrightarrow[\text{by} -1]{\text{Multiply second}}$

   $\begin{cases} 1x + 0y = 40 \\ 0x + 1y = 60 \end{cases}$

   The solution is $x = 40$, $y = 60$. There are many other possible sequences of equivalent systems that solve this problem, and all reach the same conclusion.

3.1 Exercises

Represent each pair of statements as a system of two equations using the given definitions of x and y. Verify that the values given for x and y are a solution for the system of equations.

1. "The sum of two numbers is eighteen" and "the first number is two more than the second number." Let x be the first number and y be the second number. $x = 10$, $y = 8$

2. "The sum of two numbers is twenty-five" and "twice the first number added to the second number totals thirty-two." Let x be the first number and y be the second number. $x = 7$, $y = 18$

3. "Tom has $6 more than Alice" and "together, they have $40." Let x be the amount of money that Tom has and y be the amount Alice has. $x = 23$, $y = 17$

4. "Bill and Jessica together have $25" and "Jessica has $12." Let x be the amount of money that Bill has and y be the amount that Jessica has. $x = 13$, $y = 12$

5. "An envelope of $1 and $5 bills contains thirty bills" and "the money in the envelope is worth $70." Let x be the number of $1 bills and y be the number of $5 bills. $x = 20$, $y = 10$

6. "An envelope of $10 and $20 bills contains eight bills" and "the money in the envelope is worth $110." Let x be the number of $10 bills and y be the number of $20 bills. $x = 5$, $y = 3$

7. "A small theater sold tickets for all one hundred seats" and "the box office receipts of $650 came from adult tickets at $10 and child tickets at $5." Let x be the number of adult tickets sold and y be the number of child tickets sold. $x = 30$, $y = 70$

8. "A movie theater sold tickets for three hundred seats" and "the box office receipts of $2400 came from adult tickets at $9 and child tickets at $6." Let x be the number of adult tickets sold and y be the number of child tickets sold. $x = 200$, $y = 100$

9. "A corn and beet farmer planted 225 acres of crops" and "he planted twice as many acres of corn as acres of beets." Let x be the number of acres of corn and y be the number of acres of beets. $x = 150$, $y = 75$

10. "A stock and bond speculator invested $10,000 in the market" and "she invested three times as much in stocks as in bonds." Let x be the amount in stocks and y be the amount in bonds. $x = 7500$, $y = 2500$

Solve each system by graphing. If the solution is not unique, identify the system as "inconsistent" or "dependent."

You may use a graphing calculator if permitted by your instructor.

11. $\begin{cases} x + y = 6 \\ x - y = 2 \end{cases}$

12. $\begin{cases} -x + y = 2 \\ x + y = 4 \end{cases}$

13. $\begin{cases} 2x + y = 8 \\ x = 3 \end{cases}$

14. $\begin{cases} x + 2y = 10 \\ y = 4 \end{cases}$

15. $\begin{cases} x - y = 4 \\ -x + 2y = -6 \end{cases}$

16. $\begin{cases} 2x - y = 2 \\ x + 2y = 6 \end{cases}$

17. $\begin{cases} x + y = 10 \\ -x - y = 10 \end{cases}$

18. $\begin{cases} -2x + 4y = -16 \\ x - 2y = 4 \end{cases}$

19. $\begin{cases} x + y = 10 \\ -x - y = -10 \end{cases}$

20. $\begin{cases} 2x - 4y = 16 \\ -x + 2y = -8 \end{cases}$

Solve each system by the substitution method. If the solution is not unique, identify the system as "inconsistent" or "dependent."

21. $\begin{cases} x + 2y = 10 \\ y = 3 \end{cases}$

22. $\begin{cases} 2x + y = 8 \\ x = 2 \end{cases}$

23. $\begin{cases} 2x + y = 20 \\ x + y = 15 \end{cases}$

24. $\begin{cases} x + y = 12 \\ x + 2y = 14 \end{cases}$

25. $\begin{cases} 5x + 2y = 30 \\ 2x + y = 10 \end{cases}$

26. $\begin{cases} 2x + 3y = 25 \\ x - y = 5 \end{cases}$

27. $\begin{cases} 3x + 2y = 30 \\ x - y = -5 \end{cases}$

28. $\begin{cases} 2x + y = 20 \\ x + 3y = 15 \end{cases}$

29. $\begin{cases} -2x + 2y = -20 \\ x - y = 10 \end{cases}$

30. $\begin{cases} 3x - 2y = 30 \\ -6x + 4y = 30 \end{cases}$

Solve each system by the elimination method. If the solution is not unique, identify the system as "inconsistent" or "dependent."

31. $\begin{cases} x + y = 11 \\ 2x + 3y = 30 \end{cases}$

32. $\begin{cases} 3x + 2y = 30 \\ x + y = 13 \end{cases}$

33. $\begin{cases} 3x + y = 15 \\ x + 2y = 10 \end{cases}$

34. $\begin{cases} x + 3y = 30 \\ 2x + y = 10 \end{cases}$

35. $\begin{cases} x + 2y = 14 \\ 3x + 4y = 36 \end{cases}$

36. $\begin{cases} 2x + 3y = 30 \\ x - y = 10 \end{cases}$

37. $\begin{cases} 2x + 5y = 60 \\ 2x + 3y = 48 \end{cases}$ 38. $\begin{cases} 4x + 3y = 36 \\ x + 3y = 18 \end{cases}$ 39. $\begin{cases} 3x + 4y = -24 \\ 6x + 8y = 24 \end{cases}$ 40. $\begin{cases} x - 4y = 20 \\ -2x + 8y = -40 \end{cases}$

APPLIED EXERCISES

Formulate each situation as a system of two linear equations in two variables. Be sure to state clearly the meaning of your x- and y-variables. Solve the system by the elimination method. Be sure to state your final answer in terms of the original question.

41. **BUSINESS: Apartment Ownership** A lawyer has found 60 investors for a limited partnership to purchase an inner-city apartment building, with each contributing either $5000 or $10,000. If the partnership raised $430,000, how many investors contributed $5000 and how many contributed $10,000?

42. **BUSINESS: Violin Syndication** Ninety supporters of a musical prodigy raised $1.2 million to purchase a violin by Peter Guarneri for her use. If each supporter contributed either $5000 or $20,000, how many contributed each amount?

43. **GENERAL: Coins in a Jar** A jar contains 60 nickels and dimes worth $4.30. How many of each are in the jar?

44. **GENERAL: Bills in an Envelope** An envelope found in a safe deposit box contains a total of ninety $5 and $20 bills worth $1200. How many of each are in the envelope?

45. **PERSONAL FINANCE: Financial Planning** A retired couple wish to invest their nest egg of $10,000 in a money market account paying 6% and in a stock mutual fund returning 11%. If their income tax and Social Security situation requires that they earn $1000 from these investments, how much should they invest in each?

46. **PERSONAL FINANCE: Home Financing** A young couple needs to borrow $168,000 to finance their first house. They could borrow the whole amount at 12% from their bank, but her father offers to lend them enough of the money at 5% (with the same terms as the bank loan) to reduce the overall interest rate to just 8%. How much does he lend them and how much do they borrow from the bank?

47. **BUSINESS: Ice Hockey Concession Receipts** The concession stand at an ice hockey rink had receipts of $7200 from selling a total of 3000 sodas and hot dogs. If each soda sold for $2 and each hot dog sold for $3, how many of each were sold?

48. **BUSINESS: Baseball Tickets** A college baseball game generated box office receipts of $4800 from 600 ticket sales. If general admission tickets were $12 and student tickets were half price, how many of each were sold?

49. **PERSONAL FINANCE: Income Taxes** Find the federal and state taxes on a taxable income of $49,900 if the federal tax is 10% of the taxable income after first deducting the state tax and the state tax is 2% of the taxable income after first deducting the federal tax.

50. **GENERAL: Estate Division** A will specifies that each of two sons receive one half of the $3 million estate after first deducting the other's share, and that any remainder is then to be given to their sister. How much does each son receive? How much is left for the sister?

51. **ATHLETICS: Sports Nutrition** The dietician at a sports training facility has determined that one of her athletes needs an additional 600 mg each of calcium and phosphorus daily. Two supplements are available containing the milligrams of calcium and phosphorus per tablet as given by the table. How many tablets of each supplement will provide the required calcium and phosphorus?

	Calcium	Phosphorus
Supplement A	150	100
Supplement B	120	120

52. BUSINESS: Bicycle Shop Management
A bicycle shop has $10,500 to spend on new bikes and 390 hours of assembly time to put them together. Each mountain bike costs $50 wholesale and takes two hours to assemble. Each racing bike costs $70 wholesale and takes two-and-one-half hours to assemble. How many of each can the shop buy and assemble to use all the available money and time?

Explorations and Excursions

The following problems extend and augment the material presented in the text.

More About Parameterizations

53. In Example 2a we sketched the line
$2x + 3y = 12$, and after Example 3 we parameterized this line as $x = 6 - \frac{3}{2}t$, $y = t$. To graph these *parametric equations* on your calculator, change the MODE from FUNCtion to PARametric and enter the pair of equations $x_{1T} = 6 - (3/2)T$, $y_{1T} = T$. Set the WINDOW with $Tmin = -2$, $Tmax = 2$, $Tstep = 0.1$, $Xmin = -10$, $Xmax = 10$, $Ymin = -10$, and $Ymax = 10$, and watch as your calculator draws the graph. Change $Tmin$ to -3, $Tmax$ to 3, and GRAPH it again. Experiment with different values for $Tmin$ and $Tmax$ until you see how to get a "complete" picture of this line in this window.

54. Check that the line $2x - 5y = 8$ can be parameterized as $x = 4 + \frac{5}{2}t$, $y = t$. Set the WINDOW with $Tmin = -2$, $Tmax = 2$, $Tstep = 0.1$, $Xmin = -10$, $Xmax = 10$, $Ymin = -10$, and $Ymax = 10$. GRAPH these equations (be sure to first change the MODE from FUNCtion to PARametric) and use TRACE to explore this line segment. Experiment with different values for $Tmin$ and $Tmax$ until you see how to get a "complete" picture of this line in this window.

55. Find a parameterization for the line $x + 3y = 15$ and use your calculator to verify that your parametric equations determine a line with x-intercept $(15, 0)$ and y-intercept $(0, 5)$.

56. Use the parametric equations from Exercises 54 and 55 to GRAPH Example 4 on the WINDOW with $Tmin = -5$, $Tmax = 10$, $Tstep = 0.1$, $Xmin = -5$, $Xmax = 20$, $Ymin = -10$, and $Ymax = 10$. Use TRACE to explore both lines and verify that the intersection point is $(9, 2)$.

Round-off Errors

Round-off errors can completely misrepresent the true nature of a system of equations. (If you wish to remove fractions from a problem, multiply each equation through by the least common denominator instead of rounding off.)

57. Use the elimination method to show that the
system $\begin{cases} x + \frac{1}{3}y = 39 \\ 2x + \frac{2}{3}y = 84 \end{cases}$ is inconsistent.

58. Rounding to one decimal place, the system in
Exercise 57 becomes $\begin{cases} x + 0.3y = 39 \\ 2x + 0.7y = 84 \end{cases}$. Use the elimination method to show that this new system has the unique solution $x = 21$, $y = 60$.

59. Use the elimination method to show that the
system $\begin{cases} x + \frac{2}{9}y = 9.79 \\ 4x + \frac{8}{9}y = 39.16 \end{cases}$ is dependent.

60. Rounding to two decimal places, the system in
Exercise 59 becomes $\begin{cases} x + 0.22y = 9.79 \\ 4x + 0.89y = 39.16 \end{cases}$. Use the elimination method to show that this new system has the unique solution $x = 9.79$, $y = 0$.

61. Use the elimination method to show that the
system $\begin{cases} 2.1x + \frac{1}{7}y = 157 \\ 3x + \frac{1}{5}y = 224 \end{cases}$ has the unique solution $x = 70$, $y = 70$. Rounding to two decimal places, this system becomes
$\begin{cases} 2.10x + 0.14y = 157 \\ 3.00x + 0.20y = 224 \end{cases}$. Use the elimination method to show that this new system is inconsistent.

3.2 MATRICES AND LINEAR EQUATIONS IN TWO VARIABLES

Introduction

In this section we will use matrix notation to streamline the elimination method used in the previous section to solve systems of two linear equations in two variables. In the next section we shall extend this method to the solutions of systems of many linear equations in many variables.

Matrices

A *matrix* is a rectangular array of numbers called *elements*. This rectangular array has *rows* (with the first at the top, the second below the first, and so on) and *columns* (with the first on the left, the second to the right of the first, and so on). The *dimension* of a matrix with m rows and n columns is written $m \times n$ (rows $\times$ columns). Thus a 5×2 matrix is "tall and thin," while a 2×5 matrix is "short and wide." A *square matrix* has the same number of rows as columns. A *row matrix* has just one row, and a *column matrix* has just one column.

$$(1 \quad 2 \quad 3) \qquad \begin{pmatrix} 1 \\ 2 \\ 3 \end{pmatrix} \qquad \begin{pmatrix} 1 & 2 & 3 \\ 4 & 5 & 6 \\ 7 & 8 & 9 \end{pmatrix}$$

Row matrix   Column matrix   Square matrix
(dimension 1×3)   (dimension 3×1)   (dimension 3×3)

We name matrices with capital letters (A, B, C, . . .) and then the elements are named by the corresponding lowercase letter together with subscripts indicating the row and then the column. For instance, if

$$A = \begin{pmatrix} 1 & 2 & 3 & 4 \\ 5 & 6 & 7 & 8 \\ 9 & 10 & 11 & 12 \end{pmatrix}$$

$a_{2,3}$ is $\begin{pmatrix} \downarrow 2 \\ \\ \xrightarrow{3} a_{2,3} \end{pmatrix}$

then the dimension of A is 3×4 and $a_{2,3} = 7$ because 7 is the element in the second row and third column. This *double subscript* notation is sometimes used without the comma so that a_{23} means $a_{2,3}$. The elements on the *main diagonal* are those with the same row number as column number. For the matrix A above, the main diagonal consists of the elements $a_{1,1} = 1$, $a_{2,2} = 6$, and $a_{3,3} = 11$.

$\begin{pmatrix} \diagdown \\ \diagdown \end{pmatrix}$
└ main diagonal

Augmented Matrices from Systems of Equations

An *augmented matrix* is a matrix created from two "smaller" matrices having the same number of rows by placing them beside each other and joining them into one "larger" matrix. The system of equations

$$\begin{cases} ax + by = h \\ cx + dy = k \end{cases}$$ has a *coefficient matrix* $\begin{pmatrix} a & b \\ c & d \end{pmatrix}$ and a *constant term matrix* $\begin{pmatrix} h \\ k \end{pmatrix}$. Taken together, these form the augmented matrix $\begin{pmatrix} a & b & h \\ c & d & k \end{pmatrix}$, separating the original matrices by a vertical line. The augmented matrix represents the system of equations with the vertical bar representing the equals signs. The first column gives the coefficients of the *x*-variable, the second gives those of the *y*-variable, and the last column (after the bar) contains the constant terms after the equals signs.

Augmented Matrix of a System of Equations

The augmented matrix $\begin{pmatrix} a & b & h \\ c & d & k \end{pmatrix}$ represents the system of equations $\begin{cases} ax + by = h \\ cx + dy = k \end{cases}$.

EXAMPLE 1

AUGMENTED MATRICES AND SYSTEMS OF EQUATIONS

a. Find the augmented matrix representing the system $\begin{cases} 2x + 3y = 24 \\ 4x + 5y = 60 \end{cases}$.

b. Find the system represented by the augmented matrix $\begin{pmatrix} 6 & 8 & 84 \\ 4 & 5 & 60 \end{pmatrix}$.

Solution

a. The system $\begin{cases} 2x + 3y = 24 \\ 4x + 5y = 60 \end{cases}$ is represented by $\begin{pmatrix} 2 & 3 & 24 \\ 4 & 5 & 60 \end{pmatrix}$.

b. The augmented matrix $\begin{pmatrix} 6 & 8 & 84 \\ 4 & 5 & 60 \end{pmatrix}$ represents $\begin{cases} 6x + 8y = 84 \\ 4x + 5y = 60 \end{cases}$.

Practice Problem 1

a. Find the augmented matrix representing the system $\begin{cases} 2x - y = 14 \\ x + 3y = 21 \end{cases}$.

b. Find the system represented by the augmented matrix $\begin{pmatrix} 3 & 2 & 35 \\ 1 & 3 & 21 \end{pmatrix}$.

➤ Solutions on page 175

Row Operations

We will solve a system of equations by operating on its augmented matrix, using the following *matrix row operations* to make the solution obvious. These row operations correspond to the steps that we used on page 157 to solve systems of equations, and result in a succession of *equivalent* matrices, each representing a system that has exactly the same solution as the original system.

Matrix Row Operations

1. Switch any two rows.
2. Multiply or divide one of the rows by a nonzero number.
3. Replace a row by its sum or difference with a multiple of another row.

We may apply row operations to matrices of any size. We write short formulas for row operations using R for "row" and either a double arrow $\leftrightarrow$ for "is switched with" or an arrow $\rightarrow$ for "becomes." When we write the new matrix, the formula for each new row ends with an arrow $\rightarrow$ pointing to the row that was changed.

EXAMPLE 2

PERFORMING MATRIX ROW OPERATIONS

Carry out the indicated row operation on the given matrix.

a. $R_1 \leftrightarrow R_2$ on $\begin{pmatrix} 2 & 3 & | & 25 \\ 3 & 4 & | & 36 \end{pmatrix}$

b. $3R_2 \rightarrow R_2$ on $\begin{pmatrix} 2 & 3 & | & 30 \\ 1 & 1 & | & 13 \end{pmatrix}$

c. $R_1 - R_2 \rightarrow R_1$ on $\begin{pmatrix} 2 & 3 & | & 30 \\ 2 & -2 & | & 20 \end{pmatrix}$

Solution

a. $R_1 \leftrightarrow R_2$ says that row 1 and row 2 are switched:

$$\begin{matrix} R_2 \rightarrow \\ R_1 \rightarrow \end{matrix} \begin{pmatrix} 3 & 4 & | & 36 \\ 2 & 3 & | & 25 \end{pmatrix} \qquad \text{From } \begin{pmatrix} 2 & 3 & | & 25 \\ 3 & 4 & | & 36 \end{pmatrix}$$

b. $3R_2 \rightarrow R_2$ says that 3 times row 2 becomes R_2:

$$\begin{matrix} \\ 3R_2 \rightarrow \end{matrix} \begin{pmatrix} 2 & 3 & | & 30 \\ 3 & 3 & | & 39 \end{pmatrix} \qquad \text{From } \begin{pmatrix} 2 & 3 & | & 30 \\ 1 & 1 & | & 13 \end{pmatrix}$$

c. $R_1 - R_2 \rightarrow R_1$ says that row 1 minus row 2 becomes row 1:

$$R_1 - R_2 \rightarrow \begin{pmatrix} 0 & 5 & | & 10 \\ 2 & -2 & | & 20 \end{pmatrix} \qquad \text{From} \quad \begin{pmatrix} 2 & 3 & | & 30 \\ 2 & -2 & | & 20 \end{pmatrix}$$

Practice Problem 2

a. Carry out $R_1 \leftrightarrow R_2$ on $\begin{pmatrix} 2 & 1 & | & 14 \\ 1 & -3 & | & 21 \end{pmatrix}$.

b. Carry out $R_1 + R_2 \rightarrow R_1$ on $\begin{pmatrix} 6 & 3 & | & 42 \\ 1 & -3 & | & 21 \end{pmatrix}$.

c. Carry out $\frac{1}{7}R_1 \rightarrow R_1$ on $\begin{pmatrix} 7 & 0 & | & 63 \\ 1 & -3 & | & 21 \end{pmatrix}$.

d. Is $0R_1 \rightarrow R_1$ a valid row operation?

e. Is $5R_1 \rightarrow R_2$ a valid row operation?

➤ **Solutions on page 176**

Solving Equations by Row Reduction

Two matrices are *equivalent* if one can be transformed into the other by a sequence of row operations. Since an augmented matrix represents a system of equations and the row operations correspond to the steps used to solve the system, we can solve the system by *row-reducing* the augmented matrix to an equivalent matrix that displays the solution. There are three possibilities. If we can obtain the form

$$\begin{pmatrix} 1 & 0 & | & p \\ 0 & 1 & | & q \end{pmatrix} \qquad \begin{cases} 1x + 0y = p \\ 0x + 1y = q \end{cases} \quad \text{or} \quad \begin{cases} x = p \\ y = q \end{cases}$$

then the unique solution is $x = p$, $y = q$. However, if we obtain a row of zeros ending with a nonzero number:

$$0 \quad 0 \quad | \quad m \qquad\qquad 0x + 0y = m, \quad m \neq 0$$

then the system is *inconsistent* and has *no solution* because the equation $0x + 0y = m$ makes the impossible claim that zero equals a nonzero number. On the other hand, if there is no such "inconsistent" row but there is a row consisting entirely of zeros:

$$0 \quad 0 \quad | \quad 0 \qquad\qquad 0x + 0y = 0$$

then the system is *dependent*, and there are *infinitely many solutions* because the equation $0x + 0y = 0$ is *always* true and represents no restriction at all. These important observations will be extended in the next section.

EXAMPLE 3

SOLVING EQUATIONS BY ROW REDUCTION

Solve $\begin{cases} x + 3y = 15 \\ 2x - 5y = 8 \end{cases}$ by row-reducing an augmented matrix.

Solution

The augmented matrix for this system is

$$\begin{pmatrix} 1 & 3 & | & 15 \\ 2 & -5 & | & 8 \end{pmatrix}$$

We hope to change the first column from $\begin{smallmatrix}1\\2\end{smallmatrix}$ to $\begin{smallmatrix}1\\0\end{smallmatrix}$ (which is the same as removing the x-variable from the second equation) and then the second column to $\begin{smallmatrix}0\\1\end{smallmatrix}$ (removing the y-variable from the first equation). Therefore we first want a 0 in the bottom-left corner. Subtracting twice the 1 in the top row from the 2 in the second row will achieve this 0, suggesting the row operation $R_2 - 2R_1 \rightarrow R_2$. We write "twice row 1" in small numbers above row 1 to make the subtraction of "row 2 minus twice row 1" easier:

$$\begin{array}{ccc} 2 & 6 & 30 \end{array}$$

$$\boxed{\text{Want a 0 here}} \qquad \begin{pmatrix} 1 & 3 & | & 15 \\ 2 & -5 & | & 8 \end{pmatrix} \qquad R_2 - 2R_1 \rightarrow R_2 \text{ gives}$$

$$R_2 - 2R_1 \rightarrow \begin{pmatrix} 1 & 3 & | & 15 \\ 0 & -11 & | & -22 \end{pmatrix}$$

The bottom row has a common factor of -11, which we divide out, obtaining the simpler matrix below. We then want a 0 where the 3 is. Subtracting three times the 1 in the second row from the 3 will achieve this, suggesting the row operation $R_1 - 3R_2 \rightarrow R_1$. To help with the subtraction we write "three times row 2" in small numbers under row 2:

$$\boxed{\text{Want a 0 here}} \qquad \tfrac{1}{-11}R_2 \rightarrow \begin{pmatrix} 1 & 3 & | & 15 \\ 0 & 1 & | & 2 \end{pmatrix} \qquad R_1 - 3R_2 \rightarrow R_1 \text{ gives}$$

$$\begin{array}{ccc} 0 & 3 & 6 \end{array}$$

$$R_1 - 3R_2 \rightarrow \begin{pmatrix} 1 & 0 & | & 9 \\ 0 & 1 & | & 2 \end{pmatrix} \qquad \text{Equivalent to} \quad \begin{cases} 1x + 0y = 9 \\ 0x + 1y = 2 \end{cases}$$

The unique solution is $x = 9$, $y = 2$. (We solved this same system by the substitution method on pages 155–156 and found the same answer.)

There are many other sequences of row operations to reduce this augmented matrix, and all lead to the same solution.

Practice Problem 3 Solve $\begin{cases} 2x + y = 14 \\ x - 3y = 21 \end{cases}$ by row-reducing an augmented matrix.

➤ Solution on page 176

Graphing Calculator Exploration

Some graphing calculators can row-reduce matrices. If your calculator has a RREF command (for "reduced row-echelon form"), you can easily check your row reduction. For the system of equations in Example 3:

a. Enter $\begin{pmatrix} 1 & 3 & 15 \\ 2 & -5 & 8 \end{pmatrix}$ as matrix [A] using MATRX EDIT.

b. QUIT and select the RREF command from the MATRX MATH menu.

c. Apply RREF to the matrix [A].

Although this serves as a useful check of your answer, do not rely on it completely because the calculator sometimes returns an answer with round-off errors. Furthermore, this calculator command may not work for matrices with more rows than columns. We will be interested in such matrices in the next section.

EXAMPLE **4**

SOLVING EQUATIONS BY ROW REDUCTION

Solve $\begin{cases} 6x - 3y = 30 \\ -8x + 4y = -40 \end{cases}$ by row-reducing an augmented matrix.

Solution

The augmented matrix is

$$\begin{pmatrix} 6 & -3 & | & 30 \\ -8 & 4 & | & -40 \end{pmatrix}$$

When the numbers in a row have an obvious common factor, removing that factor can sometimes simplify the reduction.

$$\begin{matrix} \frac{1}{3}R_1 \to \\ \frac{1}{4}R_2 \to \end{matrix} \begin{pmatrix} 2 & -1 & | & 10 \\ -2 & 1 & | & -10 \end{pmatrix} \quad \text{(Want a 0 here)}$$

Adding the first row to the second, we get a zero at the bottom of the first column.

$$\begin{matrix} \\ R_2 + R_1 \to \end{matrix} \begin{pmatrix} 2 & -1 & 10 \\ 0 & 0 & 0 \end{pmatrix} \quad \text{(Want a 1 here)}$$

To finish, we divide the first row by 2 to make the row begin with a 1 on the left.

$$\frac{1}{2}R_1 \to \begin{pmatrix} 1 & -\frac{1}{2} & 5 \\ 0 & 0 & 0 \end{pmatrix} \quad \leftarrow \text{a zero row}$$

The zero row means that the system is *dependent*, so there are *infinitely many solutions*. The first row of this final matrix says that

$$x - \tfrac{1}{2}y = 5$$

or, solving for x,

$$x = 5 + \tfrac{1}{2}y$$

We may let y be *any* number t and then determine x from this equation. That is, the solutions may be parameterized as $x = 5 + \tfrac{1}{2}t$, $y = t$, where t is *any* number. The following table lists some of these solutions for various values of the parameter t.

t	$x = 5 + \frac{1}{2}t$	$y = t$
-20	-5	-20
-10	0	-10
0	5	0
10	10	10
20	15	20

Parameterized solution

$\left.\vphantom{\begin{array}{c}a\\a\\a\\a\end{array}}\right\}$ Evaluating x and y at t

There are many other sequences of row operations to reduce this augmented matrix and all reach the same conclusion.

Practice Problem 4

Verify that $x = -5$, $y = -20$, and $x = 15$, $y = 20$ from the preceding table solve $\begin{cases} 6x - 3y = 30 \\ -8x + 4y = -40 \end{cases}$.

➤ **Solution on page 176**

EXAMPLE 5

PRODUCTION MANAGEMENT

A worker in a plastics factory breaks apart sheets of component A and strips of component B and then snaps one of each together to make a finished item. If the worker can break off 20 A's or 30 B's per minute and snap together 10 pairs of A's and B's per minute, how should the worker's 440-minute workday be divided so as to complete as many items as possible with no unused pieces left over?

Solution

Let x be the number of minutes spent breaking apart sheets of component A and y be the number of minutes breaking apart strips of component B. The remainder of the worker's time, $440 - (x + y)$, will then be spent snapping A's and B's together. Because the worker needs as many A's as B's,

$$20x = 30y$$

Rate × time gives number finished

As many finished items as A's will be completed:

$$20x = 10(440 - (x + y))$$

Simplifies to $30x + 10y = 4400$

Rewriting these as a system of equations, the problem becomes

$$\begin{cases} 20x - 30y = 0 \\ 30x + 10y = 4400 \end{cases}$$

The augmented matrix is

$$\begin{pmatrix} 20 & -30 & | & 0 \\ 30 & 10 & | & 4400 \end{pmatrix}$$

We begin the row reduction by removing from both rows the common factor of 10.

$$\begin{matrix} \frac{1}{10}R_1 \to \\ \frac{1}{10}R_2 \to \end{matrix} \begin{pmatrix} 2 & -3 & | & 0 \\ 3 & 1 & | & 440 \end{pmatrix} \quad \text{(Want a 1 here)}$$

To get a one in the upper left-hand corner:

$$R_2 - R_1 \to \begin{pmatrix} 1 & 4 & | & 440 \\ 3 & 1 & | & 440 \end{pmatrix} \quad \text{(Want a 0 here)}$$

To get a zero in the lower left-hand corner:

$$3R_1 - R_2 \to \begin{pmatrix} 1 & 4 & 440 \\ 0 & 11 & 880 \end{pmatrix}$$

Removing the common factor in row 2:

$$\frac{1}{11}R_2 \to \begin{pmatrix} 1 & 4 & 440 \\ 0 & 1 & 80 \end{pmatrix} \quad \text{(Want a 0 here)}$$

To get a zero above the 1 in the second row:

$$R_1 - 4R_2 \to \begin{pmatrix} 1 & 0 & 120 \\ 0 & 1 & 80 \end{pmatrix} \qquad\qquad x = 120$$

The solution is $x = 120$, $y = 80$. In terms of the original question, the worker should break apart sheets of component A for 120 minutes, break apart strips of component B for 80 minutes, and snap A's and B's together for the remaining 240 minutes.

 If you have a graphing calculator with a RREF command, you may use it to check the result of the row reduction in Example 5.

3.2 Section Summary

The *augmented matrix* $\begin{pmatrix} a & b & | & h \\ c & d & | & k \end{pmatrix}$ represents the system of equations $\begin{cases} ax + by = h \\ cx + dy = k \end{cases}$. The rows correspond to the equations, and the first column corresponds to the x-variable, the second column to the y-variable, and the column after the bar to the constant terms of the equations.

The three basic row operations are

1. Switch any two rows.

2. Multiply or divide one of the rows by a nonzero number.

3. Replace a row by its sum or difference with a multiple of another row.

If these row operations lead to a row of zeros ending with a one, $0 \quad 0 \mid 1$, then the system is *inconsistent* and has *no* solutions. If there is no such "inconsistent" row but there is a row of *all* zeros, $0 \quad 0 \mid 0$, then the system is *dependent* and has *infinitely many* solutions. Otherwise, the system has a unique solution. In fact, any 2×3 augmented matrix is equivalent to exactly one of the five types below.

The augmented matrix $\begin{pmatrix} a & b & \mid h \\ c & d & \mid k \end{pmatrix}$ of $\begin{cases} ax + by = h \\ cx + dy = k \end{cases}$ is equivalent to exactly one of the following (where p and q stand for any numbers).

Matrix	Equations	Solution
$\begin{pmatrix} 1 & 0 & \mid p \\ 0 & 1 & \mid q \end{pmatrix}$	Unique solution	$x = p, \quad y = q$
$\begin{pmatrix} 1 & p & \mid 0 \\ 0 & 0 & \mid 1 \end{pmatrix}$	Inconsistent	No solution
$\begin{pmatrix} 0 & 1 & \mid 0 \\ 0 & 0 & \mid 1 \end{pmatrix}$	Inconsistent	No solution
$\begin{pmatrix} 1 & p & \mid q \\ 0 & 0 & \mid 0 \end{pmatrix}$	Dependent	Infinitely many solutions: $x = q - pt, \quad y = t$
$\begin{pmatrix} 0 & 1 & \mid p \\ 0 & 0 & \mid 0 \end{pmatrix}$	Dependent	Infinitely many solutions: $x = t, \quad y = p$

➤ **Solutions to Practice Problems**

1. a. $\begin{pmatrix} 2 & -1 & \mid 14 \\ 1 & 3 & \mid 21 \end{pmatrix}$ 　　　**b.** $\begin{cases} 3x + 2y = 35 \\ x + 3y = 21 \end{cases}$

2. a. $\begin{matrix} R_2 \to \\ R_1 \to \end{matrix} \begin{pmatrix} 1 & -3 & | & 21 \\ 2 & 1 & | & 14 \end{pmatrix}$ **b.** $R_1 + R_2 \to \begin{pmatrix} 7 & 0 & | & 63 \\ 1 & -3 & | & 21 \end{pmatrix}$

c. $\frac{1}{7}R_1 \to \begin{pmatrix} 1 & 0 & | & 9 \\ 1 & -3 & | & 21 \end{pmatrix}$

d. No; you can multiply only by a *nonzero* number.

e. No; multiplying row 1 by 5 must still give row 1 (*not* row 2).

3. Starting with the augmented matrix $\begin{pmatrix} 2 & 1 & | & 14 \\ 1 & -3 & | & 21 \end{pmatrix}$, one possible sequence of row operations is

$$R_1 - R_2 \to \begin{pmatrix} 1 & 4 & | & -7 \\ 1 & -3 & | & 21 \end{pmatrix}$$

$$R_1 - R_2 \to \begin{pmatrix} 1 & 4 & | & -7 \\ 0 & 7 & | & -28 \end{pmatrix}$$

$$\tfrac{1}{7}R_2 \to \begin{pmatrix} 1 & 4 & | & -7 \\ 0 & 1 & | & -4 \end{pmatrix}$$

$$R_1 - 4R_2 \to \begin{pmatrix} 1 & 0 & | & 9 \\ 0 & 1 & | & -4 \end{pmatrix}$$

The unique solution is $x = 9, \ y = -4$. There are many other sequences of row operations to reduce this augmented matrix and all reach the same conclusion.

4. For $x = -5, \ y = -20,$ the equations become

$$\begin{cases} 6(-5) - 3(-20) = -30 + 60 = 30 \ ✔ \\ -8(-5) + 4(-20) = 40 - 80 = -40 \ ✔ \end{cases}$$

For $x = 15, \ y = 20,$ the equations become

$$\begin{cases} 6(15) - 3(20) = 90 - 60 = 30 \ ✔ \\ -8(15) + 4(20) = -120 + 80 = -40 \ ✔ \end{cases}$$

Both check!

3.2 Exercises

Find the dimension of each matrix and the value(s) of the specified element(s).

1. $\begin{pmatrix} 1 & 2 \\ 2 & 5 \\ 3 & 6 \end{pmatrix}; \ a_{1,1}, a_{3,2}$ **2.** $\begin{pmatrix} 6 & 3 \\ 5 & 2 \\ 4 & 1 \end{pmatrix}; \ a_{1,1}, a_{3,2}$

3. $\begin{pmatrix} 1 & -1 & 2 & -2 \\ -3 & 3 & -4 & 4 \\ 5 & -5 & 6 & -6 \end{pmatrix}; \ a_{2,2}, a_{3,4}$

4. $\begin{pmatrix} 1 & -2 & -1 \\ -3 & 4 & 3 \\ 5 & -6 & -5 \\ -7 & 8 & 7 \end{pmatrix}; \ a_{2,2}, a_{4,3}$

5. $\begin{pmatrix} 1 & 0 & 0 & 0 \\ 0 & 1 & 0 & 0 \\ 0 & 0 & 1 & 0 \\ 0 & 0 & 0 & 1 \end{pmatrix}; \ a_{1,1}, a_{2,2}, a_{3,3}, a_{4,4}, a_{1,4}$

6. $\begin{pmatrix} 1 & 0 & 0 \\ 0 & 1 & 0 \\ 0 & 0 & 1 \end{pmatrix}$; $a_{1,1}, a_{2,2}, a_{3,3}, a_{2,1}$

7. $(4 \quad 5 \quad 6 \quad 7)$; $a_{1,3}$ **8.** $(2 \quad 3 \quad 4 \quad 5 \quad 6)$; $a_{1,4}$

9. $\begin{pmatrix} 9 \\ 8 \\ 7 \\ 6 \\ 5 \end{pmatrix}$; $a_{2,1}, a_{4,1}$ **10.** $\begin{pmatrix} 2 \\ 8 \\ 3 \\ 7 \end{pmatrix}$; $a_{2,1}, a_{3,1}$

Find the augmented matrix representing the system of equations.

11. $\begin{cases} x + 2y = 2 \\ 3x + 4y = 12 \end{cases}$ **12.** $\begin{cases} 2x - y = 10 \\ -3x + 4y = 60 \end{cases}$

13. $\begin{cases} -4x + 3y = 84 \\ 5x - 2y = 70 \end{cases}$ **14.** $\begin{cases} -x + 2y = 2 \\ 2x - 3y = 6 \end{cases}$

15. $\begin{cases} 3x - 2y = 24 \\ x = 6 \end{cases}$ **16.** $\begin{cases} 4x - 3y = 24 \\ y = 8 \end{cases}$

17. $\begin{cases} 5x - 15y = 30 \\ -4x + 12y = 24 \end{cases}$ **18.** $\begin{cases} 12x - 4y = 36 \\ -15x + 5y = 45 \end{cases}$

19. $\begin{cases} x = 20 \\ y = 30 \end{cases}$ **20.** $\begin{cases} y = 18 \\ x = 12 \end{cases}$

Find the system of equations represented by the augmented matrix.

21. $\begin{pmatrix} 1 & 1 & | & 9 \\ 0 & 1 & | & 4 \end{pmatrix}$ **22.** $\begin{pmatrix} 1 & 0 & | & -3 \\ 1 & 1 & | & 5 \end{pmatrix}$

23. $\begin{pmatrix} -4 & 3 & | & -60 \\ 1 & -2 & | & 20 \end{pmatrix}$ **24.** $\begin{pmatrix} 3 & 4 & | & 24 \\ 1 & 2 & | & 6 \end{pmatrix}$

25. $\begin{pmatrix} 1 & -3 & | & -70 \\ 1 & 1 & | & 10 \end{pmatrix}$ **26.** $\begin{pmatrix} 1 & 1 & | & 5 \\ 2 & 3 & | & 7 \end{pmatrix}$

27. $\begin{pmatrix} 2 & 1 & | & 6 \\ 1 & 2 & | & -6 \end{pmatrix}$ **28.** $\begin{pmatrix} 3 & 1 & | & -24 \\ 1 & 3 & | & 24 \end{pmatrix}$

29. $\begin{pmatrix} 20 & -15 & | & 60 \\ -16 & 12 & | & -48 \end{pmatrix}$ **30.** $\begin{pmatrix} 20 & -15 & | & 60 \\ -16 & 12 & | & 48 \end{pmatrix}$

Carry out the row operation on the matrix.

31. $R_1 \leftrightarrow R_2$ on $\begin{pmatrix} 3 & 4 & | & 24 \\ 5 & 6 & | & 30 \end{pmatrix}$

32. $R_1 \leftrightarrow R_2$ on $\begin{pmatrix} 8 & 7 & | & 56 \\ 6 & 5 & | & 60 \end{pmatrix}$

33. $R_1 - R_2 \rightarrow R_1$ on $\begin{pmatrix} 8 & 7 & | & 56 \\ 6 & 5 & | & 60 \end{pmatrix}$

34. $2R_1 \rightarrow R_1$ on $\begin{pmatrix} 3 & 4 & | & 24 \\ 5 & 6 & | & 30 \end{pmatrix}$

35. $R_1 - 3R_2 \rightarrow R_1$ on $\begin{pmatrix} 5 & 6 & | & 30 \\ 1 & 2 & | & 18 \end{pmatrix}$

36. $R_1 - 3R_2 \rightarrow R_1$ on $\begin{pmatrix} 6 & 5 & | & 60 \\ 2 & 2 & | & -4 \end{pmatrix}$

37. $R_1 - R_2 \rightarrow R_2$ on $\begin{pmatrix} 6 & 6 & | & -12 \\ 6 & 5 & | & 60 \end{pmatrix}$

38. $R_1 - R_2 \rightarrow R_1$ on $\begin{pmatrix} 5 & 6 & | & 30 \\ 4 & 8 & | & 72 \end{pmatrix}$

39. $\frac{1}{8}R_2 \rightarrow R_2$ on $\begin{pmatrix} 1 & -2 & | & -42 \\ 0 & 8 & | & 120 \end{pmatrix}$

40. $\frac{1}{6}R_1 \rightarrow R_1$ on $\begin{pmatrix} 6 & 0 & | & 420 \\ 0 & 1 & | & -72 \end{pmatrix}$

Interpret each augmented matrix as the solution of a system of equations. State the solution or identify the system as "inconsistent" or "dependent."

41. $\begin{pmatrix} 1 & 0 & | & 7 \\ 0 & 1 & | & -3 \end{pmatrix}$ **42.** $\begin{pmatrix} 1 & 0 & | & -5 \\ 0 & 1 & | & 8 \end{pmatrix}$

43. $\begin{pmatrix} 1 & 1 & | & 0 \\ 0 & 0 & | & 1 \end{pmatrix}$ **44.** $\begin{pmatrix} 1 & -1 & | & 0 \\ 0 & 0 & | & 1 \end{pmatrix}$

45. $\begin{pmatrix} 0 & 1 & | & 0 \\ 0 & 0 & | & 1 \end{pmatrix}$ **46.** $\begin{pmatrix} 1 & 0 & | & 0 \\ 0 & 1 & | & 0 \end{pmatrix}$

47. $\begin{pmatrix} 1 & 2 & | & 3 \\ 0 & 0 & | & 0 \end{pmatrix}$ **48.** $\begin{pmatrix} 1 & -4 & | & 6 \\ 0 & 0 & | & 0 \end{pmatrix}$

49. $\begin{pmatrix} 0 & 1 & | & -3 \\ 0 & 0 & | & 0 \end{pmatrix}$ **50.** $\begin{pmatrix} 0 & 1 & | & 9 \\ 0 & 0 & | & 0 \end{pmatrix}$

Solve each system by row-reducing the corresponding augmented matrix. State the solution or identify the system as "inconsistent" or "dependent."

If you have a graphing calculator with a RREF command, use it to check your row reduction.

51. $\begin{cases} x + y = 5 \\ x = 3 \end{cases}$ **52.** $\begin{cases} x + y = 8 \\ y = 6 \end{cases}$

53. $\begin{cases} x + y = 4 \\ x - y = 2 \end{cases}$ **54.** $\begin{cases} x - 2y = 6 \\ x + y = 3 \end{cases}$

55. $\begin{cases} 2x + y = 4 \\ x + y = 3 \end{cases}$

56. $\begin{cases} x + y = 4 \\ 3x + y = 6 \end{cases}$

57. $\begin{cases} x + y = 5 \\ 2x + 3y = 12 \end{cases}$

58. $\begin{cases} 3x + y = 9 \\ 2x + y = 4 \end{cases}$

59. $\begin{cases} 2x + y = 20 \\ x + 3y = 15 \end{cases}$

60. $\begin{cases} x + y = 8 \\ 3x + 5y = 30 \end{cases}$

61. $\begin{cases} -x + 2y = 4 \\ x - 2y = 6 \end{cases}$

62. $\begin{cases} x - 3y = 12 \\ -x + 3y = 9 \end{cases}$

63. $\begin{cases} 6x + 2y = 18 \\ 5x + 2y = 10 \end{cases}$

64. $\begin{cases} 5x + 3y = 15 \\ 4x + 2y = 8 \end{cases}$

65. $\begin{cases} -3x + y = 3 \\ 5x - 2y = -10 \end{cases}$

66. $\begin{cases} 3x - y = 3 \\ -2x + y = 2 \end{cases}$

67. $\begin{cases} 2x - 6y = 18 \\ -3x + 9y = -27 \end{cases}$

68. $\begin{cases} 3x - 6y = 24 \\ -5x + 10y = -40 \end{cases}$

69. $\begin{cases} 4x + 7y = 56 \\ 2x + 3y = 30 \end{cases}$

70. $\begin{cases} 4x + 3y = 24 \\ 6x + 5y = 30 \end{cases}$

APPLIED EXERCISES

Express each situation as a system of two equations in two variables. Be sure to state clearly the meaning of your x- and y-variables. Solve the system by row-reducing the corresponding augmented matrix. State your final answer in terms of the original question.

 If you have a graphing calculator with a RREF command, use it to check your row reduction.

71. BUSINESS: Commodity Futures A corn and soybean commodities speculator invested $15,000 yesterday with twice as much in soybean futures as in corn futures. How much did she invest in each?

72. BUSINESS: Stamps and Coins A stamp and coin dealer spent $8000 at a numismatics auction last weekend. If he spent three times as much on coins as on stamps, how much did he spend on each?

73. GENERAL: Estate Division A will specifies that the older brother is to receive one half of the $12 million estate after first deducting the younger brother's share, the younger brother is to receive one third of the estate after first deducting the older brother's share, and the remainder is to be given to their sister. How much does each brother receive? How much is left over for their sister?

74. BUSINESS: Real Estate Taxes Find the state and city property taxes on an apartment building assessed at $833,000 if the state tax is 4% of the assessed value after first deducting the city tax and the city tax is 1% of the assessed value after first deducting the state tax.

75. BUSINESS: Racehorse Syndication The stud rights for a recent Triple-Crown winner were purchased by a syndicate for $3,675,000 and each member of the syndicate invested either $5000 or $25,000. If twice as many of the investors had invested $5000 but only half as many had invested $25,000, the syndicate would have raised only $3,150,000. How many investors contributed $5000 and how many contributed $25,000?

76. PERSONAL FINANCE: Financial Planning Last year, a retired couple's investments in a money market account yielding 5% and in a stock mutual fund yielding 14% paid a total of $16,800. If the stock mutual fund can return 17% this year and the money market account can continue to yield 5%, the couple will receive $17,850. How much is invested in each?

77. BIOMEDICAL: Geriatrics Nutrition The dietician at a senior care facility has decided to supplement the weekly menu with "NutraDrink" and "VitaPills," which contain calcium and vitamin C as listed in the table. If each resident needs an additional 300 mg of calcium and 240 mg of vitamin C per week, how many cans of NutraDrink and tablets of VitaPills should each resident be given per week?

	Calcium (mg)	Vitamin C (mg)
NutraDrink	25	20
VitaPills	20	16

78. GENERAL: Plant Fertilizer A garden field needs 308 pounds of potash and 330 pounds of nitrogen. Two brands of fertilizer, GrowRite and GreatGreen, are available and contain the amounts of potash and nitrogen per bag listed in the table. How many bags of each brand should be used to provide the required potash and nitrogen?

(Pounds per Bag)	GrowRite	GreatGreen
Potash	4	7
Nitrogen	5	6

79. SOCIAL SCIENCE: Political Advertising For the final days before the election, the campaign manager has a total of $36,000 to spend on TV and radio campaign advertisements. Each TV ad costs $3000 and is seen by 10,000 voters, while each radio ad costs $500 and is heard by 2000 voters. Ignoring repeated exposures to the same voter, how many TV and radio ads will contact 130,000 voters using the allocated funds?

80. MANAGEMENT SCIENCE: Consumer Preferences A marketing company wants to gather information on consumer preferences for laundry detergents using telephone interviews and direct mail questionnaires. Each attempted telephone interview costs $2.25 whether successful or not, and each mailed questionnaire costs $0.75 whether returned or not. The research director estimates that 21% of the telephone calls will result in usable interviews while only 7% of the questionnaires will be returned in usable condition. If the director's budget is $15,750 and she needs 1470 usable responses, how many telephone calls should she attempt and how many questionnaires should she mail out?

Explorations and Excursions

The following problems extend and augment the material presented in the text.

More about 2 × 3 Row-Reduced Matrices

The table on page 175 lists exactly five possible final matrices for the solution of a system of equations by reducing the corresponding augmented matrix. The following problems provide examples of each possibility.

Sketch each system of equations and then solve it by row-reducing the corresponding augmented matrix. State the solution or identify the system as "inconsistent" or "dependent."

If permitted by your instructor, graph each system on the window $[-30, 50]$ by $[-30, 50]$, use the RREF command to row-reduce the augmented matrix, and graph the parametric solution if the system is dependent.

81. $\begin{cases} x - 2y = -14 \\ 2x + 3y = 84 \end{cases}$ **82.** $\begin{cases} x + 2y = 56 \\ 2x - 3y = 42 \end{cases}$

83. $\begin{cases} 3x - 6y = 18 \\ -2x + 4y = 16 \end{cases}$ **84.** $\begin{cases} 4x - 6y = 36 \\ -2x + 3y = 30 \end{cases}$

85. $\begin{cases} 2y = 8 \\ 3y = 18 \end{cases}$ **86.** $\begin{cases} 5y = 20 \\ 3y = 30 \end{cases}$

87. $\begin{cases} 4x + 14y = 56 \\ 6x + 21y = 84 \end{cases}$ **88.** $\begin{cases} 6x + 14y = 42 \\ 9x + 21y = 63 \end{cases}$

89. $\begin{cases} 5y = 30 \\ 3y = 18 \end{cases}$ **90.** $\begin{cases} 2y = 18 \\ 3y = 27 \end{cases}$

Row Operations Are Reversible

Carry out the row operation on the augmented matrix $\begin{pmatrix} 3 & -10 & | & -65 \\ -4 & 13 & | & 84 \end{pmatrix}$ and then find a row operation to perform on your new matrix that will return it to the original $\begin{pmatrix} 3 & -10 & | & -65 \\ -4 & 13 & | & 84 \end{pmatrix}$.

91. $R_1 \leftrightarrow R_2$ **92.** $R_2 \leftrightarrow R_1$

93. $5R_1 \rightarrow R_1$ **94.** $3R_2 \rightarrow R_2$

95. $\frac{1}{4}R_2 \rightarrow R_2$ **96.** $\frac{1}{3}R_1 \rightarrow R_1$

97. $R_1 + R_2 \rightarrow R_1$ **98.** $R_2 + R_1 \rightarrow R_2$

99. $R_1 - R_2 \rightarrow R_1$ **100.** $R_2 - R_1 \rightarrow R_2$

3.3 SYSTEMS OF LINEAR EQUATIONS AND THE GAUSS–JORDAN METHOD

Introduction

In this section we use augmented matrices to solve larger systems of linear equations. We simply enlarge the augmented matrix to allow for more equations (rows) and more variables (columns) and then apply row operations to find an equivalent matrix that displays the solution.

Names for Many Variables

To deal with many variables, we now distinguish them by subscripts instead of different letters: x_1 ("x sub one"), x_2 ("x sub two"), and so on for as many as we need. For example:

$$\begin{cases} 3x_1 + 2x_2 + x_3 = 39 \\ 2x_1 + 3x_2 + x_3 = 34 \\ x_1 + 2x_2 + 3x_3 = 26 \end{cases}$$

We form the augmented matrix exactly as before, so for this system we have

$$\begin{array}{ccc} x_1 & x_2 & x_3 \end{array}$$
$$\left(\begin{array}{ccc|c} 3 & 2 & 1 & 39 \\ 2 & 3 & 1 & 34 \\ 1 & 2 & 3 & 26 \end{array} \right)$$

Variables corresponding to columns (last column represents constant terms)

Each row represents an equation and the bar represents the equals sign

Row-Reduced Form

We continue to use row operations to "solve" the augmented matrix by finding an equivalent matrix that displays the solution. With our list of 2×3 matrices on page 175 as a guide, we make the following definition for matrices of any dimension. In this definition, a *zero row* is a row containing only zeros, and a *nonzero row* is a row with at least one nonzero element.

Row-Reduced Form

A matrix is in *row-reduced form* if it satisfies the following four conditions.

1. All zero rows are grouped below the nonzero rows.
2. The first nonzero element in each nonzero row is a 1. We call these particular 1s "leftmost 1s."

3. If a column contains a leftmost 1, all other entries in the column are 0s.

4. Each leftmost 1 appears to the *right* of the leftmost 1 in the row above it.

The following matrices are in row-reduced form:

$$\begin{pmatrix} 0 & 1 & 2 & 0 & 1 \\ 0 & 0 & 0 & 1 & 4 \end{pmatrix}, \quad \begin{pmatrix} 1 & 2 & 3 & 0 \\ 0 & 0 & 0 & 1 \\ 0 & 0 & 0 & 0 \end{pmatrix}, \quad \text{and} \quad \begin{pmatrix} 1 & 0 & 0 & 0 & 1 \\ 0 & 1 & 0 & 0 & 2 \\ 0 & 0 & 1 & 0 & 3 \\ 0 & 0 & 0 & 1 & 4 \end{pmatrix}$$

Leftmost 1s   Zero row

Notice that the 1s in the upper right corners of the first and last matrices above are not *leftmost* 1s and do not need to have 0s below them. The following matrix is *not* in row-reduced form because although conditions (1), (2), and (3) are satisfied, condition (4) fails for the third row.

$$\begin{pmatrix} 1 & 0 & 0 & 2 \\ 0 & 0 & 1 & 4 \\ 0 & 1 & 0 & 6 \end{pmatrix}$$

Not row-reduced:

The 1 in the third row is not to the right of the 1 in the second row

Practice Problem 1   Find a row operation that will correct this defect and result in a matrix in row-reduced form. ➤ Solution on page 192

In many augmented matrices the row-reduced form will be particularly simple: to the left of the bar will be a square matrix with 1s down the main diagonal and 0s elsewhere, as shown below. In such cases the system has a *unique solution* and we may immediately read off the values of the variables from the numbers to the right of the bar (top to bottom):

$$\left(\begin{array}{ccc|c} 1 & 0 & 0 & 5 \\ 0 & 1 & 0 & -2 \\ 0 & 0 & 1 & 3 \end{array} \right) \text{ gives } \begin{cases} x_1 = 5 \\ x_2 = -2 \\ x_3 = 3 \end{cases} \quad \text{From} \quad \begin{array}{l} 1x_1 + 0x_2 + 0x_3 = 5 \\ 0x_1 + 1x_2 + 0x_3 = -2 \end{array}$$

This is the pattern of 1s and 0s that we seek, since it leads to a unique solution. If another pattern occurs, we will see how to interpret it as indicating either no solution or infinitely many solutions. We will row-reduce augmented matrices by following a procedure called the

Gauss–Jordan method.* It operates on one column at a time, first obtaining a 1 and then 0s above and below it.

Gauss–Jordan Method

To row-reduce a matrix:

For instance, you might get:

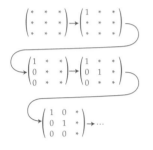

1. If any rows have leading 0s, switch the rows so that the ones with the *fewest* leading 0s are at the top, down to the ones with the *most* leading 0s at the bottom.
2. In the first row, find the leftmost nonzero entry and divide the row through by that number. This gives a *leftmost 1* in that row.
3. Add or subtract multiples of the first row to each other row to obtain 0s in the rest of the column above and below the leftmost 1 found in step 2.
4. Repeat steps 1, 2, and 3 but replacing "first row" by "second row" and then by "third row" and so on. Stop when you reach the bottom row or a row consisting entirely of zeros—when this happens the matrix is *row-reduced*.

Just as in the last section, there are three possibilities: a unique solution, no solution, or infinitely many solutions that can be parameterized.

EXAMPLE 1

SOLVING A SYSTEM USING THE GAUSS–JORDAN METHOD

Solve by the Gauss–Jordan method: $\begin{cases} 2x_1 + 4x_2 + 2x_3 = 2 \\ 3x_1 + 7x_2 + 3x_3 = 0. \\ x_1 + 2x_2 + 4x_3 = 7 \end{cases}$

Solution

Beginning with the augmented matrix, we carry out steps 1–4 of the Gauss–Jordan method.

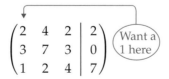

Step 1 concerns zero rows, and there are none. Step 2 says to divide row 1 by the first nonzero element, which is 2, to get a leftmost 1. The following row operation achieves this.

---

* Named after the mathematician Carl Friedrich Gauss (1777–1855) and the geodesist Wilhelm Jordan (1842–1899).

$\frac{1}{2}R_1 \rightarrow$ $\begin{pmatrix} 1 & 2 & 1 & | & 1 \\ 3 & 7 & 3 & | & 0 \\ 1 & 2 & 4 & | & 7 \end{pmatrix}$ Want 0s here

Step 3 says to subtract multiples of row 1 from the other rows to get 0s below the leftmost 1. The following two row operations do this.

$R_2 - 3R_1 \rightarrow$ $\begin{pmatrix} 1 & 2 & 1 & | & 1 \\ 0 & 1 & 0 & | & -3 \\ 0 & 0 & 3 & | & 6 \end{pmatrix}$ Want a 0 here
$R_3 - R_1 \rightarrow$

Row 2 already has a leftmost 1, so to achieve the 0 above it, in the next step we subtract twice row 2.

$R_1 - 2R_2 \rightarrow$ $\begin{pmatrix} 1 & 0 & 1 & | & 7 \\ 0 & 1 & 0 & | & -3 \\ 0 & 0 & 3 & | & 6 \end{pmatrix}$ Want a 1 here

Step 2 then says to divide row 3 by its first nonzero element, 3, to obtain a leftmost 1 as follows.

$\begin{pmatrix} 1 & 0 & 1 & | & 7 \\ 0 & 1 & 0 & | & -3 \\ 0 & 0 & 1 & | & 2 \end{pmatrix}$ Want a 0 here
$\frac{1}{3}R_3 \rightarrow$

The next step subtracts row 3 from row 1, achieving the final zero and completing the row-reduction of the augmented matrix.

$R_1 - R_3 \rightarrow$ $\begin{pmatrix} 1 & 0 & 0 & | & 5 \\ 0 & 1 & 0 & | & -3 \\ 0 & 0 & 1 & | & 2 \end{pmatrix}$

The solution is $\begin{cases} x_1 = 5 \\ x_2 = -3. \\ x_3 = 2 \end{cases}$

Reading the numbers from the last column

Graphing Calculator Exploration

The program* ROWOPS carries out the arithmetic for the type of row operation you select from a menu. To have your calculator perform the row operations used in Example 1 (above), proceed as follows:

a. Enter the augmented matrix as matrix [A]. Because this matrix is too large to fit on the calculator screen, it will scroll from left to right and back again as you enter your numbers.

* See the Preface for information on how to obtain this and other programs.

```
┌─ROW OPERATIONS─┐
│1: SWITCH       │
│2:  MULTIPLY    │
│3:  DIVIDE      │
│4: ADD TOGETHER │
│5: SUBTRACT FROM│
│6:  ADD MULTIPLE│
│7: QUIT         │
└────────────────┘
```

```
┌────────────────────┐
│       [1 2 4 7]]▪   │
│      DIVIDE         │
│DIVIDE ROW 1         │
│  BY 2              │
│     [[1 2 1 1]     │
│     [3 7 3 0]      │
│     [1 2 4 7]]     │
└────────────────────┘
```

b. Run the program ROWOPS. It will display the current values in the matrix [A] and you can use the arrows to scroll the screen to see the rest of it. Press (ENTER) to select the type of row operation you want to use.

c. Choose the type of row operation you want by using the arrows to move to its number and pressing (ENTER), or just press the number. Enter the specific details for the operation and the program will carry out your request.

d. Press (ENTER) to select another row operation or to choose 7 and QUIT the program.

The program ROWOPS allows you to multiply rows by fractions and displays them in the usual 3/5 notation. The reduction of this matrix took six row operations to complete in Example 1. Can you find other ways to achieve the same result?

The system in Example 1 had a unique solution. If there is an "inconsistent" row, the system will have *no solutions*, just as in the previous section.

No Solutions

If a row-reduced matrix has a row of 0s ending in a 1 (such as 0 0 0 | 1), then the system of equations from which it came is *inconsistent* and has *no solutions*.

EXAMPLE **2**

A SYSTEM WITH NO SOLUTION

Solve by the Gauss–Jordan method: $\begin{cases} 2x_1 - 4x_2 = 2 \\ -3x_1 + 6x_2 = 4 \end{cases}$.

Solution

The augmented matrix is

$$\begin{pmatrix} 2 & -4 & | & 2 \\ -3 & 6 & | & 4 \end{pmatrix} \quad \text{Want a } 1 \text{ here}$$

There are no zero rows, so we go to step 2, dividing row 1 by the first nonzero element, 2, as follows.

$$\tfrac{1}{2}R_1 \rightarrow \begin{pmatrix} 1 & -2 & | & 1 \\ -3 & 6 & | & 4 \end{pmatrix}$$

(Want a 0 here)

Step 3 says to add or subtract a multiple of row 1 to row 2 to get a zero, so in the next matrix we add three times row 1.

$$R_2 + 3R_1 \rightarrow \begin{pmatrix} 1 & -2 & | & 1 \\ 0 & 0 & | & 7 \end{pmatrix}$$

(Want a 1 here)

Step 2 says to get a leading 1 in row 2, divide the row by 7 leading to the following matrix.

$$\tfrac{1}{7}R_2 \rightarrow \begin{pmatrix} 1 & -2 & | & 1 \\ 0 & 0 & | & 1 \end{pmatrix}$$

An "inconsistent" row

Row 2 is of the form $\quad 0 \quad 0 \ | \ 1$ saying that $\quad 0 = 1, \quad$ which means the system has *no solutions*.

The system is inconsistent and has no solutions.

Once we encountered the line $\quad 0 \quad 0 \ | \ 7 \quad$ it was clear that by dividing by 7 we would obtain the "inconsistent" row $\quad 0 \quad 0 \ | \ 1$. Therefore, in practice, whenever we encounter a row of 0s ending with any nonzero number after the bar we may stop at that point and declare that the system is inconsistent and has no solutions.

Practice Problem 2   Solve by the Gauss–Jordan method: $\begin{cases} 3x_1 - 6x_2 = 12 \\ -5x_1 + 10x_2 = -14 \end{cases}$

➤ Solution on page 192

The rule for infinitely many solutions is a little more complicated than in the case of two equations. It depends on the number of leftmost 1s compared to the number of variables.

Infinitely Many Solutions

For a row-reduced augmented matrix with no "inconsistent" rows, if the number of leading 1s is *less than* the number of variables, then the original system is *dependent* and has *infinitely many solutions*.

If there are infinitely many solutions, we may *parameterize* them as we did on pages 172–173. First we classify each variable as *determined* or *free* by looking at the column in the row-reduced augmented matrix corresponding to that variable:

The variable is *determined* if its column has a leftmost 1.

The variable is *free* if its column does *not* have a leftmost 1.

For example, given the following row-reduced matrix (with two columns to the left of the bar corresponding to two variables x_1 and x_2), the first variable x_1 is *determined* and the second variable x_2 is *free*.

x_1 is *determined* (its column has a leftmost 1)
x_2 is *free* (its column does not have a leftmost 1)

$$\begin{pmatrix} 1 & 2 & | & 3 \\ 0 & 0 & | & 0 \end{pmatrix}$$

$x_1 + 2x_2 = 3$
so $x_1 = 3 - 2x_2$

That is, x_2 is *free to take any value t*, while the value of x_1 is *determined* by the equation $x_1 = 3 - 2x_2$, giving the parameterized solution (replacing x_2 by t):

$$\begin{cases} x_1 = 3 - 2t \\ x_2 = t \end{cases}$$

From $x_1 = 3 - 2x_2$ with x_2 replaced by t

Free variables are sometimes called *independent* and determined variables are then called *dependent*.

EXAMPLE 3

A SYSTEM WITH INFINITELY MANY SOLUTIONS

Solve by the Gauss–Jordan method: $\begin{cases} 2x_1 + 6x_2 + 10x_3 = 8 \\ 3x_1 + 9x_2 + 15x_3 = 12. \\ 2x_1 + 5x_2 + 8x_3 = 7 \end{cases}$

Solution

The augmented matrix is

$$\begin{pmatrix} 2 & 6 & 10 & | & 8 \\ 3 & 9 & 15 & | & 12 \\ 2 & 5 & 8 & | & 7 \end{pmatrix}$$

Want a 1 here

Since there are no leading 0s, we go to step 2, which says to divide the first row by 2, as follows.

$$\frac{1}{2}R_1 \rightarrow \begin{pmatrix} 1 & 3 & 5 & | & 4 \\ 3 & 9 & 15 & | & 12 \\ 2 & 5 & 8 & | & 7 \end{pmatrix}$$

Want 0s here

To "zero-out" the rest of the column, in the next matrix we subtract appropriate multiples of the first row.

$$\begin{array}{c} \\ R_2 - 3R_1 \rightarrow \\ R_3 - 2R_1 \rightarrow \end{array} \begin{pmatrix} 1 & 3 & 5 & | & 4 \\ 0 & 0 & 0 & | & 0 \\ 0 & -1 & -2 & | & -1 \end{pmatrix}$$

Has more leading 0s than

Since row 2 has more leading 0s than row 3, step 1 says to switch rows 2 and 3, as we do on the next page.

$$\begin{array}{c} R_3 \to \\ R_2 \to \end{array} \begin{pmatrix} 1 & 3 & 5 & 4 \\ 0 & -1 & -2 & -1 \\ 0 & 0 & 0 & 0 \end{pmatrix}$$
(Want a 1 here)

Now divide by -1 (which is equivalent to multiplying by -1) to get a leftmost 1 in the second row, as follows.

$$-1R_2 \to \begin{pmatrix} 1 & 3 & 5 & 4 \\ 0 & 1 & 2 & 1 \\ 0 & 0 & 0 & 0 \end{pmatrix}$$
(Want a 0 here)

Using the 1 to "zero-out" the rest of the column gives the matrix below.

$$R_1 - 3R_2 \to \begin{pmatrix} 1 & 0 & -1 & 1 \\ 0 & 1 & 2 & 1 \\ 0 & 0 & 0 & 0 \end{pmatrix}$$

Row-reduced!

The corresponding equations are $\begin{cases} x_1 & -x_3 = 1 \\ & x_2 + 2x_3 = 1 \end{cases}$.

Since the variables x_1 and x_2 are *determined* (their columns *do* have leftmost 1s) and the variable x_3 is *free* (column 3 does not have a leading 1), the free variable may take *any* value, $x_3 = t$, and we solve the other equations for the *determined* variables x_1 and x_2. The solution is:

$$\begin{cases} x_1 = 1 + t \\ x_2 = 1 - 2t \\ x_3 = t \end{cases}$$
From solving $x_1 - x_3 = 1$ for x_1
From solving $x_2 + 2x_3 = 1$ for x_2
and then replacing x_3 by t

We obtain all solutions by taking different values for t. For example:

$$t = 3 \text{ gives } \begin{cases} x_1 = 4 \\ x_2 = -5 \\ x_3 = 3 \end{cases} \text{ and } t = -2 \text{ gives } \begin{cases} x_1 = -1 \\ x_2 = 5 \\ x_3 = -2 \end{cases}.$$

It is easily checked that these are solutions to the original equations.

Practice Problem 3

Find the solution of a system of equations with augmented matrix equivalent to the row-reduced matrix

$$\begin{pmatrix} 1 & 2 & 0 & 0 & 0 \\ 0 & 0 & 1 & 0 & 3 \\ 0 & 0 & 0 & 1 & 4 \\ 0 & 0 & 0 & 0 & 0 \end{pmatrix}$$

➤ Solution on page 192

EXAMPLE 4 **MANAGING PRODUCTION**

A hand-thrown pottery shop manufactures plates, cups, and vases. Each plate requires 4 ounces of clay, 6 minutes of shaping, and 5 minutes of painting; each cup requires 4 ounces of clay, 5 minutes of shaping, and 3 minutes of painting; and each vase requires 3 ounces of clay, 4 minutes of shaping, and 4 minutes of painting. This week the shop has 165 pounds of clay, 59 hours of skilled labor for shaping, and 46 hours of skilled labor for painting. If the shop manager wishes to use all these resources fully, how many of each product should the shop produce?

Solution

Suggestion: List the information in a table.

	Plates	Cups	Vases
Clay	4	4	3
Shaping	6	5	4
Painting	5	3	4

Let x_1, x_2, and x_3 be the numbers of plates, cups, and vases produced. The number of ounces of clay required is then $4x_1 + 4x_2 + 3x_3$ and this must match the 165 pounds available:

$$4x_1 + 4x_2 + 3x_3 = 2640$$

Use ounces on both sides of the equation

Similarly, for the time in minutes required for shaping and painting,

$$6x_1 + 5x_2 + 4x_3 = 3540$$
$$5x_1 + 3x_2 + 4x_3 = 2760$$

Use minutes on both sides of the equation

Therefore, the augmented matrix is

$$\begin{pmatrix} 4 & 4 & 3 & | & 2640 \\ 6 & 5 & 4 & | & 3540 \\ 5 & 3 & 4 & | & 2760 \end{pmatrix}$$

We now solve this problem by row-reducing the augmented matrix using the Gauss–Jordan method.

$$\tfrac{1}{4}R_1 \rightarrow \begin{pmatrix} 1 & 1 & \tfrac{3}{4} & | & 660 \\ 6 & 5 & 4 & | & 3540 \\ 5 & 3 & 4 & | & 2760 \end{pmatrix}$$

Get a 1 in the first column of the first row

$$\begin{matrix} \\ R_2 - 6R_1 \rightarrow \\ R_3 - 5R_1 \rightarrow \end{matrix} \begin{pmatrix} 1 & 1 & \tfrac{3}{4} & | & 660 \\ 0 & -1 & -\tfrac{1}{2} & | & -420 \\ 0 & -2 & \tfrac{1}{4} & | & -540 \end{pmatrix}$$

"Zero-out" the rest of the first column

$$-R_2 \rightarrow \begin{pmatrix} 1 & 1 & \tfrac{3}{4} & | & 660 \\ 0 & 1 & \tfrac{1}{2} & | & 420 \\ 0 & -2 & \tfrac{1}{4} & | & -540 \end{pmatrix}$$

Get a 1 in the second column of the second row

$$\begin{matrix} R_1 - R_2 \rightarrow \\ \\ R_3 + 2R_2 \rightarrow \end{matrix} \begin{pmatrix} 1 & 0 & \tfrac{1}{4} & | & 240 \\ 0 & 1 & \tfrac{1}{2} & | & 420 \\ 0 & 0 & \tfrac{5}{4} & | & 300 \end{pmatrix}$$

"Zero-out" the rest of the second column

$$\frac{4}{5}R_3 \to \begin{pmatrix} 1 & 0 & \frac{1}{4} & 240 \\ 0 & 1 & \frac{1}{2} & 420 \\ 0 & 0 & 1 & 240 \end{pmatrix}$$

Get a 1 in the third column of the third row

$$\begin{matrix} R_1 - \frac{1}{4}R_3 \to \\ R_2 - \frac{1}{2}R_3 \to \end{matrix} \begin{pmatrix} 1 & 0 & 0 & 180 \\ 0 & 1 & 0 & 300 \\ 0 & 0 & 1 & 240 \end{pmatrix}$$

"Zero-out" the rest of the third column

The system has a unique solution: $x_1 = 180$, $x_2 = 300$, $x_3 = 240$. You should check that these values satisfy the original equations and that the clay and skilled labor resources are fully used. In terms of the original question, the shop should produce 180 plates, 300 cups, and 240 vases this week.

EXAMPLE 5

MODELING A COMPUTER NETWORK

The office manager for the accounting division of a large company is writing a report to her supervisor about the demands placed on the computers serving her division. Besides meeting the needs of the accounting division, her four "file servers" are expected to accept and relay messages to and from other areas. The diagram shows the four file servers (I, II, III, IV), and the arrows and numbers indicate the data packets per minute passing from the senders to the receivers. Computers A through H are outside the accounting division. Connections passing an unknown number of data packets per minute are marked with variables (for instance, x_1 is the number of data packets per minute sent from server I to server II, while x_5 is the number leaving server IV to outside computer G). For the network to function, the number of data packets per minute arriving at each computer must match the number leaving it. How many data packets per minute must leave IV for G? Is it possible for each connection within the accounting division to carry no more than 1500 data packets per minute?

Solution

In terms of the notations from the diagram, the requirements that the data packets per minute arriving at each file server match the number leaving are represented by the equations

$$1500 + 1500 = x_1 + x_2$$
$$x_1 + 900 = x_3 + 1200$$
$$900 + x_2 = x_4 + 1200$$
$$x_3 + x_4 = 1200 + x_5$$

"In" equals "out"
Server I
Server II
Server III
Server IV

Rewriting to put the variables on the left, we have the system

$$\begin{cases} x_1 + x_2 = 3000 \\ x_1 - x_3 = 300 \\ x_2 - x_4 = 300 \\ x_3 + x_4 - x_5 = 1200 \end{cases}$$

The following augmented matrix represents these equations, with its row-reduced form shown on the right. (You may wish to carry out the reduction—it can be done easily because of the many zeros and ones—or use your calculator's RREF command to check the reduction.)

$$\left(\begin{array}{ccccc|c} 1 & 1 & 0 & 0 & 0 & 3000 \\ 1 & 0 & -1 & 0 & 0 & 300 \\ 0 & 1 & 0 & -1 & 0 & 300 \\ 0 & 0 & 1 & 1 & -1 & 1200 \end{array}\right) \xrightarrow{\text{Leads to}} \left(\begin{array}{ccccc|c} 1 & 0 & 0 & 1 & 0 & 2700 \\ 0 & 1 & 0 & -1 & 0 & 300 \\ 0 & 0 & 1 & 1 & 0 & 2400 \\ 0 & 0 & 0 & 0 & 1 & 1200 \end{array}\right)$$

The variables x_1, x_2, x_3, and x_5 are *determined* (since their columns have leftmost 1s), and x_4 is a free variable (its column does *not* have a leftmost 1). We may parameterize the solution as follows:

$$\begin{aligned} x_1 &= 2700 - t \\ x_2 &= 300 + t \\ x_3 &= 2400 - t \\ x_4 &= t \qquad\qquad \text{The free variable} \\ x_5 &= 1200 \end{aligned}$$

To prevent a "back flow" along the connections, each variable must stay nonnegative, which means that $t \geq 0$ (since $x_4 = t$) and $t \leq 2400$ (since $x_3 = 2400 - t$). For example, the choice $t = 1200$ gives the solution

$$x_1 = 1500, \quad x_2 = 1500, \quad x_3 = 1200, \quad x_4 = 1200, \quad x_5 = 1200$$

In terms of the original questions, 1200 data packets per minute must leave computer IV for G, and the example with $t = 1200$ shows that it *is* possible to arrange the transmissions so that each connection within the accounting division carries no more than 1500 data packets per minute.

Section Summary

The solution of any system of linear equations can be found by representing it as an augmented matrix, applying row operations (page 168) using the Gauss–Jordan method (page 182) to find the equivalent row-reduced matrix (pages 180–181), and interpreting this final matrix as a statement about the original system of equations (pages 181 and 184–185). Each step in this method is a direct extension of the method for solving systems of two linear equations in two variables presented in the previous section.

For three equations in three unknowns with a unique solution, this would look like:

$$\begin{pmatrix} * & * & * & | & * \\ * & * & * & | & * \\ * & * & * & | & * \end{pmatrix} \xrightarrow{\text{Row-reduces to}} \begin{pmatrix} 1 & 0 & 0 & | & * \\ 0 & 1 & 0 & | & * \\ 0 & 0 & 1 & | & * \end{pmatrix}$$

$$\underbrace{}_{\text{solution}}$$

A row of zeros that ends in a nonzero number (such as $0 \; 0 \; 0 \, | \, 1$) means that the system is *inconsistent* and has *no solutions*. If there is no such "inconsistent row" and if the number of leftmost 1s *is less* than the number of variables, then there are *infinitely many* solutions, which may be parameterized as on page 190.

A system of equations may have more than one free variable. The number of free variables will be equal to the total number of variables minus the number of leftmost 1s in the row-reduced form.

The three possibilities—unique solution, no solution, or infinitely many solutions—may be understood geometrically as follows. On page 153 we saw that equations in *two* variables represent *lines in the plane* and that two lines may intersect in a single point (a unique solution), not at all (parallel lines), or at infinitely many points (the same line). Analogously, equations in *three* variables represent *planes in three-dimensional space,* and three planes can intersect at a single point, not at all, or at infinitely many points, as shown below.

| Unique solution | No solutions | Infinitely many solutions |

➤ **Solutions to Practice Problems**

1. $R_2 \leftrightarrow R_3$

2.
$$\begin{pmatrix} 3 & -6 & | & 12 \\ -5 & 10 & | & -14 \end{pmatrix}$$

$$\tfrac{1}{3}R_1 \rightarrow \begin{pmatrix} 1 & -2 & | & 4 \\ -5 & 10 & | & -14 \end{pmatrix}$$

$$R_2 + 5R_1 \rightarrow \begin{pmatrix} 1 & -2 & | & 4 \\ 0 & 0 & | & 6 \end{pmatrix}$$

Inconsistent, so *no* solution.

3. x_1, x_3, and x_4 are determined while x_2 is free. The solution is
$$\begin{cases} x_1 = -2t \\ x_2 = t \\ x_3 = 3 \\ x_4 = 4 \end{cases}.$$

3.3 Exercises

Find the augmented matrix representing the system of equations.

1. $\begin{cases} x_1 + x_2 + x_3 = 4 \\ x_1 + 2x_2 + x_3 = 3 \\ x_1 + 2x_2 + 2x_3 = 5 \end{cases}$

2. $\begin{cases} x_1 + 5x_2 + 4x_3 = 6 \\ x_1 + x_2 + x_3 = 4 \\ 2x_1 + 3x_2 + 3x_3 = 9 \end{cases}$

3. $\begin{cases} 2x_1 - x_2 + 2x_3 = 11 \\ -x_1 + x_2 - 3x_3 = -12 \\ 2x_1 - 2x_2 + 7x_3 = 27 \end{cases}$

4. $\begin{cases} 4x_1 + 3x_2 - x_3 = 2 \\ 3x_1 + 3x_2 + 2x_3 = 9 \\ 2x_1 + x_2 - 3x_3 = -6 \end{cases}$

5. $\begin{cases} 2x_1 + x_2 + 5x_3 + 4x_4 + 5x_5 = 2 \\ x_1 + x_2 + 3x_3 + 3x_4 + 3x_5 = -1 \end{cases}$

6. $\begin{cases} 5x_1 + 2x_2 - 4x_3 + x_4 + 5x_5 = 7 \\ 3x_1 + x_2 - 3x_3 + x_4 + 3x_5 = 5 \end{cases}$

Find the system of equations represented by the augmented matrix.

7. $\begin{pmatrix} 4 & 3 & 2 & | & 11 \\ 3 & 3 & 1 & | & 6 \\ 1 & -2 & 3 & | & 13 \end{pmatrix}$

8. $\begin{pmatrix} 3 & -2 & 5 & | & 23 \\ -1 & 1 & -3 & | & -12 \\ 2 & -2 & 7 & | & 27 \end{pmatrix}$

9. $\begin{pmatrix} 2 & 1 & 1 & | & 7 \\ 2 & 2 & 1 & | & 6 \\ 3 & 3 & 2 & | & 10 \end{pmatrix}$

10. $\begin{pmatrix} 5 & 1 & 3 & | & 20 \\ 1 & 1 & 2 & | & 6 \\ 4 & 1 & 3 & | & 17 \end{pmatrix}$

11. $\begin{pmatrix} 8 & 3 & -2 & 19 & | & 15 \\ 3 & 1 & -1 & 7 & | & 6 \end{pmatrix}$

12. $\begin{pmatrix} 6 & -2 & -4 & -2 & | & 36 \\ 2 & -1 & -10 & 5 & | & 6 \end{pmatrix}$

Interpret each row-reduced matrix as the solution of a system of equations.

13. $\begin{pmatrix} 1 & 0 & 0 & | & 4 \\ 0 & 1 & 0 & | & 5 \\ 0 & 0 & 1 & | & -4 \end{pmatrix}$
14. $\begin{pmatrix} 1 & 0 & 0 & | & 0 \\ 0 & 1 & 0 & | & 0 \\ 0 & 0 & 1 & | & 1 \end{pmatrix}$

15. $\begin{pmatrix} 1 & 0 & 1 & | & 0 \\ 0 & 1 & 0 & | & 0 \\ 0 & 0 & 0 & | & 1 \end{pmatrix}$
16. $\begin{pmatrix} 1 & 1 & | & 0 \\ 0 & 0 & | & 1 \end{pmatrix}$

17. $\begin{pmatrix} 1 & 0 & -1 & | & -5 \\ 0 & 1 & 1 & | & 5 \\ 0 & 0 & 0 & | & 0 \end{pmatrix}$

18. $\begin{pmatrix} 1 & 1 & 0 & 0 & | & 2 \\ 0 & 0 & 1 & 0 & | & -1 \\ 0 & 0 & 0 & 1 & | & 3 \\ 0 & 0 & 0 & 0 & | & 0 \end{pmatrix}$

Use an appropriate row operation or sequence of row operations to find the equivalent row-reduced matrix.

If your instructor permits, you may carry out the calculations on a graphing calculator using the ROWOPS program.

19. $\begin{pmatrix} 0 & 1 & 0 & | & 2 \\ 1 & 0 & 0 & | & 1 \\ 0 & 0 & 1 & | & 3 \end{pmatrix}$
20. $\begin{pmatrix} 1 & 0 & 0 & | & 1 \\ 0 & 0 & 1 & | & 3 \\ 0 & 1 & 0 & | & 2 \end{pmatrix}$

21. $\begin{pmatrix} 1 & 0 & 1 & | & 4 \\ 0 & 1 & 0 & | & 2 \\ 0 & 0 & 1 & | & 3 \end{pmatrix}$
22. $\begin{pmatrix} 1 & 0 & 0 & | & 1 \\ 1 & 1 & 0 & | & 3 \\ 0 & 0 & 1 & | & 3 \end{pmatrix}$

23. $\begin{pmatrix} 2 & 4 & 0 & | & 6 \\ 0 & 0 & 1 & | & -3 \\ 0 & 0 & 0 & | & 0 \end{pmatrix}$
24. $\begin{pmatrix} 1 & 0 & 2 & | & 5 \\ 0 & 3 & -6 & | & 3 \\ 0 & 0 & 0 & | & 0 \end{pmatrix}$

Solve each system of equations by the Gauss–Jordan method. If the solution is not unique, identify the system as "dependent" or "inconsistent."

If you have a graphing calculator with a RREF command, use it to check your row reduction.

25. $\begin{cases} x_1 + x_2 + x_3 = 2 \\ x_1 + 2x_2 + 2x_3 = 3 \\ x_1 + 3x_2 + 2x_3 = 1 \end{cases}$

26. $\begin{cases} 2x_1 + 3x_2 + x_3 = 4 \\ 3x_1 + 3x_2 + 2x_3 = 12 \\ x_1 + 2x_2 + x_3 = 4 \end{cases}$

27. $\begin{cases} 2x_1 + 2x_2 - x_3 = -5 \\ -2x_1 - x_2 + x_3 = 3 \\ 3x_1 + 4x_2 - x_3 = -8 \end{cases}$

28. $\begin{cases} 3x_1 - 4x_2 + 2x_3 = -15 \\ -x_1 + 2x_2 - x_3 = 6 \\ 4x_1 - 3x_2 + 2x_3 = -16 \end{cases}$

29. $\begin{cases} 2x_1 + x_2 - 2x_3 + x_4 = 2 \\ x_1 + x_2 + 2x_3 + x_4 = 5 \\ x_1 + x_2 + x_3 + x_4 = 4 \\ 2x_1 + 2x_2 + 3x_3 + x_4 = 8 \end{cases}$

30. $\begin{cases} x_1 - x_2 + x_3 + 2x_4 = 7 \\ x_1 - 2x_2 + x_3 + 2x_4 = 8 \\ 2x_1 + 2x_2 + x_3 + 3x_4 = 7 \\ x_1 - x_2 + x_3 + x_4 = 6 \end{cases}$

31. $\begin{cases} 2x_1 + 3x_2 + x_3 = 4 \\ 3x_1 + 5x_2 + 2x_3 = 12 \\ x_1 + 2x_2 + x_3 = 3 \end{cases}$

32. $\begin{cases} x_1 + 3x_2 + 2x_3 = 6 \\ x_1 + 2x_2 + 2x_3 = 3 \\ 2x_1 + 5x_2 + 4x_3 = 8 \end{cases}$

33. $\begin{cases} 4x_1 + 3x_2 + 2x_3 = 24 \\ x_1 + x_2 + 3x_3 = 7 \\ 5x_1 + 4x_2 + 5x_3 = 31 \end{cases}$

34. $\begin{cases} 5x_1 - 7x_2 + 3x_3 = -9 \\ -x_1 + x_2 - x_3 = 1 \\ 4x_1 - 5x_2 + 3x_3 = -6 \end{cases}$

35. $\begin{cases} x_1 + x_2 + 2x_3 + x_4 = 2 \\ 2x_1 + 2x_2 + 3x_3 + 3x_4 = 9 \\ 2x_1 + x_2 + 2x_3 + 2x_4 = 7 \\ x_1 + x_2 + x_3 + x_4 = 4 \end{cases}$

36. $\begin{cases} 3x_1 + 7x_2 + 4x_3 + 4x_4 = -7 \\ 7x_1 + 7x_2 + 4x_3 + 7x_4 = 3 \\ 4x_1 + 3x_2 + 2x_3 + 3x_4 = 2 \\ 3x_1 + 2x_2 + x_3 + 3x_4 = 4 \end{cases}$

37. $\begin{cases} x_1 - x_2 + x_4 = 1 \\ 2x_1 - 2x_2 + 2x_4 = 2 \\ x_1 - x_2 - x_3 - x_4 = 1 \\ 2x_1 - 2x_2 - x_3 = 1 \end{cases}$

39. $\begin{cases} x_1 + x_2 + x_3 + 2x_4 = 3 \\ x_1 - x_3 + x_4 = 2 \\ x_1 + 2x_2 + 3x_3 + 3x_4 = 4 \\ x_2 + 2x_3 + x_4 = 1 \end{cases}$

38. $\begin{cases} x_1 - x_3 + x_4 = 2 \\ x_2 + 2x_3 + x_4 = 0 \\ 2x_1 - 2x_2 - 6x_3 = 5 \\ x_1 + x_2 + x_3 + 2x_4 = 2 \end{cases}$

40. $\begin{cases} x_1 + 2x_2 + x_3 - 2x_4 = -3 \\ 2x_1 + 4x_2 + 2x_3 - x_4 = 0 \\ x_1 + 2x_2 + x_3 + x_4 = 3 \\ x_1 + 2x_2 + x_3 = 1 \end{cases}$

APPLIED EXERCISES

Formulate each situation as a system of linear equations. Be sure to state clearly the meaning of each variable. Solve using the Gauss–Jordan method. State your final answer in terms of the original question.

If you have a graphing calculator with a RREF command, use it to check your row reduction.

41. GENERAL: Plant Fertilizer A backyard garden needs 35 pounds of potash, 68 pounds of nitrogen, and 25 pounds of phosphoric acid. Three brands of fertilizer, GrowRite, MiracleMix, and GreatGreen, are available and contain the amounts of potash, nitrogen, and phosphoric acid per bag listed in the table. How many bags of each brand should be used to provide the required potash, nitrogen, and phosphoric acid?

(Pounds per Bag)	GrowRite	MiracleMix	GreatGreen
Potash	4	6	7
Nitrogen	5	10	16
Phosphoric acid	3	4	5

42. ATHLETICS: Nutrition A student athlete has decided to "bulk up" before the wrestling season by supplementing his weekly diet with an additional 325 grams of protein, 185 grams of fiber, and 110 grams of fat. If a hamburger contains 20 grams of protein, 10 grams of fiber, and 5 grams of fat; a cheeseburger contains 25 grams of protein, 10 grams of fiber, and 5 grams of fat; and a "sloppy-joe" contains 20 grams of protein, 15 grams of fiber, and

10 grams of fat, how many of each should he eat this week to meet his goal?

43. BUSINESS: Oil Well Ownership An accountant manages a limited-partnership that exploits oil-depletion tax advantages for wealthy clients, and sells memberships in the partnership for $5000, $10,000 or $25,000. If the partnership has seven hundred members and an investment value of $6 million, how many memberships of each amount are there? What is the greatest possible number of $25,000 members?

44. PERSONAL FINANCE: Financial Planning An international investment banker wishes to invest $150,000 in U.S. and German stocks and bonds. Stocks are not as secure as bonds, so he plans to spread his investments so that he has three times as much in U.S. bonds as in U.S. stocks and twice as much in German bonds as in German stocks. Furthermore, he intends to invest just $45,000 in stocks altogether. How much does he invest in each?

45. PERSONAL FINANCE: Income Taxes
a. Find the federal, state, and city income taxes on a taxable income of $180,000 if the federal tax is 50% of the taxable income after first deducting the state and city taxes, the state tax is 17% of the taxable income after first deducting the federal and city taxes, and the city tax is 3% of the taxable income after first deducting the federal and state taxes. **b.** Although it appears that the taxpayer is facing a 70% nominal tax rate, use the actual taxes to determine the effective combined tax rate for this situation.

46. GENERAL: Estate Division As the old patriarch lay dying, he spoke to his four sons: "I leave you my fortune of 11,100 pieces of gold, which you must divide as I command. The oldest is to take one half, the second oldest is to take one third, the third oldest is to take one quarter, and the youngest is to take one fifth. But each of you must take your part from what remains after the others have taken theirs and then you must give any remainder to your mother and sister." The old woman was frantic: "How can there be any left for our daughter when you have given away more than everything?" "Don't worry, Mother," said the daughter, who had gotten an MBA while her brothers had been off fighting at Troy. How much did each son receive and how much was left over for the mother and daughter?

47. BUSINESS: Apparel Production a. A "limited edition" ladies fashion shop has a 240-yard supply of a silk fabric suitable for scarves, dresses, blouses, and skirts. Each scarf requires 1 yard of material, 3 minutes of cutting, and 5 minutes of sewing; each dress requires 3 yards of material, 14 minutes of cutting, and 40 minutes of sewing; each blouse requires 1.5 yards of material, 9 minutes of cutting, and 30 minutes of sewing; and each skirt requires 2 yards of material, 8 minutes of cutting, and 20 minutes of sewing. If the shop has 17 hours of skilled pattern cutter labor and 45 hours of skilled seamstress labor available, how many of each can be made? **b.** If 20 blouses and 10 skirts are made, how many scarves and dresses can be made?

48. BUSINESS: Agriculture Management A farmer grows wheat, barley, and oats on his 360-acre farm. The labor (for planting, tending, and harvesting) and capital (for seeds, fertilizers, and pesticides) requirements for each crop are given in the table. If the farmer can contract for 1050 days of migrant worker labor and has $18,600 set aside for capital expenses, how many acres of each crop can he grow?

(Per Acre)	Wheat	Barley	Oats
Days of labor	2	4	3
Capital expenses	$50	$40	$60

49. BUSINESS: Advertising The promotional director for a new movie has a total of $110,000 to spend on TV, radio, and newspaper advertisements in the metropolitan area. Each TV ad costs $2500 and is seen by 10,000 moviegoers, each radio ad costs $500 and is heard by 2000 moviegoers, and each newspaper ad costs $1000 and is read by 5000 moviegoers. Ignoring repeated exposures to the same person, how many of each ad will contact 500,000 moviegoers using the allocated funds?

50. MANAGEMENT SCIENCE: Mass Transit Part of a subway system is shown in the diagram. The numbers of subway cars per hour arriving at and leaving stations I, II, III, and IV are indicated by arrows with numbers or variables. In order for the subway to function, the numbers of cars arriving and leaving per hour must match at each station. To prevent collisions, the subway cars must travel in the directions indicated by the arrows, and so none of the variables may be negative. A group of concerned citizens has petitioned the city council to build an express service from station I to IV (indicated on the diagram as x_5) capable of handling 40 cars per hour. Is this a good idea or is it a waste of money?

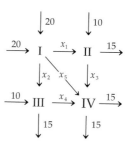

Explorations and Excursions

The following problems extend and augment the material presented in the text.

Row-Echelon Form

Row-echelon form has the same requirements as row-reduced form (pages 180–181) except that condition (3) is relaxed to require only that *the column of each leftmost 1 contains zeros below it.* Such a matrix is also called a *triangular matrix,* because the nonzero elements can only be on and above the main diagonal.

Find the system of equations represented by each row-echelon form matrix:

51. $\begin{pmatrix} 1 & 2 & | & 3 \\ 0 & 1 & | & 1 \end{pmatrix}$ **52.** $\begin{pmatrix} 1 & 3 & | & 9 \\ 0 & 1 & | & 2 \end{pmatrix}$

Unlike row-reduced matrices, there are many row-echelon form matrices equivalent to a given augmented matrix.

Row-reduce each group of matrices to show that they are all equivalent because they are all equivalent to the same row-reduced matrix.

53. $\begin{pmatrix} 1 & 2 & | & 3 \\ 2 & 3 & | & 5 \end{pmatrix}$, $\begin{pmatrix} 1 & 2 & | & 3 \\ 0 & 1 & | & 1 \end{pmatrix}$, $\begin{pmatrix} 1 & 3 & | & 4 \\ 0 & 1 & | & 1 \end{pmatrix}$

54. $\begin{pmatrix} 2 & 5 & | & 16 \\ 1 & 4 & | & 11 \end{pmatrix}$, $\begin{pmatrix} 1 & 3 & | & 9 \\ 0 & 1 & | & 2 \end{pmatrix}$, $\begin{pmatrix} 1 & -1 & | & 1 \\ 0 & 1 & | & 2 \end{pmatrix}$

Find an equivalent matrix in row-echelon form using the REF matrix command.

55. $\begin{pmatrix} 1 & 2 & | & 3 \\ 2 & 3 & | & 5 \end{pmatrix}$ **56.** $\begin{pmatrix} 1 & 2 & 2 & | & 7 \\ 1 & 1 & -1 & | & 0 \\ 2 & 2 & 1 & | & 6 \end{pmatrix}$

Back Substitution

If the augmented matrix for a system of equations is in row-echelon form, the last equation gives the value for the last variable. Substituting this value into the second-to-last equation, the value of the second-to-last variable is easy to find. Continuing upward through the equations, the values of all the remaining variables are easily found.

Solve each system of equations by back substitution.

57. $\begin{cases} x_1 + 2x_2 = 3 \\ x_2 = 1 \end{cases}$ **58.** $\begin{cases} x_1 + 3x_2 = 9 \\ x_2 = 2 \end{cases}$

59. $\begin{cases} x_1 + x_2 + x_3 = 4 \\ x_2 + x_3 = 3 \\ x_3 = 2 \end{cases}$ **60.** $\begin{cases} x_1 + 2x_2 + 2x_3 = 7 \\ x_2 + 2x_3 = 5 \\ x_3 = 2 \end{cases}$

61. $\begin{cases} x_1 + x_3 = 7 \\ x_2 + x_4 = 3 \\ x_3 = 4 \\ x_4 = 1 \end{cases}$ **62.** $\begin{cases} x_1 - x_2 + x_3 - x_4 = 4 \\ x_2 - x_3 + x_4 = -1 \\ x_3 - x_4 = 3 \\ x_4 = 1 \end{cases}$

Gauss–Jordan Elimination

Gauss–Jordan Elimination solves a system of equations by row-reducing the augmented matrix. It consists of two steps: First use row operations to find an equivalent row-echelon form matrix, and then use more row operations to back-substitute and complete the solution. Of course, if the row-echelon form matrix contains a row of zeros ending in a one, the corresponding system of equations is inconsistent, there can be no solution, and the method stops *without* attempting to carry out the back-substitution.

Solve each system of equations by Gauss–Jordan elimination.

63. $\begin{cases} x_1 + 3x_2 + x_3 = 2 \\ 2x_1 + 5x_2 + 2x_3 = 5 \\ x_1 + 2x_2 + 2x_3 = 5 \end{cases}$

64. $\begin{cases} 2x_1 + 2x_2 + x_3 + 3x_4 = 8 \\ x_1 + 2x_2 + x_3 + x_4 = 4 \\ 2x_1 + 3x_2 + x_3 + 2x_4 = 6 \\ 3x_1 + 2x_2 - x_3 + 2x_4 = 3 \end{cases}$

3.4 MATRIX ARITHMETIC

Introduction

Our focus thus far has been on the augmented matrix representation of a system of equations. This representation is "unnatural" in the sense that it somehow "contains" but does not show the equals signs from the equations. We shall now define the arithmetic of matrices so that we may write an entire system of equations as a single *matrix equation.*

In the next section, the solution of such a matrix equation will deepen our understanding of what it means to solve a system of equations.

Equality of Matrices

Two matrices are *equal* if they have the same dimension and if elements in corresponding locations are equal; that is, $A = B$ if $a_{i,j} = b_{i,j}$ for every row i and column j. For instance, the following matrices are equal:

$$\begin{pmatrix} 1 & 2 \\ 3 & 4 \end{pmatrix} = \begin{pmatrix} 1 & 1 + 1 \\ 4 - 1 & 2^2 \end{pmatrix}$$ Equal: same dimension and corresponding values are equal

The following matrices are *not* equal:

$$\begin{pmatrix} 1 & 2 \\ 3 & 4 \end{pmatrix} \neq \begin{pmatrix} 1 \\ 2 \\ 3 \\ 4 \end{pmatrix}$$ Not equal because their dimensions are different

$$\begin{pmatrix} 1 & 2 \\ 3 & 4 \end{pmatrix} \neq \begin{pmatrix} 1 & 2 \\ 4 & 3 \end{pmatrix}$$ Not equal because some corresponding values differ

Transpose of a Matrix

The *transpose* of a matrix A is the matrix A^t formed by turning each row into a column (or, equivalently, each column into a row). For example,

$$\begin{pmatrix} 1 & 2 & 3 \\ 4 & 5 & 6 \end{pmatrix}^t = \begin{pmatrix} 1 & 4 \\ 2 & 5 \\ 3 & 6 \end{pmatrix}$$ Transpose: First row becomes first column Second row becomes second column

Identity Matrix

An *identity matrix* is a square matrix with 1s on the main diagonal and 0s elsewhere. We write $I = I_n$ for the $n \times n$ identity matrix. The identity matrices $I_1, I_2, I_3,$ and I_4 are written below.

$$I_1 = (1), \quad I_2 = \begin{pmatrix} 1 & 0 \\ 0 & 1 \end{pmatrix}, \quad I_3 = \begin{pmatrix} 1 & 0 & 0 \\ 0 & 1 & 0 \\ 0 & 0 & 1 \end{pmatrix}, \quad I_4 = \begin{pmatrix} 1 & 0 & 0 & 0 \\ 0 & 1 & 0 & 0 \\ 0 & 0 & 1 & 0 \\ 0 & 0 & 0 & 1 \end{pmatrix}$$

Scalar Multiplication

Scalar multiplication of a matrix simply means multiplying each element of a matrix by the same number. ("Scalar" is just another word for "number" and is used to distinguish a number from a matrix.) In particular, the *negative* of a matrix is $-A = -1 \cdot A$ and is found by multiplying each element by -1. For example, if

$$A = \begin{pmatrix} 1 & 2 & 3 \\ 4 & 5 & 6 \end{pmatrix}$$

then

$$3A = \begin{pmatrix} 3 & 6 & 9 \\ 12 & 15 & 18 \end{pmatrix} \quad \text{and} \quad -A = \begin{pmatrix} -1 & -2 & -3 \\ -4 & -5 & -6 \end{pmatrix}$$

Scalar Product

The product of a scalar (a number) times a matrix, kA, is the matrix A with each element multiplied by k.

Matrix Addition and Subtraction

For two matrices with the same dimensions, *matrix addition* means adding elements in corresponding locations.

$$\begin{pmatrix} 1 & 4 \\ 2 & 5 \\ 3 & 6 \end{pmatrix} + \begin{pmatrix} 11 & 12 \\ 10 & 9 \\ 7 & 8 \end{pmatrix} = \begin{pmatrix} 12 & 16 \\ 12 & 14 \\ 10 & 14 \end{pmatrix}$$

Matrices with *different* dimensions cannot be added.

$$\begin{pmatrix} 1 & 4 \\ 2 & 5 \\ 3 & 6 \end{pmatrix} + \begin{pmatrix} 7 & 8 & 9 \\ 10 & 11 & 12 \end{pmatrix} \qquad \text{Is not possible}$$

Similarly, two matrices with the same dimensions can be *subtracted* by subtracting corresponding elements (or equivalently, adding the negative of the second matrix).

$$\begin{pmatrix} 11 & 12 \\ 10 & 9 \\ 7 & 8 \end{pmatrix} - \begin{pmatrix} 1 & 4 \\ 2 & 5 \\ 3 & 6 \end{pmatrix} = \begin{pmatrix} 10 & 8 \\ 8 & 4 \\ 4 & 2 \end{pmatrix}$$

Addition and Subtraction of Matrices

For two matrices A and B of the same dimensions, their sum $A + B$ and difference $A - B$ are found by adding or subtracting corresponding elements. Matrices of different dimensions *cannot* be added or subtracted.

A *zero matrix* has all elements equal to zero. We write 0 for the zero matrix of whatever size is appropriate for the situation. If two matrices are equal, their difference is a zero matrix: that is, if $A = B$, then $A - B = 0$ means that 0 is the zero matrix of the same dimension as A and B.

$$\begin{pmatrix} 1 & 2 \\ 3 & 4 \\ 5 & 6 \end{pmatrix} - \begin{pmatrix} 1 & 2 \\ 3 & 4 \\ 5 & 6 \end{pmatrix} = 0 \quad \text{means} \quad \begin{pmatrix} 1 & 2 \\ 3 & 4 \\ 5 & 6 \end{pmatrix} - \begin{pmatrix} 1 & 2 \\ 3 & 4 \\ 5 & 6 \end{pmatrix} = \begin{pmatrix} 0 & 0 \\ 0 & 0 \\ 0 & 0 \end{pmatrix}$$

Graphing Calculator Exploration

The basic matrix operations are found in the MATRX MATH menu. To have the calculator demonstrate the preceding concepts, proceed as follows:

```
MATRIX[A] 2 ×3
[1     2     3   ]
[4     5     6   ]




2,3=6
```

a. Use MATRX EDIT to enter $\begin{pmatrix} 1 & 2 & 3 \\ 4 & 5 & 6 \end{pmatrix}$ in [A].

```
[A]
          [[1 2 3]
           [4 5 6]]
[A]ᵀ
          [[1 4]
           [2 5]
           [3 6]]
```

b. Find the transpose of A.

```
3[A]
      [[3   6   9 ]
       [12 15 18]]

(1/2)[A]
      [[.5 1   1.5]
       [2  2.5 3  ]]
```

c. Find the scalar multiples $3A$ and $(1/2)A$. Check that attempting the division $[A]/2$ gives a "data type" error.

```
[A]+[A]
      [[2  4  6 ]
       [8 10 12]]

[A]-[A]
      [[0 0 0]
       [0 0 0]]
```

d. Find $A + A$ and $A - A$.

```
identity(4)
        [[1 0 0 0]
         [0 1 0 0]
         [0 0 1 0]
         [0 0 0 1]]
```

e. Find the 4×4 identity matrix.

EXAMPLE 1 **USING MATRIX ARITHMETIC**

A McBurger restaurant sells hamburgers for \$1, cheeseburgers for \$1.50, and fries for 75¢, while the BurgerQueen across the street sells hamburgers for \$1.25, cheeseburgers for \$1.75, and fries for 50¢. At McBurger the preparation costs are 30¢ per hamburger, 40¢ per cheeseburger, and 25¢ per order of fries, while at BurgerQueen the costs are 35¢ per hamburger, 45¢ per cheeseburger, and 20¢ per order of fries.

a. Represent the selling prices and preparation costs as matrices.

b. Using the matrices from (a), find the profit matrix for the items at the restaurants.

c. Using the matrix from (b), find the franchise fee charged on each item at each restaurant if this fee is 30% of the profit.

Solution

a. Let R be the revenue matrix of selling prices and let C be the cost matrix of preparation costs. We have 3 items from 2 restaurants, so we can choose the matrices to be either 2×3 or 3×2. We choose the matrices to be 2×3 with the rows corresponding to the different restaurants and the columns to the different menu items:

$$
\begin{array}{ccc}
\text{Hamburger} & \text{Cheeseburger} & \text{Fries} \\
\downarrow & \downarrow & \downarrow
\end{array}
$$

$$
\begin{array}{c}
\text{McBurger} \rightarrow \\
\text{BurgerQueen} \rightarrow
\end{array}
\left(
\begin{array}{ccc}
\underline{\quad} & \underline{\quad} & \underline{\quad} \\
\underline{\quad} & \underline{\quad} & \underline{\quad}
\end{array}
\right)
$$

From the given information, we obtain the revenue and cost matrices

$$
R = \begin{pmatrix} 1.00 & 1.50 & 0.75 \\ 1.25 & 1.75 & 0.50 \end{pmatrix} \quad \text{and} \quad C = \begin{pmatrix} 0.30 & 0.40 & 0.25 \\ 0.35 & 0.45 & 0.20 \end{pmatrix}
$$

b. Profit is revenue minus cost, so we find $P = R - C$ by matrix subtraction:

$$\overbrace{\begin{pmatrix} 1.00 & 1.50 & 0.75 \\ 1.25 & 1.75 & 0.50 \end{pmatrix}}^{R} - \overbrace{\begin{pmatrix} 0.30 & 0.40 & 0.25 \\ 0.35 & 0.45 & 0.20 \end{pmatrix}}^{C} = \overbrace{\begin{pmatrix} 0.70 & 1.10 & 0.50 \\ 0.90 & 1.30 & 0.30 \end{pmatrix}}^{P}$$

c. The franchise fee matrix is 30% of the profit matrix P, which we find by scalar multiplication:

$$0.30P = 0.30 \cdot \begin{pmatrix} 0.70 & 1.10 & 0.50 \\ 0.90 & 1.30 & 0.30 \end{pmatrix} = \begin{pmatrix} 0.21 & 0.33 & 0.15 \\ 0.27 & 0.39 & 0.09 \end{pmatrix}$$

Matrix Multiplication as Evaluation

To evaluate an expression like: $\qquad 5x_1 + 6x_2 + 7x_3$

when x_1, x_2, and x_3 are equal to: $\qquad 2, \qquad 3,$ and 4

we simply multiply and add: $\qquad 5 \cdot 2 + 6 \cdot 3 + 7 \cdot 4 = 56$

That is, evaluating linear expressions amounts to multiplying pairs of numbers and adding. With this kind of evaluation in mind, we define matrix multiplication as *multiplying in order the numbers from a row by the numbers from a column and adding the results*. For example, the matrix product of a row matrix by a column matrix is a 1×1 matrix:

$$(5 \quad 6 \quad 7) \cdot \begin{pmatrix} 2 \\ 3 \\ 4 \end{pmatrix} = (5 \cdot 2 + 6 \cdot 3 + 7 \cdot 4) = (56)$$

Of course, there must be exactly the same number of elements in the row as in the column. In general,

$$(a_1 \quad a_2 \quad \cdots \quad a_n) \cdot \begin{pmatrix} b_1 \\ b_2 \\ \vdots \\ b_n \end{pmatrix} = (a_1 b_1 + a_2 b_2 + \cdots + a_n b_n)$$

Practice Problem 1   Find $(1 \quad 2) \cdot \begin{pmatrix} 3 \\ 4 \end{pmatrix}$.

➤ Solution on page 207

If there are several rows on the left and several columns on the right, then we multiply *each row* times *each column*, with each answer placed in the product matrix at the row and column position from which it came. This means, for example, that when multiplying the *first row* of a matrix by the *second column* of another, place the result in the *first row* and *second column* of the answer. Note that we always take *rows* (from the left) times *columns* (on the right).

First row times second column goes here, in first row, second column

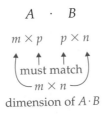

Go from here

Try to multiply each row times column "in your head"

directly to here

In general, to multiply two matrices, the row-length of the first must match the column-length of the second, with the other numbers giving the dimension of the product:

$$A \quad \cdot \quad B$$
$$m \times p \quad p \times n$$

must match

$m \times n$

dimension of $A \cdot B$

Outside numbers give dimension

Inside numbers get "absorbed"

Matrix Multiplication

If A is an $m \times p$ matrix and B is a $p \times n$ matrix, the matrix product $A \cdot B$ is the $m \times n$ matrix whose element in row i and column j is the product of row i from A times column j from B. If the row-length of A does not equal the column-length of B, then the product $A \cdot B$ is not defined.

Even when both products $A \cdot B$ and $B \cdot A$ are defined, the different orders may give very different results. For example:

$$(5 \quad 6 \quad 7)\begin{pmatrix} 2 \\ 3 \\ 4 \end{pmatrix} = (5 \cdot 2 + 6 \cdot 3 + 7 \cdot 4) = (56)$$

Multiplying in one order gives a 1×1 matrix

$$\begin{pmatrix} 2 \\ 3 \\ 4 \end{pmatrix}(5 \quad 6 \quad 7) = \begin{pmatrix} 2 \cdot 5 & 2 \cdot 6 & 2 \cdot 7 \\ 3 \cdot 5 & 3 \cdot 6 & 3 \cdot 7 \\ 4 \cdot 5 & 4 \cdot 6 & 4 \cdot 7 \end{pmatrix} = \begin{pmatrix} 10 & 12 & 14 \\ 15 & 18 & 21 \\ 20 & 24 & 28 \end{pmatrix}$$

Multiplying in the other order gives a 3×3 matrix

That is:

> Matrix multiplication is not commutative.
>
> In general: $A \cdot B \neq B \cdot A$

Spreadsheet Exploration

The spreadsheet* below shows the sum and product of the matrices
$$A = \begin{pmatrix} 1 & 2 & 3 \\ 4 & -2 & 4 \\ 3 & 2 & 1 \end{pmatrix} \quad \text{and} \quad B = \begin{pmatrix} 6 & 3 & 8 \\ 2 & 4 & 1 \\ 5 & 7 & 9 \end{pmatrix}.$$ The sum $A + B$ is
found by adding the range of cells **a3:c5** (containing the entries of matrix A) to the range of cells **e3:g5** (containing the entries of matrix B, while the product $A \cdot B$ uses the matrix multiplication function **MMULT** applied to the two ranges of cells representing the matrices.

	A	B	C	D	E	F	G
1		Matrix A				Matrix B	
2							
3	1	2	3		6	3	8
4	4	-2	4		2	4	1
5	3	2	1		5	7	9
6							
7		Sum A+B				Product A*B	
8							
9	7	5	11		25	32	37
10	6	2	5		40	32	66
11	8	9	10		27	24	35
12							

If you modify this spreadsheet to calculate $B + A$ and $B \cdot A$, which answer will change and which will remain the same?

* See the Preface for how to obtain this and other Excel spreadsheets.

EXAMPLE **2** **MATRIX MULTIPLICATION**

Use the selling price ("revenue") matrix from Example 1 (pages 200–201) and matrix multiplication to find the total selling price for a meal of three hamburgers, a cheeseburger, and two fries at each of the restaurants.

Solution

The revenue matrix R from Example 1 gives the selling prices in rows, one row for each restaurant. If we write the three hamburgers, one cheeseburger, and two fries as a *column*, we can use matrix multiplication to find the total price at each restaurant.

$$\begin{pmatrix} 1.00 & 1.50 & 0.75 \\ 1.25 & 1.75 & 0.50 \end{pmatrix} \begin{pmatrix} 3 \\ 1 \\ 2 \end{pmatrix} = \begin{pmatrix} 1.00\cdot 3 + 1.50\cdot 1 + 0.75\cdot 2 \\ 1.25\cdot 3 + 1.75\cdot 1 + 0.50\cdot 2 \end{pmatrix} = \begin{pmatrix} 6.00 \\ 6.50 \end{pmatrix}$$

$\uparrow$ $\uparrow$ $\uparrow$

R Order   Prices

(from Example 1)

The total selling prices are $6.00 at McBurger and $6.50 at Burger-Queen.

Practice Problem 2   Modify Example 2 to find the total selling price for a meal of four hamburgers, two cheeseburgers, and five fries at each of the restaurants.

➤ Solution on page 207

Identity Matrices

Multiplication by an identity matrix (of the proper size) does not change the matrix:

$$\begin{pmatrix} 1 & 0 & 0 \\ 0 & 1 & 0 \\ 0 & 0 & 1 \end{pmatrix} \begin{pmatrix} 2 & 7 \\ 3 & 6 \\ 4 & 5 \end{pmatrix} = \begin{pmatrix} 2 & 7 \\ 3 & 6 \\ 4 & 5 \end{pmatrix}$$

Same matrix on the right as on the left

The first row (1   0   0) picks out just the first value, the second (0   1   0) picks out the second value, and so on. (You should carefully verify this multiplication to see how the identity works.) To multiply by the identity matrix on the other side, we need to use an identity matrix of a different size:

$$\begin{pmatrix} 2 & 7 \\ 3 & 6 \\ 4 & 5 \end{pmatrix} \begin{pmatrix} 1 & 0 \\ 0 & 1 \end{pmatrix} = \begin{pmatrix} 2 & 7 \\ 3 & 6 \\ 4 & 5 \end{pmatrix}$$

Again, the matrix is duplicated

Writing I for an identity matrix of the appropriate size, we have for any matrix A:

Multiplication by I

$$I \cdot A = A \cdot I = A$$

That is, the matrix I plays the role in matrix arithmetic that the number 1 plays in ordinary arithmetic—multiplying by it gives back exactly what you started with.

Matrix Multiplication and Systems of Equations

Because we defined matrix multiplication as evaluation of linear expressions (page 201), we may use it to write an entire system of equations as a single matrix equation. Consider the system of equations

$$\begin{cases} 2x_1 + 3x_2 + 3x_3 = 15 \\ 3x_1 + 4x_2 + 4x_3 = 22 \\ 3x_1 + 4x_2 + 5x_3 = 30 \end{cases}$$

The "coefficient" matrix A and the "constant term" matrix B are

$$A = \begin{pmatrix} 2 & 3 & 3 \\ 3 & 4 & 4 \\ 3 & 4 & 5 \end{pmatrix} \quad \text{and} \quad B = \begin{pmatrix} 15 \\ 22 \\ 30 \end{pmatrix}$$

Let X be the column matrix of the variables:

$$X = \begin{pmatrix} x_1 \\ x_2 \\ x_3 \end{pmatrix}$$

Then the matrix product AX is precisely the left sides of the original equations (as you should check):

$$\underbrace{\begin{pmatrix} 2 & 3 & 3 \\ 3 & 4 & 4 \\ 3 & 4 & 5 \end{pmatrix}}_{A} \underbrace{\begin{pmatrix} x_1 \\ x_2 \\ x_3 \end{pmatrix}}_{X} = \begin{pmatrix} 2x_1 + 3x_2 + 3x_3 \\ 3x_1 + 4x_2 + 4x_3 \\ 3x_1 + 4x_2 + 5x_3 \end{pmatrix}$$

The system of equations is the same as the matrix equation $AX = B$:

$$\begin{cases} 2x_1 + 3x_2 + 3x_3 = 15 \\ 3x_1 + 4x_2 + 4x_3 = 22 \\ 3x_1 + 4x_2 + 5x_3 = 30 \end{cases} \quad \text{is the same as} \quad \underbrace{\begin{pmatrix} 2 & 3 & 3 \\ 3 & 4 & 4 \\ 3 & 4 & 5 \end{pmatrix}}_{A} \underbrace{\begin{pmatrix} x_1 \\ x_2 \\ x_3 \end{pmatrix}}_{X} = \underbrace{\begin{pmatrix} 15 \\ 22 \\ 30 \end{pmatrix}}_{B}$$

We can therefore represent the entire system of equations as just *one matrix equation* $AX = B$, as on the bottom of the previous page. This, at the very least, saves writing out the variables in each equation, a saving that is even greater for larger systems.

Matrix Multiplication and Row Operations

Because matrix multiplication of $I \cdot A$ picks out the elements of A in the correct order, switching the rows of I before multiplying will pick out the elements of A in a different order. For example:

$$\text{\textit{I} with rows 1 and 2 switched} \longrightarrow \begin{pmatrix} 0 & 1 & 0 \\ 1 & 0 & 0 \\ 0 & 0 & 1 \end{pmatrix} \begin{pmatrix} 2 & 7 \\ 3 & 6 \\ 4 & 5 \end{pmatrix} = \begin{pmatrix} 3 & 6 \\ 2 & 7 \\ 4 & 5 \end{pmatrix} \longleftarrow \text{\textit{A} with rows 1 and 2 switched}$$

Replacing one of the 1s in I with a different value before multiplying will pick out the corresponding element of A that many times:

$$\text{\textit{I} with row 3 multiplied by 8} \longrightarrow \begin{pmatrix} 1 & 0 & 0 \\ 0 & 1 & 0 \\ 0 & 0 & 8 \end{pmatrix} \begin{pmatrix} 2 & 7 \\ 3 & 6 \\ 4 & 5 \end{pmatrix} = \begin{pmatrix} 2 & 7 \\ 3 & 6 \\ 32 & 40 \end{pmatrix} \longleftarrow \text{\textit{A} with row 3 multiplied by 8}$$

Replacing one of the 0s in I with another number before multiplying will add that multiple of the elements of another row to the current row of A:

$$\text{\textit{I} with twice row 3 added to row 1} \longrightarrow \begin{pmatrix} 1 & 0 & 2 \\ 0 & 1 & 0 \\ 0 & 0 & 1 \end{pmatrix} \begin{pmatrix} 2 & 7 \\ 3 & 6 \\ 4 & 5 \end{pmatrix} = \begin{pmatrix} 10 & 17 \\ 3 & 6 \\ 4 & 5 \end{pmatrix} \longleftarrow \text{\textit{A} with twice row 3 added to row 1}$$

Thus the row operations of switching two rows, multiplying a row by a constant, and adding a multiple of one row to another can be accomplished by matrix multiplications. Furthermore, to find the matrix that performs a given row operation, we need only apply the same row operation to the identity matrix (of the appropriate size).

> Any row operation can be accomplished by a matrix multiplication.

Furthermore, any *sequence* of row operations can be carried out by multiplying by a single matrix. This is because multiplying by several matrices on the left is equivalent to multiplying just once by the *product* of all of the matrices, which can also be found by applying the same sequence of row operations to the identity matrix. Therefore, if a matrix A is equivalent to the identity matrix, then there is a matrix R (corresponding to the row operations performed on the identity matrix) such that $R \cdot A = I$. This observation will be very important on page 212 of the next section.

3.4 Section Summary

Matrix arithmetic uses the following terms and operations.

Equal matrices	$A = B$	Same dimension and same values in same positions
Transpose of a matrix	A^t	Row and column positions reversed
Identity matrix	I	Square, 0s except for 1s on diagonal
Scalar multiple	kA	Multiply every element by k
Matrix addition	$A + B$	Add elements in same positions
Matrix subtraction	$A - B$	Subtract elements in same positions
Zero matrix	0	All elements are zeros
Matrix multiplication	$A \cdot B$	Rows on left times columns on right (row length of A = column length of B)

Any system of linear equations may be written as a matrix equation $AX = B$ where A is the coefficient matrix, X is the column matrix of variables, and B is the constant term matrix.

➤ **Solutions to Practice Problems**

1. $(1 \quad 2) \cdot \begin{pmatrix} 3 \\ 4 \end{pmatrix} = (1 \cdot 3 + 2 \cdot 4) = (11)$

2. $\begin{pmatrix} 1.00 & 1.50 & 0.75 \\ 1.25 & 1.75 & 0.50 \end{pmatrix} \begin{pmatrix} 4 \\ 2 \\ 5 \end{pmatrix} = \begin{pmatrix} 1.00 \cdot 4 + 1.50 \cdot 2 + 0.75 \cdot 5 \\ 1.25 \cdot 4 + 1.75 \cdot 2 + 0.50 \cdot 5 \end{pmatrix} = \begin{pmatrix} 10.75 \\ 11.00 \end{pmatrix}$

The prices are \$10.75 at McBurger and \$11 at BurgerQueen.

3.4 Exercises

Use the given matrices to find each expression.

$$A = \begin{pmatrix} 1 & 2 & 3 \\ 4 & 5 & 6 \\ 7 & 8 & 9 \end{pmatrix} \quad B = \begin{pmatrix} 9 & 8 & 7 \\ 6 & 5 & 4 \\ 3 & 2 & 1 \end{pmatrix}$$

$$C = \begin{pmatrix} 1 & 6 & 8 \\ 4 & 2 & 7 \\ 9 & 5 & 3 \end{pmatrix}$$

1. A^t

2. B^t

3. $3C$

4. $2A$

5. $-B$

6. $-C$

7. $A + C$

8. $B + C$

9. $C - (A + I)$

10. $(A + I) - B$

Find each matrix product.

11. $(1 \quad -1 \quad 1)\begin{pmatrix} 4 \\ 3 \\ 5 \end{pmatrix}$

12. $(1 \quad 1 \quad -1 \quad -1)\begin{pmatrix} 3 \\ 2 \\ 4 \\ 1 \end{pmatrix}$

13. $\begin{pmatrix} 1 \\ 2 \end{pmatrix} \cdot (3 \quad 4)$

14. $\begin{pmatrix} 2 \\ -1 \end{pmatrix} \cdot (4 \quad 0)$

15. $\begin{pmatrix} 1 & 2 & 1 \\ 2 & 1 & 2 \end{pmatrix}\begin{pmatrix} 2 \\ -3 \\ 2 \end{pmatrix}$

16. $\begin{pmatrix} 1 & 3 & 1 \\ 3 & 1 & 3 \end{pmatrix}\begin{pmatrix} 4 \\ -2 \\ 4 \end{pmatrix}$

17. $\begin{pmatrix} 1 & 3 & 1 \\ 2 & 1 & 2 \end{pmatrix}\begin{pmatrix} 1 & 2 \\ 1 & 1 \\ 2 & 1 \end{pmatrix}$

18. $\begin{pmatrix} 1 & 2 & 1 \\ 3 & 1 & 3 \end{pmatrix}\begin{pmatrix} 2 & 3 \\ 3 & 1 \\ 2 & -1 \end{pmatrix}$

19. $\begin{pmatrix} 2 & 3 \\ 3 & 1 \\ 2 & -1 \end{pmatrix}\begin{pmatrix} 1 & 2 & 1 \\ 3 & 1 & 3 \end{pmatrix}$

20. $\begin{pmatrix} 1 & 2 \\ 1 & 1 \\ 2 & 1 \end{pmatrix}\begin{pmatrix} 1 & 3 & 1 \\ 2 & 1 & 2 \end{pmatrix}$

Use the given matrices to find each expression.

If your instructor permits, you may carry out the calculations on a graphing calculator.

$A = \begin{pmatrix} 2 & -1 & 1 \\ 1 & 0 & 1 \end{pmatrix}$ $B = \begin{pmatrix} 1 & 2 & 1 \\ 3 & -2 & 0 \end{pmatrix}$

$C = \begin{pmatrix} 2 & -1 & 2 \\ 1 & 1 & 1 \\ 0 & 2 & 1 \end{pmatrix}$

21. $A \cdot C$

22. $B \cdot C$

23. $C \cdot B^t$

24. $C \cdot A^t$

25. $(A - B) \cdot C$

26. $(B - A) \cdot C$

27. $A^t \cdot B + C$

28. $B^t \cdot A - C$

29. $B \cdot (C + I)$

30. $A \cdot (C - I)$

Rewrite each system of linear equations as a matrix equation $AX = B$.

31. $\begin{cases} x_1 + 5x_2 + 4x_3 = 6 \\ x_1 + x_2 + x_3 = 4 \\ 2x_1 + 3x_2 + 3x_3 = 9 \end{cases}$

32. $\begin{cases} x_1 + x_2 + x_3 = 4 \\ x_1 + 2x_2 + x_3 = 3 \\ x_1 + 2x_2 + 2x_3 = 5 \end{cases}$

33. $\begin{cases} 4x_1 + 3x_2 - x_3 = 2 \\ 3x_1 + 3x_2 + 2x_3 = 9 \\ 2x_1 + x_2 - 3x_3 = -6 \end{cases}$

34. $\begin{cases} 2x_1 - x_2 + 2x_3 = 11 \\ -x_1 + x_2 - 3x_3 = -12 \\ 2x_1 - 2x_2 + 7x_3 = 27 \end{cases}$

35. $\begin{cases} 5x_1 + 2x_2 - 4x_3 + x_4 + 5x_5 = 7 \\ 3x_1 + x_2 - 3x_3 + x_4 + 3x_5 = 5 \end{cases}$

36. $\begin{cases} 2x_1 + x_2 + 5x_3 + 4x_4 + 5x_5 = 2 \\ x_1 + x_2 + 3x_3 + 3x_4 + 3x_5 = -1 \end{cases}$

Rewrite each matrix equation $AX = B$ as a system of linear equations.

37. $\begin{pmatrix} 5 & 9 & 9 \\ 4 & 7 & 6 \\ 3 & 5 & 3 \\ 4 & 7 & 5 \end{pmatrix}\begin{pmatrix} x_1 \\ x_2 \\ x_3 \end{pmatrix} = \begin{pmatrix} 11 \\ 9 \\ 8 \\ 10 \end{pmatrix}$

38. $\begin{pmatrix} 6 & 3 & 5 \\ 1 & 2 & 2 \\ 4 & 3 & 4 \\ 5 & 1 & 3 \end{pmatrix}\begin{pmatrix} x_1 \\ x_2 \\ x_3 \end{pmatrix} = \begin{pmatrix} 8 \\ 1 \\ 5 \\ 7 \end{pmatrix}$

39. $\begin{pmatrix} 5 & 4 & 7 & 6 \\ 2 & 2 & 3 & 3 \\ 4 & 3 & 5 & 5 \\ 3 & 2 & 3 & 3 \end{pmatrix}\begin{pmatrix} x_1 \\ x_2 \\ x_3 \\ x_4 \end{pmatrix} = \begin{pmatrix} 18 \\ 9 \\ 16 \\ 11 \end{pmatrix}$

40. $\begin{pmatrix} 3 & 4 & 2 & 4 \\ 1 & 2 & 1 & 1 \\ 4 & 5 & 2 & 5 \\ 6 & 6 & 1 & 6 \end{pmatrix}\begin{pmatrix} x_1 \\ x_2 \\ x_3 \\ x_4 \end{pmatrix} = \begin{pmatrix} 12 \\ 4 \\ 14 \\ 15 \end{pmatrix}$

APPLIED EXERCISES

Formulate each situation in matrix form. Be sure to indicate the meaning of your rows and columns. Find the requested quantities using the appropriate matrix arithmetic.

41. BUSINESS: Sales Commissions A salesman at a furniture store sells mattresses manufactured by SlumberKing, DreamOn, and RestEasy. The selling prices for the three models of SlumberKing mattresses are $300 for the economy, $350 for the best, and $500 for the deluxe; DreamOn mattresses are $350 for the economy, $400 for the best, and $550 for the deluxe; and RestEasy mattresses are $400 for the economy, $500 for the best, and $700 for the deluxe. Represent these selling prices as a price matrix. Use this matrix to find the salesperson's commission matrix for these mattresses from these manufacturers if the commission is 15% of the selling price.

42. BUSINESS: Sales Taxes The Toys4U stores in Rockland and Martinville sell "Little Tykes" baseballs, bats, gloves, and caps. At the Rockland store, the retail prices are $1 per baseball, $8 per bat, $7 per glove, and $10 per cap while at the Martinville store, the prices are $1 per baseball, $9 per bat, $8 per glove, and $11 per cap. Represent these retail prices as a price matrix. Use this matrix to find the sales tax matrix for these items at these stores if the state sales tax is 5% of the retail price.

43. BUSINESS: Automobile Sales A car dealer sells sedans, station wagons, vans, and pickup trucks at sales lots in Oakdale and Roanoke. The "dealer markup" is the difference between the sticker price and the dealer invoice price. The dealer invoice prices at both locations are the same: $15,000 per sedan, $19,000 per wagon, $23,000 per van, and $25,000 per pickup. The sticker prices at the Oakdale lot are $18,900 per sedan, $22,900 per wagon, $26,900 per van, and $29,900 per pickup, while at the Roanoke lot the sticker prices are $19,900 per sedan, $21,900 per wagon, $27,900 per van, and $28,900 per pickup. Represent these prices as a dealer invoice matrix and a sticker price matrix.

Use these matrices to find the dealer markup matrix for these vehicles at these sales lots.

44. BUSINESS: Fuel Prices An oil refinery in Louisiana produces gasoline, kerosene, and diesel fuel for sale at service stations in Tennessee, Alabama, and Florida. In Tennessee, the pump prices per gallon are $1.18 for gasoline, $0.87 for kerosene, and $1.09 for diesel fuel while the combined federal and state taxes per gallon are 52¢ for gasoline, 35¢ for kerosene, and 46¢ for diesel fuel. In Alabama, the pump prices per gallon are $1.15 for gasoline, $0.88 for kerosene, and $1.04 for diesel fuel, while the combined federal and state taxes per gallon are 48¢ for gasoline, 33¢ for kerosene, and 41¢ for diesel fuel. And in Florida, the pump prices per gallon are $1.27 for gasoline, $0.93 for kerosene, and $1.16 for diesel fuel while the combined federal and state taxes per gallon are 56¢ for gasoline, 41¢ for kerosene, and 48¢ for diesel fuel. Represent these prices and taxes as matrices. Use these matrices to find the pretax price matrix for these fuels in these states.

45. MANAGEMENT SCIENCE: Overseas Manufacturing A sports apparel company manufactures shorts, tee shirts, and caps in Costa Rica and Honduras for importation and sale in the United States. In Costa Rica, the labor costs per item are 75¢ per pair of shorts, 25¢ per tee shirt, and 45¢ per cap, while the costs of the necessary materials are $1.60 per pair of shorts, 95¢ per tee shirt, and $1.15 per cap. In Honduras, the labor costs per item are 80¢ per pair of shorts, 20¢ per tee shirt, and 55¢ per cap, while the costs of the necessary materials are $1.50 per pair of shorts, 80¢ per tee shirt, and $1.10 per cap. Represent these costs as a labor cost matrix and a materials cost matrix. Use these matrices to find the total cost matrix for these products in these countries.

46. SOCIAL SCIENCE: Retirement Income A study of retired Chicago municipal workers now living in the Ozark Plateau found that in Missouri the average monthly pension benefits were $2700 for former police officers, $2500 for former mass transit workers, and $2800 for former firefighters, while in Arkansas the averages were $2750 for former police officers, $2300 for former mass transit workers, and $2900 for former firefighters. The average monthly Social Security benefits received by the same individuals in Missouri were $1100 for former police officers, $800 for former mass transit workers, and $1300 for former firefighters, while in Arkansas the averages were $1000 for former police officers, $900 for former mass transit workers, and $1200 for former firefighters. Represent these average monthly incomes as a Chicago pension matrix and a Social Security matrix. Use these matrices to find the average monthly retirement income matrix for these groups of retired employees in these states.

47. **GENERAL: Picnic Supplies** A supermarket advertises a "summer picnic sale" with 2-liter bottles of soda for 89¢, a large bottle of pickles for $1.29, packages of hot dogs for $2.39, and large bags of chips for $1.69. The Culbert family wants 12 sodas, 2 bottles of pickles, 3 packages of hot dogs, and 4 bags of chips. Represent the sale prices as a row matrix and the Culbert's shopping list as a column matrix. Use these matrices to find the total cost of these items at these prices.

48. **PERSONAL FINANCE: Part-Time Jobs** A college student makes money during the semester shelving books in the library at $8.50 per hour, tutoring freshmen at $30 per hour, and pumping gas on weekends at $6.75 per hour. Last week this student shelved books for 10 hours, tutored for 3 hours, and pumped gas for 16 hours. Represent the hourly pay rates as a row matrix and the hours worked as a column matrix. Use these matrices to find the total amount this student earned last week.

49. **MANAGEMENT SCIENCE: Furniture Production** A furniture company manufactures pine tables, chairs, and desks at factories in Wytheville and Andersen. Each table requires 2 hours of cutting and milling, 1 hour of assembly, and 2 hours of finishing; each chair requires 1.5 hours of cutting and milling, 1 hour of assembly, and 0.5 hours of finishing; and each desk requires 3 hours of cutting and milling, 2 hours of assembly, and 3 hours of finishing. At the Wytheville factory, the per-hour labor costs are $9 for cutting and milling work, $14 for assembly work, and $13 for finishing work, while at the Andersen factory, the per-hour labor costs are $10 for cutting and milling work, $13 for assembly work, and $12 for finishing work. Represent the table, chair, and desk labor requirements as a time matrix and the factory per-hour labor costs as a cost matrix. Use these matrices to find the production costs of these pieces of furniture at these factories.

50. **MANAGEMENT SCIENCE: Municipal Management** The Hollins County business manager has received a request from the police department for 3 new cars, 5 new motorcycles, and 1 van as well as a request from the rescue squad for 1 new car, 2 new vans, and 4 new ambulances. The local Ford dealer's prices are $20,000 per car, $8000 per motorcycle, $27,000 per van, and $52,000 per ambulance, while the local GM dealer's prices are $19,000 per car, $6000 per motorcycle, $29,000 per van, and $58,000 per ambulance. Represent the police and rescue requests as a vehicle matrix and the Ford and GM bids as a price matrix. Use these matrices to find the costs of these requests at these dealers.

Explorations and Excursions

The following problem extends and augments the material presented in the text.

51. Show that the solution of $\begin{cases} ax + by = h \\ cx + dy = k \end{cases}$ is

$$\begin{pmatrix} x \\ y \end{pmatrix} = \frac{1}{ad - bc} \begin{pmatrix} hd - bk \\ ak - hc \end{pmatrix}$$ for any values of

$a, b, c, d, h,$ and k such that $ad - bc \neq 0$.

3.5 INVERSE MATRICES AND SYSTEMS OF LINEAR EQUATIONS

Introduction

We can write a system of linear equations as a single matrix equation $AX = B$. Just as we solved the simple equation $2x = 10$ by multiplying each side by $\frac{1}{2}$ (the inverse of 2) to obtain $1x = 5$, we will solve the matrix equation $AX = B$ by multiplying by an "inverse" matrix.

Inverse Matrices

The inverse of the number 2 is $\frac{1}{2}$ because $\frac{1}{2}$ times 2 (in either order) is 1. Similarly, the *inverse* of a matrix is defined as another matrix whose product with the first (in either order) gives I, the identity matrix. The inverse of the matrix A is denoted A^{-1} just as the inverse of 2 is $2^{-1} = \frac{1}{2}$.

Inverse Matrix

A square matrix A has an *inverse matrix*, denoted A^{-1}, if and only if

$$A \cdot A^{-1} = I \quad \text{and} \quad A^{-1} \cdot A = I$$

If the inverse A^{-1} exists, then A is *invertible*, and otherwise A is *singular*. The inverse A^{-1} will exist if and only if A is row-equivalent to the identity matrix I.

We read A^{-1} as "A inverse." The I in the equations stands for the identity matrix of the *same size* as A. Since the product in either order must give the identity matrix, if one matrix is the inverse of another, then the second is also the inverse of the first. Note that only square matrices can have inverses. Recall that a matrix being "row-equivalent to the identity matrix" means that it can be row-reduced to the identity by a sequence of row operations.

EXAMPLE 1

CHECKING INVERSE MATRICES

Verify that the inverse of $A = \begin{pmatrix} \frac{1}{2} & 3 \\ 1 & 5 \end{pmatrix}$ is $A^{-1} = \begin{pmatrix} -10 & 6 \\ 2 & -1 \end{pmatrix}$.

Solution

We must show that $A^{-1} \cdot A = I$ and $A \cdot A^{-1} = I$.

$$A^{-1} \cdot A = \begin{pmatrix} -10 & 6 \\ 2 & -1 \end{pmatrix} \begin{pmatrix} \frac{1}{2} & 3 \\ 1 & 5 \end{pmatrix} = \begin{pmatrix} 1 & 0 \\ 0 & 1 \end{pmatrix} = I$$

and

Multiplication in
either order gives I

$$A \cdot A^{-1} = \begin{pmatrix} \frac{1}{2} & 3 \\ 1 & 5 \end{pmatrix} \begin{pmatrix} -10 & 6 \\ 2 & -1 \end{pmatrix} = \begin{pmatrix} 1 & 0 \\ 0 & 1 \end{pmatrix} = I$$

as required. (You should check the arithmetic in both multiplications.)

To verify that two matrices are inverses of each other, it is enough to multiply them in either order to get I (see Exercise 55 on page 224).

How to Find Inverse Matrices

Not all matrices have inverses. We saw on pages 206–207 that if a matrix A is equivalent to the identity matrix I, then the row reduction can be accomplished by multiplying A by the product R of the row operation matrices: $RA = I$, making R the inverse of A. Thus the inverse of an invertible matrix A can be found by adjoining an identity matrix, $(A \mid I)$, and row reducing to obtain $(I \mid A^{-1})$, with the inverse matrix on the right. This means we can calculate A^{-1} as follows:

Calculating an Inverse Matrix

If the square matrix A is invertible, then row-reducing the augmented matrix $(A \mid I)$ gives $(I \mid A^{-1})$, so that A^{-1} appears to the right of the bar.

If row-reducing $(A \mid I)$ concludes with a matrix *other* than I on the left of the bar, then A is *singular* and does not have an inverse.

EXAMPLE **2**

CALCULATING AN INVERSE MATRIX

Find the inverse of $A = \begin{pmatrix} 1 & 0 & 2 \\ 1 & 1 & 1 \\ 1 & 1 & 2 \end{pmatrix}$.

Solution

A is 3×3, so we augment it by I_3 and row-reduce $(A \mid I)$.

$$\left(\begin{array}{ccc|ccc} 1 & 0 & 2 & 1 & 0 & 0 \\ 1 & 1 & 1 & 0 & 1 & 0 \\ 1 & 1 & 2 & 0 & 0 & 1 \end{array} \right) \qquad (A \mid I)$$

$$\underbrace{}_{A} \quad \underbrace{}_{I}$$

There are many different sequences of row operations to reduce this augmented matrix, and all reach the same conclusion. One way is as follows:

$$R_3 - R_2 \rightarrow \left(\begin{array}{ccc|ccc} 1 & 0 & 2 & 1 & 0 & 0 \\ 1 & 1 & 1 & 0 & 1 & 0 \\ 0 & 0 & 1 & 0 & -1 & 1 \end{array} \right)$$

$$\begin{array}{c} R_1 - 2R_3 \rightarrow \\ R_2 - R_3 \rightarrow \end{array} \left(\begin{array}{ccc|ccc} 1 & 0 & 0 & 1 & 2 & -2 \\ 1 & 1 & 0 & 0 & 2 & -1 \\ 0 & 0 & 1 & 0 & -1 & 1 \end{array} \right)$$

$$R_2 - R_1 \rightarrow \left(\begin{array}{ccc|ccc} 1 & 0 & 0 & 1 & 2 & -2 \\ 0 & 1 & 0 & -1 & 0 & 1 \\ 0 & 0 & 1 & 0 & -1 & 1 \end{array} \right) \quad \begin{array}{l} \text{Since } I \text{ is on} \\ \text{the left, } A^{-1} \\ \text{is on the right} \end{array}$$

$$\underbrace{}_{I} \quad \underbrace{}_{A^{-1}}$$

Therefore, $A^{-1} = \begin{pmatrix} 1 & 2 & -2 \\ -1 & 0 & 1 \\ 0 & -1 & 1 \end{pmatrix}$ is the inverse of $A = \begin{pmatrix} 1 & 0 & 2 \\ 1 & 1 & 1 \\ 1 & 1 & 2 \end{pmatrix}$.

Practice Problem 1 For the matrices A and A^{-1} in Example 2, verify that A^{-1} is the inverse of A by showing that $A^{-1}A = I$. ➤ Solution on page 220

Spreadsheet Exploration

The following spreadsheet* uses the matrix inverse function **MINVERSE** to calculate inverses of

$$\begin{pmatrix} 1 & 0 & 2 \\ 1 & 1 & 1 \\ 1 & 1 & 2 \end{pmatrix} \quad \text{and} \quad \begin{pmatrix} 1 & -6 & 4 & 6 \\ 2 & 5 & 4 & 0 \\ 2 & 2 & 3 & 2 \\ 1 & 4 & 2 & -1 \end{pmatrix}.$$

* See the Preface for how to obtain this and other Excel spreadsheets.

	A	B	C	D	E	F	G	H	I	
1		Matrix A					Inverse of A			
2										
3	1	0	2			1	2	-2		
4	1	1	1			-1	0	1		
5	1	1	2			0	-1	1		
6										
7	1	-6	4	6		-5	34	-8	-46	
8	2	5	4	0		2	-15	4	20	
9	2	2	3	2		-3.25E-17	2	-1	-2	
10	1	4	2	-1		3	-22	6	29	
11										

However, the first answer is correct and the second is not. Spreadsheet programs sometimes produce round-off errors (such as the small number -3.25×10^{-17} instead of zero). The correct inverse

of $\begin{pmatrix} 1 & -6 & 4 & 6 \\ 2 & 5 & 4 & 0 \\ 2 & 2 & 3 & 2 \\ 1 & 4 & 2 & -1 \end{pmatrix}$ is $\begin{pmatrix} -5 & 34 & -8 & -46 \\ 2 & -15 & 4 & 20 \\ 0 & 2 & -1 & -2 \\ 3 & -22 & 6 & 29 \end{pmatrix}$

as you can easily check.

Be careful! Because apparent round-off errors might also indicate that A is invertible when it is in fact singular, *always* verify that the spreadsheet (or computer or calculator) answer does indeed satisfy $A^{-1}A = I$.

Solving $AX = B$ Using A^{-1}

Just as the equation $2x = 10$ is solved simply by multiplying both sides by 2^{-1} or $\frac{1}{2}$, we can solve $AX = B$ by multiplying both sides (on the left) by A^{-1}.

$AX = B$	Original equation
$A^{-1} \cdot A \cdot X = A^{-1} \cdot B$	Left-multiplying by A^{-1}
$I \cdot X = A^{-1} \cdot B$	Since $A^{-1}A = I$
$X = A^{-1}B$	Since $IX = X$

Notice that because matrix multiplication is *not* commutative, we must multiply both sides of $AX = B$ on the *left* by A^{-1}. (And besides, the product BA^{-1} is not defined because B is a column matrix and A^{-1} is a square matrix.)

Solving $AX = B$ Using A^{-1}

If A is invertible, the solution of $AX = B$ is

$$X = A^{-1}B$$

If A is singular, then we must use the Gauss–Jordan method (from Section 3.3) and we will then find that there are either no solutions or infinitely many solutions.

EXAMPLE 3

SOLVING A SYSTEM USING AN INVERSE MATRIX

Use an inverse matrix to solve $\begin{cases} x_1 + 2x_3 = 22 \\ x_1 + x_2 + x_3 = 11. \\ x_1 + x_2 + 2x_3 = 20 \end{cases}$

Solution

Writing this system as $AX = B,$ we have

$$\underbrace{\begin{pmatrix} 1 & 0 & 2 \\ 1 & 1 & 1 \\ 1 & 1 & 2 \end{pmatrix}}_{A} \underbrace{\begin{pmatrix} x_1 \\ x_2 \\ x_3 \end{pmatrix}}_{X} = \underbrace{\begin{pmatrix} 22 \\ 11 \\ 20 \end{pmatrix}}_{B} \qquad AX = B$$

Ordinarily, we would now find the inverse of matrix A by row-reducing $(A \mid I)$. However, this is exactly the matrix that we just row-reduced in Example 2, so we will simply use the A^{-1} found on page 213 and write the solution as $X = A^{-1}B$.

$$\underbrace{\begin{pmatrix} x_1 \\ x_2 \\ x_3 \end{pmatrix}}_{X} = \underbrace{\begin{pmatrix} 1 & 2 & -2 \\ -1 & 0 & 1 \\ 0 & -1 & 1 \end{pmatrix}}_{A^{-1}} \underbrace{\begin{pmatrix} 22 \\ 11 \\ 20 \end{pmatrix}}_{B} = \begin{pmatrix} 4 \\ -2 \\ 9 \end{pmatrix} \qquad \text{Using } A^{-1} \text{ from Example 2}$$

The solution is $x_1 = 4,\ x_2 = -2,\ x_3 = 9$.

 We may check this solution by substituting into the original equations.

$$4 + 2 \cdot 9 = 22$$
$$4 - 2 + 9 = 11$$
$$4 - 2 + 2 \cdot 9 = 20 \qquad \text{It checks!}$$

Graphing Calculator Exploration

If A is invertible, you can solve $AX = B$ by entering the matrices A and B and then calculating $A^{-1}B$. To solve the problem from Example 3, enter $\begin{pmatrix} 1 & 0 & 2 \\ 1 & 1 & 1 \\ 1 & 1 & 2 \end{pmatrix}$ in [A], enter $\begin{pmatrix} 22 \\ 11 \\ 20 \end{pmatrix}$ in [B], and calculate $[A]^{-1}[B]$:

```
[A]
       [[1  0  2]
        [1  1  1]
        [1  1  2]]
```

```
[B]
       [[22]
        [11]
        [20]]
```

```
[A]⁻¹[B]
       [[4 ]
        [-2]
        [9 ]]
```

Now choose three other integers, enter them in [B], and find $[A]^{-1}[B]$. Do these values satisfy your new equations?

Practice Problem 2   Solve $\begin{cases} x_1 + 2x_3 = -5 \\ x_1 + x_2 + x_3 = 10. \\ x_1 + x_2 + 2x_3 = 0 \end{cases}$

➤ Solution on page 220

[*Hint:* Write this system as $AX = B$ and notice that A is the same as in Example 3 but B is different, and so the solution can be found by matrix multiplication using the inverse found in Example 2.]

Solving $AX = B$ by finding the inverse and then using $X = A^{-1}B$ is only slightly more difficult than row-reducing the augmented matrix for the system of equations. However, if you need to solve $AX = B$ with the same A but with several different matrices B, using the inverse A^{-1} means that you only do the row reduction *once* and then the solution for any new matrix B is given by a simple matrix multiplication. Such problems, where A remains the same but B changes, occur frequently in applications.

Solving $AX = B$ for Many Different B's

If A is invertible, the solutions of the matrix equations

$$AX = B_1, \quad AX = B_2, \quad \dots, \quad AX = B_n$$

may all be found by calculating A^{-1} once and then finding the solutions as the products

$$X = A^{-1}B_1, \quad X = A^{-1}B_2 \quad , \ldots, \quad X = A^{-1}B_n$$

EXAMPLE **4**   **JEWELRY PRODUCTION**

An employee-owned jewelry company fabricates enameled gold rings, pendants, and bracelets. Each ring requires 3 grams of gold, 1 gram of enameling compound, and 2 hours of labor; each pendant requires 6 grams of gold, 2 grams of enameling compound, and 3 hours of labor; and each bracelet requires 8 grams of gold, 3 grams of enameling compound, and 2 hours of labor. Each of the five employee–owners works 160 hours each month, and the company has contracts guaranteeing the delivery of the grams of gold and enameling compound shown in the table on the first day of the months shown. How many rings, pendants, and bracelets should the company fabricate each month to use all the available materials and time?

	March	April	May	June
Gold	1720	2620	2460	2220
Enamel	600	960	900	800

Solution

Summarizing

	Rings	Pendants	Bracelets
Gold	3	6	8
Enamel	1	2	3
Labor	2	3	2

Let $x_1, x_2,$ and x_3 be the numbers of rings, pendants, and bracelets produced in one month. Then the required grams of gold, grams of enameling compound, and hours of labor are

Gold	$3x_1 + 6x_2 + 8x_3$
Enamel	$x_1 + 2x_2 + 3x_3$
Labor	$2x_1 + 3x_2 + 2x_3$

For the month of March, these quantities must match the amounts available:

$$\begin{cases} 3x_1 + 6x_2 + 8x_3 = 1720 \\ x_1 + 2x_2 + 3x_3 = 600 \\ 2x_1 + 3x_2 + 2x_3 = 800 \end{cases}$$

From the table and 5 workers at 160 hours each

The systems of equations for the other months follow in a similar manner, and we have four problems to solve:

March	April
$\begin{cases} 3x_1 + 6x_2 + 8x_3 = 1720 \\ x_1 + 2x_2 + 3x_3 = 600 \\ 2x_1 + 3x_2 + 2x_3 = 800 \end{cases}$	$\begin{cases} 3x_1 + 6x_2 + 8x_3 = 2620 \\ x_1 + 2x_2 + 3x_3 = 960 \\ 2x_1 + 3x_2 + 2x_3 = 800 \end{cases}$
May	**June**
$\begin{cases} 3x_1 + 6x_2 + 8x_3 = 2460 \\ x_1 + 2x_2 + 3x_3 = 900 \\ 2x_1 + 3x_2 + 2x_3 = 800 \end{cases}$	$\begin{cases} 3x_1 + 6x_2 + 8x_3 = 2220 \\ x_1 + 2x_2 + 3x_3 = 800 \\ 2x_1 + 3x_2 + 2x_3 = 800 \end{cases}$

The coefficient matrices for these four problems are the same, so if we find the inverse matrix for the March problem by row-reducing $(A \mid I)$, we will only have to carry out the row operations to reduce A once, and then the four problems can be solved with just four matrix multiplications.

We first find A^{-1} by row-reducing $(A \mid I)$. There are many different sequences of row operations to reduce this augmented matrix, and all reach the same conclusion. One possible way is as follows:

$$\left(\begin{array}{ccc|ccc} 3 & 6 & 8 & 1 & 0 & 0 \\ 1 & 2 & 3 & 0 & 1 & 0 \\ 2 & 3 & 2 & 0 & 0 & 1 \end{array}\right) \quad (A \mid I)$$

$$\begin{array}{c} 3R_2 - R_1 \rightarrow \\ \\ 2R_2 - R_3 \rightarrow \end{array} \left(\begin{array}{ccc|ccc} 0 & 0 & 1 & -1 & 3 & 0 \\ 1 & 2 & 3 & 0 & 1 & 0 \\ 0 & 1 & 4 & 0 & 2 & -1 \end{array}\right)$$

Using 1 in the first column to zero-out the rest of that column

$$\begin{array}{c} \\ R_2 - 2R_3 \rightarrow \\ \\ \end{array} \left(\begin{array}{ccc|ccc} 0 & 0 & 1 & -1 & 3 & 0 \\ 1 & 0 & -5 & 0 & -3 & 2 \\ 0 & 1 & 4 & 0 & 2 & -1 \end{array}\right)$$

Using 1 in the second column to zero-out the rest of that column

$$\begin{array}{c} \\ R_2 + 5R_1 \rightarrow \\ R_3 - 4R_1 \rightarrow \end{array} \left(\begin{array}{ccc|ccc} 0 & 0 & 1 & -1 & 3 & 0 \\ 1 & 0 & 0 & -5 & 12 & 2 \\ 0 & 1 & 0 & 4 & -10 & -1 \end{array}\right)$$

Using 1 in third column to zero-out the rest of that column

$$\begin{array}{c} R_2 \rightarrow \\ R_3 \rightarrow \\ R_1 \rightarrow \end{array} \left(\begin{array}{ccc|ccc} 1 & 0 & 0 & -5 & 12 & 2 \\ 0 & 1 & 0 & 4 & -10 & -1 \\ 0 & 0 & 1 & -1 & 3 & 0 \end{array}\right)$$

Switching rows to achieve the correct order

Since the left side is I, the right side is the inverse matrix

$$A^{-1} = \left(\begin{array}{ccc} -5 & 12 & 2 \\ 4 & -10 & -1 \\ -1 & 3 & 0 \end{array}\right)$$

The solution for each month is simply the product of this inverse matrix with the column matrix of the constant terms for that month.

Month	$X = A^{-1}B$	*Solution*
March	$\begin{pmatrix} x_1 \\ x_2 \\ x_3 \end{pmatrix} = \begin{pmatrix} -5 & 12 & 2 \\ 4 & -10 & -1 \\ -1 & 3 & 0 \end{pmatrix} \begin{pmatrix} 1720 \\ 600 \\ 800 \end{pmatrix} = \begin{pmatrix} 200 \\ 80 \\ 80 \end{pmatrix}$	200 rings 80 pendants 80 bracelets
April	$\begin{pmatrix} x_1 \\ x_2 \\ x_3 \end{pmatrix} = \begin{pmatrix} -5 & 12 & 2 \\ 4 & -10 & -1 \\ -1 & 3 & 0 \end{pmatrix} \begin{pmatrix} 2620 \\ 960 \\ 800 \end{pmatrix} = \begin{pmatrix} 20 \\ 80 \\ 260 \end{pmatrix}$	20 rings 80 pendants 260 bracelets
May	$\begin{pmatrix} x_1 \\ x_2 \\ x_3 \end{pmatrix} = \begin{pmatrix} -5 & 12 & 2 \\ 4 & -10 & -1 \\ -1 & 3 & 0 \end{pmatrix} \begin{pmatrix} 2460 \\ 900 \\ 800 \end{pmatrix} = \begin{pmatrix} 100 \\ 40 \\ 240 \end{pmatrix}$	100 rings 40 pendants 240 bracelets
June	$\begin{pmatrix} x_1 \\ x_2 \\ x_3 \end{pmatrix} = \begin{pmatrix} -5 & 12 & 2 \\ 4 & -10 & -1 \\ -1 & 3 & 0 \end{pmatrix} \begin{pmatrix} 2220 \\ 800 \\ 800 \end{pmatrix} = \begin{pmatrix} 100 \\ 80 \\ 180 \end{pmatrix}$	100 rings 80 pendants 180 bracelets

3.5 Section Summary

A square matrix A is *invertible* if there is an *inverse matrix* A^{-1} such that $A^{-1}A = I$ and $AA^{-1} = I$. A square matrix is *singular* if it is not invertible.

The inverse of an invertible matrix A may be found by row-reducing the augmented matrix $(A \mid I)$ to obtain $(I \mid A^{-1})$.

The solution of the matrix equation $AX = B$ is $X = A^{-1}B$, provided the matrix A is invertible.

The solutions to a succession of problems

$$AX = B_1, \quad AX = B_2, \quad \ldots, \quad AX = B_n$$

may be found from A^{-1} as

$$X = A^{-1}B_1, \quad X = A^{-1}B_2, \quad \ldots, \quad X = A^{-1}B_n$$

➤ **Solutions to Practice Problems**

1. $A^{-1}A = \begin{pmatrix} 1 & 2 & -2 \\ -1 & 0 & 1 \\ 0 & -1 & 1 \end{pmatrix} \begin{pmatrix} 1 & 0 & 2 \\ 1 & 1 & 1 \\ 1 & 1 & 2 \end{pmatrix}$

$= \begin{pmatrix} 1+2-2 & 0+2-2 & 2+2-4 \\ -1+0+1 & 0+0+1 & -2+0+2 \\ 0-1+1 & 0-1+1 & 0-1+2 \end{pmatrix} = \begin{pmatrix} 1 & 0 & 0 \\ 0 & 1 & 0 \\ 0 & 0 & 1 \end{pmatrix} = I$

2. Using $X = A^{-1}B$:

$\begin{pmatrix} x_1 \\ x_2 \\ x_3 \end{pmatrix} = \begin{pmatrix} 1 & 2 & -2 \\ -1 & 0 & 1 \\ 0 & -1 & 1 \end{pmatrix} \begin{pmatrix} -5 \\ 10 \\ 0 \end{pmatrix} = \begin{pmatrix} 15 \\ 5 \\ -10 \end{pmatrix}$ so $\begin{cases} x_1 = 15 \\ x_2 = 5 \\ x_3 = -10 \end{cases}$

3.5 Exercises

Find each matrix product. Identify each pair of matrices as "a matrix and its inverse" or "not a matrix and its inverse."

1. $\begin{pmatrix} 1 & 2 \\ -1 & -1 \end{pmatrix}$ and $\begin{pmatrix} -1 & -2 \\ 1 & 1 \end{pmatrix}$

2. $\begin{pmatrix} 5 & 3 \\ -3 & -2 \end{pmatrix}$ and $\begin{pmatrix} 2 & 3 \\ -3 & -5 \end{pmatrix}$

3. $\begin{pmatrix} 1 & 1 & 0 \\ 2 & 1 & 1 \\ 1 & 0 & 0 \end{pmatrix}$ and $\begin{pmatrix} 0 & 0 & 1 \\ 1 & 0 & -1 \\ -1 & 1 & -1 \end{pmatrix}$

4. $\begin{pmatrix} 2 & 1 & -1 \\ -2 & 0 & 1 \\ -3 & -1 & 2 \end{pmatrix}$ and $\begin{pmatrix} 1 & -1 & 1 \\ 1 & 1 & 0 \\ 2 & -1 & 2 \end{pmatrix}$

5. $\begin{pmatrix} 4 & 6 & 3 \\ 3 & 4 & 1 \\ 5 & 7 & 3 \end{pmatrix}$ and $\begin{pmatrix} -5 & -3 & 6 \\ 4 & 3 & -5 \\ -1 & -2 & 2 \end{pmatrix}$

6. $\begin{pmatrix} 3 & 2 & 3 \\ 5 & 2 & 6 \\ 2 & 3 & 1 \end{pmatrix}$ and $\begin{pmatrix} 16 & -7 & -6 \\ -7 & 3 & 3 \\ -11 & 5 & 4 \end{pmatrix}$

7. $\begin{pmatrix} 10 & -4 & -7 \\ -7 & 3 & 5 \\ 4 & -1 & -3 \end{pmatrix}$ and $\begin{pmatrix} 4 & 5 & -1 \\ 1 & 2 & 1 \\ 5 & 6 & 2 \end{pmatrix}$

8. $\begin{pmatrix} 4 & 1 & 5 \\ -1 & 1 & -2 \\ 3 & 4 & 2 \end{pmatrix}$ and $\begin{pmatrix} 10 & 18 & -7 \\ -4 & -7 & -3 \\ -7 & -13 & 5 \end{pmatrix}$

For each matrix A, row-reduce $(A \mid I)$ to find the inverse matrix A^{-1} or to identify A as a singular matrix.

9. $\begin{pmatrix} 1 & 3 \\ 0 & 1 \end{pmatrix}$

10. $\begin{pmatrix} 1 & 4 \\ 0 & 1 \end{pmatrix}$

11. $\begin{pmatrix} 11 & 2 \\ 6 & 1 \end{pmatrix}$

12. $\begin{pmatrix} 5 & 7 \\ 2 & 3 \end{pmatrix}$

13. $\begin{pmatrix} 1 & 1 & 0 \\ 3 & 0 & 2 \\ 1 & 0 & 1 \end{pmatrix}$

14. $\begin{pmatrix} 1 & 0 & 1 \\ 1 & 1 & 0 \\ 5 & 0 & 4 \end{pmatrix}$

15. $\begin{pmatrix} 1 & 1 & 0 \\ 0 & 1 & 1 \\ 1 & 2 & 1 \end{pmatrix}$

16. $\begin{pmatrix} 1 & 0 & 1 \\ 2 & 1 & 3 \\ 0 & 1 & 1 \end{pmatrix}$

17. $\begin{pmatrix} 1 & 1 & 0 & 1 \\ 0 & 1 & 0 & 0 \\ 1 & 0 & 1 & 0 \\ 0 & 1 & 0 & 1 \end{pmatrix}$

18. $\begin{pmatrix} 1 & 0 & 0 & 1 \\ 0 & 1 & 1 & 1 \\ 2 & 0 & 0 & 1 \\ 0 & 1 & 0 & 1 \end{pmatrix}$

19. $\begin{pmatrix} 2 & 3 & 1 \\ 1 & 2 & 1 \\ 2 & 3 & 2 \end{pmatrix}$

20. $\begin{pmatrix} 1 & 2 & 2 \\ 1 & 1 & 1 \\ 1 & 3 & 2 \end{pmatrix}$

21. $\begin{pmatrix} 3 & -4 & 2 \\ -1 & 2 & -1 \\ 4 & -3 & 2 \end{pmatrix}$

22. $\begin{pmatrix} 2 & 2 & -1 \\ -2 & -1 & 1 \\ 3 & 4 & -1 \end{pmatrix}$

23. $\begin{pmatrix} 1 & 0 & -1 & 1 \\ 1 & 1 & 1 & 2 \\ 2 & -2 & -6 & 0 \\ 1 & 0 & 2 & 1 \end{pmatrix}$

24. $\begin{pmatrix} 1 & -1 & 0 & 1 \\ 2 & -2 & -1 & 0 \\ 1 & -1 & -1 & -1 \\ 2 & -2 & 0 & 2 \end{pmatrix}$

Rewrite each system of equations as a matrix equation $AX = B$ and use the inverse of A to find the solution. Be sure to check your solution in the original system of equations.

25. $\begin{cases} 11x_1 + 2x_2 = 9 \\ 6x_1 + x_2 = 5 \end{cases}$ 26. $\begin{cases} 5x_1 + 7x_2 = 2 \\ 2x_1 + 3x_2 = 1 \end{cases}$

27. $\begin{cases} x_1 + x_2 = 2 \\ 3x_1 + 2x_3 = 5 \\ x_1 + x_3 = 2 \end{cases}$ 28. $\begin{cases} x_1 + x_3 = 1 \\ x_1 + x_2 = 3 \\ 5x_1 + 4x_3 = 6 \end{cases}$

29. $\begin{cases} 2x_1 + 3x_2 + x_3 = 6 \\ x_1 + 2x_2 + x_3 = 4 \\ 2x_1 + 3x_2 + 2x_3 = 7 \end{cases}$

30. $\begin{cases} x_1 + 2x_2 + 2x_3 = 5 \\ x_1 + x_2 + x_3 = 3 \\ x_1 + 3x_2 + 2x_3 = 6 \end{cases}$

31. $\begin{cases} 3x_1 - 4x_2 + 2x_3 = 12 \\ -x_1 + 2x_2 - x_3 = -4 \\ 4x_1 - 3x_2 + 2x_3 = 15 \end{cases}$

32. $\begin{cases} 2x_1 + 2x_2 - x_3 = 1 \\ -2x_1 - x_2 + x_3 = 0 \\ 3x_1 + 4x_2 - x_3 = 2 \end{cases}$

33. $\begin{cases} x_1 - 2x_2 + x_3 + 2x_4 = 8 \\ x_1 - x_2 + x_3 + 2x_4 = 7 \\ 2x_1 + 2x_2 + x_3 + 3x_4 = 7 \\ x_1 - x_2 + x_3 + x_4 = 6 \end{cases}$

34. $\begin{cases} 2x_1 + x_2 - 2x_3 + x_4 = 2 \\ x_1 + x_2 + 2x_3 + x_4 = 5 \\ x_1 + x_2 + x_3 + x_4 = 4 \\ 2x_1 + 2x_2 + 3x_3 + x_4 = 8 \end{cases}$

35. $\begin{cases} x_1 - 2x_2 + x_3 + 2x_4 = 10 \\ x_1 - x_2 + x_3 + 2x_4 = 8 \\ 2x_1 + 2x_2 + x_3 + 3x_4 = 5 \\ x_1 - x_2 + x_3 + x_4 = 6 \end{cases}$

36. $\begin{cases} 2x_1 + x_2 - 2x_3 + x_4 = 8 \\ x_1 + x_2 + 2x_3 + x_4 = 3 \\ x_1 + x_2 + x_3 + x_4 = 4 \\ 2x_1 + 2x_2 + 3x_3 + x_4 = 5 \end{cases}$

37. $\begin{cases} x_1 + 2x_2 + x_3 = 11 \\ x_1 + 4x_2 + x_3 = 19 \\ 2x_1 + 2x_2 + x_3 = 13 \end{cases}$

38. $\begin{cases} 2x_1 + 3x_2 + x_3 = 4 \\ 3x_1 + 3x_2 + 2x_3 = 12 \\ x_1 + 2x_2 + x_3 = 4 \end{cases}$

39. $\begin{cases} 4x_1 + 6x_2 + 5x_3 + 9x_4 = 75 \\ 4x_1 + x_2 + 2x_3 + 4x_4 = 10 \\ x_1 + 4x_2 + 3x_3 + 6x_4 = 65 \\ 4x_1 + 3x_2 + 3x_3 + 4x_4 = 10 \end{cases}$

40. $\begin{cases} 3x_1 + 2x_2 + x_3 + 3x_4 = 10 \\ 4x_1 + 3x_2 + 3x_3 + 7x_4 = 15 \\ 6x_1 + 7x_2 + 3x_3 + 8x_4 = 50 \\ 5x_1 + 3x_2 + 3x_3 + 7x_4 = 10 \end{cases}$

APPLIED EXERCISES

Formulate each situation as a collection of systems of linear equations. Be sure to state clearly the meaning of each variable. Solve each collection using an inverse matrix. State your final answers in terms of the original questions.

If permitted by your instructor, you may use a graphing calculator to find the necessary inverse matrices and matrix products.

41. **BUSINESS: Movie Tickets** A five-screen multiplex cinema charges $10 for adults and $5 for children under twelve. The number of tickets sold for each of today's shows and the corresponding gross receipts are given in the table. How many tickets of each kind were sold for each film?

	Film No. 1	Film No. 2	Film No. 3	Film No. 4	Film No. 5
Tickets sold	500	400	450	500	600
Gross receipts	$3250	$3000	$3500	$4500	$6000

42. GENERAL: Summer Day Care An inner-city antipoverty foundation staffs summer day care sites serving children aged 6 to 12 at Hollis Avenue, Beaverton Boulevard, Gramson Park, and Riverside Street. Certified instructors earn $350 per week and supervise 8 children, and college student group leaders earn $250 per week and supervise 6 children. The number of children and the weekly payroll at each site are given in the table. How many instructors and group leaders work at each site?

	Children	Payroll
Hollis Avenue	92	$3900
Beaverton Boulevard	130	$5500
Gramson Park	124	$5300
Riverside Street	152	$6500

43. MANAGEMENT SCIENCE: Speculative Partnerships A broker offers limited partnerships in the amounts of $1000, $5000, and $10,000 in three organizations speculating in gold mines, oil wells, and modern art. The capitalization values and number of members in each are given in the table, together with the "altered value" of the capitalization should all of the $1000 memberships be replaced by the same number of $5000 memberships. How many memberships of each amount are in each partnership?

	Gold Mines	Oil Wells	Modern Art
Value	$2.0 million	$3.9 million	$2.5 million
Number of Members	500	700	600
"Altered Value"	$3.0 million	$4.5 million	$3.8 million

44. GENERAL: Plant Fertilizer The manager of a garden supply store has received the soil test results for the gardens of Mr. Smith, Mrs. Jones, Miss Roberts, and Mr. Wheeler. Using the size of each garden plot, the manager calculated the ounces of nutrients needed for each garden as listed in the first table. The store sells three brands of fertilizer, Great-Green, MiracleMix, and GrowRite, which contain the amounts of potash, nitrogen, and phosphoric acid per box listed in the second table. How many boxes of each fertilizer should be sold to each customer?

Ounces Needed	Mr. Smith	Mrs. Jones	Miss Roberts	Mr. Wheeler
Potash	26	31	25	25
Nitrogen	31	33	31	29
Phosphoric acid	19	20	19	18

Ounces per Box	GreatGreen	MiracleMix	GrowRite
Potash	4	5	2
Nitrogen	5	5	3
Phosphoric acid	3	3	2

45. BIOMEDICAL: Infant Nutrition A pediatric dietician at an inner-city foundling hospital needs to supplement each bottle of baby formula given to three infants in her ward with the units of vitamin A, vitamin D, calcium, and iron given in the first table. If four diet supplements are available with the nutrient content per drop given in the second table, how many drops of each supplement per bottle of formula should each infant receive?

Units Needed	Vitamin A	Vitamin D	Calcium	Iron
Billy	48	26	19	49
Susie	46	26	27	65
Jimmy	47	26	19	49

Units per Drop	Vitamin A	Vitamin D	Calcium	Iron
Supplement 1	5	3	0	1
Supplement 2	0	1	3	6
Supplement 3	4	2	3	7
Supplement 4	4	2	2	5

46. PERSONAL FINANCE: International Investments An international investment advisor recommends industrial stocks in Nigeria, Bolivia, Thailand, and Hungary, with four times as much in Thailand as in Bolivia and three times as much in Hungary as in Nigeria. Furthermore, he suggests that more than 20% but less than 25% of the investor's portfolio be Bolivian and Nigerian stocks. Acting on this advice, four investors commit the amounts given in the table. How much does each invest in each country?

	Total Investment	Bolivia and Nigeria
Mr. Croft	$100,000	$23,000
Mrs. Fredericks	$150,000	$33,000
Mr. Spencer	$90,000	$22,000
Ms. Winpeace	$135,000	$28,000

47. PERSONAL FINANCE: Financial Planning A retirement planning counselor recommends investing in a stock fund yielding 18%, a money market fund returning 6%, and a bond fund paying 8%, with twice as much in the bond and money market funds together as in the stock fund. Mr. and Mrs. Jordan have $300,000 to invest and need an annual return of $31,000; Mr. and Mrs. French have $234,900 to invest and need an annual return of $25,600; and Mr. and Mrs. Daimen have $270,000 to invest and need an annual return of $28,500. How much should each elderly couple place in each investment to receive their desired income?

48. BUSINESS: Year-End Bonuses At the end of each year, the owner of a small company gives himself, his salesman, and his secretary a bonus by dividing up whatever remains in the "office supplies" account. He takes one third for himself, gives one quarter to his salesman, and gives one fifth to his secretary, but each share is taken after the others have been given out. Any remaining money is then spent on the company's New Year's Eve dinner at the best restaurant in town. The ending balances in the office supplies account for four years are given in the table. How much did each person receive each year and how much was left over each year for the dinner party?

2000	2001	2002	2003
$2500	$2000	$3000	$2800

49. MANAGEMENT SCIENCE: Mass Transit The metropolitan area mass transit manager is revising the subway, bus, and jitney service to the suburbs of Brighton, Conway, Longwood, and Oakley. To meet federal clean air mandates, the mayor's office demands that twice as many electric subway cars as diesel buses and jitneys combined be used for each suburb. The transit workers' union contract requires that like the buses and jitneys, each subway car must have a driver/ticket taker whether or not that subway car is part of a longer train. The number of commuters from each suburb using mass transit and the number of transit workers assigned to each area are given in the table. If each subway car carries 70 commuters, each bus carries 60 commuters, and each jitney carries 10 commuters, how many of each vehicle must be assigned to each suburb?

	Commuters	Transit Workers
Brighton	11,500	180
Conway	9000	150
Longwood	9500	150
Oakley	10,250	165

50. GENERAL: Community Food Pantry A town-wide food drive to aid the interfaith ministries' food pantry for the needy was supported by the Boy Scouts, the Girl Scouts, and the Lions Club. Both dry food (in 10-ounce boxes) and canned food, in small (15-ounce) and large (40-ounce) sizes, were collected. The Boy Scouts collected 240 items weighing 355 pounds, of which 315 pounds were canned goods; the Girl Scouts collected 240 items weighing 320 pounds, of which 260 pounds were canned goods; and the Lions Club collected 416 items weighing 565 pounds, of which 465 pounds were canned goods. How many boxes, small cans, and large cans did each group collect?

Explorations and Excursions

The following problems extend and augment the material presented in the text.

More About 2 × 2 Matrices

51. Show that the inverse of a 2×2 matrix $\begin{pmatrix} a & b \\ c & d \end{pmatrix}$ is $\dfrac{1}{ad - bc}\begin{pmatrix} d & -b \\ -c & a \end{pmatrix}$ provided that the *determinant* $ad - bc$ is not zero.

52. Solve $\begin{cases} ax + by = h \\ cx + dy = k \end{cases}$ for any values of $a, b, c, d, h,$ and k such that $ad - bc \neq 0$.

    [*Hint:* Use the preceding exercise and the formula $X = A^{-1}B$.]

More About Inverses

53. Show that the inverse of a product is the product of the inverses in the reverse order; that is, show that $(A \cdot B)^{-1} = B^{-1} \cdot A^{-1}$ for any invertible matrices A and B of the same dimension.

54. Show that the $n \times n$ identity matrix is unique; that is, given that $J \cdot A = A$ for every A, establish that $J = I$. [*Hint:* Consider $A = I$.]

55. Show that for an invertible matrix A, if $B \cdot A = I$, then $A \cdot B = I$ and $B = A^{-1}$.

56. Show that the inverse of a transposed matrix is the transpose of the inverse matrix; that is, show that $(A^t)^{-1} = (A^{-1})^t$ for any invertible matrix A.

Matrix Inverses and Geometric Series

Exercises 57–60 develop material that will be used in Chapter 7 on Markov Chains.

57. Show that the geometric series formula on page 121 can be written as

$$1 + x + x^2 + \cdots + x^{n-1} = \frac{1 - x^n}{1 - x}$$

and that the corresponding matrix equation is then

$$I + A + A^2 + \cdots + A^{n-1} = (I - A^n) \cdot (I - A)^{-1}$$

where A is a square matrix such that $I - A$ is invertible.

A square matrix with nonnegative entries is a *substochastic* matrix of size w (where $0 \leq w < 1$) if the sum of the entries in each row is no more than w.

58. Show that for square matrices A and B of the same dimension, if A is substochastic of size u and B is substochastic of size v, then $A \cdot B$ is substochastic of size $u \cdot v$.

59. Show that if A is a substochastic matrix of size w, then A^n is a substochastic matrix of size w^n for any positive integer n.

60. Use Exercises 59 and 57 to show that if A is a substochastic matrix of size w, then the inverse matrix $(I - A)^{-1}$ is the same as the infinite geometric series $I + A + A^2 + \cdots$.

3.6 INTRODUCTION TO MODELING: LEONTIEF MODELS AND LEAST SQUARES

Introduction

We conclude this chapter with two applications of matrix methods: the Leontief input–output model of an economy and least squares estimation. Because of the computational nature of these topics, you will need a graphing calculator or a computer.

Leontief "Open" Input–Output Models

Input–output analysis was developed by Wassily Leontief* to study the flow of goods and services among different sectors of an economy. In a "closed" model all the goods produced are used by the producers, while in an "open" model the economy produces more than is needed by the producers, with the extra output available to consumers. In the global economy of today, each country may be considered an open system. Given the relationships among the different sectors of an economy, we can calculate the extra production from the economic activity within each sector. And, conversely, we can determine the level of economic activity necessary within each sector to achieve a desired level of excess production. The Leontief "open" input–output economic model performs these two computational tasks by expressing the relation between the activity levels of the sectors and the extra production as a matrix equation.

We begin by considering the simple economy shown in the following diagram, consisting of a blacksmith (B), who makes nails, plows, and other tools; a carpenter (C), who builds barns and other useful buildings; and a farmer (F), who grows food.

The connections between the sectors of the economy are the values of materials needed to produce one dollar of output from each.

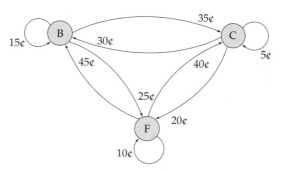

A diagram of a simple economy.

Each sector uses some of the output of the other sectors for its own production (for example, the carpenter uses nails from the blacksmith), and the arrows indicate the amount of output of one sector that is used by the other sector. To be precise, the number on an arrow from a sector indicates *the value of output from that sector needed to produce $1 of output of the target sector*. For example,

* Russian-born American economist (1906–1999), winner of the 1973 Nobel Memorial Prize for Economics.

means that 35¢ of the blacksmith's output is needed to produce $1 of the carpenter's output. Since each sector uses as input some of its own output (for example, the carpenter builds his own workshop), there are circular arrows from a sector to itself. These values of input materials from each sector needed to produce $1 of output from each sector are the connections between the sectors of the economy, and are the basic data needed to construct the Leontief model of the economy.

Let x_1, x_2, and x_3 be the values of the goods and services produced by the blacksmith, the carpenter, and the farmer, and let y_1, y_2, and y_3 be the values of the *excess* production from each that can be sold outside this economic system. The value produced by the blacksmith is the same as the values of the blacksmith's products used by the blacksmith, the carpenter, and the farmer, together with the amount sold outside this economy. Using the numbers from the diagram, we have

$$x_1 = \underbrace{0.15x_1}_{\substack{\text{Value of blacksmith's} \\ \text{products used} \\ \text{by the blacksmith}}} + \underbrace{0.35x_2}_{\substack{\text{Value of blacksmith's} \\ \text{products used} \\ \text{by the carpenter}}} + \underbrace{0.25x_3}_{\substack{\text{Value of blacksmith's} \\ \text{products used} \\ \text{by the farmer}}} + \underbrace{y_1}_{\substack{\text{Value of blacksmith's} \\ \text{products sold} \\ \text{to the "outside"}}}$$

Similarly, for the carpenter and the blacksmith,

$$x_2 = \underbrace{0.30x_1}_{\substack{\text{Value of carpenter's} \\ \text{products used} \\ \text{by the blacksmith}}} + \underbrace{0.05x_2}_{\substack{\text{Value of carpenter's} \\ \text{products used} \\ \text{by the carpenter}}} + \underbrace{0.20x_3}_{\substack{\text{Value of carpenter's} \\ \text{products used} \\ \text{by the farmer}}} + \underbrace{y_2}_{\substack{\text{Value of carpenter's} \\ \text{products sold} \\ \text{to the "outside"}}}$$

$$x_3 = \underbrace{0.45x_1}_{\substack{\text{Value of farmer's} \\ \text{products used} \\ \text{by the blacksmith}}} + \underbrace{0.40x_2}_{\substack{\text{Value of farmer's} \\ \text{products used} \\ \text{by the carpenter}}} + \underbrace{0.10x_3}_{\substack{\text{Value of farmer's} \\ \text{products used} \\ \text{by the farmer}}} + \underbrace{y_3}_{\substack{\text{Value of farmer's} \\ \text{products sold} \\ \text{to the "outside"}}}$$

In matrix form, this system of equations becomes

$$\begin{pmatrix} x_1 \\ x_2 \\ x_3 \end{pmatrix} = \begin{pmatrix} 0.15 & 0.35 & 0.25 \\ 0.30 & 0.05 & 0.20 \\ 0.45 & 0.40 & 0.10 \end{pmatrix} \begin{pmatrix} x_1 \\ x_2 \\ x_3 \end{pmatrix} + \begin{pmatrix} y_1 \\ y_2 \\ y_3 \end{pmatrix}$$

The numerical matrix in this equation was called the "interindustry matrix of technical coefficients" by Leontief and is now usually referred to as the "technology matrix" of the economy. The columns of this matrix are the input values for each sector as shown in the economy diagram. For example, the first column lists the amounts of the various outputs that are consumed by the blacksmith. In general:

Leontief "Open" Input–Output Model

A Leontief "open" input–output model is a matrix equation

$$X = AX + Y$$

where the column matrix X lists the values produced by each of the n economic sectors, the column matrix Y lists the values of the excess production of these same sectors, and the element in row i and column j of the technology matrix A represents the value of sector i goods and services needed to produce one dollar of sector j output.

Solving this matrix equation for the excess productions Y in terms of the sector production values X, we have

$$X - AX = Y$$

so that

$$Y = (I - A)X.$$

Alternatively, solving for the sector production values X required for given excess productions Y, we have

$$X = (I - A)^{-1}Y.$$

That is:

Excess and Sector Productions

In a Leontief "open" input–output model economy, the excess productions Y from given sector production values X is

$$Y = (I - A)X \qquad \text{Excess production}$$

The sector production values X necessary to provide required excess productions Y is

$$X = (I - A)^{-1}Y \qquad \text{Sector production}$$

Returning to our example economy, if a government survey of economic activity establishes that both the blacksmith and the farmer each produce \$200 of value each year and the carpenter produces \$160, so that X is

$$\begin{pmatrix} x_1 \\ x_2 \\ x_3 \end{pmatrix} = \begin{pmatrix} 200 \\ 160 \\ 200 \end{pmatrix} \qquad \begin{matrix} \text{Blacksmith} \\ \text{Carpenter} \\ \text{Farmer} \end{matrix}$$

Then the amount of extra production is given by $Y = (I - A)X$:

$$\begin{pmatrix} y_1 \\ y_2 \\ y_3 \end{pmatrix} = \left[\begin{pmatrix} 1 & 0 & 0 \\ 0 & 1 & 0 \\ 0 & 0 & 1 \end{pmatrix} - \begin{pmatrix} 0.15 & 0.35 & 0.25 \\ 0.30 & 0.05 & 0.20 \\ 0.45 & 0.40 & 0.10 \end{pmatrix} \right] \begin{pmatrix} 200 \\ 160 \\ 200 \end{pmatrix} = \begin{pmatrix} 64 \\ 52 \\ 26 \end{pmatrix}$$

Therefore, the amount of extra production will be $64 from the blacksmith, $52 from the carpenter, and $26 from the farmer, for a total of $142.

How much will the production of each sector have to change in order to provide $151 of excess production, specifically $59 from the blacksmith, $48 from the carpenter, and $44 from the farmer? So Y is

$$\begin{pmatrix} y_1 \\ y_2 \\ y_3 \end{pmatrix} = \begin{pmatrix} 59 \\ 48 \\ 44 \end{pmatrix}$$

The formula $X = (I - A)^{-1}Y$ becomes (using a calculator)

$$\begin{pmatrix} x_1 \\ x_2 \\ x_3 \end{pmatrix} = \left[\begin{pmatrix} 1 & 0 & 0 \\ 0 & 1 & 0 \\ 0 & 0 & 1 \end{pmatrix} - \begin{pmatrix} 0.15 & 0.35 & 0.25 \\ 0.30 & 0.05 & 0.20 \\ 0.45 & 0.40 & 0.10 \end{pmatrix} \right]^{-1} \begin{pmatrix} 59 \\ 48 \\ 44 \end{pmatrix} = \begin{pmatrix} 200 \\ 160 \\ 220 \end{pmatrix}$$

Thus if the blacksmith and the carpenter continue to produce the same value as before but the farmer increases production by $20 to $220, the economy can generate an additional $9 in excess production (but notice that the respective sources of this excess change drastically).

Graphing Calculator Exploration

Leontief "open" input–output model calculations are simple to do on your calculator. To verify the above results for the blacksmith–carpenter–farmer economy example:

```
[A]
 [[.15 .35 .25]
  [.3  .05 .2 ]
  [.45 .4  .1 ]]
```

a. Store the technology matrix A in [A].

```
[B]
         [[200]
         [160]
         [200]]
```

b. To find the excess productions Y for productions $X = \begin{pmatrix} 200 \\ 160 \\ 200 \end{pmatrix}$, store these values in [B].

```
(identity(3)-[A]
)*[B]
         [[64]
         [52]
         [26]]
```

c. Then use the MATRX MATH IDENTITY(3) command to find $Y = (I - A)X$.

```
(identity(3)-[A]
)⁻¹*[C]
         [[200]
         [160]
         [220]]
```

d. To find the sector productions X necessary to provide $Y = \begin{pmatrix} 59 \\ 48 \\ 44 \end{pmatrix}$ excess productions, store these values in [C] and use the $\boxed{x^{-1}}$ button to find $X = (I - A)^{-1}Y$.

There are many reasons to alter the production levels or the excess production levels of the sectors, such as raising the standard of living or supporting government programs through taxes. The Leontief "open" input–output model provides a tool to evaluate the effects of such possible changes.

Least Squares

The method of least squares was invented in 1794 by Carl Friedrich Gauss to find the best "compromise solution" to problems with inconsistent data. He became world famous as a scientist in 1801 when he used his method to predict when and where the asteroid Ceres could next be seen after it had been lost in the sun's glare on February 11, 1801 shortly after its discovery on January 1, 1801 by the Italian astronomer Piazzi. The sensation created by the accuracy of Gauss's prediction ultimately led to his appointment as astronomer at the Gottingen Observatory, assuring him the financial security that allowed him to pursue his many other ideas.

To understand how the least squares method is used, suppose we believe that some quantity y depends linearly on some other quantity x (that is, $y = mx + b$). Since "real world" measurements are not exact and always have errors, suppose we have data for x and y as follows:

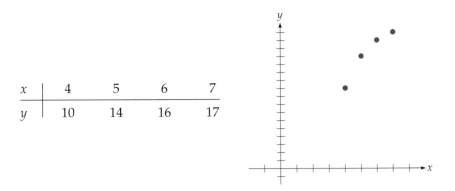

x	4	5	6	7
y	10	14	16	17

Clearly, these four data points are *not* collinear, so it is impossible to find a line $y = mx + b$ passing through all four of them. If we were to write the problems of finding m and b as a system of linear equations,

$$\begin{cases} 4m + b = 10 \\ 5m + b = 14 \\ 6m + b = 16 \\ 7m + b = 17 \end{cases} \longrightarrow \begin{pmatrix} 4 & 1 \\ 5 & 1 \\ 6 & 1 \\ 7 & 1 \end{pmatrix} \begin{pmatrix} m \\ b \end{pmatrix} = \begin{pmatrix} 10 \\ 14 \\ 16 \\ 17 \end{pmatrix} \qquad AX = B$$

then the system would be *inconsistent* since there is no solution. However, instead of an *exact* solution we can look for the best *compromise* line passing *closest* to the points. We multiply both sides of the above matrix equation by A^t.

$$\begin{pmatrix} 4 & 5 & 6 & 7 \\ 1 & 1 & 1 & 1 \end{pmatrix} \begin{pmatrix} 4 & 1 \\ 5 & 1 \\ 6 & 1 \\ 7 & 1 \end{pmatrix} \begin{pmatrix} m \\ b \end{pmatrix} = \begin{pmatrix} 4 & 5 & 6 & 7 \\ 1 & 1 & 1 & 1 \end{pmatrix} \begin{pmatrix} 10 \\ 14 \\ 16 \\ 17 \end{pmatrix} \qquad A^tAX = A^tB$$

Multiplying out yields

$$\begin{pmatrix} 126 & 22 \\ 22 & 4 \end{pmatrix} \begin{pmatrix} m \\ b \end{pmatrix} = \begin{pmatrix} 325 \\ 57 \end{pmatrix}$$

so

$$\begin{pmatrix} m \\ b \end{pmatrix} = \begin{pmatrix} 126 & 22 \\ 22 & 4 \end{pmatrix}^{-1} \begin{pmatrix} 325 \\ 57 \end{pmatrix} = \begin{pmatrix} \frac{1}{5} & -\frac{11}{10} \\ -\frac{11}{10} & \frac{63}{10} \end{pmatrix} \begin{pmatrix} 325 \\ 57 \end{pmatrix} = \begin{pmatrix} 2.3 \\ 1.6 \end{pmatrix}$$

These values, $m = 2.3$ and $b = 1.6$, give the line $y = 2.3x + 1.6$, which is called the "least squares" line, providing the best fit to the four points in that it minimizes the sum of the squared vertical distances.* The following graph shows the four points together with the least squares line.

---

* For a proof, see the section on least squares in *Applied Calculus, Brief Applied Calculus*, or *Finite Mathematics and Applied Calculus* by the authors of this text.

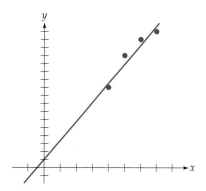

If the four points *did* lie on a line, then the *m* and *b* given by this procedure would be the *exact* values for the correct line. The least squares line is used widely in business and the sciences to predict future trends from current but imperfect data.

Least Squares Line

For the data points $(x_1, y_1), (x_2, y_2), \ldots, (x_n, y_n)$, form the matrices

$$A = \begin{pmatrix} x_1 & 1 \\ x_2 & 1 \\ \vdots & \vdots \\ x_n & 1 \end{pmatrix} \quad \text{and} \quad B = \begin{pmatrix} y_1 \\ y_2 \\ \vdots \\ y_n \end{pmatrix}. \quad \text{The coefficients } m \text{ and } b \text{ for}$$

the least squares line $y = mx + b$ are then given by

$$\begin{pmatrix} m \\ b \end{pmatrix} = (A^t A)^{-1}(A^t B)$$

provided that the numbers $x_1, x_2, \ldots, x_n$ are distinct.

EXAMPLE **1**

FINDING A LEAST SQUARES LINE

The numbers of paid subscribers to an internet service provider at the end of its first four years of operations are given in the following table. Find the least squares line for these data and use it to predict sales at the end of year 5.

Year	1	2	3	4
Subscribers (1000s)	9	13	14	18

Solution

Using $A = \begin{pmatrix} 1 & 1 \\ 2 & 1 \\ 3 & 1 \\ 4 & 1 \end{pmatrix}$ and $B = \begin{pmatrix} 9 \\ 13 \\ 14 \\ 18 \end{pmatrix}$ with the above formula,

we find (see the following Graphing Calculator Exploration) that

$$\begin{pmatrix} m \\ b \end{pmatrix} = (A^t A)^{-1}(A^t B) = \begin{pmatrix} 2.8 \\ 6.5 \end{pmatrix}$$

Therefore, the least squares line is $y = 2.8x + 6.5$, as shown in the graph on the left. The predicted number of subscribers at the end of year 5 is found by substituting $x = 5$:

$$y = 2.8 \cdot 5 + 6.5 = 20.5 \qquad \text{(in thousands)}$$

The predicted number of subscribers at the end of the fifth year is 20,500.

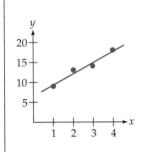

Graphing Calculator Exploration

The matrix calculations for the least squares line in the above example are easily carried out on a graphing calculator.

```
[A]
        [[1  1]
         [2  1]
         [3  1]
         [4  1]]
```

a. Define a 4×2 matrix [A] containing the *x*-values.

```
[B]
        [[9 ]
         [13]
         [14]
         [18]]
```

b. Define a 4×1 matrix [B] containing the *y*-values.

```
([A]ᵀ[A])⁻¹([A]ᵀ[
B])
        [[2.8]
         [6.5]]
```

c. Use the TRANSPOSE command and the $\boxed{x^{-1}}$ key to find $(A^t A)^{-1}(A^t B)$.

The least squares line can also be found by using the LinReg command on your graphing calculator.

3.6 Section Summary

A *Leontief "open" input–output model* is a matrix equation $X = AX + Y$, where the column matrix X lists the values produced by each of the n economic sectors, the column matrix Y lists the values of the excess production of these same sectors, and the element in row i and column j of the technology matrix A represents the value of sector i goods and services needed to produce one dollar of sector j output. Solving this model for the excess productions Y in terms of the sector production values X gives that $Y = (I - A)X$, while the sector production values X necessary to provide excess productions Y is $X = (I - A)^{-1}Y$.

The least squares line of a collection of data points can be found by the formula on page 231. It is widely used to analyze linear trends and predict future values from such trends.

3.6 Exercises (will be helpful.)

LEONTIEF "OPEN" INPUT–OUTPUT MODELS

Let A denote agriculture, C denote construction, E denote electronics, F denote fishing, H denote heavy industry, L denote light industry, M denote mining, R denote railroads, and T denote tourism.

Find the technology matrix for each economy diagram.

1.

2.

3.

4.

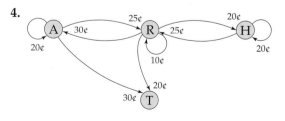

Draw an economy diagram for each technology matrix.

5. $\begin{pmatrix} 0.15 & 0.10 \\ 0.25 & 0.30 \end{pmatrix}$ for sectors M and H

6. $\begin{pmatrix} 0.10 & 0.30 & 0.10 \\ 0.20 & 0.40 & 0.20 \\ 0.10 & 0.30 & 0.10 \end{pmatrix}$ for sectors A, R, and M

7. $\begin{pmatrix} 0.20 & 0.30 & 0.10 \\ 0.50 & 0.20 & 0.20 \\ 0.30 & 0.10 & 0.20 \end{pmatrix}$ for sectors C, A, and L

8. $\begin{pmatrix} 0.10 & 0 & 0.20 \\ 0 & 0.20 & 0.10 \\ 0.10 & 0.40 & 0 \end{pmatrix}$ for sectors A, C, and R

Find the excess production Y of each economy with technology matrix A and economic activity level X.

9. $A = \begin{pmatrix} 0.20 & 0.30 \\ 0.35 & 0.25 \end{pmatrix}$ and $X = \begin{pmatrix} 130 \\ 110 \end{pmatrix}$

10. $A = \begin{pmatrix} 0.45 & 0.20 \\ 0.30 & 0.35 \end{pmatrix}$ and $X = \begin{pmatrix} 140 \\ 100 \end{pmatrix}$

11. $A = \begin{pmatrix} 0.05 & 0.15 & 0.20 \\ 0.15 & 0.05 & 0.15 \\ 0.10 & 0.10 & 0.05 \end{pmatrix}$ and $X = \begin{pmatrix} 150 \\ 170 \\ 140 \end{pmatrix}$

12. $A = \begin{pmatrix} 0.10 & 0.05 & 0.10 & 0.15 \\ 0.15 & 0.10 & 0.10 & 0.05 \\ 0 & 0.05 & 0.10 & 0 \\ 0.10 & 0 & 0.10 & 0.05 \end{pmatrix}$

and $X = \begin{pmatrix} 120 \\ 100 \\ 110 \\ 80 \end{pmatrix}$

Find, for each economy with technology matrix A, the economic activity level X necessary to generate excess production Y.

13. $A = \begin{pmatrix} 0.20 & 0.30 \\ 0.30 & 0.20 \end{pmatrix}$ and $Y = \begin{pmatrix} 84 \\ 51 \end{pmatrix}$

14. $A = \begin{pmatrix} 0.25 & 0.35 \\ 0.45 & 0.15 \end{pmatrix}$ and $Y = \begin{pmatrix} 33 \\ 57 \end{pmatrix}$

15. $A = \begin{pmatrix} 0.10 & 0.20 & 0 \\ 0 & 0.15 & 0.20 \\ 0.30 & 0.10 & 0.20 \end{pmatrix}$ and $Y = \begin{pmatrix} 60 \\ 31 \\ 50 \end{pmatrix}$

16. $A = \begin{pmatrix} 0.10 & 0.15 & 0.10 & 0.05 \\ 0.15 & 0.15 & 0.05 & 0.10 \\ 0 & 0.20 & 0.10 & 0.10 \\ 0.10 & 0 & 0.15 & 0.05 \end{pmatrix}$

and $Y = \begin{pmatrix} 60 \\ 55 \\ 60 \\ 70 \end{pmatrix}$

Represent each situation as a Leontief "open" input–output model by constructing an economy diagram and the corresponding technology matrix. Find the required excess production or level of economic activity and state your final answer in terms of the original question.

17. ECONOMICS: Industrial Production
The heavy and light industry sectors of the Birmingham economy depend on each other in the following way: Each dollar of production from the heavy industry sector requires $0.25 of heavy industry produce and $0.15 of light industry produce, while each dollar of production from the light industry sector requires $0.35 of heavy industry produce and $0.05 of light industry produce. The city leaders decide to increase production in order to generate tax revenues for a new sports stadium. How much must each type of industry produce to yield an excess production of $127 million of heavy industry produce and $221 million of light industry produce?

18. ECONOMICS: County Production Planning
An analysis of the mining, railroad, construction, and light industry sectors of the Hanover County economy revealed that each dollar produced by the mining sector requires $0.20 of mining products, $0.10 of railroad services, $0.10 of construction, and $0.10 of light industry products. Each dollar produced by the railroad sector requires $0.10 of mining products, $0.20 of railroad services, and $0.10 of light industry products. Each dollar produced by the construction sector requires $0.30 of mining products, $0.20 of construction, and $0.10 of light industry products. Each dollar produced by the light industry sector requires $0.10 of mining products, $0.20 of railroad services, and $0.10 of light industry products. If these

industries in Hanover County presently produce excess productions valued at $8 million from mining, $30 million from railroads, $32 million from construction, and $31 million from light industry, what is the current production level of each of these economic sectors?

19. ECONOMICS: Developing Productivity
In an effort to raise the standard of living, the new government of a developing country wants to increase its excess production by stimulating the heavy industry sector of the national economy. An analysis of the relationships between its heavy industry, light industry, and railroad sectors found that the current production levels for these sectors are $100 million from heavy industry, $150 million from light industry, and $100 million from the railroads, with the technology matrix given in the table. Find the current excess production from these sectors of the economy. If the light industry and railroad productions remain the same, how much does this excess production increase with each $10 million increase in heavy industry production? What is the greatest heavy industry production level that this part of the national economy can tolerate?

$From \backslash ^{To}$	Heavy Industry	Light Industry	Railroads
Heavy industry	0.20	0.20	0.30
Light industry	0.50	0.20	0.20
Railroads	0.20	0.10	0.20

20. ECONOMICS: Island Economy An economist is studying the agriculture-, fishing-, and tourism-based economy of a small Pacific island nation. Although the country's excess production was $158 million last year, she has found that the economy actually produced $300 million, of which $142 million was consumed in the course of production. Her breakdown of each sector's production is given in the table. (The first line shows that the total value of agriculture production was $100 million, with $20 million consumed by the agriculture sector, $8 million consumed by the fishing sector, and $36 million consumed by the tourism sector, leaving an excess of $36 million.) If the same relations persist among the sectors of the economy, what will happen to the excess production next year if the fishing sector declines to $60 million while the other sectors produce the same amounts as before?

In Millions of $	Agriculture	Fishing	Tourism	Excess Production
Agriculture	$20	$8	$36	$36
Fishing	$10	$8	$36	$26
Tourism	$0	$0	$24	$96

LEAST SQUARES

Find the least squares best approximation line $y = mx + b$ for each collection of x and y data pairs.

21.

x	-1	0	1
y	24	36	42

22.

x	-1	0	1
y	8	8	14

23.

x	-1	0	1	2
y	120	90	70	30

24.

x	-1	0	1	3
y	761	656	691	551

25.

x	1	4	5	8	12
y	185	195	195	245	275

26.

x	1	4	5	7	8	11
y	50	80	110	140	170	200

Use the least squares best approximation line to make each prediction. Be sure to state your final answer in terms of the original question.

27. BUSINESS: Real Estate The new salesman at Abbott Associates, Real Estate Brokers, had an impressive first three months with sales of $300,000, $480,000, and $600,000. If he can keep improving this much every month, how much will he sell in his fifth month?

28. **BUSINESS: Commodity Futures** A soybean speculator has been nervously watching the December delivery price per bushel drop from $6.59 on Monday to $6.41 on Tuesday and then to $6.29 on Wednesday. If this trend continues, what will the price be on Friday?

29. **BEHAVIORAL SCIENCE: Price–Demand** The market research division of a large candy manufacturer has test-marketed the new treat YummieCrunchies in five different markets at five different prices per 8-ounce bag to determine the relation between the selling price and the sales volume. The results of its study are presented in the table and are normalized to give the weekly sales per 20,000 consumers. How many weekly sales per 20,000 consumers can the manufacturer expect when it begins national distribution next week with an introductory price of 79¢ per 8-ounce bag?

Selling Price	Sales Volume
70¢	2050
55¢	2675
75¢	1975
90¢	1250
85¢	1425

30. **MANAGEMENT SCIENCE: Advertising** The accounts manager at television station WXXB claims that every dollar spent on commercials by local retailers generates $23 of sales. As proof, she shows prospective advertisers the following table of advertising budgets and gross receipts for five area companies that aired commercials last month. Is her claim justified?

Dollars Spent on WXXB Commercials	Gross Receipts
$8000	$245,000
$5000	$215,000
$2000	$125,000
$9000	$305,000
$6000	$230,000

Chapter Summary with Hints and Suggestions

Reading the text and doing the exercises in this chapter have helped you to master the following skills, which are listed by section (in case you need to review them) and are keyed to particular Review Exercises. Answers for all Review Exercises are given at the back of the book, and full solutions can be found in the Student Solutions Manual.

3.1 Systems of Two Linear Equations in Two Variables

■ Represent a pair of statements as a system of two linear equations in two variables. *(Review Exercises 1–2.)*

$$\begin{cases} ax + by = h \\ cx + dy = k \end{cases}$$

■ Solve a system of two linear equations in two variables by graphing. *(Review Exercises 3–4.)*

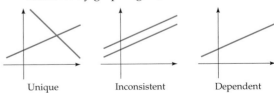

Unique   Inconsistent   Dependent

- Solve a system of two linear equations in two variables by the substitution method. *(Review Exercises 5–6.)*

- Solve a system of two linear equations in two variables by the elimination method. *(Review Exercises 7–8.)*

- Solve an applied problem by using a system of two equations in two variables and the elimination method. *(Review Exercises 9–10.)*

3.2 Matrices and Linear Equations in Two Variables

- Find the dimension of a matrix and identify a particular element using double subscript notation. *(Review Exercises 11–12.)*

$$A = \begin{pmatrix} a_{1,1} & \cdots & a_{1,n} \\ \vdots & & \\ a_{m,1} & \cdots & a_{m,n} \end{pmatrix}$$

- Carry out a given row operation on an augmented matrix. *(Review Exercises 13–14.)*

$$\begin{cases} ax + by = h \\ cx + dy = k \end{cases} \longleftrightarrow \left(\begin{array}{cc|c} a & b & h \\ c & d & k \end{array} \right)$$

- Solve a system of two linear equations in two variables by row-reducing an augmented matrix. *(Review Exercises 15–18.)*

- Solve an applied problem by using a system of two linear equations in two variable and row-reducing an augmented matrix. *(Review Exercises 19–20.)*

3.3 Systems of Linear Equations and the Gauss–Jordan Method

- Interpret a row-reduced matrix as the solution of a system of equations. If the solution is not unique, identify the system as "inconsistent" or "dependent." *(Review Exercises 21–22.)*

$$\begin{array}{ccccc|c} 0 & 0 & \cdots & 0 & 1 \end{array} \quad \text{means inconsistent.}$$
$$\begin{array}{ccccc|c} 0 & 0 & \cdots & 0 & 0 \end{array} \quad \text{means dependent.}$$

- Use the Gauss–Jordan method to find the equivalent row-reduced matrix. *(Review Exercises 23–24.)*

- Solve a system of equations by the Gauss–Jordan method. If the solution is not unique, identify the system as "inconsistent" or "dependent." *(Review Exercises 25–28.)*

- Formulate an application as a system of linear equations. Solve the system by the Gauss–Jordan method. State the final answer in terms of the original question. *(Review Exercises 29–30.)*

3.4 Matrix Arithmetic

- Find a matrix product or determine that the product is not defined. *(Review Exercises 31–32.)*

$$\begin{array}{ccccc} A & \cdot & B & = & C \\ m \times p & & p \times n & & m \times n \end{array}$$

- Find the value of a matrix expression involving scalar multiplication, matrix addition, subtraction, transposition, or multiplication. *(Review Exercises 33–34.)*

$$A \cdot I = I \cdot A = A \qquad A + 0 = 0 + A = A$$

- Rewrite a system of linear equations as a matrix equation $AX = B$. *(Review Exercises 35–36.)*

- Formulate an application in matrix form, and indicate the meaning of each row and column. Find the requested quantity using matrix arithmetic. *(Review Exercises 37–38.)*

3.5 Inverse Matrices and Systems of Linear Equations

- Find the product of a pair of matrices to identify the pair as "a matrix and its inverse" or "not a matrix and its inverse." *(Review Exercises 39–40.)*

$$AA^{-1} = A^{-1}A = I$$

- Find the inverse of a matrix or identify it as singular. *(Review Exercises 41–42.)*

$$(A|I) \rightarrow (I|A^{-1})$$

- Rewrite a system of equations as a matrix equation $AX = B$ and use the inverse of A to find the solution. *(Review Exercises 43–46.)*

$$X = A^{-1}B$$

- Formulate an application in terms of systems of linear equations. Solve the collection of systems of equations by finding the inverse of a common coefficient matrix and using matrix multiplications of this inverse times the various constant term matrices. State the final answer in terms of the original question. (*Review Exercises 47–48.*)

$$X = A^{-1}B_1, \ldots, X = A^{-1}B_n$$

3.6 Introduction to Modeling

Leontief "Open" Input–Output Models

- Find the technology matrix from an economy diagram. (*Review Exercises 49–50.*)

- Draw an economy diagram from a technology matrix. (*Review Exercises 51–52.*)

- Find the excess production of an economy with a given technology matrix and economic activity level. (*Review Exercises 53–54.*)

$$Y = (I - A)X$$

- Find, for an economy with a given technology matrix, the economic activity level necessary to generate a specified excess production. (*Review Exercises 55–56.*)

$$X = (I - A)^{-1}Y$$

- Represent an application as a Leontief "open" input–output model by constructing an economy diagram and the corresponding technology matrix. Find the required excess production or level of economic activity, and state the final answer in terms of the original question. (*Review Exercises 57–58.*)

Least Squares

- Find the least squares best approximation line $y = mx + b$ for a collection of x and y data pairs such that all the x-values are distinct. (*Review Exercises 59–62.*)

$$\binom{m}{b} = (A^tA)^{-1}(A^tB)$$

- Use the least squares best approximation line to make a prediction in an application. (*Review Exercises 63–64.*)

Hints and Suggestions

- (*Overview*) Row operations on matrices are a generalization of the elimination method of finding the intersection of two lines, extending the technique to problems with many equations in many variables. Matrix arithmetic is similar to real number operations except that matrix multiplication is not commutative and many matrices do not have inverses. However, if a square matrix A *does* have an inverse, then the equation $AX = B$ can be solved as $X = A^{-1}B$, just as a real number linear equation can be solved by dividing. Matrices are used to represent and solve many large and complicated problems important to both science and society.

- Although not every system of linear equations has a solution, they can all be identified as "having a unique solution," "inconsistent," or "dependent" by row-reducing the corresponding augmented matrix.

- When setting up a word problem, look first for the questions "how many" or "how much" to help identify the variables. Be sure that finding values for your variables will answer the question stated in the problem. Use the rest of the given information to build equations describing facts about your variables.

- Many row-reduction problems require many steps to solve, so don't give up; keep improving the matrix until it meets all the requirements to be row-reduced. Make sure that each row operation you choose will move you toward your goal without undoing the parts you already have gotten the way you need.

- **Practice for Test:** Review Exercises 4, 6, 8, 11, 17, 20, 25, 30, 33, 35, 38, 41, 47, 49, 51, 55, 57, 59, and 63.

Review Exercises for Chapter 3

Practice test exercise numbers are in green.

3.1 Systems of Two Linear Equations in Two Variables

Represent each pair of statements as a system of two linear equations in two variables. Be sure to state clearly the meaning of your x- and y-variables.

1. "A small commuter airplane has thirty passengers" and "the ticket receipts of $3970 come from 30-day advance sale tickets at $79 and full fare tickets at $159."

2. "A cow and horse rancher has four hundred twenty animals" and "there are twice as many cows as horses."

Solve each system by graphing. If the solution is not unique, identify the system "inconsistent" or "dependent."

3. $\begin{cases} x + y = 18 \\ x - y = 8 \end{cases}$ **4.** $\begin{cases} 2x - 4y = 36 \\ -3x + 6y = -54 \end{cases}$

Solve each system by the substitution method. If the solution is not unique, identify the system as "inconsistent" or "dependent."

5. $\begin{cases} 5x - 2y = 10 \\ -2x + y = 2 \end{cases}$ **6.** $\begin{cases} -4x + 2y = 12 \\ 2x - y = 12 \end{cases}$

Solve each system by the elimination method. If the solution is not unique, identify the system as "inconsistent" or "dependent."

7. $\begin{cases} x + y = 12 \\ x + 3y = 18 \end{cases}$ **8.** $\begin{cases} 4x + 5y = 60 \\ 2x + 3y = 42 \end{cases}$

Formulate each situation as a system of equations. Be sure to state clearly the meaning of your x- and y-variables. Solve the system by the elimination method. State your final answer in terms of the original question.

9. GENERAL: Garden Plants A retired investment broker has rosebushes in his flower garden and tomato plants in his vegetable garden. He spends one hour each day tending his twenty-five plants. If each rosebush takes three minutes of care and each tomato plant takes two minutes, how many of each does he have?

10. GENERAL: Fraternity Convention Twenty-six members of the Alpha Alpha Alpha fraternity want to go to the national convention in Orlando and each has put in $10 for gas. If each car holds 5 people and uses $45 worth of gas and each van holds 8 people and uses $85 worth of gas, how many of each vehicle do they need for the trip?

3.2 Matrices and Linear Equations in Two Variables

Find the dimension of each matrix and the values of the specified elements.

11. $\begin{pmatrix} 8 & 3 & 4 \\ 1 & 5 & 9 \\ 6 & 7 & 2 \end{pmatrix}$; $a_{2,2}, a_{3,1}, a_{1,3}$

12. $\begin{pmatrix} 1 & 15 & 14 & 4 \\ 12 & 6 & 7 & 9 \\ 8 & 10 & 11 & 5 \\ 13 & 3 & 2 & 16 \end{pmatrix}$; $a_{2,3}, a_{3,2}, a_{4,1}$

Carry out the row operation on the matrix.

13. $3R_2 \to R_2$ on $\begin{pmatrix} 3 & 4 & | & 12 \\ 1 & 2 & | & 2 \end{pmatrix}$

14. $R_1 + R_2 \to R_1$ on $\begin{pmatrix} -1 & 2 & | & 2 \\ 2 & -3 & | & 6 \end{pmatrix}$

Solve each system by row-reducing an augmented matrix. If the solution is not unique, identify the system as "inconsistent" or "dependent."

15. $\begin{cases} x + 3y = 63 \\ 4x + 5y = 140 \end{cases}$ **16.** $\begin{cases} -3x + 4y = 60 \\ 2x - y = 10 \end{cases}$

17. $\begin{cases} 12x - 4y = 36 \\ -15x + 5y = 45 \end{cases}$ **18.** $\begin{cases} -16x + 12y = -48 \\ 20x - 15y = 60 \end{cases}$

Formulate each situation as a system of equations. Be sure to state clearly the meaning of your x- and y-variables. Solve the system by row-reducing the corresponding augmented matrix. State your final answer in terms of the original question.

19. **BIOMEDICAL: Pharmaceuticals** The pharmacist at the Charter Drug Shop filled 92 prescriptions today for antibiotics and cough suppressants. If there were 34 more prescriptions for antibiotics than for cough suppressants, how many prescriptions for each were filled?

20. **BUSINESS: Classic Magazines** A used book store offers grab bag packages of old *Life* and *The New Yorker* magazines from the 1940s containing 4 copies of *Life* and 3 copies of *The New Yorker* for $39 and larger bags of 12 copies of *Life* and 10 copies of *The New Yorker* for $122. What is the price of one copy of each old magazine?

3.3 Systems of Linear Equations and the Gauss–Jordan Method

Interpret each row-reduced matrix as the solution of a system of equations. If the solution is not unique, identify the system as "inconsistent" or "dependent."

21. $\begin{pmatrix} 1 & 0 & 0 & | & 3 \\ 0 & 1 & 0 & | & -3 \\ 0 & 0 & 1 & | & 6 \end{pmatrix}$ 22. $\begin{pmatrix} 1 & 1 & 0 & | & 4 \\ 0 & 0 & 1 & | & 2 \\ 0 & 0 & 0 & | & 0 \end{pmatrix}$

Use the Gauss–Jordan method to find the equivalent row-reduced matrix.

23. $\begin{pmatrix} 0 & 1 & 1 & | & -1 \\ 1 & 0 & 1 & | & 6 \\ 2 & 0 & 1 & | & 10 \end{pmatrix}$ 24. $\begin{pmatrix} 1 & 0 & 1 & 1 & | & 4 \\ 5 & 1 & 1 & 4 & | & 12 \\ 2 & 1 & 0 & 1 & | & 5 \\ 3 & 0 & 1 & 3 & | & 8 \end{pmatrix}$

Solve each system of equations by using the Gauss–Jordan method. If the solution is not unique, identify the system as "inconsistent" or "dependent."

If permitted by your instructor, you may use a graphing calculator.

25. $\begin{cases} x_1 - 2x_3 = 2 \\ -x_1 + x_2 + 2x_3 = 1 \\ -x_1 + 2x_2 + 3x_3 + x_4 = 7 \\ x_1 - 2x_3 + x_4 = 4 \end{cases}$

26. $\begin{cases} x_1 - x_2 + x_4 = 3 \\ x_2 + x_3 = 3 \\ 2x_1 - x_2 + x_3 + 2x_4 = 9 \\ 2x_1 - x_2 + x_3 + x_4 = 5 \end{cases}$

27. $\begin{cases} x_1 + 2x_3 + x_4 = 3 \\ x_1 + x_2 + 3x_3 + x_4 = 2 \\ 3x_1 + 3x_2 + 9x_3 + 4x_4 = 7 \\ 2x_1 + 4x_3 + 2x_4 = 7 \end{cases}$

28. $\begin{cases} x_1 + x_2 + x_3 + x_4 = 2 \\ x_1 + x_2 + x_3 + x_4 = 1 \\ x_1 + x_2 + x_3 + x_4 = 0 \\ x_1 + x_2 + x_3 + x_4 = 3 \end{cases}$

Formulate each situation as a system of linear equations. Be sure to state clearly the meaning of each variable. Solve using the Gauss–Jordan method. State your final answer in terms of the original question.

If permitted by your instructor, you may use a graphing calculator.

29. **BUSINESS: Nursery Management** The Nyack Nursery starts plants from seeds and sells potted plants to garden stores for resale. Each dahlia costs 16¢ to start and needs a 10¢ flowerpot and 5 ounces of soil; each chrysanthemum costs 11¢ to start and needs a 12¢ flowerpot and 6 ounces of soil; and each daisy costs 13¢ to start and needs an 8¢ flowerpot and 5 ounces of soil. If the nursery has $50 to spend on starting the plants, $48 to spend on flowerpots, and 153 pounds of potting soil, how many of each plant can it raise using all of the available resources?

30. **GENERAL: Family Entertainment** The Family Fun Center in Asheville offers the package specials listed in the table for an afternoon of fun at its go-kart track, miniature golf course, house of funny mirrors, and snack stand serving hot dogs and sodas. How much does one hot dog cost? How much does a go-kart ride cost if the mirror house is $1.50 and sodas are $1?

Go-kart	Golf	Mirror House	Hot Dog	Soda	Package Price
2	1	1	1	1	$15
3	2	'1	2	1	$23
4	2	2	3	2	$31

3.4 Matrix Arithmetic

Find each matrix product.

31. $\begin{pmatrix} 1 & 2 & 3 & 4 \\ 4 & 3 & 2 & 1 \end{pmatrix} \begin{pmatrix} 1 & 0 \\ 0 & -1 \\ 1 & 0 \\ 0 & -1 \end{pmatrix}$

32. $\begin{pmatrix} 1 & 0 \\ 0 & -1 \\ 1 & 0 \\ 0 & -1 \end{pmatrix} \begin{pmatrix} 1 & 2 & 3 & 4 \\ 4 & 3 & 2 & 1 \end{pmatrix}$

Use the given matrices to find each matrix expression.

$$A = \begin{pmatrix} -1 & 2 & -1 \\ 2 & -1 & 2 \end{pmatrix} \qquad B = \begin{pmatrix} 3 & 6 & 8 \\ 7 & 5 & 4 \end{pmatrix}$$

$$C = \begin{pmatrix} 2 & 1 & 2 \\ 1 & -2 & 1 \\ 2 & 1 & 2 \end{pmatrix} \qquad D = \begin{pmatrix} 5 & 8 \\ 7 & 6 \end{pmatrix}$$

33. $3D - A \cdot B^t + I$ **34.** $3I + A^t \cdot B - C$

Rewrite each system of linear equations as a matrix equation $AX = B$.

35. $\begin{cases} x_1 + 4x_2 + x_3 = 15 \\ 2x_1 + 8x_2 + 3x_3 = 26 \\ x_1 + 5x_2 + 2x_3 = 17 \end{cases}$

36. $\begin{cases} 2x_1 + 3x_2 - x_3 + x_4 = 20 \\ 5x_1 + 4x_2 + x_3 + 2x_4 = 35 \\ 2x_1 + x_2 + x_3 + x_4 = 12 \end{cases}$

Formulate each situation in matrix form. Be sure to indicate the meaning of your rows and columns. Find the requested quantities using the appropriate matrix arithmetic.

37. GENERAL: Growing Grandchildren
A proud grandmother's record book shows that her grandson Thomas is now 61 inches tall and weighs 90 pounds, while last year he was 58 inches tall and weighed 80 pounds; her grandson Richard is now 54 inches tall and weighs 75 pounds, while last year he was 52 inches tall and weighed 70 pounds; and her granddaughter Harriet is now 47 inches tall and weighs 60 pounds, while last year she was 46 inches tall and weighed 55 pounds. Represent these facts as a "this year" matrix and a "last

year" matrix. Use these matrices to find how much each grandchild grew.

38. MANAGEMENT SCIENCE: Spring Fashions
The buyer for the ladies sportswear division of a large department store needs 200 jackets, 300 blouses, 250 skirts, and 175 pairs of slacks. The spring lines shown by both an East Coast designer and an Italian team are acceptable to her, but the East Coast designer wants $195 for each jacket, $85 for each blouse, $145 for each skirt, and $130 for each pair of slacks, while the Italian company wants $190 for each jacket, $90 for each blouse, $150 for each skirt, and $125 for each pair of slacks. Represent her needs as a column matrix and the prices as a price matrix. Use these matrices to find the cost of her order from each source.

3.5 Inverse Matrices and Systems of Linear Equations

Find each matrix product. Identify each pair of matrices as "a matrix and its inverse" or "not a matrix and its inverse."

39. $\begin{pmatrix} 1 & 2 & 3 \\ 1 & 1 & 1 \\ 0 & 1 & 3 \end{pmatrix}$ and $\begin{pmatrix} -2 & 3 & 1 \\ 3 & -3 & -2 \\ -1 & 1 & 1 \end{pmatrix}$

40. $\begin{pmatrix} -3 & 0 & 1 \\ 1 & 3 & 1 \\ -3 & 2 & 2 \end{pmatrix}$ and $\begin{pmatrix} -4 & -2 & 3 \\ 5 & 3 & -4 \\ -11 & -6 & 8 \end{pmatrix}$

Find the inverse of each matrix or identify it as singular.

41. $\begin{pmatrix} 1 & 1 & 0 \\ 0 & -3 & 1 \\ 2 & 3 & 0 \end{pmatrix}$ **42.** $\begin{pmatrix} 1 & 0 & 1 \\ 1 & 1 & 0 \\ 2 & 1 & 1 \end{pmatrix}$

Rewrite each system of equations as a matrix equation $AX = B$ and use the inverse of A to find the solution. Be sure to check the solution in the original system of equations.

43. $\begin{cases} 13x_1 + 4x_2 = 33 \\ 3x_1 + x_2 = 8 \end{cases}$ **44.** $\begin{cases} 3x_1 + 8x_2 = 25 \\ 2x_1 + 5x_2 = 16 \end{cases}$

45. $\begin{cases} 5x_1 + x_2 + 2x_3 = 11 \\ 2x_1 + 2x_2 + x_3 = 7 \\ 2x_1 + x_2 + x_3 = 5 \end{cases}$

46. $\begin{cases} 3x_1 + 2x_2 + x_3 + 2x_4 = 7 \\ 2x_1 + 5x_2 + 2x_3 + 2x_4 = 10 \\ x_1 + 2x_2 + x_3 + x_4 = 4 \\ 2x_1 + 2x_2 + x_3 + 2x_4 = 6 \end{cases}$

Formulate each situation as a collection of systems of linear equations. Be sure to state clearly the meaning of each variable. Solve each collection using an inverse matrix. State your final answer in terms of the original question.

If permitted by your instructor, you may use a graphing calculator.

47. MANAGEMENT SCIENCE: Retail Displays
A chain of furniture stores sells living room and bedroom suites and displays them in the store windows and inside on the showroom floor. A living room suite window display requires 3 square yards of window space and then the same furniture group is allowed 7 square yards of showroom space, while a bedroom suite window display requires 4 square yards of window space and then the same furniture group is allowed 9 square yards of showroom space. For furniture styles not given window space but shown only on the showroom floor, each living room suite is allowed 8 square yards and each bedroom suite is allowed 10 square yards. The setup costs per square yard are $50 for furniture shown in the window and on the showroom floor, while for furniture shown only on the showroom floor, $60 for living room suites and $70 for bedroom suites. The table lists the available space (in square yards) at the chain's stores in Kingman, Prescott, and Holbrook, together with the approved setup budgets and the amount of the total showroom space that must be given to furniture not displayed in the window. How many different living room and bedroom suites can be shown at each location?

	Window Space	Showroom Floor	Floor Space for Nonwindow Suites	Setup Budget
Kingman	18	131	90	$ 8,850
Prescott	21	182	134	$12,190
Holbrook	17	187	148	$12,680

48. PERSONAL FINANCE: Income Taxes Find the federal, state, and city income taxes on the taxable incomes of the individuals listed in the table if the federal income tax is 20% of the taxable income after first deducting the state and city taxes; the state income tax is 10% of the taxable income after first deducting the federal and city taxes; and the city income tax is 5% of the taxable income after first deducting the federal and state taxes.

	Mr. Dahlman	Mrs. Farrell	Ms. Mazlin	Mr. Seidner
Taxable income	$96,700	$48,350	$77,360	$145,050

3.6 Introduction to Modeling

Leontief "Open" Input–Output Models

Find the technology matrix for each economy diagram.

49.

50.

Draw an economy diagram for each technology matrix.

51. $\begin{pmatrix} 0.10 & 0 & 0.20 \\ 0.10 & 0.15 & 0 \\ 0 & 0.15 & 0.20 \end{pmatrix}$

52. $\begin{pmatrix} 0.15 & 0.10 & 0 & 0.15 \\ 0 & 0.10 & 0.20 & 0 \\ 0.15 & 0 & 0.10 & 0 \\ 0 & 0.10 & 0 & 0.15 \end{pmatrix}$

Find the excess production Y of each economy with technology matrix A and economic activity level X.

53. $A = \begin{pmatrix} 0.10 & 0.05 & 0.15 \\ 0.15 & 0.10 & 0.05 \\ 0.10 & 0.05 & 0.10 \end{pmatrix}$ and $X = \begin{pmatrix} 540 \\ 620 \\ 560 \end{pmatrix}$

54. $A = \begin{pmatrix} 0.15 & 0.15 & 0.10 & 0.05 \\ 0.10 & 0.10 & 0.05 & 0.10 \\ 0.10 & 0.05 & 0.10 & 0.15 \\ 0.05 & 0.10 & 0.15 & 0.10 \end{pmatrix}$

and $X = \begin{pmatrix} 280 \\ 200 \\ 260 \\ 240 \end{pmatrix}$

Find, for each economy with technology matrix A, the economic activity level X necessary to generate excess production Y.

55. $A = \begin{pmatrix} 0.10 & 0.35 \\ 0.40 & 0.15 \end{pmatrix}$ and $Y = \begin{pmatrix} 175 \\ 225 \end{pmatrix}$

56. $A = \begin{pmatrix} 0.10 & 0.10 & 0.20 \\ 0.30 & 0.20 & 0.10 \\ 0.10 & 0.30 & 0.20 \end{pmatrix}$ and $Y = \begin{pmatrix} 245 \\ 294 \\ 196 \end{pmatrix}$

Represent each situation as a Leontief "open" input–output model by constructing an economy diagram and the corresponding technology matrix. Find the required excess production or level of economic activity and state your final answer in terms of the original question.

57. BUSINESS: A 5&10 Problem Each of the four divisions of the Woolworth Corporation depends on the others in the following way: Each dollar of production from each division requires 10¢ of production from that division and 5¢ of production from each of the others. How much must each division produce to yield an excess production of $3 million from each?

58. SOCIAL SCIENCE: Energy Dependence As a result of treaty obligations, the domestic and foreign oil consumption of an industrialized nation is linked to the protection provided by its military forces. Each dollar of domestic oil produced requires 5¢ of domestic oil and 10¢ of military protection; each dollar of foreign oil requires 20¢ of military protec-

tion; and each dollar of military protection consumes 10¢ of domestic oil, 10¢ of foreign oil, and 5¢ of military protection. If the domestic oil production level is $940 million, the foreign production is $2000 million, and the military protection budget is $520 million, how much of each can be used elsewhere in the nation's economy?

Least Squares

Find the least squares best approximation line $y = mx + b$ for each collection of x and y data pairs.

59.

x	2	3	4
y	22	32	36

60.

x	3	5	6
y	238	364	462

61.

x	2	3	5	6
y	40	50	80	90

62.

x	10	11	12	13	14
y	140	150	180	200	210

Use the least squares best approximation line to make each prediction. Be sure to state your final answer in terms of the original question.

63. MANAGEMENT SCIENCE: Store Hours The owners of a "Mom and Pop" corner store have had their store open various numbers of hours on recent Mondays, the usual day off. The table shows the number of hours the store was open and the sales receipts for the day. How much could they expect to sell next Monday if they keep their store open for 12 hours?

Hours open	6	8	10	14
Total sales	$2230	$3035	$3770	$5065

64. MANAGEMENT SCIENCE: Production Accidents The supervisor of a factory assembly line gathered the data in the table over the last year to compare the number of minor accidents each month with the number of extra five-minute mini-breaks she allows during the day. How many minor accidents could she expect next month if she allows four mini-breaks during the day?

Mini-breaks	0	2	6	8
Minor accidents	21	16	8	3

4 Linear Programming

*These two sections provide *alternate* methods for solving the *same* types of problems. The choice of which one to cover is left to the instructor.

Application Preview

Managing an Investment Portfolio

The adage "never put all your eggs in one basket" particularly applies to investing money. There is no such thing as a low-risk yet high-return investment, so each investor must reconcile desire for high returns with fear of losing the principal. An "investment strategy" balances greed against safety by determining the investments to be made. Given the many different opportunities in today's global economy, the actual solution of this "portfolio management problem" can be very difficult.

Let us imagine a simplified situation in which there are only two possible investments: one with low risk and low return and the other with high risk and high return. Suppose you have $10,000 to invest and wish to protect yourself by investing at least as much in the low-risk investment as in the high-risk investment. We will express these restrictions as inequalities and then draw their graphs just as we drew the graphs of equations. The portfolio management problem is to find the combination of investments that gives the greatest possible return but still satisfies the restrictions. This will correspond to finding the point on our graph that maximizes the return subject to satisfying the inequalities. In this chapter we will develop ways to find this maximal point, first geometrically for simple problems and then numerically for larger problems. We will then use these methods to solve problems like the portfolio problem described above (see pages 318 and 334) as well as many other problems of practical importance.

4.1 LINEAR INEQUALITIES

Introduction

This chapter describes a method for solving a large and important class of problems called *linear programming problems.* A typical linear programming problem, such as the portfolio problem just described, asks for the maximum value of a linear function subject to a collection of linear inequalities. Such problems often arise in modern business, sometimes involving many variables, and are usually solved by a procedure known as the *simplex method.* Although the simplex method is a numerical algorithm, it is based on the geometry of linear inequalities such as those we studied on page 3. We will first consider linear programming problems in *two* variables since they can be solved graphically. In this section we will graph inequalities in two variables, restricting ourselves to inequalities using $\geq$ and $\leq$ signs since these are the ones that occur in practice.

Inequalities in Two Variables

To graph an inequality in two variables means to graph all of the points that satisfy the inequality. The *boundary* of the inequality is the corresponding *equality,* that is, an *equation.* For example, the boundary of the inequality $3x - 7y \leq 42$ is the equation $3x - 7y = 42$. The graph of the inequality is then one side or the other of the boundary, the correct side being determined by a *test point.* The correct side of the boundary is called the *feasible region* for the inequality.

EXAMPLE 1

GRAPHING A LINEAR INEQUALITY IN THE PLANE

Graph the linear inequality $2x + 3y \geq 12$.

Solution

First we graph its boundary, $2x + 3y = 12$, by plotting the intercepts. We find the y-intercept from $2x + 3y = 12$ with $x = 0$:

$$0 + 3y = 12 \quad \text{so} \quad y = 4 \qquad \text{For the point } (0, 4)$$

then the x-intercept from $2x + 3y = 12$ with $y = 0$:

$$2x + 0 = 12 \quad \text{so} \quad x = 6 \qquad \text{For the point } (6, 0)$$

Using these two points, we draw the boundary line shown on the left.

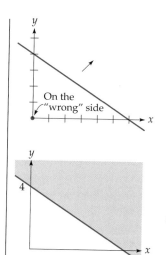

To determine the correct side we use the test point $(0, 0)$. (We may use any point *not on* the line.) Substituting $x = 0$, $y = 0$ into the inequality $2x + 3y \geq 12$ gives $0 \geq 12$, which is *false* (zero is *not* greater than twelve), meaning that the correct side of the line is the *opposite* side from the point $(0, 0)$, as indicated by the arrow in the graph.

Finally, we shade the correct side of the boundary to show the feasible region, as in the third graph.

The general procedure for graphing an inequality is as follows.

To Graph a Linear Inequality

1. Draw the boundary line (possibly using the intercepts).
2. Use a test point not on the line (possibly the origin) to determine the correct side. If this point satisfies the inequality, the point lies on the correct side; if not, the *other* side is the correct side.
3. Shade the correct side to show the feasible region.

Graphing Calculator Exploration

You can view any nonvertical boundary line $ax + by = h$ by entering it as $y = (h - ax)/b$. To see the boundary line $2x + 3y = 12$ from Example 1, enter it as $y_1 = (12 - 2x)/3$ and graph it on the window $[-5, 10]$ by $[-5, 10]$. Some calculators will also "shade" above or below the line provided you change the \ marker to the left of the y_1.

Practice Problem **1**   Graph the linear inequality $3x - 5y \le 60$.   ▶ **Solution on page 255**

Two or more inequalities joined by a brace { gives a *system* of inequalities, meaning that *each* of the inequalities must hold. To graph the feasible region for the system we graph each boundary, determine its correct side (which we mark with an arrow), and then shade the region that lies on the correct side of *all* of the lines. The system is *feasible* if there is *at least one point* in the feasible region, and *infeasible* if there are *no* points in the feasible region.

EXAMPLE **2**

GRAPHING A SYSTEM OF INEQUALITIES

a. Graph the system $\begin{cases} x + 2y \le 12 \\ x \ge 0 \\ y \ge 0 \end{cases}$

b. Is the system feasible?

c. What happens if the last inequality, $y \ge 0$, is replaced by $y \ge 7$?

Solution

a. We graph the boundary just as before, plotting the intercepts of $x + 2y = 12$: Substituting $x = 0$ gives $2y = 12$ or $y = 6$, for the point $(0, 6)$; substituting $y = 0$ into $x + 2y = 12$ gives $x = 12$ for the point $(12, 0)$. These two points give the line in the first of the following three graphs.

Feasible region

6

12

No feasible points

The point $(0, 0)$ *does* satisfy $x + 2y \leq 12$ (since $0 + 0 \leq 12$), so the origin is on the *correct* side, giving the arrow in the second graph. We add the arrows on the axes because $x \geq 0$ means from the y-axis to the *right*, and $y \geq 0$ means from the x-axis *upwards*. From the three arrows we get the shaded region in the third graph.

b. Since there *are* points in the feasible region, the system *is* feasible.

c. Replacing $y \geq 0$ by $y \geq 7$ gives the graph shown on the left. There are no points on the correct side of all of the lines, so the system with $y \geq 7$ becomes *infeasible*.

Practice Problem 2   Sketch the system $\begin{cases} x + 2y \leq 20 \\ x \geq 0 \\ y \geq 0 \end{cases}$ Is this system feasible?

➤ Solution on page 255

Vertices of Feasible Regions

The "corner points" of the feasible region are called *vertices*. More precisely:

> **Vertex**
>
> A *vertex* of a system of linear inequalities is an intersection point of two (or more) of the boundaries that satisfies all of the inequalities.

Each vertex may be found by solving for an intersection of two of the boundaries; if this point satisfies all of the inequalities, then it is a vertex of the region.

EXAMPLE 3

FINDING THE VERTICES OF A FEASIBLE REGION

Graph and find the vertices of the feasible region for $\begin{cases} x + 2y \leq 12 \\ x - y \geq 0 \\ y \geq 0 \end{cases}$

Solution

We graph the boundaries in the usual way, by plotting their intercepts. In Example 2 we found that the boundary $x + 2y = 12$ had intercepts $(0, 6)$ and $(12, 0)$. For the boundary $x - y = 0$ we use the intercept $(0, 0)$ and any other point satisfying $x - y = 0$, such as $(1, 1)$, leading to the graph on the left. The arrows are determined by test points just as in Examples 1 and 2. The inequality $y \geq 0$ means "upwards from the *x*-axis."

The only vertex that is not obvious from the graph is the intersection of $x + 2y = 12$ and $x - y = 0$. Subtracting:

$$\begin{array}{r} x + 2y = 12 \\ x - y = 0 \\ \hline 3y = 12 \end{array}$$

⟵ Top minus bottom

This last equation gives $y = 4$, which, when substituted into $x - y = 0$, gives $x - 4 = 0$ or $x = 4$, so the intersection point is $(4, 4)$. The feasible region is shown on the left, with vertices $(0, 0)$, $(4, 4)$, and $(12, 0)$.

Notice that the point $(0, 6)$ is *not* a vertex of the feasible region since it is on the *wrong* side of the line $x - y = 0$ (it does not satisfy $x - y \geq 0$). Vertices of the feasible region must be on the correct side of *all* of the lines (that is, their coordinates must satisfy *all* of the inequalities).

Bounded and Unbounded Regions

A region is *bounded* if it is enclosed on all sides by boundary lines. For example, the feasible region in Example 3 is bounded. However, if in Example 3 the direction of the inequality $x + 2y \leq 12$ were reversed to become $x + 2y \geq 12$, the feasible region would be on the *other* side of the boundary $x + 2y = 12$, giving the *unbounded* region shown on the right below.

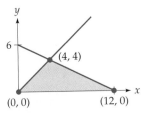

The bounded region
from Example 3.

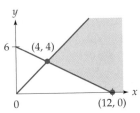

An unbounded region.

Practice Problem **3**   Graph the feasible region $\begin{cases} x + 2y \geq 20 \\ x \geq 0 \\ y \geq 0 \end{cases}$ Is the region bounded?

[*Hint:* Modify your graph from Practice Problem 2.]

➤ Solution on page 256

Applications

Since most physical quantities are nonnegative, applied problems usually include the two *nonnegativity inequalities* $x \geq 0$ and $y \geq 0$, which mean that the feasible region will be in the first quadrant.

EXAMPLE **4**   **JEWELRY PRODUCTION**

A small jewelry company prepares and mounts semiprecious stones. There are 10 lapidaries (who cut and polish the stones) and 12 jewelers (who mount the stones in gold settings). Each employee works 7 hours each day. Each tray of agates requires 5 hours of cutting and polishing and 4 hours of mounting, while each tray of onyxes requires 2 hours of cutting and polishing and 3 hours of mounting. How many trays of each stone can be processed each day?

Formulate this situation as a system of linear inequalities, sketch the feasible region, and find the vertices.

Solution

Clearly, there are many different ways the company could operate each day: It could choose to do nothing and give the employees the day off, it could process just one kind of stone, or it could process some combination. But it could not exceed the amount of time the lapidaries could work (10 workers at 7 hours each is 70 work-hours) or the amount of time the jewelers could work (12 workers at 7 hours each is 84 work-hours). Nor could it process a negative number of trays.

Let

$$x = \begin{pmatrix} \text{Number of trays} \\ \text{of agates} \end{pmatrix} \quad \text{and} \quad y = \begin{pmatrix} \text{Number of trays} \\ \text{of onyxes} \end{pmatrix}$$

For the lapidaries,

$$\underbrace{5x}_{\substack{x \text{ trays} \\ @ 5 \text{ hours} \\ \text{each}}} + \underbrace{2y}_{\substack{y \text{ trays} \\ @ 2 \text{ hours} \\ \text{each}}} \underset{\substack{\text{No} \\ \text{more} \\ \text{than}}}{\leq} \underbrace{70}_{\substack{10 \text{ workers} \\ @ 7 \text{ hours} \\ \text{each}}} \qquad \begin{pmatrix} \text{Time} \\ \text{for} \\ \text{agates} \end{pmatrix} + \begin{pmatrix} \text{Time} \\ \text{for} \\ \text{onyxes} \end{pmatrix} \leq \begin{pmatrix} \text{Total} \\ \text{time for} \\ \text{lapidaries} \end{pmatrix}$$

Summarizing

$$\left. \begin{array}{l} \textbf{10 Lapidaries} \\ \textbf{12 Jewelers} \end{array} \right\} \times \textbf{7 hours}$$

(hours)	Agates	Onyxes
Cut & Polish	5	2
Mount	4	3

For the jewelers,

$$4x \quad + \quad 3y \quad \leq \quad 84 \qquad \left(\begin{array}{c}\text{Time} \\ \text{for} \\ \text{agates}\end{array}\right) + \left(\begin{array}{c}\text{Time} \\ \text{for} \\ \text{onyxes}\end{array}\right) \leq \left(\begin{array}{c}\text{Total} \\ \text{time for} \\ \text{jewelers}\end{array}\right)$$

x trays   y trays   No   12 workers
@ 4 hours   @ 3 hours   more   @ 7 hours
each   each   than   each

Combining these constraints with the nonnegativity conditions $x \geq 0$ and $y \geq 0$, we can represent the problem by the system of linear inequalities

$$\begin{cases} 5x + 2y \leq 70 \\ 4x + 3y \leq 84 \\ x \geq 0 \\ y \geq 0 \end{cases}$$

We can now proceed as usual. The intercepts of the boundary line $5x + 2y = 70$ are $(14, 0)$ and $(0, 35)$. The intercepts of the boundary line $4x + 3y = 84$ are $(21, 0)$ and $(0, 28)$. The nonnegativity conditions $x \geq 0$ and $y \geq 0$ place the feasible region in the first quadrant. Since the test point $(0, 0)$ satisfies the constraints, this region has the origin for one of its vertices.

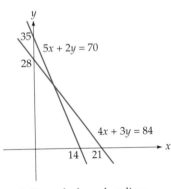

1. Draw the boundary lines

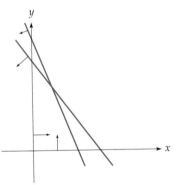

2. Mark the correct sides

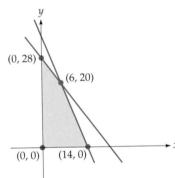

3. Shade the feasible region

Three of the four vertices of this bounded region are already known from the x- and y-intercepts of the boundary lines. The fourth is the intersection of $5x + 2y = 70$ and $4x + 3y = 84$:

$$\begin{array}{rcl} 5x + 2y = 70 & \times 3 \rightarrow & 15x + 6y = 210 \\ 4x + 3y = 84 & \times 2 \rightarrow & 8x + 6y = 168 \\ \hline & & 7x \qquad = 42 \quad \leftarrow \text{Top minus bottom} \end{array}$$

The solution to this last equation is $x = 6$ which, substituted into either equation, gives $y = 20$, for the intersection point $(6, 20)$. The feasible region has vertices $(0, 0)$, $(14, 0)$, $(6, 20)$, and $(0, 28)$, as shown on the right above.

Graphing Calculator Exploration

Intersection
X=6   Y=20

You can graph the boundary lines from Example 4 on your graphing calculator by entering them as $y_1 = (70 - 5x)/2$ and $y_2 = (84 - 4x)/3$.

a. Set the window to $[-10, 60]$ by $[-10, 40]$ and graph your lines. Remember that the nonnegativity conditions $x \geq 0$ and $y \geq 0$ put this region in the first quadrant.

b. You can use TRACE or INTERSECT to find the intersection point of the lines y_1 and y_2.

EXAMPLE 5   **WASTE MANAGEMENT**

The Marshall County trash incinerator in Norton burns 10 tons of trash per hour and co-generates 6 kilowatts (kW) of electricity, while the Wiseburg incinerator burns 5 tons per hour and co-generates 4 kilowatts. If the county needs to burn at least 70 tons of trash and co-generate at least 48 kilowatts of electricity each day, how many hours should each plant operate?

Formulate this situation as a system of linear inequalities, sketch the feasible region, and find the vertices.

Solution

Again, there are many different schedules of operating times that could burn all the trash and co-generate enough electricity. But there must be at least enough time to get the job done, and neither plant can operate a negative number of hours. Let

	Tons	kW
Summarizing		
Norton	10	6
Wiseburg	5	4
	≥ 70	≥ 48

$$x = \begin{pmatrix} \text{Number of hours} \\ \text{Norton operates} \end{pmatrix} \quad \text{and} \quad y = \begin{pmatrix} \text{Number of hours} \\ \text{Wiseburg operates} \end{pmatrix}$$

For the amount of trash to be burned, we have the inequality

$$\underbrace{10x}_{\substack{x \text{ hours} \\ @\ 10 \text{ tons} \\ \text{each}}} + \underbrace{5y}_{\substack{y \text{ hours} \\ @\ 5 \text{ tons} \\ \text{each}}} \underbrace{\geq}_{\substack{\text{At} \\ \text{least}}} \underbrace{70}_{\substack{\text{Tons of} \\ \text{trash} \\ \text{to burn}}} \qquad \begin{pmatrix} \text{Tons} \\ \text{burned} \\ \text{at Norton} \end{pmatrix} + \begin{pmatrix} \text{Tons} \\ \text{burned at} \\ \text{Wiseburg} \end{pmatrix} \geq \begin{pmatrix} \text{Tons} \\ \text{to} \\ \text{burn} \end{pmatrix}$$

For the electricity to be produced,

$$\underbrace{6x}_{\substack{x \text{ hours} \\ @\ 6 \text{ kWs} \\ \text{each}}} + \underbrace{4y}_{\substack{y \text{ hours} \\ @\ 4 \text{ kWs} \\ \text{each}}} \underbrace{\geq}_{\substack{\text{At} \\ \text{least}}} \underbrace{84}_{\substack{\text{kWs} \\ \text{needed}}} \qquad \begin{pmatrix} \text{Norton} \\ \text{electricity} \end{pmatrix} + \begin{pmatrix} \text{Wiseburg} \\ \text{electricity} \end{pmatrix} \geq \begin{pmatrix} \text{Electricity} \\ \text{needed} \end{pmatrix}$$

Combining these constraints with the nonnegativity conditions $x \geq 0$ and $y \geq 0$, we can represent the problem by the system of linear inequalities

$$\begin{cases} 10x + 5y \geq 70 \\ 6x + 4y \geq 48 \\ x \geq 0 \\ y \geq 0 \end{cases}$$

The intercepts of $10x + 5y = 70$ are $(7, 0)$ and $(0, 14)$. The intercepts of $6x + 4y = 48$ are $(8, 0)$ and $(0, 12)$. The nonnegativity conditions place the feasible region in the first quadrant. Since the origin does not satisfy either of the first two inequalities, this region does *not* have the origin for one of its vertices.

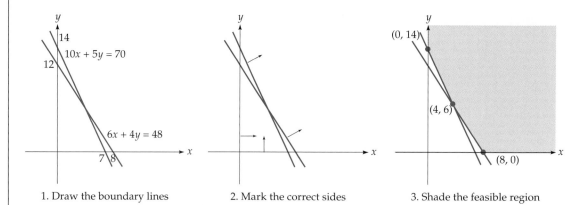

1. Draw the boundary lines   2. Mark the correct sides   3. Shade the feasible region

Two of the three vertices of this unbounded region are already known from the *x*- and *y*-intercepts of the boundary lines. The third is the intersection of $10x + 5y = 70$ and $6x + 4y = 48$, which, found in the usual way (omitting the details), is $(4, 6)$. The three vertices of this region are $(8, 0)$, $(4, 6)$, and $(0, 14)$, as shown in the third diagram above.

4.1 **Section Summary**

The *feasible region* of a system of linear inequalities $\begin{cases} ax + by \leq c \\ \vdots \end{cases}$ consists of all the (x, y) points on the correct sides of the *boundary lines* $\begin{cases} ax + by = c \\ \vdots \end{cases}$. We usually graph the boundary lines by plotting their

For a, b, and c all positive

intercepts. The correct side of the boundary can be found by trying a "test point" in the inequality [the origin $(0, 0)$ is usually the easiest to use, but any point *not* on the line will do].

The *vertices* are the "corner points" of the region and are the intersections of two (or more) boundary lines that satisfy the inequalities. These intersections can be found by solving the equations representing the lines.

A feasible region is *bounded* if it is enclosed on all sides by boundary lines, and *unbounded* if it is not.

Two special inequalities are the *nonnegativity conditions* $x \geq 0$ and $y \geq 0$, which place the feasible region in the first quadrant and frequently appear in word problems when the variables represent amounts of materials or other real objects.

➤ **Solutions to Practice Problems**

1. The boundary is the line $3x - 5y = 60$. From $x = 60/3 = 20$, the x-intercept is $(20, 0)$. From $y = 60/(-5) = -12$, the y-intercept is $(0, -12)$. Because $3 \cdot 0 - 5 \cdot 0$ is ≤ 60, the origin $(0, 0)$ is on the correct side of the boundary line.

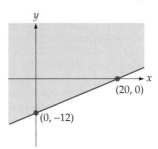

2. The boundary line is $x + 2y = 20$ with intercepts $(20, 0)$ and $(0, 10)$. The origin $(0, 0)$ is on the correct side of $x + 2y \leq 20$ (since $0 \leq 20$). $x \geq 0$ means "to the right of the y-axis" and $y \geq 0$ means "above the x-axis."

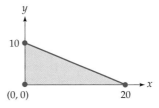

This system *is* feasible.

3. The "reversed" inequality $x + 2y \geq 20$ means the other side of the boundary line, so the region is:

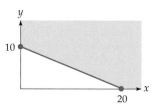

The region is *unbounded*.

4.1 Exercises

For each region, select the inequality it represents.

1. **a.** $5x + 8y \geq 40$
   b. $8x + 5y \geq 40$
   c. $5x + 8y \leq 40$
   d. $8x + 5y \leq 40$

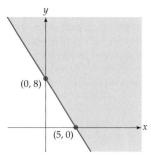

2. **a.** $7x + 6y \geq 42$
   b. $6x + 7y \geq 42$
   c. $7x + 6y \leq 42$
   d. $6x + 7y \leq 42$

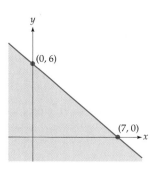

3. **a.** $-12x + 5y \geq -60$
   b. $5x - 12y \geq -60$
   c. $5x - 12y \leq 60$
   d. $10x + 24y \leq 60$

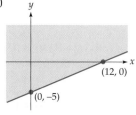

4. **a.** $5x - 3y \leq -15$
   b. $5x - 3y \geq -15$
   c. $-3x + 5y \leq 15$
   d. $-5x + 3y \leq 15$

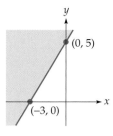

For each region, select the system of inequalities it represents.

5. **a.** $\begin{cases} 5x + 2y \leq 20 \\ y \geq 0 \end{cases}$

   b. $\begin{cases} 4x + 10y \leq 40 \\ x \leq 0 \end{cases}$

   c. $\begin{cases} 10x - 4y \leq 40 \\ y \leq 0 \end{cases}$

   d. $\begin{cases} -2x + 5y \geq -20 \\ y \geq 0 \end{cases}$

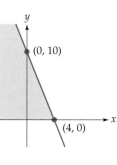

6. **a.** $\begin{cases} 5x - 3y \leq -30 \\ x \geq 0 \end{cases}$

   b. $\begin{cases} -6x + 10y \leq -60 \\ x \geq 0 \end{cases}$

   c. $\begin{cases} -5x + 3y \leq 30 \\ x \leq 0 \end{cases}$

   d. $\begin{cases} 5x + 3y \leq -30 \\ x \geq 0 \end{cases}$

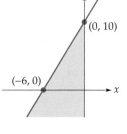

7. a. $\begin{cases} x - 3y \le -60 \\ x \ge 0 \\ y \ge 0 \end{cases}$

b. $\begin{cases} 2x - 6y \ge -120 \\ x \ge 0 \\ y \ge 0 \end{cases}$

c. $\begin{cases} -60x + 20y \le 240 \\ y \ge 0 \end{cases}$

d. $\begin{cases} 3x - y \ge -60 \\ x \le 0 \\ y \ge 0 \end{cases}$

8. a. $\begin{cases} 2x + y \ge 16 \\ x \ge 0 \end{cases}$

b. $\begin{cases} 8x + 16y \le 32 \\ y \le 0 \\ x \ge 0 \end{cases}$

c. $\begin{cases} x + 2y \ge 16 \\ x \ge 0 \end{cases}$

d. $\begin{cases} 2x + y \ge -16 \\ y \ge 0 \end{cases}$

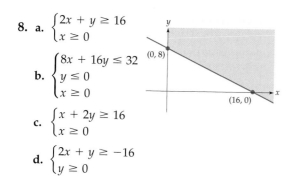

Sketch each system of inequalities. List all vertices and identify the region as "bounded" or "unbounded."

9. $\begin{cases} x + 2y \le 40 \\ x \ge 0, \ y \ge 0 \end{cases}$ **10.** $\begin{cases} 3x + y \le 90 \\ x \ge 0, \ y \ge 0 \end{cases}$

11. $\begin{cases} -2x + y \le 10 \\ x \le 10 \\ x \ge 0, \ y \ge 0 \end{cases}$ **12.** $\begin{cases} 2x - y \le 20 \\ y \le 40 \\ x \ge 0, \ y \ge 0 \end{cases}$

13. $\begin{cases} x + 2y \le 8 \\ x + y \le 6 \\ x \ge 0, \ y \ge 0 \end{cases}$ **14.** $\begin{cases} 2x + y \le 10 \\ x + y \le 8 \\ x \ge 0, \ y \ge 0 \end{cases}$

15. $\begin{cases} 5x + 2y \ge 20 \\ x \ge 0, \ y \ge 0 \end{cases}$ **16.** $\begin{cases} 4x + 5y \ge 20 \\ x \ge 0, \ y \ge 0 \end{cases}$

17. $\begin{cases} 4x + 3y \ge 24 \\ y \ge 4 \\ x \ge 0, \ y \ge 0 \end{cases}$ **18.** $\begin{cases} 3x + 4y \ge 12 \\ x \le 8 \\ x \ge 0, \ y \ge 0 \end{cases}$

19. $\begin{cases} 3x + y \ge 12 \\ x + y \ge 8 \\ x \ge 0, \ y \ge 0 \end{cases}$ **20.** $\begin{cases} x + 3y \ge 15 \\ x + y \ge 9 \\ x \ge 0, \ y \ge 0 \end{cases}$

21. $\begin{cases} 2x + y \le 80 \\ x + 3y \ge 30 \\ x \ge 0, \ y \ge 0 \end{cases}$ **22.** $\begin{cases} 3x + y \le 90 \\ x + 2y \ge 20 \\ x \ge 0, \ y \ge 0 \end{cases}$

23. $\begin{cases} x - 3y \le 15 \\ 2x + y \ge 30 \\ x \ge 0, \ y \ge 0 \end{cases}$ **24.** $\begin{cases} 3x + y \ge 30 \\ x - 2y \ge -60 \\ x \ge 0, \ y \ge 0 \end{cases}$

APPLIED EXERCISES

Formulate each situation as a system of inequalities, sketch the feasible region, and find the vertices. Be sure to state clearly the meaning of your x- and y-variables.

25. BUSINESS: Livestock Management A rancher raises goats and llamas on his 400-acre ranch. Each goat needs 2 acres of land and requires $100 of veterinary care per year, while each llama needs 5 acres of land and requires $80 of veterinary care per year. If the rancher can afford no more than $13,200 for veterinary care this year, how many of each animal can he raise?

26. BUSINESS: Agriculture Management A farmer grows wheat and barley on her 500-acre farm. Each acre of wheat requires 3 days of labor to plant, tend, and harvest, while each acre of barley requires 2 days of labor. If the farmer and her hired field hands can provide no more than 1200 days of labor this year, how many acres of each crop can she grow?

27. BUSINESS: Production Planning A boat company manufactures aluminum dinghies and rowboats. The amounts of metal work and painting needed for each are shown in the table, together with the number of hours of skilled labor available for each task. How many of each kind of boat can the company manufacture?

	Dinghy	Rowboat	Labor Available
Metal work	2 hours	3 hours	120 hours
Painting	2 hours	2 hours	100 hours

28. BUSINESS: Resource Allocation A sailboat company manufactures fiberglass prams and yawls. The amounts of molding, painting, and finishing needed for each are shown in the table, together with the number of hours of skilled labor available for each task. How many of each kind of sailboat can the company manufacture?

	Pram	Yawl	Labor Available
Molding	3 hours	6 hours	150 hours
Painting	3 hours	2 hours	114 hours
Finishing	2 hours	6 hours	132 hours

29. BIOMEDICAL: Nutrition Joshua loves "junk food" but wants to stay within the recommended daily limits of 80 grams of fat and 2250 calories. If each serving of Sugar-Snaks contains 5 grams of fat and 125 calories and each bag of Gobbl'Ems contains 8 grams of fat and 250 calories, how much of each may he eat today?

30. BIOMEDICAL: Diet Planning Justin plays a lot of sports and wants to make sure he gets at least 70 grams of protein and 20 milligrams of iron each day. If each serving of Pro-Team Power Bars has 10 grams of protein and 2 milligrams of iron and each glass of Bulk-Up-Delight has 5 grams of protein and 2 milligrams of iron, how much of each should he eat each day?

31. ENVIRONMENTAL SCIENCES: Pollution Control A smelting company refines metals at two factories located in Ohio and in Pennsylvania. The smokestacks release both sulfur dioxide (which combines with water vapor to form "acid rain") and particulates (solid mat-

ter such as soot that can cause respiratory problems) at the rates shown in the table. The EPA has obtained an injunction against the company preventing it from releasing more than 64 pounds of sulfur dioxide and 60 pounds of particulates into the atmosphere each day. How many hours can the company operate these factories each day?

(Pounds Per Hour)	Sulfur Dioxide	Particulates
Ohio factory	4	5
Pennsylvania factory	4	3

32. ENVIRONMENTAL SCIENCES: Pollution Control A chemical company manufactures batteries at two factories located in Connecticut and in Alabama. The factories discharge both heavy metals (such as mercury and cadmium, which are very toxic) and nitric acid into the local river systems at the rates shown in the table. An environmental organization has obtained an injunction against the company preventing it from discharging more than 54 pounds of heavy metals and 60 pounds of nitric acid each day. How many hours can the company operate these factories each day?

(Pounds Per Hour)	Heavy Metals	Nitric Acid
Connecticut factory	6	4
Alabama factory	3	4

33. PERSONAL FINANCE: Investment Strategy An investment portfolio manager has $8 million to invest in stock and bond funds. If the amount invested in stocks can be no more than the amount invested in bonds, how much can be invested in each type of fund?

34. PERSONAL FINANCE: Financial Planning A retired couple want to invest their $20,000 life savings in bank certificates of deposit and Treasury bonds. If they want at least $5000 in each type of investment, how much can they invest in each?

4.2 TWO-VARIABLE LINEAR PROGRAMMING PROBLEMS

Introduction

In this section we will explain what a linear programming problem is and then use the geometry of feasible regions from the previous section to show that the solution of a linear programming problem occurs at a vertex of the region.

Linear Programming Problems

We begin by formulating an example of a linear programming problem, which we will then solve in Example 2.

EXAMPLE 1

FARM MANAGEMENT

A farmer grows soybeans and corn on his 300-acre farm. To maintain soil fertility, the farmer rotates the crops and always plants at least as many acres of soybeans as acres of corn. If each acre of soybeans yields a profit of $100 and each acre of corn yields a profit of $200, how many acres of each crop should the farmer plant to obtain the greatest possible profit?

Formulate this situation as a linear programming problem by identifying the variables, the objective function, and the constraints.

Solution

Since the question asks "how many acres of each crop," we let

$$x = \begin{pmatrix} \text{Number of} \\ \text{acres of soybeans} \end{pmatrix} \quad \text{and} \quad y = \begin{pmatrix} \text{Number of} \\ \text{acres of corn} \end{pmatrix}$$

The objective is to maximize the farmer's profit, and this profit is $P = 100x + 200y$ because the profits per acre are $100 for soybeans and $200 for corn. The 300-acre size of the farm leads to the constraint $x + y \leq 300$, while the crop rotation requirement that $x \geq y$ can be written as $x - y \geq 0$. Since x and y cannot be negative, we also have the nonnegativity constraints $x \geq 0$ and $y \geq 0$. This maximum linear programming problem may be written as

Objective function

Maximize $P = 100x + 200y$ Corn and soybean profits

Subject to $\begin{cases} x + y \leq 300 \\ x - y \geq 0 \\ x \geq 0 \ \text{and} \ y \geq 0 \end{cases}$

Size of farm
Crop rotation
Nonnegativity

Constraints

We call $P = 100x + 200y$ the "objective function" because the objective of the entire procedure is to maximize it. The inequalities are called "constraints" because they constrain the profit by, for example, the limiting the crops to 300 acres.

In general, a *linear programming problem* consists of a linear *objective function* to be maximized or minimized subject to *constraints* in the form of a system of linear inequalities. Since an inequality with a $\geq$ can be changed into one with a $\leq$ by multiplying by -1 (for example, $x - y \geq 0$ can be written as $-x + y \leq 0$), every linear programming problem in two variables may be written in one of the following forms:

Linear Programming Problems

Maximize $P = Mx + Ny$

Subject to $\begin{cases} ax + by \leq c \\ \vdots \end{cases}$

or

Minimize $C = Mx + Ny$

Subject to $\begin{cases} ax + by \geq c \\ \vdots \end{cases}$

We call the objective function in the maximum problem P for profit and in the minimum problem C for cost, but problems with different goals are perfectly acceptable. Nonnegativity conditions should be included only when appropriate.

Fundamental Theorem of Linear Programming

A *solution* of a *maximum* linear programming problem is a point that satisfies the system of linear inequalities (a "feasible" point) and that gives the *largest* possible value of the objective function. Similarly, a solution of a *minimum* linear programming problem is a feasible point that gives the *smallest* possible value of the objective function. With either problem, if the constraints have no solutions (they are "infeasible") then the problem has no solution.

How can we solve such a problem? Consider the objective function $P = 100x + 200y$ from Example 1. If we solve this equation for y, we obtain:

$$200y = -100x + P \qquad \text{Switching sides}$$
$$y = -\tfrac{1}{2}x + \tfrac{1}{200}P \qquad \text{Dividing through by 200}$$

The $y = mx + b$ form of this equation shows that it represents a line of slope $-\tfrac{1}{2}$. Different values of P will give parallel "objective function lines," all with slope $-\tfrac{1}{2}$. In Example 2 we will see that the feasible region for Example 1 is the triangular region in the following graph. On this graph are drawn several of these objective function lines for various values of P. Higher lines correspond to larger value of P, so *maximizing P* means choosing the uppermost line that still intersects the feasible region.

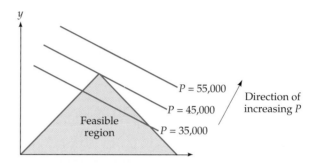

As the diagram shows, the uppermost line that intersects the region does so at a *vertex* of the region. (Any line with a higher value of P, such as $P = 55,000$, does not intersect the feasible region at all.) It should be clear that the situation will be the same for *any* linear objective function and *any* feasible region bounded by linear inequalities: *For a linear programming problem with a bounded region, the solution occurs at a vertex of the region.*

For an *unbounded* region the solution might not exist: The objective function lines might move in an unbounded direction as P increases, allowing higher and higher values of P, so there would never be a largest value. However, if increasing the value of P always means moving toward a boundary, then the solution *will* exist and again will occur at a vertex.

Similar reasoning applies to a *minimum* linear programming problem by finding the objective function line with the *minimum* value of P that intersects the feasible region. Combining these observations, we have established the following important result about the solution of a linear programming problem.

Fundamental Theorem of Linear Programming

If a linear programming problem has a solution,
then it occurs at a *vertex* of the feasible region.

The Fundamental Theorem of Linear Programming gives us the following procedure to find a solution.

> **How to Solve a Linear Programming Problem**
>
> If the region is bounded, list the vertices and calculate the value of the objective function at each. The solution occurs at the vertex that gives the largest (or smallest) value.
>
> If the region is unbounded, first check whether the objective function "improves" in an unbounded direction. If it does, there is *no solution*; if it does not, list the vertices and calculate the value of the objective function at each; the solution occurs at the vertex that gives the largest (or smallest) value.

EXAMPLE **2**

SOLUTION OF EXAMPLE 1

Solve the linear programming problem from Example 1 on page 259:

$$\text{Maximize} \quad P = 100x + 200y \qquad \text{Objective function}$$

$$\text{Subject to} \quad \begin{cases} x + y \le 300 \\ x - y \ge 0 \\ x \ge 0 \quad \text{and} \quad y \ge 0 \end{cases} \qquad \text{Constraints}$$

Solution

We begin by sketching the region determined by the constraints. The nonnegativity conditions $x \ge 0$ and $y \ge 0$ place the region in the first quadrant. The boundary line $x + y = 300$ has intercepts $(300, 0)$ and $(0, 300)$, and the boundary line $x - y = 0$ passes through the origin $(0, 0)$ and the point $(1, 1)$. These points give the first of the following graphs. For test points, the origin satisfies the inequality $x + y \le 300$, and the point $(300, 0)$ satisfies the inequality $x - y \ge 0$.

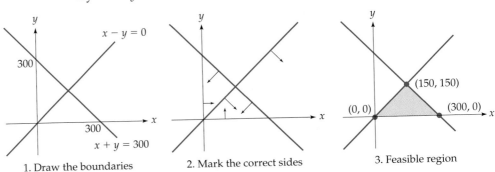

1. Draw the boundaries　　2. Mark the correct sides　　3. Feasible region

The region has three vertices, and two are already known from the x- and y-intercepts of the boundary lines. The third is the intersection of $x - y = 0$ and $x + y = 300$:

$$x + y = 300$$
$$\underline{x - y = \quad 0}$$
$$2x \quad = 300 \qquad \longleftarrow \text{Adding}$$

The last equation gives $x = 150$, and from $x + y = 300$, we find $y = 150$, for the vertex $(150, 150)$. The vertices are $(0, 0)$, $(300, 0)$, and $(150, 150)$, as shown in the third graph.

Since the region is bounded, this problem *does* have a solution, and it occurs at a vertex. Evaluating the objective function at the vertices, we find the following values:

Vertex	Value of $P = 100x + 200y$	
$(0, 0)$	0	$0 = 100 \cdot 0 + 200 \cdot 0$
$(300, 0)$	30,000	$30,000 = 100 \cdot 300 + 200 \cdot 0$
$(150, 150)$	45,000 $\longleftarrow$ Largest	$45,000 = 100 \cdot 150 + 200 \cdot 150$

Since the largest value of the objective function occurs at the vertex $(150, 150)$, the solution of this problem is:

The maximum value of P is 45,000 at the vertex $(150, 150)$.

In terms of the original word problem in Example 1, the maximum profit is $45,000 when the farmer plants 150 acres of corn and 150 acres of soybeans.

Practice Problem 1   Solve the linear programming problem:

Maximize $P = 5x + 2y$

Subject to $\begin{cases} 3x + y \le 60 \\ y \le 36 \\ x \ge 0 \ \text{ and } \ y \ge 0 \end{cases}$

➤ Solution on page 269

EXAMPLE **3**

A MINIMUM PROBLEM ON AN UNBOUNDED REGION

Solve the linear programming problem:

$$\text{Minimize} \quad C = 3x + 5y$$

$$\text{Subject to} \quad \begin{cases} 2x + y \geq 12 \\ x + y \geq 8 \\ x \geq 0 \quad \text{and} \quad y \geq 0 \end{cases}$$

Solution

We begin by sketching the region determined by the constraints. The nonnegativity conditions $x \geq 0$ and $y \geq 0$ place the region in the first quadrant. The boundary line $2x + y = 12$ has intercepts $(6, 0)$ and $(0, 12)$, while the boundary line $x + y = 8$ has intercepts $(8, 0)$ and $(0, 8)$. Since the test point $(0, 0)$ does *not* satisfy $2x + y \geq 12$ and $x + y \geq 8$, the region is on the sides of these boundaries *away* from the origin.

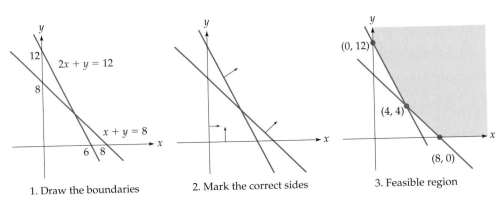

1. Draw the boundaries   2. Mark the correct sides   3. Feasible region

This unbounded region has three vertices, and two are already known from the x- and y-intercepts of the boundary lines. The third is the intersection of $2x + y = 12$ and $x + y = 8$, which we solve together:

$$\begin{array}{r} 2x + y = 12 \\ x + y = 8 \\ \hline x = 4 \end{array}$$ ← Top minus bottom

Substituting $x = 4$ into $x + y = 8$ gives $y = 4$, for the intersection point $(4, 4)$. The vertices are $(0, 12)$, $(4, 4)$, and $(8, 0)$, as shown in the third graph above.

    The region is unbounded, so the solution may not exist. The third graph above shows that the region is unbounded in the *positive x direc-*

tion and also in the *positive y direction*, which means that we could increase x as much as we choose, and we could increase y as much as we choose. However, increasing either x or y will *increase* the objective function $C = 3x + 5y$ (since x and y are multiplied by *positive* numbers) which is the *opposite* of what we should do to minimize C. Therefore, the objective function does not "improve" in an unbounded direction, so the solution *does* exist. Evaluating the objective function at the vertices, we find the following values.

Vertex	$C = 3x + 5y$		
(8, 0)	24	⟵ Smallest	$24 = 3 \cdot 8 + 5 \cdot 0$
(4, 4)	32		$32 = 3 \cdot 4 + 5 \cdot 4$
(0, 12)	60		$60 = 3 \cdot 0 + 5 \cdot 12$

Since the smallest value of the objective function occurs at the vertex (8, 0), the solution of this problem is:

The minimum value of C is 24 at the vertex (8, 0).

A linear programming problem with an unbounded feasible region may or may not have a solution. For example, in a maximum problem if the objective function can be made arbitrarily large by choosing x and y values far out in the unbounded direction, then it will not have a largest value, so the problem will have no solution. The general situation is described in the following box, where the phrase *the objective function "improves"* means that it *increases* for a *maximum* problem and that it *decreases* for a *minimum* problem.

Determining Whether a Solution Exists

If the feasible region is unbounded, the linear programming problem may not have a solution. First determine the direction(s) in which the feasible region is unbounded.

1. If the objective function *improves* when the x and y variables take values further out an unbounded direction, then the solution *does not exist*.

2. If the objective function does *not* improve when the x and y variables take values further out in any of the unbounded directions, then the solution *does* exist.

Practice Problem 2   Solve the linear programming problem:

$$\text{Maximize} \quad P = 5x + 11y$$

$$\text{Subject to} \quad \begin{cases} x + 3y \geq 60 \\ x \geq 0 \text{ and } y \geq 0 \end{cases}$$

➤ **Solution on page 269**

EXAMPLE 4   **A MANUFACTURING PROBLEM**

A fully automated plastics factory produces two toys, a racing car and a jet airplane, in three stages: molding, painting, and packaging. After allowing for routine maintenance, the equipment for each stage can operate no more than 150 hours per week. Each batch of racing cars requires 6 hours of molding, 2.5 hours of painting, and 5 hours of packaging, while each batch of jet airplanes requires 3 hours of molding, 7.5 hours of painting, and 5 hours of packaging. If the profit per batch of toys is $120 for cars and $100 for airplanes, how many batches of each toy should be produced each week to obtain the greatest possible profit?

Solution

Summarizing

(hours)	Car	Plane
Molding	6	3
Painting	2.5	7.5
Packaging	5	5
Profit	$120	$100

Let

$$x = \begin{pmatrix} \text{Number of} \\ \text{batches of cars} \end{pmatrix} \quad \text{and} \quad y = \begin{pmatrix} \text{Number of} \\ \text{batches of jets} \end{pmatrix}$$

made during the week. The profit is $P = 120x + 100y$ and the problem is to maximize P subject to the constraints that the molding, painting, and packaging processes can each take no more than 150 hours:

$$6x + 3y \leq 150 \qquad \text{Molding time}$$
$$2.5x + 7.5y \leq 150 \qquad \text{Painting time}$$
$$5x + 5y \leq 150 \qquad \text{Packaging time}$$

and the nonnegativity conditions $x \geq 0$ and $y \geq 0$. Simplifying the constraints by removing common factors (dividing the molding constraint by 3, the painting constraint by 2.5, and the packaging constraint by 5), we can rewrite this problem as the linear programming problem

$$\text{Maximize} \quad P = 120x + 100y \qquad \text{Objective function}$$

$$\text{Subject to} \quad \begin{cases} 2x + y \leq 50 \\ x + 3y \leq 60 \\ x + y \leq 30 \\ x \geq 0 \text{ and } y \geq 0 \end{cases} \qquad \text{Constraints}$$

Proceeding as usual, we graph the boundary lines by plotting their intercepts (the first graph), determine the correct sides by using the origin as a test point (the second graph), and finally shade the feasible region (the third graph).

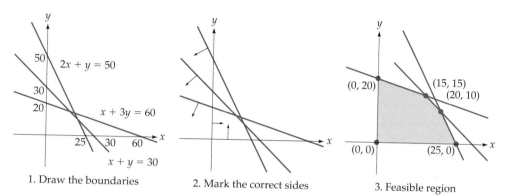

1. Draw the boundaries   2. Mark the correct sides   3. Feasible region

The third graph shows that the region has five vertices, three of which are already known from the x- and y-intercepts. The other two come from solving pairs of equations in the usual way, giving the following vertices:

$$
\begin{array}{rr}
2x + y = 50 & x + 3y = 60 \\
\underline{x + y = 30} & \underline{x + y = 30} \\
x \qquad = 20 & 2y = 30 \\
\text{Vertex: } (20, 10) & \text{Vertex: } (15, 15)
\end{array}
$$

[The intersection of $2x + y = 50$ and $x + 3y = 60$ is the point $(18, 14)$, but this is *not* a vertex of the region because it does not also satisfy $x + y \leq 30$: $18 + 14$ is not ≤ 30.] The vertices are $(0, 0)$, $(25, 0)$, $(20, 10)$, $(15, 15)$, and $(0, 20)$ as shown in the preceding diagram.

The region is bounded, so this problem *has* a solution, and it occurs at a vertex. Evaluating the objective function at the vertices, we find the following values:

Vertex	$P = 120x + 100y$
$(0, 0)$	0
$(25, 0)$	3000
$(20, 10)$	3400 ←—— Largest
$(15, 15)$	3300
$(0, 20)$	2000

Since the largest value of the objective function occurs at the vertex $(20, 10)$, the solution is:

The maximum value is 3400 at the vertex $(20, 10)$.

In terms of the original question, the maximum profit is $3400 when the factory produces 20 batches of racing car toys and 10 batches of jet airplane toys each week.

Be careful! Except in very simple problems, some boundary line intersections are *not* vertices of the feasible region because they violate at least one of the other constraints. Particularly when sketching regions by hand, you may want to find all the intersection points anyway and then verify which are feasible and which are not.

Extensions to Larger Problems

Our geometric method is limited to two variables and therefore to two products, which is unrealistic for a large company. In the next two sections we will develop a numerical method that applies to problems with *any* number of variables, and that also eliminates the need to find and check *all* of the vertices.

4.2 Section Summary

A *linear programming problem* consists of a linear *objective function* to be *maximized* or *minimized* subject to *constraints*, written as a system of linear inequalities. Every such problem can be written in the form

Maximize $P = Mx + Ny$

Subject to $\begin{cases} ax + by \leq c \\ \vdots \end{cases}$

or

Minimize $C = Mx + Ny$

Subject to $\begin{cases} ax + by \geq c \\ \vdots \end{cases}$

A *solution* is a feasible point that gives the largest (or smallest) possible value for the objective function. The *Fundamental Theorem of Linear Programming* says that if there is a solution, it occurs at a *vertex* of the feasible region. Thus the problem can be solved by graphing the feasible region of the constraints, evaluating the objective function at the vertices, and selecting the vertex with the optimal value. If the region is unbounded, you must check that a solution exists before selecting the "best" vertex.

➤ **Solutions to Practice Problems**

1.

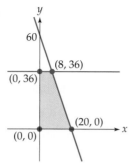

Vertex	$P = 5x + 2y$
(0, 0)	0
(20, 0)	100
(8, 36)	112
(0, 36)	72

The maximum value is 112 when $x = 8$ and $y = 36$.

2.

The region is unbounded in the positive x and positive y directions. Since the objective function $P = 5x + 11y$ increases ("improves") as x and y take values in those directions, *the solution does not exist.*

4.2 **Exercises**

For Exercises 1–5: Use the region below to find each maximum or minimum value. If such a value does not exist, explain why not.

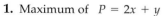

1. Maximum of $P = 2x + y$

2. Maximum of $P = x + 2y$

3. Minimum of $C = 3x + 4y$

4. Minimum of $C = 4x + 3y$

5. Maximum of $P = x - y$

For Exercises 6–10: Use the region below to find each maximum or minimum value. If such a value does not exist, explain why not.

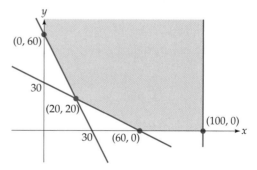

6. Minimum of $C = x + y$

7. Minimum of $C = 3x + y$

8. Maximum of $P = x + 3y$

9. Maximum of $P = 5x + 8y$

10. Maximum of $P = -x - y$

Solve each linear programming problem by sketching the region and labeling the vertices, deciding whether a solution exists, and then finding it if it does exist.

11. Maximize $P = 30x + 40y$

Subject to $\begin{cases} 2x + y \le 16 \\ x + y \le 10 \\ x \ge 0, \ y \ge 0 \end{cases}$

12. Maximize $P = 80x + 70y$

Subject to $\begin{cases} x + 2y \le 18 \\ x + y \le 10 \\ x \ge 0, \ y \ge 0 \end{cases}$

13. Minimize $C = 15x + 45y$

Subject to $\begin{cases} 2x + 5y \ge 20 \\ x \ge 0, \ y \ge 0 \end{cases}$

14. Minimize $C = 35x + 25y$

Subject to $\begin{cases} 5x + 3y \ge 60 \\ x \ge 0, \ y \ge 0 \end{cases}$

15. Maximize $P = 4x + 5y$

Subject to $\begin{cases} 2x + y \le 50 \\ x + 3y \le 75 \\ x \ge 0, \ y \ge 0 \end{cases}$

16. Maximize $P = 7x + 8y$

Subject to $\begin{cases} 3x + y \le 90 \\ x + 2y \le 60 \\ x \ge 0, \ y \ge 0 \end{cases}$

17. Minimize $C = 12x + 10y$

Subject to $\begin{cases} 4x + y \ge 40 \\ 2x + 3y \ge 60 \\ x \ge 0, \ y \ge 0 \end{cases}$

18. Minimize $C = 20x + 30y$

Subject to $\begin{cases} 3x + 2y \ge 120 \\ x + 4y \ge 80 \\ x \ge 0, \ y \ge 0 \end{cases}$

19. Maximize $P = 5x + 3y$

Subject to $\begin{cases} 2x + y \le 90 \\ x + y \le 50 \\ x + 2y \le 90 \\ x \ge 0, \ y \ge 0 \end{cases}$

20. Maximize $P = 6x + 5y$

Subject to $\begin{cases} x + 2y \le 96 \\ x + y \le 54 \\ 2x + y \le 96 \\ x \ge 0, \ y \ge 0 \end{cases}$

21. Maximize $P = 10x + 12y$

Subject to $\begin{cases} 3x + 2y \le 180 \\ 4x + y \le 120 \\ 3x + y \le 105 \\ x \ge 0, \ y \ge 0 \end{cases}$

22. Maximize $P = 20x + 15y$

Subject to $\begin{cases} 2x + 3y \le 60 \\ x + 4y \le 40 \\ x + 3y \le 33 \\ x \ge 0, \ y \ge 0 \end{cases}$

23. Minimize $C = 20x + 25y$

Subject to $\begin{cases} 3x + y \ge 60 \\ x + y \ge 42 \\ x + 3y \ge 60 \\ x \ge 0, \ y \ge 0 \end{cases}$

24. Minimize $C = 50x + 35y$

Subject to $\begin{cases} x + 3y \ge 72 \\ x + y \ge 48 \\ 3x + y \ge 72 \\ x \ge 0, \ y \ge 0 \end{cases}$

APPLIED EXERCISES

Formulate each situation as a linear programming problem by identifying the variables, the objective function, and the constraints. Be sure to state clearly the meaning of each variable. Determine whether a solution exists, and if it does, find it. State your final answer in terms of the original question.

25. **BUSINESS: Livestock Management** A rancher raises goats and llamas on his 400-acre ranch. Each goat needs 2 acres of land and requires $100 of veterinary care per year, while each llama needs 5 acres of land and requires $80 of veterinary care per year. The rancher can afford no more than $13,200 for veterinary care this year. If the expected profit is $60 for each goat and $90 for each llama, how many of each animal should he raise to obtain the greatest possible profit?

26. **BUSINESS: Agriculture Management** A farmer grows wheat and barley on her 500-acre farm. Each acre of wheat requires 3 days of labor to plant, tend, and harvest, while each acre of barley requires 2 days of labor. The farmer and her hired field hands can provide no more than 1200 days of labor this year. If the expected profit is $50 for each acre of wheat and $40 for each acre of barley, how many acres of each crop should she grow to obtain the greatest possible profit?

27. **BUSINESS: Resource Allocation** A sailboat company manufactures fiberglass prams and yawls. The amount of molding, painting, and finishing needed for each is shown in the table, together with the number of hours of skilled labor available for each task. If the expected profit is $150 for each pram and $180 for each yawl, how many of each kind of sailboat should the company manufacture to obtain the greatest possible profit?

	Pram	Yawl	Labor Available
Molding	3 hours	6 hours	150 hours
Painting	3 hours	2 hours	114 hours
Finishing	2 hours	6 hours	132 hours

28. **BUSINESS: Production Planning** A small jewelry company prepares and mounts semiprecious stones. There are 10 lapidaries (who cut and polish the stones) and 12 jewelers (who mount the stones in gold settings). Each employee works 7 hours each day. Each tray of agates requires 5 hours of cutting and polishing and 4 hours of mounting, while each tray of onyxes requires 2 hours of cutting and polishing and 3 hours of mounting. If the profit is $15 for each tray of agates and $10 for each tray of onyxes, how many trays of each stone should be processed each day to obtain the greatest possible profit?

29. **ENVIRONMENTAL SCIENCE: Waste Management** The Marshall County trash incinerator in Norton burns 10 tons of trash per hour and co-generates 6 kilowatts of electricity, while the Wiseburg incinerator burns 5 tons per hour and co-generates 4 kilowatts. The county needs to burn at least 70 tons of trash and co-generate at least 48 kilowatts of electricity every day. If the Norton incinerator costs $80 per hour to operate and the Wiseburg incinerator costs $50, how many hours should each incinerator operate each day with the least cost to the county?

30. **GENERAL: Disaster Relief** An international relief agency has been asked to provide medical support to a Caribbean island devastated by a recent hurricane. The agency estimates that it must be able to perform at least 50 major surgeries, 78 minor surgeries, and 130 outpatient services each day. The daily capacities of its portable field hospitals and clinics are given in the table, along with the daily operating costs. How many field hospitals and clinics should the agency airlift to the island to provide the needed help at the least daily cost?

	Major Surgeries	Minor Surgeries	Outpatient Services	Daily Cost
Field hospitals	5	3	2	$2000
Clinics	0	2	10	$500

31. BIOMEDICAL: Nutrition A pet store owner raises baby bunnies and feeds them leftover salad greens from a nearby restaurant. The weekly nutritional needs of each bunny are given in the table, as well as the nutrition provided by the salad greens and nutritional supplement drops. If the salad greens cost 2¢ per handful and the nutritional supplement costs 1¢ per drop, how many handfuls of greens and drops of supplement should each bunny receive weekly to minimize the owner's costs?

	Salad Greens (Per Handful)	Nutritional Supplement (Per Drop)	Minimum Weekly Requirement
Fiber (grams)	3	0	90
Vitamins (mg)	2	5	140
Minerals (mg)	1	6	84

32. BIOMEDICAL: Diet Planning The director of a school district's hot lunch program estimates that each lunch should contain at least 1000 calories and 15 grams of protein and no more than 30 grams of fat. The nutritional contents per ounce of the cafeteria's famous mystery foods X and Y are given in the table. If each ounce of X costs 5¢ and each ounce of Y costs 6¢, how much of each food should be served to meet the director's nutritional goals at the least cost?

	Calories	Protein	Fat
Mystery food X	200	1	3
Mystery food Y	100	3	3

33. ENVIRONMENTAL SCIENCE: Land Reclamation A state government included $1.2 million in its current appropriations bill to reclaim some of the land at a 3000-acre strip mine site. It will cost $800 per acre to return the land to productive grassland suitable for livestock and $500 per acre to return the land to forest suitable for commercial timber production. The appropriations bill also requires that at least 1800 acres be reclaimed immediately and then the income from leasing it to ranchers and/or paper companies be used to reclaim the rest of the site in the coming years. If long-term-use agreements yield $200 per acre of grassland and $150 per acre of forest, how many acres of each should be reclaimed this year to raise the greatest amount for next year's reclamation efforts?

34. SOCIAL SCIENCE: Urban Renewal A nonprofit urban development corporation has agreed to rebuild at least 24 city blocks in the south side of Megatropolis in which at least 15 blocks will be semidetached single family homes and at least 3 but not more than 10 blocks will be commercial buildings (retail stores and/or light industry). If it will cost $6 million to rebuild 1 block with homes and $7 million to rebuild 1 block for commercial use, how many blocks of each should be rebuilt to meet the goals at the least cost?

4.3 **THE SIMPLEX METHOD FOR STANDARD MAXIMUM PROBLEMS**

Introduction

Using the geometric method from the previous section as our guide, we shall explain the *simplex method* for solving linear programming problems. This method finds a *path of vertices* leading from an initial vertex to a solution without finding all of the intersection points of the boundary lines or even all of the vertices of the feasible region. We begin by defining a *standard maximum problem,* the type of problem that we will solve in this section. Other types of problems will be solved in succeeding sections.

Standard Maximum Problems

A standard maximum problem is a linear programming problem in which the objective function is to be maximized, the variables $x_1, x_2, \ldots, x_n$ are all nonnegative, and the other constraints have $\leq$ inequalities with nonnegative numbers on the right-hand sides.

Standard Maximum Problem

A problem is a *standard maximum problem* if it can be written:

Maximize an objective function

$$P = c_1 x_1 + c_2 x_2 + \cdots + c_n x_n$$

Subject to inequalities of the form

$$a_1 x_1 + a_2 x_2 + \cdots + a_n x_n \leq b \quad \text{with} \quad b \geq 0$$

and

$$x_1 \geq 0, \, x_2 \geq 0, \ldots, \, x_n \geq 0$$

Observe that if we substitute 0 for each variable, the inequalities are *all* satisfied, so for a standard maximum problem the origin *is* a vertex of the feasible region.

EXAMPLE 1

VERIFYING STANDARD MAXIMUM FORM

Is the following a standard maximum problem?

$$\text{Maximize} \quad P = 9x_1 + 16x_2$$

$$\text{Subject to} \quad \begin{cases} 3x_1 + 4x_2 \leq 24 \\ x_1 + 2x_2 \leq 10 \\ x_1 \geq 0 \quad \text{and} \quad x_2 \geq 0 \end{cases}$$

Solution

The objective function is to be maximized, the variables are both non-negative, and the other inequalities all have $\leq$ with nonnegative numbers on the right, so this *is* a standard maximum problem.

Practice Problem 1   Is the following a standard maximum problem?

$$\text{Maximize} \quad P = 7x_1 + 5x_2$$

$$\text{Subject to} \quad \begin{cases} 2x_1 - 3x_2 \le 24 \\ 5x_1 + 2x_2 \le -4 \\ x_1 \ge 0 \quad \text{and} \quad x_2 \ge 0 \end{cases}$$

➤ Solution on page 289

Matrix Form of a Standard Maximum Problem

We may express a standard maximum problem in the matrix notation of the preceding chapter. Just as two matrices are *equal* if they have the same dimension and corresponding elements are equal, two matrices obey an *inequality* if they have the same dimension and corresponding elements obey that inequality. For example:

$$\begin{pmatrix} 1 & 2 \\ 3 & 4 \end{pmatrix} \le \begin{pmatrix} 1 & 2 \\ 6 & 8 \end{pmatrix}$$

Matrices are the same size and corresponding elements obey $\le$

Using matrix inequalities, the problem written on the left below (from Example 1) may be expressed in the matrix form on the right:

$$\text{Maximize} \quad P = 9x_1 + 16x_2$$

$$\text{Subject to} \quad \begin{cases} 3x_1 + 4x_2 \le 24 \\ x_1 + 2x_2 \le 10 \\ x_1 \ge 0 \\ x_2 \ge 0 \end{cases}$$

$$\text{Maximize} \quad P = \begin{pmatrix} 9 & 16 \end{pmatrix} \begin{pmatrix} x_1 \\ x_2 \end{pmatrix}$$

$$\text{Subject to} \quad \begin{cases} \begin{pmatrix} 3 & 4 \\ 1 & 2 \end{pmatrix} \begin{pmatrix} x_1 \\ x_2 \end{pmatrix} \le \begin{pmatrix} 24 \\ 10 \end{pmatrix} \\ \begin{pmatrix} x_1 \\ x_2 \end{pmatrix} \ge 0 \end{cases}$$

In this way, any standard maximum problem may be expressed in the following matrix form, using $X = \begin{pmatrix} x_1 \\ \vdots \\ x_n \end{pmatrix}$ for a column matrix of variables:

Standard Maximum Problem in Matrix Form*

$$\text{Maximize} \quad P = c^t X$$

$$\text{Subject to} \quad \begin{cases} AX \le b \\ X \ge 0 \end{cases} \quad \text{where} \quad b \ge 0$$

---

* Writing c^t for the row matrix of coefficients of the objective function is for historical reasons only. The mathematicians who developed these methods preferred *column* matrices, so row matrices were represented as *transposes* of column matrices, hence c^t.

The Initial Simplex Tableau

Equations are easier to solve than inequalities, so we first simplify the problem by changing the inequalities to equations by adding to each a *slack variable* that is nonnegative and represents the "amount not used." For example, in the inequality $4 \le 7$ the two sides differ by 3, so adding 3 to the left side gives an *equation* $4 + 3 = 7$. Similarly, the inequality $x \le 7$ can be rewritten as the *equation* $x + s = 7$ by adding a slack variable $s \ge 0$ to the left side to make up the difference between the sides. We do this for each inequality, using a different slack variable for each inequality.

EXAMPLE 2

INTRODUCING SLACK VARIABLES

Write the following inequalities (from Example 1) as equations with slack variables.

$$\begin{cases} 3x_1 + 4x_2 \le 24 \\ x_1 + 2x_2 \le 10 \end{cases}$$

Solution

We introduce a slack variable $s_1 \ge 0$ into the first inequality and another slack variable $s_2 \ge 0$ into the second to obtain the *equations* shown in the middle below.

$$\begin{cases} 3x_1 + 4x_2 \le 24 \\ x_1 + 2x_2 \le 10 \end{cases} \quad \text{become} \quad \begin{cases} 3x_1 + 4x_2 + s_1 \quad\;\; = 24 \\ x_1 + 2x_2 + \quad\;\; s_2 = 10 \end{cases} \quad \text{or} \quad \begin{cases} 3x_1 + 4x_2 + 1s_1 + 0s_2 = 24 \\ x_1 + 2x_2 + 0s_1 + 1s_2 = 10 \end{cases}$$

The version on the right above has *all* of the variables included, with the unnecessary ones multiplied by zeros.

Slack variables are sometimes called simply *slacks*.

We can write the objective function $P = 9x_1 + 16x_2$ (from Example 1) with all variables moved to the left of the equals sign, including the slacks multiplied by zeros:

$$P = 9x_1 + 16x_2 \qquad \text{becomes} \qquad P - 9x_1 - 16x_2 + 0s_1 + 0s_2 = 0$$

Since both the objective function and the constraints from Example 1 are now written as equations, with the variables in the same order, the entire problem can be summarized in a table, called the "simplex tableau" (using the French word for *table*, whose plural is *tableaux*). We state the general form and then give an example.

Initial Tableau

The initial tableau for the linear programming problem

$$\text{Maximize} \quad P = c^t X$$
$$\text{Subject to} \quad \begin{cases} AX \leq b \\ X \geq 0 \end{cases} \quad \text{is}$$

$$\begin{array}{c|cc|c} & X & S & \\ \hline S & A & I & b \\ \hline P & -c^t & 0 & 0 \end{array}$$

where I stands for the identity matrix and S stands for the slack variables.

EXAMPLE 3

WRITING A SIMPLEX TABLEAU

Write the initial simplex tableau for the following linear programming problem (from Example 1):

$$\text{Maximize} \quad P = 9x_1 + 16x_2$$
$$\text{Subject to} \quad \begin{cases} 3x_1 + 4x_2 \leq 24 \\ x_1 + 2x_2 \leq 10 \\ x_1 \geq 0, \text{ and } x_2 \geq 0 \end{cases}$$

Solution

We write the constraints as:

$$\begin{cases} 3x_1 + 4x_2 + 1s_1 + 0s_2 = 24 \\ x_1 + 2x_2 + 0s_1 + 1s_2 = 10 \end{cases} \qquad \text{From Example 2}$$

and the objective function as:

$$P - 9x_1 - 16x_2 + 0s_1 + 0s_2 = 0 \qquad \begin{array}{l}\text{From} \\ P = 9x_1 + 16x_2\end{array}$$

For the initial simplex tableau we write the variables on the top, the constraint numbers in the middle, and the objective numbers on the bottom (all in the correct columns), and finally the slack variables on the left:

	x_1	x_2	s_1	s_2	
s_1	3	4	1	0	24
s_2	1	2	0	1	10
P	−9	−16	0	0	0

Column variables

Constraints

Objective function (signs changed)

Represents =

From the original problem:

$$\text{Maximize} \quad P = 9x_1 + 16x_2$$
$$\text{Subject to} \quad \begin{cases} 3x_1 + 4x_2 \leq 24 \\ x_1 + 2x_2 \leq 10 \\ x_1 \geq 0, x_2 \geq 0 \end{cases}$$

Practice Problem 2   Write the initial simplex tableau for the problem

$$\text{Maximize} \quad P = 3x_1 + 5x_2$$

$$\text{Subject to} \quad \begin{cases} 2x_1 - 3x_2 \le 24 \\ 5x_1 + 2x_2 \le 20 \\ 3x_1 + 2x_2 \le 12 \\ x_1 \ge 0 \quad \text{and} \quad x_2 \ge 0 \end{cases}$$

➤ **Solution on page 289**

Basic and Nonbasic Variables

In an initial simplex tableau like the one below (from Example 3), the slack variables s_1 and s_2 have two special properties.

	x_1	x_2	s_1	s_2	
s_1	3	4	1	0	24
s_2	1	2	0	1	10
P	-9	-16	0	0	0

1. Their columns contain all zeros except for exactly one 1, which occurs above the bottom row.

2. Each of the rows above the bottom row has exactly one of these special 1s.

These special variables determine a feasible point in the region without any further calculation: Just give the variable the value that is at the right-hand end of its row, and set all other variables equal to 0. For the above tableau, this means:

	x_1	x_2	s_1	s_2		
s_1	3	4	1	0	24	← $s_1 = 24$
s_2	1	2	0	1	10	← $s_2 = 10$
P	-9	-16	0	0	0	and $x_1 = x_2 = 0$ Other variables set equal to 0

The general procedure for determining basic and nonbasic variables in a tableau is as follows.

> **Basic and Nonbasic Variables in a Simplex Tableau**
>
> Count the number of slack variables (or, equivalently, the number of constraints). This number, m, will be the number of basic variables throughout the solution.
>
> 1. For basic variables, choose any m variables whose columns have all zeros except for one 1 that must appear above the

bottom row and such that each of the rows above the bottom row has exactly one of these special 1s; assign to each basic variable the value at the right-hand end of the row containing this 1.

2. All *other* variables are *nonbasic* variables; assign to each the value 0.

3. At these values, the objective function equals the number in the bottom right corner.

Notice that statements 1 and 2 agree with the values we assigned to the variables in the tableau above the box. Furthermore, the objective function $P = 9x_1 + 16x_2$ evaluated at the assigned $x_1 = 0$ and $x_2 = 0$ is $P = 0$, the number in the bottom right corner of the tableau, agreeing with statement 3 in the box. Because of their importance, we list the basic variables on the left side of the tableau, each in the row corresponding to its special 1. We interpret the tableau as specifying a basic feasible point (a vertex) of the feasible region along with the value of the objective function at that point.

The Pivot Element

The simplex method begins by taking the slack variables as the basic variables and then successively finds better basic variables to increase the value of the objective function until the solution is found. The method mimics the geometric procedure that we used on pages 262–263 to find vertices, adding and subtracting equations as we did on page 170 to solve systems of equations, a process we now call *pivoting*. We will explain the process by referring to the problem in Example 1, but explaining how to carry out the steps on the simplex tableau since we use the tableau to represent the problem.

In the objective function from Example 1, $P = 9x_1 + 16x_2$, x_1 is multiplied by 9 and x_2 is multiplied by a larger number, 16, so increasing x_2 should do more to increase P than increasing x_1. In general, to increase *any* objective function, we should increase the variable with the *largest positive* coefficient. Since the bottom row of the tableau contains the *negatives* of these coefficients, this means choosing the column with the *smallest negative* number in the bottom row. This is the *pivot column*.

Pivot Column

The *pivot column* is the column with the smallest negative entry in the bottom row (omitting the right column). If there is a tie, choose the leftmost such column.

Be careful! The smallest negative number among 12, -8, and -5 is -8. (Students sometimes call this choosing *the "most negative" number*).

Having chosen the column and therefore the variable to increase, how much can we increase it without violating a constraint? The boundary lines for the constraints in Example 1 are on the left below. Recall from our geometric method that we cannot increase x_2 beyond the *intercepts* of these boundaries, which are calculated on the right below.

Boundary line	x_2-intercept	
$3x_1 + 4x_2 = 24$	$x_2 = \frac{24}{4} = 6$	$\longleftarrow x_2$ cannot exceed 6
$x_1 + 2x_2 = 10$	$x_2 = \frac{10}{2} = 5$	$\longleftarrow x_2$ cannot exceed 5

To stay below *two* numbers means staying below the *smaller* of them, so x_2 must not exceed the *smaller* of the two numbers, each number being the right-hand side of the inequality divided by the number in the pivot column. This gives the rule for choosing the *pivot row* from the simplex tableau:

Pivot Row

For each row (except the bottom), divide the rightmost entry by the pivot column entry (omitting any row with a zero or negative pivot column entry). The row with the smallest ratio is the *pivot row*. If there is a tie, choose the uppermost such row.

We ignore rows with zeros in the pivot column (because the corresponding inequality would not include that variable), and with negative numbers (because such an inequality would not restrict the variable from increasing).

Pivot Element

The *pivot element* is the entry in the pivot column and the pivot row.

The Pivot Operation

Once we have found the pivot element, we want to increase the selected variable as much as possible (up to the boundary of the feasible region) and to increase the objective function correspondingly. This is accomplished by the pivot operation, which is just the tableau version of the elimination method that we used for solving equations on pages 157–158.

Pivot Operation

1. Divide every entry in the pivot row by the pivot element to obtain a *new* pivot row, which will then have a 1 where the pivot element was. Replace the variable at the left-hand end of this row by the variable corresponding to the pivot column. The variables listed on the left will be the new *basic* variables.

2. Subtract multiples of the new pivot row from all other rows of the tableau to get zeros in the pivot column above and below the 1.

   Specifically, if R_p is the new pivot row found from step 1 and R is any *other* row, and if we denote the entry in that row and in the pivot column by p, then the row operation in step 2 for that row is $R - p \cdot R_p \rightarrow R$.

EXAMPLE 4

FINDING THE PIVOT ELEMENT AND PIVOTING

Find the pivot element in the following simplex tableau (from Example 3) and carry out the pivot operation.

	x_1	x_2	s_1	s_2	
s_1	3	4	1	0	24
s_2	1	2	0	1	10
P	-9	-16	0	0	0

Solution

The smallest negative number in the bottom row is -16, so the pivot column is column 2 (as shown below). To find the pivot row, we divide the last entry of each row by the pivot column entry for that row (skipping the bottom row and any row with a zero or a negative pivot column entry). We choose the row with the smallest nonnegative ratio, as shown by the calculations to the right of the following tableau. The pivot row is then row 2.

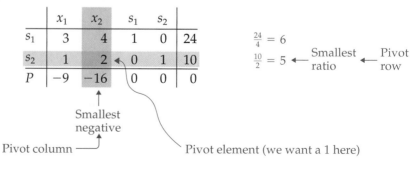

Having found the pivot element, we carry out the pivot operation. The first step is to divide the pivot row by the pivot element, so we divide the second row by 2, which changes the pivot element to a 1. We also update the basis on the left—the pivot column variable x_2 replaces the pivot row variable s_2, so the new basic variables are s_1 and x_2.

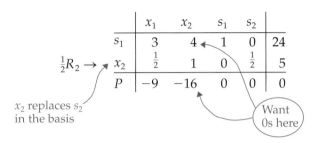

The second step in the pivot operation is to subtract multiples of the *new* pivot row from the other rows to "zero out" the rest of the pivot column (the 4 and the -16). This is accomplished by the row operations listed on the left of the following tableau.

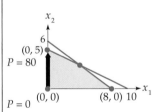

Geometrically, we have moved from the initial vertex $(0, 0)$ to the vertex $(0, 5)$, increasing P from 0 to 80.

This completes the pivot operation. The basic variables have the values shown on the right, with the other variables set equal to zero (in particular, $x_1 = 0$). Observe that when $x_1 = 0$ and $x_2 = 5$ the objective function $P = 9x_1 + 16x_2$ *does* take the value of 80 as given in the lower right corner of the tableau.

All of the steps we have just carried out—finding the pivot element, updating the basic variables on the left of the tableau, dividing to replace the pivot element by a 1, and zeroing out the rest of the column by subtracting multiples of the pivot row—constitute *one* pivot operation. You may want to carry out some of the calculations on scratch paper.

Notice that the tableau above still has a negative number (the -1) in the bottom row of the above tableau, meaning that there is more pivoting to be done.

EXAMPLE 5

PIVOTING A SECOND TIME

Perform the pivot operation on the following tableau (from the end of Example 4):

	x_1	x_2	s_1	s_2	
s_1	1	0	1	-2	4
x_2	$\frac{1}{2}$	1	0	$\frac{1}{2}$	5
P	-1	0	0	8	80

Solution

The -1 in the bottom row means that the pivot column is column 1. The calculations on the right below show that the pivot row is row 1.

	x_1	x_2	s_1	s_2	
s_1	1	0	1	-2	4
x_2	$\frac{1}{2}$	1	0	$\frac{1}{2}$	5
P	-1	0	0	8	80

$\frac{4}{1} = 4$ ⟵ Smallest ⟵ Pivot row

$\frac{5}{1/2} = 10$

Pivot column   Pivot element

The pivot entry is already a 1, so no division is needed (but we update the basis by replacing s_1 by x_1). The two row operations listed on the left below complete the pivot operation.

		x_1	x_2	s_1	s_2		
x_1 replaces s_1 ⟶	x_1	1	0	1	-2	4	⟵ $x_1 = 4$
$R_2 - \frac{1}{2}R_1$ ⟶	x_2	0	1	$-\frac{1}{2}$	$\frac{3}{2}$	3	⟵ $x_2 = 3$
$R_3 + R_1$ ⟶	P	0	0	1	6	84	⟵ $P = 84$

There are no negative numbers in the bottom row, so no more pivoting is possible—the objective function has been maximized. According to the box on pages 277–278, we take the values from this final tableau: P from the bottom right corner, the basic variables from the right side, with the nonbasic variables set to 0:

The maximum value of P is 84, when $x_1 = 4$ and $x_2 = 3$.

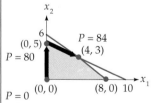

Geometrically, we have moved from (0, 5) to the vertex (4, 3), increasing P from 80 to 84.

Usually only the x-values are given in the solution, since only they were stated in the original problem. The nonbasic variables here take the values $s_1 = 0$ and $s_2 = 0$. In Example 7 we will see how to interpret the slack variables as "unused resources."

Graphing Calculator Exploration

The program* PIVOT carries out the pivot operation after you specify the pivot column and pivot row. To carry out the pivot operation with your calculator on the simplex tableau from Example 5, proceed as follows:

```
MATRIX [A] 3 X5
[1    0       1   -
[.5   1       0   -
[-1   0       0   -


2, 1=1/2
```

a. Enter the simplex tableau as one large matrix [A] having 3 rows and 5 columns. Because this matrix is too large to fit on the calculator screen, it will scroll from left to right and back again as you enter your numbers.

```
...1    0 1 -2   4 ...
...1/2  1 0 1/2  5 ...
...-1   0 0 8    80...
P C?  1
P R?  1
```

b. Run the program PIVOT. It will display the current tableau in matrix [A] and you can use the arrows to scroll the screen to see the rest of it. Press ENTER to enter the **P**ivot **C**olumn and then again to enter the **P**ivot **R**ow of your pivot element.

```
...1/2  1 0 1/2  5 ...:
...-1   0 0 8    80...
P C?  1
P R?  1
[[1 0  1     -2 ...
 [0 1  -1/2 3/2 ...
 [0 0  1    6   ...
```

c. After you ENTER the pivot row, the calculator will perform the pivot operation and display the new tableau. You can use the arrows to scroll the screen to see the rest of it. Press ENTER to exit the program.

If your problem requires several pivot operations, you can rerun the program with the new tableau by pressing ENTER again.

In general, we continue to pivot until no more pivoting is possible, that is, until we arrive at a *final tableau*.

How do we recognize and interpret a final tableau? Recall that choosing the smallest negative number in the bottom row was equivalent to choosing the variable that will increase the objective function most quickly. Therefore, if there is *no* negative number in the bottom row, then there is no variable that will increase the objective function, so P must be at its maximum value, and we read the solution from the tableau as we did in the preceding example. On the other hand, if there *is* a negative number in the bottom row (so the objective function

* See the Preface for information on how to obtain this and other programs.

can be increased) but there is *no* pivot row, then there are no constraints limiting the objective function, which can therefore be increased arbitrarily, meaning that there is *no solution*.

Interpreting a Final Tableau

1. If all the numbers in the bottom row are positive or zero (so there is no pivot column), then the original problem *does* have a solution:

   a. The (basic) variables listed on the left of the rows take the values at the right ends of those rows.

   b. The (nonbasic) variables not listed on the left take the value 0.

   c. The maximum value of the objective function is in the lower right corner of the tableau.

2. If there *is* a negative number in the bottom row but every other element in its column is negative or zero (so there is a pivot column but no pivot row), then the original problem has *no solution*.

EXAMPLE 6

INTERPRETING A FINAL TABLEAU

Interpret the following final tableaux:

a.

	x_1	x_2	s_1	s_2	
s_1	2	0	1	3	4
x_2	5	1	0	6	7
P	8	0	0	9	10

b.

	x_1	x_2	s_1	s_2	
s_1	-2	0	1	3	4
x_2	0	1	0	6	7
P	-8	0	0	9	10

Solution

a. The first tableau has no negative numbers in the bottom row, so the original problem *does* have a solution. The basic variables listed on the left, s_1 and x_2, take the values on the right, 4 and 7; the other variables, s_2 and x_1, take the value 0. The value of P is in the lower right corner:

   The maximum value of P is 10, when $x_1 = 0$ and $x_2 = 7$.

b. The second tableau *does* have a negative number in the bottom row, -8, so there *is* a pivot column (column 1). However, there is no pivot *row* (the ratios are $\frac{4}{-2}$ which is negative and $\frac{7}{0}$ which is undefined). Therefore, there is *no solution*.

The Simplex Method

The simplex method proceeds by pivoting until a final tableau is reached, and then interpreting it.

Simplex Method

To solve a standard maximum problem (page 273) by the simplex method:

1. Construct the simplex tableau (see page 276).

2. Locate the pivot element (see pages 278–279) and go to step 3. If the tableau does not have a pivot element, go to step 4.

3. Perform the pivot operation (see page 280) using the pivot element and return to step 2.

4. If the final tableau does not have a pivot column (see page 278), then the solution occurs at the vertex where the (basic) variables listed on the left of the tableau take the values at the right-hand ends of those rows; the other (nonbasic) variables take the value 0. The maximum value of the objective function is in the bottom right corner of the tableau.

   If the final tableau has a pivot *column* but no pivot *row* (see page 279), then the problem has *no solution*.

EXAMPLE 7   **A MANUFACTURING PROBLEM WITH MANY VARIABLES**

A pottery shop manufactures dinnerware in four different patterns by shaping the clay, decorating it, and then kiln-firing it. The numbers of hours required per place setting for the four designs are given in the following table, together with the expected profits. The shop employs two skilled workers to do the initial shaping and three artists to do the decorating, none of whom will work more than 40 hours each week. The kiln can be used no more than 55 hours per week. How many place settings of each pattern should be made this week to obtain the greatest possible profit?

	Classic	Modern	Art Deco	Floral
Shaping	2 hours	1 hour	4 hours	2 hours
Decorating	3 hours	1 hour	6 hours	4 hours
Kiln-firing	1 hour	1 hour	1 hour	1 hour
Expected profit	$10	$6	$9	$8

Solution

Let

$$x_1 = \begin{pmatrix} \text{Number of} \\ \text{Classic settings} \end{pmatrix}, \qquad x_2 = \begin{pmatrix} \text{Number of} \\ \text{Modern settings} \end{pmatrix},$$

$$x_3 = \begin{pmatrix} \text{Number of} \\ \text{Art Deco settings} \end{pmatrix}, \quad \text{and} \quad x_4 = \begin{pmatrix} \text{Number of} \\ \text{Floral settings} \end{pmatrix}$$

made during the week, so $x_1 \geq 0$, $x_2 \geq 0$, $x_3 \geq 0$, and $x_4 \geq 0$. From the last line in the table, the profit to be maximized is

$$P = 10x_1 + 6x_2 + 9x_3 + 8x_4.$$

The constraints on the time spent shaping, decorating, and firing come from the other numbers in the table, along with the time available for each:

$$2x_1 + x_2 + 4x_3 + 2x_4 \leq 80 \qquad \text{2 shapers @ 40 hours each}$$
$$3x_1 + x_2 + 6x_3 + 4x_4 \leq 120 \qquad \text{3 decorators @ 40 hours each}$$
$$x_1 + x_2 + x_3 + x_4 \leq 55 \qquad \text{Time kiln can be used}$$

The initial simplex tableau is

	x_1	x_2	x_3	x_4	s_1	s_2	s_3	
s_1	②←	1	4	2	1	0	0	80
s_2	3	1	6	4	0	1	0	120
s_3	1	1	1	1	0	0	1	55
P	−10	−6	−9	−8	0	0	0	0

$\frac{80}{2} = 40 \longleftarrow$ Smallest ratio $\longleftarrow$ Pivot row

$\frac{120}{3} = 40$ (Tie for smallest ratio, so take top row)

$\frac{55}{1} = 55$

↑ Smallest negative

Pivot element

Pivot column ↗

When we pivot (on the 2 in column 1, row 1), the tableau becomes

x_1 replaces s_1 in basis →

	x_1	x_2	x_3	x_4	s_1	s_2	s_3	
x_1	1	1/2	2	1	1/2	0	0	40
s_2	0	−1/2	0	1	−3/2	1	0	0
s_3	0	①/2←	−1	0	−1/2	0	1	15
P	0	−1	11	2	5	0	0	400

$\frac{40}{1/2} = 80$

(Cannot use since $-\frac{1}{2}$ is < 0)

$\frac{15}{1/2} = 30 \longleftarrow$ Smallest ratio

Pivot row ↗

↑ Smallest negative

Pivot element

Pivot column ↗

Pivoting again (on the $1/2$ in column 2, row 3), we reach the final tableau:

	x_1	x_2	x_3	x_4	s_1	s_2	s_3	
x_1	1	0	3	1	1	0	-1	25
s_2	0	0	-1	1	-2	1	1	15
x_2	0	1	-2	0	-1	0	2	30
P	0	0	9	2	4	0	2	430

x_2 replaces s_3 in the basis →

No negatives in bottom row
so final tableau

The basic variables take the values at the right ends of their rows, the nonbasic variables are zero, and the objective function value appears in the bottom right corner. Therefore, the maximum value of P is 430 when $x_1 = 25$, $x_2 = 30$, $x_3 = 0$, and $x_4 = 0$. The fact that $s_2 = 15$ means that 15 hours of the available decorating time will not be needed.

Answer: The pottery shop should manufacture 25 place settings of Classic, 30 of Modern, and none of Art Deco and Floral.

How Good Is the Simplex Method?

Will the simplex method always find the solution? Even if it does find the solution, will it get to the answer after a reasonable number of tableaux?

In 1955, E. M. I. Beale published a linear programming problem with the awful property that the simplex method found the same sequence of tableaux over and over again. This repetition of tableaux is known as "cycling" because the tableaux repeat in a cyclic pattern. If a computer were solving this problem, it would never stop. Several examples of cycling and changes to the simplex method that prevent cycling are given in Exercises 42–46.

In 1972, V. Klee and G. J. Minty described a class of problems that take many more tableaux than their small size would suggest: With each new variable, the number of tableaux doubles. This means that a Klee–Minty problem with 2 variables takes 4 tableaux to solve, one with 3 variables takes 8 tableaux, one with 4 variables takes 16 tableaux, and so on. The 10-variable problem in this sequence will take 1024 tableaux with larger problems becoming even more unmanageable. Examples of these kinds of problems are given in Exercises 47–50.

Most "real world" computer programs for solving linear programming problems include subroutines to protect against cycling and

Klee–Minty situations. Stephen Smale has shown that "... the number of pivots required to solve a linear programming problem grows in proportion to the number of variables on average." Several other methods (by L. Khachiyan and N. Karmarkar) to solve linear programming problems guarantee that the number of steps required does not grow too fast for every problem. However, the simplex method continues to provide the fastest solutions to commercial problems.

4.3 Section Summary

The maximum linear programming problem

$$\text{Maximize} \quad P = c^t X$$

$$\text{Subject to} \quad \begin{cases} AX \le b \\ X \ge 0 \end{cases}$$

is a *standard maximum problem* if $b \ge 0$. The *initial simplex tableau* represents this problem in the form

	X	S	
S	A	I	b
P	$-c^t$	0	0

The variables listed on the far left form a *basis*, and setting them equal to the values on the far right of their rows and setting the nonbasic variables equal to zero gives a *basic feasible point* that is a vertex of the region. The pivot column is the column with the smallest negative entry in the bottom row (omitting the right column), and the pivot row is the row with the smallest ratio between the last entry and the pivot column entry (omitting any row with a zero or negative pivot column entry, and the bottom row). The *pivot element* is the entry in the pivot column and pivot row.

The *pivot operation* divides the pivot row by the pivot element and then subtracts from each of the other rows the pivot column entry of that row times the new pivot row. The pivot column variable enters the basis, replacing the variable of the pivot row. The *simplex method* solves a standard maximum problem by pivoting until the tableau does not have a pivot element. If there is no pivot *column*, then the solution occurs at the vertex given by the basic variables, and the maximum value of the objective function appears in the bottom right corner. If there is a pivot column but no pivot *row*, then there is no solution.

The simplex method uses the pivot operation to make the bottom row "less negative" while keeping the right column nonnegative. That is, *the simplex method makes the tableau "more nearly optimal" while keep-*

ing it feasible. When the bottom row contains no negative entries, the tableau is *optimal* because the largest value of the objective function has been found.

➤ **Solutions to Practice Problems**

1. No. The -4 on the right of the second inequality violates the $b \geq 0$ condition in the box on page 273. (The -3 in the first inequality is not a violation: only the right-hand sides need to be nonnegative.)

2.

	x_1	x_2	s_1	s_2	s_3	
s_1	2	-3	1	0	0	24
s_2	5	2	0	1	0	20
s_3	3	2	0	0	1	12
P	-3	-5	0	0	0	0

4.3 Exercises

Construct the initial simplex tableau for each standard maximum problem.

1. Maximize $P = 8x_1 + 9x_2$

Subject to $\begin{cases} 3x_1 + 2x_2 \leq 12 \\ 6x_1 + x_2 \leq 15 \\ x_1 \geq 0, \quad x_2 \geq 0 \end{cases}$

2. Maximize $P = 10x_1 + 15x_2$

Subject to $\begin{cases} 5x_1 + 2x_2 \leq 30 \\ 3x_1 + 4x_2 \leq 36 \\ x_1 \geq 0, \quad x_2 \geq 0 \end{cases}$

3. Maximize $P = 13x_1 + 7x_2$

Subject to $\begin{cases} 4x_1 + 3x_2 \leq 12 \\ 5x_1 + 2x_2 \leq 20 \\ x_1 + 6x_2 \leq 12 \\ x_1 \geq 0, \quad x_2 \geq 0 \end{cases}$

4. Maximize $P = 8x_1 + 33x_2$

Subject to $\begin{cases} 3x_1 + 2x_2 \leq 30 \\ x_1 + 4x_2 \leq 20 \\ 5x_1 + 6x_2 \leq 60 \\ x_1 \geq 0, \quad x_2 \geq 0 \end{cases}$

5. Maximize $P = 5x_1 - 2x_2 + 10x_3 - 5x_4$

Subject to $\begin{cases} 2x_1 + x_2 + x_3 + 3x_4 \leq 6 \\ x_1 + 4x_2 - 2x_3 + x_4 \leq 8 \\ x_1 \geq 0, \quad x_2 \geq 0, \quad x_3 \geq 0, \quad x_4 \geq 0 \end{cases}$

6. Maximize $P = 5x_1 + 10x_2 - 30x_3 + 5x_4$

Subject to $\begin{cases} 3x_1 + x_2 - 2x_3 + 4x_4 \leq 12 \\ 2x_1 - x_2 + 3x_3 + x_4 \leq 12 \\ x_1 \geq 0, \quad x_2 \geq 0, \quad x_3 \geq 0, \quad x_4 \geq 0 \end{cases}$

7. Maximize $P = 10x_1 + 20x_2 + 15x_3$

Subject to $\begin{cases} 8x_1 + x_2 + 4x_3 \leq 32 \\ 3x_1 + 5x_2 + 7x_3 \leq 30 \\ 6x_1 + 2x_2 + 9x_3 \leq 28 \\ x_1 \geq 0, \quad x_2 \geq 0, \quad x_3 \geq 0 \end{cases}$

8. Maximize $P = 10x_1 + 15x_2 + 5x_3$

Subject to $\begin{cases} 7x_1 + 3x_2 + x_3 \leq 21 \\ 6x_1 + 5x_2 + 4x_3 \leq 20 \\ 8x_1 + 9x_2 + 2x_3 \leq 45 \\ x_1 \geq 0, \quad x_2 \geq 0, \quad x_3 \geq 0 \end{cases}$

For each simplex tableau, find the pivot element and carry out one complete pivot operation. If there is no pivot element, explain what the tableau shows about the solution of the original standard maximum problem.

9.

	x_1	x_2	s_1	s_2	
s_1	2	0	1	0	4
s_2	1	1	0	1	5
P	-7	-8	0	0	0

10.

	x_1	x_2	s_1	s_2	
s_1	1	-2	1	0	5
s_2	3	1	0	1	6
P	-4	-5	0	0	0

11.

	x_1	x_2	x_3	x_4	s_1	s_2	
s_1	4	2	6	2	1	0	12
s_2	3	1	2	1	0	1	8
P	-4	-5	6	-3	0	0	0

12.

	x_1	x_2	x_3	x_4	s_1	s_2	
s_1	5	3	1	3	1	0	30
s_2	6	2	2	4	0	1	16
P	-4	-6	-5	7	0	0	0

13.

	x_1	x_2	x_3	x_4	s_1	s_2	s_3	
s_1	0	0	2	1	1	1	-1	10
x_2	0	1	1	1	0	1	0	15
x_1	1	0	-2	0	0	-1	1	10
P	0	0	0	2	0	1	3	90

14.

	x_1	x_2	x_3	x_4	s_1	s_2	s_3	
x_2	0	1	-1	0	1	-1	0	20
x_4	1	0	1	1	0	1	0	40
s_3	1	0	2	0	-1	1	1	30
P	2	0	8	0	3	3	0	300

15.

	x_1	x_2	s_1	s_2	s_3	
s_1	0	0	1	-1	0	10
x_1	1	0	0	1	-1	10
x_2	0	1	0	1	0	20
P	0	0	0	25	-10	400

16.

	x_1	x_2	s_1	s_2	s_3	
s_1	0	0	1	0	-1	5
x_2	0	1	0	1	-1	10
x_1	1	0	0	1	0	20
P	0	0	0	35	-15	550

Solve each problem by the simplex method. (Exercises 17, 18, 25, and 26 can also be solved by the graphical method.)

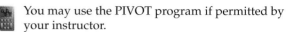 You may use the PIVOT program if permitted by your instructor.

17. Maximize $P = x_1 + 2x_2$

Subject to $\begin{cases} 3x_1 + x_2 \le 24 \\ x_1 + x_2 \le 14 \\ x_1 \ge 0, \ x_2 \ge 0 \end{cases}$

18. Maximize $P = x_1 + 3x_2$

Subject to $\begin{cases} 2x_1 + x_2 \le 24 \\ x_1 + x_2 \le 15 \\ x_1 \ge 0, \ x_2 \ge 0 \end{cases}$

19. Maximize $P = 30x_1 + 40x_2 + 15x_3$

Subject to $\begin{cases} 2x_1 + x_2 + 3x_3 \le 150 \\ 3x_1 + 2x_2 + x_3 \le 100 \\ x_1 \ge 0, \ x_2 \ge 0, \ x_3 \ge 0 \end{cases}$

20. Maximize $P = 30x_1 + 60x_2 + 15x_3$

Subject to $\begin{cases} 2x_1 + 3x_2 + x_3 \le 75 \\ 3x_1 + x_2 + 2x_3 \le 50 \\ x_1 \ge 0, \ x_2 \ge 0, \ x_3 \ge 0 \end{cases}$

21. Maximize $P = 6x_1 + 4x_2 + 5x_3$

Subject to $\begin{cases} x_1 - x_2 + x_3 \le 20 \\ x_1 + 2x_3 \le 10 \\ x_1 \ge 0, \ x_2 \ge 0, \ x_3 \ge 0 \end{cases}$

22. Maximize $P = 6x_1 + 5x_2 + 8x_3$

Subject to $\begin{cases} 2x_1 + x_3 \le 40 \\ x_1 - x_2 + x_3 \le 30 \\ x_1 \ge 0, \ x_2 \ge 0, \ x_3 \ge 0 \end{cases}$

23. Maximize $P = 4x_1 + 3x_2 - 5x_3 + 6x_4$

Subject to $\begin{cases} x_1 + x_2 + x_4 \le 60 \\ x_1 + x_3 + x_4 \le 40 \\ x_1 + x_2 + x_3 \le 50 \\ x_1 \ge 0, \ x_2 \ge 0, \ x_3 \ge 0, \ x_4 \ge 0 \end{cases}$

24. Maximize $P = 2x_1 + 3x_2 + 4x_3 + x_4$

Subject to $\begin{cases} x_1 + x_2 + x_4 \le 20 \\ x_1 + x_3 + x_4 \le 15 \\ x_1 + x_2 + x_3 \le 25 \\ x_1 \ge 0, \ x_2 \ge 0, \ x_3 \ge 0, \ x_4 \ge 0 \end{cases}$

25. Maximize $P = 4x_1 + 2x_2$

Subject to $\begin{cases} x_1 - x_2 \le 1 \\ -x_1 + x_2 \le 3 \\ x_1 + x_2 \le 5 \\ x_1 \ge 0, \quad x_2 \ge 0 \end{cases}$

26. Maximize $P = 12x_1 + 6x_2$

Subject to $\begin{cases} x_1 + x_2 \le 8 \\ x_1 - x_2 \le 2 \\ -x_1 + 2x_2 \le 10 \\ x_1 \ge 0, \quad x_2 \ge 0 \end{cases}$

27. Maximize $P = 40x_1 + 60x_2 + 50x_3$

Subject to $\begin{cases} 2x_1 + x_2 + 4x_3 \le 400 \\ 4x_1 + 3x_2 + 2x_3 \le 600 \\ x_1 \ge 0, \quad x_2 \ge 0, \quad x_3 \ge 0 \end{cases}$

28. Maximize $P = 140x_1 + 80x_2 + 100x_3$

Subject to $\begin{cases} 4x_1 + 2x_2 + x_3 \le 100 \\ 2x_1 + 3x_2 + 3x_3 \le 150 \\ x_1 \ge 0, \quad x_2 \ge 0, \quad x_3 \ge 0 \end{cases}$

29. Maximize $P = x_1 + 2x_2 + 3x_3$

Subject to $\begin{cases} x_1 + x_2 + x_3 \le 15 \\ x_2 + x_3 \le 10 \\ x_3 \le 5 \\ x_1 \ge 0, \quad x_2 \ge 0, \quad x_3 \ge 0 \end{cases}$

30. Maximize $P = 15x_1 + 10x_2 + 5x_3$

Subject to $\begin{cases} x_1 + x_2 + x_3 \le 3 \\ x_1 + x_2 \le 2 \\ x_1 \le 1 \\ x_1 \ge 0, \quad x_2 \ge 0, \quad x_3 \ge 0 \end{cases}$

APPLIED EXERCISES

Formulate each situation as a linear programming problem by identifying the variables, the objective function, and the constraints. Be sure to state clearly the meaning of each variable. Check that the problem is a standard maximum problem and then solve it by the simplex method. State your final answer in terms of the original question.

You may use the PIVOT program if permitted by your instructor.

31. BUSINESS: Production Planning An automotive parts shop rebuilds carburetors, fuel pumps, and alternators. The numbers of hours to rebuild and then inspect and pack each part are shown in the following table, together with the number of hours of skilled labor available for each task. If the profit is $12 for each carburetor, $14 for each fuel pump, and $10 for each alternator, how many of each should the shop rebuild to obtain the greatest possible profit?

	Rebuilding	Inspection & Packaging
Carburetor	5 hours	1 hour
Fuel Pump	4 hours	1 hour
Alternator	3 hours	0.5 hour
Labor Available	200 hours	45 hours

32. ENVIRONMENTAL SCIENCE: Pollution Control A storage yard at a coal-burning electric power plant can hold no more than 100,000 tons of coal. Two grades of coal are available: low-sulfur (1%) with an energy content of 20 million British thermal units (BTU) per ton and high-sulfur (2%) with an energy content of 30 million BTU per ton. If the next coal purchase may contain no more than 1400 tons of sulfur, how many tons of each type of coal should be purchased to obtain the most energy?

33. ENVIRONMENTAL SCIENCE: Recycling Management A volunteer recycle center accepts both used paper and empty glass bottles, which are then sorted and sold to a reprocessing company. The center has room to accept 800 crates of paper and glass each week and 50 hours of volunteer help to do the sorting. Each crate of paper products takes 5 minutes to sort and sells for 8¢, while each crate of bottles takes 3 minutes to sort and sells for 7¢. How many crates of each should the center accept each week to raise the most money for its ecology scholarship fund?

34. BUSINESS: Production Planning A small jewelry company prepares and mounts semiprecious stones. There are 20 lapidaries (who cut and polish the stones) and 24 jewelers (who mount the stones in gold settings). Each employee works 7 hours each day. Each tray of agates requires 5 hours of cutting and polishing and 4 hours to mount, each tray of onyxes requires 2 hours of cutting and polishing and 3 hours to mount, and each tray of garnets requires 6 hours of cutting and polishing and 3 hours to mount. If the profit is $15 for each tray of agates, $10 for each tray of onyxes, and $12 for each tray of garnets, how many trays of each stone should be processed each day to obtain the greatest possible profit?

35. BUSINESS: Production Planning Repeat the previous problem but with a profit of $13 per tray of garnets.

36. BUSINESS: Resource Allocation A furniture shop manufactures wooden desks, tables, and chairs. The numbers of hours to assemble and finish each piece are shown in the table, together with the numbers of hours of skilled labor available for each task. If the profit is $80 for each desk, $84 for each table, and $68 for each chair, how many of each should the shop manufacture to obtain the greatest possible profit?

	Assembly	**Finishing**
Desk	2 hours	2 hours
Table	1 hour	3 hours
Chair	2 hours	1 hour
Labor Available	200 hours	150 hours

37. SOCIAL SCIENCE: Political Advertising In the last few days before the election, a politician can afford to spend no more than $27,000 on TV advertisements and can arrange for no more than 10 ads. Each daytime ad costs $2000 and reaches 4000 viewers, each prime-time ad costs $3000 and reaches 5000 viewers, and each late-night ad costs $1000 and reaches 2000 viewers. Ignoring repeated viewings by the same person and assuming every viewer can vote, how many ads in each of the time periods will reach the most voters?

38. BUSINESS: Advertising The manager of a new mall may spend up to $18,000 on grand opening announcements in newspapers, on radio, and on TV. Each newspaper ad costs $300 and reaches 6000 readers, each one-minute radio commercial costs $800 and is heard by 10,000 listeners, and each 15-second TV spot costs $900 and is seen by 11,000 viewers. If there is time to arrange for no more than 5 newspaper ads and no more than 20 radio commercials and TV spots combined, how many of each should be placed to reach the largest number of potential customers? (Ignore multiple exposures to the same consumer.)

39. BUSINESS: Agriculture A farmer grows corn, peanuts, and soybeans on his 240-acre farm. To maintain soil fertility, the farmer rotates the crops and always plants at least as many acres of soybeans as the total acres of the other crops. Each acre of corn requires 2 days of labor and yields a profit of $150, each acre of peanuts requires 5 days of labor and yields a profit of $300, and each acre of soybeans requires 1 day of labor and yields a profit of $100. If the farmer and his family can put in at most 630 days of labor, how many acres of each crop should the farmer plant to obtain the greatest possible profit?

40. BUSINESS: Agriculture A farmer grows wheat, barley, and oats on her 500-acre farm. Each acre of wheat requires 3 days of labor (to plant, tend, and harvest) and costs $21 (for seed, fertilizer, and pesticides), each acre of barley requires 2 days of labor and costs $27, and each acre of oats requires 3 days of labor and costs $24. The farmer and her hired field hands can provide no more than 1200 days of labor this year, and she can afford to spend no more than $15,120. If the expected profit is $50 for each acre of wheat, $40 for each acre of barley, and $45 for each acre of oats, how many acres of each crop should she grow to obtain the greatest possible profit?

Explorations and Excursions

The following problems extend and augment the material presented in the text.

 The PIVOT program may be helpful (if permitted by your instructor).

More About the Pivot Operation

41. Show that pivoting on a tableau obeys the diagram

$$
\begin{matrix}
\vdots & \vdots & & \vdots & \vdots \\
\cdots \; c \;\; q \;\; \cdots & & \cdots \; 0 \;\; q-\dfrac{rc}{p} \;\; \cdots \\
\xrightarrow{} & & \\
\cdots \; p \;\; r \;\; \cdots & & \cdots \; 1 \;\; r/p \;\; \cdots \\
\vdots & \vdots & & \vdots & \vdots
\end{matrix}
$$

where p is the pivot element, c is any other entry in the pivot column, r is any other entry in the pivot row, and q is any entry in the tableau not in the pivot row or pivot column. In words, this diagram defines the pivot operation in four steps: "(1) the pivot element becomes 1, (2) the rest of the pivot column becomes 0, (3) the rest of the pivot row is divided by the pivot element, and (4) every entry not in the pivot row or pivot column is decreased by the product of the pivot column entry in the same row with the pivot row entry in the same column divided by the pivot element." This form of the pivot operation makes it easy to check any particular number in a new tableau without repeating the row operations.

Cycling

42. Beale's Cycling Example Attempt to solve the following problem by the simplex method:

$$
\text{Maximize} \quad P = (3/4 \;\; -20 \;\; 1/2 \;\; -6)\begin{pmatrix} x_1 \\ x_2 \\ x_3 \\ x_4 \end{pmatrix}
$$

Subject to
$$
\begin{cases}
\begin{pmatrix} 1/4 & -8 & -1 & 9 \\ 1/2 & -12 & -1/2 & 3 \\ 0 & 0 & 1 & 0 \end{pmatrix}\begin{pmatrix} x_1 \\ x_2 \\ x_3 \\ x_4 \end{pmatrix} \le \begin{pmatrix} 0 \\ 0 \\ 1 \end{pmatrix} \\[20pt]
\text{and} \quad \begin{pmatrix} x_1 \\ x_2 \\ x_3 \\ x_4 \end{pmatrix} \ge 0
\end{cases}
$$

You should return to the initial simplex tableau after pivoting at [column 1, row 1], [column 2, row 2], [column 3, row 1], [column 4, row 2], [column 5, row 1], and [column 6, row 2].

43. Rescaling Changing the size of the numbers in a problem without altering the feasible region can prevent cycling. Multiply the first constraint in Beale's cycling example (Exercise 42) by 4 to get the equivalent problem

$$
\text{Maximize} \quad P = (3/4 \;\; -20 \;\; 1/2 \;\; -6)\begin{pmatrix} x_1 \\ x_2 \\ x_3 \\ x_4 \end{pmatrix}
$$

Subject to
$$
\begin{cases}
\begin{pmatrix} 1 & -32 & -4 & 36 \\ 1/2 & -12 & -1/2 & 3 \\ 0 & 0 & 1 & 0 \end{pmatrix}\begin{pmatrix} x_1 \\ x_2 \\ x_3 \\ x_4 \end{pmatrix} \le \begin{pmatrix} 0 \\ 0 \\ 1 \end{pmatrix} \\[20pt]
\text{and} \quad \begin{pmatrix} x_1 \\ x_2 \\ x_3 \\ x_4 \end{pmatrix} \ge 0
\end{cases}
$$

You should reach the final simplex tableau after pivoting at [column 1, row 1], [column 2, row 2], [column 3, row 1], [column 4, row 2], [column 1, row 3], and [column 5, row 2].

44. Bland's Rule Eliminate cycling by changing the choice of the pivot column to be *the leftmost column with a negative entry in the bottom row* (that is, ignore the size of the numbers and just look for the first negative). Use Bland's rule to solve Beale's cycling example (Exercise 42). You should reach the final simplex tableau after pivoting at [column 1, row 1], [column 2, row 2], [column 3, row 1], [column 4, row 2], [column 1, row 3], and [column 5, row 2].

45. A Nondeterministic Simplex Algorithm The simplex method (page 285) is a *deterministic* procedure to solve standard maximum problems because once the process is started, the rules *always* select the *same* sequence of pivot elements. If we allow *chance* to play a role in the selection of the pivot column or pivot row, the resulting solution method is nondeterministic in that the sequence of pivot elements in the solution is no longer determined but rather may change from one solution attempt to another. Let us change the rule on page 279 for the selection of the pivot row to state that *if there is a tie for the smallest ratio between two of the rows, the pivot row is to be selected by a coin flip.* Solve Beale's cycling example (Exercise 42) several times using this modified procedure. Do your solutions all use the same number of pivot operations?

46. **More Cycling Examples** A large class of cyclic examples was found by K. T. Marshall and J. W. Suurballe in 1969.

   a. Try to solve this typical Marshall–Suurballe problem by the simplex method.

$$\text{Maximize} \quad P = (1 \quad -28 \quad -2 \quad -2)\begin{pmatrix} x_1 \\ x_2 \\ x_3 \\ x_4 \end{pmatrix}$$

Subject to
$$\begin{cases} \begin{pmatrix} 1/2 & -18 & -2 & 4 \\ 1/4 & -5 & -1/2 & 1/2 \end{pmatrix}\begin{pmatrix} x_1 \\ x_2 \\ x_3 \\ x_4 \end{pmatrix} \le \begin{pmatrix} 0 \\ 0 \end{pmatrix} \\ \\ \text{and} \quad \begin{pmatrix} x_1 \\ x_2 \\ x_3 \\ x_4 \end{pmatrix} \ge 0 \end{cases}$$

   You should return to the initial simplex tableau after pivoting at [column 1, row 1], [column 2, row 2], [column 3, row 1], [column 4, row 2], [column 5, row 1], and [column 6, row 2].

   b. Rescale the constraints of the problem in part (a) by multiplying the first by 2 and the second by 4. How many pivots does it now take to solve this rescaled problem?

   c. Solve the problem in part (a) using Bland's rule (see Exercise 44).
   d. Solve the problem in part (a) using the nondeterministic simplex algorithm given in Exercise 45.

Klee–Minty Problems

47. Solve the following two-variable problem by the simplex method. (Your fourth tableau should be the final tableau.)

$$\text{Maximize} \quad P = (1 \quad 10)\begin{pmatrix} x_1 \\ x_2 \end{pmatrix}$$

Subject to
$$\begin{cases} \begin{pmatrix} 0 & 1 \\ 1 & 20 \end{pmatrix}\begin{pmatrix} x_1 \\ x_2 \end{pmatrix} \le \begin{pmatrix} 1 \\ 100 \end{pmatrix} \\ \\ \text{and} \quad \begin{pmatrix} x_1 \\ x_2 \end{pmatrix} \ge 0 \end{cases}$$

48. Solve the following three-variable problem by the simplex method. (Your eighth tableau should be the final tableau.)

$$\text{Maximize} \quad P = (1 \quad 10 \quad 100)\begin{pmatrix} x_1 \\ x_2 \\ x_3 \end{pmatrix}$$

Subject to
$$\begin{cases} \begin{pmatrix} 0 & 0 & 1 \\ 0 & 1 & 20 \\ 1 & 20 & 200 \end{pmatrix}\begin{pmatrix} x_1 \\ x_2 \\ x_3 \end{pmatrix} \le \begin{pmatrix} 1 \\ 100 \\ 10{,}000 \end{pmatrix} \\ \\ \text{and} \quad \begin{pmatrix} x_1 \\ x_2 \\ x_3 \end{pmatrix} \ge 0 \end{cases}$$

49. Following the form of Exercises 47 and 48, write down a problem using four variables that will continue the pattern and require sixteen tableaux to solve.

50. Using the problems from Exercises 47–49, check that each of this family of Klee–Minty problems can be solved in one pivot by using Bland's rule (see Exercise 44) to choose the pivot column.

4.4 STANDARD MINIMUM PROBLEMS AND DUALITY

Introduction

In this section we will define and solve *standard minimum* linear programming problems. Rather than develop a new technique for minimum problems, we will find a relationship between minimum and maximum problems that allows us to solve one by solving the other.

Standard Minimum Problems

A *standard minimum problem* is a linear programming problem in which the objective function is to be minimized, all of the coefficients in the objective function are nonnegative, the variables $y_1, y_2, \ldots, y_n$ are all nonnegative, and the other constraints have $\geq$ inequalities.

Standard Minimum Problem

A problem is a *standard minimum problem* if it can be written:

Minimize an objective function

$$C = b_1 y_1 + b_2 y_2 + \cdots + b_n y_n \quad \text{with } nonnegative \ \ b_1, b_2, \ldots, b_n$$

Subject to inequalities of the form

$$a_1 y_1 + a_2 y_2 + \cdots + a_n y_n \geq c$$

and

$$y_1 \geq 0, \ y_2 \geq 0, \ldots, \ y_n \geq 0$$

We use C for the objective function to suggest cost, which is usually minimized, and y-variables to distinguish them from the x-variables used in maximum problems. Notice that in a minimum problem it is the *coefficients of the objective function* that must be nonnegative—the right-hand sides of the constraints may be of either sign.

EXAMPLE 1

VERIFYING STANDARD MINIMUM FORM

Is the following a standard minimum problem?

$$\text{Minimize} \quad C = 24y_1 + 10y_2$$

$$\text{Subject to} \quad \begin{cases} 3y_1 + y_2 \geq 9 \\ 4y_1 + 2y_2 \geq 16 \\ y_1 \geq 0 \ \text{ and } \ y_2 \geq 0 \end{cases}$$

Solution

The objective function is to be minimized, the coefficients 24 and 10 in the objective function are nonnegative, the variables are nonnegative, and the other inequalities all have $\geq$ inequalities, so this *is* a standard minimum problem.

Practice Problem 1

Is the following a standard minimum problem?

Minimize $C = 24y_1 - 4y_2$

Subject to $\begin{cases} 2y_1 - 5y_2 \geq -10 \\ 3y_1 + 2y_2 \geq 6 \\ y_1 \geq 0 \quad \text{and} \quad y_2 \geq 0 \end{cases}$ ➤ **Solution on page 308**

The Dual of a Standard Minimum Problem

Given a standard minimum problem, we want to create from it a particular maximum problem, called the *dual problem* of the original, which will help us to solve the original minimum problem. For this purpose, recall that the *transpose* of a matrix turns every row into a column, or equivalently, every column into a row (see page 197). Denoting a transpose by a superscripted *t*, we have, for example:

$$\begin{pmatrix} 1 & 2 & 3 \\ 4 & 5 & 6 \end{pmatrix}^t = \begin{pmatrix} 1 & 4 \\ 2 & 5 \\ 3 & 6 \end{pmatrix}$$

To create the dual, we summarize the minimum problem by writing its numbers in a matrix (with the objective function numbers in the bottom row), transpose the matrix, and then reinterpret the matrix as a maximum problem (with $\leq$ inequalities).

EXAMPLE 2

FINDING A DUAL PROBLEM

Find the dual of the following problem:

Minimize $C = 24y_1 + 10y_2$

Subject to $\begin{cases} 3y_1 + y_2 \geq 9 \\ 4y_1 + 2y_2 \geq 16 \\ y_1 \geq 0 \quad \text{and} \quad y_2 \geq 0 \end{cases}$

Solution

Beginning with the original problem on the left below, we summarize, transpose, and rewrite as a maximum problem as shown directly below the original.

Minimize $C = 24y_1 + 10y_2$

Subject to $\begin{cases} 3y_1 + y_2 \geq 9 \\ 4y_1 + 2y_2 \geq 16 \\ y_1 \geq 0 \text{ and } y_2 \geq 0 \end{cases}$

Original

$\xrightarrow{\text{Summarize}}$

$\begin{pmatrix} 3 & 1 & | & 9 \\ 4 & 2 & | & 16 \\ \hline 24 & 10 & | & 0 \end{pmatrix}$ $\left.\begin{matrix} \\ \\ \end{matrix}\right\}$ Constraints

$\leftarrow$ Objective function

Dual

$\big\downarrow$ Transpose

Maximize $P = 9x_1 + 16x_2$

Subject to $\begin{cases} 3x_1 + 4x_2 \leq 24 \\ x_1 + 2x_2 \leq 10 \\ x_1 \geq 0 \text{ and } x_2 \geq 0 \end{cases}$

$\xleftarrow{\text{Rewrite}}$

$\begin{pmatrix} 3 & 4 & | & 24 \\ 1 & 2 & | & 10 \\ \hline 9 & 16 & | & 0 \end{pmatrix}$ $\left.\begin{matrix} \\ \\ \end{matrix}\right\}$ *New* constraints

$\leftarrow$ *New* objective function

Be careful! The matrices on the right are *not* tableaux so do not change the signs of the objective function numbers.

The dual of a standard minimum problem is a standard maximum problem. Since transposing the numbers a second time would just recover the original, we say that *either problem is the dual of the other*.

Practice Problem 2

Find the dual of the following problem:

Minimize $C = 24y_1 + 20y_2 + 12y_3$

Subject to $\begin{cases} 2y_1 + 5y_2 + 3y_3 \geq 3 \\ -3y_1 + 2y_2 + 2y_3 \geq 5 \\ y_1 \geq 0, \ y_2 \geq 0, \text{ and } y_3 \geq 0 \end{cases}$ ➤ Solution on page 308

Practice Problem 2 shows that a problem and its dual need not have the same number of variables or the same number of constraints.

The two problems in Example 2 use only two variables and so can be solved by the graphical method of Section 4.2: drawing graphs, shading regions, finding vertices, and determining which vertex maximizes or minimizes the objective function. The graphical solutions of these two problems are shown below, with the solution to each being where the two lines intersect.

Minimize $C = 24y_1 + 10y_2$

$C = 84$ at $(1, 6)$

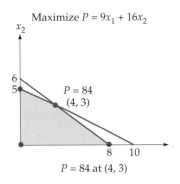

Maximize $P = 9x_1 + 16x_2$

$P = 84$ at $(4, 3)$

Notice that although the problems are quite different, the maximum value of P in one is the same as the minimum value of C in the dual (as you can easily check by substituting the intersection points into the objective functions). The fact that the minimum value of C in a standard minimum problem will *always* be the same as the maximum value of P in the dual maximum problem was a remarkable discovery, and is called the Duality Theorem. The following statement of the theorem also shows how to read the solution of the minimum problem from the final tableau of the dual problem. The Duality Theorem is proved in Exercise 37 at the end of this section.

Duality Theorem

If a maximum problem has a solution, then its final tableau also displays the solution to the dual minimum problem in its bottom row: first the slack variables, then the variables, and finally the objective function:

Even if this is not a final tableau, the values in the bottom row satisfy the constraints of the minimum problem with the exception of the nonnegativity conditions and yield the given value of the objective function.

The last sentence is not necessary for the purposes of this section but will be used in the following two sections. Note that for a minimum problem we use t's for the slack variables to distinguish them from the slacks in a maximum problem. Since dual problems are related by transposition, it is not surprising that the solution to the *minimum* problem appear in the last *row*, just as the solution to the *maximum* problem appears in the last *column*. If the maximum problem has no solution, then neither does the dual minimum problem.

For example, on pages 296–297 we began with the following minimum problem and found its dual.

$$\text{Minimize} \quad C = 24y_1 + 10y_2$$
$$\text{Subject to} \quad \begin{cases} 3y_1 + y_2 \geq 9 \\ 4y_1 + 2y_2 \geq 16 \\ y_1 \geq 0 \quad \text{and} \quad y_2 \geq 0 \end{cases}$$

The dual maximum problem is the same one we solved in the previous section, obtaining the following final tableau on page 282:

	x_1	x_2	s_1	s_2	
x_1	1	0	1	-2	4
x_2	0	1	$-\frac{1}{2}$	$\frac{3}{2}$	3
P	0	0	1	6	84

Slacks y-variables   Objective function

From the bottom row of this tableau we may read off the solution to the original minimum problem:

The minimum value of C is 84, when $y_1 = 1$ and $y_2 = 6$.

The slacks, which we usually ignore, are $t_1 = 0$ and $t_2 = 0$

This is the same as the graphical solution shown on the previous page.

Practice Problem 3

Find the solution of the linear programming problem

$$\text{Minimize} \quad C = 255y_1 + 435y_2 + 300y_3 + 465y_4$$
$$\text{Subject to} \quad \begin{cases} 2y_1 + 3y_2 + 2y_3 + 4y_4 \geq 25 \\ y_1 + 3y_2 + 2y_3 + 2y_4 \geq 18 \\ 3y_1 + 4y_2 + 3y_3 + 2y_4 \geq 36 \\ y_1 \geq 0, \quad y_2 \geq 0, \quad y_3 \geq 0, \quad \text{and} \quad y_4 \geq 0 \end{cases}$$

from the fact that the final tableau of its dual maximum problem is

	x_1	x_2	x_3	s_1	s_2	s_3	s_4	
x_3	0	0	1	0	-2	3	0	30
x_1	1	0	0	1	3	-5	0	60
x_2	0	1	0	-1	0	1	0	45
s_4	0	0	0	-2	-8	12	1	75
P	0	0	0	7	3	1	0	3390

➤ Solution on page 308

We may summarize the entire process as follows.

Solution of a Standard Minimum Problem

To solve a standard minimum problem:

1. Construct the dual maximum problem (pages 296–297).
2. Solve the dual maximum problem by the simplex method (page 285).
3. If there is a solution to the dual maximum problem, then there is a solution to the minimum problem: The values of the slacks, the variables, and the objective function appear (in that order) in the bottom row of the final tableau. If the maximum problem has no solution, then neither does the minimum problem.

EXAMPLE 3

SOLVING A STANDARD MINIMUM PROBLEM

$$\text{Minimize} \quad C = 50y_1 + 60y_2$$

$$\text{Subject to} \quad \begin{cases} y_1 + 4y_2 \geq 20 \\ 2y_1 + 3y_2 \geq 30 \\ y_1 - y_2 \geq 5 \\ y_1 \geq 0 \quad \text{and} \quad y_2 \geq 0 \end{cases}$$

Solution

First we summarize, transpose, and write the dual:

$$\begin{pmatrix} 1 & 4 & | & 20 \\ 2 & 3 & | & 30 \\ 1 & -1 & | & 5 \\ \hline 50 & 60 & | & 0 \end{pmatrix} \rightarrow \begin{pmatrix} 1 & 2 & 1 & | & 50 \\ 4 & 3 & -1 & | & 60 \\ \hline 20 & 30 & 5 & | & 0 \end{pmatrix} \rightarrow$$

$$\text{Maximize} \quad P = 20x_1 + 30x_2 + 5x_3$$

$$\text{Subject to} \quad \begin{cases} x_1 + 2x_2 + x_3 \leq 50 \\ 4x_1 + 3x_2 - x_3 \leq 60 \\ x_1 \geq 0, \quad x_2 \geq 0, \quad x_3 \geq 0 \end{cases}$$

The initial simplex tableau for the dual maximum problem is

	x_1	x_2	x_3	s_1	s_2	
s_1	1	2	1	1	0	50
s_2	4	3	-1	0	1	60
P	-20	-30	-5	0	0	0

Bottom row states that
$t_1 = -20, \ t_2 = -30, \ t_3 = -5,$
$y_1 = 0, \ y_2 = 0, \ $ and $\ C = 0$

To solve the dual maximum problem, we first pivot on the 3 in column 2 and row 2 to find the tableau

	x_1	x_2	x_3	s_1	s_2	
s_1	-5/3	0	5/3	1	-2/3	10
x_2	4/3	1	-1/3	0	1/3	20
P	20	0	-15	0	10	600

Bottom row states that
$t_1 = 20, \ t_2 = 0, \ t_3 = -15,$
$y_1 = 0, \ y_2 = 10,$
and $\ C = 600$

Then we pivot again on the 5/3 in column 3 and row 1 to reach the final tableau:

	x_1	x_2	x_3	s_1	s_2	
x_3	-1	0	1	3/5	-2/5	6
x_2	1	1	0	1/5	1/5	22
P	5	0	0	9	4	690

Bottom row states that
$t_1 = 5, \ t_2 = 0, \ t_3 = 0,$
$y_1 = 9, \ y_2 = 4, \ $ and
$C = 690$

Answer: The minimum value is 690 when $y_1 = 9$ and $y_2 = 4$.

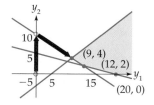

To compare this solution with the graphical method, we graph the feasible region of the minimum problem with y_1 on the horizontal axis and y_2 on the vertical axis. The path of nonfeasible points visited by the simplex tableaux of the dual maximum problem and leading to a feasible vertex is marked in bold.

The simplex method always begins at the origin and "pivots" to successively better vertices. Notice from the graph that in a minimum problem the origin is optimal (minimizing C to 0) but not feasible (not in the shaded region). This is in contrast to a maximum problem, where the origin is feasible but not optimal. That is, the simplex method *keeps the minimum problem optimal while making it feasible and keeps the maximum problem feasible while making it optimal.*

Shadow Prices and Marginal Values

The variables of the dual problem provide information about the original problem. Recall the dual minimum and maximum problems from pages 296–297 and 298, with their solutions written below them.

Minimize $C = 24y_1 + 10y_2$

Subject to $\begin{cases} 3y_1 + y_2 \geq 9 \\ 4y_1 + 2y_2 \geq 16 \\ y_1 \geq 0 \text{ and } y_2 \geq 0 \end{cases}$

$C = 84$ at $y_1 = 1$, $y_2 = 6$

Maximize $P = 9x_1 + 16x_2$

Subject to $\begin{cases} 3x_1 + 4x_2 \leq 24 \\ x_1 + 2x_2 \leq 10 \\ x_1 \geq 0 \text{ and } x_2 \geq 0 \end{cases}$

$P = 84$ at $x_1 = 4$, $x_2 = 3$

To illustrate their relationship as dual problems, suppose that the first problem involves minimizing the cost of using amounts y_1 and y_2 of two nutritional supplements that cost 24¢ and 10¢ per ounce (the objective function) subject to the daily requirements that you need at least 9 units of vitamins (the first constraint) and at least 16 units of minerals (the second constraint). The numbers of units of vitamins and minerals in each ounce of supplements A and B are given in the table on the left.

	A	B
Vitamins	3	4
Minerals	1	2

The solution says that 84¢ provides enough to satisfy both constraints, so it should be possible to see how much of this cost is for the vitamin constraint and how much is for the mineral constraint, leading to, in some sense, a *cost for the vitamins* and a *cost for the minerals*. Letting x_1 stand for this per-unit cost of vitamins and x_2 for the per-unit cost of minerals, we may interpret the two inequalities in the *dual* problem (on the right above) as saying that, for each supplement, the costs allocated to the ingredients cannot exceed the cost of the supplement.

$$3x_1 + 4x_2 \leq 24 \quad \begin{pmatrix} \text{Vitamins} \\ \text{in A} \end{pmatrix}\begin{pmatrix} \text{Cost of} \\ \text{vitamins} \end{pmatrix} + \begin{pmatrix} \text{Minerals} \\ \text{in A} \end{pmatrix}\begin{pmatrix} \text{Cost of} \\ \text{minerals} \end{pmatrix} \leq \begin{pmatrix} \text{Cost} \\ \text{of A} \end{pmatrix}$$

$$x_1 + 2x_2 \leq 10 \quad \begin{pmatrix} \text{Vitamins} \\ \text{in B} \end{pmatrix}\begin{pmatrix} \text{Cost of} \\ \text{vitamins} \end{pmatrix} + \begin{pmatrix} \text{Minerals} \\ \text{in B} \end{pmatrix}\begin{pmatrix} \text{Cost of} \\ \text{minerals} \end{pmatrix} \leq \begin{pmatrix} \text{Cost} \\ \text{of B} \end{pmatrix}$$

The values that we found, $x_1 = 4$ and $x_2 = 3$, then tell us that the cost of the vitamins is 4¢ per ounce and the cost of the minerals is 3¢ per ounce. These are called the *shadow prices* of vitamins and minerals. As an example of how such shadow prices are used, the 3¢ cost of minerals in this example means that if the mineral requirement were reduced by one unit (from 8 units to 7), the cost saving would be 3¢.

A similar analysis applied to the maximum problem on the right above, which we may interpret as maximizing the profit of a pottery company (the objective function) that sells x_1 bowls for $9 each and x_2 plates for $16, each made by "shapers" and "decorators," subject to the requirements that the shaping time cannot exceed 24 hours per day

	Shaping	Decorating
Bowls	3	1
Plates	4	2

(the first constraint) and the decorating time cannot exceed 10 hours per day (the second constraint), with the numbers of hours of shaping and decorating per bowl or plate given by the table on the left.

The profit is based on two processes, shaping and decorating, so it can be allocated between them. If we let y_1 stand for the *profit per hour based on shaping* and y_2 for the *profit per hour based on decorating*, then multiplying by the numbers in the table we would obtain constraints like those in the dual (minimum) problem on the left above. The solution $y_1 = 1$ then means that the shaping process contributes $1 per hour to the profit and $y_2 = 6$ means that the decorating process contributes $6 per hour to the profit. These numbers are called the *marginal values* of the shaping and decorating processes. For an example of how they are used, the $6 per hour contribution of decorating means that increasing the *decorating* time by 1 hour would raise the profit by $6. Further increases would be proportional (for example, an additional 5 hours would bring an additional $30) provided that none of the constraints are violated.

In general, for either a maximum or minimum problem, the *dual* variable with subscript i measures the contribution to the objective function (profit P or cost C) made by the quantity that is constrained by inequality number i, expressed in *objective units per constraint unit.*

Matrix Form

The relationships between a minimum problem and its dual maximum problem and the initial and final tableaux may be stated simply in matrix form. For a minimum problem, we use Y for a column of y-variables and A^t in the coefficients in the constraints (since we used A in the maximum problem).

Standard Minimum Problem

Minimize $C = b^t Y$ $(b \geq 0)$

Subject to $\begin{cases} A^t Y \geq c \\ Y \geq 0 \end{cases}$

Dual Maximum Problem

Maximize $P = c^t X$

Subject to $\begin{cases} AX \leq b \quad (b \geq 0) \\ X \geq 0 \end{cases}$

Notice that the *objective* numbers and the *right-hand sides of the constraints* become interchanged when writing the dual, so the requirement $b \geq 0$ applies to the objective numbers in the minimum problem and the right-hand sides of the constraints in the maximum problem. If the initial tableau for the maximum problem (on the left below) leads to a final tableau that solves the maximum problem (on the right below), then the minimum value is C with variables Y and slacks T taking values as shown in the bottom row.

$$
\begin{array}{cc}
\textbf{Initial Tableau} & \textbf{Final Tableau} \\[6pt]
\left(
\begin{array}{c|cc|c}
 & X & S & \\ \hline
S & A & I & b \\ \hline
P & -c^t & 0 & 0
\end{array}
\right)
& \quad
\left(
\begin{array}{c|cc|c}
 & X & S & \\ \hline
 & \vdots & \vdots & \vdots \\ \hline
 & T & Y & C
\end{array}
\right)
\end{array}
$$

$\Longrightarrow$ If the solution exists

Mixed Constraints: A Transportation Problem

Although we have been careful to write our standard minimum problem with $\geq$ constraints, we do not mean to exclude the possibility of mixed constraints, some $\geq$ and some $\leq$. We simply multiply each $\leq$ inequality by -1 so it becomes a $\geq$ inequality. (Recall from page 295 that the minimum problem does not require that the right-hand sides of the inequalities be nonnegative.) The following example belongs to a general type of minimization problem with mixed constraints having the nice property that the pivot elements will all be 1s.

EXAMPLE **4**

A TRANSPORTATION PROBLEM

A retail store chain has cartons of goods stored at warehouses in Maryland and Washington that must be distributed to its stores in Ohio and Louisiana. The cost to ship each carton from Maryland to Ohio is $6, from Maryland to Louisiana is $7, from Washington to Ohio is $8, and from Washington to Louisiana is $9. There are 300 cartons at the Maryland warehouse and 300 at the warehouse in Washington. If the Ohio stores need 200 cartons and the Louisiana stores need 300 cartons, how many cartons should be shipped from each warehouse to each state to incur the smallest shipping costs?

Solution

This is a minimization problem that requires four variables: one for the amount shipped from each warehouse to each state. Let

$$
y_1 = \left(\begin{array}{c}\text{Cartons shipped} \\ \text{from MD to OH}\end{array}\right), \qquad
y_2 = \left(\begin{array}{c}\text{Cartons shipped} \\ \text{from MD to LA}\end{array}\right),
$$

$$
y_3 = \left(\begin{array}{c}\text{Cartons shipped} \\ \text{from WA to OH}\end{array}\right), \quad \text{and} \quad
y_4 = \left(\begin{array}{c}\text{Cartons shipped} \\ \text{from WA to LA}\end{array}\right)
$$

Because each warehouse can ship no more than the number of cartons stored there, and each state must receive at least the required number of cartons, this problem is the linear programming problem

Minimize $C = 6y_1 + 7y_2 + 8y_3 + 9y_4$ Total shipping costs

Subject to
$$\begin{cases} y_1 + y_2 & \leq 300 & \text{Have 300 in MD} \\ y_3 + y_4 & \leq 300 & \text{Have 300 in WA} \\ y_1 + y_3 & \geq 200 & \text{Need 200 in OH} \\ y_2 + y_4 & \geq 300 & \text{Need 300 in LA} \\ y_1 \geq 0, \quad y_2 \geq 0, \quad y_3 \geq 0, \quad y_4 \geq 0 & & \text{Nonnegativity} \end{cases}$$

We multiply the first two constraints by -1 to give them $\geq$ inequalities. This problem is then simply stated in matrix form:

$$\text{Minimize} \quad C = (6 \quad 7 \quad 8 \quad 9) \begin{pmatrix} y_1 \\ y_2 \\ y_3 \\ y_4 \end{pmatrix}$$

Subject to
$$\begin{cases} \begin{pmatrix} -1 & -1 & 0 & 0 \\ 0 & 0 & -1 & -1 \\ 1 & 0 & 1 & 0 \\ 0 & 1 & 0 & 1 \end{pmatrix} \begin{pmatrix} y_1 \\ y_2 \\ y_3 \\ y_4 \end{pmatrix} \geq \begin{pmatrix} -300 \\ -300 \\ 200 \\ 300 \end{pmatrix} \\[40pt] \text{and} \quad \begin{pmatrix} y_1 \\ y_2 \\ y_3 \\ y_4 \end{pmatrix} \geq 0 \end{cases}$$

The dual maximum problem is

$$\text{Maximize} \quad P = (-300 \quad -300 \quad 200 \quad 300) \begin{pmatrix} x_1 \\ x_2 \\ x_3 \\ x_4 \end{pmatrix}$$

Subject to
$$\begin{cases} \begin{pmatrix} -1 & 0 & 1 & 0 \\ -1 & 0 & 0 & 1 \\ 0 & -1 & 1 & 0 \\ 0 & -1 & 0 & 1 \end{pmatrix} \begin{pmatrix} x_1 \\ x_2 \\ x_3 \\ x_4 \end{pmatrix} \leq \begin{pmatrix} 6 \\ 7 \\ 8 \\ 9 \end{pmatrix} \\[40pt] \text{and} \quad \begin{pmatrix} x_1 \\ x_2 \\ x_3 \\ x_4 \end{pmatrix} \geq 0 \end{cases}$$

The initial simplex tableau is

	x_1	x_2	x_3	x_4	s_1	s_2	s_3	s_4	
s_1	−1	0	1	0	1	0	0	0	6
s_2	−1	0	0	1	0	1	0	0	7
s_3	0	−1	1	0	0	0	1	0	8
s_4	0	−1	0	1	0	0	0	1	9
P	300	300	−200	−300	0	0	0	0	0

We pivot at [column 4, row 2], [column 3, row 1], and [column 1, row 3] to reach the final tableau:

	x_1	x_2	x_3	x_4	s_1	s_2	s_3	s_4	
x_3	0	−1	1	0	0	0	1	0	8
x_4	0	−1	0	1	−1	1	1	0	9
x_1	1	−1	0	0	−1	0	1	0	2
s_4	0	0	0	0	1	−1	−1	1	0
P	0	100	0	0	0	300	200	0	3700
					y_1	y_2	y_3	y_4	C

Bottom row states that $t_1 = 0$, $t_2 = 100$, $t_3 = 0$, $t_4 = 0$, $y_1 = 0$, $y_2 = 300$, $y_3 = 200$, $y_4 = 0$, and $C = 3700$

The Maryland warehouse should send nothing to Ohio and 300 cartons to Louisiana, while the Washington warehouse should send 200 cartons to Ohio and nothing to Louisiana to achieve the least shipping cost of $3700. The second slack shows that there will be 100 cartons unused in the Washington warehouse.

Spreadsheet Exploration

The spreadsheet* on the next page shows the solution of Example 4 using the **Solver** command. The entries in column E combine the coefficients and the current values of the variables to find the values that the **Solver** compares to the bounds and the goal of obtaining the minimum of the objective function.

* See the Preface for how to obtain this and other Excel spreadsheets.

The answer, sensitivity, and limits reports generated by the Solver provide additional information about the solution. Do you see where the values in the bottom row and right column of the final tableau in the solution of Example 4 on the previous page appear in this spreadsheet solution?

4.4 Section Summary

We solve a standard minimum problem (page 295) by transposing to find its dual (pages 296–297), which we solve by the simplex method, and then interpret the solution according to the Duality Theorem (page 298):

Standard Minimum Problem

Minimize $C = b^t Y$

Subject to $\begin{cases} A^t Y \geq c \\ Y \geq 0 \end{cases}$

Dual Maximum Problem

Maximize $P = c^t X$

Subject to $\begin{cases} AX \leq b \\ X \geq 0 \end{cases}$

with $b \geq 0$ in both.

If the initial tableau for the dual problem (on the left below) leads to a tableau that solves the maximum problem (on the right), then the solution to the minimum problem may be read from the bottom row of the final tableau: the slack variables T, then the variables Y, followed by the minimum value C.

Initial Tableau

$$\left(\begin{array}{c|cc|c} & X & S & \\ \hline S & A & I & b \\ P & -c^t & 0 & 0 \end{array} \right)$$

$\xrightarrow{\text{If the}}$ solution exists

Final Tableau

$$\left(\begin{array}{c|cc|c} & X & S & \\ \hline & \vdots & \vdots & \vdots \\ & T & Y & C \end{array} \right)$$

> ### Solutions to Practice Problems

1. No. The negative coefficient -4 in the objective function violates the condition in the box on page 295. (The -10 on the right-hand side of the first inequality is *not* a violation—only the objective coefficients need to be nonnegative.)

2.
$$\left(\begin{array}{ccc|c} 2 & 5 & 3 & 3 \\ -3 & 2 & 2 & 5 \\ \hline 24 & 20 & 12 & 0 \end{array} \right)^t \rightarrow \left(\begin{array}{cc|c} 2 & -3 & 24 \\ 5 & 2 & 20 \\ 3 & 2 & 12 \\ \hline 3 & 5 & 0 \end{array} \right) \rightarrow$$

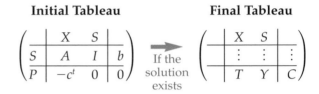

Maximize $P = 3x_1 + 5x_2$

Subject to $\begin{cases} 2x_1 - 3x_2 \leq 24 \\ 5x_1 + 2x_2 \leq 20 \\ 3x_1 + 2x_2 \leq 12 \\ x_1 \geq 0 \text{ and } x_2 \geq 0 \end{cases}$

Notice that the original problem had three variables and two constraints, while the dual has two variables and three constraints. The numbers of variables and constraints will always interchange in this way because of the transposition.

3. The bottom row of the final tableau for the dual maximum problem displays the values for the slack variables $(t_1 = 0, \; t_2 = 0, \; t_3 = 0)$, the variables $(y_1 = 7, \; y_2 = 3, \; y_3 = 1, \; y_4 = 0)$, and the objective function $(C = 3390)$. The minimum value is 3390 when $y_1 = 7, \; y_2 = 3, \; y_3 = 1,$ and $y_4 = 0$.

4.4 Exercises

For each standard minimum problem construct the dual maximum problem and its initial simplex tableau.

1. Minimize $C = 84y_1 + 21y_2$

Subject to $\begin{cases} 3y_1 + y_2 \geq 21 \\ 4y_1 - y_2 \geq 0 \\ y_1 \geq 0, \ y_2 \geq 0 \end{cases}$

2. Minimize $C = 7y_1 + 7y_2$

Subject to $\begin{cases} 2y_1 + 3y_2 \geq 42 \\ 3y_1 + y_2 \geq 21 \\ y_1 \geq 0, \ y_2 \geq 0 \end{cases}$

3. Minimize $C = 60y_1 + 100y_2 + 300y_3$

Subject to $\begin{cases} y_1 + 2y_2 + 3y_3 \geq 180 \\ 4y_1 + 5y_2 + 6y_3 \geq 120 \\ y_1 \geq 0, \ y_2 \geq 0, \ y_3 \geq 0 \end{cases}$

4. Minimize $C = 100y_1 + 60y_2 + 280y_3$

Subject to $\begin{cases} 5y_1 + 9y_2 + 7y_3 \geq 315 \\ 2y_1 + 6y_2 + 4y_3 \geq 480 \\ y_1 \geq 0, \ y_2 \geq 0, \ y_3 \geq 0 \end{cases}$

5. Minimize $C = 3y_1 + 20y_2$

Subject to $\begin{cases} 3y_1 + 2y_2 \geq 150 \\ y_1 + 4y_2 \geq 100 \\ 3y_1 + 4y_2 \geq 228 \\ y_1 \geq 0, \ y_2 \geq 0 \end{cases}$

6. Minimize $C = 60y_1 + 16y_2$

Subject to $\begin{cases} y_1 + 4y_2 \geq 20 \\ 3y_1 + 2y_2 \geq 30 \\ 3y_1 + 4y_2 \geq 48 \\ y_1 \geq 0, \ y_2 \geq 0 \end{cases}$

7. Minimize $C = 15y_1 + 20y_2 + 5y_3$

Subject to $\begin{cases} y_1 - y_2 - 2y_3 \leq 30 \\ y_1 + 2y_2 + y_3 \geq 30 \\ y_1 \geq 0, \ y_2 \geq 0, \ y_3 \geq 0 \end{cases}$

8. Minimize $C = 30y_1 + 40y_2 + 80y_3$

Subject to $\begin{cases} -3y_1 + y_2 - y_3 \leq 60 \\ y_1 + y_2 + 2y_3 \geq 60 \\ y_1 \geq 0, \ y_2 \geq 0, \ y_3 \geq 0 \end{cases}$

Solve each standard minimum problem by finding the dual maximum problem and using the simplex method. (Exercises 9–16 can also be solved by the graphical method.)

You may use the PIVOT program if permitted by your instructor.

9. Minimize $C = 15y_1 + 10y_2$

Subject to $\begin{cases} y_1 + 2y_2 \geq 20 \\ 3y_1 - y_2 \geq 60 \\ y_1 \geq 0, \ y_2 \geq 0 \end{cases}$

10. Minimize $C = 40y_1 + y_2$

Subject to $\begin{cases} 5y_1 - y_2 \geq 5 \\ 4y_1 + y_2 \geq 4 \\ y_1 \geq 0, \ y_2 \geq 0 \end{cases}$

11. Minimize $C = 4y_1 + 3y_2$

Subject to $\begin{cases} y_1 + y_2 \geq 15 \\ 3y_1 + y_2 \leq 60 \\ y_1 \geq 0, \ y_2 \geq 0 \end{cases}$

12. Minimize $C = 4y_1 + 5y_2$

Subject to $\begin{cases} y_1 + y_2 \geq 20 \\ 2y_1 + y_2 \leq 50 \\ y_1 \geq 0, \ y_2 \geq 0 \end{cases}$

13. Minimize $C = 4y_1 + 5y_2$

Subject to $\begin{cases} y_1 + y_2 \geq 10 \\ y_1 \geq 2 \\ y_2 \geq 3 \\ y_1 \geq 0, \ y_2 \geq 0 \end{cases}$

14. Minimize $C = 3y_1 + 2y_2$

Subject to $\begin{cases} y_1 + y_2 \geq 20 \\ y_1 \geq 4 \\ y_2 \geq 5 \\ y_1 \geq 0, \quad y_2 \geq 0 \end{cases}$

15. Minimize $C = 2y_1 + y_2$

Subject to $\begin{cases} -y_1 + y_2 \leq 20 \\ y_1 - y_2 \leq 20 \\ y_1 + y_2 \geq 10 \\ y_1 \geq 0, \quad y_2 \geq 0 \end{cases}$

16. Minimize $C = 2y_1 + 3y_2$

Subject to $\begin{cases} -y_1 + y_2 \leq 30 \\ y_1 - y_2 \leq 30 \\ y_1 + y_2 \geq 10 \\ y_1 \geq 0, \quad y_2 \geq 0 \end{cases}$

17. Minimize $C = 10y_1 + 20y_2 + 10y_3$

Subject to $\begin{cases} -y_1 + y_2 + y_3 \geq 50 \\ y_1 + y_2 - y_3 \geq 30 \\ y_1 \geq 0, \quad y_2 \geq 0, \quad y_3 \geq 0 \end{cases}$

18. Minimize $C = 30y_1 + 50y_2 + 30y_3$

Subject to $\begin{cases} -y_1 + y_2 + y_3 \geq 10 \\ y_1 + y_2 - y_3 \geq 20 \\ y_1 \geq 0, \quad y_2 \geq 0, \quad y_3 \geq 0 \end{cases}$

19. Minimize $C = 30y_1 + 19y_2 + 30y_3$

Subject to $\begin{cases} 2y_1 + y_2 + y_3 \geq 6 \\ y_1 + y_2 + 2y_3 \geq 4 \\ y_1 \geq 0, \quad y_2 \geq 0, \quad y_3 \geq 0 \end{cases}$

20. Minimize $C = 60y_1 + 39y_2 + 60y_3$

Subject to $\begin{cases} 2y_1 + y_2 + y_3 \geq 6 \\ y_1 + y_2 + 2y_3 \geq 8 \\ y_1 \geq 0, \quad y_2 \geq 0, \quad y_3 \geq 0 \end{cases}$

21. Minimize $C = 20y_1 + 30y_2 + 40y_3$

Subject to $\begin{cases} y_1 - y_2 + y_3 \geq 15 \\ y_1 + y_2 + y_3 \geq 20 \\ y_1 - y_2 + y_3 \leq 10 \\ y_1 \geq 0, \quad y_2 \geq 0, \quad y_3 \geq 0 \end{cases}$

22. Minimize $C = 20y_1 + 50y_2 + 30y_3$

Subject to $\begin{cases} 2y_1 - y_2 + y_3 \leq 10 \\ y_1 + y_2 + y_3 \geq 30 \\ 2y_1 - y_2 + y_3 \geq 20 \\ y_1 \geq 0, \quad y_2 \geq 0, \quad y_3 \geq 0 \end{cases}$

23. Minimize $C = 132y_1 + 102y_2 + 60y_3$

Subject to $\begin{cases} 3y_1 + 2y_2 + y_3 \geq 48 \\ 4y_1 + 3y_2 + 2y_3 \geq 72 \\ 2y_1 + 2y_2 + y_3 \geq 42 \\ y_1 \geq 0, \quad y_2 \geq 0, \quad y_3 \geq 0 \end{cases}$

24. Minimize $C = 96y_1 + 144y_2 + 84y_3$

Subject to $\begin{cases} 2y_1 + 3y_2 + 2y_3 \geq 102 \\ 3y_1 + 4y_2 + 2y_3 \geq 132 \\ y_1 + 2y_2 + y_3 \geq 60 \\ y_1 \geq 0, \quad y_2 \geq 0, \quad y_3 \geq 0 \end{cases}$

APPLIED EXERCISES

Formulate each situation as a standard minimum linear programming problem by identifying the variables, the objective function, and the constraints. Be sure to state clearly the meaning of each variable. Solve it by finding the dual maximum problem and using the simplex method. State your final answer in terms of the original question.

 You may use the PIVOT program if permitted by your instructor.

25. BIOMEDICAL: Nutrition An athlete's training diet needs at least 44 more grams of carbohydrates, 12 more grams of fat, and 16 more grams of protein each day. A dietician recommends two food supplements, Bulk-Up

Bars (costing 48¢ each) and Power Drink (costing 45¢ per can), with nutritional contents (in grams) as given in the table. How much of each food supplement will provide the extra needed nutrition at the least cost?

	Bulk-Up Bar	Power Drink
Carbohydrates	4	3
Fat	1	1
Protein	2	1

26. **BIOMEDICAL: Diet Planning** The residents of an elder care facility need at least 156 more milligrams (mg) of calcium, 180 more micrograms (μg) of folate, and 66 more grams (g) of protein in their weekly diets. The staff chef has created a new fish entree and a salad that the residents should find appealing, and the nutritional contents of the new menu items are given in the accompanying table. If the fish entree costs 21¢ per ounce and the salad costs 15¢ per ounce, how many ounces of each should be added to every resident's weekly menu to provide the additional nutrition at the least cost?

(per ounce)	Fish Entree	Salad
Calcium (mg)	6	4
Folate (μg)	6	5
Protein (g)	2	2

27. **MANAGEMENT SCIENCE: Office Supplies** The office manager of a large accounting firm needs at least 315 more boxes of pens and 120 more boxes of pencils but has only $1500 left in his budget. Jack's Office Supplies has packages of 5 boxes of pens with 2 boxes of pencils on sale for $20, while John's Discount offers packages of 3 boxes of pens and 1 box of pencils for $11. How many packages from each store should he buy to restock the store room at the least cost?

28. **MANAGEMENT SCIENCE: Highway Construction** The project engineer for a highway construction company needs to cut through a small hill to make way for a new road. She estimates that at least 3500 cubic yards of dirt and at least 2400 cubic yards of crushed rock will have to be hauled away. A heavy-duty dump truck can haul either 10 cubic yards of dirt or 6 cubic yards of crushed rock, while a regular dump truck can haul either 5 cubic yards of dirt or 4 cubic yards of crushed rock. The dirt and crushed rock cannot be mixed, because they go to different dumping sites. The union contract demands that this job use at least 400 loads carried in heavy-duty dump trucks. If each heavy-duty dump truck load costs $90 and each regular dump truck load costs $50, how many truckloads with each type of truck and cargo will be needed to complete the job at the least cost?

29. **MANAGEMENT SCIENCE: Highway Construction** Repeat Exercise 28 with the additional requirement that there can be no more than 800 truckloads hauled from the job site in order to get the project finished on time.

30. **MANAGEMENT SCIENCE: Office Supplies** Repeat Problem 27 with the additional requirement that the office manager feels he must purchase at least $1200 worth of pens and pencils from Jack's Office Supplies because it gave him such a good deal on some filing cabinets last month.

31. **BUSINESS: Agriculture** A soil analysis of a farmer's field showed that he needs to apply at least 3000 pounds of nitrogen, 2400 pounds of phosphoric acid, and 2100 pounds of potash. Plant fertilizer is labeled with three numbers giving the percentages of nitrogen, phosphoric acid, and potash. The local farm supply store sells 15-30-15 Miracle Mix for 15¢ per pound and a 10-5-5 store brand for 8¢ per pound. How many pounds of each fertilizer should the farmer buy to meet the needs of the field at the least cost?

32. **GENERAL: Gardening** A weekend gardener's vegetable patch needs at least 10.2 pounds of nitrogen, 7.8 pounds of phosphoric acid, and 6.6 pounds of potash. Plant fertilizer is labeled with three numbers giving the percentages of nitrogen, phosphoric acid, and

potash. The local garden center sells 15-30-15 Miracle Mix for 16¢ per pound, 15-10-10 Grow Great for 13¢ per pound, and a 10-5-5 store brand for 8¢ per pound. How many pounds of each fertilizer should the gardener buy to meet the needs of the vegetable patch at the least cost?

33. MANAGEMENT SCIENCE: Transportation A retail store chain has cartons of goods stored at warehouses in Kentucky and Utah that must be distributed to its stores in Kansas, Texas, and Oregon. Each carton shipped from the Utah warehouse costs $2 whether it goes to Kansas, Texas, or Oregon. However, the cost to ship one carton from the Kentucky warehouse to Kansas is $2, to Texas is $4, and to Oregon is $5. There are 200 cartons at the Utah warehouse and 400 at the warehouse in Kentucky. If the Kansas stores need 200 cartons, the Texas stores need 300 cartons, and the Oregon stores need 100 cartons, how many cartons should be shipped from each warehouse to each state to incur the smallest shipping costs?

34. MANAGEMENT SCIENCE: Transportation A soda distributor has warehouses in Seaford and Centerville and needs to supply stores in Huntington and Towson. The costs of shipping one case of sodas are 8¢ from Seaford to Huntington, 5¢ from Seaford to Towson, 6¢ from Centerville to Huntington, and 4¢ from Centerville to Towson. There are 800 cases in the Seaford warehouse and 1200 in the Centerville warehouse. If the Huntington stores need at least 1000 cases and the Towson stores need at least 600 cases, how many cases should be shipped from each warehouse to each city to incur the smallest shipping costs?

Explorations and Excursions

The following problems extend and augment the material presented in the text.

Why $P \le C$ for Dual Problems

35. *(Requires matrix algebra)* For the pair of dual problems

Maximize $P = c^t X$

Subject to $\begin{cases} AX \le b \\ X \ge 0 \end{cases}$

Minimize $C = b^t Y$

Subject to $\begin{cases} A^t Y \ge c \\ Y \ge 0 \end{cases}$

show that $P \le C$ for any feasible X and Y by justifying each $=$ and $\le$ in the chain of statements

$$P = c^t \cdot X = X^t \cdot c \le X^t \cdot A^t Y$$
$$= (X^t A^t) \cdot Y = (AX)^t \cdot Y \le b^t \cdot Y = C$$

36. *(Requires matrix algebra)* Check the chain of matrix statements in Exercise 35 for the matrices

$$A = \begin{pmatrix} 3 & 4 \\ 1 & 2 \end{pmatrix}, \quad b = \begin{pmatrix} 24 \\ 10 \end{pmatrix}, \quad c = \begin{pmatrix} 9 \\ 16 \end{pmatrix}, \quad X = \begin{pmatrix} x_1 \\ x_2 \end{pmatrix},$$

and $Y = \begin{pmatrix} y_1 \\ y_2 \end{pmatrix}$ from Example 2 on pages 296–297.

A Proof of the Duality Theorem

37. *(Requires matrix algebra)* The following sequence of statements provides a proof of the Duality Theorem on page 298. Justify each statement to verify this proof.

a. If $mx + b$ is the same as $0x + v$ for every value of x, then $m = 0$ and $b = v$.

b. The dual problems in Exercise 35 may be rewritten as

Maximize $P = c^t X$

Subject to $\begin{cases} AX + S = b \\ X, S \ge 0 \end{cases}$

and

Minimize $C = b^t Y$

Subject to $\begin{cases} A^t Y - T = c \\ Y, T \ge 0 \end{cases}$

to clearly show both the variables (X and Y) and the slack variables (S and T).

c. The simplex tableau for the maximum problem (both the initial tableau and after *any* sequence of pivot operations) has numbers in the bottom row:

	x_1	$\cdots$	x_n	s_1	$\cdots$	s_m	
		$\vdots$			$\vdots$		
P	w_1	$\cdots$	w_n	z_1	$\cdots$	z_m	V

To prove the Duality Theorem, we must show that these values can be used for the values of Y and T in the minimum problem. That is, if we set $y_1 = z_1, \ldots, y_m = z_m,$ and $t_1 = w_1, \ldots, t_n = w_n,$ then the number V is the same as the value $C = b^tY$ and the constraint $A^tY - T = c$ is satisfied.

d. If we use the notations $W = \begin{pmatrix} w_1 \\ \vdots \\ w_n \end{pmatrix}$ and

$Z = \begin{pmatrix} z_1 \\ \vdots \\ z_m \end{pmatrix},$ the bottom row of the tableau

is the same as the matrix equation $P + W^tX + Z^tS = V,$ and this equation holds for *any* values of X and S such that $AX + S = b.$

e. Since $P = c^tX$ and $S = b - AX,$ we may rewrite the matrix equation $P + W^tX + Z^tS = V$ as

$$c^tX + W^tX + Z^t(b - AX) = V$$

f. That is, $X^tc + X^tW + b^tZ - (AX)^tZ = V$ and so

$$X^t \cdot (c + W - A^tZ) + b^tZ = V$$

Thus $(c + W - A^tZ) = 0$ and $b^tZ = V.$ So the number V is the value of the minimum problem objective function $C = b^tY$ when Y takes the value Z and this value for Y satisfies the constraint $A^tY - T = c$ when the slack T takes the value $W.$

Complementary Slackness

38. Use the Duality Theorem for the simplex method (see page 298) to show that *at least* one of *every* pair of variables x_1 and $t_1,$ x_2 and $t_2,$ $\ldots,$ x_n and $t_n,$ y_1 and $s_1,$ y_2 and $s_2,$ $\ldots,$ y_m and s_m is zero for the solution points of a linear programming problem and its dual. [*Hint:* What can you say about the bottoms of the basic variable columns in the simplex tableau of a maximum problem?]

39. Verify the Complementary Slackness Theorem (Exercise 38) for the final tableau given in Practice Problem 3 (see pages 299–300).

NONSTANDARD PROBLEMS: THE DUAL PIVOT ELEMENT AND THE TWO-STAGE METHOD

This section and the next are optional and present alternative methods for solving nonstandard linear programming problems. The choice of which one to cover is left to the instructor. The examples and exercises duplicate those in Section 4.6.

Introduction

In this section we extend the simplex method to maximum problems that include constraints with $\geq$ inequalities as well as $\leq$ inequalities. Because a $\geq$ inequality can be changed into a $\leq$ inequality by multiplying through by -1 (for example, changing $5 \geq -3$ into $-5 \leq 3$), this extension is equivalent to dropping the requirement that $b \geq 0$ in the constraints for the standard maximum problem on page 273.

The Dual Pivot Element

The origin is not a vertex for a nonstandard maximum problem because an inequality like $x_1 \geq 3$ separates the feasible region from the origin. The initial tableau begins at the origin so it is not feasible as well as not optimal. The question becomes: How can we move to a feasible point so that we can then work on making it optimal? On page 301 we saw that the pivot operations for a standard maximum problem make the dual minimum problem feasible. This gives the key to the answer to our question: Pivoting on "dual pivot elements" will make the problem feasible, and then pivoting on (regular) pivot elements will make it optimal.

Since dual problems interchange the roles of rows and columns, the following definition is just the definition of the pivot element with the roles of the rows and columns reversed, and we must take the *largest* ratio instead of the smallest because we are now dividing by negative numbers.

Dual Pivot Element

The *dual pivot row* is the row with the smallest negative entry in the rightmost column of the tableau (omitting the bottom row). If there is a tie, choose the uppermost such row.

The *dual pivot column* is the column with the largest ratio found by dividing the bottom entry by the dual pivot row entry, omitting any column with a zero or positive dual pivot row entry and omitting the rightmost column. If there is a tie, choose the leftmost such column.

The *dual pivot element* is the entry in the dual pivot row and the dual pivot column.

EXAMPLE 1

THE DUAL PIVOT ELEMENT

Find the dual pivot element in the following simplex tableau or explain why the tableau does not have a dual pivot element.

	x_1	x_2	x_3	x_4	s_1	s_2	s_3	
s_1	2	1	1	2	1	0	0	30
s_2	-1	-2	-1	-1	0	1	0	-5
s_3	-1	-1	-1	2	0	0	1	-10
P	-8	12	-10	14	0	0	0	0

Solution

The dual pivot row is row 3 because -10 is the smallest negative entry on the right (omitting the bottom row). The dual pivot column is column 3 because the ratio $\frac{-10}{-1} = 10$ is greater than the others (the remaining columns are not considered because their dual pivot row entries are zero or positive and the rightmost column is never considered).

	x_1	x_2	x_3	x_4	s_1	s_2	s_3	
s_1	2	1	1	2	1	0	0	30
s_2	−1	−2	−1	−1	0	1	0	−5
s_3	−1	−1	−1	2	0	0	1	−10
P	−8	12	−10	14	0	0	0	0

s_3 ← Smallest negative

↑ Dual pivot row

$\frac{-8}{-1} = 8$ $\frac{12}{-1} = -12$ $\frac{-10}{-1} = 10$ (Omit)   (Omit)   (Omit)   (Omit)

↑ Largest ratio

↑ Dual pivot column

Dual pivot element

Choose the dual pivot row first then the dual pivot column

The dual pivot element is the −1 in row 3 and column 3.

Practice Problem 1 For each of the following simplex tableaux, find the dual pivot element or explain why the tableau does not have a dual pivot element.

a.

	x_1	x_2	x_3	x_4	s_1	s_2	s_3	
s_1	−1	−1	2	−2	1	0	0	−10
s_2	1	1	−1	−1	0	1	0	20
s_3	0	−1	3	2	0	0	1	−5
P	−5	−6	14	8	0	0	0	0

b.

	x_1	x_2	x_3	s_1	s_2	s_3	
s_1	1	5	1	1	0	0	25
s_2	−1	−2	−1	0	1	0	−15
s_3	2	1	4	0	0	1	−20
P	−6	10	−10	0	0	0	0

➤ Solutions on page 321

The Two-Stage Simplex Method

Pivoting on dual pivot elements will make the tableau feasible. After the tableau becomes feasible, pivoting on *regular* pivot elements will make it optimal, leading to the solution of the problem.

Two-Stage Simplex Method

To solve any maximum problem by the simplex method:

1. Write the constraints with $\leq$ inequalities (keeping the variables nonnegative), and construct the initial simplex tableau.

2. If at least one basic variable has a negative value in the rightmost column (the tableau is not feasible), go to step 3. If all the basic variables have nonnegative values in the rightmost column (the tableau is feasible), go to step 4.

3. Locate the dual pivot element (page 314), perform the pivot operation (page 280), and return to step 2. If the tableau has a dual pivot row but no dual pivot column, then there is *no solution* to the problem (the constraints are infeasible).

4. Locate the pivot element (pages 278–279), perform the pivot operation (page 280), and return to step 2. If the tableau does not have a pivot column, then the solution occurs at the vertex given by the basic variables, and the maximum value of the objective function appears in the bottom right corner of the tableau. If the tableau has a pivot column but no pivot row, then there is *no solution* to the problem.

EXAMPLE 2

THE TWO-STAGE SIMPLEX METHOD

Solve the following nonstandard linear programming problem by the two-stage simplex method.

$$\text{Maximize } P = 5x_1 + 7x_2$$

$$\text{Subject to } \begin{cases} x_1 + x_2 \leq 8 \\ x_1 + x_2 \geq 4 \\ 2x_1 + x_2 \geq 6 \\ x_1 \geq 0 \text{ and } x_2 \geq 0 \end{cases}$$

"Mixed constraints" because they have $\leq$ and $\geq$ inequalities

Solution

We change the second and third inequalities from $\geq$ to $\leq$ by multiplying by -1, obtaining $-x_1 - x_2 \leq -4$ and $-2x_1 - x_2 \leq -6$.

The initial simplex tableau is

	x_1	x_2	s_1	s_2	s_3		
s_1	1	1	1	0	0	8	
s_2	-1	-1	0	1	0	-4	
s_3	-2	-1	0	0	1	-6	
P	-5	-7	0	0	0	0	

$\left.\begin{array}{l} s_1 = 8 \\ s_2 = -4 \\ s_3 = -6 \end{array}\right\}$ Some basic variables are negative, so *not* feasible

Since s_2 and s_3 are negative, the tableau is not feasible and we look for a dual pivot element. The dual pivot row is row 3 (-6 is the smallest negative entry in the right column, omitting the bottom row). The dual pivot column is column 2 (the ratio $\frac{-7}{-1} = 7$ is greater than $\frac{-5}{-2} = 2.5$, and none of the other columns may be considered). The dual pivot element is the -1 in row 3 and column 2. Pivoting on this dual pivot element, the tableau becomes:

	x_1	x_2	s_1	s_2	s_3		
s_1	-1	0	1	0	1	2	
s_2	1	0	0	1	-1	2	
x_2	2	1	0	0	-1	6	
P	9	0	0	0	-7	42	

$\left.\begin{array}{l} s_1 = 2 \\ s_2 = 2 \\ x_2 = 6 \end{array}\right\}$ Feasible because *all* basic variables are nonnegative

The bottom row has
a negative entry, so
not optimal

(If some basic variables were still negative, we would look for another dual pivot element.) The tableau is now feasible but not optimal, so we look for a (regular) pivot element. The pivot column is column 5 (-7 is the only negative entry in the bottom row). The pivot row is row 1 (the other rows cannot be considered because their pivot column entries are negative). The pivot element is the 1 in column 5 and row 1. Pivoting on this pivot element, the tableau becomes:

	x_1	x_2	s_1	s_2	s_3		
s_3	-1	0	1	0	1	2	
s_2	0	0	1	1	0	4	
x_2	1	1	1	0	0	8	
P	2	0	7	0	0	56	

$\left.\begin{array}{l} \\ \\ \end{array}\right\}$ Feasible: *all* are nonnegative

Optimal: *all* are *nonnegative*

This is the final tableau because it is both feasible and optimal. The maximum is $P = 56$ when $x_1 = 0$ and $x_2 = 8$.

To compare this solution with the graphical method, we graph the feasible region with x_1 on the horizontal axis and x_2 on the vertical axis. The five vertices and the values of the objective function are given in the table. The path of vertices visited by the two-stage simplex method is marked in bold.

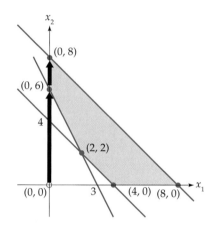

Vertex (x_1, x_2)	$P = 5x_1 + 7x_2$
(8, 0)	40
(4, 0)	20
(2, 2)	24
(0, 6)	42
(0, 8)	56

Notice that the first pivot reaches a feasible point and the second makes it optimal.

Practice Problem 2 Solve the following nonstandard linear programming problem by the two-stage simplex method.

$$\text{Maximize} \quad P = 4x_1 + x_2 + 3x_3$$
$$\text{Subject to} \quad \begin{cases} x_1 + 2x_2 + x_3 \leq 50 \\ 2x_1 + x_2 + 2x_3 \geq 10 \\ x_1 \geq 0, \quad x_2 \geq 0, \quad \text{and} \quad x_3 \geq 0 \end{cases}$$ ➤ Solution on page 321

EXAMPLE 3

MANAGING AN INVESTMENT PORTFOLIO

The manager of an $80 million mutual fund has three investment possibilities: secure government bonds yielding 4%, a blue chip growth stock paying 6%, and a biotechnology start-up company returning 12%. If the manager wants to invest at least $50 million in a combination of safe government bonds and blue chip stock, and at least as much in the bonds as the total in blue chip and biotechnology stocks, how much should be invested in each to obtain the greatest possible return?

Solution

Let

$$x_1 = \begin{pmatrix} \text{Amount in} \\ \text{government bonds} \end{pmatrix}$$

$$x_2 = \begin{pmatrix} \text{Amount in} \\ \text{blue chip stock} \end{pmatrix} \qquad \text{In millions of dollars}$$

$$x_3 = \begin{pmatrix} \text{Amount in} \\ \text{biotechnology stock} \end{pmatrix}$$

The objective is to maximize the return

$$P = 0.04x_1 + 0.06x_2 + 0.12x_3 \qquad \text{Rate} \times \text{amount for each investment}$$

subject to the restrictions

$$
\begin{aligned}
x_1 + x_2 + x_3 &\le 80 && \text{\$80 million available to invest} \\
x_1 + x_2 &\ge 50 && \text{\$50 million in safe investments} \\
x_1 &\ge x_2 + x_3 && \text{Bonds} \ge \text{total in stocks} \\
x_1 \ge 0, \quad x_2 \ge 0, \quad x_3 &\ge 0 && \text{Nonnegativity}
\end{aligned}
$$

Multiplying the second constraint by -1 to get $-x_1 - x_2 \le -50$ and rewriting the third as $-x_1 + x_2 + x_3 \le 0$, the initial simplex tableau is:

	x_1	x_2	x_3	s_1	s_2	s_3	
s_1	1	1	1	1	0	0	80
s_2	-1	-1	0	0	1	0	-50
s_3	-1	1	1	0	0	1	0
P	-0.04	-0.06	-0.12	0	0	0	0

$s_2 < 0 \longleftarrow$ Dual pivot row

Since this is not feasible (because $s_2 < 0$), we look for a dual pivot element. The dual pivot row is row 2 and the dual pivot column is column 2 (because the ratio $\frac{-0.06}{-1} = 0.06$ is greater than $\frac{-0.04}{-1} = 0.04$ and none of the other columns may be considered). Pivoting on the -1 in row 2 and column 2, the tableau becomes:

	x_1	x_2	x_3	s_1	s_2	s_3	
s_1	0	0	1	1	1	0	30
x_2	1	1	0	0	-1	0	50
s_3	-2	0	1	0	1	1	-50
P	0.02	0	-0.12	0	-0.06	0	3

$s_3 < 0 \longleftarrow$ Dual pivot row

This is still not feasible (because $s_3 < 0$). Pivoting on the dual pivot element in row 3 and column 1, the tableau becomes:

	x_1	x_2	x_3	s_1	s_2	s_3	
s_1	0	0	1	1	1	0	30
x_2	0	1	1/2	0	−1/2	1/2	25
x_1	1	0	−1/2	0	−1/2	−1/2	25
P	0	0	−0.11	0	−0.05	0.01	2.5

Feasible: all are ≥ 0

Not optimal: some are < 0

This is now feasible but not optimal (because the bottom row has some negative entries). Pivoting on the (regular) pivot element in column 3 and row 1, we obtain:

	x_1	x_2	x_3	s_1	s_2	s_3	
x_3	0	0	1	1	1	0	30
x_2	0	1	0	−1/2	−1	1/2	10
x_1	1	0	0	1/2	0	−1/2	40
P	0	0	0	0.11	0.06	0.01	5.8

Feasible: all are ≥ 0

Optimal: all are ≥ 0

This is the final tableau because it is both feasible and optimal.

The maximum is $P = 5.8$ when $x_1 = 40$, $x_2 = 10$, and $x_3 = 30$.

In terms of the original situation, the manager should invest $40 million in the government bonds, $10 million in the blue chip stock, and $30 million in the biotechnology start-up company for a maximum return of $5.8 million.

Any Linear Programming Problem Can Now Be Solved

The two-stage simplex method allows us to solve *any* maximum linear programming problem, no matter what mixture of $\leq$ and $\geq$ inequalities are present in the constraints, provided that the variables are non-negative. A direct way to change a minimum problem into a maximum problem is described in Exercises 35–36. Should the initial formulation include equality constraints, Exercises 37–38 demonstrate two ways to change the problem into a form that we can solve. Exercises 39–41 explain how to reformulate a problem so that all the variables are non-negative. Taken together, these techniques allow us to solve *any* linear programming problem.

4.5 ## Section Summary

A *nonstandard* maximum problem has both $\geq$ and $\leq$ constraints. Change the $\geq$ inequalities into $\leq$ form by multiplying by -1, and then construct the initial simplex tableau. In general, this tableau will have negative numbers in the rightmost column (making it not feasible) as well as negative numbers in the bottom row (making it not optimal).

The dual pivot *row* is the row with the smallest negative entry in the rightmost column of the tableau (omitting the bottom row). The dual pivot *column* is the column with the largest ratio of the bottom entry divided by the dual pivot row entry, omitting any column with a zero or positive dual pivot row entry and omitting the rightmost column.

The two-stage simplex method pivots on dual pivot elements until the tableau is feasible (the right-hand column is nonnegative) and then continues to pivot on (regular) pivot elements until the tableau is optimal (the bottom row is nonnegative). The problem has no solution if the dual pivot column or the (regular) pivot row does not exist.

➤ **Solutions to Practice Problems**

1. a. The dual pivot row is row 1 because -10 is the smallest negative entry in the rightmost column (omitting the bottom row). The dual pivot column is column 2 because the ratio $\frac{-6}{-1} = 6$ is greater than the ratio $\frac{-5}{-1} = 5$ for the first column and the ratio $\frac{8}{-2} = -4$ for the fourth column; the other columns may not be considered since their dual pivot row entries are zero or positive, and the rightmost column is never considered. The dual pivot element is the -1 in row 1 and column 2.

b. The tableau does not have a dual pivot element. The dual pivot row is row 3 because the smallest negative entry in the rightmost column (omitting the bottom row) is -20. The other entries in row 3 are either zero or positive, so there is no dual pivot column. (This means that the constraints are infeasible. The third row represents the inequality $2x_1 + x_2 + 4x_3 \leq -20$ and this is impossible because the variables are nonnegative. The problem has no solution.)

2. Rewriting the second inequlity as $-2x_1 - x_2 - 2x_3 \leq -10$, the initial simplex tableau is

	x_1	x_2	x_3	s_1	s_2	
s_1	1	2	1	1	0	50
s_2	-2	-1	-2	0	1	-10
P	-4	-1	-3	0	0	0

This is not feasible because $s_2 = -10$. Pivoting on the dual pivot element in row 2 and column 1, the tableau becomes feasible (no negatives on the right-hand column):

	x_1	x_2	x_3	s_1	s_2	
s_1	0	3/2	0	1	1/2	45
x_1	1	1/2	1	0	−1/2	5
P	0	1	1	0	−2	20

Pivoting on the (regular) pivot element in column 5 and row 1, the tableau becomes optimal (no negatives in the bottom row):

	x_1	x_2	x_3	s_1	s_2	
s_2	0	3	0	2	1	90
x_1	1	2	1	1	0	50
P	0	7	1	4	0	200

This is the final tableau because it is both feasible and optimal. The maximum is $P = 200$ when $x_1 = 50$ and $x_2 = 0$.

4.5 Exercises

For each nonstandard linear programming problem, construct the initial simplex tableau and locate the dual pivot element. (You do not need to carry out the pivot operation.)

1. Maximize $P = 15x_1 + 20x_2 + 18x_3$

Subject to $\begin{cases} 3x_1 + 2x_2 + 8x_3 \leq 96 \\ 5x_1 + x_2 + 6x_3 \geq 30 \\ x_1 \geq 0, \ x_2 \geq 0, \ x_3 \geq 0 \end{cases}$

2. Maximize $P = 60x_1 + 90x_2 + 30x_3$

Subject to $\begin{cases} 4x_1 + 5x_2 + x_3 \geq 40 \\ 10x_1 + 18x_2 + 3x_3 \leq 150 \\ x_1 \geq 0, \ x_2 \geq 0, \ x_3 \geq 0 \end{cases}$

3. Maximize $P = 6x_1 + 4x_2 + 6x_3$

Subject to $\begin{cases} 2x_1 + x_2 + 3x_3 \geq 30 \\ x_1 + x_2 + 2x_3 \geq 20 \\ x_1 \geq 0, \ x_2 \geq 0, \ x_3 \geq 0 \end{cases}$

4. Maximize $P = 9x_1 + 5x_2 + 4x_3$

Subject to $\begin{cases} 3x_1 + x_2 + x_3 \geq 40 \\ 2x_1 + 2x_2 + x_3 \geq 30 \\ x_1 \geq 0, \ x_2 \geq 0, \ x_3 \geq 0 \end{cases}$

5. Maximize $P = 20x_1 + 30x_2 + 10x_3$

Subject to $\begin{cases} x_1 + x_2 + x_3 \geq 8 \\ x_1 + 2x_2 + 3x_3 \leq 30 \\ x_1 + 2x_2 + x_3 \leq 18 \\ x_1 \geq 0, \ x_2 \geq 0, \ x_3 \geq 0 \end{cases}$

6. Maximize $P = 24x_1 + 18x_2 + 30x_3$

Subject to $\begin{cases} x_1 + x_2 + x_3 \geq 4 \\ x_1 + 2x_2 + 3x_3 \leq 15 \\ x_1 + 2x_2 + x_3 \leq 9 \\ x_1 \geq 0, \ x_2 \geq 0, \ x_3 \geq 0 \end{cases}$

7. Maximize $P = 3x_1 + 2x_2 + 5x_3 + 4x_4$

Subject to $\begin{cases} x_1 + x_2 + x_3 + x_4 \geq 30 \\ 2x_1 + 3x_2 + 2x_3 + x_4 \geq 20 \\ 4x_1 + 2x_2 + x_3 + 2x_4 \leq 80 \\ x_1 \geq 0, \ x_2 \geq 0, \ x_3 \geq 0, \ x_4 \geq 0 \end{cases}$

8. Maximize $P = 6x_1 + 8x_2 + 3x_3 + 5x_4$

Subject to $\begin{cases} x_1 + x_2 + x_3 + x_4 \geq 70 \\ 3x_1 + 2x_2 + 4x_3 + 3x_4 \geq 60 \\ 4x_1 + x_2 + 2x_3 + 5x_4 \leq 100 \\ x_1 \geq 0, \ x_2 \geq 0, \ x_3 \geq 0, \ x_4 \geq 0 \end{cases}$

Solve each nonstandard linear programming problem by the two-stage simplex method. (Exercises 9, 10, 13, 14, 17, and 18 can also be solved by the graphical method.)

 You may use the PIVOT program if permitted by your instructor.

9. Maximize $P = 4x_1 + x_2$

Subject to $\begin{cases} 3x_1 + 2x_2 \leq 120 \\ x_1 + x_2 \geq 50 \\ x_1 \geq 0, \; x_2 \geq 0 \end{cases}$

10. Maximize $P = x_1 + 4x_2$

Subject to $\begin{cases} 3x_1 + 2x_2 \leq 120 \\ 2x_1 + x_2 \geq 60 \\ x_1 \geq 0, \; x_2 \geq 0 \end{cases}$

11. Maximize $P = 30x_1 + 20x_2 + 28x_3$

Subject to $\begin{cases} 3x_1 + x_2 + 2x_3 \geq 30 \\ 4x_1 + x_2 + 3x_3 \leq 60 \\ x_1 \geq 0, \; x_2 \geq 0, \; x_3 \geq 0 \end{cases}$

12. Maximize $P = 18x_1 + 10x_2 + 20x_3$

Subject to $\begin{cases} 3x_1 + x_2 + 4x_3 \geq 30 \\ 2x_1 + x_2 + 5x_3 \leq 50 \\ x_1 \geq 0, \; x_2 \geq 0, \; x_3 \geq 0 \end{cases}$

13. Maximize $P = 4x_1 + 5x_2$

Subject to $\begin{cases} x_1 + 2x_2 \leq 12 \\ x_1 + x_2 \geq 15 \\ 2x_1 + x_2 \leq 12 \\ x_1 \geq 0, \; x_2 \geq 0 \end{cases}$

14. Maximize $P = 5x_1 + 4x_2$

Subject to $\begin{cases} 2x_1 + 3x_2 \leq 12 \\ x_1 + x_2 \geq 10 \\ 3x_1 + 2x_2 \leq 12 \\ x_1 \geq 0, \; x_2 \geq 0 \end{cases}$

15. Maximize $P = 15x_1 + 12x_2 + 18x_3$

Subject to $\begin{cases} 5x_1 + x_2 + 2x_3 \geq 30 \\ 2x_1 + x_2 + 3x_3 \leq 24 \\ x_1 \geq 0, \; x_2 \geq 0, \; x_3 \geq 0 \end{cases}$

16. Maximize $P = 4x_1 + 5x_2 + 6x_3$

Subject to $\begin{cases} 5x_1 + x_2 + 3x_3 \leq 30 \\ 2x_1 + x_2 + 2x_3 \geq 24 \\ x_1 \geq 0, \; x_2 \geq 0, \; x_3 \geq 0 \end{cases}$

17. Maximize $P = x_1 + x_2$

Subject to $\begin{cases} x_1 \leq 8 \\ x_2 \leq 5 \\ x_1 + 2x_2 \geq 6 \\ x_1 \geq 0, \; x_2 \geq 0 \end{cases}$

18. Maximize $P = x_1 + x_2$

Subject to $\begin{cases} x_1 \leq 6 \\ x_2 \leq 9 \\ 3x_1 + x_2 \geq 6 \\ x_1 \geq 0, \; x_2 \geq 0 \end{cases}$

19. Maximize $P = 4x_1 + 6x_2 + 12x_3 + 10x_4$

Subject to $\begin{cases} x_1 + x_2 + x_3 + x_4 \leq 60 \\ 2x_1 + x_2 + x_3 + 2x_4 \geq 10 \\ 2x_1 + 2x_2 + x_3 + x_4 \leq 100 \\ x_1 \geq 0, \; x_2 \geq 0, \; x_3 \geq 0, \; x_4 \geq 0 \end{cases}$

20. Maximize $P = 8x_1 - x_2 + 5x_3 + 14x_4$

Subject to $\begin{cases} x_1 + 2x_2 + 3x_4 \leq 55 \\ x_1 + x_2 + x_3 + 2x_4 \geq 25 \\ x_1 + 3x_3 + 3x_4 \leq 45 \\ x_1 \geq 0, \; x_2 \geq 0, \; x_3 \geq 0, \; x_4 \geq 0 \end{cases}$

21. Maximize $P = 5x_1 + 32x_2 + 3x_3 + 15x_4$

Subject to $\begin{cases} 5x_1 + 8x_2 + x_3 + 3x_4 \geq 18 \\ 2x_1 - x_2 + x_3 + x_4 \leq 10 \\ 3x_1 - 3x_2 + x_3 + 2x_4 \geq 6 \\ x_1 \geq 0, \; x_2 \geq 0, \; x_3 \geq 0, \; x_4 \geq 0 \end{cases}$

22. Maximize $P = 70x_1 + 12x_2 + 60x_3 + 20x_4$

Subject to $\begin{cases} 2x_1 + x_2 + x_3 + x_4 \leq 14 \\ 3x_1 + 3x_2 + x_3 + 2x_4 \leq 24 \\ 5x_1 + 3x_2 + 4x_3 + 2x_4 \geq 44 \\ x_1 \geq 0, \; x_2 \geq 0, \; x_3 \geq 0, \; x_4 \geq 0 \end{cases}$

23. Maximize $P = -x_1 + x_2 - 2x_3$

Subject to $\begin{cases} x_1 + x_2 + x_3 \leq 50 \\ x_1 + x_3 \geq 10 \\ x_2 + x_3 \geq 20 \\ x_1 \geq 0, \; x_2 \geq 0, \; x_3 \geq 0 \end{cases}$

24. Maximize $P = x_1 + 2x_2 + x_3$

Subject to $\begin{cases} x_1 + x_2 + x_3 \leq 30 \\ x_1 + x_2 \leq 25 \\ x_2 + x_3 \geq 15 \\ x_1 \geq 0, \; x_2 \geq 0, \; x_3 \geq 0 \end{cases}$

APPLIED EXERCISES

Express each situation as a nonstandard maximum problem. Be sure to state the meaning of each variable. Solve the problem by the two-stage simplex method. State your final answer in terms of the original question.

You may use the PIVOT program if permitted by your instructor.

25. PERSONAL FINANCE: Financial Planning A retired couple want to invest their $20,000 life savings in bank certificates of deposit yielding 6% and Treasury bonds yielding 5%. If they want at least $5000 in each type of investment, how much should they invest in each to receive the greatest possible income?

26. ENVIRONMENTAL SCIENCE: Recycling Management A volunteer recycling center accepts both used paper and empty glass bottles, which it then sorts and sells to a reprocessing company. The center has room to accept a total of 800 crates of paper and glass each week and has 50 hours of volunteer help to do the sorting. Each crate of paper products takes 5 minutes to sort and sells for 8¢, while each crate of bottles takes 3 minutes to sort and sells for 7¢. To support the city's "grab that glass" recycle theme this week, the center wants to accept at least 600 crates of glass. How many crates of each should the center accept this week to raise the most money for its ecology scholarship fund?

27. BUSINESS: Advertising The manager of a new mall may spend up to $18,000 on "grand opening" announcements in newspapers, on radio, and on TV. Each newspaper ad costs $300 and reaches 5000 readers, each one-minute radio commercial costs $800 and is heard by 13,000 listeners, and each 15-second TV spot costs $900 and is seen by 15,000 viewers. If the manager wants at least 5 newspaper ads and at least 20 radio commercials and TV spots combined, how many of each should be placed to reach the largest number of potential customers? (Ignore multiple exposures to the same customer.)

28. SOCIAL SCIENCE: Political Advertising In the last few days before the election, a politician can afford to spend no more than $27,000 on TV advertisements and can arrange for no more than 10 ads. Each daytime ad costs $2000 and reaches 4000 viewers, each prime time ad costs $3000 and reaches 5000 viewers, and each late night ad costs $1000 and reaches 2000 viewers. To be sure to reach the widest variety of voters, the politician's advisors insist on at least 5 ads scheduled during daytime and late night combined. Ignoring repeated viewings by the same person and assuming every viewer can vote, how many ads in each of the time periods will reach the most voters?

29. BUSINESS: Resource Allocation A furniture shop manufactures wooden desks, tables, and chairs. The numbers of hours to assemble and finish each piece are shown in the table, together with the number of hours of skilled labor available for each task. To meet expected demand, a total of at least 30 desks and tables combined must be made. If the profit is $75 for each desk, $84 for each table, and $66 for each chair, how many of each should the company manufacture to obtain the greatest possible profit?

	Desk	Table	Chair	Labor Available
Assembly	2 hours	1 hour	2 hours	210 hours
Finishing	2 hours	3 hours	1 hour	150 hours

30. BUSINESS: Production Planning An automotive parts shop rebuilds carburetors, fuel pumps, and alternators. The numbers of hours to rebuild and then inspect and pack each part are shown in the table, together with the number of hours of skilled labor available for each task. At least 12 carburetors must be rebuilt. If the profit is $12 for each carburetor, $14 for each fuel pump, and $10 for each alternator, how many of each should the shop rebuild to obtain the greatest possible profit?

	Carburetor	**Fuel Pump**	**Alternator**	**Labor Available**
Rebuilding	5 hours	4 hours	3 hours	200 hours
Inspection & packaging	1 hour	1 hour	0.5 hour	45 hours

31. **BUSINESS: Agriculture** A farmer grows wheat, barley, and oats on her 500-acre farm. Each acre of wheat requires 3 days of labor (to plant, tend, and harvest) and costs $21 (for seed, fertilizer, and pesticides), each acre of barley requires 2 days of labor and costs $27, and each acre of oats requires 3 days of labor and costs $24. The farmer and her hired field hands can provide no more than 1200 days of labor this year. She wants to grow at least 100 acres of oats and can afford to spend no more than $15,120. If the profit is $50 for each acre of wheat, $40 for each acre of barley, and $45 for each acre of oats, how many acres of each crop should she grow to obtain the greatest possible profit?

32. **BUSINESS: Agriculture** A farmer grows corn, peanuts, and soybeans on his 240-acre farm. To maintain soil fertility, the farmer rotates the crops and always plants at least as many acres of soybeans as the total acres of the other crops. Because he has promised to sell some of his corn to a neighbor who raises cattle, he must plant at least 42 acres of corn. Each acre of corn requires 2 days of labor and yields a profit of $150, each acre of peanuts requires 5 days of labor and yields a profit of $300, and each acre of soybeans requires 1 day of labor and yields a profit of $100. If the farmer and his children can put in at most 630 days of labor, how many acres of each crop should the farmer plant to obtain the greatest possible profit?

33. **ENVIRONMENTAL SCIENCE: Pollution Control** A storage yard at a coal burning electric power plant can hold no more than 100,000 tons of coal. Two grades of coal are available: low sulfur (1%) with an energy content of 20 million BTU per ton and high sulfur (2%) with an energy content of 30 million BTU per ton. If existing contracts with the high-sulfur coal mine operator require that at least 50,000 tons of high sulfur coal be purchased,

and if the next coal purchase may contain no more than 1400 tons of sulfur, how many tons of each type of coal should be purchased to obtain the most energy?

34. **BUSINESS: Production Planning** A small jewelry company prepares and mounts semi-precious stones. There are 20 lapidaries (who cut and polish the stones) and 24 jewelers (who mount the stones in gold settings). Each employee works 7 hours each day, 5 days each week. Each tray of agates requires 5 hours of cutting and polishing and 4 hours to mount, each tray of onyxes requires 2 hours of cutting and polishing and 3 hours to mount, and each tray of garnets requires 6 hours of cutting and polishing and 3 hours to mount. Furthermore, the company's owner has decided that the company will process at least 12 trays of agates each week. If the profit is $15 for each tray of agates, $10 for each tray of onyxes, and $13 for each tray of garnets, how many trays of each stone should be processed each week to obtain the greatest possible profit?

Explorations and Excursions

The following problems extend and augment the material presented in the text.

The PIVOT program may be helpful (if permitted by your instructor).

Minimum Problems and the Dual Pivot Element

35. **a.** Solve both of the following problems by the graphical method from Section 4.2. Is it true that the minimum of C is the same as -1 times the maximum of $P = -C$?

$$\text{Minimize}\quad C = 3x + 4y$$
$$\text{Subject to}\quad \begin{cases} x + y \le 10 \\ 2x + y \ge 8 \\ x \ge 0 \;\text{ and }\; y \ge 0 \end{cases}$$

$$\text{Maximize}\quad P = -3x - 4y$$
$$\text{Subject to}\quad \begin{cases} x + y \le 10 \\ 2x + y \ge 8 \\ x \ge 0 \;\text{ and }\; y \ge 0 \end{cases}$$

(continues)

b. Solve the following problem by finding the dual maximum problem and using the (regular) simplex method as we did in Section 4.4.

$$\text{Minimize} \quad C = 3y_1 + 4y_2$$

$$\text{Subject to} \quad \begin{cases} -y_1 - y_2 \ge -10 \\ 2y_1 + y_2 \ge 8 \\ y_1 \ge 0 \text{ and } y_2 \ge 0 \end{cases}$$

c. Solve the following nonstandard problem by the two-stage simplex method.

$$\text{Maximize} \quad P = -3x_1 - 4x_2$$

$$\text{Subject to} \quad \begin{cases} x_1 + x_2 \le 10 \\ -2x_1 - x_2 \le -8 \\ x_1 \ge 0 \text{ and } x_2 \ge 0 \end{cases}$$

d. Compare the tableaux and pivot elements in parts (b) and (c). Are the pivot elements the same numbers? Is the arithmetic to choose each pair of pivot elements the same? Is the "dual pivot element" just the "dual" of the (regular) pivot element?

36. Repeat the process of Exercise 35 to find "dual" solutions to the following problems.

$$\text{Minimize} \quad C = 20y_1 + 18y_2$$

$$\text{Subject to} \quad \begin{cases} 2y_1 + y_2 \ge 18 \\ y_1 + y_2 \ge 14 \\ 4y_1 + 3y_2 \ge 48 \\ y_1 \ge 0 \text{ and } y_2 \ge 0 \end{cases}$$

$$\text{Maximize} \quad P = -20x_1 - 18x_2$$

$$\text{Subject to} \quad \begin{cases} -2x_1 - x_2 \le -18 \\ -x_1 - x_2 \le -14 \\ -4x_1 - 3x_2 \le -48 \\ x_1 \ge 0 \text{ and } x_2 \ge 0 \end{cases}$$

Equality Constraints

Equality constraints may either be rewritten as a pair of inequalities (one with $\le$ and the other with $\ge$) or used to eliminate some of the variables from the problem.

37. Replace the equality constraint $2x_1 + x_2 = 6$ in the following problem by the two inequalities $2x_1 + x_2 \le 6$ and $2x_1 + x_2 \ge 6$ and then solve it by the two-stage simplex method.

Check your answer by solving the original problem by the graphical method of Section 4.2.

$$\text{Maximize} \quad P = 5x_1 + 3x_2$$

$$\text{Subject to} \quad \begin{cases} x_1 + x_2 \le 4 \\ 2x_1 + x_2 = 6 \\ x_1 \ge 0 \text{ and } x_2 \ge 0 \end{cases}$$

38. Solve the equality constraint $x_1 + x_3 = 20$ in the following problem for x_3 and then eliminate that variable from the objective function and the constraints (including the nonnegativity requirement that $x_3 \ge 0$). Solve the resulting two-variable problem by the two-stage simplex method. Be sure to state your solution in terms of the original problem.

$$\text{Maximize} \quad P = 5x_1 + 6x_2 + 4x_3$$

$$\text{Subject to} \quad \begin{cases} 2x_1 + x_2 + x_3 \le 60 \\ x_1 + x_2 + x_3 \ge 40 \\ x_1 + x_3 = 20 \\ x_1 \ge 0, \ x_2 \ge 0, \text{ and } x_3 \ge 0 \end{cases}$$

More About Nonnegativity Conditions

The following problems show how to reformulate linear programming problems to have nonnegativity conditions on *all* the variables when the original versions do not.

39. Show that the first problem may be rewritten as the second by replacing the variable x_1 by a new variable u_1 where $u_1 = x_1 + 8$ and $u_1 \ge 0$. Solve the first problem by the graphical method from Section 4.2, solve the second by the simplex method from Section 4.3, and check that your answers agree.

$$\text{Maximize} \quad P = 5x_1 + 3x_2$$

$$\text{Subject to} \quad \begin{cases} 2x_1 + x_2 \le 6 \\ x_1 + x_2 \le 4 \\ x_1 \ge -8 \text{ and } x_2 \ge 0 \end{cases}$$

$$\text{Maximize} \quad P = 5u_1 + 3x_2 - 40$$

$$\text{Subject to} \quad \begin{cases} 2u_1 + x_2 \le 22 \\ u_1 + x_2 \le 12 \\ u_1 \ge 0 \text{ and } x_2 \ge 0 \end{cases}$$

40. Show that the first problem may be rewritten as the second by replacing the variable x_1 by a new variable u_1 where $u_1 = -x_1$ and $u_1 \geq 0$. Solve the first problem by the graphical method from Section 4.2, solve the second by the simplex method from Section 4.3, and check that your answers agree.

$$\text{Maximize} \quad P = 5x_1 + 3x_2$$

$$\text{Subject to} \quad \begin{cases} 2x_1 + x_2 \leq 10 \\ -x_1 + x_2 \leq 4 \\ x_1 \leq 0 \quad \text{and} \quad x_2 \geq 0 \end{cases}$$

$$\text{Maximize} \quad P = -5u_1 + 3x_2$$

$$\text{Subject to} \quad \begin{cases} -2u_1 + x_2 \leq 10 \\ u_1 + x_2 \leq 4 \\ u_1 \geq 0 \quad \text{and} \quad x_2 \geq 0 \end{cases}$$

Unrestricted Variables

41. A variable is *unrestricted* if it may take *any* value (positive, negative, or zero). Show that the first problem may be rewritten as the sec-

ond by replacing the variable x_1 by two new variables u_1 and u_2 where $u_1 - u_2 = x_1$ and $u_1, u_2 \geq 0$. Solve the first problem by the graphical method from Section 4.2, solve the second by the simplex method from Section 4.3, and check that your answers agree.

$$\text{Maximize} \quad P = 3x_1 + 5x_2$$

$$\text{Subject to} \quad \begin{cases} 2x_1 + x_2 \leq 6 \\ x_2 \leq 4 \\ x_1 \text{ unrestricted} \quad \text{and} \quad x_2 \geq 0 \end{cases}$$

$$\text{Maximize} \quad P = 3u_1 - 3u_2 + 5x_2$$

$$\text{Subject to} \quad \begin{cases} 2u_1 - 2u_2 + x_2 \leq 6 \\ x_2 \leq 4 \\ u_1 \geq 0, \ u_2 \geq 0, \ \text{and} \ x_2 \geq 0 \end{cases}$$

[*Hint:* Can any real number be written as the difference of two nonnegative numbers?]

4.6 # NONSTANDARD PROBLEMS: ARTIFICIAL VARIABLES AND THE BIG-M METHOD

This and the previous section are optional and present alternative methods for solving nonstandard linear programming problems. The choice of which one to cover is left to the instructor. The examples and exercises duplicate those in Section 4.5.

Introduction

In this section we extend the simplex method to maximum problems with constraints involving $\geq$ and $=$ as well as the usual $\leq$. Because a basis for such a problem is not necessarily feasible or even immediately apparent, we introduce new "artificial" variables to provide an initial basis and then force the simplex method to seek a solution involving just the original variables.

Artificial Variables

The method requires that the right-hand side of each inequality be nonnegative. We multiply constraints by -1 as necessary to bring this about, remembering that such multiplication reverses the sense of the inequality (changing, for example $5 \geq -3$ into $-5 \leq 3$). We then introduce new variables into the constraints to make them into equations, but in different ways for the different types of constraints. Each of these new variables is required to be *nonnegative.*

- For $\leq$ constraints, we add slack variables in the usual way, so that, for example, $2x_1 + x_2 \leq 30$ becomes $2x_1 + x_2 + s_1 = 30$, with s_1 in the basis as $s_1 = 30$.

- For $\geq$ constraints, meaning that the left side may be *larger* than the right, we must *subtract* a slack variable* to obtain an equation, so that, for example, $x_1 + 2x_2 \geq 6$ becomes $x_1 + 2x_2 - s_2 = 6$. However, we cannot include s_2 in the basis because $-s_2 = 6$ means $s_2 = -6$, violating the nonnegativity condition of the variables. Therefore, we add an *artificial variable* a_1, giving $x_1 + 2x_2 - s_2 + a_1 = 6$. Such variables are called "artificial" because they have no interpretation in the original problem and are used only to obtain variables for the initial basis, in this case $a_1 = 6$.

- For *equality* constraints, which need no slack variables, and therefore do not give basic variables, we again introduce an artificial variable a_2 so that, for example, $x_1 + x_2 = 10$ becomes $x_1 + x_2 + a_2 = 10$, with $a_2 = 10$ in the basis.

The artificial variables a_1 and a_2 were added merely to create a feasible basis to begin with. However, since they have no real role in the problem, we must ensure that the simplex method removes them permanently from the basis. To do this, we subtract a large multiple M of each artificial variable from the objective function, so that, for example, $P = 8x_1 + 12x_2$ becomes $P = 8x_1 + 12x_2 - Ma_1 - Ma_2$. It is from this large number that the big-M method gets its name.

How do we know that there is a sufficiently large number M, and if so, how do we find it? Because a linear programming problem has a finite number of constraints, variables, and vertices, M need only be larger than the absolute values of a finite list of numbers, and so certainly exists. If the solution is to be calculated on a computer, a number near the largest possible number that the computer can represent will do (and if this doesn't suffice, the solution can't be calculated on that computer anyway). When calculating by hand, we will keep M as a *variable*, remembering that it stands for a large positive number.

The new variables gives us an "extended" maximum problem.

---

* Subtracted slack variables are sometimes called *surplus* variables.

Extended Problem

To form the extended problem for a nonstandard maximum problem:

1. Multiple constraints by -1 as necessary so that their right-hand sides are nonnegative.
2. In each $\leq$ constraint, add a slack variable.
3. In each $\geq$ constraint, subtract a slack variable and add an artificial variable.
4. In each $=$ constraint, add an artificial variable.
5. In the objective function, subtract M times each artificial variable.

Although it is not necessary, we will order constraints according to their type: first $\leq$, then $\geq$, and then $=$.

From the extended problem we write a *preliminary* tableau in the usual way, which we then modify to bring the artificial variables into the initial basis, giving the *initial* tableau.

EXAMPLE **1**

CONSTRUCTING AN INITIAL TABLEAU

Construct the extended problem and write the preliminary and initial tableaux for the following nonstandard linear programming problem.

Maximize $P = 8x_1 + 12x_2$

Subject to $\begin{cases} 2x_1 + x_2 \leq 30 \\ x_1 + 2x_2 \geq 6 \\ x_1 + x_2 = 10 \\ x_1 \geq 0 \text{ and } x_2 \geq 0 \end{cases}$ 　　"Mixed constraints"
　　　　$\leq, \ \geq, \ =$

Solution

The right-hand sides of the constraints are nonnegative, so we do not need to multiply by -1. Adding a slack variable to the $\leq$ constraint, subtracting a slack and adding an artificial variable to the $\geq$ constraint, and adding an artificial variable to the $=$ constraint, and finally subtracting M times the artificial variables from the objective function gives the extended problem:

Maximize $P = 8x_1 + 12x_2 - Ma_1 - Ma_2$ M is a large positive number

Subject to
$$\begin{cases} 2x_1 + x_2 + s_1 = 30 \\ x_1 + 2x_2 - s_2 + a_1 = 6 \\ x_1 + x_2 + a_2 = 10 \\ x_1 \geq 0, \; x_2 \geq 0, \; s_1 \geq 0, \\ s_2 \geq 0, \; a_1 \geq 0, \; a_2 \geq 0 \end{cases}$$

s_1 and s_2 are slack variables
a_1 and a_2 are artificial variables

All variables are nonnegative

From the extended problem we construct a *preliminary* tableau. Be sure you understand why the 1s and -1s are where they are, since later you may want to go directly from the original problem to the preliminary tableau (or even to the *initial* tableau given in the next step).

	x_1	x_2	s_1	s_2	a_1	a_2	
s_1	2	1	1	0	0	0	30
(a_1)	1	2	0	-1	1	0	6
(a_2)	1	1	0	0	0	1	10
P	-8	-12	0	0	M	M	0

"Preliminary" because the artificial variables a_1 and a_2 do not have 0s at the bottoms of their columns

To obtain zeros where the M's are we pivot on the 1s in their columns, which is equivalent to subtracting M times row 2 and M times row 3 from row 4, obtaining the following *initial tableau*, with the values of the basic variables written on the right.

	x_1	x_2	s_1	s_2	a_1	a_2		
s_1	2	1	1	0	0	0	30	$\leftarrow s_1 = 30$
a_1	1	2	0	-1	1	0	6	$\leftarrow a_1 = 6$
a_2	1	1	0	0	0	1	10	$\leftarrow a_2 = 10$
P	$-2M - 8$	$-3M - 12$	0	M	0	0	$-16M$	$\leftarrow P = -16M$

The basic solution on the right of the above tableau, with the remaining variables all set equal to zero, is *feasible* (all variables are nonnegative) but is far from optimal (P is negative). To solve the problem we would now pivot as usual on this tableau to try to make P optimal, in the process removing a_1 and a_2 from the basis.

Initial Tableau

The tableau written directly from the extended problem is the *preliminary* tableau. Pivoting on the 1s in the columns of the artificial variables gives the *initial* tableau.

Practice Problem

For the following nonstandard maximum problem, write:

a. the extended problem

b. the preliminary tableau

c. the initial tableau.

$$\text{Maximize} \quad P = 5x_1 + 6x_2 - 3x_3$$

$$\text{Subject to} \quad \begin{cases} x_1 - x_2 + 2x_3 \leq 10 \\ x_1 + 2x_2 - x_3 \geq 20 \\ x_1 \geq 0, \quad x_2 \geq 0, \quad x_3 \geq 0 \end{cases}$$

➤ Solution on page 337

The Big-M Method

The big-M method solves a nonstandard maximum problem by introducing artificial variables to form the *extended* problem, writing the preliminary tableau to obtain the initial tableau that displays an initial basic feasible point, and then pivoting as usual to find the solution.

> **Big-M Method**
>
> To solve a nonstandard maximum problem by the big-M method:
>
> 1. Write the extended problem (page 329).
>
> 2. Construct the initial tableau for the extended problem (page 330).
>
> 3. Solve the extended problem by the simplex method (page 285).
>
> 4. If there is a solution to the extended problem in which every artificial variable is 0, then this is also the solution to the original problem. Otherwise, there is no solution to the original problem.

EXAMPLE 2

USING THE BIG-M METHOD

Solve the following nonstandard maximum problem by the big-M method.

$$\text{Maximize} \quad P = x_1 - x_2 + 3x_3$$

$$\text{Subject to} \quad \begin{cases} x_1 + x_2 \quad\quad \leq 12 \\ \quad\quad x_2 + x_3 \geq 6 \\ x_1 \quad\quad + x_3 = 3 \\ x_1 \geq 0, \quad x_2 \geq 0, \quad x_3 \geq 0 \end{cases}$$

Solution

The right-hand sides of the constraints are all nonnegative so we do not need to multiply any by -1. With slack and artificial variables the problem becomes:

Maximize $P = x_1 - x_2 + 3x_3 - Ma_1 - Ma_2$

Subject to
$$\begin{cases} x_1 + x_2 \quad\ + s_1 \qquad\qquad\qquad = 12 \\ \qquad x_2 + x_3 \quad\ - s_2 + a_1 \qquad = 6 \\ x_1 \qquad\ + x_3 \qquad\qquad\ + a_2 = 3 \\ x_1 \ge 0, \quad x_2 \ge 0, \quad x_3 \ge 0, \quad s_1 \ge 0, \quad s_2 \ge 0, \quad a_1 \ge 0, \quad a_2 \ge 0 \end{cases}$$

The *preliminary* tableau is

	x_1	x_2	x_3	s_1	s_2	a_1	a_2	
s_1	1	1	0	1	0	0	0	12
(a_1)	0	1	1	0	-1	1	0	6
(a_2)	1	0	1	0	0	0	1	3
P	-1	1	-3	0	0	M	M	0

"Preliminary" because the artificial variables a_1 and a_2 do not have 0s at the bottoms of their columns

To obtain the *initial* tableau we need 0s where the M's are, which we accomplish by pivoting on the 1s in their columns (column 6, row 2 and column 7, row 3), or equivalently, subtracting M times row 2 and M times row 3 from the bottom row.

	x_1	x_2	x_3	s_1	s_2	a_1	a_2	
s_1	1	1	0	1	0	0	0	12
a_1	0	1	1	0	-1	1	0	6
a_2	1	0	1	0	0	0	1	3
P	$-M-1$	$-M+1$	$-2M-3$	0	M	0	0	$-9M$

New pivot element

The pivot column is column 3 (because $-2M - 3$ is smaller than both $-M - 1$ and $-M + 1$ for large positive values of M) and the pivot row is row 3 (because $\frac{3}{1} = 3$ is smaller than $\frac{6}{1} = 6$ and we don't consider the first row because of the 0), and pivoting on the 1 in column 3 and row 3 gives:

	x_1	x_2	x_3	s_1	s_2	a_1	a_2	
s_1	1	1	0	1	0	0	0	12
a_1	-1	1	0	0	-1	1	-1	3
x_3	1	0	1	0	0	0	1	3
P	$M+2$	$-M+1$	0	0	M	0	$2M+3$	$-3M+9$

New pivot element

The bottom row was found by multiplying row 3 (the pivot row) by $2M + 3$ and adding it to the bottom row $\longrightarrow$

We now pivot on column 2 (because $-M + 1$ is negative) and row 2 (because $\frac{3}{1} = 3$ is less than $\frac{12}{1} = 12$ and we don't consider the third row because of the 0), and pivoting on the 1 in column 2 and row 2 gives:

	x_1	x_2	x_3	s_1	s_2	a_1	a_2		
s_1	2	0	0	1	1	-1	1	9	← $s_1 = 9$
x_2	-1	1	0	0	-1	1	-1	3	← $x_2 = 3$
x_3	1	0	1	0	0	0	1	3	← $x_3 = 3$
P	3	0	0	0	1	$M-1$	$M+4$	6	← $P = 6$

This is the *final* tableau because there are no negative numbers in the bottom row. The solution of the extended problem is given on the right of the tableau, with the nonbasic variables $x_1, s_2, a_1,$ and a_2 all given the value zero. Therefore, the solution to the original problem is the same:

The maximum value of P is 6 when $x_1 = 0,$ $x_2 = 3,$ and $x_3 = 3.$

Graphing Calculator Exploration

a. On calculators that perform only numerical calculations, the big-M method is carried out by using a particular large value for M. The TI-83 screen on the left shows the right-hand part of the final tableau in the preceding example, after pivoting on column 2 and row 2 using the program PIVOT (see page 283) with $M = 100.$ The last column shows the solution we found, and the bottom row agrees with bottom row of the preceding tableau if $M = 100.$

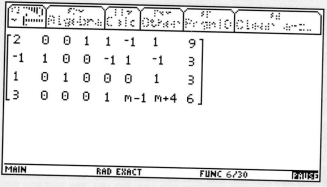

b. Some more advanced graphing calculators can perform symbolic as well as numerical calculations. Using the program* PIVOT for the TI-92 gives the same final tableau that we found by hand, with the last column again giving the solution that we found.

* See the Preface for information on how to obtain this and other programs.

EXAMPLE 3 **MANAGING AN INVESTMENT PORTFOLIO**

The manager of an $80 million mutual fund has three investment possibilities: secure government bonds yielding 4%, a blue chip growth stock paying 6%, and a biotechnology start-up company returning 12%. If the manager wants to invest at least $50 million in a combination of safe government bonds and blue chip stock, and at least as much in the bonds as the total in blue chip and biotechnology stocks, how much should be invested in each to obtain the greatest possible return?

Solution

Let

$$x_1 = \begin{pmatrix} \text{Amount in} \\ \text{government bonds} \end{pmatrix}$$

$$x_2 = \begin{pmatrix} \text{Amount in} \\ \text{blue chip stock} \end{pmatrix} \qquad \text{In millions of dollars}$$

$$x_3 = \begin{pmatrix} \text{Amount in} \\ \text{biotechnology stock} \end{pmatrix}$$

The objective is to maximize the return

$$P = 0.04x_1 + 0.06x_2 + 0.12x_3 \qquad \begin{array}{l} \text{Rate} \times \text{amount} \\ \text{for each investment} \end{array}$$

subject to the restrictions

$$
\begin{array}{ll}
x_1 + x_2 + x_3 \le 80 & \text{\$80 million available to invest} \\
x_1 + x_2 \ge 50 & \text{\$50 million in safe investments} \\
x_1 \ge x_2 + x_3 & \text{Bonds} \ge \text{total in stocks} \\
x_1 \ge 0, \quad x_2 \ge 0, \quad x_3 \ge 0 & \text{Nonnegativity}
\end{array}
$$

Rewriting the third constraint as $x_1 - x_2 - x_3 \ge 0$, we find that the second and third constraints require artificial variables. Pivoting twice on the preliminary tableau, the initial tableau is:

	x_1	x_2	x_3	s_1	s_2	s_3	a_1	a_2	
s_1	1	1	1	1	0	0	0	0	80
a_1	1	1	0	0	-1	0	1	0	50
a_2	1	-1	-1	0	0	-1	0	1	0
P	$-2M - 0.04$	-0.06	$M - 0.12$	0	M	M	0	0	$-50M$

Pivoting on the 1 in column 1 and row 3 yields

	x_1	x_2	x_3	s_1	s_2	s_3	a_1	a_2	
s_1	0	2	2	1	0	1	0	-1	80
a_1	0	2	1	0	-1	1	1	-1	50
x_1	1	-1	-1	0	0	-1	0	1	0
P	0	$-2M - 0.10$	$-M - 0.16$	0	M	$-M - 0.04$	0	$-2M + 0.04$	$-50M$

Pivoting on the 2 in column 2 and row 2 gives:

	x_1	x_2	x_3	s_1	s_2	s_3	a_1	a_2	
s_1			1	1	1	0	-1	0	30
x_2	0	1	$1/2$	0	$-1/2$	$1/2$	$1/2$	$-1/2$	25
x_1	1	0	$-1/2$	0	$-1/2$	$-1/2$	$1/2$	$1/2$	25
P	0	0	-0.11	0	-0.05	0.01	$M + 0.05$	$M - 0.01$	2.5

There are now no artificial variables in the basis. Pivoting on the 1 in column 3 and row 1 brings us to the final tableau:

	x_1	x_2	x_3	s_1	s_2	s_3	a_1	a_2	
x_3	0	0	1	1	1	0	-1	0	30
x_2	0	1	0	$-1/2$	-1	$1/2$	1	$-1/2$	10
x_1	1	0	0	$1/2$	0	$-1/2$	0	$1/2$	40
P	0	0	0	0.11	0.06	0.01	$M - 0.06$	$M - 0.01$	5.8

Because the extended problem has a solution with the artificial variables all equal to zero, we have the solution of the original problem:

The maximum is $P = 5.8$ when $x_1 = 40$, $x_2 = 10$, and $x_3 = 30$.

In terms of the original situation, the manager should invest $40 million in the government bonds, $10 million in the blue chip stock, and $30 million in the biotechnology start-up company for a maximum return of $5.8 million.

Computer Exploration

Linear programming problems can also be solved by specialized computer programs. One of the best known, both in education and in industry, is LINDO (for linear, **in**teractive, and **d**iscrete **o**ptimizer).* The following screen shows the LINDO 6.1 solution of Example 3. Note the similarity between the problem input format and our notation for linear programming problems.

* Further information about LINDO can be found at the Internet site *http://www.lindo .com* or obtained from LINDO Systems, Inc., Chicago, Illinois.

```
LINDO
File  Edit  Solve  Reports  Window  Help

 C:\LINDO\LP-EX3.LTX
MAX  0.04 GBD + 0.06 BCS + 0.12 BIO
ST
    GBD + BCS + BIO <= 80
    GBD + BCS       >= 50
    GBD - BCS - BIO >= 0
END
```

```
 Reports Window

LP OPTIMUM FOUND AT STEP        0

          OBJECTIVE FUNCTION VALUE

    1)      5.800000

VARIABLE          VALUE          REDUCED COST
    GBD        40.000000          0.000000
    BCS        10.000000          0.000000
    BIO        30.000000          0.000000


    ROW    SLACK OR SURPLUS     DUAL PRICES
    2)        0.000000           0.110000
    3)        0.000000          -0.060000
    4)        0.000000          -0.010000

NO. ITERATIONS=        0
```

Any Linear Programming Problem Can Now Be Solved

The big-M method allows us to solve *any* maximum linear programming problem, no matter what mixture of $\leq$ and $\geq$ inequalities or $=$ equalities are present in the constraints, provided that the variables are nonnegative. A direct way to change a minimum problem into a maximum problem is described in Exercises 35–36. Exercises 37–39 explain how to reformulate a problem so that all the variables are nonnegative. Taken together, these techniques allow us to solve *any* linear programming problem.

4.6 Section Summary

A nonstandard maximum problem may have any combination of $\leq$, $\geq$, and $=$ constraints, which we write with nonnegative numbers on their right-hand sides. The big-M method (page 331) adds artificial variables to the $\geq$ and $=$ constraints and subtracts M times these variables from the objective function. The initial tableau for this extended problem then displays a basic feasible point including these variables, which the simplex method removes from the basis. If this extended problem has a solution in which every artificial variable is zero, then the original problem has the same solution; otherwise, the problem does not have a solution.

➤ **Solution to Practice Problem**

a. Maximize $P = 5x_1 + 6x_2 - 3x_3 - Ma_1$

Subject to
$$\begin{cases} x_1 - x_2 + 2x_3 + s_1 = 10 \\ x_1 + 2x_2 - x_3 - s_2 + a_1 = 20 \\ x_1 \geq 0, \quad x_2 \geq 0, \quad x_3 \geq 0 \\ s_1 \geq 0, \quad s_2 \geq 0, \quad a_1 \geq 0 \end{cases}$$

b.

	x_1	x_2	x_3	s_1	s_2	a_1	
s_1	1	-1	2	1	0	0	10
(a_1)	1	2	-1	0	-1	1	20
P	-5	-6	3	0	0	M	0

c.

	x_1	x_2	x_3	s_1	s_2	a_1	
s_1	1	-1	2	1	0	0	10
a_1	1	2	-1	0	-1	1	20
P	$-M - 5$	$-2M - 6$	$M + .3$	0	M	0	$-20M$

Construct the initial simplex tableau for each nonstandard linear programming problem.

1. Maximize $P = 15x_1 + 20x_2 + 18x_3$

Subject to
$$\begin{cases} 3x_1 + 2x_2 + 8x_3 \leq 96 \\ 5x_1 + x_2 + 6x_3 \geq 30 \\ x_1 \geq 0, \quad x_2 \geq 0, \quad x_3 \geq 0 \end{cases}$$

2. Maximize $P = 60x_1 + 90x_2 + 30x_3$

Subject to
$$\begin{cases} 10x_1 + 18x_2 + 3x_3 \leq 150 \\ 4x_1 + 5x_2 + x_3 \geq 40 \\ x_1 \geq 0, \quad x_2 \geq 0, \quad x_3 \geq 0 \end{cases}$$

3. Maximize $P = 6x_1 + 4x_2 + 6x_3$

Subject to
$$\begin{cases} 2x_1 + x_2 + 3x_3 \geq 30 \\ x_1 + x_2 + 2x_3 \geq 20 \\ x_1 \geq 0, \quad x_2 \geq 0, \quad x_3 \geq 0 \end{cases}$$

4. Maximize $P = 9x_1 + 5x_2 + 4x_3$

Subject to
$$\begin{cases} 3x_1 + x_2 + x_3 \geq 40 \\ 2x_1 + 2x_2 + x_3 \geq 30 \\ x_1 \geq 0, \quad x_2 \geq 0, \quad x_3 \geq 0 \end{cases}$$

5. Maximize $P = 20x_1 + 30x_2 + 10x_3$

Subject to $\begin{cases} x_1 + 2x_2 + 3x_3 \leq 30 \\ x_1 + 2x_2 + x_3 \leq 18 \\ x_1 + x_2 + x_3 \geq 8 \\ x_1 \geq 0, \quad x_2 \geq 0, \quad x_3 \geq 0 \end{cases}$

6. Maximize $P = 24x_1 + 18x_2 + 30x_3$

Subject to $\begin{cases} x_1 + 2x_2 + 3x_3 \leq 15 \\ x_1 + 2x_2 + x_3 \leq 9 \\ x_1 + x_2 + x_3 \geq 4 \\ x_1 \geq 0, \quad x_2 \geq 0, \quad x_3 \geq 0 \end{cases}$

7. Maximize $P = 3x_1 + 2x_2 + 5x_3 + 4x_4$

Subject to $\begin{cases} x_1 + x_2 + x_3 + x_4 \geq 30 \\ 2x_1 + 3x_2 + 2x_3 + x_4 \geq 20 \\ 4x_1 + 2x_2 + x_3 + 2x_4 = 80 \\ x_1 \geq 0, \ x_2 \geq 0, \ x_3 \geq 0, \ x_4 \geq 0 \end{cases}$

8. Maximize $P = 6x_1 + 8x_2 + 3x_3 + 5x_4$

Subject to $\begin{cases} x_1 + x_2 + x_3 + x_4 \geq 70 \\ 3x_1 + 2x_2 + 4x_3 + 3x_4 \geq 60 \\ 4x_1 + x_2 + 2x_3 + 5x_4 = 100 \\ x_1 \geq 0, \quad x_2 \geq 0, \quad x_3 \geq 0, \quad x_4 \geq 0 \end{cases}$

Solve each nonstandard linear programming problem by the big-M method. (Problems 9, 10, 13, 14, 17, and 18 can also be solved by the graphical method.)

You may use the PIVOT program if permitted by your instructor.

9. Maximize $P = 4x_1 + x_2$

Subject to $\begin{cases} 3x_1 + 2x_2 \leq 120 \\ x_1 + x_2 \geq 50 \\ x_1 \geq 0, \quad x_2 \geq 0 \end{cases}$

10. Maximize $P = x_1 + 4x_2$

Subject to $\begin{cases} 3x_1 + 2x_2 \leq 120 \\ 2x_1 + x_2 \geq 60 \\ x_1 \geq 0, \quad x_2 \geq 0 \end{cases}$

11. Maximize $P = 30x_1 + 20x_2 + 28x_3$

Subject to $\begin{cases} 4x_1 + x_2 + 3x_3 \leq 60 \\ 3x_1 + x_2 + 2x_3 \geq 30 \\ x_1 \geq 0, \quad x_2 \geq 0, \quad x_3 \geq 0 \end{cases}$

12. Maximize $P = 18x_1 + 10x_2 + 20x_3$

Subject to $\begin{cases} 2x_1 + x_2 + 5x_3 \leq 50 \\ 3x_1 + x_2 + 4x_3 \geq 30 \\ x_1 \geq 0, \quad x_2 \geq 0, \quad x_3 \geq 0 \end{cases}$

13. Maximize $P = 4x_1 + 5x_2$

Subject to $\begin{cases} x_1 + 2x_2 \leq 12 \\ 2x_1 + x_2 \leq 12 \\ x_1 + x_2 \geq 15 \\ x_1 \geq 0, \quad x_2 \geq 0 \end{cases}$

14. Maximize $P = 5x_1 + 4x_2$

Subject to $\begin{cases} 2x_1 + 3x_2 \leq 12 \\ 3x_1 + 2x_2 \leq 12 \\ x_1 + x_2 \geq 10 \\ x_1 \geq 0, \quad x_2 \geq 0 \end{cases}$

15. Maximize $P = 15x_1 + 12x_2 + 18x_3$

Subject to $\begin{cases} 2x_1 + x_2 + 3x_3 \leq 24 \\ 5x_1 + x_2 + 2x_3 \geq 30 \\ x_1 \geq 0, \quad x_2 \geq 0, \quad x_3 \geq 0 \end{cases}$

16. Maximize $P = 4x_1 + 5x_2 + 6x_3$

Subject to $\begin{cases} 5x_1 + x_2 + 3x_3 \leq 30 \\ 2x_1 + x_2 + 2x_3 \geq 24 \\ x_1 \geq 0, \quad x_2 \geq 0, \quad x_3 \geq 0 \end{cases}$

17. Maximize $P = x_1 + x_2$

Subject to $\begin{cases} x_1 \leq 8 \\ x_2 \leq 5 \\ x_1 + 2x_2 \geq 6 \\ x_1 \geq 0, \quad x_2 \geq 0 \end{cases}$

18. Maximize $P = x_1 + x_2$

Subject to $\begin{cases} x_1 \leq 6 \\ x_2 \leq 9 \\ 3x_1 + x_2 \geq 6 \\ x_1 \geq 0, \quad x_2 \geq 0 \end{cases}$

19. Maximize $P = 4x_1 + 6x_2 + 12x_3 + 10x_4$

Subject to $\begin{cases} x_1 + x_2 + x_3 + x_4 \leq 60 \\ 2x_1 + x_2 + x_3 + 2x_4 \geq 10 \\ 2x_1 + 2x_2 + x_3 + x_4 = 100 \\ x_1 \geq 0, \quad x_2 \geq 0, \quad x_3 \geq 0, \quad x_4 \geq 0 \end{cases}$

20. Maximize $P = 8x_1 - x_2 + .5x_3 + 14x_4$

Subject to $\begin{cases} x_1 + 2x_2 + 3x_4 \le 55 \\ x_1 + x_2 + x_3 + 2x_4 \ge 25 \\ x_1 + 3x_3 + 3x_4 = 45 \\ x_1 \ge 0, \quad x_2 \ge 0, \quad x_3 \ge 0, \quad x_4 \ge 0 \end{cases}$

21. Maximize $P = 5x_1 + 32x_2 + 3x_3 + 15x_4$

Subject to $\begin{cases} 2x_1 - x_2 + x_3 + x_4 \le 10 \\ 5x_1 + 8x_2 + x_3 + 3x_4 \ge 18 \\ 3x_1 - 3x_2 + x_3 + 2x_4 \ge 6 \\ x_1 \ge 0, \quad x_2 \ge 0, \quad x_3 \ge 0, \quad x_4 \ge 0 \end{cases}$

22. Maximize $P = 70x_1 + 12x_2 + 60x_3 + 20x_4$

Subject to $\begin{cases} 2x_1 + x_2 + x_3 + x_4 \le 14 \\ 3x_1 + 3x_2 + x_3 + 2x_4 \le 24 \\ 5x_1 + 3x_2 + 4x_3 + 2x_4 \ge 44 \\ x_1 \ge 0, \quad x_2 \ge 0, \quad x_3 \ge 0, \quad x_4 \ge 0 \end{cases}$

23. Maximize $P = -x_1 + x_2 - 2x_3$

Subject to $\begin{cases} x_1 + x_2 + x_3 \le 50 \\ x_1 + x_3 \ge 10 \\ x_2 + x_3 \ge 20 \\ x_1 \ge 0, \quad x_2 \ge 0, \quad x_3 \ge 0 \end{cases}$

24. Maximize $P = x_1 + 2x_2 + x_3$

Subject to $\begin{cases} x_1 + x_2 + x_3 \le 30 \\ x_1 + x_2 \le 25 \\ x_2 + x_3 \ge 15 \\ x_1 \ge 0, \quad x_2 \ge 0, \quad x_3 \ge 0 \end{cases}$

APPLIED EXERCISES

Express each situation as a nonstandard maximum problem. Be sure to state the meaning of each variable. Solve the problem by the big-M method. State your final answer in terms of the original question.

 You may use the PIVOT program if permitted by your instructor.

25. PERSONAL FINANCE: Financial Planning
A retired couple wants to invest their $20,000 life savings in bank certificates of deposit yielding 6% and Treasury bonds yielding 5%. If they want at least $5000 in each type of investment, how much should they invest in each to receive the greatest possible income?

26. ENVIRONMENTAL SCIENCE: Recycling Management A volunteer recycling center accepts both used paper and empty glass bottles which it then sorts and sells to a reprocessing company. The center has room to accept a total of 800 crates of paper and glass each week and has 50 hours of volunteer help to do the sorting. Each crate of paper products takes 5 minutes to sort and sells for 8¢, while each crate of bottles takes 3 minutes to sort and sells for 7¢. To support their city's "grab that glass"

recycle theme this week, the center wants to accept at least 600 crates of glass. How many crates of each should the center accept this week to raise the most money for their ecology scholarship fund?

27. BUSINESS: Advertising The manager of a new mall may spend up to $18,000 on "grand opening" announcements in newspapers, on radio and on TV. Each newspaper ad costs $300 and reaches 5000 readers, each one-minute radio commercial costs $800 and is heard by 13,000 listeners, and each 15-second TV spot costs $900 and is seen by 15,000 viewers. If the manager wants at least 5 newspaper ads and at least 20 radio commercials and TV spots combined, how many of each should be placed to reach the largest number of potential customers? (Ignore multiple exposures to the same customer.)

28. SOCIAL SCIENCE: Political Advertising In the last few days before the election, a politician can afford to spend no more than $27,000 on TV advertisements and can arrange for no more than 10 ads. Each day time ad costs $2000 and reaches 4000 viewers, each prime time ad costs $3000 and reaches 5000 viewers, and each late night ad costs $1000 and reaches 2000 viewers. To be sure to reach the widest variety of voters, the politician's advisors insist on at least 5 ads scheduled during day time and late night combined. Ignoring repeated viewings by the same person and assuming every viewer can vote, how many ads in each of the time periods will reach the most voters?

29. **BUSINESS: Resource Allocation** A furniture shop manufactures wooden desks, tables and chairs. The numbers of hours to assemble and finish each piece are shown in the table together with the number of hours of skilled labor available for each task. To meet expected demand, a total of at least 30 desks and tables combined must be made. If the profit is $75 for each desk, $84 for each table, and $66 for each chair, how many of each should the company manufacture to obtain the greatest possible profit?

	Desk	Table	Chair	Labor Available
Assembly	2 hours	1 hour	2 hours	210 hours
Finishing	2 hours	3 hours	1 hours	150 hours

30. **BUSINESS: Production Planning** An automotive parts shop rebuilds carburetors, fuel pumps, and alternators. The numbers of hours to rebuild and then inspect and pack each part are shown in the table, together with the number of hours of skilled labor available for each task. At least 12 carburetors must be rebuilt. If the profit is $12 for each carburetor, $14 for each fuel pump, and $10 for each alternator, how many of each should the shop rebuild to obtain the greatest possible profit?

	Carburetor	Fuel Pump	Alternator	Labor Available
Rebuilding	5 hours	4 hours	3 hours	200 hours
Inspection & packaging	1 hour	1 hour	0.5 hour	45 hours

31. **BUSINESS: Agriculture** A farmer grows wheat, barley, and oats on her 500-acre farm. Each acre of wheat requires 3 days of labor (to plant, tend, and harvest) and costs $21 (for seed, fertilizer, and pesticides), each acre of barley requires 2 days of labor and costs $27, and each acre of oats requires 3 days of labor and costs $24. The farmer and her hired field hands can provide no more than 1200 days of labor this year. She wants to grow at least 100 acres of oats and can afford to spend no more than $15,120. If the profit is $50 for each acre of wheat, $40 for each acre of barley, and $45 for each acre of oats, how many acres of each crop should she grow to obtain the greatest possible profit?

32. **BUSINESS: Agriculture** A farmer grows corn, peanuts, and soybeans on his 240-acre farm. To maintain soil fertility, the farmer rotates the crops and always plants at least as many acres of soybeans as the total acres of the other crops. Because he has promised to sell some of his corn to a neighbor who raises cattle, he must plant at least 42 acres of corn. Each acre of corn requires 2 days of labor and yields a profit of $150, each acre of peanuts requires 5 days of labor and yields a profit of $300, and each acre of soybeans requires 1 day of labor and yields a profit of $100. If the farmer and his children can put in at most 630 days of labor, how many acres of each crop should the farmer plant to obtain the greatest possible profit?

33. **ENVIRONMENTAL SCIENCE: Pollution Control** A storage yard at a coal burning electric power plant can hold no more than 100,000 tons of coal. Two grades of coal are available: low sulfur (1%) with an energy content of 20 million BTU per ton and high sulfur (2%) with an energy content of 30 million BTU per ton. If existing contracts with the high-sulfur coal mine operator require that at least 50,000 tons of high-sulfur coal be purchased, and if the next coal purchase may contain no more than 1400 tons of sulfur, how many tons of each type of coal should be purchased to obtain the most energy?

34. **BUSINESS: Production Planning** A small jewelry company prepares and mounts semi-precious stones. There are 20 lapidaries (who cut and polish the stones) and 24 jewelers (who

mount the stones in gold settings). Each employee works 7 hours each day, 5 days each week. Each tray of agates requires 5 hours of cutting and polishing and 4 hours to mount, each tray of onyxes requires 2 hours of cutting and polishing and 3 hours to mount, and each tray of garnets requires 6 hours of cutting and polishing and 3 hours to mount. Furthermore, the company's owner has decided that they will process at least 12 trays of agates each week. If the profit is $15 for each tray of agates, $10 for each tray of onyxes, and $13 for each tray of garnets, how many trays of each stone should be processed each week to obtain the greatest possible profit?

Explorations and Excursions

The following problems extend and augment the material presented in the text.

The PIVOT program may be helpful (if permitted by your instructor).

Minimum Problems and the Big-M Method

35. a. Solve both of the following problems by the graphical method from Section 4.2. Is it true that the minimum of C is the same as -1 times the maximum of $P = -C$?

$$\text{Minimize} \quad C = 3x + 4y$$
$$\text{Subject to} \quad \begin{cases} x + y \leq 10 \\ 2x + y \geq 8 \\ x \geq 0 \quad \text{and} \quad y \geq 0 \end{cases}$$

$$\text{Maximize} \quad P = -3x - 4y$$
$$\text{Subject to} \quad \begin{cases} x + y \leq 10 \\ 2x + y \geq 8 \\ x \geq 0 \quad \text{and} \quad y \geq 0 \end{cases}$$

b. Solve the following problem by finding the dual maximum problem and using the (regular) simplex method as we did in Section 4.4.

$$\text{Minimize} \quad C = 3y_1 + 4y_2$$
$$\text{Subject to} \quad \begin{cases} -y_1 - y_2 \geq -10 \\ 2y_1 + y_2 \geq 8 \\ y_1 \geq 0 \quad \text{and} \quad y_2 \geq 0 \end{cases}$$

c. Solve the following nonstandard problem by the big-M method.

$$\text{Maximize} \quad P = -3x_1 - 4x_2$$
$$\text{Subject to} \quad \begin{cases} x_1 + x_2 \leq 10 \\ 2x_1 + x_2 \geq 8 \\ x_1 \geq 0 \quad \text{and} \quad x_2 \geq 0 \end{cases}$$

d. For each sequence of tableaux from parts (b) and (c), trace the basic points on the graphs from part (a). Which solution method is more efficient?

36. Repeat the process of Exercise 35 to compare the "dual" solution to the following minimum problem with the big-M solution of the corresponding maximum problem.

$$\text{Minimize} \quad C = 20y_1 + 18y_2$$
$$\text{Subject to} \quad \begin{cases} 2y_1 + y_2 \geq 18 \\ y_1 + y_2 \geq 14 \\ 4y_1 + 3y_2 \geq 48 \\ y_1 \geq 0 \quad \text{and} \quad y_2 \geq 0 \end{cases}$$

$$\text{Maximize} \quad P = -20x_1 - 18x_2$$
$$\text{Subject to} \quad \begin{cases} 2x_1 + x_2 \geq 18 \\ x_1 + x_2 \geq 14 \\ 4x_1 + 3x_2 \geq 48 \\ x_1 \geq 0 \quad \text{and} \quad x_2 \geq 0 \end{cases}$$

More About Nonnegativity Conditions

The following problems show how to reformulate linear programming problems to have nonnegativity conditions on *all* the variables when the original versions do not.

37. Show that the first problem may be rewritten as the second by replacing the variable x_1 by a new variable u_1 where $u_1 = x_1 + 8$ and $u_1 \geq 0$. Solve the first problem by the graphical method from Section 4.2, solve the second by the simplex method from Section 4.3, and check that your answers agree.

$$\text{Maximize} \quad P = 5x_1 + 3x_2$$
$$\text{Subject to} \quad \begin{cases} 2x_1 + x_2 \leq 6 \\ x_1 + x_2 \leq 4 \\ x_1 \geq -8 \quad \text{and} \quad x_2 \geq 0 \end{cases}$$

Maximize $P = 5u_1 + 3x_2 - 40$

Subject to $\begin{cases} 2u_1 + x_2 \le 22 \\ u_1 + x_2 \le 12 \\ u_1 \ge 0 \quad \text{and} \quad x_2 \ge 0 \end{cases}$

38. Show that the first problem may be rewritten as the second by replacing the variable x_1 by a new variable u_1 where $u_1 = -x_1$ and $u_1 \ge 0$. Solve the first problem by the graphical method from Section 4.2, solve the second by the simplex method from Section 4.3, and check that your answers agree.

Maximize $P = 5x_1 + 3x_2$

Subject to $\begin{cases} 2x_1 + x_2 \le 10 \\ -x_1 + x_2 \le 4 \\ x_1 \le 0 \quad \text{and} \quad x_2 \ge 0 \end{cases}$

Maximize $P = -5u_1 + 3x_2$

Subject to $\begin{cases} -2u_1 + x_2 \le 10 \\ u_1 + x_2 \le 4 \\ u_1 \ge 0 \quad \text{and} \quad x_2 \ge 0 \end{cases}$

Unrestricted Variables

39. A variable is *unrestricted* if it may take *any* value (positive, negative, or zero). Show that the first problem may be rewritten as the second by replacing the variable x_1 by two new variables u_1 and u_2 where $u_1 - u_2 = x_1$ and $u_1, u_2 \ge 0$. Solve the first problem by the graphical method from Section 4.2, solve the second by the simplex method from Section 4.3, and check that your answers agree.

Maximize $P = 3x_1 + 5x_2$

Subject to $\begin{cases} 2x_1 + x_2 \le 6 \\ x_2 \le 4 \\ x_1 \text{ unrestricted} \quad \text{and} \quad x_2 \ge 0 \end{cases}$

Maximize $P = 3u_1 - 3u_2 + 5x_2$

Subject to $\begin{cases} 2u_1 - 2u_2 + x_2 \le 6 \\ x_2 \le 4 \\ u_1 \ge 0, \quad u_2 \ge 0, \quad \text{and} \quad x_2 \ge 0 \end{cases}$

[*Hint:* Can any real number be written as the difference of two nonnegative numbers?]

Chapter Summary with Hints and Suggestions

Reading the text and doing the exercises in this chapter have helped you to master the following skills, which are listed by section (in case you need to review them) and are keyed to particular Review Exercises. Answers for all Review Exercises are given at the back of the book, and full solutions can be found in the Student Solutions Manual.

4.1 Linear Inequalities

■ Graph a linear inequality $ax + by \le c$ by drawing the boundary line $ax + by = c$ from its intercepts, using a test point to choose the correct side, and shading in the feasible region. *(Review Exercises 1–2.)*

■ Graph a system of linear inequalities $\begin{cases} ax + by \le c \\ \vdots \end{cases}$ by drawing the boundary lines, finding the correct sides of the boundaries, and shading in the feasible region. List the vertices or "corners" of the region and identify the region as bounded or unbounded. *(Review Exercises 3–8.)*

■ Formulate an applied situation as a system of linear inequalities and sketch the feasible region with the vertices. *(Review Exercises 9–10.)*

4.2 Two-Variable Linear Programming Problems

■ Find the maximum or minimum of a linear function on a given region or explain why

such a value does not exist.
(Review Exercises 11–14.)

■ Solve a two-variable linear programming problem by sketching the feasible region, determining whether a solution exists, and then finding it by evaluating the objective function at the vertices. *(Review Exercises 15–18.)*

■ Solve an applied linear programming problem by identifying the variables, the objective function, and the constraints, and then using the vertices of the feasible region.
(Review Exercises 19–20.)

4.3 The Simplex Method for Standard Maximum Problems

■ Check that a maximum linear programming problem is a standard problem and construct the initial simplex tableau.
(Review Exercises 21–22.)

$$\text{Maximize} \quad P = c^t X$$

$$\text{Subject to} \quad \begin{cases} AX \le b & (b \ge 0) \\ X \ge 0 \end{cases}$$

	X	S	
S	A	I	b
P	$-c^t$	0	0

■ Find the pivot element in a tableau and carry out one complete pivot operation. If there is no pivot element, explain what this means about the original problem. *(Review Exercises 23–24.)*

■ Solve a standard maximum problem by the simplex method (construct the initial simplex tableau, pivot until the tableau does not have a pivot element, and interpret this final tableau).
(Review Exercises 25–28.)

■ Solve an applied linear programming problem by the simplex method.
(Review Exercises 29–30.)

4.4 Standard Minimum Problems and Duality

■ Check that a minimum linear programming problem is a standard problem, and construct

the dual maximum problem and the initial simplex tableau. *(Review Exercises 31–34.)*

$$\text{Minimize} \quad C = b^t Y \qquad \text{Maximize} \quad P = c^t X$$

$$\text{Subject to} \quad \begin{cases} A^t Y \ge c \\ Y \ge 0 \end{cases} \qquad \text{Subject to} \quad \begin{cases} AX \le b \\ X \ge 0 \end{cases}$$

$$\text{where } b \ge 0 \qquad\qquad \text{where } b \ge 0$$

■ Solve a standard minimum problem by finding the dual maximum problem and using the simplex method. *(Review Exercises 35–38.)*

■ Solve an applied minimum problem by finding the dual maximum problem and using the simplex method. *(Review Exercises 39–40.)*

4.5 Nonstandard Problems: The Dual Pivot Element and the Two-Stage Method

■ For a nonstandard maximum problem, construct the initial simplex tableau and locate the dual pivot element. *(Review Exercises 41–42.)*

■ Solve a nonstandard maximum problem by the two-stage simplex method.
(Review Exercises 43–48.)

■ Solve an applied nonstandard maximum problem by the two-stage simplex method.
(Review Exercises 49–50.)

4.6 Nonstandard Problems: Artificial Variables and the Big-M Method

■ Construct the initial simplex tableau for a nonstandard maximum problem by using artificial variables. *(Review Exercises 51–52.)*

■ Solve a nonstandard maximum problem by the big-M method. *(Review Exercises 53–58.)*

■ Solve an applied nonstandard maximum problem by the big-M method.
(Review Exercises 59–60.)

Hints and Suggestions

■ *(Overview)* A linear programming problem asks for the maximum or minimum of a linear objective function subject to constraints in the form of linear inequalities. If there is a solution to the problem, it occurs at a *vertex* of the feasible region determined by the constraints. A problem with two variables may be solved graphically, but one with more than two variables must be solved algebraically using the simplex method, duality (for a standard mini-mum problem), and either the two-stage simplex or the big-M method (for a nonstandard problem).

■ While a problem with a *bounded* region always has a solution, a problem with an *unbounded* region may or may not have a solution (if it *does*, then the solution occurs at a vertex).

■ When setting up a word problem, look first for the question "how many" or "how much" to help identify the variables. Be sure that finding values for your variables will answer the question stated in the problem. Use the rest of the given information to find the objective function and constraints. Include nonnegativity conditions for your variables as appropriate (for example, a farmer *cannot* raise a negative number of cows).

■ A simplex tableau is *feasible* if it has no negative numbers in the rightmost *column* and is *optimal* if it has no negative numbers in the bottom *row*. If a simplex tableau is both feasible and optimal, it displays the solution of the problem.

■ The pivot operation is not completed until *all* the rows of the tableau have been recalculated.

■ If you are solving a standard maximum problem by the simplex method and you have a negative number in the rightmost column after pivoting, you've made an error in your choice of the pivot element and/or your arithmetic.

■ Many simplex tableaux require several pivots to solve, so don't give up: Keep pivoting until the tableau does not have a pivot element.

■ **Practice for Test:** Review Exercises 1, 7, 9, 11, 13, 16, 17, 20, 25, 29, 35, 39, and (41, 45, 50) or (51, 55, 60).

Review Exercises for Chapter 4

Practice test exercise numbers are in green.

4.1 Linear Inequalities

Graph each inequality.

1. $3x - 2y \le 24$ 2. $x + y \ge -6$

Graph each system of inequalities. List all vertices and identify the region as "bounded" or "unbounded."

3. $\begin{cases} x + y \le 20 \\ x \ge 0, \ y \ge 0 \end{cases}$ 4. $\begin{cases} x + 2y \ge 10 \\ x \ge 0, \ y \ge 0 \end{cases}$

5. $\begin{cases} 2x + y \le 20 \\ x + y \le 15 \\ x \ge 0, \ y \ge 0 \end{cases}$ 6. $\begin{cases} 2x + y \ge 12 \\ x + y \ge 8 \\ x \ge 0, \ y \ge 0 \end{cases}$

7. $\begin{cases} 3x + 2y \le 24 \\ x + y \le 10 \\ x \ge 2 \\ x \ge 0, \ y \ge 0 \end{cases}$ 8. $\begin{cases} x + y \le 10 \\ -x + y \le 4 \\ x \le 9 \\ x \ge 0, \ y \ge 0 \end{cases}$

Formulate each application as a system of linear inequalities, sketch the feasible region, and find the vertices.

9. **GENERAL: Dog Training** A dog trainer works with Irish setters and Labrador retrievers for no more than 6 hours each day. Each training session with a "setter" takes half an hour and requires 8 dog treats, while each session with a "lab" takes three-quarters of an hour and also requires 8 dog treats. If the trainer has only 80 dog treats left, how many of each breed can he train today?

10. **GENERAL: Blending Bird Seed** A pet store owner blends custom bird seed by mixing SongBird and MeadowMix brands together. Each pound of SongBird contains 2 ounces of sunflower hearts and 3 ounces of crushed peanuts (along with a lot of cheap filler seed),

while each pound of MeadowMix contains 4 ounces of sunflower hearts and 2 ounces of crushed peanuts. If each bag of the custom blend is labeled "contains at least 104 ounces of sunflower hearts and 84 ounces of crushed peanuts," how many pounds of each brand must be put in each bag?

4.2 Two-Variable Linear Programming Problems

Use the region below to find each maximum or minimum value. If such a value does not exist, explain why not.

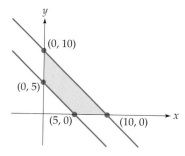

11. Maximum of $P = 3x + 4y$

12. Minimum of $C = 3x + 2y$

Use the region below to find each maximum or minimum value. If such a value does not exist, explain why not.

13. Maximum of $P = 4x + 3y$

14. Minimum of $C = 2x + 5y$

Solve each linear programming problem by sketching the region, labeling the vertices, checking whether a solution exists, and then finding it if it does.

15. Maximize $P = 20x + 15y$

Subject to $\begin{cases} x + 3y \le 18 \\ x + y \le 12 \\ x \ge 0, \ y \ge 0 \end{cases}$

16. Minimize $C = 30x + 40y$

Subject to $\begin{cases} 2x + y \ge 120 \\ x + 3y \ge 120 \\ x \ge 0, \ y \ge 0 \end{cases}$

17. Maximize $P = 20x + 30y$

Subject to $\begin{cases} 2x + y \le 24 \\ x + 4y \le 40 \\ x + y \le 13 \\ x \ge 0, \ y \ge 0 \end{cases}$

18. Minimize $C = 7x + 11y$

Subject to $\begin{cases} x + y \ge 20 \\ x - y \ge -10 \\ x \ge 0, \ y \ge 0 \end{cases}$

Formulate each application as a linear programming problem by identifying the variables, the objective function, and the constraints. Determine whether a solution exists, and if it does, find it using the vertices of the feasible region. State your final answer in terms of the original question.

19. BUSINESS: Production Planning A woodworking shop makes antique reproduction cases for wall and mantel clocks. The amount of cutting and finishing needed for each is shown in the table, together with the number of hours of skilled labor available for each task. If the profit for each case is $250 for wall clocks and $200 for mantel clocks, how many of each kind of clock case should the shop make to obtain the greatest possible profit?

	Wall Clock	Mantel Clock	Labor Available
Cutting	5 hours	3 hours	165 hours
Finishing	3 hours	4 hours	132 hours

20. **BIOMEDICAL: Nutrition** A large and active cat needs at least 70 grams of fat and 134.4 grams of protein each week. Each ounce of canned cat food contains 2 grams of fat and 3.2 grams of protein, while each ounce of dry food contains 2.5 grams of fat and 9.6 grams of protein. If canned cat food costs 5¢ per ounce and dry food costs 8¢ per ounce, how many ounces of each type of food will meet the cat's nutritional needs for the week at the least cost?

4.3 The Simplex Method for Standard Maximum Problems

Construct the initial simplex tableau for each standard maximum problem. (Do not solve the problem any further.)

21. Maximize $P = 4x_1 + 10x_2 + 9x_3$

   Subject to $\begin{cases} 5x_1 + 2x_2 + 3x_3 \le 30 \\ 2x_1 + 3x_2 + 4x_3 \le 24 \\ x_1 \ge 0, \quad x_2 \ge 0, \quad x_3 \ge 0 \end{cases}$

22. Maximize $P = 5x_1 + 7x_2$

   Subject to $\begin{cases} x_1 + x_2 \le 10 \\ 2x_1 + x_2 \le 14 \\ x_1 - x_2 \le 4 \\ x_1 \ge 0, \quad x_2 \ge 0 \end{cases}$

For each simplex tableau, find the pivot element and carry out one complete pivot operation. If there is no pivot element, explain what this means about the original problem.

23.

	x_1	x_2	x_3	s_1	s_2	s_3	
s_1	1	1	1	1	0	0	6
s_2	3	1	2	0	1	0	8
s_3	1	1	-3	0	0	1	6
P	-6	4	-8	0	0	0	0

24.

	x_1	x_2	s_1	s_2	s_3	
s_1	2	0	1	0	-2	8
s_2	6	0	0	1	3	42
x_2	1	1	0	0	1	10
P	5	0	0	0	15	150

Solve each problem by the simplex method.

25. Maximize $P = 7x_1 + 8x_2 + 7x_3$

   Subject to $\begin{cases} 2x_1 + x_2 + 4x_3 \le 16 \\ x_1 + 2x_2 - x_3 \le 30 \\ x_1 + x_2 + 3x_3 \le 12 \\ x_1 \ge 0, \quad x_2 \ge 0, \quad x_3 \ge 0 \end{cases}$

26. Maximize $P = 8x_1 + 10x_2 + 9x_3$

   Subject to $\begin{cases} 3x_1 + 2x_2 + 3x_3 \le 18 \\ 2x_1 + x_2 + 4x_3 \le 14 \\ x_1 \ge 0, \quad x_2 \ge 0, \quad x_3 \ge 0 \end{cases}$

27. Maximize $P = 8x_1 + 7x_2 + 5x_3 + 7x_4$

   Subject to $\begin{cases} 3x_1 + 2x_2 + 4x_3 + 3x_4 \le 12 \\ x_1 + x_2 + x_3 + x_4 \le 5 \\ 3x_1 + x_2 + 5x_3 + 3x_4 \le 15 \\ x_1 \ge 0, \quad x_2 \ge 0, \quad x_3 \ge 0, \quad x_4 \ge 0 \end{cases}$

28. Maximize $P = 5x_1 + 4x_2 + 3x_3$

   Subject to $\begin{cases} x_1 + x_2 \le 8 \\ x_2 + x_3 \le 7 \\ x_1 + x_2 - x_3 \le 6 \\ x_1 \ge 0, \quad x_2 \ge 0, \quad x_3 \ge 0 \end{cases}$

Formulate each application as a linear programming problem by identifying the variables, the objective function, and the constraints. Check that the problem is a standard maximum problem and then solve it by the simplex method. State your final answer in terms of the original question.

29. **BUSINESS: Book Publishing** A book publisher needs more copies of its latest best seller. By rescheduling several other projects over the next two weeks, it can arrange up to 300 hours of printing and 250 hours of binding. For each thousand copies, the trade edition requires 3 hours of printing and 2 hours of binding, the book club edition requires 2 hours of printing and 2 hours of binding, and the paperback edition requires 1 hour of printing and 1 hour of binding. If the profit per book is $7 for the trade edition, $6 for the book club edition, and $4 for the paperback, how many of each edition should be printed to obtain the greatest possible profit?

30. **BUSINESS: Production Planning** A fully automated plastics factory produces three toys—a racing car, a jet airplane, and a speed boat—in three stages: molding, painting, and packaging. After allowing for routine maintenance, the equipment can operate no more than 150 hours per week. Each batch of racing cars requires 6 hours of molding, 2.5 hours of painting, and 5 hours of packaging; each batch of jet airplanes requires 3 hours of molding, 7.5 hours of painting, and 5 hours of packaging; and each batch of speed boats requires 3 hours of molding, 5 hours of painting, and 5 hours of packaging. If the profits per batch of toys are $120 for cars, $100 for airplanes, and $110 for boats, how many batches of each toy should be produced each week to obtain the greatest possible profit?

4.4 Duality and Standard Minimum Problems

For each minimum linear programming problem check that it is a standard problem and then construct the dual maximum problem and the initial simplex tableau. (Do not solve the problem any further.)

31. Minimize $C = 10y_1 + 15y_2 + 20y_3$

    Subject to $\begin{cases} 2y_1 - y_2 + y_3 \geq 40 \\ y_1 + y_2 + 3y_3 \geq 30 \\ y_1 \geq 0, \quad y_2 \geq 0, \quad y_3 \geq 0 \end{cases}$

32. Minimize $C = 30y_1 + 20y_2$

    Subject to $\begin{cases} 5y_1 + 2y_2 \geq 210 \\ 3y_1 + 4y_2 \geq 252 \\ 5y_1 + 4y_2 \leq 380 \\ y_1 \geq 0, \quad y_2 \geq 0 \end{cases}$

33. Minimize $C = 42y_1 + 36y_2$

    Subject to $\begin{cases} 7y_1 + 4y_2 \geq 84 \\ y_1 + y_2 \geq 18 \\ y_1 \geq 0, \quad y_2 \geq 0 \end{cases}$

34. Minimize $C = 130y_1 + 40y_2 + 98y_3$

    Subject to $\begin{cases} 3y_1 + y_2 + 2y_3 \geq 51 \\ 4y_1 + y_2 + 2y_3 \geq 60 \\ 3y_1 + y_2 + 3y_3 \geq 57 \\ y_1 \geq 0, \quad y_2 \geq 0, \quad y_3 \geq 0 \end{cases}$

Solve each standard minimum problem by finding its dual and using the simplex method.

35. Minimize $C = 9y_1 + 24y_2 + 6y_3$

    Subject to $\begin{cases} y_1 + 2y_2 \geq 18 \\ 3y_2 + y_3 \geq 15 \\ y_1 \geq 0, \quad y_2 \geq 0, \quad y_3 \geq 0 \end{cases}$

36. Minimize $C = 30y_1 + 36y_2$

    Subject to $\begin{cases} 5y_1 + 3y_2 \geq 60 \\ y_1 + y_2 \geq 16 \\ y_2 \geq 5 \\ y_1 \geq 0, \quad y_2 \geq 0 \end{cases}$

37. Minimize $C = 3y_1 + 4y_2 + 10y_3 + 8y_4$

    Subject to $\begin{cases} -y_1 + y_2 + y_3 - 2y_4 \geq 10 \\ y_1 - y_2 - 2y_3 + y_4 \geq 20 \\ y_1 \geq 0, \quad y_2 \geq 0, \quad y_3 \geq 0, \quad y_4 \geq 0 \end{cases}$

38. Minimize $C = 12y_1 + 5y_2 + 6y_3$

    Subject to $\begin{cases} 3y_1 + y_2 + 3y_3 \geq 9 \\ 2y_1 + y_2 - 3y_3 \geq 12 \\ 4y_1 + 2y_2 + 2y_3 \geq 8 \\ y_1 \geq 0, \quad y_2 \geq 0, \quad y_3 \geq 0 \end{cases}$

Formulate each application as a standard minimum problem and then solve it by finding its dual and using the simplex method. State your final answer in terms of the original question.

39. **GENERAL: School Supplies** A student needs at least 3 new pens and 2 dozen ink cartridges. The bookstore sells single pens for 99¢, packages of 6 ink cartridges for 89¢, and a $1.19 "writer's combo" package containing 1 pen and 2 ink cartridges. How many of each will meet the student's needs at the least cost?

40. **BUSINESS: Production Planning** A pasta company is expanding its linguini production facility. Two machines are available: a small-capacity machine costing $5000 that produces 20 pounds per minute and needs 1 operator, and a large-capacity machine costing $6000 that produces 30 pounds per minute and needs 2 operators. The company wants to hire no more than 34 additional employees yet increase production by at least 600 pounds per minute. How many of each machine should

the company buy to expand its production at the least cost?

4.5 Nonstandard Problems: The Dual Pivot Element and the Two-Stage Method

For each nonstandard maximum problem construct the initial simplex tableau and locate the dual pivot element. (Do not carry out the pivot operation.)

41. Maximize $P = 80x_1 + 30x_2$

Subject to $\begin{cases} 4x_1 + x_2 \leq 40 \\ 2x_1 + 3x_2 \leq 60 \\ x_1 + x_2 \geq 10 \\ x_1 \geq 0, \;\; x_2 \geq 0 \end{cases}$

42. Maximize $P = 4x_1 + 8x_2 + 6x_3 + 10x_4$

Subject to $\begin{cases} 2x_1 + 3x_2 - 3x_3 + x_4 \leq 30 \\ 7x_1 - 5x_2 + x_3 + 5x_4 \geq 35 \\ x_1 \geq 0, \;\; x_2 \geq 0, \;\; x_3 \geq 0, \;\; x_4 \geq 0 \end{cases}$

Solve each nonstandard maximum problem by the two-stage simplex method.

43. Maximize $P = 3x_1 + 4x_2$

Subject to $\begin{cases} x_1 + x_2 \leq 10 \\ x_1 + x_2 \geq 5 \\ x_1 \geq 0, \;\; x_2 \geq 0 \end{cases}$

44. Maximize $P = 8x_1 - 20x_2 - 18x_3$

Subject to $\begin{cases} x_1 + 3x_2 + 2x_3 \leq 55 \\ x_1 + x_2 + x_3 \geq 25 \\ 2x_1 + 3x_2 + 3x_3 \leq 70 \\ x_1 \geq 0, \;\; x_2 \geq 0, \;\; x_3 \geq 0 \end{cases}$

45. Maximize $P = 6x_1 - 9x_2$

Subject to $\begin{cases} 3x_1 + x_2 \leq 24 \\ x_1 + 3x_2 \leq 24 \\ x_1 + x_2 \geq 6 \\ x_1 \geq 0, \;\; x_2 \geq 0 \end{cases}$

46. Maximize $P = 12x_1 - 21x_2 + 2x_3$

Subject to $\begin{cases} 2x_1 - 3x_2 + x_3 \geq 20 \\ x_1 - x_2 + x_3 \leq 25 \\ x_1 - 3x_2 \geq 5 \\ x_1 \geq 0, \;\; x_2 \geq 0, \;\; x_3 \geq 0 \end{cases}$

47. Maximize $P = 6x_1 + 7x_2$

Subject to $\begin{cases} x_1 \geq 2 \\ x_2 \geq 3 \\ x_1 \leq 4 \\ x_2 \leq 5 \\ x_1 \geq 0, \;\; x_2 \geq 0 \end{cases}$

48. Maximize $P = 8x_1 + 2x_2 + 10x_3$

Subject to $\begin{cases} 2x_1 + x_2 + x_3 \leq 30 \\ x_1 + 2x_2 + x_3 \geq 5 \\ x_1 - x_2 + x_3 \geq 10 \\ x_1 \geq 0, \;\; x_2 \geq 0, \;\; x_3 \geq 0 \end{cases}$

Formulate each application as a nonstandard maximum problem and solve it by the two-stage simplex method. State your final answer in terms of the original question.

49. MANAGEMENT SCIENCE: Lumber Production A saw mill rough-cuts lumber and planes some of it to make finished-grade boards. The saw cuts 1 thousand board-feet per hour and can operate up to 10 hours each day, while the plane finishes 1 thousand board-feet per hour and can operate up to 8 hours each day. The profit per thousand board-feet is $120 for rough-cut lumber and $100 for finished-grade boards. If at least 4 thousand board-feet of finished-grade boards must be produced each day, how many board-feet of each type of lumber should be produced daily to obtain the greatest possible profit?

50. PERSONAL FINANCE: Financial Planning An international money manager wishes to invest no more than $150,000 in United States and Canadian stocks and bonds, with at least twice as much in bonds as in stocks. The current market yields are 10% on U.S. stocks, 5% on U.S. bonds, 11% on Canadian stocks, and 4% on Canadian bonds. If at least $30,000 must be invested in Canadian stocks, how much should be invested in each to obtain the greatest possible return?

4.6 Nonstandard Problems: Artificial Variables and the Big-M Method

Construct the initial simplex tableau for each non-standard maximum problem.

51. Maximize $P = 80x_1 + 30x_2$

Subject to $\begin{cases} 4x_1 + x_2 \le 40 \\ 2x_1 + 3x_2 \le 60 \\ x_1 + x_2 \ge 10 \\ x_1 \ge 0, \quad x_2 \ge 0 \end{cases}$

52. Maximize $P = 4x_1 + 8x_2 + 6x_3 + 10x_4$

Subject to $\begin{cases} 2x_1 + 3x_2 - 3x_3 + x_4 \le 30 \\ 7x_1 - 5x_2 + x_3 + 5x_4 \ge 35 \\ x_1 \ge 0, \quad x_2 \ge 0, \quad x_3 \ge 0, \quad x_4 \ge 0 \end{cases}$

Solve each nonstandard maximum problem by the big-M method.

53. Maximize $P = 3x_1 + 4x_2$

Subject to $\begin{cases} x_1 + x_2 \le 10 \\ x_1 + x_2 \ge 5 \\ x_1 \ge 0, \quad x_2 \ge 0 \end{cases}$

54. Maximize $P = 8x_1 - 20x_2 - 18x_3$

Subject to $\begin{cases} x_1 + 3x_2 + 2x_3 \le 55 \\ 2x_1 + 3x_2 + 3x_3 \le 70 \\ x_1 + x_2 + x_3 \ge 25 \\ x_1 \ge 0, \quad x_2 \ge 0, \quad x_3 \ge 0 \end{cases}$

55. Maximize $P = 6x_1 - 9x_2$

Subject to $\begin{cases} 3x_1 + x_2 \le 24 \\ x_1 + 3x_2 \le 24 \\ x_1 + x_2 \ge 6 \\ x_1 \ge 0, \quad x_2 \ge 0 \end{cases}$

56. Maximize $P = 12x_1 - 21x_2 + 2x_3$

Subject to $\begin{cases} x_1 - x_2 + x_3 \le 25 \\ x_1 - 3x_2 \ge 5 \\ 2x_1 - 3x_2 + x_3 \ge 20 \\ x_1 \ge 0, \quad x_2 \ge 0, \quad x_3 \ge 0 \end{cases}$

57. Maximize $P = 6x_1 + 7x_2$

Subject to $\begin{cases} x_1 \le 4 \\ x_2 \le 5 \\ x_1 \ge 2 \\ x_2 \ge 3 \\ x_1 \ge 0, \quad x_2 \ge 0 \end{cases}$

58. Maximize $P = 8x_1 + 2x_2 + 10x_3$

Subject to $\begin{cases} 2x_1 + x_2 + x_3 \le 30 \\ x_1 + 2x_2 + x_3 \ge 5 \\ x_1 - x_2 + x_3 \ge 10 \\ x_1 \ge 0, \quad x_2 \ge 0, \quad x_3 \ge 0 \end{cases}$

Formulate each application as a nonstandard maximum problem and solve it by the big-M method. State your final answer in terms of the original question.

59. MANAGEMENT SCIENCE: Lumber Production A saw mill rough-cuts lumber and planes some of it to make finished-grade boards. The saw cuts 1 thousand board-feet per hour and can operate up to 10 hours each day, while the plane finishes 1 thousand board-feet per hour and can operate up to 8 hours each day. The profit per thousand board-feet is $120 for rough-cut lumber and $100 for finished-grade boards. If at least 4 thousand board-feet of finished-grade boards must be produced each day, how many board-feet of each type of lumber should be produced daily to obtain the greatest possible profit?

60. PERSONAL FINANCE: Financial Planning An international money manager wishes to invest no more than $150,000 in United States and Canadian stocks and bonds, with at least twice as much in bonds as in stocks. The current market yields are 10% on U.S. stocks, 5% on U.S. bonds, 11% on Canadian stocks, and 4% on Canadian bonds. If at least $30,000 must be invested in Canadian stocks, how much should be invested in each to obtain the greatest possible return?

5 Probability

Application Preview

The Chevalier de Meré

Probability as a mathematical subject began when a French nobleman, Antoine Gombaude (whose title was the Chevalier de Meré), wanted to understand why he was losing money in his dice games. The Chevalier's first bet was that in four rolls of a die (the singular of *dice*) he could roll at least one 6, and he was winning so often that he had no more takers. The Chevalier then reasoned that in six times as many rolls he should get a *second* six, so he bet that in twenty-four rolls of *two* dice he could roll at least one *double*-6. Much to his surprise, he was losing these bets. Rather than just giving up, he had the good sense to seek the advice of one of the foremost mathematicians of the era, Blaise Pascal (1623–1662). Pascal corresponded with the other great French mathematician of the time, Pierre de Fermat (1601–1665), about the Chevalier's difficulties, and together they found that an analysis of the situation depended on a careful listing of all possible outcomes from throwing the dice. This analysis and the calculation of the corresponding probabilities is considered to be the beginning of the mathematical study of probability. We will analyze the Chevalier's bets on page 395 and see why he won one and lost the other.

Introduction

Probability is used frequently in everyday conversation, as in "you will probably get the job" or "it will probably rain today." In mathematics, however, we need to give a precise definition for probability, and we will see that it depends on a careful counting all of the possible events that can occur. With this in mind, we begin by reviewing sets and some ways of counting their members.

Sets and Set Operations

A *set* is any well-defined collection of objects (also called *elements* or *members*). By "well-defined" we mean you can tell whether an object is in the set or not. For example, we may speak of *the set of American citizens*, since there are specific conditions for being an American citizen. On the other hand, we cannot speak of *the set of all thin people*, since the word "thin" is not precisely defined.

We will often use diagrams to represent sets. We use a rectangle for the *universal set*, denoted U, which is the set of all of the elements that we are discussing. For example, depending on the question, the universal set might consist of all people, all customers of a company, all cars on the road, or any other collection of objects of interest. To represent sets within the universal set, we draw ellipses inside the rectangle containing the elements of the sets. Such drawings are called *Venn diagrams*.* There are several operations on sets, which we illustrate with Venn diagrams.

The *intersection* of sets A and B, denoted $A \cap B$, is the set of all elements that are in *both A and B*.

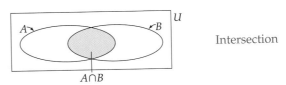

Intersection

$A \cap B$

Two sets A and B are *disjoint* if their intersection is empty (that is, if they have no elements in common).

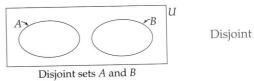

Disjoint

Disjoint sets A and B

* After the English logician John Venn (1843–1923), author of *Logic of Chance*.

The *union* of sets A and B, denoted $A \cup B$, is the set of all elements that are in *either A or B* (or both).

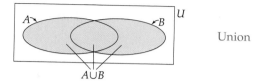

Union

$A \cup B$

The *complement* of a set A, denoted A^c, is the set of all elements *not* in A.

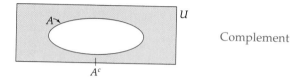

Complement

A^c

In general, the words "and," "or," and "not" translate into "intersection," "union," and "complement." We use the word "or" in the *inclusive* sense, meaning one possibility or the other *or both*. The complement of the universal set is the *empty* set, also called the *null* set, denoted $\varnothing$, which has no members. A *subset* of a set is a set (possibly empty) of elements from the original set. For a finite set A, we use the symbol $n(A)$ to mean the number of elements in A:

$$n(A) = \left(\begin{array}{c} \text{Number of} \\ \text{elements in } A \end{array} \right)$$

For example, $n(\varnothing) = 0$ because the empty set has no elements. From the previous Venn diagram, it is clear that the elements of the set A^c are precisely the elements of the universal set U that are *not* in A. Therefore:

Complementary Principle of Counting

$$n(A^c) = n(U) - n(A)$$

The number of elements in the *complement* of a set is the number of elements in the universal set minus the number of elements in the original set

We will find this principle useful when it is easier to count the elements in the complement rather than the elements in the set itself.

Addition Principle for Counting

How can we count the number of elements in the *union* of two sets without counting them one by one? In the following Venn diagram there are $n(A) = 7$ elements (dots) in A and $n(B) = 5$ elements in B.

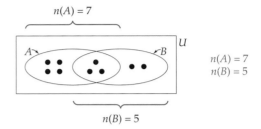

Altogether there are 9 elements in $A \cup B$, but adding $n(A) = 7$ and $n(B) = 5$ gives 12, not the correct total of 9 for the union. The diagram shows the trouble: Adding $n(A)$ and $n(B)$ counts the elements in the middle (in $A \cap B$) *twice*—once in A and a second time in B. To correct this double-counting, we must subtract the number of elements in $A \cap B$ from the total, obtaining the general rule:

Addition Principle of Counting

$$n(A \cup B) = n(A) + n(B) - n(A \cap B)$$

The number of elements in the *union* of two sets is the number of elements in one plus the number of elements in the other minus the number of elements in *both*

Of course, if A and B are disjoint, then $A \cap B$ is empty, so $n(A \cap B) = 0$, giving a simpler addition principle for disjoint sets:

$$n(A \cup B) = n(A) + n(B) \qquad \text{for } A, B \text{ disjoint}$$

EXAMPLE **1**   **CARS IN A PARKING LOT**

A mall parking lot has 300 cars, of which 150 have alarm systems, 200 have sound systems, and 90 have both alarm and sound systems.

a. How many cars have an alarm system or a sound system?

b. How many have neither?

Solution

Let A be the set of cars with alarm systems and S be the set of cars with sound systems. Starting with the Venn diagram on the left, we enter the numbers successively, beginning with the intersection:

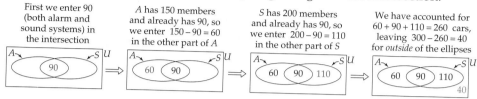

First we enter 90 (both alarm and sound systems) in the intersection

A has 150 members and already has 90, so we enter $150-90=60$ in the other part of A

S has 200 members and already has 90, so we enter $200-90=110$ in the other part of S

We have accounted for $60+90+110=260$ cars, leaving $300-260=40$ for *outside* of the ellipses

The 40 *outside* of the ellipses means that 40 cars have *neither* system. Answer:

260 cars have either an alarm system or a sound system. $60+90+110$
40 cars have neither an alarm system nor a sound system. $300-260$

Alternatively, since "or" means "union," we could find $n(A \cup S)$ by the Addition Principal of Counting, using $n(A)=150$, $n(S)=200$, and $n(A \cap S)=90$:

$$n(A \cup S) = \underbrace{n(A)}_{150} + \underbrace{n(S)}_{200} - \underbrace{n(A \cap S)}_{90} = 150 + 200 - 90 = 260$$

Addition principle of counting

Again we see that 260 cars have an alarm or a sound system, so that $300-260=40$ have neither.

Hint: With Venn diagrams you usually begin at the "inside" and work "out."

Practice Problem 1

A survey of insurance coverage in 300 metropolitan businesses revealed that 150 offer their employees dental insurance, 150 offer vision coverage, and 100 offer both dental and vision coverage. How many of these businesses offer their employees dental or vision insurance? How many offer neither? ➤ Solution on page 359

The Multiplication Principle for Counting

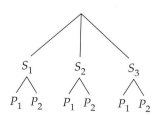

Suppose that you are choosing an outfit to wear, and you may choose any one of 3 shirts, S_1, S_2, or S_3, and either of 2 pairs of pants, P_1 or P_2. If the 3 shirts can be combined freely with the 2 pairs of pants, there are $3 \cdot 2 = 6$ different possible outfits, namely S_1P_1, S_1P_2, S_2P_1, S_2P_2, S_3P_1, and S_3P_2. The possible choices are shown in the *tree diagram* on the left, indicating a first choice of a shirt, then "branching"

from each shirt to the second choice of a pair of pants, ending in 6 "leaves" at the bottom. More generally, we have the following *multiplication principle*, which will be very useful throughout this chapter.

Multiplication Principle for Counting

If two choices are to be made, and there are m possibilities for the first choice and n possibilities for the second choice, and if any first choice can be combined with any second choice, then the two choices together can be made in $m \cdot n$ ways.

The multiplication principle can also be proved by enumerating all of the possibilities, as in the following example.

EXAMPLE 2

COUNTING DIFFERENT PRODUCTS

A toy company makes red, green, blue, and yellow plastic cars, trucks, and planes. How many different kinds of toys do they make?

Solution

Let the set of colors be $C = \{$red, green, blue, yellow$\}$, and let the set of shapes be $S = \{$car, truck, plane$\}$. Each possible toy is described by its color and shape:

	Red	**Green**	**Blue**	**Yellow**
Car	(red, car)	(green, car)	(blue, car)	(yellow, car)
Truck	(red, truck)	(green, truck)	(blue, truck)	(yellow, truck)
Plane	(red, plane)	(green, plane)	(blue, plane)	(yellow, plane)

This rectangular table contains all possible combination of colors and shapes, and the number of boxes is found by multiplying length times width: There are $4 \cdot 3 = 12$ possible toys that can be made.

A convenient way of using the multiplication principle is to imagine making up a typical combination of the two choices. For instance, in the preceding example there were 4 choices for the color and 3 choices for the shape, giving

$4 \cdot 3 = 12$ possible toys

The Multiplication Principle for Counting generalizes to *more* than two choices.

Generalized Multiplication Principle for Counting

If k choices are to be made, and there are m_1 possibilities for the first choice, m_2 possibilities for the second choice, m_3 possibilities for the third choice, and so on down to m_k possibilities for the kth choice, and if the choices can be combined in any way, then the k choices can be made together in $m_1 \cdot m_2 \cdot \cdots \cdot m_k$ ways.

EXAMPLE 3

COUNTING PARKING PERMITS

A parking permit displays an identification code consisting of a letter (A to Z) followed by two digits (0 to 9). How many different permits can be issued?

Solution

Since each identification code consists of three symbols, we need to fill three blanks:

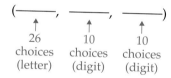

$$26 \cdot 10 \cdot 10 = 2600 \quad \text{ways}$$

Therefore, 2600 different permits can be issued.

Practice Problem 2

A computer password consists of two letters (A to Z) followed by three digits (0 to 9). How many different passwords are there?

➤ Solution on page 360

The Number of Subsets of a Set

Given a set, how many subsets does it have? For example, the set $\{a, b\}$ has 4 subsets: $\{a\}$, $\{b\}$, $\{a,b\}$, and $\varnothing$. (We consider the empty set $\varnothing$ and the set itself to be subsets.) To form these subsets we had two choices: *include or exclude a* and *include or exclude b*. Two choices (for *a* and *b*) with each made in two possible ways (*include* or *exclude*) gives a total of $2 \cdot 2$ choices, for 4 subsets (as we saw).

$$(\underline{\hspace{1.5cm}}, \underline{\hspace{1.5cm}}) \qquad 2 \cdot 2 = 4 \text{ subsets}$$

$\uparrow$ Include *a*? $\qquad$ $\uparrow$ Include *b*?
(yes/no) $\qquad$ (yes/no)

In this same way we can count the number of subsets of *any* set. A set with *n* members means *n* choices: *include* or *exclude the first member*, *include* or *exclude the second member*, and so on. Making *n* choices, each in 2 possible ways (*include* or *exclude*) means a total of 2^n choices, for 2^n subsets.

Number of Subsets of a Set

A set with *n* elements has 2^n subsets.

EXAMPLE 4

FINDING THE SUBSETS OF A SET

List all subsets of the set $\{a,b,c\}$ and verify that there are $2^3 = 8$ of them.

Solution

The subsets are

$$\{a\}, \ \{b\}, \ \{c\}, \ \{a,b\}, \ \{a,c\}, \ \{b,c\}, \ \{a,b,c\}, \text{ and } \varnothing$$

Indeed, there are 8 subsets (including the set itself and the empty set).

EXAMPLE 5

COUNTING SUBSETS

A restaurant offers pizza with mushrooms, peppers, onions, pepperoni, and sausage. How many different types of pizza can be ordered?

Solution

The set of toppings has 5 members, so there are $2^5 = 32$ possible subsets, and so 32 different pizzas. (Do you see which of these subsets corresponds to "plain" pizza?)

5.1 **Section Summary**

For a set A, its *complement* A^c is the set of all elements *not* in A. For sets A and B their *intersection* $A \cap B$ is the set of elements that are in *both* A and B, and their *union* $A \cup B$ is the set of elements that are in *either A or B or both*. In the Venn diagrams below, U stands for the *universal* set of all elements being considered.

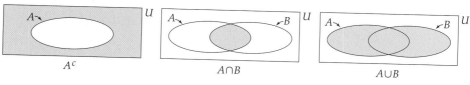

For a finite set A, $n(A)$ means the number of elements in A. The following formulas enable us to find the number of elements in one set by counting the elements in other sets:

$$n(A^c) = n(U) - n(A)$$
Complementary principle of counting

$$n(A \cup B) = n(A) + n(B) - n(A \cap B)$$
Addition principle of counting

The multiplication principle says that if two choices are to be made, and one can be made in m ways and the other can be made in n ways, and if the ways can be freely combined, then the two choices *together* can be made in $m \cdot n$ ways. The multiplication principle generalizes to more than two choices.

A set with n members has 2^n subsets.

➤ Solutions to Practice Problems

1. Let D be the set of businesses offering dental insurance, and let V be those offering vision insurance. Then, using a Venn diagram:

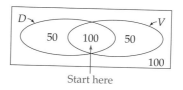

Start here

(continues)

Or, using the Addition Principle:

$$n(D \cup V) = n(D) + n(V) - n(D \cap V)$$
$$= 150 + 150 - 100 = 200$$

Two hundred businesses offer dental or vision insurance. One hundred offer neither.

2. Since $26 \cdot 26 \cdot 10 \cdot 10 \cdot 10 = 676{,}000$, there are 676,000 different passwords.

2 letters 3 digits

5.1 Exercises

Find each number using the following Venn diagram.

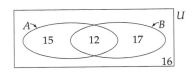

1. $n(A)$

2. $n(B)$

3. $n(U)$

4. $n(A \cap B)$

5. $n(A \cup B)$

6. $n(A^c)$

7. $n(B^c)$

8. $n(A \cap B^c)$

9. $n(A^c \cap B)$

10. $n(A \cup B^c)$

11. Given that $n(A) = 20$, $n(B) = 10$, $n(A \cap B) = 6$, and $n(U) = 40$, fill in the four regions in the Venn diagram.

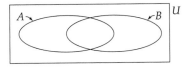

12. Given that $n(A) = 15$, $n(B) = 30$, $n(A \cap B) = 5$, and $n(U) = 65$, fill in the four regions in the Venn diagram.

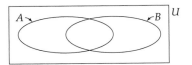

From your Venn diagram in Exercise 11 (on the left), find each number.

From your Venn diagram in Exercise 12 (above), find each number.

13. $n(A \cup B)$

14. $n(A \cup B)$

15. $n(A^c)$

16. $n(A^c)$

17. $n(B^c)$

18. $n(B^c)$

APPLIED EXERCISES (may be helpful.)

For Exercises 19–22, draw a tree diagram for each situation to find the number of possible pairs.

19. GENERAL: Parking Permits A parking permit sticker displays an identification code consisting of a letter (from A, B, and C) followed by a digit (from 1, 2, 3, and 4). How many different permits can be issued?

20. GENERAL: Computer Passwords A computer network password consists of two letters, the first being X, Y, or Z and the second being A, B, C, D, or E. How many different passwords are there?

21. **GENERAL: Wardrobes** A clothing store sells windbreakers, ski jackets, and overcoats in your choice of red or blue. How many different kinds of coats do they sell?

22. **GENERAL: Dorm Chores** Ted, Bob, Fred, Jim, Bill, and Sam share a student dorm suite. How many ways can they choose who takes out the garbage and who sweeps the floor if they can choose the same person to do both? What if different men must do the chores?

23. **ATHLETICS: Lacrosse** The sophomore lacrosse team has 24 players, of whom 10 played defense last year, 12 played offense, and 5 played both defense and offense, while the rest of the players did not play last year. How many members of the team played last year?

24. **SOCIAL SCIENCE: Political Fund Raising** If 12,300 individuals contributed to the governor's first election campaign, 15,200 contributed to her second, and 7800 contributed to both, how many individuals made contributions?

25. **SOCIAL SCIENCE: Current Events** A survey of 1200 residents of New Orleans revealed that 760 get information on international events by watching television news programs, 530 from listening to radio news shows, and 290 from both television and radio, while the rest have no interest in international events. How many have no interest in international events?

26. **BUSINESS: International Marketing** Of the 30 freshman majoring in international marketing, 12 speak French, 8 speak German, and 5 speak both languages. How many do not speak either French or German?

27. **GENERAL: Parking Permits** Look back at Example 3 on page 357 and find the number of parking permits that can be issued if the letter O and the number 0 are omitted to avoid confusion.

28. **GENERAL: Computer Passwords** A computer network password is made up of letters A to Z and may be either four or five letters long. How many such passwords are there? Find the answer in two different ways:

    a. Find the number of four-letter passwords and add it to the number of five-letter passwords.

    b. Think of a four-letter password as really having length 5, where in the fifth position is a "blank" that can be thought of as one more member of an extended "alphabet" for the last position.

29. **GENERAL: Computer Passwords** How many eight-symbol computer passwords can be formed using the letters A to J and the digits 2 to 6?

30. **GENERAL: Bagels** A bagel shop offers 12 different kinds of bagels and 18 flavors of cream cheese. How many different orders for a bagel and cream cheese can customers place?

31. **GENERAL: Combination Locks** A suitcase combination lock has four wheels, each labeled with the digits 0 to 9. How many lock combinations are possible?

32. **GENERAL: Awards** A small community consists of 25 families, each consisting of two parents and three children. If one parent and one of his or her children are to be honored as "parent and child of the year," in how many different ways can the award be made?

33. **GENERAL: Committees** A college student governance committee is made up of 3 freshmen, 4 sophomores, 4 juniors, and 5 seniors. A subcommittee of 4 consisting of one person from each class is to go to a national convention. How many different subcommittees are there?

34. **GENERAL: Mix & Match Fashions** A designer has created a completely mixable line of five shirts, nine ties, and four pairs of slacks and advertises that "the possibilities are endless." Exactly how "endless" are the possible different outfits consisting of a shirt, a tie, and a pair of slacks?

35. **SOCIAL SCIENCE: False Testimony** A witness in a trial testified that he searched 20 cars, of which 12 were convertibles, 15 had out-of-state plates, and 5 were convertibles with

out-of-state plates. Explain why the witness was not telling the truth.

36. **SOCIAL SCIENCE: False Complaints** Your neighbor complained to the police that during the last month you played your music too loud on 22 days, your car blocked his driveway on 15 days, and you did both on 4 days. How would you prove to the police that your neighbor was not telling the truth?

37. **GENERAL: Cable TV Channels** A cable TV company offers regular service plus 6 premium channels that must be ordered separately. How many different options are there?

38. **GENERAL: Book Clubs** You join a book club that each month offers five books, of which any number (or none) can be ordered. How many possibilities are there? What if at least one book must be ordered?

39. **GENERAL: Classes** There are 10 students who have the prerequisites to take a math course. How many different possible classes are there? (Assume that a class must have at least one student.)

40. **GENERAL: Making Change** You have a penny, a nickel, a dime, a quarter, a half-dollar, and a dollar in your pocket. How many different amounts of money can you make?

Explorations and Excursions

The following problems extend and augment the material presented in the text.

De Morgan's Laws

De Morgan's Laws* show the relationship between the operations of complement, union, and intersection. De Morgan's Laws are:

$$(A \cup B)^c = A^c \cap B^c \quad \text{De Morgan's First Law}$$
$$(A \cap B)^c = A^c \cup B^c \quad \text{De Morgan's Second Law}$$

41. Verify De Morgan's First Law (above) for the sets A and B shown with their members in the following Venn diagram.

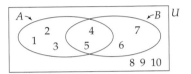

42. Verify De Morgan's Second Law (above Exercise 41) for the sets A and B shown with their members in the above Venn diagram.

43. Show De Morgan's First Law (above Exercise 41) by making Venn diagrams for $(A \cup B)^c$ and for $A^c \cap B^c$.

44. Show De Morgan's Second Law (above Exercise 41) by making Venn diagrams for $(A \cap B)^c$ and for $A^c \cup B^c$.

*After Augustus de Morgan (1806–1871), British mathematician and logician.

Introduction

In this section we will develop two very useful formulas for counting various types of choices, known as *permutations* and *combinations*, and then apply these formulas to a wide variety of problems. We begin by describing *factorial notation*.

Factorials

Products of successive integers from a number down to 1, such as $5 \cdot 4 \cdot 3 \cdot 2 \cdot 1$, are called *factorials*. We denote factorials by exclamation points, so the preceding product would be written 5! (read "5 factorial"). Formally:

Factorials

For any positive integer n,

$$n! = n(n - 1) \cdot \ldots \cdot 1 \qquad \text{n factorial is the product of the integers from n down to 1}$$

$$0! = 1 \qquad \text{Zero factorial is 1}$$

The next example shows that some factorial expressions are most easily found by using cancellation *before* evaluating the factorials.

EXAMPLE 1

CALCULATING FACTORIALS

Find: **a.** 4!   **b.** $\dfrac{7!}{6!}$ **c.** $\dfrac{6!}{3!}$ **d.** $\dfrac{100!}{99!}$

Solution

a. $4! = 4 \cdot 3 \cdot 2 \cdot 1 = 24$

b. $\dfrac{7!}{6!} = \dfrac{7 \cdot 6 \cdot 5 \cdot 4 \cdot 3 \cdot 2 \cdot 1}{6 \cdot 5 \cdot 4 \cdot 3 \cdot 2 \cdot 1} = \dfrac{7 \cdot \cancel{6 \cdot 5 \cdot 4 \cdot 3 \cdot 2 \cdot 1}}{\cancel{6 \cdot 5 \cdot 4 \cdot 3 \cdot 2 \cdot 1}} = 7$

    When finding quotients of factorials, look for cancellation

c. $\dfrac{6!}{3!} = \dfrac{6 \cdot 5 \cdot 4 \cdot 3 \cdot 2 \cdot 1}{3 \cdot 2 \cdot 1} = \dfrac{6 \cdot 5 \cdot 4 \cdot \cancel{3 \cdot 2 \cdot 1}}{\cancel{3 \cdot 2 \cdot 1}} = 6 \cdot 5 \cdot 4 = 120$

d. $\dfrac{100!}{99!} = \dfrac{100 \cdot 99 \cdot \ldots \cdot 1}{99 \cdot \ldots \cdot 1} = 100$

    Canceling $99 \cdot \ldots \cdot 1$

Parts (b) and (d) of the above example are instances of the formula $\dfrac{n!}{(n - 1)!} = n$. For $n = 1$ this becomes $\dfrac{1!}{0!} = 1$, and multiplying each side by 0! gives $1 = 0!$, which is why we define 0! to be 1.

We will use factorials to count the number of ways that objects can be *ordered*.

Permutations

How many different orderings are there for the letters a, b, and c? We may list the orderings as *abc, acb, bac, bca, cab,* and *cba,* so there are 6. Each of these orderings is called a *permutation*. Instead of listing them all, we could observe that there are 3 ways of choosing the first letter, 2 ways of choosing the second (because one letter was "used up" in the first choice), and 1 way of choosing the last (whichever is left), so by the multiplication principle there are $3 \cdot 2 \cdot 1 = 6$ possible orderings, just as we found before. How many permutations are there of n distinct objects? By the same reasoning, the answer is $n(n-1) \cdot \cdots \cdot 1 = n!$.

EXAMPLE 2   **COUNTING BATTING ORDERS**

How many different batting orders are there for a 9-player baseball team?

Solution

The first batter can be any one of the 9 players, the second batter can be any one of the remaining 8, next can be any one of the remaining 7, and so on down to the last batter who will be the only one left. By the multiplication principle we should multiply together the numbers of ways of making each choice, so the number of batting orderings will be

$$9! = 9 \cdot 8 \cdot 7 \cdot 6 \cdot 5 \cdot 4 \cdot 3 \cdot 2 \cdot 1 = 362{,}880$$

Practice Problem 1   You have six different tasks to do today. In how many different orders can they be done?   ➤ Solution on page 371

EXAMPLE 3   **COUNTING ORDERINGS OF BOOKS**

You have 3 math books, 4 history books, and 5 English books. In how many ways can these 12 books be arranged on the shelf if all books of the same subject are together?

Solution

There are 3! ways to arrange the math books among themselves, 4! ways to arrange the history books, and 5! ways to arrange the English books, so by the multiplication principle there are $3! \cdot 4! \cdot 5!$ ways if the order of the subjects is math then history then English. However, these three subjects can *themselves* be arranged in any of $3 \cdot 2 \cdot 1 = 3!$ orderings. For each of the 3! orderings of the subjects, the books can be arranged in $3! \cdot 4! \cdot 5!$ ways, so the total number of such arrangements is

$$3! \cdot 3! \cdot 4! \cdot 5! = 6 \cdot 6 \cdot 24 \cdot 120 = 103{,}680$$

To return to our baseball example, how many different orderings are there for just *the first 3 batters* on the 9-player team? Clearly any one of the 9 players can bat first, then any one of the remaining 8, then any one of the remaining 7, for a total of $9 \cdot 8 \cdot 7$ orderings (by the multiplication principle). In general, if we have n distinct objects and want to count all possible orderings of some r of them, there are n choices for the first, $n - 1$ choices for the second, $n - 2$ choices for the third, down to $n - r + 1$ choices for the rth. (Notice that taking r of them means leaving $n - r$, so the last one kept is the preceding one, $n - r + 1$.) Such orderings are called *permutations* and can be counted using the following formula.

Permutations*

The number of permutations (ordered arrangements) of n distinct objects taken r at a time is

$$\overbrace{{}_nP_r = n \cdot (n - 1) \cdot \cdots \cdot (n - r + 1)}^{r \text{ factors}}$$

Product of r numbers from n down

We define ${}_nP_0 = 1$ since there is exactly one way of taking zero objects from n, namely, taking nothing.

| EXAMPLE 4 | **COUNTING NONSENSE WORDS** |

How many five letter "nonsense" words (that is, strings of letters without regard to meaning) can be made from the letters A to Z with no letter repeated?

Solution

Words are ordered arrangements of letters, so we use the permutation formula with $n = 26$ and $r = 5$.

$$_{26}P_5 = 26 \cdot 25 \cdot 24 \cdot 23 \cdot 22 = 7{,}893{,}600 \qquad \text{Product of 5 numbers from 26 down}$$

There are 7,893,600 five-letter nonsense words with distinct letters.

---

* Alternative notations for ${}_nP_r$ are $P_{n,r}$, P_r^n, and $P(n, r)$.

Instead of using the permutation formula, we may think of building each word by filling in 5 blanks with distinct letters and multiplying the numbers of choices:

This gives $26 \cdot 25 \cdot 24 \cdot 23 \cdot 22 = 7{,}893{,}600$ nonsense words, just as before.

Graphing Calculator Exploration

Values of ${}_nP_r$ are easy to find on some graphing calculators. The following displays show ${}_nP_r$ for $n = 26$ and several values of r, showing that ${}_nP_r$ gets large quickly as r increases.

```
MATH NUM CPX PRB
1:rand
2:nPr
3:nCr
4:!
5:randInt(
6:randNorm(
7:randBin(
```

```
26 nPr 1
                26
26 nPr 3
             15600
26 nPr 5
           7893600
```

EXAMPLE 5

COUNTING LICENSE PLATES

If a car license plate consists of three distinct letters followed by three distinct digits, how many different license plates are possible?

Solution

For three-letters with no repeats we have ${}_{26}P_3 = 26 \cdot 25 \cdot 24$, and for three-digits with no repeats we have ${}_{10}P_3 = 10 \cdot 9 \cdot 8$. By the multiplication principle, the number of possible license plates is

$$\underbrace{26 \cdot 25 \cdot 24}_{{}_{26}P_3} \cdot \underbrace{10 \cdot 9 \cdot 8}_{{}_{10}P_3} = 11{,}232{,}000$$

Practice Problem 2 Jurors at an art exhibition must select the first-, second-, and third-place winners from an exhibition of 35 paintings. In how many different ways can the winning paintings be chosen? ➤ Solution on page 371

Combinations

Sometimes we want to count choices *where order does not matter*. Such choices are called *combinations*. For example, a committee of three people is the same committee regardless of order in which they were chosen. How do we count the number of combinations? We begin with an example.

EXAMPLE 6

COUNTING COMMITTEES

A group of 8 student representatives wants to choose a committee of 3 to write a petition to the administration asking for better internet access. How many different committees are possible?

Solution

If we were considering the order in which the committee members were chosen, then there would be $8 \cdot 7 \cdot 6$ ways. However, a committee consisting of Alice, Bob, and Cathy is exactly the same as the committee of Bob, Cathy, and Alice—the order in which they were chosen makes no difference. Any particular committee of 3 people can be chosen in $3 \cdot 2 \cdot 1 = 6$ different orders, all making the same committee. Therefore, to eliminate multiple countings of the same committee we divide the $8 \cdot 7 \cdot 6$ orderings by $3 \cdot 2 \cdot 1$, obtaining for the number of different committees:

$$\frac{8 \cdot 7 \cdot 6}{3 \cdot 2 \cdot 1} = 56$$

Number of committees
from 8 of size 3

This number of ways of choosing from 8 a committee of 3 is called the *number of combinations of 8 things taken 3 at a time*, written $_8C_3$, sometimes read "8 choose 3." If we were to multiply the numerator of the above fraction by $5 \cdot 4 \cdot 3 \cdot 2 \cdot 1$ it would "complete" the factorial into 8!. Therefore, multiplying numerator and denominator by $5 \cdot 4 \cdot 3 \cdot 2 \cdot 1$ gives an alternate way expression for $_8C_3$:

$$\frac{8 \cdot 7 \cdot 6}{3 \cdot 2 \cdot 1} = \frac{\overbrace{8 \cdot 7 \cdot 6 \cdot 5 \cdot 4 \cdot 3 \cdot 2 \cdot 1}^{8!}}{\underbrace{3 \cdot 2 \cdot 1}_{3!} \cdot \underbrace{5 \cdot 4 \cdot 3 \cdot 2 \cdot 1}_{5!}} = \frac{8!}{3! \cdot 5!}$$

Multiplying top
and bottom by 5!

The expressions on the left and the right above are each equal to $_8C_3$: the left side is easier to calculate, but the right side simpler to write (using factorials). Replacing the 8 and 3 by *any* positive integers, we obtain the following general formulas.

Combinations*

The number of combinations (*un*ordered arrangements) of n distinct objects taken r at a time is

$$_nC_r = \frac{n \cdot (n-1) \cdot \cdots \cdot (n-r+1)}{r \cdot (r-1) \cdot \cdots \cdot 1} \qquad \begin{matrix} \leftarrow r \text{ numbers beginning with } n \\ \leftarrow r! \end{matrix}$$

$$= \frac{n!}{r!(n-r)!} \qquad \text{In factorial form}$$

The second formula comes from the first by multiplying the numerator and denominator by $(n-r) \cdot \cdots \cdot 1 = (n-r)!$. We define $_nC_0 = 1$ since there is clearly one way to choose zero objects from n, namely, taking nothing.

Notice that 8 things taken 3 at a time is the same as 8 things taken 5 at a time:

$$_8C_3 = \frac{8 \cdot 7 \cdot 6}{3 \cdot 2 \cdot 1} = 56$$

Same value

$$_8C_5 = \frac{8 \cdot 7 \cdot 6 \cdot 5 \cdot 4}{5 \cdot 4 \cdot 3 \cdot 2 \cdot 1} = 56$$

There is a simple reason for this, which may be explained in terms of committees: Whenever you choose a committee of 3 from 8 people you are, in a sense, also selecting another committee, *the 5 left behind.* Any different choice of the 3 results in a different 5 being left behind. Therefore, from 8 people there must be exactly as many committees of 3 as committees of 5. This is also clear from the factorial formulas, since $_8C_3 = \frac{8!}{3!\,5!}$ and $_8C_5 = \frac{8!}{5!\,3!}$ differ only in the order of the denominators. As before, replacing 8 and 3 by *any* positive integers n and r shows that from n people there are exactly as many committees of r people as committees of $n-r$ people (in symbols: $_nC_r = {}_nC_{n-r}$). This relationship is useful in reducing calculation; for example, given $_8C_5$ it is easier to calculate it as $_8C_3$, as we saw in the calculations above this paragraph.

---

* Alternative notations for $_nC_r$ are $C_{n,r}$, C_r^n, $C(n, r)$, and $\binom{n}{r}$.

EXAMPLE **7**   **COUNTING PERMUTATIONS AND COMBINATIONS**

A student club has 15 members.

a. How many ways can a president, vice president, and treasurer be chosen?

b. How many ways can a committee of three members be chosen?

Solution

Each question involves choosing three members from the club, but for the officers we want *ordered* arrangements (the order determines the offices: the president is listed first, the vice president second, and the treasurer third), while for the committee we want *unordered* arrangements. Thus part (a) asks for permutations, part (b) for combinations.

a. For the officers: $_{15}P_3 = 15 \cdot 14 \cdot 13 = 2730$ 3 numbers from 15 down

b. For the committee: $_{15}C_3 = \dfrac{15 \cdot 14 \cdot 13}{3 \cdot 2 \cdot 1} = 455$ Permutations divided by 3!

There are 2730 different ways of choosing the president, vice president, and treasurer, and 455 ways of choosing the committee.

Practice Problem 3

A college business major can also minor in computer science by taking any 7 courses from an approved list of 10 courses. How many different collections of courses will satisfy the requirements for the computer science minor?

➤ Solution on page 371

 Graphing Calculator Exploration

Values of $_nC_r$ are easy to find if your calculator includes this command. The following screens show several values of $_nC_r$ for $n = 15$. Notice that $_nC_r$ gets larger and then smaller as r increases.

```
MATH NUM CPX PRB
1:rand
2:nPr
3:nCr
4:!
5:randInt(
6:randNorm(
7:randBin(
```

```
8 nCr 3
              56
8 nCr 4
              70
8 nCr 5
              56
```

A *standard deck* of 52 playing cards contains four *suits* (spades, hearts, diamonds, and clubs), each of which contains three *face cards* (jack, queen, and king), cards numbered 2 through 10, and an *ace*. A *hand* of cards is a selection of cards from the deck.

EXAMPLE 8

COUNTING 5-CARD HANDS

How many different 5-card hands are there?

Solution

Since the order in which the cards are dealt makes no difference to the final hand, we want the number of *combinations* of 52 cards taken 5 at a time.

$$_{52}C_5 = \frac{52 \cdot 51 \cdot 50 \cdot 49 \cdot 48}{5 \cdot 4 \cdot 3 \cdot 2 \cdot 1} = 2{,}598{,}960$$

There are 2,598,960 possible 5-card hands.

EXAMPLE 9

WHY FOUR OF A KIND BEATS A FULL HOUSE

In cards, "kind" means *value* (ace through king), so "four of a kind" means, for example, 4 aces or 4 sevens. Among all 5-card hands, how many have:

a. four of a kind?

b. three of a kind and two of a kind (called a *full house*)?

Solution

a. To get four of a kind, there are 13 ways to choose the "kind" (ace through king) and 48 ways to choose the fifth card (52 minus the 4 already chosen), giving:

$$13 \cdot 48 = 624$$

Number of hands
with *four of a kind*

b. For a full house (three of a kind and two of a kind, such as 3 kings and 2 sevens), there are 13 ways to choose the first kind, and then 3 of the 4 cards of this kind can be chosen in $_4C_3 = {}_4C_1 = \frac{4}{1} = 4$ ways. For the "kind" in the two of a kind there are 12 choices (since one kind has been "used up"), and choosing 2 of these 4 cards can be done in $_4C_2 = \frac{4 \cdot 3}{2 \cdot 1} = 6$ ways, giving:

$$13 \cdot 4 \cdot 12 \cdot 6 = 3744$$

Number of hands with a *full house*

These numbers show why four of a kind beats a full house in poker: Four of a kind is much more rare, occurring only about a sixth as often as a full house.

5.2 Section Summary

Permutations are *ordered* choices; *combinations* are choices *without regard to order*. That is, in *permutations* a different order means a *different object* (*abc* and *bca* are different), but in combinations changing the order does *not* represent a new object (*abc* and *bca* are counted as the same). If order matters (as with letters in words or listings of people for president, vice president, and treasurer), use *permutations*; if order does not matter (as with committees or hands of cards), use *combinations*.

r factors

$$_nP_r = \overbrace{n \cdot (n-1) \cdot \cdots \cdot (n-r+1)}$$

Permutations (*orderings*)

$$_nC_r = \frac{n \cdot (n-1) \cdot \cdots \cdot (n-r+1)}{r \cdot (r-1) \cdot \cdots \cdot 1} = \frac{n!}{r!(n-r)!}$$

Combinations (*collections*)

➤ Solutions to Practice Problems

1. There are $6! = 6 \cdot 5 \cdot 4 \cdot 3 \cdot 2 \cdot 1 = 720$ different orders.

2. Since the winners must be selected in order (first, second, and third place), there are $_{35}P_3 = 35 \cdot 34 \cdot 33 = 39{,}270$ different ways of choosing the winners.

3. Since the courses can be taken in any order, there are $_{10}C_7$ different ways. It is easier to calculate this as $_{10}C_3 = \frac{10 \cdot 9 \cdot 8}{3 \cdot 2 \cdot 1} = 120$ different ways to fulfill a minor in computer science.

5.2 Exercises

Calculate each factorial or quotient of factorials.

1. a. 2! **b.** 6! **2. a.** 1! **b.** 5!

3. $\dfrac{10!}{7!}$ **4.** $\dfrac{12!}{10!}$

5. $\dfrac{200!}{198!}$ **6.** $\dfrac{1000!}{998!}$

Determine each number of permutations.

7. $_6P_3$ **8.** $_6P_2$

9. $_8P_1$ **10.** $_9P_8$

11. a. $_{13}P_4$ **b.** $_{13}P_5$ **c.** $_{13}P_6$

12. a. $_{11}P_5$ **b.** $_{11}P_6$ **c.** $_{11}P_7$

13. a. $_{10}P_1$ **b.** $_{10}P_5$ **c.** $_{10}P_{10}$

14. a. $_{12}P_1$ **b.** $_{12}P_6$ **c.** $_{12}P_{12}$

Find each number of combinations.

15. $_6C_2$ **16.** $_5C_3$

17. $_7C_3$ **18.** $_8C_1$

19. a. $_{11}C_4$ **b.** $_{11}C_5$ **c.** $_{11}C_6$ **d.** $_{11}C_7$

20. a. $_{13}C_5$ **b.** $_{13}C_6$ **c.** $_{13}C_7$ **d.** $_{13}C_8$

21. a. $_{12}C_1$ **b.** $_{12}C_2$ **c.** $_{12}C_6$ **d.** $_{12}C_{11}$ **e.** $_{12}C_{12}$

22. a. $_{10}C_1$ **b.** $_{10}C_2$ **c.** $_{10}C_5$ **d.** $_{10}C_9$ **e.** $_{10}C_{10}$

APPLIED EXERCISES

23. SOCIAL SCIENCE: Election Ballots How many different ways can the eight candidates for the school board election be listed on the ballot?

24. MANAGEMENT SCIENCE: Employee Training Six candidates in an employee training program are ranked according to the time it takes them to perform a task. Assuming that all of the times are different, how many different rankings are possible?

25. BUSINESS: Commercial Art A bank lobby wall has space to display 4 paintings by local artists. If 12 paintings are available, how many different ways can they be displayed on the wall?

26. GENERAL: Postal Codes How many five-digit ZIP codes have no repeated digits?

27. GENERAL: Telephone Numbers A telephone number consists of seven digits and the first digit cannot be a zero or a one. How many telephone numbers have no repeated digits?

28. GENERAL: Horse Racing How many different ways can the 14 horses in a race finish first, second, and third?

29. GENERAL: Computer Passwords A computer password is to consist of 4 alphanumeric characters with no repeats. (An alphanumeric character is a letter from A to Z or a digit from 0 to 9.) How many such passwords are there? How many are there if the letter O and the digit 0 are excluded to avoid confusion?

30. GENERAL: Social Security Numbers How many nine-digit Social Security numbers start with three odd digits, then have two even digits, and end with four distinct digits from 0 to 9?

31. ATHLETICS: Team Sports A league has 6 college teams, and each team must host each of the others once in a season. How many games are played per season?

32. GENERAL: Books You have 2 science books, 3 language books, and 4 philosophy books. In how many ways can they be arranged on the shelf if all books of the same subject are to be together?

33. ATHLETICS: Basketball Teams A junior high girls' basketball team is to consist of 5 players. How many different teams can the manager select from a roster of 12 girls?

34. GENERAL: Novels By the end of an English class you are to have read 3 novels from a list of 21. How many choices are possible?

35. GENERAL: Playing Cards How many five-card hands from a standard deck will contain only spades? How many will contain only cards of the same suit?

36. GENERAL: Playing Cards How many five-card hands from a standard deck will contain only face cards? How many will contain no face cards?

37. MANAGEMENT SCIENCE: Labor Negotiations The union leadership and the management strike team have agreed to each select 4 representatives to enter into a "closed-room, around-the-clock" marathon in a final attempt to reach a settlement. If there are 10 members of the union leadership and 12 members of the management team, how many ways can the marathon negotiators be chosen?

38. GENERAL: Lines and Circles Ten distinct points have been marked on the circumference of a circle. How many lines can be drawn through pairs of these points?

39. GENERAL: Test Taking A test consists of 12 questions, from which each student chooses 10 to answer. How many choices are possible?

40. GENERAL: Homework Grading A sociology professor assigned 5 problems from Unit One, 10 problems from Unit Two, and 8 problems from Unit Three for next Wednesday

and announced that she will grade only 2 of the assigned problems from Unit One, 4 of those from Unit Two, and 3 from Unit Three. How many different ways can she choose the problems that she will grade?

41. SOCIAL SCIENCE: Senate Committees The U.S. Senate consists of 100 members, two from each state. A committee of 5 members is to be chosen. How many such committees are possible? How many if the committee cannot have more than one senator from the same state?

42. SOCIAL SCIENCE: Jury Verdicts In how many ways can a civil jury of 6 members split 4 to 2 on a decision?

43. ATHLETICS: Baseball Lineups A sports broadcaster, irritated that a baseball manager kept changing his lineup, once said that the manager should "choose his most talented nine players, try them in all different batting orders, and then stick with the best one." To see whether this advice makes sense answer the following questions.

a. Find the number of possible orderings for nine players (or see Example 2 on page 364).

b. If the manager were to try each batting order for just 15 minutes, holding tryouts 8 hours a day 365 days a year, how long would it take to try them all?

44. ATHLETICS: Lacrosse A women's lacrosse team plays 12 games in a season, winning 7, losing 3, and tying 2. Show that the number of ways that this can occur is $(_{12}C_7)(_5C_3) = \frac{12!}{7!3!2!}$.

Explorations and Excursions

The following problems extend and augment the material presented in the text.

The Binomial Theorem

The numbers $_nC_r$ are often called *binomial coefficients* because of their connection with the *binomial theorem*:

Binomial Theorem

$$(x + y)^n = {}_nC_0x^n + {}_nC_1x^{n-1}y + {}_nC_2x^{n-2}y^2 + \cdots + {}_nC_ny^n \qquad \text{For any positive integer } n$$

---

45. Prove the binomial theorem (above) by justifying the following equation.

$$\underbrace{(x + y)(x + y) \cdots (x + y)}_{n \text{ factors}} =$$

$$\underbrace{x^n}_{\substack{x \text{ from} \\ \text{each} \\ \text{factor}}} + \underbrace{{}_nC_1x^{n-1}y}_{\substack{y \text{ from} \\ \text{one factor,} \\ x\text{'s from} \\ \text{the others}}} + \underbrace{{}_nC_2x^{n-2}y^2}_{\substack{y\text{'s from} \\ \text{two factors,} \\ x\text{'s from} \\ \text{the other}}} + \cdots + \underbrace{y^n}_{\substack{y \text{ from} \\ \text{each} \\ \text{factor}}}$$

(*Hint:* $(x + y)(x + y) \cdots (x + y)$ is found by adding all products consisting of one term (either x or y) from each of the factors and combining similar terms: x^n comes from taking the x from each factor; $x^{n-1}y$ comes from taking y from one factor and x's from the others (explain why this can be done in ${}_nC_1$ ways); $x^{n-2}y^2$ takes y from two factors and x's from the others (explain why this can be done in ${}_nC_2$ ways); and so on. Then explain why the above equation is the same as the binomial theorem.)

46. Use the binomial theorem to expand $(x + y)^4$.

Pascal's Triangle

The coefficients of $(x + y)^n$ can also be found from *Pascal's triangle,** in which each number is the sum of the two closest numbers in the row just above it and each row begins and ends with a 1.

$\leftarrow$ Row 2 gives 1st power
$\leftarrow$ Row 3 gives 2nd power
$\leftarrow$ Row 4 gives 3rd power
$\leftarrow$ Row 5 gives 4th power

---

* Blaise Pascal, *Treatise on the Arithmetic Triangle*, (*Traité du Triangle Arithmétique*), published in 1665. "Pascal's" triangle dates back to the Chinese in about 1100, and certain parts of it to the Pythagoreans in 540 BC. For more information, see *Pascal's Arithmetical Triangle* by A.W.F. Edwards, Griffin, 1987.

47. Use the appropriate row of Pascal's triangle to expand $(x + y)^4$.

48. The connection between the binomial coefficients and Pascal's triangle is given by the formula ${}_nC_k + {}_nC_{k-1} = {}_{n+1}C_k$. Explain why this formula is the rule used to find the numbers in each row of Pascal's triangle and then show that it is correct in two ways:

   a. By algebra.

   b. By interpreting the two sides in terms of committees. [*Hint:* Think of forming committees of k people from n people plus yourself, and interpret the left-hand side as the number of committees excluding you plus the number of committees including you.]

Interpreting the Binomial Theorem in Terms of Committees

49. a. Show that evaluating the binomial theorem at $x = 1$ and $y = 1$ gives

$$2^n = {}_nC_0 + {}_nC_1 + {}_nC_2 + \cdots + {}_nC_n.$$

   b. Interpret the above equation in terms of committees. [*Hint:* To interpret the left-hand side see page 358, and to interpret the right-hand side think of committees of different sizes.]

50. a. Show that evaluating the binomial theorem at $x = 1$ and $y = -1$ gives

$$0 = {}_nC_0 - {}_nC_1 + {}_nC_2 - \cdots + (-1)^n{}_nC_n$$

   (with alternating signs).

   b. In the above equation, move the negative terms to the left-hand side and interpret the result in terms of committees with odd numbers of members compared to committees with even numbers of members.

PROBABILITY SPACES

Introduction

Some experiments always have the same outcome, while others involve "chance" or "random" effects that produce a variety of outcomes. In this section we will consider such "chance" experiments and show that, in spite of their random nature, we can draw many useful conclusions about their results. We restrict ourselves to experiments that have a finite number of possible outcomes.

Random Experiments and Sample Spaces

A *random* experiment is one that, when repeated under identical conditions, may produce different outcomes. Each repetition of the experiment is a called a *trial*, and each result is an *outcome*. The set of all possible outcomes, exactly one of which must occur, is called the *sample space* for the experiment. The following table lists some simple random experiments and sample spaces.

Random Experiment	Sample Space	
a. Flip a coin	$\{H, T\}$	H means Heads, T means Tails
b. Flip a coin twice	$\{(H, H), (H, T), (T, H), (T, T)\}$	(H, T) means Heads then Tails
c. Choose a card from a deck and observe its suit	$\{\spadesuit, \heartsuit, \diamondsuit, \clubsuit\}$	
d. Roll a die	$\{1, 2, 3, 4, 5, 6\}$	

There may be more than one possible sample space for a given random experiment. For example, in the experiment of choosing a card (experiment (c) above) the sample space could be all 52 cards (the ace of spades to the king of clubs). In general, we will choose the *simplest* sample space that allows us to observe what we are interested in (which in experiment (c) is just the suit). The only restriction is that exactly one of the outcomes in the sample space must occur whenever the experiment is performed. In general, if the possible outcomes are represented as $e_1, e_2, \ldots, e_n$, then the sample space is the *set* of these possible outcomes $\{e_1, e_2, \ldots, e_n\}$.

Events

An *event* is a *subset* of the sample space.

EXAMPLE **1**

ROLLING A DIE

For the experiment of rolling a die (experiment (d) on the previous page), represent the following events as subsets of the sample space $\{1, 2, 3, 4, 5, 6\}$:

a. Event E: *Rolling evens.*   **b.** Event F: *Rolling 5 or better.*

Solution

a. $E = \{2, 4, 6\}$ Rolling evens

b. $F = \{5, 6\}$ Rolling 5 or better

Practice Problem **1**

For the experiment of flipping a coin twice (experiment (b) on the previous page), represent the event S: *same face* as a subset of the sample space $\{(H, H), (H, T), (T, H), (T, T)\}$.   ➤ Solution on page 383

Probabilities of Possible Outcomes

Having identified the possible outcomes of an experiment, we now assign to each of them a probability from 0 (impossible) to 1 (certain). How do we assign probabilities? This depends on our knowledge of the experiment. For example, if there are three possible outcomes, and each seems equally likely, then we assign probability $\frac{1}{3}$ to each; if there are 10 outcomes that seem equally likely, then we assign probability $\frac{1}{10}$ to each. In general,

Equally Likely Outcomes

If each of the n possible outcomes in the sample space S is equally likely to occur, then we assign probability $\frac{1}{n}$ to each.

EXAMPLE **2**

ASSIGNING EQUAL PROBABILITIES

a. A coin is said to be *fair* if "heads" and "tails" are equally likely. Therefore, for a fair coin we assign probabilities $P(H) = \frac{1}{2}$ and $P(T) = \frac{1}{2}$.

b. A die is said to be *fair* if each of its six faces is equally likely. Therefore, for a fair die we assign $P(1) = P(2) = P(3) = P(4) = P(5) = P(6) = \frac{1}{6}$.

c. A card randomly drawn from a standard deck is equally likely to be any one of the 52 cards. Therefore, the probability of drawing any particular card (such as the queen of hearts) is $\frac{1}{52}$.

From now on when we speak of coins or dice, we will assume that they are fair, unless stated otherwise.

EXAMPLE **3**

ASSIGNING EQUAL PROBABILITY USING COMBINATIONS

A student club has 15 members. If a 3-member fund-raising committee is selected at random, what is the probability that the committee will consist of Bob, Sue, and Tim?

Solution

Since there are $_{15}C_3 = \frac{15 \cdot 14 \cdot 13}{3 \cdot 2 \cdot 1} = 455$ different 3-member committees that can be selected from the 15 members, and each committee is equally likely, the probability that any one particular committee is selected is $\frac{1}{455}$.

Not all outcomes are equally likely. For example, long-term weather records for an area might indicate that the probability that it will rain on any given day is $P(R) = \frac{1}{10}$ so that the probability of no rain is $P(N) = \frac{9}{10}$. Probabilities must be assigned on the basis of experience and knowledge of the basic experiment, but they are subject to two conditions: Each probability must be between 0 and 1 (inclusive), and they must add to 1.

EXAMPLE **4**

ASSIGNING UNEQUAL PROBABILITIES

The arrow of the spinner on the right can point to any one of three regions labeled A, B, and C. If the probability of pointing to a region is proportional to its area, find the probability of pointing to each of the areas A, B, and C.

Solution

Let the sample space be $S = \{A, B, C\}$, representing the events that the spinner lands in regions A, B, or C. Clearly, region A is $\frac{60°}{360°} = \frac{1}{6}$ of the circle, region B is $\frac{1}{2}$ of the circle, and region C is the remaining $\frac{1}{3}$ of the circle. Therefore

$$P(A) = \tfrac{1}{6} \qquad P(B) = \tfrac{1}{2} \qquad P(C) = \tfrac{1}{3}$$

Practice Problem **2**   Suppose that the spinner on the right is spun. What are the probabilities of the three outcomes?

> ➤ **Solution on page 383**

Probabilities of Events

We have assigned probabilities to events that consist of exactly one outcome. How do we find the probabilities of other events? We simply add up the probabilities of all of the outcomes in the event.

Probability Summation Formula

$$P(E) = \sum_{\text{All } e_i \text{ in } E} P(e_i)$$

The probability of an event is the sum of the probabilities of the possible outcomes in the event

---

Σ is the Greek capital letter sigma, which is equivalent to our capital S, and stands for "sum." From this formula, since the sample space S contains *all* of the outcomes,

$$P(S) = 1 \qquad \text{Something } must \text{ happen}$$

and, since the empty set contains *no* outcomes,

$$P(\varnothing) = 0 \qquad \text{"Nothing" cannot happen}$$

EXAMPLE 5   **FINDING THE PROBABILITY OF AN EVENT**

When rolling a pair of dice, what is the probability of rolling *doubles*?

Solution

For two dice, the sample space is

$$S = \{ \ (1,1), \quad (1,2), \quad (1,3), \quad (1,4), \quad (1,5), \quad (1,6),$$
$$(2,1), \quad (2,2), \quad (2,3), \quad (2,4), \quad (2,5), \quad (2,6),$$
$$(3,1), \quad (3,2), \quad (3,3), \quad (3,4), \quad (3,5), \quad (3,6),$$
$$(4,1), \quad (4,2), \quad (4,3), \quad (4,4), \quad (4,5), \quad (4,6),$$
$$(5,1), \quad (5,2), \quad (5,3), \quad (5,4), \quad (5,5), \quad (5,6),$$
$$(6,1), \quad (6,2), \quad (6,3), \quad (6,4), \quad (6,5), \quad (6,6) \ \}$$

Each of the 36 outcomes is equally likely, so each has probability $\frac{1}{36}$. Rolling doubles means rolling one of the outcomes $(1,1), (2,2), (3,3),$

(4, 4), (5, 5), or (6, 6), each having probability $\frac{1}{36}$, so we add $\frac{1}{36}$ six times:

$$\underbrace{\frac{1}{36} + \frac{1}{36} + \cdots + \frac{1}{36}}_{6 \text{ times}} = \frac{6}{36} = \frac{1}{6} \qquad \text{Probability of rolling doubles}$$

In the preceding example, we could instead have taken the number of "favorable" outcomes (the 6 doubles) divided by the *total* number of possible outcomes (36) to get $\frac{6}{36} = \frac{1}{6}$, the same answer as before. Although this method is sometimes stated as a basic rule of probability, it is important to remember that it holds only when the outcomes are *equally likely*.

Probability for Equally-Likely Outcomes

$$P(E) = \frac{n(E)}{n(S)} \quad \begin{array}{l} \longleftarrow \text{Number of outcomes "favorable to" (or in) } E \\ \longleftarrow \text{Total number of outcomes} \end{array}$$

Probability That an Event Does *Not* Occur

An event E is a subset of the sample space, so the event that E does *not* occur, the event E^c, is the complement of E. Since E and E^c are disjoint and contain all outcomes, we must have $P(E) + P(E^c) = 1$. Solving this equation for $P(E^c)$ gives the probability of the complement:

Complementary Probability

$$P(E^c) = 1 - P(E) \qquad \begin{array}{l} \text{The probability that an event does \textit{not} occur} \\ \text{is 1 minus the probability that it \textit{does} occur} \end{array}$$

EXAMPLE 6 **QUALITY CONTROL**

A shipment of 100 memory chips contains 4 that are defective. The company that ordered them has a policy of testing 3 of the chips and, if any are defective, rejecting the entire shipment. What is the probability that the shipment is rejected?

Solution

The shipment will be rejected if among the three chosen chips there are 1, 2, or 3 defectives. It is much easier to find the probability of the *complementary* event, that the shipment is *accepted* (the number of defectives is 0). For no defectives, all three tested chips must be selected from the 96 good ones in the shipment, and this can occur in $_{96}C_3$ ways. Since the total number of ways that the sample of three can be selected is $_{100}C_3$, and all choices are equally likely, the probability that the shipment passes inspection is

$$\frac{_{96}C_3}{_{100}C_3} = \frac{96 \cdot 95 \cdot 94}{100 \cdot 99 \cdot 98} \approx 0.884$$

Numerator and denominator are both divided by 3!, which cancels and so may be omitted

This is the probability that the shipment is *accepted*, so the probability that it is *rejected* is one minus this number: The shipment will be rejected with probability $1 - 0.884 = 0.116$, or about 12%.

Practice Problem 3

For the spinner in the Example 4, the probability of the spinner pointing to region A was $\frac{1}{6}$. What is the probability of it *not* pointing to region A?

➤ Solution on page 383

Probability Space

A sample space and an assignment of probabilities to the possible outcomes make up a *probability space*. To summarize:

Probability Space

A *probability space* is a sample space $S = \{e_1, e_2, \ldots, e_n\}$ of possible outcomes $e_1, e_2, \ldots, e_n$ together with their probabilities $P(e_1), P(e_2), \ldots, P(e_n)$ satisfying two conditions:

1. $0 \le P(e_i) \le 1$ for each outcome e_i in S

Probabilities are between 0 and 1

2. $P(e_1) + P(e_2) + \cdots + P(e_n) = 1$

Probabilities sum to 1

An *event* E is a subset of S and has probability

$$P(E) = \sum_{\text{All } e_i \text{ in } E} P(e_i)$$

Sum of the probabilities of the possible outcomes

In particular, $P(S) = 1$, $P(\varnothing) = 0$, and $P(E^c) = 1 - P(E)$.

Addition Rule for Probability

On page 354 we found that the number of elements in the union of two sets is given by the formula $n(A \cup B) = n(A) + n(B) - n(A \cap B)$, with the subtraction at the end done to avoid double-counting the elements that are in both A and B. For exactly the same reason, to find the probability of a union $A \cup B$, we add the probabilities of events A and B but then must subtract the probability of the intersection (to avoid double-counting outcomes in both events):

Addition Rule for Probability

$$P(A \cup B) = P(A) + P(B) - P(A \cap B)$$

The probability of a *union* is the sum of the probabilities minus the probability of the intersection

For disjoint events (also called *mutually exclusive events*), the intersection is empty and so has probability zero, leading to a simpler addition rule:

Addition Rule for Disjoint Events

$$P(A \cup B) = P(A) + P(B)$$

For *disjoint* events, the probability of the *union* is the sum of the probabilities

EXAMPLE **7**

PROBABILITY OF A UNION

If you roll two dice, what is the probability of rolling doubles or a sum of at least 11?

Solution

Let A be the event of rolling doubles and B the event of rolling a sum of at least 11 (that is, 11 or 12). The sample space and these events may be diagrammed as follows.

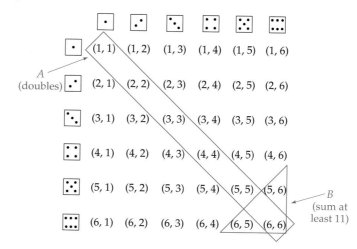

The diagram shows that the probability of doubles is $P(A) = \frac{6}{36}$, and that the probability of a sum of at least 11 is $P(B) = \frac{3}{36}$. The outcome $(6, 6)$ satisfies both conditions, so $P(A \cap B) = \frac{1}{36}$. Therefore:

$$P(A \cup B) = \underbrace{P(A)}_{6/36} + \underbrace{P(B)}_{3/36} - \underbrace{P(A \cap B)}_{1/36} \qquad \text{Addition rule}$$

$$= \frac{6}{36} + \frac{3}{36} - \frac{1}{36} = \frac{8}{36} = \frac{2}{9}$$

The probability of doubles or a sum of at least 11 is $\frac{2}{9}$, or about 22%.

Notice that the subtraction of $P(A \cap B)$ prevents the outcome $(6, 6)$ from being counted twice, once in A and once in B.

For events that are disjoint, we may use the simpler Addition Rule for Disjoint Events.

EXAMPLE **8** **FINDING THE PROBABILITY OF A DISJOINT UNION**

If you roll two dice, what is the probability of rolling a sum of 10 or 11?

Solution

Let A be the event that the sum is 10 and B the event that it is 11. Clearly the sum cannot be two different numbers at the same time, so the events A and B are *disjoint* and we may use the simpler Addition Rule for Disjoint Events. The sum is 10 for the three outcomes $(4, 6)$, $(5, 5)$, and $(6, 4)$, giving $P(A) = \frac{3}{36}$, and it is 11 for the two outcomes $(5, 6)$ and $(6, 5)$, so $P(B) = \frac{2}{36}$. Therefore:

$$P(A \cup B) = \underbrace{P(A)}_{3/36} + \underbrace{P(B)}_{2/36} = \frac{5}{36} \qquad \begin{array}{l}\text{Using the Addition Rule} \\ \text{for Disjoint Events}\end{array}$$

The probability of rolling a sum of 10 or 11 is $\frac{5}{36}$ or about 14%.

The disjoint events A: sum is 10 and B: sum is 11 are shown in the following diagram, again showing that the probability is $\frac{5}{36}$.

	$\boxed{\cdot}$	$\boxed{\cdot\,^{\cdot}}$	$\boxed{^{\cdot}\!\cdot_{\cdot}}$	$\boxed{:\;:}$	$\boxed{^{:}\!\cdot^{:}}$	$\boxed{:::}$
$\boxed{\cdot}$	$(1,1)$	$(1,2)$	$(1,3)$	$(1,4)$	$(1,5)$	$(1,6)$
$\boxed{\cdot\,^{\cdot}}$	$(2,1)$	$(2,2)$	$(2,3)$	$(2,4)$	$(2,5)$	$(2,6)$
$\boxed{^{\cdot}\!\cdot_{\cdot}}$	$(3,1)$	$(3,2)$	$(3,3)$	$(3,4)$	$(3,5)$	$(3,6)$
$\boxed{:\;:}$	$(4,1)$	$(4,2)$	$(4,3)$	$(4,4)$	$(4,5)$	$(4,6)$ ← A: sum is 10
$\boxed{^{:}\!\cdot^{:}}$	$(5,1)$	$(5,2)$	$(5,3)$	$(5,4)$	$(5,5)$	$(5,6)$ ← B: sum is 11
$\boxed{:::}$	$(6,1)$	$(6,2)$	$(6,3)$	$(6,4)$	$(6,5)$	$(6,6)$

5.3 Section Summary

For a random experiment, the *sample space* is the set S of possible outcomes. We assign to each possible outcome a probability between 0 and 1 such that the sum of the probabilities is 1. If the n possible events are equally likely, we assign each probability $\frac{1}{n}$. A *probability space* is a sample space S together with an assignment of probabilities. An *event E* is a subset of S, and the probability of the event E is

$$P(E) = \sum_{\text{All } e_i \text{ in } E} P(e_i)$$

Adding the probabilities for each outcome in the event

In particular, $P(S) = 1$, $P(\varnothing) = 0$, and $P(E^c) = 1 - P(E)$. The probability of a *union* $A \cup B$ of two events is found by the *addition rule*:

$$P(A \cup B) = P(A) + P(B) - P(A \cap B)$$

If events A and B are *mutually exclusive* or *disjoint* (that is, $A \cap B = \varnothing$), this rule simplifies to $P(A \cup B) = P(A) + P(B)$.

➤ **Solutions to Practice Problems**

1. $\{(H, H), (T, T)\}$

2. $P(A) = \frac{1}{4}$, $P(B) = \frac{1}{2}$, and $P(C) = \frac{1}{4}$

3. $P(A^c) = 1 - P(A) = 1 - \frac{1}{6} = \frac{5}{6}$

5.3 Exercises

1. **SOCIAL SCIENCE: Committees** Find the sample space for a committee of two chosen from Alice, Bill, Carol, and David. Then find the sample space if both sexes must be represented. (*Use initials.*)

2. **SOCIAL SCIENCE: Committees** Find the sample space for a committee of two formed from Alice, Bill, Carol, David, and Edgar. Then find the sample space if the committee must include at least one woman. (*Use initials.*)

3. **GENERAL: Wardrobes** Find the sample space for choosing an outfit consisting of one of four shirts $(S_1, S_2, S_3, \text{or } S_4)$ and one of two jackets $(J_1 \text{ or } J_2)$.

4. **GENERAL: Wardrobes** Find the sample space for choosing an outfit consisting of one of two shirts $(S_1 \text{ or } S_2)$, one of two jackets $(J_1 \text{ or } J_2)$, and one of two pairs of pants $(P_1 \text{ or } P_2)$.

5. **GENERAL: Marbles** A box contains three marbles, one red, one green, and one blue. A first marble is chosen, its color recorded, and then it is replaced in the box and a second marble is chosen, and its color is recorded. Find the sample space. Then find the sample space if the first marble is *not* replaced before the second is chosen.

6. **GENERAL: Coin Tossing** You toss a coin until you get the first head or until you have tossed it five times, whichever comes first. Find the sample space.

GENERAL: Dice A die is rolled, and $A = \{1, 3, 5\}$ (rolling an odd number), and $B = \{3, 6\}$ (rolling a three or a six). Specify each event as a subset of the sample space $S = \{1, 2, 3, 4, 5, 6\}$.

7. **a.** $A \cap B$ **b.** $A^c \cup B$

8. **a.** $A \cup B$ **b.** $A \cap B^c$

GENERAL: Dice Two dice are rolled. E is the event that the sum is even, F is the event of rolling at least one six, and G is the event that the sum is eight. List the outcomes for the following events.

9. **a.** $E \cap F$ **b.** $E^c \cap G$

10. **a.** $F \cap G$ **b.** $E^c \cap F \cap G$

Probabilities For the sample space $S = \{e_1, e_2, e_3, e_4, e_5\}$ with probabilities $P(e_1) = 0.10$, $P(e_2) = 0.30$, $P(e_3) = 0.40$, $P(e_4) = 0.05$, and $P(e_5) = 0.15$, and events $A = \{e_1, e_2, e_3\}$, $B = \{e_1, e_3, e_5\}$, and $C = \{e_2, e_4\}$, find each probability.

11. **a.** $P(A)$ **b.** $P(B \cup C)$

12. **a.** $P(B)$ **b.** $P(A \cup C)$

13. **a.** $P(A^c)$ **b.** $P(A^c \cap B)$

14. **a.** $P(B^c)$ **b.** $P(B^c \cap C)$

15. **GENERAL: Committees** If a committee of 3 is to be chosen at random from a class of 12 students, what is the probability of any particular committee being selected? What if the committee is to consist of a president, a vice president, and a treasurer?

16. **GENERAL: Committees** A committee of 3 is to be chosen at random from a class of 20 students. What is the probability that a particular committee will be selected? What if the 3 are to be a president, a vice president, and a treasurer?

17. **GENERAL: Marbles** One marble is selected at random from a box containing 6 red and 4 blue marbles, and its color noted, so the sample space is {R, B}. What probability should be assigned to each outcome?

GENERAL: Spinners Find the probability of the arrow landing in each numbered region. Assume that the probability of any region is proportional to its area and that areas that look the same size are the same size.

18.   **19.**

20. GENERAL: Dartboards
A circular dartboard has a radius of 12 inches with a circular bull's-eye of radius 2 inches at the center. You throw a dart and hit the dartboard. Assume that the probability of hitting any region is proportional to the area of the region.

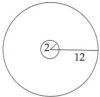

a. What is the probability of hitting the bull's-eye?

b. What is the probability of hitting the rest of the dartboard?

21. ATHLETICS: Baseball A baseball hits the 12-foot-by-10-foot wall of your house. Assume that the probability of its hitting any region is proportional to the area of the region.

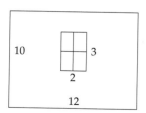

a. What is the probability that it hits the 3-foot-by-2-foot window?

b. What is the probability that it misses the window?

22. GENERAL: Choosing Colors One ball is to be chosen from a box containing red, blue, and green balls. If the probability that the chosen ball is red is $\frac{1}{5}$, and the probability that it is blue is $\frac{1}{3}$, what is the probability that it is green?

23. SOCIAL SCIENCE: Political Contributions
In a town, 38% of the citizens contributed to the Republicans, 42% contributed to the Democrats, and 12% contributed to both. What percentage contributed to neither party?

24. BUSINESS: Credit Cards A store accepts both Visa cards and Mastercards. If 61% of its customers carry Visa cards, 52% carry Mastercards, and 28% carry both, what proportion carry a card that the store will accept?

25. SOCIAL SCIENCE: Surveys A college survey claimed that 63% of students took English composition, 48% took calculus, 15% took both, and 10% took neither. Show that these figures cannot be correct.

26. GENERAL: Dice You roll a pair of dice. Find the probability of:

a. rolling a sum of 7.

b. the first die show a higher number than the second.

27. GENERAL: Marbles A box contains 4 red and 8 green marbles. You reach in and remove 3 marbles all at once. Find the probability that these 3 marbles:

a. are all red.

b. are all of the same color.

28. GENERAL: Lottery In a lottery you choose 5 numbers out of 40. Then 6 numbers are announced, and you win something if you have 5 of the 6 numbers. What is the probability that you win something?*

29. GENERAL: Guessing Numbers Someone picks a number between 1 and 10 (inclusive) and you have three guesses. What is the probability that you will get it?

---

* One strategy for such lotteries is to avoid numbers that are dates or that are likely to be chosen by others. The probability of winning is the same for any number, and you want to avoid having to share the prize with others.

30. **GENERAL: Words** Among all 5-letter nonsense words (that is, without regard to meaning), what is the probability that a word has:

    a. no vowels?
    b. at least one vowel?

31. **GENERAL: Cards** If you are dealt 5 cards at random from an ordinary deck, what is the probability that your hand contains all four aces?

32. **GENERAL: Cards** If you are dealt 5 cards at random from an ordinary deck, what is the probability of being dealt a flush (all 5 cards of the same suit)?

33. **GENERAL: Elevator Stops** An elevator has 5 people and makes 7 stops. What is the probability that no two people get off on the same floor?

34. **BUSINESS: Defective Products** A carton of 24 CD players includes 4 that are missing a part. If you choose 4 at random, what is the probability that you get those 4?

35. **BUSINESS: Defective Products** A box of 100 screws contains 10 that are defective. If you choose 10 at random, what is the probability that none are defective?

36. **SOCIAL SCIENCE: Committees** A committee of 12 is to be formed from your class of 100 students. What is the probability that you and your best friend will be on the committee?

37. **SOCIAL SCIENCE: Senate Committees** The United States Senate consists of 100 members, 2 from each state. A committee of 8 senators is formed. What is the probability that it contains at least one senator from your state?

38. **GENERAL: Light Bulbs** Your house uses fifteen light bulbs, five in each of three different wattages, and you keep one spare bulb of each wattage. Two bulbs burn out. What is the probability that you have spares for both?

39. **GENERAL: Keys** You carry six keys in your pocket, two of which are for the two locks on your front door. You lose one key. What is the probability that you can get into your house through the front door?

40. **GENERAL: Shoes** You are rushing to leave on a trip, and you randomly grab four shoes from the five pairs in your closet. What is the probability that you take at least one pair?

41. **GENERAL: Monte Hall Problem** Contestants on Monte Hall's game show "Let's Make a Deal!" were asked to choose one of three doors to win the prize behind the door. Behind one door was a car and behind the others were goats. The contestant would choose a door, say door 1. Then Monte, who knew where the car was, would open one of the doors, say door 3, that had a goat and would then ask the contestant: "Do you want to change your choice to door 2?" Is it advantageous to switch? In 1990, Marilyn vos Savant wrote about this problem in her "Ask Marilyn" column in *Parade Magazine* and said that the contestant should switch. She subsequently received more than 10,000 letters, most of them disagreeing with her answer. Was she right?

42. **ATHLETICS: Batting Averages** Suppose that one baseball player (call him Babe) has a higher batting average than another (Ty) in the first half of the season, and also has the higher batting average in the second half of the season. Can Ty then have the higher batting average over the whole season? Consider the following batting averages (hits over at-bats) for the two players and complete the table by calculating hits over at-bats for each over the whole season.

	1st Half	2nd Half	Season
Babe	$\frac{60}{200} = 0.300$	$\frac{40}{100} = 0.400$	?
Ty	$\frac{29}{100} = 0.290$	$\frac{78}{200} = 0.390$	?

Interpreting these averages as probabilities, explain how the average of two higher probabilities turns out to be lower than the average of two lower probabilities. [*Hint:* Think about whether you are really averaging the averages or taking a *weighted average*, $\frac{2}{3}$ of one and $\frac{1}{3}$ of the other, and how that can make a difference.]

Explorations and Excursions

The following problems extend and augment the material presented in the text.

Odds

If E is an event with probability $P(E)$ such that $0 < P(E) < 1$, then "the *odds for* the event E" means the ratio $P(E) : P(E^c)$ [read: "$P(E)$ to $P(E^c)$"] while "the *odds against* the event E" means the ratio $P(E^c) : P(E)$ [read: "$P(E^c)$ to $P(E)$"]. Since odds are ratios, we multiply to clear the fractions, so that, for example, $\frac{2}{3} : \frac{1}{3}$ becomes $2 : 1$. For equally likely outcomes, the odds for an event can be interpreted as the ratio of the number of *favorable* outcomes to the number of *unfavorable* outcomes.

43. Show that the odds for heads in a flip of a coin are $1 : 1$.

44. Show that the odds for a two on a roll of a die are $1 : 5$.

45. Show that if $P(E) = \frac{n}{m}$ then the odds for E are $n : (m - n)$.

46. Show that if $P(E) = \frac{n}{m}$ then the odds against E are $(m - n) : n$.

47. Show that if the odds for E are $n : m$, then $P(E) = \frac{n}{n + m}$.

48. Show that if the odds against E are $n : m$, then $P(E) = \frac{m}{n + m}$.

49. ATHLETICS: Baseball A radio announcer says the odds that the Yankees will beat the Orioles are $7 : 4$. What does the announcer believe is the probability that the Yankees will beat the Orioles?

50. GENERAL: Horse Racing A bookie has changed his odds for Win-By-A-Neck from $3 : 14$ to $5 : 24$. Does he think the probability that this horse will win has gone up or down?

51. BUSINESS: Stocks A stock exchange announces that gainers beat losers by 7 to 2. What does this mean about the probability that a stock gained value?

5.4   CONDITIONAL PROBABILITY AND INDEPENDENCE

Introduction

We often ask about the probability of one event *given another*. For example, a card player may want to know the probability of being dealt a third ace given that the first two were aces. A smoker might ask about the probability of developing cancer given that one continues to smoke. Such questions involve *conditional probability*, which is the subject of this section. Conditional probability will then lead to a discussion of independence.

Conditional Probability

When two dice (always fair unless stated otherwise) are rolled, there are $6 \cdot 6 = 36$ possible outcomes in the sample space:

$$S = \{(1, 1), (1, 2), \dots, (2, 1), (2, 2), \dots, (6, 6)\} \qquad \text{See also page 378}$$

EXAMPLE 1 **FINDING A CONDITIONAL PROBABILITY**

When rolling two dice, what is the probability that the first is a 3 given that the sum is 5?

Solution

Let B be the event that the sum is 5, so $B = \{(1, 4), (2, 3), (3, 2), (4,1)\}$.
B contains four equally likely outcomes, and the first roll is a 3 in only
one, $(3, 2)$, so the probability that the first is a 3 (event A) given that
the sum is 5 is one out of four, which we write as:

$$P(A \text{ given } B) = \frac{1}{4}$$

A: first is 3
B: sum is 5

In this example we found the conditional probability by restricting ourselves to a new *smaller* sample space corresponding to the "given"
event B. The *conditional* probability was then the probability of
A *in* B,   that is, the probability of $A \cap B$ relative to the *restricted* sample
space B. This leads to the following definition of conditional probability.

Conditional Probability

For events A and B with $P(B) > 0$,   the conditional probability
of A given B is

$$P(A \text{ given } B) = \frac{P(A \cap B)}{P(B)}$$

The probability of A given B is
the probability of *both* divided
by the probability of the *second*

We assume $P(B) > 0$ to avoid zero denominators. The conditional
probability of A given B may also be written $P(A|B)$.   If we show the
events in a Venn diagram with area representing probability, then the
conditional probability is the ratio of the intersection relative to the
"given" area.

$$P(A \text{ given } B) = \frac{P(A \cap B)}{P(B)} = \frac{\text{Area} \boxtimes}{\text{Area} \boxslash}$$

Area of the intersection
over the area of B

We may check the result of Example 1 by using the above formula:

$$P(A \cap B) = \frac{1}{36}$$

The first is 3 *and* the
sum is 5 means just $(3, 2)$

$$P(B) = \frac{4}{36} = \frac{1}{9}$$

The four outcomes were
listed in Example 1

Then, according to the formula,

$$P(A \text{ given } B) = \frac{P(A \cap B)}{P(B)} = \frac{\frac{1}{36}}{\frac{1}{9}} = \frac{9}{36} = \frac{1}{4}$$ Same answer as before

The *unconditional* probability of getting a 3 on the first roll is $\frac{1}{6}$ (all six numbers are equally likely), so conditioning on the sum being 5 changed the probability from $\frac{1}{6}$ to $\frac{1}{4}$.

EXAMPLE **2** **FINDING PROBABILITIES OF HANDS OF CARDS**

You are playing cards with a friend, and each of you has been dealt 5 cards at random. If you have no face cards, what is the probability that your friend doesn't either?

Solution

Let A be the event that your friend has no face cards, and let B be the event that you have no face cards. The event $A \cap B$ means that *neither* of you has face cards, and since the deck has 12 face cards, this means that the first 10 cards dealt were from the 40 nonface cards.

$$P(A \cap B) = \frac{{}_{40}C_{10}}{{}_{52}C_{10}} \approx 0.0536$$ ⟵ Ways of choosing 10 from the 40 nonface cards
⟵ Ways of choosing 10 from 52

— Using a calculator

$$P(B) = \frac{{}_{40}C_{5}}{{}_{52}C_{5}} \approx 0.2532$$ ⟵ Ways of choosing 5 from the 40 nonface cards
⟵ Ways of choosing 5 from 52

Therefore,

$$P(A \text{ given } B) = \frac{P(A \cap B)}{P(B)} \approx \frac{0.0536}{0.2532} \approx 0.2117$$

The probability that your friend has no face cards given that you don't have any is about 21%.

Notice that the probability of having no face cards decreases from 25% to 21% when you include the information that you have none. Can you give an intuitive reason why this information should *decrease* the probability?

   We could also solve this problem by looking directly at the restricted sample space: Given that you have 5 nonface cards, your friend's cards come from the remaining 47 cards, 12 of which are face cards and 35 of which are not. Therefore,

$$P(A \text{ given } B) = \frac{_{35}C_5}{_{47}C_5} \approx 0.2117$$

⟵ 5 from the remaining 35 nonface cards

⟵ 5 from the remaining 47

⌐ Same answer as before

We will see that this is true in general: You can solve conditional probability problems either by using the formula or by looking at the restricted sample space. You should use whichever way seems easier for that problem, but preferably use both ways to check that they agree.

The Product Rule for Probability

The conditional probability formula $P(A \text{ given } B) = \dfrac{P(A \cap B)}{P(B)}$ multiplied through by $P(B)$ gives $P(B) \cdot P(A \text{ given } B) = P(A \cap B)$. Reversing sides and interchanging A and B gives the following formula.

Product Rule for Probability

$$P(A \cap B) = P(A) \cdot P(B \text{ given } A)$$

The probability of $A \cap B$ is the probability of A times the probability of B *given* A

This rule is very useful when a conditional probability is known.

EXAMPLE 3

USING THE PRODUCT RULE

You need to be somewhere in half an hour, and your friend has borrowed your car. If your friend arrives soon (you give this a 50–50 chance), the probability that you will be on time is 90%. Otherwise, you will have to walk, with only a 60% chance of being on time. What is the probability that you arrive in your car and on time?

Solution

Much depends on whether your friend arrives soon, so that will be the "given" event, A: *Your friend arrives soon.* Then B is the other event, B: *You are on time.* The 90% is the probability of being on time *if* (or *given that*) your friend arrives soon, so $P(B \text{ given } A) = 0.90$. We are asked for the probability that you arrive in your car and on time, meaning that *both* event A and B occur. Therefore, by the Product Rule for Probabilities:

$$P(A \cap B) = P(A) \cdot P(B \text{ given } A)$$
$$= 0.5 \cdot 0.9 = 0.45$$

A: Friend arrives soon
B: You are on time

$\underbrace{}\ \underbrace{}$
$P(A)\ P(B \text{ given } A)$

The probability that you arrive on time and in your car is 45%.

Suppose that instead we were asked for the probability of *arriving on time* (either by car or on foot). We will use a *tree diagram* to represent the different ways, similar to our use of tree diagrams to find different outcomes in Section 5.1.

EXAMPLE 4

USING A TREE DIAGRAM

For the situation in Example 3, find the probability that you arrive on time.

Solution

Again, much depends upon your friend's arrival, so we first "branch" on whether your friend arrives soon (event A, having probability 0.5) or not (event A^c, also having complementary probability 0.5). The second-stage branching represents whether you arrive on time (event B) or not (B^c), with the given (conditional) probabilities depending on the first-stage branching. Be sure you understand the correct placement of the probabilities given in the problem, 0.90 and 0.60, along with their complements 0.10 and 0.40. The probabilities at the end of *two* branchings are the *products* of the probabilities along those branches.

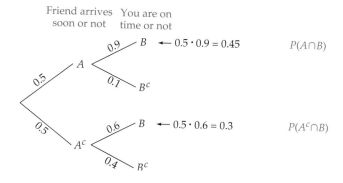

Arriving on time (by car or on foot) corresponds to the two branches ending in *B*. We multiply out those probabilities and add the results, $0.45 + 0.30 = 0.75$. Answer:

You have a 75% chance of arriving on time.

Notice that the uppermost branches give the 0.45 that we found in Example 3. In general, a tree diagram gives all possible probabilities that can be found from the given information.

Practice Problem 1

Use the tree diagram in Example 4 to answer the following questions.

a. What is the probability of arriving late and in your car?

b. What is the probability of arriving late?   ➤ Solution on page 397

Notice several things about such tree diagrams:

- The branches from any point must have probabilities that add to 1.
- The probabilities at the right-hand end of a branch are found by multiplying along the branch.
- The probabilities written on the second-stage branches are *conditional* probabilities, so we are indeed using the Product Rule for Probability.
- There can be any number of branches per stage, and any number of stages.
- There is no need to multiply out all of the products, only the ones you need.

EXAMPLE 5

SAMPLING WITHOUT REPLACEMENT

A box contains three blue marbles, two green marbles, and one red marble. A marble is randomly chosen (and not replaced) and then a second marble is chosen. What is the probability that the second marble is green?

Solution

We branch on the color of the first marble, using *B*, *G*, and *R* for the events that it is blue, green, or red. Of the six marbles, 3 are blue, 2 are green, and 1 is red, so the probabilities on the first-stage branching are $\frac{3}{6} = \frac{1}{2}$, $\frac{2}{6} = \frac{1}{3}$, and $\frac{1}{6}$ as shown in the following diagram. For the second-stage branching, if a blue is chosen first, then the remaining 5 are 2 blue, 2 green, and 1 red, leading to the probabilities $\frac{2}{5}, \frac{2}{5}, \frac{1}{5}$ for the top part of the second stage choice. The other second-stage probabilities are calculated similarly, and depend on the color that was removed in the first stage.

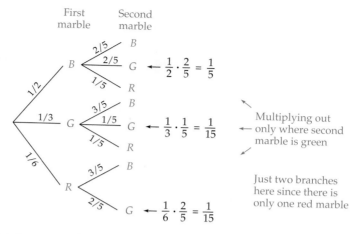

First marble / Second marble

$\frac{1}{2} \cdot \frac{2}{5} = \frac{1}{5}$

Multiplying out only where second marble is green

$\frac{1}{3} \cdot \frac{1}{5} = \frac{1}{15}$

Just two branches here since there is only one red marble

$\frac{1}{6} \cdot \frac{2}{5} = \frac{1}{15}$

Multiplying out and adding up the branches that end in green:

The probability that the second marble is green is $\frac{1}{5} + \frac{1}{15} + \frac{1}{15} = \frac{5}{15} = \frac{1}{3}$.

Practice Problem 2

Use the diagram in Example 5 to find the probability that the second marble is blue.

➤ Solution on page 397

Independent Events

Roughly speaking, two events are said to be *independent* if one has nothing to do with the other, so that the occurrence of one has no bearing on the occurrence of the other. In terms of conditional probability, this means that for independent events A and B, we have $P(A$ given $B) = P(A)$. Using the definition of conditional probability, this equation becomes

$$\frac{P(A \cap B)}{P(B)} = P(A) \qquad \text{For } P(B) \neq 0$$

Multiplying each side by $P(B)$ gives the following equivalent condition, which we take as the *definition* of independence.

Independent Events

Events A and B are *independent* if

$$P(A \cap B) = P(A) \cdot P(B)$$

The probability of the intersection is the *product* of the probabilities

Events that are not independent are *dependent*.

Be careful! Independent does not mean the same thing as *disjoint*, or *mutually exclusive*. If events A and B have positive probabilities and are disjoint, then $P(A \cap B) = 0$ but $P(A) \cdot P(B)$ is positive, so they cannot be independent. Intuitively, disjointness is a very strong kind of dependence: The occurrence of one of two disjoint events guarantees the *non*occurrence of the other, so they cannot be independent.

The definition above gives a simple test for the independence of events A and B: Find $P(A \cap B)$ and find the product $P(A) \cdot P(B)$. If the results are *equal*, then A and B are *independent*; otherwise, A and B are *dependent*.

EXAMPLE 6

INDEPENDENT COIN TOSSES

A coin is tossed twice. Let A be the event that the first toss is heads, and let B be the event that the second toss is heads. Are the events A and B independent?

Solution

The sample space consists of four equally likely events:

$$S = \{(H, H), (H, T), (T, H), (T, T)\}$$

so

$$A = \{(H, H), (H, T)\}, \quad B = \{(H, H), (T, H)\}, \quad A \cap B = \{(H, H)\}$$

$$P(A) = \frac{2}{4} = \frac{1}{2}, \quad P(B) = \frac{2}{4} = \frac{1}{2}, \quad P(A \cap B) = \frac{1}{4}$$

Since $P(A) \cdot P(B) = \frac{1}{2} \cdot \frac{1}{2} = \frac{1}{4}$ and $P(A \cap B) = \frac{1}{4}$ are equal, the events A and B *are* independent.

If $P(A \cap B)$ and $P(A) \cdot P(B)$ had *not* been equal, the events would have been *dependent*. The fact that successive coin tosses are independent should come as no surprise, and this is sometimes expressed by saying that the coin has "no memory." Thus, each toss is a new experiment with "no influence from the past."

The concept of independence can be extended to more than two events.

Many Independent Events

A collection $E_1, E_2, \ldots, E_m$ of events is *independent* if any subcollection of them satisfies the multiplication formula:

$$P(E_i \cap E_j \cap \ldots \cap E_k) = P(E_i) \cdot P(E_j) \cdot \cdots \cdot P(E_k)$$

The probability of the intersection is the *product* of the probabilities

In practice, the most important uses of independence do not involve *proving* it but *assuming* it. For example, just as successive coin tosses are independent of each other, different rolls of dice are independent of each other, so probabilities involving successive rolls can be found by multiplying together the probabilities for each roll. Using this idea, we can explain why the Chevalier de Meré won his first bet and lost his second.

EXAMPLE 7

WHY THE CHEVALIER DE MERÉ WON HIS FIRST BET

Recall from the Application Preview on page 351 that the Chevalier's first bet was that he could roll at least one six in four rolls of a die. Find the probability that he won this bet.

Solution

Instead of *at least one six* in four rolls it is easier to work with the complementary event, *no* sixes in four rolls, and subtract its probability from 1. Rolling a 6 has probability $\frac{1}{6}$, so *not* getting a six has probability $\frac{5}{6}$, and the probability that this will occur on four successive rolls is $\frac{5}{6} \cdot \frac{5}{6} \cdot \frac{5}{6} \cdot \frac{5}{6} = \left(\frac{5}{6}\right)^4$ (using independence). Subtracting this from 1 gives the probability of at least one six in four rolls:

$$1 - \left(\frac{5}{6}\right)^4 \approx 1 - 0.482 = 0.518$$

Therefore, he won his first bet about 52% of the time.

EXAMPLE 8

WHY THE CHEVALIER LOST HIS SECOND BET

The Chevalier's second bet was that he could roll at least one double-six in 24 rolls of two dice. Find the probability that he won this bet.

Solution

Again we begin with the complementary event. On one roll of two dice the probability of getting a double-six is $\frac{1}{36}$, so the probability of *not* getting a double-six is $\frac{35}{36}$, and the probability of this occurring on 24 successive tosses is $\left(\frac{35}{36}\right)^{24}$. Subtracting this from 1 gives the probability of at least one double-six in 24 rolls:

$$1 - \left(\frac{35}{36}\right)^{24} \approx 1 - 0.509 = 0.491$$

Therefore, he won his second bet only about 49% of the time.

Comparing the Chevalier's two bets, while both have probabilities close to $\frac{1}{2}$, he was right that in the long run the first is a winning bet and the second is a losing bet.

EXAMPLE 9

QUALITY CONTROL

A computer consists of a logic board, three memory chips, a screen, and an I/O (input-output) module. If these individual components work with probabilities 0.98, 0.99 (for each memory chip), 0.96, and 0.97 respectively, what percentage of the computers should be expected to work?

Solution

Assuming independence, we multiply the probabilities:

$$0.98 \cdot 0.99^3 \cdot 0.96 \cdot 0.97 \approx 0.885$$

About 89% of the assembled computers should be expected to work.

5.4 Section Summary

For events A and B, the conditional probability of A given B is

$$P(A \text{ given } B) = \frac{P(A \cap B)}{P(B)} \qquad \text{For } P(B) > 0$$

$P(A$ given $B)$ can also be found by restricting $P(A)$ to the sample space where B occurs. The above formula can be rewritten as:

$$P(A \cap B) = P(A) \cdot P(B \text{ given } A) \qquad \text{Product Rule for Probability}$$

This formula is useful in calculating probabilities along branches in *tree diagrams*.

Two events A and B are independent if

$$P(A \cap B) = P(A) \cdot P(B) \qquad \text{Probabilities multiply}$$

Be careful! When do you *add* probabilities and when do you *multiply* them? You *add* when the events are *disjoint* and you want *one or the other* to occur (subtracting the intersection if the events are not disjoint); you *multiply* if the events are *independent* and you want *both* to occur (using conditional probability if the events are not independent).

➤ **Solutions to Practice Problems**

1. **a.** From the branches corresponding to A and B^c, $0.50 \cdot 0.10 = 0.05$.

   b. From the branches ending in B^c, $0.05 + 0.50 \cdot 0.40 = 0.25$. [We could also have found this as 1 minus the probability of being on time, which we found in Example 4.]

2. Adding up the products along branches ending in blue:

$$\tfrac{1}{2} \cdot \tfrac{2}{5} + \tfrac{1}{3} \cdot \tfrac{3}{5} + \tfrac{1}{6} \cdot \tfrac{3}{5} = \tfrac{1}{2}.$$

5.4 Exercises

For Exercises 1–4, use the given values to find:
   a. $P(A \text{ given } B)$ **b.** $P(B \text{ given } A)$

1. $P(A) = 0.6$, $P(B) = 0.4$, $P(A \cap B) = 0.2$

2. $P(A) = 0.5$, $P(B) = 0.3$, $P(A \cap B) = 0.1$

3. $P(A) = 0.4$, $P(B) = 0.5$, $P(A \cup B) = 0.6$

4. $P(A) = 0.6$, $P(B) = 0.5$, $P(A \cup B) = 0.8$

5. **GENERAL: Marbles** A box contains 4 white, 2 red, and 4 black marbles. One marble is chosen at random, and it is not black. Find the probability that it is white.

6. **GENERAL: Cards** You select two cards at random from an ordinary deck. If the first card is a spade, what is the probability that the second card is a spade?

7. **BIOMEDICAL: Gender** Your friend has two children, and you know that at least one is a girl. What is the probability that both are girls? (Assume that girls and boys are equally likely.)

8. **BIOMEDICAL: Eye Color** Suppose that each of two children in a family has probability $\frac{1}{5}$ of having blue eyes, independently of each other. If at least one child has blue eyes, what is the probability that both have blue eyes?

9. **GENERAL: Cards** A deck contains three cards: One is red on both sides, one is blue on both sides, and the third is red on one side and blue on the other. One card is chosen at random from the deck, and the color on one side

is observed. If this side is blue, what is the probability that the other side is blue?*

10. **BUSINESS: Credit Cards** Looking back at Exercise 24 on page 385, if a customer has at least one credit card, what is the probability that the customer has a Visa card?

11. **SOCIAL SCIENCE: Political Contributions** Looking back at Exercise 23 on page 385, if a person is a contributor, what is the probability that the person contributes to the Republican party?

12. **GENERAL: Cards** In the game of bridge, each of four players is dealt 13 cards. If a certain player has no aces, find the probability that that person's partner has:

    a. no aces   **b.** at least two aces

13. **GENERAL: Choosing Courses** You will take either a basket weaving course or a philosophy course, depending on what your advisor decides. You estimate that the probability of your getting an A in basket weaving is 0.95, while in philosophy it is 0.70. However, the

---

* Many people mistakenly believe that if one side is blue, then the probability that the other side is blue is $\frac{1}{2}$ since it can only be the blue–blue card or the blue–red card, and these are equally likely. The error comes from not realizing that the blue side is equally likely to be any one of the *three* blue sides. More than one probability student has made money by knowing the correct conditional probabilities and offering bets on the outcome.

chances of your advisor choosing the basket weaving course is only 20%, while there is an 80% chance of his putting you in the philosophy course. What is the probability of your ending up with an A?

14. **GENERAL: Multiple-Choice Tests** On a multiple-choice test you know the answers to 70% of the questions (and so get them right), and for the remaining 30% you choose randomly among the 5 answers. What percent of the answers should you expect to get right?

15. **GENERAL: Driving** Suppose that 70% of drivers are "careful" and 30% are "reckless." Suppose further that a careful driver has a 0.1 probability of being in an accident in a given year, while for a reckless driver the probability is 0.3. What is the probability that a randomly selected driver will have an accident within a year?

16. **MANAGEMENT SCIENCE: Quality Control** A computer manufacturer has assembly plants in three states. The Delaware plant produces 25% of the company's computers, the Michigan plant produces 35%, and the California plant produces the other 40%. The probabilities that a computer will pass inspection are 93% for the Delaware plant, 89% for the Michigan plant, and 94% for the California plant. What is the probability that a randomly selected computer from this company will pass inspection?

Independence For the experiment of tossing a coin twice, find whether events A and B are independent or dependent.

17. A: heads on the first toss
    B: different results on the two tosses

18. A: heads on the second toss
    B: the same results on both tosses

Independence For the experiment of rolling two dice, find whether events A and B are independent or dependent.

19. A: odd number on the first roll
    B: sum of the numbers is 4

20. A: even number on the first roll
    B: sum of the numbers is 10

21. **GENERAL: Dice** A pair of dice is rolled three times in succession. Find the probability that each of the rolls has a sum of seven.

22. **BUSINESS: Brand Loyalty** Suppose that each time that you buy a car, you choose between Ford and General Motors. Suppose that each time after the first, you stay with the same company with probability $\frac{2}{3}$ and switch with probability $\frac{1}{3}$. If you are equally likely to choose either company for your first car, what is the probability that your first and second choices will be Ford cars and your third and fourth choices will be General Motors cars?

23. **GENERAL: Class Attendance** Two students are registered for the same class and attend independently of each other, student A 90% of the time and student B 70% of the time. The teacher remembers that on a given day, at least one of them is in class. What is the probability that student A was in class that day?

24. **GENERAL: Class Attendance** For the students in Exercise 23, what is the probability that on any day:
    a. both will be in class?
    b. neither will be in class?
    c. at least one will be in class?

25. **GENERAL: Dice** Three dice are rolled. Find the probability of getting:
    a. all sixes
    b. all the same outcomes
    c. all different outcomes

26. **GENERAL: Fair Toss from an Unfair Coin** How can you make a fair game (one with probability $\frac{1}{2}$ of winning) from tossing an unfair coin? If a coin has probability p of heads (with p not necessarily $\frac{1}{2}$), show that for two tosses the events HT and TH have the same probability. Therefore, toss the coin twice: If you get HT call the result a "win," if you get TH call the result a "lose," and if you get HH or TT ignore those tosses and toss it twice more, repeating the procedure until you get HT or TH. "Win" and "lose" are equally likely, so each has probability $\frac{1}{2}$ of occurring first.

27. GENERAL: Dice When rolling two dice, find the probability of getting *a sum of 4 before a sum of 7*. [*Hint:* Let p be the probability of rolling a sum of 4 before a sum of 7. This event can occur in either of two ways: You roll the 4 on your first try, or your first try results in *neither* a 4 nor a 7, in which case you are back where you started and must subsequently roll a 4 before a 7 (again with probability p). This leads to the formula

$$p = P(\text{sum of 4}) + P(\text{sum of neither 4 nor 7}) \cdot p.$$

Find the two stated probabilities and then solve for p.]

28. GENERAL: Dice When rolling two dice, find the probability of rolling *a sum of 7 before a sum of 4*. [*Hint:* Modify the method outlined in Exercise 27.]

29. BIOMEDICAL: Longevity Based on *Life Tables*, in a population of 100,000 females, 89.8% will live to age 60 and 57.1% will live to age 80. Given that a woman is 60, what is the probability that she will live to be 80?

30. BIOMEDICAL: Longevity Based on *Life Tables*, in a population of 100,000 males, 81.5% will live to age 60 and 36.8% will live to age 80. Given that a man is 60, what is the probability that he will live to be 80?

31. MANAGEMENT SCIENCE: Reliability Travel between Seattle and Tacoma is sometimes impossible because of snow. The numbers p and q on possible road maps below are the probabilities that the roads are passable.

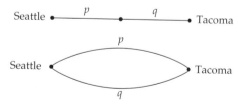

(In the language of *systems reliability* or *circuits*, the roads in the upper diagram are connected in *series* while in the lower diagram they are connected in *parallel*.)

a. For the first road map, what is the probability of being able to drive from Seattle to Tacoma?

b. For the second road map (with two possible routes), what is the probability of being able to drive from Seattle to Tacoma?

c. Evaluate your answers to parts (a) and (b) for $p = q = 0.5$ and then evaluate them for $p = q = 0.9$. Which is the more reliable highway design?

32. MANAGEMENT SCIENCE: Reliability (*continued*) Suppose that the parallel connection below has an unknown number of components (roadways), each operable (passable) with probability 80%. How many components are needed for the overall reliability to be at least 99.9%?

33. GENERAL: Shared Birthdays Assume that a person's birthday is equally likely to be on any of the 365 days of the year independently of every other person.

a. Show that the probability that a group of n people will all have different birthdays is

$$\frac{364}{365} \cdot \frac{363}{365} \cdot \frac{362}{365} \cdot \ldots \cdot \frac{366 - n}{365}$$

b. Successively multiply these fractions together until you reach a number below 0.50, and thus find the first value of n for which this occurs.

c. Explain why this value of n has the following interpretation: For a group of this size, there is a greater than 50% probability that at least two people will share the same birthday.

Explorations and Excursions

The following problems extend and augment the material presented in the text.

More About Independence

34. If event A is such that $P(A) = 0$, and B is any other event, show that events A and B are independent.

35. If events A and B are independent, show that events A and B^c are independent.

36. If events A and B are independent, show that events A^c and B^c are independent. [*Hint:* Use the result of the previous exercise.]

37. In the experiment of tossing two coins, define events as follows:

> A: heads on first toss
>
> B: heads on second coin
>
> C: outcomes agree

Show that A, B, and C are *pairwise* independent but do *not* satisfy

$$P(A \cap B \cap C) = P(A) \cdot P(B) \cdot P(C)$$

and so are not independent.

38. To see why the events in the preceding exercise are not independent, calculate $P(C)$ and also $P(C \text{ given } (A \cap B))$.

5.5 BAYES' FORMULA

Introduction

Sometimes we want to *reverse the order* in conditional probability. For example, the *probability of developing cancer given that one smokes* is different from the *probability of being a smoker given that one develops cancer*. The first might be of interest to a smoker and the second to a medical researcher. In this section we will develop *Bayes' formula** to reverse the order in conditional probability.

Bayes' Formula

In general, $P(A \text{ given } B)$ will not be the same as $P(B \text{ given } A)$.

Practice Problem 1

Explain the difference between

> $P(\text{you are rich given that you win the lottery})$

and

> $P(\text{you win the lottery given that you are rich})$

➤ Solution on page 404

Suppose that the sample space S is divided into two parts, S_1 and S_2, that are disjoint and whose union is the whole sample space $S_1 \cup S_2 = S$. These sets then partition any event A into two parts, the part in S_1 and

---

* Discovered by Thomas Bayes (1702–1761), an English nonconformist minister. Bayes was an amateur mathematician and a defender of the new subject of calculus when it was attacked as illogical and incorrect. His mathematical writings were not published during his lifetime; the one containing "Bayes' formula" was not published until three years after his death.

the part in S_2, as shown in the diagram on the left. The probability of A can then be found by adding the probabilities of the two parts:

$$P(A) = P(S_1 \cap A) + P(S_2 \cap A)$$

The definition of conditional probability, $P(S_1 \text{ given } A) = \dfrac{P(S_1 \cap A)}{P(A)}$ with the denominator replaced by the above expression gives:

$$P(S_1 \text{ given } A) = \frac{P(S_1 \cap A)}{P(S_1 \cap A) + P(S_2 \cap A)}$$

Using the Product Rule for Probability (page 390) in the numerator and twice in the denominator, we may write this in the following form:

$$P(S_1 \text{ given } A) = \frac{P(S_1) \cdot P(A \text{ given } S_1)}{P(S_1) \cdot P(A \text{ given } S_1) + P(S_2) \cdot P(A \text{ given } S_2)}$$

This formula generalizes by replacing S_1 and S_2 by *any* sets $S_1, S_2, \ldots, S_n$ that are pairwise disjoint and whose union is the entire sample space, $S_1 \cup S_2 \cup \cdots \cup S_n = S$:

Bayes' Formula

$$P(S_1 \text{ given } A) = \frac{P(S_1) \cdot P(A \text{ given } S_1)}{P(S_1) \cdot P(A \text{ given } S_1) + \cdots + P(S_n) \cdot P(A \text{ given } S_n)}$$

The sum in the denominator has n terms, one for each of the sets $S_1, S_2, \ldots, S_n$. Notice that the order in the conditional probability is reversed: S_1 *given* A on the left and A *given* S_1 on the right. Bayes' formula in this form is needed in only the most complicated cases. In fact we will use tree diagrams to carry out the calculations in Bayes' formula.

EXAMPLE 1

MEDICAL SCREENING

Suppose that a medical test for a disease is 90% accurate for both those who have the disease and for those who don't. Suppose, furthermore, that only 5% of the population has this disease. Given that a person tests positive for the disease (meaning that the test says that the person has the disease), what is the probability that the person actually has the disease? Would this test be useful for widespread medical screening?

Solution

The probabilities depend upon whether a person has the disease (event D, with probability 0.05) or not (event D^c, with probability

0.95), which represents the first-stage branching. The second-stage branching is according to whether the person tests positive (T) or not (T^c), with probabilities based on the test being 90% accurate.

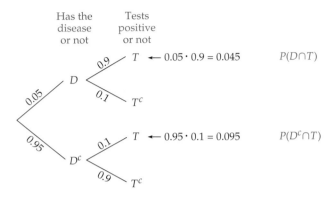

The probability that a randomly selected person has the disease given that the person tests positive is $P(D$ given $T)$, which, by the definition of conditional probability, is

$$P(D \text{ given } T) = \frac{P(D \cap T)}{P(T)}$$

For the numerator we take the branch though D and T (with probability 0.045) and for the denominator we add the two branches ending in T:

$$P(D \text{ given } T) = \frac{0.045}{0.045 + 0.095} = \frac{0.045}{0.14} \approx 0.32$$

Therefore, the probability that a person actually has the disease given that the person tests positive for it is only about 32%. In other words, 68% of those testing positive in fact do not have the disease and are being needlessly worried by a false diagnosis, so this test should *not* be used for widespread medical screening.

EXAMPLE **2** **ASSESSING VOTING PATTERNS**

Registered voters in Marin County are 45% Democratic, 30% Republican, and 25% Independent. In the last election for county supervisor, 70% of the Democrats voted, as did 80% of the Republicans and 90% of the Independents. What is the probability that a randomly selected voter in this election was a Democrat?

Solution

Since the information is stated in terms of the political parties, they represent the first-stage branching (D, R, or I), with the second-stage represented by voting or not voting (V or V^c). The probabilities come from the given values together with their complements.

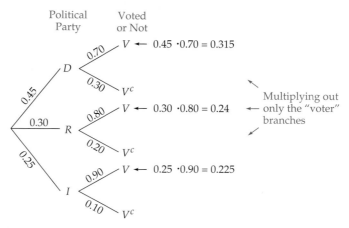

The probability that a randomly selected voter was a Democrat is $P(D$ given $V)$, which is

$$P(D \text{ given } V) = \frac{P(D \cap V)}{P(V)} \qquad \text{Using the definition of conditional probability}$$

For the numerator $P(D \cap V)$ we take the branch through D and V (the uppermost branch), and for the denominator $P(V)$, we take the sum of all of the V branches:

$$\frac{0.24}{0.315 + 0.24 + 0.225} = \frac{0.315}{0.780} \approx 0.404$$

The probability that a voter in the election was a Democrat is about 40%.

Practice Problem **2**

Using the diagram in Example 2, what is the probability that a randomly selected voter in the election was an Independent?

➤ Solution on page 404

Notice that as required for Bayes' formula, the three parties, Democrat, Republican, and Independent, do not overlap (the sets are *disjoint*) and include all voters (their probabilities *add to 1*). If the probabilities had not added to 1, we would have had to add in another group ("Others") with the remaining probability.

5.5 Section Summary

For pairwise disjoint sets $S_1, S_2, \ldots, S_n$ whose union is the entire sample space $S_1 \cup S_2 \cup \cdots \cup S_n = S$, Bayes' formula reverses the order in conditional probability:

$$P(S_1 \text{ given } A) = \frac{P(S_1) \cdot P(A \text{ given } S_1)}{P(S_1) \cdot P(A \text{ given } S_1) + \cdots + P(S_n) \cdot P(A \text{ given } S_n)}$$

In practice, we carry out this calculation using a tree diagram, where the numerator is the product along one branch and the denominator is the sum of several branches, using just the definition of conditional probability.

➤ **Solutions to Practice Problems**

1. P (you are rich given that you win the lottery) should be very high since you certainly will be rich if you win the lottery.

   P (you win the lottery given that you are rich) should be very small since your chances of winning the lottery are very small regardless of how much money you have.

2. Using the notation from the solution to Example 2, we have

   $$P(I \text{ given } V) = \frac{P(I \cap V)}{P(V)} = \frac{0.225}{0.315 + .024 + 0.225} = \frac{0.225}{0.780} \approx 0.288$$

   The probability that a voter in the last election was an Independent is about 29%.

5.5 Exercises

1. **SOCIAL SCIENCE: Voting** In a town, 60% of the citizens are Republicans and 40% are Democrats. In the last election 55% of the Republicans voted and 65% of the Democrats voted. If a voter is randomly selected, what is the probability that the person is a Republican?

2. **BIOMEDICAL: Colorblindness** An estimated 8% of men and 0.5% of women are colorblind. If a colorblind person is selected at random, what is the probability that the person is a man? (Assume that men and women occur in equal numbers.)

3. **BIOMEDICAL: Medical Testing** A new test is developed to test for a certain disease, giving "positive" or "negative" results to indicate that the person does or does not have the disease. For a person who actually *has* the disease, the test will give a positive result with probability 0.95 and a negative result with probability 0.05 (a so-called "false negative"). For a person who does *not* have the disease, the test will give a positive result with probability 0.05 (a "false positive") and a negative result with probability 0.95. Furthermore, only

one person in 1000 actually has this disease. If a randomly selected person is given the test and tests positive, what is the probability that the person actually has the disease?

4. **BIOMEDICAL: Home Pregnancy Testing** Suppose that 400 pregnant women take a home pregnancy test, and 397 of them test "positive" and the other 3 test "negative." Suppose also that 200 nonpregnant women take the test, and 184 of them test "negative" and the remaining 16 test "positive." What is the probability that a woman who tests positive is actually pregnant?

5. **MANAGEMENT SCIENCE: Manufacturing Defects** A computer chip factory has three machines, A, B, and C, for producing the memory chips. Machine A produces 50% of the factory's chips, machine B produces 30%, and machine C produces 20%. It is known that 3% of the chips produced by machine A are defective, as are 2% of chips produced by machine B and 1% of the chips from machine C. If a randomly selected chip from the factory's output is found to be defective, what is the probability that it was produced by machine B?

6. **MANAGEMENT SCIENCE: Manufacturing Defects** For the information in Exercise 5, if a randomly selected chip is found *not* to be defective, what is the probability that it came from machine B?

7. **SOCIAL SCIENCE: Airline Hijacking** The "hijacker profile" developed by the Federal Aviation Administration fits 90% of hijackers and only 0.05% of legitimate passengers. Based on historical data, assume that only 30 of 300 million passengers are hijackers. What is the probability that a person who fits the profile is actually a hijacker?

8. **SOCIAL SCIENCE: Polygraph Tests** Although seldom admissible in courts, polygraphs (sometimes called "lie detectors") are used by many businesses and branches of government. The Office of Technology Assessment has estimated that polygraphs have a "false negative" rate of 0.11 and a "false positive" rate of 0.20. Sup-

pose that you are trying to find a single thief in your company of 100 employees. What is the probability that a person who fails a polygraph test is *not* the thief?

9. **BIOMEDICAL: Random Drug Testing** In 1986 the Reagan administration issued an executive order allowing agency heads to subject all employees to urine tests for drugs. Suppose that the test is 95% accurate both in identifying drug users and in clearing nonusers. Suppose also that 1% of employees use drugs. What is the probability that a person who tests positive is *not* a drug user?

10. **GENERAL: Grades** A probability class consists of 5 math majors, 3 science majors, and 2 whose major is "undecided." If the probability of earning an A is 90% for math majors, 85% for science majors, and 70% for "undecideds," what is the probability that a randomly selected A student is a math major?

11. **GENERAL: Grades** For the information in Exercise 10, what is the probability that a randomly selected A student is "undecided"?

12. **ATHLETICS: Olympic Athletes** Suppose that 10% of olympic athletes use steroids. A blood test is 95% correct in identifying drug users and has a 2% rate of false positives (that is, of *incorrectly* indicating steroid use). What is the probability that a champion cycler who tests positive uses steroids?

13. **BIOMEDICAL: Mammograms** Many doctors recommend that women in their forties have annual mammograms to detect breast cancer. Approximately 2% of women in their forties will develop breast cancer during that decade, and the early mammograms (no longer used) for that age group had a 30% rate of false positives and a 25% rate of false negatives. What is the probability that a woman who tested positive on this test had breast cancer?

14. **SOCIAL SCIENCE: Unemployment** A city has five districts, and 30% of its citizens live in district I, 25% in district II, 20% in district III, 15% in district IV, and 10% in district V. The unemployment rate in district I is 3%, while in

districts II, III, IV, and V it is 4%, 4%, 5%, and 6%, respectively. The local newspaper interviews a randomly selected unemployed resident. What is the probability that this resident lives in district I?

15. SOCIAL SCIENCE: Unemployment For the information in Exercise 14, what is the probability that this resident lives in district V?

16. GENERAL: Automotive Advice Your automobile mechanic says "Ninety percent of new cars I see in here for repairs were made in Detroit—practically none from Nashville. When you buy your next car, make sure it was made in Nashville." Is this good advice? Suppose that 10% of all new cars (whether from Detroit or Nashville) need repairs during their first year, so that all new cars are, in fact, equally good. Suppose also that 90% of all cars are manufactured in Detroit with the other 10% in Nashville. Show that your mechanic's observation was right, but his advice was not.

Explorations and Excursions

The following problem extends and augments the material presented in the text.

How to Ask an Embarrassing Question

17. Suppose that you want to determine the proportion of people who have a certain medical condition that they may not want to disclose if asked directly. Here is a way to find out while maintaining everyone's confidentiality: Each person in the survey is given a coin, and in a private room they toss the coin. If it comes up heads, they answer truthfully about the medical condition (Yes or No, written on a piece of paper), and if it is tails, they toss the coin again and write Yes or No depending on whether the second toss is heads or tails. Each person's written Yes or No discloses nothing about them since it is equally likely to be truthful or random. Given the resulting proportion $P(Y)$ of Yes answers, show how to find the actual proportion p of people who have the medical condition as follows.

a. Use a probability tree to find a formula for $P(Y)$ in terms of p.
b. Solve the resulting equation for p in terms of $P(Y)$.
c. If 62% of answers were Yes, what proportion of people actually have the condition?

RANDOM VARIABLES AND DISTRIBUTIONS

Introduction

Often a random experiment results in a *number*, such as the sum on the faces of two dice or your winnings in a lottery. Such numerical quantities that depend on chance events are called *random variables* and are the central objects of probability and statistics. In this section (and also in the following chapter) we will introduce some of the random variables that have proved most useful in applications.

Random Variables

A random variable is an assignment of a number to each outcome in the sample space. We will use capital letters such as X and Y for random variables.

EXAMPLE 1

DEFINING A RANDOM VARIABLE

A coin is tossed four times and

$$X = \begin{pmatrix} \text{Number} \\ \text{of heads} \end{pmatrix}$$

Find the possible values for X and the outcomes corresponding to each of its possible values.

Solution

$$\left\{ \begin{array}{ll} (H,H,H,H), & (H,H,H,T), \\ (H,H,T,H), & (H,T,H,H), \\ (T,H,H,H), & (H,H,T,T), \\ (H,T,H,T), & (H,T,T,H), \\ (T,H,H,T), & (T,H,T,H), \\ (T,T,H,H), & (H,T,T,T), \\ (T,H,T,T), & (T,T,H,T), \\ (T,T,T,H), & (T,T,T,T) \end{array} \right\}$$

The number of heads in four tosses can be 0, 1, 2, 3, or 4, so these are the possible values for X. The sample space consists of the 16 sequences of H's and T's shown on the left.

Since X is the number of heads, for any particular outcome we can find its value by counting H's. For example, (H, T, H, H) gives $X = 3$. The following table lists the possible values of X and the outcomes for which it takes those values (as you should check by "counting heads").

Values of *X*	Outcomes
$X = 0$	(T, T, T, T)
$X = 1$	$(H, T, T, T), (T, H, T, T), (T, T, H, T), (T, T, T, H)$
$X = 2$	$(H, H, T, T), (H, T, H, T), (H, T, T, H), (T, H, H, T), (T, H, T, H), (T, T, H, H)$
$X = 3$	$(H, H, H, T), (H, H, T, H), (H, T, H, H), (T, H, H, H)$
$X = 4$	(H, H, H, H)

We can find the probability that X takes any particular value, such as $X = 2$, by adding up the probabilities of the outcomes corresponding to that value. The *probability distribution* of a random variable is the collection of these probabilities for its various values.

Random Variable

A *random variable* X is an assignment of a number to each element in the sample space. The *probability distribution* of the random variable X is the collection of all probabilities $P(X = x)$ for each possible value x.

EXAMPLE 2

FINDING A PROBABILITY DISTRIBUTION

A coin is tossed four times. Find and graph the probability distribution for

$$X = \begin{pmatrix} \text{Number} \\ \text{of heads} \end{pmatrix}$$

Solution

Using the table in Example 1, $X = 0$ occurs only for the outcome (T, T, T, T), which is one out of 16 equally likely outcomes, so $P(X = 0) = \frac{1}{16}$. The event $X = 1$ corresponds to 4 outcomes in the table, so $P(X = 1) = \frac{4}{16} = \frac{1}{4}$. The other probabilities $P(X = 2)$, $P(X = 3)$, and $P(X = 4)$ are similarly found by counting outcomes and dividing by 16, giving the probabilities in the following table. These probabilities are graphed on the left, the height of each bar being the probability that X takes the value at the bottom of the bar.

$P(X = x)$

3/8

1/4

1/16

   0 1 2 3 4   → x

Probability distribution for X

x	0	1	2	3	4
$P(X = x)$	$\frac{1}{16}$	$\frac{1}{4}$	$\frac{3}{8}$	$\frac{1}{4}$	$\frac{1}{16}$

Observe that the *most likely* number of heads in 4 tosses is 2 and that the least likely are the extreme values 0 and 4, just as you might expect. The probabilities add to 1: $\frac{1}{16} + \frac{1}{4} + \frac{3}{8} + \frac{1}{4} + \frac{1}{16} = \frac{16}{16} = 1$, so *the area under the graph is 1* (since each bar has width 1).

Practice Problem 1

Your dog has a litter of four pups. Assuming that each pup is equally likely to be male or female, which is the more likely distribution: two of each sex or 3–1 split (3 of one sex and 1 of the other)? [*Hint:* Use Example 2.] ➤ Solution on page 416

EXAMPLE 3

RAFFLE WINNINGS

To raise money, a children's hospital sells four hundred raffle tickets for $100 each. First prize is a $3000 Florida vacation for two, second prize is a $1000 credit at the town's supermarket, and the five third prizes are $100 "dinner for two" gift certificates at a local restaurant. Find the probability distribution of the value of a raffle ticket.

Solution

If X represents the value of a ticket, then its possible values are $3000, 1000, 100,$ and $0,$ with probabilities

$$P(X = 3000) = \frac{1}{400}$$ One \$3000 prize among 400 tickets

$$P(X = 1000) = \frac{1}{400}$$ One \$1000 prize

$$P(X = 100) = \frac{5}{400}$$ Five \$100 prizes

$$P(X = 0) = \frac{393}{400}$$ The other 393 tickets are worth nothing

Expected Value

On average, how much is each ticket worth? We divide the total value of all of the tickets, $3000 + 1000 + 100 \cdot 5 + 0 \cdot 393$, by the number of tickets, 400. The result can be written $3000 \cdot \frac{1}{400} + 1000 \cdot \frac{1}{400} + 100 \cdot \frac{5}{400} + 0 \cdot \frac{393}{400}$. Looking back at the above example, this is just the values of X multiplied by their probabilities and added. This leads to the following definition of the expected value of a random variable.

Expected Value

A random variable X taking values $x_1, x_2, \ldots, x_n$ with probabilities $p_1, p_2, \ldots, p_n$ has expected value:

$$E(X) = x_1 \cdot p_1 + x_2 \cdot p_2 + \cdots + x_n \cdot p_n$$
Sum of the possible values times their probabilities

The expected value is also called the *expectation* or the *mean* of the random variable (or of the probability distribution), and is sometimes denoted by μ (pronounced "mu," the Greek letter m). Using this definition, we may calculate the expected value of the raffle ticket in the preceding example by multiplying the possible values times their probabilities and adding:

$$E(X) = 3000 \cdot \frac{1}{400} + 1000 \cdot \frac{1}{400} + 100 \cdot \frac{5}{400} + 0 \cdot \frac{393}{400} = \$11.25$$

This expected value represents your average winnings per ticket. Since the tickets cost \$100, you are on average losing \$88.75 on each ticket. To put this in a more positive light, this latter figure represents your true generosity to the hospital.

Practice Problem 2   If the prizes in Example 3 were changed to one $3000 first prize, two $1000 second prizes, and ten $100 third prizes, what would be the expected value of a raffle ticket?   ➤ Solution on page 416

EXAMPLE 4

EXPECTED DICE WINNINGS

You roll a die and win any *even* amount and lose any *odd* amount (in dollars). (That is, if you roll 2 you win $2, and if you roll 3 you lose $3.) What is the expected value of your winnings?

Solution

The expected value of $X = \begin{pmatrix} \text{Your} \\ \text{winnings} \end{pmatrix}$ is the sum of the values times their probabilities:

$$E(X) = -1 \cdot \tfrac{1}{6} + 2 \cdot \tfrac{1}{6} - 3 \cdot \tfrac{1}{6} + 4 \cdot \tfrac{1}{6} - 5 \cdot \tfrac{1}{6} + 6 \cdot \tfrac{1}{6} = \tfrac{3}{6} = \tfrac{1}{2}$$ Expected value is half a dollar, or 50¢

This answer of $\frac{1}{2}$ means that if you play the game many times, your *average winnings per play* will be about half a dollar. The expected value is positive, so this game is *favorable* to you. If the expected value were zero, the game would be *fair*, and if it were negative, the game would be *unfavorable* to you. The expected value can be interpreted as *the fair price for playing the game*, meaning that paying that amount before each play would make the game *fair*.

Note that if the values are equally likely, then the expected value is just the arithmetic average of the values. Generally, the probabilities will not all be equal, so the expected value will be a *weighted average*, weighting each value by its probability. If none of the probabilities are zero, the expected value will lie somewhere between the highest and lowest values of the random variable.

Practice Problem 3   Find the expected value of one roll of a die.

➤ Solution on page 416

EXAMPLE 5

CONCERT INSURANCE

A concert promoter is planning an outdoor concert and estimates a profit of $100,000 if there is no rain but only $10,000 if it rains. He can take out an insurance policy costing $15,000 that will pay $100,000 if it rains. The weather bureau estimate the chances of rain at 10%. Based on expected value, should he buy the insurance?

Solution

First we calculate the profit under both conditions.

	Without Insurance	With Insurance	
No rain	100,000	100,000 − 15,000 = 85,000	← Subtracting the cost of insurance
Rain	10,000	10,000 − 15,000 + 100,000 = 95,000	← and adding the insurance payment

The expected values are:

Without insurance: $100{,}000 \cdot \frac{9}{10} + 10{,}000 \cdot \frac{1}{10} = 91{,}000$ Each value times its probability

With insurance: $85{,}000 \cdot \frac{9}{10} + 95{,}000 \cdot \frac{1}{10} = 86{,}000$

The expected value *without* insurance is higher, so on this basis he should skip the insurance. (However, in view of the relatively small difference in expected profits, he might decide to take the insurance after all to reduce the risk.)

Binomial Distribution

Suppose a biased coin has probability p of coming up heads (and therefore probability $1 - p$ of tails) on any toss. Since successive tosses are independent, the probability of a particular succession of outcomes would be just the product of the probabilities for each toss. For example, the probability of the outcome $HTHHT$ is $p \cdot (1 - p) \cdot p \cdot p \cdot (1 - p)$ which simplifies to $p^3(1 - p)^2$. Clearly, any other particular ordering of 3 heads and 2 tails would have this same probability. How many such orderings are there? Choosing the three tosses on which heads will occur (so the other two tosses are automatically tails) can be done in exactly $_5C_3 = \frac{5 \cdot 4 \cdot 3}{3 \cdot 2 \cdot 1}$ ways (using the combinations formula from page 368). Therefore, the probability of exactly 3 heads in 5 tosses is $_5C_3 p^3(1 - p)^2$.

We can generalize this result to any number n of tosses with k heads and $n - k$ tails to obtain the formula $_nC_k p^k(1 - p)^{n-k}$. Furthermore, we spoke of heads and tails only for familiarity. The same ideas apply to any experiment that has just two possible outcomes, *success* and *failure*, occurring with probabilities p and $1 - p$, respectively, with successive repetitions being independent. Such repeated experiments with fixed probabilities are called *Bernoulli trials*, and the number of successes in Bernoulli trials is called a *binomial random variable*.[*]

[*] After the Swiss mathematician James Bernoulli (1654–1705), who first recognized the importance of such random variables. His main work was *Ars Conjectandi* (*Art of Conjecturing*).

> ### Binomial Distribution
>
> For independent repetitions of an experiment with probability p of success on each trial, the probability that in n trials the number X of successes will be k is
>
> $$P(X = k) = {}_nC_kp^k(1 - p)^{n-k}$$
> $\quad n$ = number of trials
> $\quad k$ = number of successes $\quad(0 \le k \le n)$
> $\quad p$ = probability of success
>
> X is called a *binomial random variable with parameters n and p.* Its expected value or mean is $\quad E(X) = n \cdot p.$

The formula for the expected value is proven in Exercises 53–55, but may be made intuitively reasonable as follows: If a coin with probability of heads $\frac{1}{3}$ is tossed 12 times, you should expect about a third of them, $12 \cdot \frac{1}{3} = 4$, to be heads, and this is just the formula $E(X) = n \cdot p$. Recall that the number of combinations $\,{}_nC_k$ may be calculated from the formula $\,{}_nC_k = \frac{n \cdot (n-1) \cdot \,\cdots\, \cdot (n-k+1)}{k!} = \frac{n!}{k!(n-k)!}.$ We may use the binomial distribution in any situation with independent trials having probability p of success, where the word "success" may mean whatever outcome we choose.

EXAMPLE **6**

A BINOMIAL DISTRIBUTION

Let X be the number of heads in four tosses of a coin. Find the probability distribution and expected value of X.

Solution

Coin tossing is merely Bernoulli trials, so we want the probability distribution of a binomial random variable with $n = 4$ and $p = \frac{1}{2}$.

$$P(X = 0) = 1 \cdot \left(\tfrac{1}{2}\right)^0 \cdot \left(\tfrac{1}{2}\right)^4 = \tfrac{1}{16} \qquad\qquad {}_4C_0\,p^0(1 - p)^4$$

$$P(X = 1) = 4 \cdot \left(\tfrac{1}{2}\right)^1 \cdot \left(\tfrac{1}{2}\right)^3 = \tfrac{4}{16} = \tfrac{1}{4} \qquad {}_4C_1\,p^1(1 - p)^3$$

$$P(X = 2) = 6 \cdot \left(\tfrac{1}{2}\right)^2 \cdot \left(\tfrac{1}{2}\right)^2 = \tfrac{6}{16} = \tfrac{3}{8} \qquad {}_4C_2\,p^2(1 - p)^2$$

$$P(X = 3) = 4 \cdot \left(\tfrac{1}{2}\right)^3 \cdot \left(\tfrac{1}{2}\right)^1 = \tfrac{4}{16} = \tfrac{1}{4} \qquad {}_4C_3\,p^3(1 - p)^1$$

$$P(X = 4) = 1 \cdot \left(\tfrac{1}{2}\right)^4 \cdot \left(\tfrac{1}{2}\right)^0 = \tfrac{1}{16} \qquad\qquad {}_4C_4\,p^4(1 - p)^0$$

These five probabilities, graphed on the left, make up the probability distribution for X, the number of heads in four tosses. They could also have been found using a graphing calculator. For the expected value, we use the formula in the preceding box:

$$E(x) = n \cdot p = 4 \cdot \tfrac{1}{2} = 2$$

We found exactly this distribution in Example 2 (page 408). Here we found it using the binomial probability formula, and there we found it from basic principles. The results agree. The fact that the expected number of heads in four tosses of a fair coin is two should come as no surprise. This "mean" of 2 is the "balance point" of the probability distribution graph, as shown on the bottom left of the previous page.

Spreadsheet Exploration

The spreadsheet* below uses a "random number generator" to simulate 4 coin tosses repeated 400 times. The numbers of heads on each of these 4 tosses are recorded in cells `A21:T40` and the results are summarized in cells `G18:K18`. The graph below shows that the results of this experiment approximate the probability distribution found in Example 6.

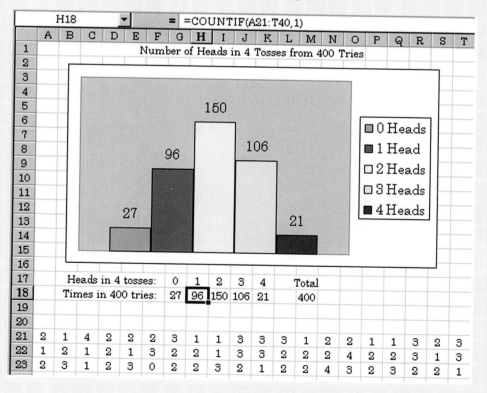

---

* See the Preface for information on how to obtain this and other Excel spreadsheets.

Practice Problem 4

Let X be the number of heads in six tosses of a coin. Find $P(X = 3)$ and $E(X)$.

➤ Solution on page 416

EXAMPLE 7

EMPLOYEE RETENTION

A restaurant manager estimates the probability that a newly hired waiter will still be working at the restaurant six months later is only 60%. For the five new waiters just hired, what is the probability that at least four of them will still be working at the restaurant in six months?

Solution

Assuming that the waiters decide independently of each other, their decisions make up five Bernoulli trials. Counting a waiter who stays as a "success," the number X of waiters who stay is a binomial random variable with $n = 5$ and $p = 0.6$. We want $P(X \geq 4)$, which means $P(X = 4 \text{ or } 5) = P(X = 4) + P(X = 5)$.

$$P(X = 4) + P(X = 5) = \underset{5}{\underbrace{{}_5C_4(0.6)^4(0.4)^1}} + \underset{1}{\underbrace{{}_5C_5(0.6)^5(0.4)^0}} \approx 0.337$$

```
binompdf(5,.6,4)
+binompdf(5,.6,5
)
        .33696
```

The probability that at least four of the new waiters will stay for six months is only about 34%.

EXAMPLE 8

MONEY-BACK GUARANTEES

A manufacturer of compact disks (CD's) for computer storage sells them in packs of 10 with a "double your money back" guarantee if more than one CD is defective. If each CD is defective with a 1% probability independently of the others, what proportion of packages will require refunds?

Solution

If X is the number defective CD's in a package, then X is a binomial random variable with $n = 10$ and $p = 0.01$. Rather than calculating the probability of a refund (2 or more defectives), we find the complementary probability:

$$P(X = 0) + P(X = 1) = \underset{1}{\underbrace{{}_{10}C_0(0.01)^0(0.99)^{10}}} + \underset{10}{\underbrace{{}_{10}C_1(0.01)^1(0.99)^9}} \approx 0.996$$

```
binompdf(10,.01,
0)+binompdf(10,.
01,1)
        .9957337998
1-Ans
        .0042662002
```

Subtracting this answer from 1 (since it is the complementary probability) gives 0.004. Therefore, the company will have to provide refunds for only about 0.4% of its packs.

EXAMPLE 9

MULTIPLE-CHOICE TESTS

A multiple-choice test consists of 25 questions, each of which has 5 possible answers. What is the expected number of correct answers that would result from random guessing?

Solution

The 25 questions are 25 Bernoulli trials, and random guessing among the 5 possible answers for each question gives a probability $p = \frac{1}{5}$ of success for each. Therefore, the expected number of correct answers is

$$E(X) = 25 \cdot \tfrac{1}{5} = 5 \qquad\qquad E(X) = n \cdot p$$

This is why multiple-choice tests are often *rescaled* by only counting the number of correct answers above a certain expected number. In this case, the grade would be based on the number of correct answers beyond the first 5.

5.6 **Section Summary**

A *random variable* X is an assignment of a number to each outcome in a sample space. The *probability distribution* of X is the collection of probabilities $P(X = x)$ for each possible value x of X. The *expected value* or *mean* of X is

$$E(X) = x_1 \cdot P(X = x_1) + \cdots + x_n \cdot P(X = x_n) \qquad \text{Sometimes denoted } \mu$$

Bernoulli trials are independent repetitions of an experiment that results in *success* with probability p (and therefore *failure* with probability $1 - p$). The number X of successes in n Bernoulli trials is called a *binomial* random variable with parameters n and p and has probability distribution and mean

$$P(X = k) = {}_nC_k p^k (1 - p)^{n-k} \qquad \text{For } k = 0, 1, \dots, n$$
$$E(X) = n \cdot p$$

The following graphs show the binomial distribution for $n = 6$ and different values of p. Notice that the middle graph is symmetric, and that values away from $p = \frac{1}{2}$ skew the graph to one side or the other.

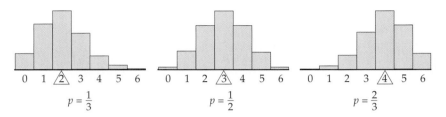

Under each graph the mean (calculated from $E(X) = n \cdot p$) is shown in a triangle to indicate that the distribution "balances" at that point.

➤ **Solutions to Practice Problems**

1. The sample space of all gender orders (such as F, M, M, F) is just like that for four coin tosses on page 408 but with different letters. Example 2 says that two of each gender has probability $\frac{3}{8}$ while a 3–1 split (which can occur in two ways) has probability $\frac{1}{4} + \frac{1}{4} = \frac{1}{2}$. Therefore, three of one sex and one of the other is more likely than two of each.

2. $E(X) = 3000 \cdot \frac{1}{400} + 1000 \cdot \frac{2}{400} + 100 \cdot \frac{10}{400} + 0 \cdot \frac{387}{400} = 15$.
   The expected value is $15.

3. $E(X) = 1 \cdot \frac{1}{6} + 2 \cdot \frac{1}{6} + 3 \cdot \frac{1}{6} + 4 \cdot \frac{1}{6} + 5 \cdot \frac{1}{6} + 6 \cdot \frac{1}{6} = \frac{21}{6} = 3.5$.
   This answer shows that the expected value may not be one of the possible values of the random variable.

4. $P(X = 3) = {}_6C_3 \left(\frac{1}{2}\right)^3 \left(\frac{1}{2}\right)^3 = \frac{6 \cdot 5 \cdot 4}{3 \cdot 2 \cdot 1} \left(\frac{1}{2}\right)^6 = 20 \cdot \frac{1}{64} = \frac{5}{16}$

   $E(x) = n \cdot p = 6 \cdot \frac{1}{2} = 3$

   ```
   binompdf(6,.5,3)
                   .3125
   Ans▶Frac
                    5/16
   ```

| 5.6 | **Exercises** (▦ will be helpful.) |

---

1. **GENERAL: Coins** A coin is tossed three times, and X is the number of heads. Find and graph the probability distribution of X.

2. **GENERAL: Dice** Two dice are rolled, and X is the sum of the faces. Find and graph the probability distribution of X.

3. **GENERAL: Coins** You toss three coins and win $11 if they all agree (all heads or all tails), and otherwise you lose $1. Find and graph the probability distribution of your winnings.

4. **GENERAL: Marbles** Two marbles are chosen without replacement from a box containing

3 red and 5 green marbles. Let X be the number of green marbles chosen. Find and graph the probability distribution of X.

5. **GENERAL: Dice** A die is rolled, and you win $2 if it comes up odd, you lose $12 if it comes up 2, and you win $3 if it comes up 4 or 6. Find and graph the probability distribution of your winnings.

6. **GENERAL: Dice** A die is rolled and you win $8 if it comes up odd, you lose $15 if it comes up 2 or 4, and you win $6 if it comes up 6. Find and graph the probability distribution of your winnings.

7. **GENERAL: Dice** Two dice are rolled, and X is the larger of the two numbers that come up. Find and graph the probability distribution of X.

8. **GENERAL: Dice** Two dice are rolled, and X is the smaller of the two numbers that come up. Find and graph the probability distribution of X.

9. **Mean** Find the mean of the random variable in Exercise 1.

10. **Mean** Find the mean of the random variable in Exercise 2.

11. **Mean** Find the mean of the random variable in Exercise 3.

12. **Mean** Find the mean of the random variable in Exercise 4.

13. **Mean** Find the mean of the random variable in Exercise 5.

14. **Mean** Find the mean of the random variable in Exercise 6.

15. **Mean** Find the mean of the random variable in Exercise 7.

16. **Mean** Find the mean of the random variable in Exercise 8.

17. **GENERAL: Raffle Tickets** One thousand raffle tickets are sold, and there is one first prize worth $2000, one second prize worth $250, and 20 third prizes worth $50 each. Find the expected value of a ticket.

18. **BUSINESS: Insurance** A concert promoter is planning an outdoor concert and estimates a profit of $120,000 if there is no rain, but only

$15,000 if it rains. She can take out an insurance policy costing $20,000 that will pay $90,000 if it rains. The weather bureau estimates the chances of rain at 20%. Based on expected value, should she take the insurance?

19. **BUSINESS: Insurance** A college is planning an outdoor fund raising event and predicts donations of $80,000 if there is no rain, but only $40,000 if it rains. They can take out an insurance policy costing $8,000 that will pay $40,000 if it rains. The weather bureau estimates the chances of rain at 15%. Based on expected value, should they take the insurance?

20. **Binomial Distribution** Find and graph the probability distribution of a binomial random variable with parameters $n = 4$ and $p = \frac{1}{5}$. Use the formula on page 412 to find its mean.

Binomial Distribution For a binomial random variable with the given parameters, find and graph its probability distribution, and use the formula on page 412 to find its mean.

21. $n = 20$ and $p = \frac{1}{2}$

22. $n = 25$ and $p = \frac{1}{2}$

23. $n = 20$ and $p = 0.8$

24. $n = 25$ and $p = 0.64$

25. **GENERAL: Coins** For 6 tosses of a coin, find the probability that the number of heads is between 2 and 4 (inclusive).

26. **GENERAL: Coins** What is the probability of getting exactly 5 heads in 10 tosses of a coin?

27. **GENERAL: Coins** What is the probability of getting exactly 4 heads in 8 tosses of a coin?

28. **GENERAL: Coins** For an unfair coin whose probability of heads is 0.7, what is the most likely number of heads in 10 tosses, and what is its probability? What is the *least* likely number of heads, and what is its probability?

29. **GENERAL: Coins** For an unfair coin whose probability of heads is $\frac{1}{3}$, what is the most likely number of heads in 6 tosses, and what is its probability? What is the *least* likely number of heads, and what is its probability?

30. GENERAL: Dice In rolling a die 10 times, what is the most likely number of sixes, and what is its probability? What is the mean number of sixes?

31. GENERAL: Dice In rolling a die 8 times, what is the most likely number of sixes, and what is its probability? What is the mean number of sixes?

32. GENERAL: Testing You know that one of three batteries is dead. If you test them one at a time until you find the defective one, what is the expected number of tests?

33. BUSINESS: Sales An automobile salesperson predicts that a customer will buy a $30,000 car with probability $\frac{1}{10}$, a $25,000 car with probability $\frac{1}{5}$, a $20,000 car with probability $\frac{3}{10}$, and otherwise buy nothing. What is the expected value of the sale?

34. BUSINESS: Insurance An insurance company estimates that on a typical policy it will have to pay out $10,000 with probability 0.05, $5000 with probability 0.1, and $1000 with probability 0.2. If the company wants to charge $200 more than the expected payout of the policy, what should it charge?

35. BUSINESS: Product Quality A company manufactures products of which 2% have hidden defects. If you buy 10 of them, then the number of defective ones in your purchase is a binomial random variable. Find its mean.

36. SOCIAL SCIENCE: Multiple-Choice Testing A multiple-choice test has 5 possible answers for each of 10 questions. For a student who guesses randomly:

a. What is the mean of the number of right answers?

b. What is the probability that this student gets 4 or more correct answers?

37. SOCIAL SCIENCE: ESP A person claims to have ESP (extrasensory perception) and calls 7 out of 10 tosses of a fair coin correctly. What is the probability of doing at least that well by guessing randomly?

38. SOCIAL SCIENCE: Juries Suppose that on a jury of 12 people it takes at least 9 votes to

convict. If each person decides correctly with probability 0.9, what is the probability that a guilty person is convicted? What is the probability that an innocent person is found innocent? What assumptions are you making about independence?

39. GENERAL: Communications A one-digit message (0 or 1) is to be transmitted over a communications line that has a $\frac{1}{10}$ probability of changing the digit. Because of this, the message is to be transmitted in triplicate, 000 or 111, with *decoding by majority*, which means that the digit that occurs two or three times will be taken to be the message. What is the probability that the message is received correctly? How does this compare with the probability of correct reception when only *one* digit is transmitted?

40. BIOMEDICAL: Family Distribution Assuming that boys and girls are equally likely, what is the probability that a family of 6 children consists of 3 boys and 3 girls?

41. GENERAL: Cards A bridge hand consists of 13 cards. What is the probability that you get no aces in three consecutive hands?

42. GENERAL: Target Practice If the probability of hitting a target is $\frac{1}{4}$ and 5 shots are fired, what is the probability of hitting the target at least once?

43. ATHLETICS: Winning a Series Suppose that in a sports contest with no ties, the stronger team has probability $\frac{3}{5}$ of winning any particular game. Find the probability that the stronger team wins:

a. A two-out-of-three series. [*Hint:* For ease of calculation, assume that all three games are played. What assumptions about independence are you making?]

b. A three-out-of-five series.

c. A four-out-of-seven series.

44. ATHLETICS: Home Runs On the basis of records from 1996 to 2002, the probability that Mark McGuire will hit a home run on any particular at-bat is $p_M = 0.110$. Model the number X of home runs by McGuire in a season by a binomial random variable with parameters

$n = 530$ (a typical number of at-bats in a season) and p_M. Find $P(X > 50)$, $P(X > 60)$, and $P(X > 70)$. Then do the same for Sammy Sosa, using his home-run-per-at-bat probability $p_S = 0.095$.

45. SOCIAL SCIENCE: Multiple-Choice Tests A multiple-choice test consists of 20 questions, each of which has 4 possible answers. What is the expected number of correct answers that would result from random guessing?

46. SOCIAL SCIENCE: Multiple-Choice Tests A multiple-choice test consists of 30 questions, each of which has 6 possible answers. What is the expected number of correct answers that would result from random guessing?

47. GENERAL: Airline Safety An airplane has four independent engines and each operates with probability 0.99. If it takes at least two engines to land safely, what is the probability of a safe landing? What if the two engines must be on opposite sides of the plane?

48. ATHLETICS: Football After scoring a touchdown, a football team has a choice of attempting to run or pass the ball into the end zone for two points (a *two-point conversion*) or kicking the ball through the goalposts for one point (a *field goal*). If the two-point conversions work 37% of the time and one-point field goals are successful 94% of the time,* which play has the better expected gain?

49. GENERAL: Odd Man Out Five people in an office decide who fetches (and pays for) midmorning coffee in the following way. Each tosses a coin, all at the same time, and if everyone gets the same outcome except for one person (such as all tails except for one head), then that person gets coffee for everyone that day. If there is no "odd man out" on the first round, then they repeat the coin tosses as many times as necessary to choose someone.

a. What is the probability of getting an "odd man out" on one round of tosses?
b. What is the probability that it takes more than four rounds to choose an odd man out?
c. More than eight rounds?

50. GENERAL: Expected Number of Trials Until a Success Let T be the number of trials up to and including the first success in Bernoulli trials. We can find the expected number of trials, $E(T)$, by reasoning about what happens after the first trial: If the first trial is a success (this happens with probability p) then $T = 1$; if the first trial is a failure (this happens with probability $1 - p$), then we begin the wait for the first success all over again, with expected number of trials $E(T)$ beyond the first trial.*

a. Explain why the reasoning in the previous sentence leads to the equation
$$E(T) = 1 \cdot p + (1 + E(T)) \cdot (1 - p).$$
b. Solve this equation for $E(T)$ to find that $E(T) = \frac{1}{p}$.
c. What is the expected number of tosses up to and including the first 5 when rolling a die?
d. What is the expected number of tosses up to and including the first double-5 when rolling two dice?

51. GENERAL: Coin Tossing A coin is tossed n times. Given that there are exactly k heads, show that the probability of a head on any given toss is $\frac{k}{n}$.

52. GENERAL: Coin Tossing A coin is tossed n times where n is an even number. Let E be the event that the first toss is heads and F_k be the event that exactly k of the tosses are heads. For which values of k are E and F_k independent events?

Explorations and Excursions

The following exercises extend and augment the material presented in the text.

---

* According to data from the National Football League.

* Technically, this problem goes beyond the finite sample spaces we have discussed, since the first success might never occur. The reasoning, however, avoids this difficulty so that it causes no trouble.

The Mean of the Binomial Distribution

The following exercises prove the formulas $E(x) = np$ for binomial random variables with parameters n (number of trials) and p (probability of success on each trial). We will use the fact that adding together n binomial random variables, each with parameters 1 and p, gives a binomial random variable with parameters n and p (since adding up the number of "successes" in each of n Bernoulli trials gives the total number of successes in all n trials).

53. If X is binomial with $n = 1$, show that $E(X) = 1 \cdot p + 0 \cdot (1 - p) = p$.

54. Justify each numbered equals sign. For X and Y independent,

$$E(X + Y) \overset{1}{=} \sum_{\text{All } x,y} (x + y) \cdot P(X = x \text{ and } Y = y)$$

$$\overset{2}{=} \sum_{\text{All } x,y} (x + y) \cdot P(X = x) \cdot P(Y = y)$$

$$\overset{3}{=} \sum_{\text{All } x,y} x \cdot P(X = x) \cdot P(Y = y) + \sum_{\text{All } x,y} y \cdot P(X = x) \cdot P(Y = y)$$

$$\overset{4}{=} \left(\sum_{\text{All } x} x \cdot P(X = x) \right) \cdot \left(\sum_{\text{All } y} P(Y = y) \right) + \left(\sum_{\text{All } y} y \cdot P(Y = y) \right) \cdot \left(\sum_{\text{All } x} P(X = x) \right)$$

$$\overset{5}{=} \left(\sum_{\text{All } x} x \cdot P(X = x) \right) \cdot 1 + \left(\sum_{\text{All } y} y \cdot P(Y = y) \right) \cdot 1 = E(X) + E(Y)$$

That is, for independent random variables, $E(X + Y) = E(X) + E(Y)$ (in words, the expectation of the sum is the sum of the expectations). This result immediately generalizes to any number of independent random variables:

$$E(X_1 + X_2 + \cdots + X_n) = E(X_1) + E(X_2) + \cdots + E(X_n).$$

55. If X is a binomial random variable with parameters n and p, we may write $X = X_1 + X_2 + \cdots + X_n$, where $X_1, X_2, \ldots, X_n$ are independent binomial random variables each with parameters 1 and p. Justify:

$$E(X) = E(X_1 + X_2 + \cdots + X_n) = E(X_1) + E(X_2) + \cdots + E(X_n) = \underbrace{p + p + \cdots + p}_{n \text{ terms}} = np$$

Therefore, a binomial random variable with parameters n and p has mean $E(x) = np$.

Chapter Summary with Hints and Suggestions

Reading the text and doing the exercises in this chapter have helped you to master the following skills, which are listed by section (in case you need to review them) and are keyed to particular Review Exercises. Answers for all Review Exercises are given at the back of the book, and full solutions can be found in the Student Solutions Manual.

5.1 Sets, Counting, and Venn Diagrams

- Read and interpret a Venn diagram. *(Review Exercise 1.)*

- Use the complementary or the addition principles of counting or Venn diagrams to solve an applied problem. *(Review Exercise 2.)*

$$n(A^c) = n(U) - n(A)$$
$$n(A \cup B) = n(A) + n(B) - n(A \cap B)$$

- Use the multiplication principle of counting to solve an applied problem. *(Review Exercise 3.)*

5.2 Permutations and Combinations

- Calculate numbers of permutations and combinations. *(Review Exercise 4.)*

$$_nP_r = n \cdot (n - 1) \cdot \cdots \cdot (n - r + 1)$$
$$_nC_r = \frac{n(n - 1) \cdot \cdots \cdot (n - r + 1)}{r \cdot (r - 1) \cdot \cdots \cdot 1} = \frac{n!}{r!(n - r)!}$$

- Use the permutation and combination formulas (possibly using ⌨) to solve an applied problem. *(Review Exercise 5–7.)*

5.3 Probability Spaces

- Find an appropriate sample space for a random experiment. *(Review Exercises 8–9.)*

- Describe an event in terms of the sample space. *(Review Exercise 10.)*

- Assign probabilities to outcomes in a sample space, and find probabilities of events. *(Review Exercises 11–15.)*

- Use the techniques of this section to find a probability in an applied problem. *(Review Exercises 16–24.)*

5.4 Conditional Probability and Independence

- Find a conditional probability using either the definition or a restricted sample space. *(Review Exercises 25–26.)*

$$P(A \text{ given } B) = \frac{P(A \cap B)}{P(B)}$$

- Use conditional probability to solve an applied problem. *(Review Exercises 27–29.)*

- Use a tree diagram to solve an applied problem. *(Review Exercises 30–32.)*

- Determine whether two events are independent or dependent. *(Review Exercise 33.)*

$$P(A \cap B) = P(A) \cdot P(B)$$

- Use independence to find a probability in an applied problem. *(Review Exercises 34–35.)*

5.5 Bayes' Formula

- Use Bayes' formula to solve an applied problem. *(Review Exercises 36–38.)*

5.6 Random Variables and Distributions

- Determine the outcomes that correspond to the values of a random variable. *(Review Exercise 39.)*

- Find and graph the probability distribution of a random variable, and find the mean. *(Review Exercises 40–44.)*

- For a binomial random variable, find and graph the probability distribution, and find its mean. *(Review Exercise 45.)*

$$P(X = k) = {_nC_k}p^k(1 - p)^{n-k}$$
$$E(X) = n \cdot p$$

- Use the binomial probability distribution to solve an applied problem. *(Review Exercises 46–51.)*

Hints and Suggestions

- Much of probability depends on counting, and there are several principles to simplify counting large numbers of objects. Roughly speaking, the *complementary* principle says that you can count the *opposite* set and then subtract; the *addition* principle says that you can add numbers from two sets but you must then subtract what was double-counted; and the *multiplication* principle says that the number of two-part choices is the product of the number of first-part choices times the number of second-part choices.

- *Permutations* are arrangements where a different order means a different object (such as letters in a word or rankings of people). *Combinations* are arrangements where reordering does *not* make a new object (such as hands of cards or committees of people).

- A set of *n* objects has 2^n subsets (counting the "empty" subset and the set itself).

- Probabilities are assigned to possible outcomes of a random experiment so that each probability is between 0 and 1 (inclusive) and they add to 1. Equally likely outcomes should be assigned equal probabilities. Probabilities of more complicated events are found by adding up the probabilities of the outcomes in the event.

- Conditional probability is the relative probability of the intersection of the events compared to the probability of the *given* event. Conditional probabilities are sometimes more easily found from the restricted sample space than from the definition.

- A complicated probability can sometimes be found by a tree diagram, branching on some event on which others depend.

- Independent events "have nothing to do with each other," and the probability of both occurring is the *product* of their probabilities. This is not the same as *disjoint* events, where one *precludes* the other, and which therefore *cannot* be independent.

- Bayes' formula is useful for finding events of the form S_1 *given A* in terms of events of the form *A given* S_i where $S_1, \ldots, S_n$ are disjoint events whose union is the entire sample space.

- The mean of a random variable gives a *representative* or *typical* value.

- Bernoulli trials are repeated independent experiments with only two outcomes, *success* and *failure*. The number of successes in several Bernoulli trials is a *binomial* random variable.

- A graphing calculator that finds permutations, combinations, and binomial and other probabilities is very useful.

- **Practice for Test:** Review Exercises 1, 2, 3, 5, 7, 9, 10, 11, 16, 17, 18, 22, 23, 26, 30, 32, 35, 37, 40, 42, 43, 44, 47, 49, 51.

Review Exercises for Chapter 5

Practice test exercise numbers are in green.

5.1 Sets, Counting, and Venn Diagrams

1. **Venn Diagrams** For the Venn diagram below, find

    a. $n(A)$ **b.** $n(A \cup B)$ **c.** $n(B^c)$ **d.** $n(A^c \cap B)$

2. **GENERAL: Purchases** A survey of 1000 homeowners found that during the last year 230 bought an automobile, 340 bought a major appliance, and 540 bought neither. How many homeowners had bought both an automobile and a major appliance?

3. **GENERAL: Initials** Find how many three-letter monograms there are if:

    a. Repeated letters are allowed.
    b. Repeated letters are not allowed.

5.2 Permutations and Combinations

4. **Permutations and Combinations** Find:

    a. $_6P_3$ **b.** $_6C_3$

5. **GENERAL: Committees** How many 4-member committees can be formed from a club consisting of 20 students? What if the committee is to consist of a president, vice-president, secretary, and treasurer?

6. **GENERAL: Cards** How many 5-card hands are there that contain only spades?

7. **GENERAL: Cards** How many 5-card hands are there containing only cards 10 or higher? [*Hint:* Remember that aces may count high.]

5.3 Probability Spaces

8. **GENERAL: Wardrobes** Find the sample space for choosing an outfit consisting of one of two coats (C_1 or C_2), one of two scarves (S_1 or S_2), and one of two hats (H_1 or H_2).

9. **GENERAL: Marbles** A box contains three marbles: one blue, one yellow, and one red. A first marble is chosen, then it is replaced, and a second marble is chosen. Find the sample space. Then find the sample space if the first marble is *not* replaced before the second is chosen.

10. **Events** For the experiment of tossing a coin twice, describe each event as a subset of the sample space $\{(H, H), (H, T), (T, H), (T, T)\}$.

    a. At least one head.
    b. At most one head.
    c. Different faces.

11. **Probabilities of Events** Consider the sample space in Exercise 10.

    a. What probability should be assigned to each outcome?
    b. What is the probability of getting at least one head?
    c. What is the probability of getting different faces?

12. **GENERAL: Committees** A committee of 2 is to be chosen at random from a class of 15 students. What probability should be assigned to any particular committee? What if one of the 2 members is to be designated committee spokesperson?

GENERAL: Spinners Find the probability of the arrow landing in each numbered region. Assume that the probability of any region is proportional to its area, and assume that areas that look the same size are the same size.

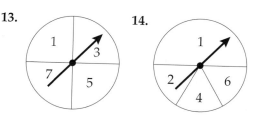

13.   14.

15. **GENERAL: Marbles** A marble is drawn at random from a box containing 3 red, 6 green, and 9 black marbles, and the color is noted. What is the probability of each outcome in the sample space $\{R, G, B\}$?

16. **BIOMEDICAL: Smoking and Weight** For a randomly selected person, the probability of being a smoker is 0.35, the probability of being overweight is 0.40, and the probability of being both a smoker and overweight is 0.20. What is the probability of being *neither* a smoker *nor* overweight?

17. **GENERAL: Dice** If you roll one die, find the probability of:

    a. Rolling at most 2.
    b. Rolling an even number.

18. **GENERAL: Lottery** In a lottery you choose 4 numbers out of 40. Then 5 numbers are announced, and you win something if you have 4 of the 5 numbers. What is the probability that you win something?

19. **GENERAL: Cards** You are dealt 5 cards at random from an ordinary deck. Find the probability of being dealt:

    a. All face cards.
    b. No face cards.

20. **GENERAL: Cards** You are dealt 13 cards at random from an ordinary deck. Find the probability of being dealt:

    a. No spades.
    b. No face cards.

21. **BUSINESS: Defective Products** You have 30 computer CDs, and 2 of them contain computer viruses. If you lend 5 to a friend, what is the probability that they are virus-free?

22. **BUSINESS: Defective Products** A store shelf has 50 light bulbs, of which 2 are defective. If you buy 4, what is the probability that you get both defective bulbs? Neither defective bulb?

23. **GENERAL: Committees** A committee of 3 is to be formed from your class of 30. What is the probability that both you and your best friend will be on it?

24. **BIOMEDICAL: Memory** You remember the names of 6 of the 10 people you just met. If you run into 2 of the 10 on the street, what is the probability that you will remember their names?

5.4 Conditional Probability and Independence

25. If $P(A) = 0.5$, $P(B) = 0.4$, and $P(A \cap B) = 0.3$, find

   a. $P(A$ given $B)$
   b. $P(B$ given $A)$

26. **GENERAL: Dice** If you roll two dice, what is the probability of at least one six given that the sum of the numbers is seven?

27. **BIOMEDICAL: Gender** If you know that a family of three children has at least one boy, what is the probability that all three are boys? Assume that boys and girls are equally likely.

28. **GENERAL: Cards** You draw three cards at random from a standard deck. If the first is an ace, what is the probability that the other two are also aces?

29. **GENERAL: Cards** If you are being dealt five cards and the first four are hearts, what is the probability that the fifth is also a heart?

30. **BUSINESS: Market Share** An automobile salesperson estimates that with a male customer she can make a sale with probability 0.3 and that with a female customer the probability is 0.4. If 80% of her customers are male and 20% female, what is the probability that for a randomly selected customer she can make a sale?

31. **GENERAL: Searches** An airplane is missing, and you estimate that with probability 0.6 it crashed in the mountains and with probability 0.4 it crashed in the valley. If it is in the mountains, a search there will find it with probability 0.8, whereas if it is in the valley, a search there will find it with probability 0.9. What is the probability that the plane will be found? (The complementary probabilities to the numbers 0.8 and 0.9 are called the *overlook probabilities* because they give the probability that it will *not* be found in a region if it is there.)

32. **BIOMEDICAL: Births** Twelve percent of all births are by cesarean (surgical) section, and of these births 96% survive. Overall, 98% of all babies survive delivery. For a randomly chosen mother who did not have a cesarean, what is the probability that her baby survives?

33. **GENERAL: Coins** For the experiment of tossing a coin twice, with sample space $\{(H, H), (H, T), (T, H), (T, T)\}$, are the following events independent or dependent?

   a. *At least one head* and *at most one head.*
   b. *Heads on first toss* and *same face on both tosses.*

34. **BUSINESS: Computer Malfunctions** An airline reservations computer breaks down with probability 0.01, at which time a second computer takes over, but it also has probability 0.01 of failing. Find the probability that the airlines will be able to take reservations.

35. **GENERAL: Coins** Five coins are tossed. Find the probability of getting

   a. All heads.
   b. All the same outcome.
   c. Alternating outcomes.

5.5 Bayes' Formula

36. **GENERAL: Coins** You have two coins in your pocket, one with two heads, and one with a head and a tail. You choose one at random and toss it, and it comes up heads. What

is the probability that it is the two-headed coin?

37. **MANAGEMENT SCIENCE: Manufacturing Defects** A soft drink bottler has three bottling machines: A, B, and C. Machine A bottles 25% of the company's output, machine B bottles 35%, and machine C bottles 40%. It is known that 4% of the bottles produced by machine A are defective, as are 3% of the bottles produced by machine B and 2% of the bottles from machine C. If a randomly selected bottle from the factory's output is found to be defective, what is the probability that it was produced by machine A?

38. **MANAGEMENT SCIENCE: Manufacturing Defects** For the information in Exercise 37, if a randomly selected bottle is *not* defective, what is the probability that it came from machine A?

5.6 Random Variables and Distributions

39. **GENERAL: Coins** Five coins are tossed, and X is the number of heads. List the outcomes (from the usual sample space) corresponding to each of the events $X = 5$, $X = 1$, and $X = 0$.

40. **GENERAL: Coins** You toss two coins. You win $34 if you get double heads, and otherwise you lose $2. Find and graph the probability distribution of your winnings.

41. **Mean** Find the mean of the random variable in Exercise 40.

42. **GENERAL: Marbles** Two marbles are chosen at random and without replacement from a box containing 5 red and 4 green marbles. If X is the number of green marbles chosen, find and graph the probability distribution of X.

43. **Mean** Find the mean of the random variable in Exercise 42.

44. **GENERAL: Raffle Tickets** Five hundred raffle tickets are sold for the following prizes: one first prize worth $1000, two second prizes worth $150, and 10 third prizes worth $25. Find the expected value of a ticket.

45. **Binomial Distribution** Find and graph the distribution of a binomial random variable with parameters $n = 4$ and $p = \frac{4}{5}$. Find its mean using the formula developed for a binomial random variable.

46. **GENERAL: Coins** For 8 tosses of a coin, find the probability that the number of heads is between 3 and 5 inclusive.

47. **BUSINESS: Sales** A television salesperson estimates that a customer will buy a $900 TV with probability $\frac{1}{10}$, a $500 TV with probability $\frac{1}{2}$, a $200 TV with probability $\frac{1}{10}$, and otherwise buy nothing. What is the expected value of the sale?

48. **GENERAL: Testing** You know that one of three keys will work in a lock. If you try them one at a time until you find the right one, what is the expected number of tries?

49. **BUSINESS: Product Quality** Of the products a company produces, 1% have hidden defects. If you buy a dozen of them, what is the mean of the number of defective ones in your purchase?

50. **BUSINESS: Product Quality** A company makes screws, and 98% of them are usable. If you buy a box of 100 screws, what is the probability of at least 97 being usable?

51. **GENERAL: Target Practice** You shoot at a target 5 times. If the probability of a hit on each shot is 75%, what is the probability that you hit it at least 4 times?

6 Statistics

Application Preview

Household Income

The following graph shows the percentages of U.S. households with 2001 incomes from $0 to $100,000 grouped in blocks of $5000 (the remaining 14% had incomes of $100,000 or more).

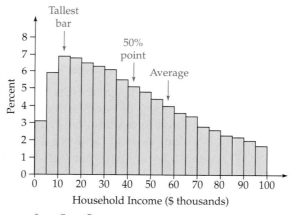

Source: Census Bureau

This is a *histogram*, which is just one of the graphical displays that we will use to exhibit data. There is a vast amount of information in this graph, and it is natural to ask whether there is some sort of "middle" income that might represent a "typical American household." There are several possible approaches. The tallest bar, representing the income block with the greatest number of households, has midpoint $12,500, but this is lower than the vast majority of households. The average of all incomes is $58,208, but if you look where this falls on the graph, it is much higher than most of the households (in fact, it's higher than 65% of them). The 50-percent point, with as many households above it as below it, is $42,228. Each of these figures has some claim to being a *representative* income, and in this chapter we will discuss the merits of each.

6.1 RANDOM SAMPLES AND DATA ORGANIZATION

Introduction

Often you have to make decisions based on incomplete information. For example, to find out how many Americans watch a television program you can't ask everyone, and to find out how long a light bulb lasts a company can't test all of its bulbs, particularly if it wants to have any left to sell. These and many other "real world" problems depend on taking samples and analyzing the data. Statistics is the branch of mathematics concerned with the collection, analysis, and interpretation of numerical data.

Random Samples

A *statistical population* is the collection of data that you are studying. Examples of statistical populations are the ages of citizens of the United States, the size of bank accounts in Illinois, and the brands of automobiles in California. Although in principle it is possible to measure the entire population, considerations of time and cost usually dictate that we measure only part of the population. For that purpose we use a *random sample* to represent the population.

Random Sample

A *random sample* is a selection of members of the population satisfying two requirements:

1. Every member of the population is equally likely to be included in the sample.
2. Every possible sample of the same size from the population is equally likely to be chosen.

A sample that does not meet these criteria is not representative and cannot be used to infer characteristics of the entire population. For example, a telephone survey to homes at 10 A.M. on a weekday would miss most employed people and so could not produce valid information about work skills. Random samples can be chosen using random number tables or by other methods, and we will assume in this chapter that the data we are analyzing are from a random sample.

Levels of Measurement

We shall consider only numerical data, which we classify into four types: *nominal, ordinal, interval,* and *ratio* data.

> *Nominal* data (*nominal* means "in name only") means numbers used only to *identify* objects. For example, the numbers on the backs of football jerseys are nominal data, since they do not mean that "number 24" can run twice as fast or throw twice as far as "number 12."
>
> *Ordinal* data (the word comes from "order") means numbers that can be arranged in order but differences between them are not meaningful. For example, you might rate the teachers you have had in college as "best" (number 1), "second best" (number 2), and so on. However, the difference between "1" and "2" may not be the same as the difference between "7" and "8."
>
> *Interval* data means numbers that can be put in order *and* differences between them can be compared, but ratios cannot. For example, it makes sense to talk about the *difference* between two temperatures, like 40° and 80°, but it makes no sense to say that 80° is *twice* as hot as 40°, since the zero point in the Fahrenheit (or Celsius) system is arbitrary and does not mean *zero heat*.
>
> *Ratio* data means numbers that can be put in order, whose differences can be compared, *and* whose *ratios* have meaning. Examples of ratio data are amounts of money and weights, since 80 of either really *is* twice 40. In general, with ratio data, the zero level *does* mean "none" of something.

When we refer to the *type* of a data collection, we mean the highest of these levels that applies, from nominal (lowest) to ratio (highest).

EXAMPLE **1**

JUDGING LEVELS OF DATA MEASUREMENT

Identify the level of measurement of each data collection.

a. Ten vehicles passing through a turnpike toll booth were classified as 1 for a car, 2 for a bus, and 3 for a truck, giving data {1, 3, 2, 3, 1, 1, 1, 1, 3, 1}.

b. A rock concert enthusiast rates the last eight performances she attended on a scale from 1 (awful) to 10 (awesome) as {8, 5, 7, 9, 7, 10, 6, 9}.

c. The temperatures of five students at the college infirmary are {99.3, 102.1, 101.8, 101.5, 100.9} degrees Fahrenheit.

d. The weights of twelve members of the cross-country team are {124, 151, 132, 153, 142, 147, 120, 127, 154, 119, 118, 116} pounds.

Solution

a. *Nominal data:* The numbers are used only to identify the type of vehicle. The number 2 does not mean more or less than the number 3, so differences between these numbers have no meaning.

b. *Ordinal data:* The rankings put the performances in relative order, but no meaning can be given to differences between the rankings. The difference between ratings 5 and 6 may be more or less than the difference between ratings 9 and 10.

c. *Interval data:* The temperatures can be put in order from "lowest" to "highest," and differences between them can be compared. However, this is not ratio data because the zero point in temperature is arbitrary—101° is not 1% hotter than 100°.

d. *Ratio data:* The weights can be put in order, "10 pounds heavier" has meaning, *and* 200 pounds really is twice as heavy as 100 pounds.

Practice Problem 1

Students filling out a course evaluation were asked for the following information: (a) gender: 1—male, 2—female; (b) course rating: 1—poor, 2—acceptable, 3—good, 4—outstanding; and (c) current grade point average. Identify the level of measurement of the data gathered from each question.   ➤ Solution on page 436

Bar Chart

A *bar chart* provides a visual summary of data with the number of times each value appears corresponding to the length of the bar for that value. The bars may be drawn vertically or horizontally, and they may have spaces between them.

EXAMPLE 2 **CONSTRUCTING A BAR CHART**

The colors of a random sample of thirty cars in the Country Corners Mall parking lot were recorded using: 1—blue, 2—green, 3—white,

4—yellow, 5—brown, and 6—black. Construct a bar chart for the data {3, 5, 2, 4, 1, 5, 1, 1, 1, 2, 4, 1, 6, 1, 6, 5, 6, 3, 1, 1, 1, 1, 1, 6, 3, 1, 2, 2, 4, 2}.

Solution

We first tally the frequency of each color and then sketch the bar chart.

Color	Tally	Frequency
1	ⅢⅢ ⅢⅢ ‖	12
2	ⅢⅢ	5
3	‖‖	3
4	‖‖	3
5	‖‖	3
6	‖‖‖	4

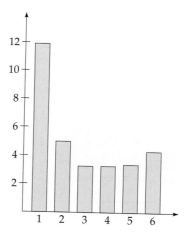

Notice that the most popular color by far is blue—about as popular as any three of the other colors combined. This observation is immediately obvious from the bar chart but is much more difficult to see from the list of data.

Spreadsheet Exploration

The following spreadsheet* shows the car colors and frequencies from Example 2, together with some of the many graphs that can be constructed from such data. The *pie chart* on the lower right is easily drawn by a computer, but it is difficult to draw accurately by hand because the size of the "slices" must be calculated first.

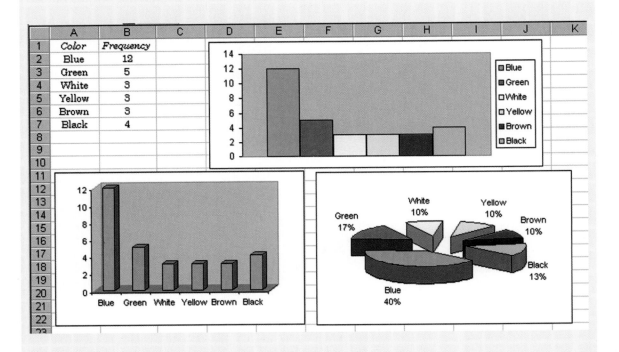

The pie chart is best for showing relative proportions because the total pie represents all of the data and the size of each slice shows its fraction of the whole.

Practice Problem 2   Construct a bar chart for the data given in Example 1(a) (page 429). Which type of vehicle was the most common? Which was the least common?

➤ Solutions on page 436

* See the Preface for information on how to obtain this and other Excel spreadsheets.

Histogram

A *histogram* is a type of bar chart in which the sides of the bars touch and the *width* of the rectangles has meaning. The data is divided into *classes*, usually numbering from 5 to 15, one for each bar, and each class has the same *width*:

$$\begin{pmatrix} \text{Class} \\ \text{width} \end{pmatrix} \approx \frac{\begin{pmatrix} \text{Largest} \\ \text{data value} \end{pmatrix} - \begin{pmatrix} \text{Smallest} \\ \text{data value} \end{pmatrix}}{\begin{pmatrix} \text{Number} \\ \text{of classes} \end{pmatrix}}$$

Round the class width *up* so that all data will be covered

We then make a tally of the number of data values that fall into each class. Some histograms use the convention that any value on the boundary between two classes belongs to the upper class, while others are designed so that no data value falls on a class boundary.

EXAMPLE **3**

CONSTRUCTING A HISTOGRAM

When FirstAmerica National Bank set up its website for online banking, they asked users to fill out a questionnaire. Those who said that they would use online banking at least twice a week listed their ages as follows: {34, 10, 53, 50, 80, 38, 39, 31, 52, 41, 46, 46, 41, 69, 73, 57, 40, 52, 47, 68, 22, 33, 51, 65, 23, 47, 64, 45, 26, 74}. Construct a histogram for the data and determine whether online banking is most popular with young, middle-aged, or elderly users.

Solution

Scanning the data, we see that the smallest and largest ages are 10 and 80. Therefore, if we choose 6 classes we obtain a class width of $\frac{80 - 10}{6} \approx 11.7$, which we round up to 12. Since we rounded up, we may choose our first class to start just below the lowest data point of 10, say at 9.5. We then add successively the class width of 12 to get the other class boundaries.

Class Boundaries	Tally	Frequency
9.5–21.5	|	1
21.5–33.5	⦀⦀	5
33.5–45.5	⦀⦀ ||	7
45.5–57.5	⦀⦀ ⦀⦀	10
57.5–69.5	||||	4
69.5–81.5	|||	3

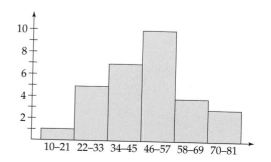

The histogram shows that FirstAmerica's website is most popular with middle-aged customers (in their 40's and 50's).

We may label the base of each bar with either the range of its data values or the class boundaries. Different numbers of bars or class boundaries will give different histograms, as the following Graphing Calculator Exploration shows.

Graphing Calculator Exploration

Histograms can be drawn on graphing calculators using the STAT PLOT command. To explore several histograms of the data from Example 3 (page 433), proceed as follows:

```
{34,10,53,50,80,
38,39,31,52,41,4
6,46,41,69,73,57
,40,52,47,68,22,
33,51,65,23,47,6
4,45,26,74}→L₁
```

a. Enter the data values as a list and store it in the list L_1.

b. Turn off the axes in the window FORMAT menu, turn on STAT PLOT 1, select the histogram icon with Xlist L_1, and set the WINDOW parameters Xmin = 9.5, Xmax = 81.5, Xscl = 12 to match the start, finish, and class width we used. GRAPH and then TRACE to explore your histogram.

Of course, different choices of the class width, number of classes, and starting value will result in slightly different histograms. The following are several possibilities:

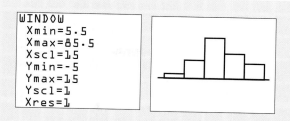

```
WINDOW
  Xmin=5.5
  Xmax=85.5
  Xscl=15
  Ymin=-5
  Ymax=15
  Yscl=1
  Xres=1
```

Notice that each histogram supports the conclusion that the website is most popular with those in the middle age range.

How do histograms compare to the graphs of probability distributions that we drew in Chapter 5? They are very similar, but in Chapter 5 the heights of the bars were *probabilities*, with the heights adding up to 1. Here the heights of the bars are *frequencies*, with the heights adding to equal the total number of data points.

6.1 Section Summary

A *statistical population* is the collection of data that you are studying. A *random sample* is a selection from the population such that every member is equally likely to be included and every possible sample of the same size is equally likely to be chosen.

There are four levels of data measurement:

Data	Uses
Nominal	Classify by name
Ordinal	Compare order (1st, 2nd, . . .)
Interval	Compare order and differences
Ratio	Compare order, differences, and ratios

Bar charts can be drawn for *any* type of data, with the length of the bars indicating the frequency of the values. Histograms involve dividing the data into adjacent intervals or *classes*, whose widths are calculated from the formula on page 433. Histograms are used only for interval and ratio data since only in these cases does the calculation of the class width have meaning.

➤ **Solutions to Practice Problems**

1. a. Nominal   **b.** Ordinal   **c.** Ratio

2.

Vehicle	Tally	Frequency
1	⊞ I	6
2	I	1
3	III	3

1 = car
2 = bus
3 = truck

Cars were most common and buses were least common.

6.1 Exercises

Identify the level of data measurement in each situation.

1. BUSINESS: Discount Brokers A survey of ten randomly selected traders at a discount broker service records the numbers of transactions involving stocks, bonds, and mutual funds last Wednesday.

2. BUSINESS: Travel Agencies An airline surveys twenty randomly selected travel agencies in and around San Francisco and records the numbers of tickets sold at each last month to Hawaii, Alaska, Mexico, and New York.

3. BUSINESS: Mortgage Banking The home office of the First & National Bank sends questionnaires to two hundred randomly selected families who applied for mortgages last summer, asking for ratings (from "poor" to "outstanding") of the courtesy of the mortgage staff at the branch offices.

4. BUSINESS: Shoe Sales The national personnel manager for the Feets-R-Us shoe store franchise reviews the supervisor reports on thirty randomly selected sales employees and counts the number of "below average," "average," and "above average" performance evaluations.

5. GENERAL: SAT Scores The test results on the verbal SAT administered by the Educational Testing Service are numerical scores from 200 to 800. A survey records the verbal SAT scores of twelve randomly selected seniors at J. T. Marshall High School.

6. SOCIAL SCIENCE: Midwest Aviation A survey records the years of birth of fifty randomly selected commercial pilots residing in Missouri.

7. SOCIAL SCIENCE: IQ Scores An educational research team working on a grant from the National Institutes of Health records the IQ scores of eighty randomly selected fifth graders in the Baltimore County Public School system.

8. BUSINESS: Quality Control A major cereal manufacturer randomly selects thirty boxes of Corn Chex Crunches from the shipping warehouse and records the actual net weights.

9. PERSONAL FINANCE: Disposable Income The Chamber of Commerce tabulates the disposable family incomes of ninety randomly selected households in Thomsonville.

10. BUSINESS: Financial Trends The *Get Rich Slowly* investment newsletter reports the price-to-earnings ratios of sixty randomly selected technology stocks traded last month on the American Stock Exchange.

Construct a bar chart of the data in each situation.

11. **SOCIAL SCIENCE: Marital Status** A random sample of shoppers in the Kingston Valley Mall counted 39 "never married," 97 "married," 16 "widowed," and 8 "divorced" individuals.

12. **SOCIAL SCIENCE: Place of Birth** A random sample of skiers in Grand Junction last February counted 12 from Colorado, 27 from California, 19 from New York, 8 from Florida, and 6 from other states or nations.

13. **SOCIAL SCIENCE: State Legislators** A random sample of candidates running for election to the Nevada legislature counted 4 business executives, 7 real estate developers, 18 lawyers, 1 doctor, 3 certified public accountants, and 2 retirees.

14. **SOCIAL SCIENCE: Voter Concerns** A random sample of voters leaving the polls identified the most important issue in the Higginsville school board election as follows: 8 for teacher tenure, 54 for budget, 11 for after-school day care, 19 for the bond proposal for a new football stand at the high school, and 3 making "no comment."

For each situation construct a histogram and use it to answer the question. (For Exercises 19–25, histograms may vary.)

You may use a graphing calculator if your instructor permits it.

15. **BUSINESS: Shipping Weights** A random sample of twenty packages shipped from the Holiday Values Corporation mail order center found the following weights (in ounces). (Use 6 classes of width 8 starting at 12.5.)

29	21	60	23	39	24	33	49	37	17
13	27	56	21	30	43	34	24	18	13

Based on your histogram, which type of boxes are in greatest demand: lighter weight, mid-weight, or heavier weight?

16. **PERSONAL FINANCE: Blue Book Values** A random sample of thirty-two cars parked at Emmett's Field last June found the following "blue book" values (in thousands of dollars) (use 6 classes of width 4 starting at 2.5):

15	5	8	17	18	17	8	9
5	13	6	9	5	19	8	13
20	9	7	13	25	16	5	25
25	17	24	10	23	14	15	4

Based on your histogram, which two adjacent classes have the highest total number of cars?

17. **GENERAL: Websites** A random sample of twenty-eight student accounts at the campus Academic Computing Center showed the following number of website "hits" between 9 and 12 P.M. last Wednesday (use 8 classes of width 10 starting at 10.5):

38	29	16	35	69	63	13
58	11	58	34	73	45	57
70	16	45	37	39	88	69
47	30	40	12	47	20	61

Based on your histogram, which two classes have the highest number of hits?

18. **ATHLETICS: Football Players** A random sample of twenty-four defensive linemen in the East Coast Division II college conference found the following weights (in pounds) (use 8 classes of width 15 starting at 140.5):

173	171	211	169	258	213	143	236
191	163	192	235	196	208	157	221
174	200	195	198	228	152	188	201

Based on your histogram, are the weights evenly distributed or more concentrated in one part of the range?

19. **BUSINESS: Airline Tickets** A random sample of forty passengers traveling economy class from New York to Los Angeles last Thanksgiving weekend found the following one-way ticket prices (in dollars):

425	236	481	559	258	473	522	480
544	451	244	440	530	565	445	569
255	515	412	377	453	518	490	301
379	439	229	531	465	252	219	510
565	477	440	551	518	525	230	437

Based on your histogram, are more ticket prices grouped at the lower end, the middle, or the upper end of the range of costs?

20. **GENERAL: Insurance Claims** A random sample of thirty accident repair claims received in December at the western division office of the Some States Insurance Company were for the following dollar amounts:

5197	6012	1859	6207	3497	1097
4785	5950	3652	6279	3639	4553
6249	5999	6601	1193	5961	4625
5918	4289	2948	5043	1329	5483
968	5297	1423	5159	5527	4880

Based on your histogram, are more claims grouped at the lower end, in the middle, or at the upper end of the range?

21. **SOCIAL SCIENCE: Law Practice** A random sample of twenty-one attorneys hired within the past year at the U.S. Department of Justice found the following numbers of hours worked last week:

48	44	50	49	44	45	51
50	57	36	53	44	48	54
60	52	52	39	47	49	57

Based on your histogram, are more hours grouped at the lower end, the middle, or the upper end of the workweeks?

22. **GENERAL: Light Bulbs** A random sample of thirty-six 75-watt light bulbs found the following lifetimes (number of hours until burn out):

1070	1210	1280	1230	1340	1280
1200	1120	1310	1100	1170	1000
1290	1420	1190	1280	1160	1290
1250	1130	1180	1150	1110	1200
1230	1160	1240	1250	1230	1240
1000	1280	1180	1280	1270	1210

Based on your histogram, are the lifetimes roughly evenly distributed or concentrated in one part of the range?

23. **SOCIAL SCIENCE: Medical Insurance** A random sample of twenty private practice physicians in the Lehigh Valley found the following numbers of insurance plans accepted by each:

16	21	9	23	19	23	20	23	18	19
19	29	20	21	20	23	20	14	11	19

Based on your histogram, are the numbers of plans roughly equally distributed over the range or grouped in one area?

24. **BUSINESS: Dining Out** A random sample of forty dinner checks at the Lotus East Chinese Restaurant last Saturday evening showed the following total costs:

43	104	45	60	25	103	33	96
108	41	67	53	43	48	99	45
35	40	66	103	27	110	60	33
46	126	112	84	42	90	74	46
102	41	48	41	54	44	82	33

Based on your histogram, are the dinner checks grouped more at the lower or the upper end of the cost range?

25. **SOCIAL SCIENCE: Automobile Accidents** A random sample of twenty drivers involved in accidents in Columbia County during August showed the following ages:

23	16	75	18	19
74	17	21	73	27
61	18	72	18	44
16	75	29	17	24

Based on your histogram, are the accidents roughly evenly distributed across age groups? Give an explanation for the general shape of the histogram.

6.2 MEASURES OF CENTRAL TENDENCY

Introduction

We often hear of *averages,* from grade point averages to batting averages. Given a collection of data, we often use the average as a representative or typical value. In this section we will see that not all types of data are appropriately summarized by an average, and sometimes another kind of "typical" value is more representative. To avoid confusion with casual meanings of "average," these different kinds of typical values are called *measures of central tendency.*

Mode

The *mode* of a collection of data is the *most frequently occurring* value. If all of the values in a data set occur the same number of times, there is no mode, while if only a few of the values occur the same maximal number of times, there are several modes. The mode can be used with every level of data measurement, and it is the *only* one appropriate for *nominal* data (such as colors of cars, where "averages" have no meaning).

EXAMPLE 1

FINDING MODES

Find the mode for each collection of numbers.

a. {1, 2, 2, 2, 2, 3, 3, 4, 4, 4, 10}
b. {1, 2, 2, 2, 2, 3, 3, 4, 4, 4, 4}
c. {2, 2, 2, 3, 3, 3, 4, 4, 4}

Solution

a. Since 2 appears more often than any other value, the mode is 2.
b. Since both 2 and 4 occur four times and every other number occurs fewer times, the modes are 2 and 4.
c. Since each number occurs the same number of times, there is no mode.

The following are bar graphs for these collections of data. Notice that the presence of one number that is very different from the others, such as the 10 in the first graph, has no effect on the mode: If the 10 were replaced by 100 or even 1000, the mode would not change.

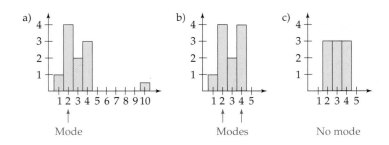

a) Mode

b) Modes

c) No mode

If a collection of data has just one mode (as in the first graph) it is said to be *unimodal,* and if it has two modes (as in the second) it is said to be *bimodal.*

Practice Problem 1   Find the mode of   {17, 20, 20, 13, 10, 17, 10, 20}.   ➤ Solution on page 446

Graphing Calculator Exploration

When finding the mode of a long list of values, it is often easier to first sort the list so that all the same values appear grouped within the list. For instance, to find the mode of   {62, 56, 52, 53, 58, 64, 68, 67, 60, 61, 57, 58, 57, 63, 55},   proceed as follows:

```
{62,56,52,53,58,
64,68,67,60,61,5
7,58,57,63,55}→L
1
```

a. Enter the data values as a list and STORE it in the list L_1 (or enter the numbers into L_1 directly using STAT and EDIT).

```
EDIT CALC TESTS
1:Edit...
2:SortA(
3:SortD(
4:ClrList
5:SetUpEditor
```

b. From the STAT or LIST menu, select SortA (to sort in *ascending* order). (You could also use SortD for descending order.)

```
{62,56,52,53,58,
64,68,67,60,61,5
7,58,57,63,55}→L
1
{62 56 52 53 58...

SortA(L1)
```

c. Complete the ascending sort command by entering the name of the list of values, L_1.

d. Examine the sorted list L_1 to identify the value(s) appearing most often. The modes of this collection of values are 57 and 58 because they appear twice and the others appear only once each.

Median

The *median* is the *middle* value of a list of data values when sorted in ascending or descending order. If there is an even number of data values, the median is the arithmetic average of the two middle values (their sum divided by 2). Since the median depends on arranging the values in order, the median is not appropriate for nominal data. For instance, it would be meaningless to find a "median color" of a car (see Example 2 on pages 430–431), since the assignment of numbers to colors was arbitrary. The median is appropriate for every level of data measurement except the nominal level.

EXAMPLE **2**

FINDING MEDIANS

Find the median of each data set.

a. {1, 3, 4, 7, 11, 18, 39}

b. {1, 3, 4, 7, 11, 18, 1000}

c. {26, 16, 10, 6, 4, 2}

d. {62, 56, 52, 53, 58, 64, 68, 67, 60, 61, 57, 58, 57, 63, 55}

Solution

a. Since these seven numbers are arranged in ascending order, the median is the fourth number, 7, because three numbers are smaller and three are larger.

b. Just as in part (a), the median is 7. Notice that replacing the largest data value in the previous data set with a much larger value did not change the median.

c. Since these six values are arranged in descending order, the median is 8, the number halfway between the two middle values of 10 and 6. Thus the median need not be one of the data values.

d. Since these fifteen values are not in ascending or descending order, we first order them (as in the Graphing Calculator Exploration on the previous page):

$$\{52, 53, 55, 56, 57, 57, 58, 58, 60, 61, 62, 63, 64, 67, 68\}$$

The median of these is the middle value, 58, since there are seven data values no larger and seven values no smaller.

Practice Problem **2**   Find the median of the values $\{4, 5, 5, 6, 7, 8\}$.   ➤ Solution on page 446

Graphing Calculator Exploration

The median of an unsorted list of values can be found with a single calculator command. To find the median of $\{62, 56, 52, 53, 58, 64, 68, 67, 60, 61, 57, 58, 57, 63, 55\}$,   the unordered values from Example 2(d) above, proceed as follows:

Enter the data values as a list and store it in the list L_1. From the LIST MATH menu, select the median command and apply it to the list L_1.

```
{62,56,52,53,58,
64,68,67,60,61,5
7,58,57,63,55}→L
₁
```

```
NAMES OPS MATH
1:min(
2:max(
3:mean(
4:median(
5:sum(
6:prod(
7↓stdDev(
```

```
median(L₁)
                58
```

Mean

The *mean* is the usual arithmetic average found by summing the values and dividing by the number of them:

Mean

The *mean* $\bar{x}$ of the n values $x_1, x_2, \ldots, x_n$ is

$$\bar{x} = \frac{1}{n}(x_1 + x_2 + \cdots + x_n)$$

The mean of a collection of numbers is equivalent to the mean of a random variable that is equally likely to be any of the numbers. The mean of a data set is sometimes called the *sample mean* to distinguish it from the mean of a probability distribution. Calculating the mean assumes that the distance between values is meaningful, so the mean is not an appropriate measure for ordinal or nominal data. For instance, it would be meaningless to find the mean of the concert ratings in Example 1(b) (page 429), from 1—awful to 10—awesome, because we do not know if a 6 and an 8 should average to a 7, since the intervals between them may not be the same. The mean is an appropriate measure of central tendency only for interval and ratio data.

EXAMPLE 3

FINDING MEANS

Find the mean of each data set.

a. {5, 16, 14, 1, 4, 20}

b. {5, 16, 14, 1, 4, 200}

```
mean({5,16,14,1,
4,20})
                10
mean({5,16,14,1,
4,200})
                40
```

Solution

a. $\bar{x} = \frac{1}{6}(5 + 16 + 14 + 1 + 4 + 20) = \frac{1}{6} \cdot 60 = 10$

b. $\bar{x} = \frac{1}{6}(5 + 16 + 14 + 1 + 4 + 200) = \frac{1}{6} \cdot 240 = 40$

The bar over x indicates *mean*

Notice that the data set in part (b) is the same as in part (a) except that the last number was changed from 20 to 200, with the result that the mean changed substantially. Thus, unlike the mode and the median, the mean is sensitive to extreme values.

The mean is where the histogram of the values would just balance.

From the data in Example 3

These calculations show that when finding the mean there is no need to arrange the numbers in any particular order and that the mean need *not* be one of the original data values.

Mean, Median, and Mode

We have three measures of central tendency, the mean, the median, and the mode. Each has advantages and disadvantages, and one may be more appropriate than the others as a "representative" value for a particular data set.

- The *mean* has the advantage of being easy to calculate, and takes all of the values into account, but it can be heavily influenced by extremes (see Example 3).

- The *median* is not influenced by extremes, and it can be used for ordinal data (that is, data that is only ranked, say from best to worst), but it is usually found by first sorting the data, which can be difficult for a large data set.

- The *mode* is easy to find, is not influenced by extremes, and can be used even for nominal data. However, a data set may not have a mode, or it may have several.

The following histograms show several possible relationships between the mean, the median, and the mode.

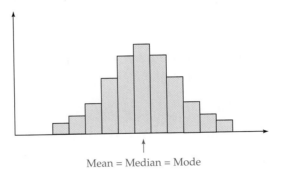

Mean = Median = Mode

For data that are symmetric about a single mode, the mean, the median, and the mode will all be the same.

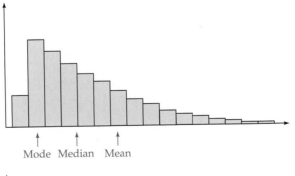

Mode Median Mean

For a *skewed* data set, the mode can occur at an early peak and the mean can be distorted by extreme values.

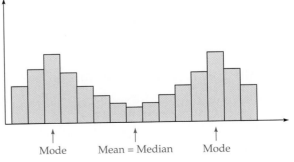

Mode   Mean = Median   Mode

For *bimodal* data, the modes may be quite far from the mean and median.

For an example of a histogram like that shown above, think of the graph of the number of automobile accidents for drivers of different ages: Teenage drivers have frequent accidents, then the number decreases with age, bottoming out in the 40s and 50s, then increasing again for elderly drivers with another peak in the top age bracket. That is, such a histogram would look like the one above but with the two peaks exactly at the ends (see Exercise 25 on page 438).

Practice Problem 3

For an extreme case of skewing, imagine that Bill Gates (the richest man in America, with an estimated wealth of $80 billion in 2001) lived in a town along with 7999 penniless people.

a. What would be the average (mean) wealth of people in the town?
b. What would be the median wealth?
c. What would be the mode?
d. Which of these seems the most "representative" measure of central tendency?

➤ Solution on page 446

EXAMPLE 4   **BASEBALL SALARIES**

In the 1995 baseball strike it was disclosed that the mean salary of major league baseball players was $1.2 million, while the median salary was only $500,000. What does this say about the distribution of baseball salaries?

Solution

If the *middle* salary is $500,000, then for the mean to be higher, there must be a few very large salaries. That is, a few baseball players must be receiving enormously high salaries to skew the mean that much. In such cases, the median is usually considered a more representative value for most players than the mean.

6.2 Section Summary

Although there are three measures of central tendency, not all of them can be used with every type of data. For nominal data only the mode is appropriate, for ordinal data either the mode or the median is appropriate, and for interval or ratio data all three are appropriate.

Measure of Central Tendency	Definition	Appropriate Data Levels
Mode	Most frequent	All levels
Median	Middle value	Ordinal, interval, and ratio
Mean	$\bar{x} = \dfrac{1}{n}(x_1 + x_2 + \cdots + x_n)$	Interval and ratio

The mean is sensitive to changes in the extremes, but the median and mode are not. The mean of a data set is sometimes called the *sample mean* to distinguish it from the mean of a probability distribution.

➤ **Solutions to Practice Problems**

1. Since 20 occurs three times and no other value occurs this often, the mode is 20.

2. Since these six values are arranged in ascending order, the median is 5.5, the number halfway between the two middle values of 5 and 6.

3. **a.** $\bar{x} = \dfrac{\$80 \text{ billion}}{8000} = \$10 \text{ million}$

   b. Median is $0.

   c. Mode is $0.

   d. The median and mode seem most representative.

6.2 Exercises (will be helpful throughout.)

Find the mode, median, and mean of each data set.

1. {6, 17, 12, 10, 15, 16, 5, 8, 9, 14, 9}

2. {10, 7, 9, 13, 11, 14, 15, 14, 8, 9}

3. {15, 14, 5, 7, 5, 14, 9, 7, 7, 14, 7, 14, 12}

4. {11, 19, 20, 11, 12, 18, 11, 20, 16, 18, 17, 10, 13, 11, 15}

5. {19, 11, 10, 19, 18, 12, 19, 10, 14, 12, 13, 17, 20, 19, 15}

6. {12, 5, 10, 6, 7, 6, 15, 7, 6, 5, 6, 5, 13, 10, 7, 13, 8, 11, 12}

7. {8, 15, 10, 14, 13, 14, 5, 10, 7, 9, 13, 14, 15, 14, 15, 19, 13, 12, 8}

8. {11, 8, 23, 21, 14, 16, 15, 11, 9, 22, 15, 11, 12, 11, 10, 15, 10, 25, 18, 6, 11}

9. {19, 9, 15, 8, 18, 15, 12, 22, 16, 19, 15, 19, 20, 24, 7, 14, 21, 19, 20, 5, 19}

10. {10, 13, 8, 15, 8, 12, 12, 12, 14, 8, 12, 11, 14, 9, 8, 14, 14, 14, 10, 10}

APPLIED EXERCISES

Find the mode, median, and mean for the data in each situation.

11. **GENERAL: Emergency Services** A random sample from the records of the Farmington Volunteer Ambulance Corps found the following twenty response times (in minutes) to 911 calls received during February:

14	9	25	14	12	16	5	20	20	25
20	24	12	7	19	12	6	13	24	18

12. **BUSINESS: Car Sales** A random sample of fifteen General Motors salesmen at Tri-State Dealers found the following earnings (in dollars) for the first week of September:

1200	1200	1100	1300	900
600	800	700	1200	500
1000	1200	600	900	900

13. **ENVIRONMENTAL SCIENCE: Volunteer Recycling** A random sample of thirty volunteers helping sort newspapers, aluminum, and glass at the Parkerville Regional Recovery Center found the following hours contributed last week:

10	7	13	11	7	11	10	5	9	14
14	12	12	5	10	12	19	10	7	10
9	21	13	10	7	18	10	15	10	12

14. **SOCIAL SCIENCE: Commuting Times** A random sample of twenty-four employees at the Grover–Smith Forge found the following morning commute times (in minutes) for last Monday:

75	60	50	55	90	53	65	71
33	81	60	74	71	75	83	53
66	75	47	36	77	54	67	65

15. **SOCIAL SCIENCE: Unemployment** A random sample of nineteen newly unemployed workers in Wilmington last August reported the following times (in weeks) until they found new jobs:

12	6	9	9	11	14	8	15	12	13
9	8	9	7	9	5	6	13	15	

16. **BIOMEDICAL: Cereal Killers** The calories per cup of several popular cold cereals, served with half a cup of milk and a peach or half a banana, are given below. Find the mode, the median, and the mean. Looking at the data, can you see why the mean is higher than the median?

Type	Calories
Banana Nut Crunch	360
Blueberry Morning	300
Cracklin Oat Bran	390
Fruit and Fibre	320
Great Grains	560
Mini Chex	360
Mother's Cream Oats	350
Müslix	420
Raspberry Granola	520
Raisin Bran Crunch	310
Raisin Nut Bran	390

17. **BIOMEDICAL: Chicken Sandwiches** *Consumer Reports* listed the calories for chicken sandwiches as follows. Find the mode, the median, and the mean. Looking at the data, can you see why the mean is higher than the median?

Type	Calories
Chick-fil-A	290
Wendy's Grilled Fillet	300
KFC Tender Roast	350
Arby's Grilled Deluxe	420
Wendy's Fillet (breaded)	430
McDonald's McGrill	450
KFC Original (breaded)	450
Burger King BK Broiler	500
McDonald's (breaded)	550
Burger King (breaded)	660
Boston Market	750

18. **SOCIAL SCIENCE: Raising IQs** When the mathematician Frederick Mosteller left Princeton for Harvard, he is reputed to have joked that he was "raising the average IQ of both places." How is this possible?

19. **ATHLETICS: Basketball Salaries** The salaries of the Los Angeles Lakers for the 2001–2002 season were as follows. Find the mode, the median, and the mean of the salaries. By looking at the data, can you see why the mean is so much higher than the median?

Player	Salary ($ million)
Kobe Bryant	11.3
Joe Crispin	0.3
Derek Fisher	3.0
Rick Fox	3.8
Devean George	0.8
Robert Horry	5.3
Lindsey Hunter	2.7
Mark Madsen	0.8
Jelani McCoy	0.5
Stanislav Medvedenkn	0.5
Shaquille O'Neal	21.4
Mitch Richmond	0.6
Brian Shaw	0.6
Samaki Walker	1.4

Source: HoopsWorld@Bskball.com

20. **SOCIAL SCIENCE: New York Income** New York State's mean income is the fourth highest in the nation, while its median income scores only 29th place, well below that of half of the states. What does this say about the incomes of New Yorkers?

21. **PERSONAL FINANCE: Household Income** The histogram in the Application Preview on page 427 shows the distribution of household incomes in the United States for the year 2001. The following is a random sample of these incomes in thousands of dollars: {38.2, 57.3, 102.1, 40.7, 47.9, 29.0, 38.1, 78.2, 33.6, 31.3, 112.8, 89.2, 67.9, 41.4, 39.6, 53.1, 113.3, 32.1, 39.7, 21.5}.

   a. Find the mode, the median, and the mean.
   b. Which is larger? Can you explain why in terms of the shape of the histogram on page 427?

Explorations and Excursions

The following problems extend and augment the material presented in the text.

Mean of Grouped Data

Sometimes the original data are not available and only the class intervals and frequencies are known. The mean of these grouped data may be estimated by assuming that each data value lies at the midpoint of its class. Therefore, we sum each midpoint value at the center of each class interval as many times as is indicated by the frequency and then divide that total by the sum of the frequencies.

22. For the grouped data

Class interval	14.5–19.5	19.5–24.5	24.5–29.5
Frequency	7	9	4

verify that the mean is

$$\bar{x} = \frac{17 \cdot 7 + 22 \cdot 9 + 27 \cdot 4}{7 + 9 + 4}$$

$$= \frac{119 + 198 + 108}{20} = \frac{425}{20} = 21\frac{1}{4}$$

23. For the grouped data

Class interval	6–12	13–19	20–26	27–33
Frequency	5	10	4	6

verify that the mean is

$$\bar{x} = \frac{9 \cdot 5 + 16 \cdot 10 + 23 \cdot 4 + 30 \cdot 6}{5 + 10 + 4 + 6}$$

$$= \frac{477}{25} = 19.08$$

Use the midpoints of the class intervals to estimate the mean of each set of grouped data.

24. **BUSINESS: Truck Repairs** The Wink-Quick package service maintains its own statewide fleet of delivery trucks. A random sample of the repair records found the following numbers of days last year that fifty trucks were not available for use because their repairs could not be completed overnight:

Number of days	0–2	3–5	6–8	9–11
Frequency	30	7	12	1

25. **BUSINESS: Overtime Pay** A random sample of payroll records for one hundred and ten craftspeople at the Old Fashion Vermont Country Furniture factory found the following amounts of overtime pay (in dollars) for the last week in November:

Overtime pay	0–19	20–39	40–59	60–79	80–99
Frequency	29	11	42	12	16

26. **GENERAL: Charity Donations** Donors at the Elktonburg Hospital Charity Ball are designated as "friends" for donations from $51 to $100, "benefactors" for $101 to $150, and "founders" for $151 to $200. A random sample of thirty hospital supporters found the following designations:

Designation	Friends	Benefactors	Founders
Frequency	9	6	15

6.3 MEASURES OF VARIATION

Introduction

It is helpful to summarize a given data set by a few numbers, as we did with random variables in the previous chapter. For a *representative* or *typical* value, we use one of the measures of central tendency (mode, median, or mean, whichever is most appropriate). But how do we measure whether the data are grouped tightly or spread widely about this central value? In this section we discuss three ways of describing the spread of the values. All three use differences among data values, so we must now restrict our discussion to data values at the interval or ratio level of measurement.

Range

The simplest measure of the spread of a data set is the *range*, which is the difference between the smallest and largest data values.

EXAMPLE 1

FINDING RANGES

Find the range of each data set.

a. $\{1, 5, 6, 7, 7, 7, 8, 15\}$

b. $\{1, 2, 10, 10, 12, 13, 15\}$

Solution

a. The largest value is 15 and the smallest is 1, so the range is $15 - 1 = 14$.

b. Again, the range is $15 - 1 = 14$. Notice that the range is *not* sensitive to the distribution of values between the two extremes.

Practice Problem 1 Find the range of the data values {15, 17, 27, 12, 30, 15, 10, 27, 25, 29}.

➤ Solution on page 457

Box-and-Whisker Plot

A *box-and-whisker plot* provides a quick way to visualize how data values are distributed between the largest and smallest by graphically displaying a *five-point summary* of the data. The *minimum* is the smallest value, the *maximum* is the largest, and we have already discussed the *median*. The *first quartile* is the median of the data values below the median (but not including the median), and the *third quartile* is the median of the data values above the median (but not including the median). The median may then be called the *second quartile*. The box extends from the first quartile to the third quartile with a vertical bar at the median. The whiskers extend from the box out to the minimum and maximum data values.

Notice that the distance between the extreme ends of the whiskers is the *range* of the data and that 50% of the data is contained in the box. The length of the box, from the first quartile to the third quartile, is sometimes called the *interquartile range*. In general, the quartiles will not be evenly spaced.

EXAMPLE 2

CONSTRUCTING A BOX-AND-WHISKER PLOT

Make a five-point summary and draw the box-and-whisker plot for the data values {20, 37, 65, 77, 78, 79, 81, 82, 83, 85, 87, 90}.

Solution

Since these values are already in order, the minimum is 20 and the maximum is 90. Since there are twelve values, the median is halfway between the sixth value (79) and the seventh (81), so the median is 80. The values below the median are the six values {20, 37, 65, 77, 78, 79}, and the first quartile is the median of these values (that is, the value halfway between 65 and 77), so the first quartile is 71. Similarly, the values above the median are the six values {81, 82, 83, 85, 87, 90}, and the third quartile is the median 84 because it is the value halfway between 83 and 85. The five-point summary of this data set and the corresponding box-and-whisker plot are shown below.

Minimum = 20
First quartile = 71
Median = 80
Third quartile = 84
Maximum = 90

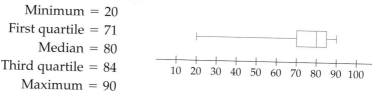

The box-and-whisker plot shows clearly that the bottom quarter of the values are spread rather thinly from 20 to 71, whereas the entire top half is concentrated between 80 and 90 (with a quarter of the values in the narrow range between 80 and 84). Such clustering is much easier to see from the box-and-whisker plot than from simply looking at the data.

Practice Problem 2

Make a five-point summary and draw the box-and-whisker plot for the data values {8, 10, 13, 14, 15, 16, 17, 19, 23, 24, 29}. From your plot, which quarter is the most tightly clustered?   ➤ Solution on page 457

Graphing Calculator Exploration

Box-and-whisker plots can be drawn on graphing calculators by using the STAT PLOT command. To explore several box-and-whisker plots, including one of the data from Example 2 on the previous page, proceed as follows.

a. Turn off the axes in the window FORMAT menu and set the WINDOW parameters to Xmin = 0 and Xmax = 100.

b. Enter the data values:

{20, 37, 65, 77, 78, 79, 81, 82, 83, 85, 87, 90} as the list L_1,

{20, 31, 39, 41, 63, 80, 80, 80, 81, 87, 88, 90} as the list L_2, and

{20, 55, 58, 60, 60, 61, 63, 64, 65, 79, 81, 90} as the list L_3.

c. Turn on STAT PLOT 1 and select the box-and-whisker plot icon with Xlist L_1, and do the same for STAT PLOT 2 with Xlist L_2 and STAT PLOT 3 with Xlist L_3.

d. GRAPH and then TRACE to explore your box-and-whisker plots. The screen shows that the first quartile of the second list of values is 40. The first plot agrees with the one we drew in Example 2.

Interpreting Box-and-Whisker Plots

How do we interpret box-and-whisker plots? Consider an example.

 EXAMPLE 3

DRAWING AND INTERPRETING BOX-AND-WHISKER PLOTS

The following five-point summaries are based on a survey of annual incomes (in thousands of dollars), for those with three different levels

of education: high school only, some college, and a bachelor's degree.*
Draw the box-and-whisker plot for each and interpret the results.

High school only:	9.6, 16.2, 23.0, 32.6, 50.1
Some college:	10.8, 19.1, 26.8, 37.5, 57.5
Bachelor's degree:	12.5, 24.9, 35.8, 51.1, 83.6

Solution

For each five-point summary (minimum, 1st quartile, median, 3rd quartile, and maximum) we plot the five points and draw the boxes and whiskers as follows.

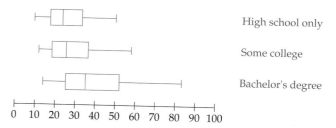

How do we interpret these plots? The first thing to look at is the median (the middle bar): Notice that it changes very little from "high school" to "some college," but then shifts significantly for "bachelor's degree," indicating that there is little income benefit from college unless you complete it. Then look at the boxes (the middle 50% of each group): As with the median, there is only a slight change from "high school" to "some college" and then a much larger change for "bachelor's degree." Notice that the minimum points (the left end of the left whisker) change very little (possibly because in any category some will leave employment), while the maximum increases much more dramatically (indicating a much greater earning potential for college graduates).

In each plot the right half of the box (the 25% above the mean) is wider than the left half (the 25% below the mean), and the right whisker is longer than the left whisker. This means that the distribution is more spread out to the right or *skewed to the right,* indicating that there are a few unusually high incomes. (If the data set were symmetric, the distances from the median to the quartiles and the extremes would be the same on either side.)

Clearly one can get much more information from comparing box-and-whisker plots than from just comparing medians.

* These and other data in this section are based on the *Current Population Survey* of the Bureau of Labor Statistics and the Bureau of the Census.

Sample Standard Deviation

A third way of measuring variation is the *sample standard deviation,* which estimates the typical variation of the data values from the (sample) mean of the data.

Sample Standard Deviation

The *sample standard deviation* of the n data values $x_1, x_2, \ldots, x_n$ is

$$s = \sqrt{\frac{(x_1 - \bar{x})^2 + \cdots + (x_n - \bar{x})^2}{n - 1}} \qquad \bar{x} = \text{mean}$$

The differences between the values and the mean are *squared,* $(x_i - \bar{x})^2$, so that positive differences don't cancel with negative ones. Since squaring gives an answer in *square units* (like "dollars squared"), we take the square root of the result to return to the original units. For n values we divide by $n - 1$. This is because the n numbers are related by the fact that their sum divided by n is the mean $\bar{x}$, so that any one value can be determined from the others together with the mean, leaving only $n - 1$ independent values in the sum. We call it the *sample* standard deviation to remind us of this.

EXAMPLE 4

CALCULATING A SAMPLE STANDARD DEVIATION

Find the sample standard deviation of the data values $\{9, 13, 16, 18\}$.

Solution

Since there are four values, $n = 4$. First we must find the mean:

$$\bar{x} = \tfrac{1}{4}(9 + 13 + 16 + 18) = \tfrac{1}{4}(56) = 14$$

Then

```
stdDev({9,13,16,
18})
         3.915780041
```

$$s = \sqrt{\frac{(9 - 14)^2 + (13 - 14)^2 + (16 - 14)^2 + (18 - 14)^2}{4 - 1}}$$

$$= \sqrt{\frac{25 + 1 + 4 + 16}{3}} = \sqrt{\frac{46}{3}} \approx 3.9 \quad s = \sqrt{\frac{(x_1 - \bar{x})^2 + \cdots + (x_n - \bar{x})^2}{n - 1}}$$

The sample standard deviation s is (approximately) 3.9.

Practice Problem 3 Find the sample standard deviation of the data values {6, 7, 12, 15}.

➤ Solution on page 457

Statistical Process Control:
A Use of the Sample Variance

An important use of the sample standard deviation arises in industry. Any production process involves variation—not every item produced will be exactly the same. *Statistical process control* is a procedure for determining whether the variation is within reasonable limits. The method is based on the idea that if data follows the *normal distribution* (described in the next section), then 95% of the time the data will fall no more than two standard deviations away from the mean, and 99.7% of the time the data will fall no more than three standard deviations away from the mean. Generally, the actual mean and standard deviation are unknown, but in statistical process control they are replaced by *averages of sample means and of sample standard deviations*. The details differ from one implementation to another, but the following example is typical.

EXAMPLE 5 **STATISTICAL PROCESS CONTROL**

Ten samples of size 3 were taken from a manufacturing process, providing the data in the following table.

Sample number	1	2	3	4	5	6	7	8	9	10
	7	7	6	8	7	9	7	8	7	9
Measurements	8	6	7	7	7	6	7	7	6	7
	7	8	10	9	7	7	6	7	6	8

a. Make a control chart for the sample mean to test whether the process is under control using $k = 2$ standard deviations.

b. If a later sample is {8, 9, 10}, is the process under control?

c. If sometime later a sample is {9, 9, 10}, is the process still under control?

Solution

a. For each of the ten samples we find the mean and standard deviation, using the formulas on pages 442 and 454. The results are listed in the two new rows at the bottom of the table.

Sample number	1	2	3	4	5	6	7	8	9	10
Measurements	7	7	6	8	7	9	7	8	7	9
	8	6	7	7	7	6	7	7	6	7
	7	8	10	9	7	7	6	7	6	8
Sample means	7.3	7	7.7	8	7	7.3	6.7	7.3	6.3	8
Sample standard deviations	0.6	1	2.1	1	0	1.5	0.6	0.6	0.6	1

We then find the average of the ten sample means (called the *grand average*), written $\bar{\bar{x}}$, and the average of the ten sample standard deviations, written $\bar{s}$. The results are:

$$\bar{\bar{x}} = 7.26 \quad \text{and} \quad \bar{s} = 0.9$$

For each, adding the 10 sample values and dividing by 10

We make a *control chart*, with the *center line* at the grand average $\bar{\bar{x}}$, the upper control limit $k = 2$ standard deviations above the center line, and the lower control limit $k = 2$ standard deviations below the center line, as shown on the left.*

Upper control limit: $7.26 + 2 \cdot 0.9 = 9.06$ $\bar{\bar{x}} + k \cdot s$

Lower control limit: $7.26 - 2 \cdot 0.9 = 5.46$ $\bar{\bar{x}} - k \cdot s$

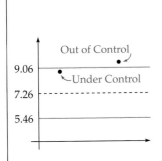

The process is considered to be *under control* if the sample mean lies between the control lines and *out of control* if not.

b. For the sample $\{8, 9, 10\}$ the sample mean is $\bar{x} = \frac{8 + 9 + 10}{3} = 9$, which lies between the control lines, so the process is under control.

c. For the later sample $\{9, 9, 10\}$ the sample mean is $\bar{x} = \frac{9 + 9 + 10}{3} = 9.3$, which is above the upper control limit, so the process is out of control. That is, after this sample the process should be stopped for repair.

One can also make analogous control tables for the sample standard deviation.

6.3 Section Summary

There are three ways to measure the spread of a data set. The simplest is the *range*, which is the largest data value minus the smallest. The

---

* The choice of k as 2 or 3 or some other value (often found from a *control chart table*) depends on the percentage of variation you choose to allow before declaring the process to be out of control. For further information, see *Introduction to Statistical Quality Control* by Douglas C. Montgomery (Wiley, 2001).

range gives no indication of the typical variation away from the mean, just the difference between the extremes.

A *box-and-whisker plot* graphically shows the range of each quarter of the data and can be drawn easily using a graphing calculator.

The *sample standard deviation* measures the typical spread of the data around the mean $\bar{x}$ as a single number and can be found easily using a calculator:

$$s = \sqrt{\frac{(x_1 - \bar{x})^2 + \cdots + (x_n - \bar{x})^2}{n - 1}}$$

$\bar{x} = \text{mean}$

➤ **Solutions to Practice Problems**

1. The largest value is 30 and the smallest is 10, so the range is $30 - 10 = 20$.

2. The minimum is 8 and the maximum is 29.
Since there are eleven values, the median is 16, the sixth value.
From the first five values, {8, 10, 13, 14, 15}, the first quartile is 13, and from the last five values, {17, 19, 23, 24, 29}, the third quartile is 23.
From this five-point summary, the box-and-whisker plot is

3. First, find the mean: $\bar{x} = \frac{1}{4}(6 + 7 + 12 + 15) = \frac{1}{4}(40) = 10$. Then

$$s = \sqrt{\frac{(6 - 10)^2 + (7 - 10)^2 + (12 - 10)^2 + (15 - 10)^2}{4 - 1}}$$

$$= \sqrt{\frac{16 + 9 + 4 + 25}{3}} = \sqrt{\frac{54}{3}} = \sqrt{18} \approx 4.2$$

6.3 **Exercises** (will be helpful throughout.)

Find the range of each data set.

1. {21, 44, 48, 52, 83}

2. {26, 35, 47, 59, 86, 92, 116}

3. {5, 25, 6, 9, 7, 7, 21, 19, 23, 16}

4. {2.0, 3.8, 5.2, 1.6, 4.3, 2.3, 5.4, 3.5, 2.5, 3.1}

5. {121, 115, 147, 163, 171, 116, 167, 147, 169, 131}

Make a five-point summary and draw the box-and-whisker plot for each data set.

6. {5, 8, 11, 13, 16, 20, 26}

7. {3, 10, 12, 14, 17, 21, 23}

8. {2, 3, 5, 9, 10, 12, 15, 17, 20}

9. {8, 9, 11, 12, 13, 14, 15, 19, 26}

10. {12, 8, 19, 17, 20, 8, 13, 14, 23, 5, 6, 9, 23, 12, 9}

11. {21, 13, 18, 14, 24, 8, 24, 24, 9, 12, 25, 20, 16, 17, 12}

12. {14, 17, 17, 13, 11, 6, 10, 21, 20, 18, 14, 11, 19, 14, 14}

13. {18, 7, 8, 18, 6, 15, 12, 15, 18, 24, 6, 18, 16, 17, 24}

14. {16, 9, 14, 12, 18, 15, 21, 19, 14, 18, 8, 4, 17, 15, 21, 14, 23, 13, 20, 17}

15. {19, 18, 7, 16, 5, 8, 9, 16, 18, 11, 20, 16, 17, 17, 8, 10, 15, 10, 8, 20}

Find the sample standard deviation of each data set.

16. {8, 13, 14, 17, 13}

17. {9, 2, 17, 13, 14}

18. {18, 11, 18, 1, 5, 19}

19. {22, 19, 11, 2, 20, 4}

20. {14, 11, 11, 9, 2, 13}

APPLIED EXERCISES

In Exercises 21–25, analyze the data by finding the range, the five-point summary, and the sample standard deviation and then drawing the box-and-whisker plot.

21. BUSINESS: Stock Prices Last Thursday was a very active trading day for shares of the new DiNextron technology stock. A random sample of twenty purchase transactions found the following share selling prices (in dollars):

16	12	17	11	12	7	12	17	11	20
10	11	14	17	16	14	25	9	17	5

22. GENERAL: Kitchen Cabinets A random sample of fifteen electric bills at Rick's Always Cooking cabinet factory found the following monthly electricity consumptions (in thousands of kilowatt hours):

10	12	6	15	13
15	12	9	14	6
5	8	9	12	9

23. BUSINESS: Airline Cancellations A random sample of thirty airplane departures from Dallas International found the following numbers of "no shows" at the boarding gates:

2	6	8	4	10	5	10	8	7	5
10	10	9	6	11	9	6	6	10	8
11	6	7	13	12	8	9	5	9	10

24. SOCIAL SCIENCE: Malpractice Settlements A random sample of twenty-four malpractice claims against a midwestern health maintenance organization (HMO) that were settled out of court found the following settlement amounts (in tens of thousands of dollars):

16	22	16	18	29	22	30	18
19	28	18	37	17	32	24	31
19	16	16	6	14	15	8	18

25. SOCIAL SCIENCE: Trial Schedules A random sample from the clerk's files at the Marksburg District Court found the following jail waiting times (in weeks) between arraignment and trial (or plea bargain) for twenty individuals charged with felonies:

11	35	15	33	20	21	2	21	23	24
18	16	9	25	12	20	20	25	21	33

26. SOCIAL SCIENCE: Income and Education The following five-point summary is based on a survey of annual incomes (in thousands of dollars) for those whose education includes an advanced degree (beyond the bachelor's). Draw the box-and-whisker plot and interpret the results in comparison with the plots in Example 3 for people without an advanced degree.

17.4, 30.2, 40.7, 54.4, 89.3

27. SOCIAL SCIENCE: Income and Gender The following five-point summaries are based on a survey of annual incomes (in thousands of dollars). The first summary is for men and the second is for women. Draw the box-and-whisker plots for each. Interpret the results.

Men's incomes: 9.6, 18.4, 28.8, 40.6, 74.1
Women's income: 8.4, 15.0, 21.5, 31.7, 53.5

28. ATHLETICS: Home Run Records In 1998, a baseball record that had stood for 37 years, Roger Maris's record of 61 home runs in a season, was broken by Sammy Sosa (with 66) and by Mark McGwire (with 70). To compare their records, make a five-point summary and draw the box-and-whisker plot for the following lists of home runs per season for each player during their major league careers. Interpret the results.

Sammy Sosa: {4, 15, 10, 8, 33, 25, 36, 40, 36, 66, 63, 50, 64, 49}

Mark McGwire: {3, 49, 32, 33, 39, 22, 42, 9, 9, 39, 58, 70, 65, 32, 29}

Roger Maris: {14, 9, 19, 16, 39, 61, 33, 23, 26, 8, 13, 9, 5}

29. GENERAL: Fuel Efficiency The following tables list the fuel economy in miles per gallon for model year 2003 family sedans and for sport utility vehicles (SUVs). For each, make a five-point summary and draw the box-and-whisker plot. Give an interpretation of the results.

Family Sedans	mpg
Volkswagen Passat	23
Toyota Camry	22
Honda Accord	24
Nissan Maxima	21
Mazda 6	22
Volvo S40	22
Subaru Legacy	21
Mitsubishi Galant	22
Chevrolet Malibu	22
Ford Taurus	21
Mercury Sable	21
Hyundai XG350	19
Dodge Intrepid	19
Chrysler Sebring	21
Pontiac Grand Am	23
Saturn L200	23

SUVs	mpg
Audi Allroad	18
BMW X5	15
Honda Pilot	19
Accura MDX	18
Lexus RX300	18
Subaru Outback	21
Volvo XC70	18
Toyota Land Cruiser	14
Mercedes-Benz ML500	15
Chevrolet Suburban	13
GMC Yukon XL	13
Nissan Pathfinder	16
Buick Rendezvous	16
Dodge Durango	13
Jeep Grand Cherokee	16
Ford Excursion	10

Source: Consumer Reports

30. MANAGEMENT SCIENCE: Process Control Suppose that ten samples of size 3 taken from a manufacturing process provided the following data.

Sample	1	2	3	4	5	6	7	8	9	10
Measurements	4	4	3	5	4	6	4	5	4	6
	1	3	4	4	5	5	4	4	5	4
	4	5	7	6	4	4	3	4	3	6

a. Find the upper and lower control limits for the sample mean (use $k = 2$).
b. If a later sample were {5, 6, 7}, would the process be under control?
c. If a later sample were {7, 7, 6}, would the process be under control?

31. MANAGEMENT SCIENCE: Process Control Suppose that ten samples of size 3 taken from a manufacturing process provided the following data.

Sample	1	2	3	4	5	6	7	8	9	10
Measurements	3	3	2	4	3	5	3	4	3	5
	4	2	3	3	4	4	3	3	2	3
	3	4	5	5	3	3	2	3	2	4

a. Find the upper and lower control limits for the sample mean (use $k = 2$).
b. If a later sample were {4, 4, 6}, would the process be under control?
c. If a later sample were {5, 6, 5}, would the process be under control?

Explorations and Excursions

The following problems extend and augment the material presented in the text.

Calculating Standard Deviations

Although the formula on page 454 defines the sample standard deviation, there is a simpler way to calculate its value that is used in most computer programs and calculators.

32. Verify each step in the following transformation of the definition formula for the sample standard deviation

$$s = \sqrt{\frac{1}{n-1} \sum (x_k - \bar{x})^2}$$

The symbol Σ means sum, and the sum inside the square root can be written as follows:

$$\sum (x_k - \bar{x})^2$$
$$= \sum (x_k^2 - 2x_k\bar{x} + \bar{x}^2)$$
$$= \left(\sum x_k^2\right) - 2\bar{x} \cdot \left(\sum x_k\right) + \bar{x}^2 \cdot \left(\sum 1\right)$$
$$= \left(\sum x_k^2\right) - 2 \cdot \left(\frac{1}{n} \cdot \sum x_k\right) \cdot \left(\sum x_k\right) + \left(\frac{1}{n} \cdot \sum x_k\right)^2 \cdot n$$
$$= \left(\sum x_k^2\right) - \frac{1}{n} \cdot \left(\sum x_k\right)^2$$

Then

$$s = \sqrt{\frac{\left(\sum x_k^2\right) - \frac{1}{n} \cdot \left(\sum x_k\right)^2}{n - 1}}$$

That is, s can be found from the sum of the data values and the sum of their squares without calculating the mean $\bar{x}$.

33. Verify the following calculation of the sample standard deviation of the data {19, 18, 27, 5, 11}.

$$\sum x_k = 19 + 18 + 27 + 5 + 11 = 80$$

$$\sum x_k^2 = 19^2 + 18^2 + 27^2 + 5^2 + 11^2 = 1560$$

so

$$s = \sqrt{\frac{1560 - \frac{1}{5}(80)^2}{5 - 1}} = \sqrt{70} \approx 8.37$$

34. Verify the following calculation of the sample standard deviation of the data {11, 18, 29, 14, 24, 6}.

$$\sum x_k = 102 \qquad \text{and} \qquad \sum x_k^2 = 2094$$

so

$$s = \sqrt{72} \approx 8.49$$

35. Use the final formula from Exercise 32 to find the sample standard deviation of the data {9, 2, 17, 13, 14} and then compare your value with your answer to Exercise 17.

36. Use the final formula from Exercise 32 to find the sample standard deviation of the data {22, 19, 11, 2, 20, 4} and then compare your value with your answer to Exercise 19.

6.4 NORMAL DISTRIBUTIONS AND BINOMIAL APPROXIMATION

Introduction

The normal distribution, with its famous bell-shaped curve, is perhaps the most important of all probability distributions. The *central limit theorems* proved by Carl F. Gauss (1777–1855) and other mathematicians during the nineteenth century showed that many quantities, such as errors in measurement or means of random samples, follow the normal distribution. Today it is used to predict everything from sizes of newborn babies to stock market fluctuations to retirement benefits. After we examine some of its most important features, we will see how it is used in applications and examine its relation to the binomial distribution.

Discrete and Continuous Random Variables

The random variables we considered in Chapter 5 take values like 0, 1, 2, . . .   that are *separated* or *discrete*, and so are called *discrete random variables*. There are other random variables whose values can be *any number in an interval*, and are called *continuous random variables*.

Examples of continuous random variables are your (exact) weight at birth and the time until a light bulb burns out, since these will most likely not be whole numbers but may be any number (in some reasonable interval). For discrete random variables, we found probabilities of events by adding probabilities of outcomes, which can be interpreted geometrically as adding areas of bars of probability distributions that change in jumps (see page 408). Analogously, for continuous random variables, we will find probabilities by calculating areas under curves.

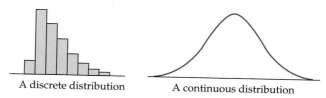

A discrete distribution　　　　A continuous distribution

The most famous of all continuous distributions is the *normal distribution*, which we now discuss.

Normal Distribution

The normal distribution depends on two constants or *parameters*, its *mean* μ, which may be any number, its *standard deviation* σ (pronounced "sigma," the Greek letter "s"), which must be positive. The parameter σ measures the *spread* of the distribution away from its mean, and it is analogous to the sample standard deviation of the previous section. The square of the standard deviation is called the *variance*. The normal distribution is given by the following formula, although we will not use it explicitly in what follows.

Normal Probability Distribution

The normal probability distribution with mean μ and standard deviation σ is given by the curve

$$f(x) = \frac{1}{\sigma\sqrt{2\pi}}e^{-\frac{1}{2}(\frac{x-\mu}{\sigma})^2} \qquad \text{for } -\infty < x < \infty$$

A random variable with this distribution is called a *normal random variable* with mean μ and standard deviation σ.

The curve has a central peak at μ and then falls back symmetrically on either side to approach the *x*-axis. Different values of μ change the location of the peak, as is shown below by three normal distributions with different means.

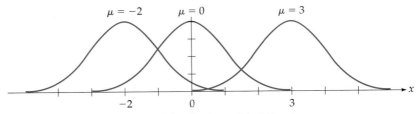

Three normal distributions with different means

The standard deviation σ measures how the values spread away from the mean. The distribution will be higher and more peaked if σ is small, and lower and more rounded if σ is large.

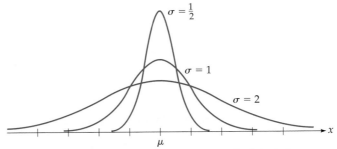

Three normal distributions with different standard deviations

As the distribution becomes "shorter" it also becomes "wider," and the area under the curve always stays at 1, although we will not prove this fact. The peaked shape of the normal distribution concentrates most of the probability (or area) near the center at μ. About 68% of the area under the normal curve is within one standard deviation of the mean, about 95% is within two standard deviations, and more than 99% is within three standard deviations. Notice also that the curve rises or falls most steeply at one standard deviation from the mean.

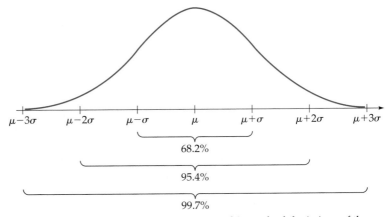

Area (probability) under the curve within 1, 2, and 3 standard deviations of the mean

The probability that a normal random variable has its value between any two given x-values corresponds to the *area* under the normal curve between those x-values. The probability is most easily found using a graphing calculator, but it can also be found using tables of values of the normal distribution.

Readers who have graphing calculators that find normal probabilities should continue reading this section. Readers who do not have such calculators should now turn to page A-1 to read the appendix near the end of this book.

Graphing Calculator Exploration

```
DISTR DRAW
1:normalpdf(
2:normalcdf(
3:invNorm(
4:tpdf(
5:tcdf(
6:X²pdf(
7↓X²cdf(
```

Areas under normal distribution curves can be found easily on a graphing calculator. To find the area under the normal distribution curve with mean $\mu = 20$ and standard deviation $\sigma = 5$ from 10 (which is $\mu - 2\sigma$) to 30 (which is $\mu + 2\sigma$), proceed as follows:

a. From the DISTRIBUTION menu, select the normalcdf command to find the cumulative values of the normal distribution (that is, the area under the curve) between two given values.

```
normalcdf(10,30,
20,5)
          .954499876
```

b. Enter the (left) starting value and the (right) ending value as well as the mean and standard deviation. This area is the same as the probability that the normal random variable is between the two given values 10 and 30.

c. To *see* this area shaded under the normal curve, use the Shade-Norm command from the DISTRIBUTION DRAW menu with an appropriate window.

```
WINDOW
 Xmin=0
 Xmax=40
 Xscl=5
 Ymin=-.05
 Ymax=.1
 Yscl=1
 Xres=1
```

```
DISTR DRAW
1:ShadeNorm(
2:Shade_t(
3:ShadeX²(
4:ShadeF(
```

```
ShadeNorm(10,30,
20,5)
```

Area=.9545
low=10   up=30

EXAMPLE **1** **FINDING A PROBABILITY FOR A NORMAL RANDOM VARIABLE**

The heights of American men are approximately normally distributed with mean $\mu = 68.1$ inches and standard deviation $\sigma = 2.7$ inches.* Find the proportion of men who are between 5 feet 9 inches and 6 feet tall.

```
normalcdf(69,72,
68.1,2.7)
        .2951343633
```

Solution

Converting to inches, we want the probability of a man's height being between 69 inches and 72 inches. From the graphing calculator screen on the left, about 30% of American men are between 5 feet 9 inches and 6 feet tall.

Practice Problem **1**   The weights of American women are approximately normally distributed with mean 134.7 pounds and standard deviation 30.4 pounds. Find the proportion of women who weigh between 130 and 150 pounds.
➤ **Solution on page 470**

z-Scores

We have seen that graphing calculators quickly calculate normal probabilities for *any* mean and standard deviation. The normal distribution with *mean* 0 *and standard deviation* 1 has special significance and is called the *standard* normal distribution (the word *standard* indicating mean 0 and standard deviation 1). The letter z is traditionally used for the variable in standard normal calculations. Since the mean is zero, the highest point on the curve is at $z = 0$, and the curve is symmetric about this value.

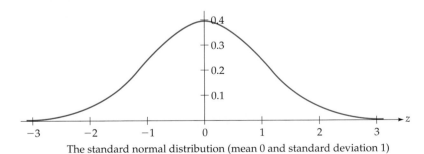

The standard normal distribution (mean 0 and standard deviation 1)

---

* These and other data in this section are from *Handbook of Human Factors and Ergonomics* by Gavriel Salvendy (Wiley-Interscience, 1997).

From the diagram on page 462 with $\mu = 0$ and $\sigma = 1$ we have the following facts about the standard normal distribution: More than

68% of its probability is between -1 and $+1$

95% of its probability is between -2 and $+2$

99% of its probability is between -3 and $+3$

To change any x-value for a normal distribution with mean μ and standard deviation σ into the corresponding z-value for the *standard normal distribution*, we subtract the mean and then divide by the standard deviation. The resulting number is called the *z-score* and this process is called *standardizing*.

---

z-Score

$$z = \frac{x - \mu}{\sigma}$$

Subtracting the mean μ and dividing by the standard deviation σ

The z-score gives the number of standard deviations the value is from the mean.

EXAMPLE 2

FINDING z-SCORES FOR HEIGHTS OF MEN

Using the mean and standard deviation for men's heights given in Example 1, find the z-scores corresponding to the following heights (in inches) and interpret the results:

a. 73.5   **b.** 62.7.

Solution

a. $z = \dfrac{73.5 - 68.1}{2.7} = \dfrac{5.4}{2.7} = 2$ Using $z = (x - \mu)/\sigma$ with $\mu = 68.1$ and $\sigma = 2.7$ from Example 1

Therefore, a height of 73.5 inches is two standard deviations *above* the mean.

b. $z = \dfrac{62.7 - 68.1}{2.7} = \dfrac{-5.4}{2.7} = -2$

Therefore, a height of 62.7 inches is two standard deviations *below* the mean.

Since the standard normal distribution has more than 95% of its probability between -2 and $+2$, these results mean that more than 95% of men have heights between 62.7 inches and 73.5 inches (5 feet 2.7 inches and 6 feet 1.5 inches).

Practice Problem 2 Using the mean and standard deviation for women's weights given in Practice Problem 1 on the previous page, find the z-scores for the following weights, and interpret your answers:

a. 165.1 lbs.   **b.** 104.3 lbs.   ➤ Solutions on page 470

The Normal and Binomial Distributions

As we saw on page 412, the binomial distribution has a kind of "bell" shape, with a peak at the expected value $\mu = np$ and falling on both sides to very small probabilities further away from the mean.

$$n = 20, p = 0.4 \qquad n = 15, p = 0.5 \qquad n = 25, p = 0.7$$

Several binomial distributions with different values for n and p

Since both the binomial and the normal distributions have similar "mound" shapes, could they be related? In the eighteenth century, Abraham de Moivre (1667–1754) and Pierre-Simon Laplace (1749–1827) discovered and proved that for any choice of p between 0 and 1, the binomial distribution *approaches* the normal distribution as n becomes large. This fundamental fact is known as the de Moivre–Laplace theorem.

Graphing Calculator Exploration

The program* SEENORML draws binomial distributions for $p = \frac{1}{2}$ with n from 2 to 30. Running this program shows how the "boxlike" binomial distributions change into "bell-shaped" curves as n

---

* See the Preface for information on how to obtain this and other graphing calculator programs.

increases. The window for each graph is different so that each fills as much as possible of the viewing area.

We may use the de Moivre–Laplace theorem to approximate binomial distributions by the normal distribution.

Normal Approximation to the Binomial

Let X be a binomial random variable with parameters n and p. If $np > 5$ and $n(1 - p) > 5$, then the distribution of X is approximately normal with mean $\mu = np$ and standard deviation $\sigma = \sqrt{np(1 - p)}$.

Approximating the *discrete* binomial distribution by the *continuous* normal distribution requires an adjustment: To have the same width for a "slice" of the normal distribution as for a slice of the binomial, we adopt the convention that the binomial probability $P(X = x)$ corresponds to the area under the normal distribution curve from $x - \frac{1}{2}$ to $x + \frac{1}{2}$. This is called the *continuous correction*.

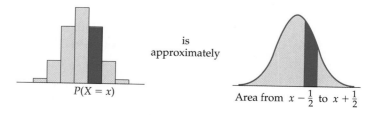

EXAMPLE 3

NORMAL APPROXIMATION OF A BINOMIAL PROBABILITY

Estimate $P(X = 12)$ for the binomial random variable X with $n = 25$ and $p = 0.6$ using the corresponding normal distribution.

Solution

Since

$$np = 25 \cdot 0.6 = 15 \quad \text{and} \quad n(1 - p) = 25 \cdot 0.4 = 10$$

are both greater than 5, we may use the normal distribution with

$$\mu = np = 25 \cdot 0.6 = 15$$

and

$$\sigma = \sqrt{np(1 - p)} = \sqrt{25 \cdot 0.6 \cdot 0.4} = \sqrt{6}$$

By the continuous correction, we interpret the event $X = 12$ as $11\frac{1}{2} \leq X < 12\frac{1}{2}$ to include numbers that would round to 12. Using a graphing calculator, we could find this probability as we did in Example 1 (page 464), but the following calculator commands also show the picture.

```
WINDOW
 Xmin=0
 Xmax=25
 Xscl=1
 Ymin=-.1
 Ymax=.25
 Yscl=1
 Xres=1
```

```
ShadeNorm(11.5,1
2.5,25*.6,√(25*.
6*.4))
```

```
Area=.077196
low=11.5 up=12.5
```

The required probability is approximately 0.077.

The normal approximation of 0.077 is indeed very close to the actual value calculated from the binomial probability formula (see page 412): $_{25}C_{12}(0.6)^{12}(1 - 0.6)^{13} \approx 0.076$.

EXAMPLE 4

MANAGEMENT MBA's

At a major Los Angeles accounting firm, 73% of the managers have MBA degrees. In a random sample of 40 of these managers, what is the probability that between 27 and 32 will have MBA's?

Solution

Because each manager either has or does not have an MBA, presumably independently of each other, the question asks for the probability that a binomial random variable X with $n = 40$ and $p = 0.73$ satisfies $27 \le X \le 32$. Since $np = 40 \cdot 0.73 = 29.2$ and $n(1 - p) = 40 \cdot 0.27 = 10.8$ are both greater than 5, this probability can be approximated as the area under the normal distribution curve with mean $\mu = np = 40 \cdot 0.73 = 29.2$ and standard deviation $\sigma = \sqrt{np(1 - p)} = \sqrt{40 \cdot 0.73 \cdot 0.27}$ from $26\frac{1}{2}$ to $32\frac{1}{2}$ (again using the continuous correction to include values that would round to between 27 and 32):

The probability is (about) 0.712, or about 71%.

The exact answer to the above problem, from summing the binomial distribution from 27 to 32, is 71.5%, so the normal approximation is very accurate.

Practice Problem 3

A brand of imported VCR is known to have defective tape rewind mechanisms in 8% of the units imported last April. If Jerry's Discount Electronics received a shipment of 80 of these VCR's, what is the probability that 10 or more are defective? ➤ Solution on page 470

6.4 Section Summary

The graph of the normal distribution with mean μ and standard deviation σ is a bell-shaped curve that peaks at μ in the center and then falls back symmetrically on either side to approach the x-axis. The area under this curve between any two x-values is the probability that the normal random variable lies between these values. These areas and probabilities can be found easily with a graphing calculator or tables.

Any value x of a normal random variable with mean μ and standard deviation σ can be converted into a z-score using the formula

$$z = \frac{x - \mu}{\sigma}$$

This z-score indicates the number of standard deviations the x-value is away from the mean.

The normal distribution with mean $\mu = np$ and standard deviation $\sigma = \sqrt{np(1-p)}$ is a good approximation to the binomial distribution provided that both np and $n(1-p)$ are greater than 5.

➤ **Solutions to Practice Problems (for those using graphing calculators)**

```
normalcdf(130,15
0,134.7,30.4)
         .2540533807
```

1. About 25% of American women weigh between 130 and 150 pounds.

2. Using $\mu = 134.7$ and $\sigma = 30.4$ from Practice Problem 1:

   a. $z = \dfrac{165.1 - 134.7}{30.4} = \dfrac{30.4}{30.4} = 1$

   b. $z = \dfrac{104.3 - 134.7}{30.4} = \dfrac{-30.4}{30.4} = -1$

   Interpretation: More than 68% of women have weights between about 104 and 165 pounds.

```
normalcdf(9.5,80
.5,80*.08,√(80*.
08*.92))
         .1007041942
```

3. Since $np = 80 \cdot 0.08 = 6.4$ and $n(1-p) = 80 \cdot 0.92 = 73.6$ are both greater than 5, we may use the normal distribution with $\mu = np = 80 \cdot 0.08$ and $\sigma = \sqrt{np(1-p)} = \sqrt{80 \cdot 0.08 \cdot 0.92}$ to approximate $P(X \geq 10)$ as the area under the normal curve from 9.5 to 80.5 (corresponding to all 80 being defective). The probability of at least ten defective VCR's is (about) 0.10, or 10%.

6.4 Exercises

Let X be a normal random variable with mean $\mu = 12$ and standard deviation $\sigma = 3$. Find each probability as an area under the normal curve.

1. $P(9 \leq X \leq 15)$
2. $P(6 \leq X \leq 18)$
3. $P(3 \leq X \leq 21)$
4. $P(0 \leq X \leq 24)$
5. $P(12 \leq X \leq 15)$
6. $P(9 \leq X \leq 12)$
7. $P(15 \leq X \leq 16)$
8. $P(7 \leq X \leq 10)$
9. $P(6\frac{1}{2} \leq X \leq 11)$
10. $P(10\frac{1}{2} \leq X \leq 14)$

Find the z-score corresponding to each x-value.

11. $x = 6$ with $\mu = 4$ and $\sigma = 2$

12. $x = 20$ with $\mu = 15$ and $\sigma = 5$
13. $x = 10$ with $\mu = 10$ and $\sigma = 3$
14. $x = 6$ with $\mu = 6$ and $\sigma = 5$
15. $x = 15$ with $\mu = 20$ and $\sigma = 5$
16. $x = 6$ with $\mu = 8$ and $\sigma = 2$
17. $x = 15$ with $\mu = 20$ and $\sigma = 1$
18. $x = 6$ with $\mu = 8$ and $\sigma = 1$
19. $x = 130$ with $\mu = 100$ and $\sigma = 15$
20. $x = 90$ with $\mu = 100$ and $\sigma = 10$

Use the normal approximation to the binomial random variable X with $n = 20$ and $p = 0.7$ to find each probability.

21. $P(X = 10)$

22. $P(X = 12)$

23. $P(11 \leq X \leq 12)$

24. $P(8 \leq X \leq 10)$

25. $P(3 \leq X \leq 8)$

26. $P(13 \leq X \leq 19)$

27. $P(15 \leq X \leq 20)$

28. $P(2 \leq X \leq 7)$

29. $P(0 \leq X \leq 20)$

30. $P(14 \leq X \leq 20)$

APPLIED EXERCISES

Answer each question using an appropriate normal probability.

31. BIOMEDICAL: Weights Weights of men are approximately normally distributed with mean 163 pounds and standard deviation 28 pounds. If the minimum and maximum weights to be a volunteer fireman in Martinsville are 128 and 254, what proportion of men meet this qualification?

32. BUSINESS: Advertising The percentage of early afternoon television viewers who watch episodes of "As the World Spins" is a normal random variable with mean 22% and standard deviation 3%. The producers guarantee advertisers that the viewer percentage will be better than 20% or else the advertisers will receive free air time. What is the probability that the producers will have to give free air time after any particular episode?

33. BUSINESS: Airline Overbooking To avoid empty seats, airlines generally sell more tickets than there are seats. Suppose that the number of ticketed passengers who show up for a flight with 100 seats is a normal random variable with mean 97 and standard deviation 6. What is the probability that at least one person will have to be "bumped" from the flight?

34. GENERAL: SAT Scores The scores at Centerville High School on last year's mathematics SAT test were approximately normally distributed with mean 490 and standard deviation 140. What proportion of the scores were between 550 and 750?

35. BIOMEDICAL: Smoking In a large-scale study carried out in the 1960s of male smokers 35–45 years old, the number of cigarettes smoked daily was approximately normally distributed with mean 28 and standard deviation

10. What proportion of the smokers smoked between 30 and 40 (two packs!) each day?

36. GENERAL: Coin Tosses A fair coin is tossed fifty times. What is the probability that it comes up heads at least thirty times?

37. GENERAL: Actuarial Exams The probability of passing the first actuarial examination is 60%. If 940 college students take the exam in March, what is the probability that the number that pass is between 575 and 725?

38. BIOMEDICAL: Flu Shots Although getting a flu shot in October is a good way of planning ahead for the winter flu season, approximately 1% of those getting the shots develop side effects. If 800 students at Micheles College get flu shots, what is the probability that between 6 and 12 of them will develop side effects?

39. BUSINESS: Real Estate Nationwide, 40% of the sales force at the Greener Pastures real estate franchises are college graduates. If 200 of these sales personnel are chosen at random, what is the probability that between 85 and 100 of them are college graduates?

40. SOCIAL SCIENCE: Old Age If 3% of the population lives to age 90, what is the probability that between 10 and 15 of the 280 business majors at Henderson College will live to be that old?

Explorations and Excursions

The following exercises extend and augment the material presented in the text.

Chebychev's Theorem

(*Requires Section 5.6*) On page 409 we defined the *mean* or expected value of a random variable X as

$$\mu = E(X) = x_1 \cdot P(X = x_1) + \cdots + x_n \cdot P(X = x_n)$$

The standard deviation of a random variable X is defined as

$$\sigma = \sqrt{E(X-\mu)^2}$$
$$= \sqrt{(x_1-\mu)^2 \cdot P(X=x_1) + \cdots + (x_n-\mu)^2 \cdot P(X=x_n)}$$

The standard deviation measures the *spread* of the random variable away from its mean just as the sample standard deviation measures the spread of a data set away from its sample mean. The relationship between the standard deviation and the values of a random variable is given by *Chebychev's Theorem*:

> The probability that the values of a random variable lie within k standard deviations from its mean is at least $1 - \frac{1}{k^2}$.

For example, taking $k = 2$, Chebychev's Theorem says that the probability that the values of a random variable lie within 2 standard deviations from its mean is at least $1 - \frac{1}{2^2} = 1 - \frac{1}{4} = \frac{3}{4}$.

Use Chebychev's Theorem to find the probability that the values of a random variable lie within:

41. 3 standard deviations from its mean.

42. 4 standard deviations from its mean.

Suppose that a random variable has mean $\mu = 100$ and standard deviation $\sigma = 5$. Then, again taking $k = 2$, Chebychev's Theorem says

that the probability that the values of the random variable lie between $\mu \pm 2\sigma = 100 \pm 2 \cdot 5$, that is, between 90 and 110, is at least $1 - \frac{1}{2^2} = \frac{3}{4}$.

43. State the result of Chebychev's Theorem with $k = 3$ for a random variable with $\mu = 100$ and $\sigma = 5$.

44. State the result of Chebychev's Theorem with $k = 4$ for a random variable with $\mu = 100$ and $\sigma = 5$.

While Chebychev's Theorem applies to all random variables, stronger results are available for *particular* random variables, such as normal random variables. For example, the probability that the values of a normal random variable with $\mu = 100$ and $\sigma = 5$ lie between 90 and 110 is greater than 95%, a stronger result than the $\frac{3}{4} = 75\%$ from Chebychev's Theorem.

45. For a normal random variable with $\mu = 100$ and $\sigma = 5$, find the probability that its values lies between 85 and 115. Compare the answer with that of Exercise 43, which gives the probability for any random variable.

46. For a normal random variable with $\mu = 100$ and $\sigma = 5$, find the probability that its values lie between 80 and 120. Compare the answer with that of Exercise 44, which gives the probability for any random variable.

Chapter Summary with Hints and Suggestions

Reading the text and doing the exercises in this chapter have helped you to master the following skills, which are listed by section (in case you need to review them) and are keyed to particular Review Exercises. Answers for all Review Exercises are given at the back of the book, and full solutions can be found in the Student Solutions Manual.

6.1 Random Samples and Data Organization

- Identify levels of data measurement (nominal, ordinal, interval, and ratio). (*Review Exercises 1–4.*)

- Construct a bar chart or a histogram for a given data set. (*Review Exercises 5–8.*)

6.2 Measures of Central Tendency

- Find the mode, median, and mean of a data set. (*Review Exercises 9–13.*)

    Mode: most frequent

    Median: middle

    Mean: $\bar{x} = \dfrac{x_1 + \cdots + x_n}{n}$

6.3 Measures of Variation

- Find the range of a data set.
  (Review Exercise 14.)

- Find the five-point summary of a data set and draw the box-and-whisker plot.
  (Review Exercise 15.)

- Find the sample standard deviation of a data set. *(Review Exercise 16.)*

$$s = \sqrt{\frac{(x_1 - \bar{x})^2 + \cdots + (x_n - \bar{x})^2}{n - 1}}$$

- Analyze the data in an application by finding the range, the five-point summary, and the sample standard deviation and then drawing the box-and-whisker plot.
  (Review Exercises 17–18.)

6.4 Normal Distributions and Binomial Approximation

- Find the probability that the normal random variable with mean μ and standard deviation σ is between two given values.
  (Review Exercises 19–20.)

- Find the z-score of an x-value of the normal random variable with mean μ and standard deviation σ. *(Review Exercises 21–22.)*

$$z = \frac{x - \mu}{\sigma}$$

- Use the normal approximation to the binomial random variable X to find the probability that X takes given values.
  (Review Exercises 23–24.)

- Solve an applied problem involving probabilities using the normal distribution.
  (Review Exercises 25–28.)

Hints and Suggestions

- *(Overview)* Properties of "many" can be inferred from just "some" only if the "some" are a random sample from the population. There are four different levels of data measurement: nominal, ordinal, interval, and ratio.

Data may be organized into bar charts or histograms. Measures of central tendency (mode, median, and mean) summarize all the data as one typical value, while measures of variation (range, box-and-whisker plot, and sample standard deviation) show how closely the values cluster about the center. The bell-shaped curve of the normal distribution applies to many situations, and z-scores give the number of standard deviations the values are from the mean. The binomial distribution is approximated by the normal distribution, provided both np and $n(1 - p)$ are sufficiently large.

- A bar chart may have its bars separated or touching. Histograms are for interval or ratio data and the bars touch each other on both sides.

- To find the mode or the median of data values, it is helpful to begin by sorting them in order.

- For measures of central tendency, with nominal ("in name only") data, only the mode is appropriate; with ordinal data, both the mode and the median are appropriate; with interval or ratio data, all three (mode, median, and mean) are appropriate.

- All three measures of variation are appropriate only for interval or ratio data.

- The mean is sensitive to changes in a few extreme values, while the mode and median are not.

- Box-and-whisker plots graphically show how the data are distributed among the four quartiles.

- Areas under the normal distribution curve are probabilities for the normal random variable.

- When using the normal distribution to approximate the binomial distribution, be sure to increase the largest value by $\frac{1}{2}$ and to decrease the smallest by $\frac{1}{2}$ (the continuous correction) to cover all the area represented by the boxes making up the binomial distribution.

- **Practice for Test:** Review Exercises 1, 3, 5, 7, 9, 17, 19, 21, 24, 26.

Review Exercises for Chapter 6

Practice test exercise numbers are in green.

(will be helpful throughout.)

6.1 Random Samples and Data Organization

Identify the level of data measurement in each situation.

1. **GENERAL: Chicago Commuters** A random sample of seventy Chicago area commuters counts the number traveling to work last Thursday by car, taxi, bus, and train.

2. **GENERAL: Gourmet Coffee** At the House of Java Coffee Emporium, a random sample of eighteen customers rates the new Turkish Sultan flavor on a scale from 1 ("I'll try something else") to 5 ("Love at first sip").

3. **GENERAL: Fourth of July** A random sample of five hundred suburban households records the backyard temperatures at 3 P.M. on Independence Day.

4. **ENVIRONMENTAL SCIENCE: Drinking Water** A random sample of eighty small municipalities (less than 50,000 residents) lists the lead content (in parts per billion) of the tap water.

Construct a bar chart of the data in each situation.

5. **GENERAL: Veterans of Foreign Wars** A random sample of forty members of the Hattiesburg V.F.W. identifies their branches of military service as 14 in the Army, 12 in the Navy, 6 in the Marines, and 8 in the Air Force.

6. **SOCIAL SCIENCE: Homicides** A random sample of twenty homicides investigated last year by the Macon County Coroner's Office lists the causes as 12 by handguns, 3 by firearms other than handguns, 2 by knives, 2 by blunt instruments, and 1 by use of hands.

Construct a histogram of the data in each situation.

7. **BUSINESS: Manufacturer's Rebates** A random sample of twenty requests for the $5 rebate offered last summer on two pairs of Ultra-Cool Shades sunglasses found that the checks were received by the consumers after waits of the following numbers of days:

45	53	60	34	60	37	52	65	44	49
54	51	61	43	49	63	53	53	56	55

8. **PERSONAL FINANCE: Checking Accounts** A random sample of thirty checking account balances on March 2 at the Lobsterman Trust Company branch in Bangor found the following amounts (rounded to the nearest dollar):

906	858	1168	1397	925	797
1036	1398	912	1059	931	815
698	787	711	1485	1048	937
1272	727	1339	1458	842	1264
1045	960	802	1204	1370	841

6.2 Measures of Central Tendency

Find the mode, median, and mean of each data set.

9. {14, 11, 15, 8, 5, 4, 10, 3, 12, 11, 6}

10. {5, 6, 15, 13, 15, 6, 11, 13, 13, 6, 13, 6, 8}

11. {12, 13, 13, 5, 9, 14, 13, 8, 9, 14}

12. **BUSINESS: Stock Earnings** The third-quarter earnings per share for twelve randomly selected textile stocks on the New York Stock Exchange were (in dollars):

12	8	8	11	4	7
8	7	5	13	10	9

13. **BUSINESS: Personal Calls** A random sample of the telephone records for fifteen office clerks at the southern regional office of the Marston Glassware Products Company

counted the following numbers of personal
calls last week:

$$4 \quad 15 \quad 23 \quad 26 \quad 2$$
$$22 \quad 25 \quad 4 \quad 27 \quad 18$$
$$30 \quad 28 \quad 11 \quad 20 \quad 6$$

6.3 Measures of Variation

14. Find the range of the data {38, 28, 32, 43, 41,
25, 35, 37, 43, 36}.

15. Find the five-point summary and draw the
box-and-whisker plot for the data {4, 5, 7, 9,
10, 13, 14, 15, 20}.

16. Find the sample standard deviation of the data
{5, 7, 9, 10, 13, 16}.

Analyze the data in each situation by finding the
range, the five-point summary, and the sample
standard deviation, and then drawing the box-and-
whisker plot.

17. BUSINESS: Bank Lines A random sample
of fifteen customers at the Middletown branch
of the Peoples & First National Bank found the
following waiting times (in minutes):

$$5 \quad 6 \quad 6 \quad 5 \quad 3$$
$$5 \quad 8 \quad 3 \quad 9 \quad 6$$
$$7 \quad 9 \quad 5 \quad 9 \quad 6$$

18. BUSINESS: Mortgages A random sample of
twenty mortgages filed last month at the
Seaford Town Clerk office had the following
values (in thousands of dollars):

$$132 \quad 141 \quad 154 \quad 143 \quad 115$$
$$141 \quad 125 \quad 132 \quad 123 \quad 132$$
$$119 \quad 126 \quad 123 \quad 114 \quad 126$$
$$132 \quad 128 \quad 131 \quad 136 \quad 131$$

6.4 Normal Distributions and Binomial Approximation

Let X be a normal random variable with mean
$\mu = 50$ and standard deviation $\sigma = 10$. Find
each probability.

19. $P(30 \leq X \leq 60)$ **20.** $P(55 \leq X \leq 75)$

Find the z-score corresponding to each x-value.

21. $x = 12$ with $\mu = 10$ and $\sigma = 1$

22. $x = 18$ with $\mu = 24$ and $\sigma = 3$

Use the normal approximation to the binomial random variable X with $n = 25$ and $p = 0.35$ to
find each probability.

23. $P(X = 14)$ **24.** $P(6 \leq X \leq 10)$

For exercises 25–28, answer each question using an
appropriate normal distribution.

25. BIOMEDICAL: Heights Women's heights
are approximately normally distributed with
mean 63.2 inches and standard deviation
2.6 inches. Find the proportion of women with
heights between 62 inches and 69 inches.

26. GENERAL: Tire Wear The life of Wear-Ever
Super Tread automobile tires is normally distributed with mean 50,000 miles and standard
deviation 5000 miles. What proportion of these
tires last between 55,000 and 65,000 miles?

27. GENERAL: Lost Luggage During the Christ-
mas rush, the chance that an airline will lose
your suitcase increases to 1%. What is the
probability that between 4 and 8 of the 690
suitcases checked in at Bellemeade Regional
Airport this holiday season will be lost?

28. BIOMEDICAL: Heart Disease The leading
cause of death in the United States is heart disease, which causes 32% of all deaths. In a random sample of 150 death certificates, what is
the probability that between 50 and 60 list
heart disease as the cause of death?

7 Markov Chains

Application Preview

Weather in Sri Lanka*

The island nation of Sri Lanka, off the southeast coast of India, produces tea, coconuts, and rice. Only 16% of the land is arable and weather conditions vary from monsoon to drought, so accurate weather projections would do much to improve the country's economic output. Even if a month's total rainfall is adequate, crops can fail if the rain comes in a downpour after several dry weeks. A recent study modeled the weekly rainfall as a "Markov chain" (to be explained in this chapter), using 50 years of weather data to estimate the probabilities that a "wet" week will be followed by a "dry" week, a "wet" week will be followed by a "wet" week, and so on. The resulting probabilities were then used to simulate the rainfall in different areas over a year, with results such as the accompanying graph.

The model, which clearly tracks the actual rainfall quite well, can be used to predict the success of planting marginal areas with various crops whose growth rates and water requirements are known.

* Adapted in part from "On Development and Comparative Study of Two Markov Models of Rainfall in the Dry Zone of Sri Lanka" by B. V. R. Punyawardena and D. Kulasiri, Research Report 96/11, Centre for Computing and Biometrics, Lincoln University (Canterbury, New Zealand).

7.1 STATES AND TRANSITIONS

Introduction

When you read the morning newspaper or watch the evening TV news, much of the information is presented as the *change* since yesterday: The stock market is up, the Dodgers won again, and the weather is moderating. Sometimes the change is given in terms of probabilities ("There is a 70% probability that the rain will end tomorrow"). In this chapter we will explore the behavior of repeated changes that depend on probabilities.

States and Transitions

The Russell 2000 Index is an average of the prices of 2000 stocks traded on the New York Stock Exchange. Each trading day this index *gains*, *loses*, or remains *unchanged* compared with the previous day, and we can denote these three possible *states* by the letters G, L, and U. Watching the market for several successive trading days might result in a sequence of states such as

$$U, G, G, L, G, L, L, L, U, G, \ldots$$

Each *transition* from one day's state to the next begins from one of the three states G, L, and U and ends in one of them, so there are $3 \cdot 3 = 9$ possible transitions. From past stock market records, it is possible to estimate the probabilities of these transitions, and we can specify them by a *state-transition diagram* or, equivalently, by a *transition matrix*.

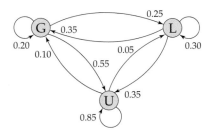

$$\begin{array}{c} \text{Next state} \\ \begin{array}{ccc} G & L & U \end{array} \\ \begin{array}{c} \text{Current} \\ \text{state} \end{array} \begin{array}{c} G \\ L \\ U \end{array} \left(\begin{array}{ccc} 0.20 & 0.25 & 0.55 \\ 0.35 & 0.30 & 0.35 \\ 0.10 & 0.05 & 0.85 \end{array} \right) \end{array}$$

State-Transition Diagram   Transition Matrix

In the state-transition diagram, the number near the arrowhead gives the probability of the transition from the starting state to the ending state. For example, the transition from state G to state L (which we denote as $G \to L$) has probability 0.25, while the transition $U \to U$

$$
\begin{array}{c}
\text{Next state} \\
\begin{array}{cc}
& \begin{array}{ccc} G & L & U \end{array} \\
\begin{array}{c} \text{C} \\ \text{u} \\ \text{r} \\ \text{r} \\ \text{e} \\ \text{n} \\ \text{t} \end{array}
\begin{array}{c} \text{s} \\ \text{t} \\ \text{a} \\ \text{t} \\ \text{e} \end{array}
\begin{array}{c} G \\ L \\ U \end{array}
&
\left(\begin{array}{ccc}
0.20 & 0.25 & 0.55 \\
0.35 & 0.30 & 0.35 \\
0.10 & 0.05 & 0.85
\end{array}\right)
\end{array}
\end{array}
$$

(that is, the market will remain unchanged) has probability 0.85. In the transition matrix, the number in a particular row and column gives the probability of the transition *from the row state to the column state*. We always list the row states from top to bottom in the same order as the column states from left to right. The matrix shows that the transition $U \to L$ has probability 0.05 and that the transition $L \to L$ (the market will continue to lose) has probability 0.30.

Practice Problem 1

What is the probability that the market will continue to gain?

➤ **Solution on page 487**

Notice that the sum of the entries in each row is 1, since any given state must be followed by one of the states G, L, or U. Such a matrix of probabilities is called a *transition matrix*.

Transition Matrix

A transition matrix is a square matrix such that the entries are nonnegative and the sum of each row is 1.

Practice Problem 2

Which of the following are transition matrices?

a. $\begin{pmatrix} 0.3 & 0.7 \\ 0.4 & 0.6 \end{pmatrix}$
b. $\begin{pmatrix} 0.8 & 0.1 \\ 0.6 & 0.4 \end{pmatrix}$
c. $\begin{pmatrix} 0.5 & 0.5 \\ 1.1 & -0.1 \end{pmatrix}$

➤ **Solution on page 487**

Markov Chains

Any collection of states (such as G, L, and U) together with the probabilities of passing from any state to any state (including from a state to itself) is called a *Markov chain*.*

Markov Chain

A *Markov chain* is a collection of *states* $S_1, S_2, \ldots, S_n$ together with a *transition matrix* T whose entry in row i and column j is the probability of the transition $S_i \to S_j$.

---

* After the Russian mathematician A. A. Markov (1856–1922) who first studied them.

A transition matrix alone may be considered a Markov chain by calling the states $S_1, S_2, \ldots, S_n$, corresponding to the rows (from top to bottom) and the columns (from left to right). Notice that the transition probabilities depend only on the present state and not on any earlier state. For this reason, a Markov chain is sometimes said to have *no memory*.

Types of Transition Matrices

What can we say about the effect of several successive transitions? Transition matrices may represent or contain several of the three basic types of transitions: *oscillating* (switching back and forth), *mixing* (moving among all possibilities), or *absorbing* (never leaving a state once it is reached).

EXAMPLE 1

TYPES OF TRANSITIONS

Characterize the transition corresponding to each transition matrix as "oscillating," "mixing," or "absorbing."

a. $\begin{pmatrix} 0 & 1 \\ 1 & 0 \end{pmatrix}$
b. $\begin{pmatrix} \frac{1}{3} & \frac{1}{3} & \frac{1}{3} \\ \frac{1}{3} & \frac{1}{3} & \frac{1}{3} \\ \frac{1}{3} & \frac{1}{3} & \frac{1}{3} \end{pmatrix}$
c. $\begin{pmatrix} 1 & 0 & 0 \\ \frac{1}{2} & \frac{1}{2} & 0 \\ \frac{1}{2} & 0 & \frac{1}{2} \end{pmatrix}$

Solution

We make a state-transition diagram for each transition matrix.

a. Since states S_1 and S_2 alternate, this is an *oscillating* transition. Repeating it several times would just flip back and forth between the two states.

b. Since each state may lead to itself or to any of the other states, this is a *mixing* transition. Repeating it several times would give some combination of the states, and any mixture is possible.

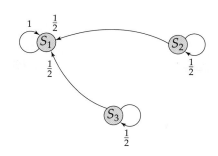

$$\begin{pmatrix} 1 & 0 & 0 \\ \frac{1}{2} & \frac{1}{2} & 0 \\ \frac{1}{2} & 0 & \frac{1}{2} \end{pmatrix}$$

Means: Once in S_1, it stays in S_1

c. Since each of states S_2 and S_3 can reach state S_1 but then can never leave, this is an *absorbing* transition. Repeating it several times might lead to several occurrences of S_2 or S_3, but once the chain reached state S_1, it would remain there forever.

A state that can never be left once it is entered is called an *absorbing state*.

Practice Problem 3

Describe the action of the following transition matrix and draw its transition diagram: $\begin{pmatrix} 1 & 0 & 0 \\ 0 & 1 & 0 \\ 0 & 0 & 1 \end{pmatrix}$.

➤ Solution on page 487

State Distribution Vectors

Given the transition matrix for a Markov chain together with information about its current state, what can we say about the chain after a transition? Consider an example.

EXAMPLE 2

Next state

$$\begin{array}{c} \text{Current state} \end{array} \begin{array}{c} \\ G \\ L \\ U \end{array} \begin{pmatrix} G & L & U \\ 0.20 & 0.25 & 0.55 \\ 0.35 & 0.30 & 0.35 \\ 0.10 & 0.05 & 0.85 \end{pmatrix}$$

$$(50\ 20\ 30)\begin{pmatrix} 0.20 \\ 0.35 \\ 0.10 \end{pmatrix} = 20$$

CALCULATING THE RESULT OF A TRANSITION

We return to the stock market situation with states G, L, and U (for *gaining, losing,* and *unchanged*) and the transition matrix shown on the left. If we have a portfolio of 100 stocks of which 50 gained, 20 lost, and 30 remained unchanged yesterday, how many may we expect to gain, lose, and remain unchanged in today's market?

Solution

We are given the numbers of stocks in the states G, L, and U, and the first column of the transition matrix gives the probabilities of transitions from these states into the state G, so multiplying and adding will give the number of gaining stocks expected today:

$$50 \cdot 0.20 + 20 \cdot 0.35 + 30 \cdot 0.10 = 20$$

Gained $P(G \to G)$ Lost $P(L \to G)$ Unchanged $P(U \to G)$
(G)   (L)   (U)
yesterday   yesterday   yesterday

Expected value is the sum of all the $n \cdot p$

Thus we can expect 20 of the stocks to gain today. Similarly, multiplying the given numbers by the probabilities in the second column (the probabilities of transitions into the state L) gives today's expected number of losers:

$$
(50 \quad 20 \quad 30) \begin{pmatrix} 0.25 \\ 0.30 \\ 0.05 \end{pmatrix}^{L} = 20
$$

$$
\underbrace{50 \cdot 0.25}_{\substack{\text{Gained } P(G \to L) \\ \text{yesterday}}} + \underbrace{20 \cdot 0.30}_{\substack{\text{Lost } P(L \to L) \\ \text{yesterday}}} + \underbrace{30 \cdot 0.05}_{\substack{\text{Unchanged } P(U \to L) \\ \text{yesterday}}} = 20 \qquad \begin{array}{l} \text{Expected} \\ \text{number of} \\ \text{losing stocks} \end{array}
$$

The remaining 60 stocks should remain unchanged today, and multiplying the given numbers by the *third* column of the transition matrix (the probabilities of transitions into the state U) gives exactly 60, as you should check.

Notice that these calculations are just the matrix product using the usual row and column multiplication described on page 201:

$$
\underbrace{(50 \quad 20 \quad 30)}_{\text{Portfolio yesterday}} \cdot \underbrace{\begin{pmatrix} 0.20 & 0.25 & 0.55 \\ 0.35 & 0.30 & 0.35 \\ 0.10 & 0.05 & 0.85 \end{pmatrix}}_{\text{Transition matrix}} = \underbrace{(20 \quad 20 \quad 60)}_{\text{Portfolio today}}
$$

Rephrasing our calculation in Example 2 in terms of probabilities for the 100 stocks in the portfolio, yesterday there were 50% gaining, 20% losing, and 30% unchanged, while after the transition to today, they became 20% gaining, 20% losing, and 60% unchanged. A row matrix consisting of such probabilities is called a *state distribution vector*, and multiplying it by the transition matrix gives the state distribution vector after one transition.

State Distribution Vector

A *state distribution vector* is a row matrix $D = (d_1 \quad d_2 \quad \ldots \quad d_n)$ of nonnegative numbers whose sum is 1. If D represents the current probability distribution for a Markov chain with transition matrix T, then the state distribution vector one step later is $D \cdot T$.

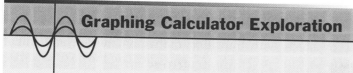

Graphing Calculator Exploration

$$T = \begin{pmatrix} 0.20 & 0.25 & 0.55 \\ 0.35 & 0.30 & 0.35 \\ 0.10 & 0.05 & 0.85 \end{pmatrix}$$

To verify that the transition matrix from Example 2 (shown on the left) transforms the state distribution vector $D = (0.50 \quad 0.20 \quad 0.30)$ into $(0.20 \quad 0.20 \quad 0.60)$, enter T and D into matrices [A] and [B] and find their product as follows.

What happens if we repeat this calculation with the state distribution vector $D = (0.140 \quad 0.104 \quad 0.756)$?

The *k*th State Distribution Vector

What can we say about the future of a Markov chain given some information about the present? If we have an *initial* state distribution vector D_0 and transition matrix T, then we have just seen that the state distribution vector after the first transition is

$$D_1 = D_0 \cdot T \qquad \text{\textit{T} gives one transition}$$

After another transition we obtain

$$D_2 = (D_0 \cdot T) \cdot T = D_0 \cdot T^2 \qquad \text{\textit{Multiplying by T again gives two transitions}}$$

Multiplying again by T (that is, $D_0 \cdot T^3$) would give the third state distribution vector, and so on.

---

kth State Distribution Vector

$$D_k = D_0 \cdot T^k \qquad \text{\textit{Multiplying by T^k gives the kth state distribution vector}}$$

To put this another way, T^k gives the probabilities for k successive transitions, so the entry in row i and column j of T^k gives the probability of going from state S_i to state S_j in exactly k transitions.

EXAMPLE 3

CALCULATING STATE DISTRIBUTION VECTORS

$$T = \begin{pmatrix} 1 & 0 & 0 \\ \frac{1}{2} & \frac{1}{2} & 0 \\ \frac{1}{2} & 0 & \frac{1}{2} \end{pmatrix}$$

For the Markov chain with transition matrix on the left and initial state distribution vector $D_0 = \begin{pmatrix} \frac{1}{5} & \frac{2}{5} & \frac{2}{5} \end{pmatrix}$, calculate the next two state distribution vectors D_1 and D_2.

Solution

We obtain the next state distribution vector by multiplying the previous one by the transition matrix:

$$D_1 = D_0 \cdot T = \begin{pmatrix} \frac{1}{5} & \frac{2}{5} & \frac{2}{5} \end{pmatrix} \cdot \begin{pmatrix} 1 & 0 & 0 \\ \frac{1}{2} & \frac{1}{2} & 0 \\ \frac{1}{2} & 0 & \frac{1}{2} \end{pmatrix}$$

$$= \begin{pmatrix} \frac{1}{5} \cdot 1 + \frac{2}{5} \cdot \frac{1}{2} + \frac{2}{5} \cdot \frac{1}{2} & 0 + \frac{2}{5} \cdot \frac{1}{2} + 0 & 0 + 0 + \frac{2}{5} \cdot \frac{1}{2} \end{pmatrix} \quad \text{Row} \times \text{columns}$$

$$= \begin{pmatrix} \frac{3}{5} & \frac{1}{5} & \frac{1}{5} \end{pmatrix}$$

and

$$D_2 = D_1 \cdot T = \begin{pmatrix} \frac{3}{5} & \frac{1}{5} & \frac{1}{5} \end{pmatrix} \cdot \begin{pmatrix} 1 & 0 & 0 \\ \frac{1}{2} & \frac{1}{2} & 0 \\ \frac{1}{2} & 0 & \frac{1}{2} \end{pmatrix} = \begin{pmatrix} \frac{4}{5} & \frac{1}{10} & \frac{1}{10} \end{pmatrix} \quad \begin{matrix} \text{Omitting} \\ \text{the details} \end{matrix}$$

$$D_0 = \begin{pmatrix} \frac{1}{5} & \frac{2}{5} & \frac{2}{5} \end{pmatrix}$$
$$D_2 = \begin{pmatrix} \frac{4}{5} & \frac{1}{10} & \frac{1}{10} \end{pmatrix}$$

Comparing D_0 with D_2, we see that the first entry has increased from $\frac{1}{5}$ to $\frac{4}{5}$. That is, in only two transitions, the probability of being in state S_1 has increased from 20% to 80%. Since we saw in Example 1c on pages 480–481 that the state S_1 is absorbing for this transition matrix, the probability of being in S_1 should increase with each transition.

Graphing Calculator Exploration

For the absorbing Markov chain in Example 3, we may easily check that the probability of being in state S_1 increases with each transition. Entering T and D_0 into matrices [A] and [B] and calculating $D_k = D_0 \cdot T^k$ for successively higher values of k gives:

```
MATRIX[A]  3 ×3      MATRIX[B]  1 ×3    [B]*[A]
[1    0    0    ]    [.2    .4    .4    ]    [[.60  .20  .20]]  ←D₁ = D₀ · A
[.5   .5   0    ]                            Ans*[A]
[.5   0    .5   ]                            [[.80  .10  .10]]  ←D₂ = D₀ · A²
                                             Ans*[A]
                                             [[.90  .05  .05]]  ←D₃ = D₀ · A³
```

↑ Probability of being in S_1

Notice that the first entry of these state distribution vectors increases from 0.6 to 0.8 and then to 0.9, showing that the probability of being in state S_1 increases with each transition. These probabilities will continue to increase, even if we begin with a different initial distribution D_0.

Duration in a Given State

In general, how long can we expect a Markov chain to remain in its current state before moving to another state? The formula in the following box is proved on pages 490–491.

Expected Duration in a Given State

For a Markov chain in a state S_i, let p be the element in row i and column i of the transition matrix. If $p = 1$, then S_i is an absorbing state, so the Markov chain will remain in S_i forever. If $p < 1$, then the expected number of times that the chain will be in that state before moving to another state is

$$E = \frac{1}{1 - p}$$

EXAMPLE 4

FINDING AN EXPECTED TIME

	H	M	I
H	0.75	0.20	0.05
M	0.20	0.70	0.10
I	0.25	0.20	0.55

The EasyDotCom Internet service provider classifies its residential customers by the number of connection hours used per week: H for high usage of more than 20 hours, M for moderate usage of between 5 and 20 hours, and I for infrequent usage of less than 5 hours. The company has found that the usage behavior of their subscribers can be modeled as a Markov chain with the transition matrix shown on the left.

If the Davidson household is presently a high user, how many weeks can it be expected to be a high user before changing to some other level of Internet use?

Solution

Since the probability that state H stays state H for one transition of this Markov chain is 0.75 (from the upper left entry in the transition matrix), the number of weeks (including the present week) that the Davidsons are expected to be heavy users before changing to some other level is

$$E = \frac{1}{1 - 0.75} = \frac{1}{0.25} = 4 \text{ weeks} \qquad E = \frac{1}{1 - p} \text{ with } p = 0.75$$

Spreadsheet Exploration

The following spreadsheet* shows 500 observations of the Markov chain from Example 4 but labels the states 1 (yellow) for H, 2 (green) for M, and 3 (blue) for I. The transition matrix is stored in the upper left-hand corner, and the counters below it summarize the number of times each state occurred in this observation.

	A	B	C
1	0.75	0.2	0.05
2	0.2	0.7	0.1
3	0.25	0.2	0.55
4			
5	State #1 :		
6	232	0.464	
7			
8	State #2 :		
9	204	0.408	
10			
11	State #3 :		
12	64	0.128	
13			
14			
15			
16			
17			
18			
19			
20			
21			
22			
23			

```
Columns D–AB (500 observations of states 1, 2, 3):

1 1 2 2 2 2 2 2 3 3 2 2 1 2 3 2 3 3 3 3 3 1 1 1
1 1 1 1 1 1 1 1 1 2 2 2 2 2 2 2 2 2 2 2 1 1 1 1 1
1 2 2 2 1 1 1 1 2 2 2 2 2 1 1 2 1 1 1 1 1 2 2 2
2 1 1 1 1 1 1 1 2 2 2 2 2 1 1 1 1 1 1 1 1 1 1 1
1 2 2 2 2 1 2 2 2 2 2 2 3 3 3 3 3 2 2 2 2 2
2 2 2 3 3 3 1 1 1 2 2 1 1 1 1 1 1 1 1 2 2 2 3 2
1 1 1 1 2 2 1 1 1 1 1 2 2 2 2 2 1 2 2 2 2 2 2 2
3 2 2 2 2 2 2 1 1 1 3 3 3 3 2 1 1 1 1 2 2 2 2 1
3 3 3 3 3 1 1 1 1 1 2 2 2 2 2 2 2 2 2 2 1 1 1 2
2 1 1 1 1 1 2 2 3 3 3 3 3 3 3 3 3 1 2 2 1 1 1 2
2 2 1 2 2 2 1 2 2 2 1 2 1 1 1 3 2 1 1 1 1 1 1
2 2 2 2 2 2 2 2 2 2 1 1 1 1 3 3 2 2 2 1 1 1 1 1
1 1 1 1 1 2 2 2 2 2 2 2 2 3 1 1 2 2 2 3 2 2 2 2 2
1 1 1 2 2 2 1 1 1 2 2 2 1 1 1 1 2 2 1 1 1 3 3 1 1
1 1 1 1 1 2 2 3 3 1 1 1 1 2 1 2 2 2 2 2 1 1 1 1 1
1 1 1 1 1 3 2 2 1 1 2 1 1 1 1 1 1 1 2 1 1 1 3 2
2 1 1 1 2 2 2 1 1 2 2 2 1 1 1 1 1 1 1 1 1 1 1 1
1 1 1 1 2 2 2 2 3 3 2 2 2 1 2 2 2 2 3 3 3 1 1 1 1
1 1 1 1 1 1 1 3 3 2 3 3 3 3 3 3 3 1 1 1 2 2 2 2
2 1 1 2 2 1 1 1 1 1 1 1 1 2 2 2 1 1 1 1 1 1 1 3
```

* See the Preface for how to obtain this and other Excel spreadsheets.

State H (yellow) sometimes appears just once and sometimes appears a dozen or more times in succession. Find the average length of these successions of H's, either by setting up your own modification of this spreadsheet or by counting from the screen image above. Is your answer consistent with our answer to Example 4?

7.1   Section Summary

A *transition matrix* is a square matrix of nonnegative entries with each row summing to 1. The entries are the transition probabilities between the states of the *Markov chain*. Transition matrices exhibit three basic transition behaviors:

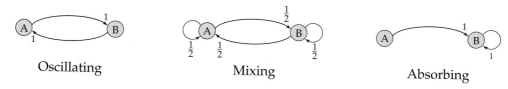

Oscillating   Mixing   Absorbing

An initial state distribution vector D_0 represents the probabilities of beginning in the various states, and the subsequent state distribution vectors $D_k = D_0 \cdot T^k$ are found by repeated multiplications by the transition matrix.

If the probability p that a Markov chain will stay in its present state for one transition is 1, then the chain will remain in that state forever. If it is less than 1, then the expected number of times it will be in this state before switching to some other state is $1/(1 - p)$.

➤ **Solutions to Practice Problems**

1. The probability of the transition $G \to G$ is 0.20.

2. Only matrix (a). Matrix (b) fails because the first row does not sum to 1, and matrix (c) fails because one of the entries is negative.

3. The identity matrix sends each state to itself, so every state is absorbing.

7.1 Exercises

Construct a transition matrix for the state-transition diagram and identify the transition as "oscillating," "mixing," or "absorbing."

1.

2.

3.

4.

5.

6.

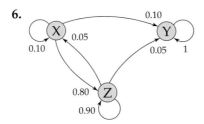

Construct a state-transition diagram for the transition matrix and identify the transition as "oscillating," "mixing," or "absorbing."

7. $\begin{pmatrix} 0 & 1 & 0 \\ 0.2 & 0 & 0.8 \\ 0 & 1 & 0 \end{pmatrix}$

8. $\begin{pmatrix} 0 & 1 & 0 & 0 \\ 0 & 0 & 0.3 & 0.7 \\ 1 & 0 & 0 & 0 \\ 1 & 0 & 0 & 0 \end{pmatrix}$

9. $\begin{pmatrix} 0.10 & 0.90 \\ 0.35 & 0.65 \end{pmatrix}$

10. $\begin{pmatrix} 0 & 1 & 0 & 0 \\ 0 & 0 & 0.3 & 0.7 \\ 0 & 1 & 0 & 0 \\ 1 & 0 & 0 & 0 \end{pmatrix}$

11. $\begin{pmatrix} 1 & 0 & 0 \\ 0.1 & 0.2 & 0.7 \\ 0.1 & 0.2 & 0.7 \end{pmatrix}$

12. $\begin{pmatrix} 1 & 0 & 0 & 0 \\ 0 & 0 & 0.3 & 0.7 \\ 0 & 1 & 0 & 0 \\ 0 & 0 & 0 & 1 \end{pmatrix}$

For each Markov chain transition matrix T and initial state distribution vector D_0, calculate the next two state distribution vectors D_1 and D_2.

13. $T = \begin{pmatrix} \frac{1}{4} & \frac{3}{4} \\ \frac{1}{2} & \frac{1}{2} \end{pmatrix}$, $D_0 = \begin{pmatrix} \frac{3}{5} & \frac{2}{5} \end{pmatrix}$

14. $T = \begin{pmatrix} \frac{1}{2} & \frac{1}{2} \\ \frac{1}{3} & \frac{2}{3} \end{pmatrix}$, $D_0 = \begin{pmatrix} \frac{2}{5} & \frac{3}{5} \end{pmatrix}$

15. $T = \begin{pmatrix} 0 & 0.2 & 0.8 \\ 0.6 & 0.4 & 0 \\ 0.4 & 0 & 0.6 \end{pmatrix}$,

$D_0 = (0.35 \quad 0.45 \quad 0.20)$

16. $T = \begin{pmatrix} 0.4 & 0.4 & 0.2 \\ 0.4 & 0 & 0.6 \\ 0.1 & 0.3 & 0.6 \end{pmatrix}$,

$D_0 = (0.25 \quad 0.60 \quad 0.15)$

For each Markov chain transition matrix T and present state, find the expected number of times the chain will be in that state before moving to some other state.

17. $T = \begin{pmatrix} 0.6 & 0.2 & 0.2 \\ 0.1 & 0.8 & 0.1 \\ 0.6 & 0.2 & 0.2 \end{pmatrix}$, presently in state S_2

18. $T = \begin{pmatrix} 0.8 & 0.2 & 0 \\ 0.3 & 0.6 & 0.1 \\ 0.1 & 0 & 0.9 \end{pmatrix}$, presently in state S_3

19. $T = \begin{pmatrix} 0.8 & 0.1 & 0.1 & 0 \\ 0.2 & 0.5 & 0.1 & 0.2 \\ 0 & 0 & 1 & 0 \\ 0 & 0.3 & 0.1 & 0.6 \end{pmatrix}$, presently in state S_4

20. $T = \begin{pmatrix} 1 & 0 & 0 & 0 \\ 0 & 0.20 & 0.70 & 0.10 \\ 0.05 & 0.10 & 0.85 & 0 \\ 0 & 0 & 0 & 1 \end{pmatrix}$, presently in state S_3

Represent each situation as a Markov chain by constructing a state-transition diagram and a transition matrix. Be sure to state your final answer in terms of the original question.

21. GENERAL: Traffic Flow A traffic light on Main Street operates on a one-minute timer. If it is green (to the traffic), at the end of the minute it will change to red only if a pedestrian has pressed a button at the crosswalk. The probability of a pedestrian pressing the button during any one-minute interval is 0.25. If the light is red, it turns back to green at the end of the minute. If the light is green now for the traffic, find the expected number of minutes that the light is green.

22. SOCIAL SCIENCE: Learning Theory The initial phase of a learning experiment is to establish that a rat will randomly wander in a maze prior to the introduction of food and other stimuli. The following maze has three connecting rooms, A, B, and C, and the location of the rat is observed at one-minute intervals. For both rooms A and B, the probability of remaining there is 50%, while the probability of remaining in room C is 75%. When the rat leaves a room, it is equally likely to choose either of the other two rooms. If the rat is in room B, find the expected number of minutes that the rat will remain there.

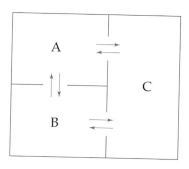

23. GENERAL: Breakfast Habits A survey of weekly breakfast eating habits found that although many people eat cereal, boredom and other reasons cause 5% to switch to something else (eggs, muffins, and so on) each week, while among those who are not eating cereal for breakfast, 15% will switch to cereal each week. If 70% of the residents of Cincinnati are eating cereal for breakfast this week, what percentage will be eating cereal next week? In two weeks?

 24. BUSINESS: Market Research A telephone survey of households in the Atlanta metropolitan area found that 85% of those using a DVD player planned to rent movies or otherwise use it next month, while 10% of those not using a DVD player this month planned on using one next month. If 2% of the households currently use a DVD player, what percentage will be using a DVD player after one year?

25. ATHLETICS: Retirement Sports A survey of retirees in Arizona who golf or fish every day found that 60% of those who played golf one day switched to fishing the next and that 90% of those who went fishing switched to golf the next day. If 30% of these retirees are golfing today, what percentage will be fishing tomorrow? Golfing the day after tomorrow?

26. MANAGEMENT SCIENCE: Emergency Medical Care Authorization specialists at a national health maintenance organization spend twelve minutes to evaluate and process each request for emergency care, and their work loads are managed by a computerized distribution system. This system checks the work queue of each specialist every twelve minutes, transfers finished work to the data storage center, and adds new requests to the queue, but never allows any queue to contain more than four requests (one "active" and three "pending review"). The number of new requests the system has to add to any particular queue is variable: 15% of the time there are none, 55% there is one, 25% there are two, and 5% there are three new requests. If a particular specialist does not have any "pending review" requests in her queue, how many minutes can we expect this situation to continue? [*Hint:* Define four states by the number of unfinished requests in the queue at the end of each twelve-minute block of time, just before the system checks it to remove finished work and add new work.]

27. BUSINESS: Portfolio Management An investment banker estimates the financial stability of mid-sized manufacturing companies as "secure," "doubtful," and "at risk." He has noticed that of the "secure" companies he follows, each year 5% decline to "doubtful" and the rest remain as they are; 10% of the "doubtful" companies improve to "secure," 5% decline to "at risk," and the rest remain as they are; and 5% of the "at risk" companies become bankrupt and never recover, 10% improve to "doubtful," and the rest remain as they are. If his current portfolio of investments is 80% "secure," 15% "doubtful," and 5% "at risk," what percentage will be "secure" in two years? (Assume that the present trends continue and that no changes are made to the portfolio.) In the long run, how many of the companies in the portfolio will become bankrupt?

28. ATHLETICS: Baseball Players A baseball fan tracks the careers of interesting players in the major and minor leagues and has found that from year to year, 80% of the major league players tracked stay in the majors, 10% drop down to the minor leagues, and 10% quit baseball or retire, while 20% of the minor league players tracked move up to the majors, 40% stay in the minors, and the rest quit or retire. If his current roster of interesting players is three-quarters major and one-quarter minor league players, what percentage of these players will

have quit or retired in two years? What will happen to all the players in the long run?

29. GENERAL: Working Mothers A working mother of three teenagers expects them to help around the house and rates the laundry room each weekend as "empty," "manageable," or "you're all grounded!" She has noticed that if it is empty one weekend, then the following weekend with probability 0.4 it is empty, with probability 0.4 it is manageable, and with probability 0.2 it is so bad that "you're all grounded!" If it is manageable one weekend, then the following weekend with probability 0.5 it is empty and with probability 0.5 it is manageable. If it is so bad that she has to ground everybody, then the following weekend the laundry room is empty. If the laundry room is manageable this weekend, how many weekends will it be manageable before changing to either "you're all grounded!" or "empty"?

30. ATHLETICS: Jogging Routes Each morning a jogger runs either at the local high school track, around the neighborhood, or in the park. If she runs at the school one day, then the next day with probability 0.7 she runs there again and with probability 0.3 she switches to the park; if she runs around the neighborhood, then the next day with probability 0.9 she runs there again and with probability 0.1 she switches to the park; and if she runs in the park, then the next day she switches with probability 0.9 to the school and with probability 0.1 to around the neighborhood. If she runs around the neighborhood today, how many days can she be expected to run around the neighborhood before switching to the park?

Explorations and Excursions

The following problems extend and augment the material presented in the text.

Proof of the Formula for the Duration in a Given State

31. For a Markov chain in a nonabsorbing state S_i, let p be the element in row i and column i of the transition matrix T. Explain why the probability of the chain being in state S_i exactly n

times and then switching to some other state is $p^{n-1}(1 - p)$.

32. Using the formula for expected value (page 409), explain why the expected number of times that the chain will be in state S_i is

$$E = 1 \cdot (1 - p) + 2 \cdot p(1 - p) + 3 \cdot p^2(1 - p) + \cdots$$

33. Multiply out this series and combine similar terms to obtain

$$E = 1 + p^2 + p^3 + \cdots$$

34. Use the result of Exercise 32 on pages 129–130 to obtain $E = \dfrac{1}{1 - p}$, which is the formula in the box on page 485.

7.2 REGULAR MARKOV CHAINS

Introduction

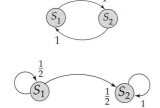

Everyone who takes their coffee or tea with cream and sugar knows that it makes no difference which is added first nor on which side of the cup it is done—a few stirs mixes everything uniformly. In this section we will study Markov chains that exhibit this sort of mixing—that is, ones that move among all of their states towards some sort of equilibrium.

Not all Markov chains exhibit this kind of mixing. The first chain on the left (from Example 1a on page 480) is an *oscillating* Markov chain, which bounces back and forth and never settles down to an equilibrium behavior. The second is an *absorbing* Markov chain, which eventually gets stuck in a single state rather than mixing around in all of them. In this section we begin by developing a condition on the transition matrix that guarantees mixing by avoiding oscillation and absorption, and then we show that the eventual equilibrium behavior of such a chain is independent of how it began.

Regular Markov Chains

Consider a Markov chain for which, in some number of moves, it is possible to go from any state to any state. Such a chain cannot rigidly oscillate nor be absorbing. Since we saw on page 483 that powers of the transition matrix give the probabilities of transitions in several moves, this means that some power of the transition matrix has all positive entries. Such chains are called *regular*.

> **Regular Markov Chains**
>
> A transition matrix is *regular* if some power of it has only positive entries. A Markov chain is regular if its transition matrix is regular.

We may determine whether a Markov chain is regular either from its transition matrix or its state-transition diagram.*

EXAMPLE 1

REGULAR MATRICES

Which of the following transition matrices are regular?

a. $A = \begin{pmatrix} \frac{1}{2} & \frac{1}{2} \\ 1 & 0 \end{pmatrix}$ **b.** $B = \begin{pmatrix} \frac{1}{2} & \frac{1}{2} \\ 0 & 1 \end{pmatrix}$ **c.** $C = \begin{pmatrix} 1 & 0 \\ 0 & 1 \end{pmatrix}$

Solution

We will first find the answer by calculating powers of the matrix and then check it from the state-transition diagram.

a. Since all entries of A^2 (calculated below) are positive, A is regular.

$$A^2 = \begin{pmatrix} \frac{1}{2} & \frac{1}{2} \\ 1 & 0 \end{pmatrix} \cdot \begin{pmatrix} \frac{1}{2} & \frac{1}{2} \\ 1 & 0 \end{pmatrix} = \begin{pmatrix} \frac{3}{4} & \frac{1}{4} \\ \frac{1}{2} & \frac{1}{2} \end{pmatrix}$$

The corresponding state-transition diagram shows that this is a mixing transition: Any state can be reached from every state in two steps. (Note that $S_2 \to S_2$ takes two steps.)

b. Since the lower left entry in B^k remains zero with each multiplication, B is not regular.

$$B = \begin{pmatrix} \frac{1}{2} & \frac{1}{2} \\ 0 & 1 \end{pmatrix}, \quad B^2 = \begin{pmatrix} \frac{1}{4} & \frac{3}{4} \\ 0 & 1 \end{pmatrix}, \quad B^3 = \begin{pmatrix} \frac{1}{8} & \frac{7}{8} \\ 0 & 1 \end{pmatrix}, \dots \quad \text{Do you see why this entry will be 0 for }any\text{ power of }B?$$

The corresponding state-transition diagram shows that S_2 is an absorbing state for this transition, so S_1 is not reachable from S_2 in any number of steps.

c. Since every power of the identity matrix is still the identity matrix, $C = I$ is not regular.

$$C = \begin{pmatrix} 1 & 0 \\ 0 & 1 \end{pmatrix}, \quad C^2 = \begin{pmatrix} 1 & 0 \\ 0 & 1 \end{pmatrix}, \quad C^3 = \begin{pmatrix} 1 & 0 \\ 0 & 1 \end{pmatrix}, \dots$$

The state-transition diagram for C shows that both S_1 and S_2 are absorbing states.

Given a particular Markov chain, we can use the methods of the previous section to calculate the long-term distribution to see whether it is affected by the initial distribution.

* A transition matrix with only positive entries is called an *ergodic* matrix. Therefore, a transition matrix is *regular* if some power of it is ergodic. The term was coined by the physicist Ludwig Boltzmann (1844–1906) from the Greek words *erg* (work) and *ode* (path) when formulating his *ergodic hypothesis* to explain how the individual paths of the molecules in a gas lead to the uniform equilibrium that we experience.

EXAMPLE 2

CALCULATING A LONG-TERM DISTRIBUTION

A rental truck company has branches in Springfield, Tulsa, and Little Rock, and each rents trucks by the week. The trucks may be returned to any of the branches. The return location probabilities are shown in the following state-transition diagram. Find the distribution of the company's 180 trucks among the three cities after one year if the trucks are initially distributed as follows:

a. 36 in Springfield and 72 each in Tulsa and in Little Rock;

b. 90 in Springfield, 30 in Tulsa, and 60 in Little Rock;

c. all 180 in Tulsa.

 Solution

We will describe what happens in terms of the *proportion* of trucks in each city, with S_1, S_2, and S_3 standing for Springfield, Tulsa, and Little Rock, respectively. From the state-transition diagram, the transition matrix is

$$T = \begin{pmatrix} \overset{\text{Springfield}}{0.6} & \overset{\text{Tulsa}}{0.1} & \overset{\text{Little Rock}}{0.3} \\ 0.2 & 0.7 & 0.1 \\ 0 & 0.5 & 0.5 \end{pmatrix} \begin{matrix} \text{Springfield} \\ \text{Tulsa} \\ \text{Little Rock} \end{matrix}$$

Since T is regular (the state-transition diagram shows that it is possible to go from any state to any state in two steps), the distribution of trucks among the rental offices over successive weeks should show that they "mix around" in some predictable fashion rather than "bunching up" in one city. For each initial state distribution D_0, the distribution after one year is the result of 52 weekly transitions:

$$D_{52} = D_0 \cdot T^{52}$$

a. For the numbers of trucks given in part (a), the initial state distribution vector is

$$D_0 = \left(\tfrac{36}{180} \quad \tfrac{72}{180} \quad \tfrac{72}{180} \right) = \left(\tfrac{1}{5} \quad \tfrac{2}{5} \quad \tfrac{2}{5} \right) \qquad \text{Each initial number divided by 180 trucks}$$

Then, after 52 weekly transitions, the distribution will be

$$D_{52} = \left(\tfrac{1}{5} \quad \tfrac{2}{5} \quad \tfrac{2}{5} \right) \cdot \begin{pmatrix} 0.6 & 0.1 & 0.3 \\ 0.2 & 0.7 & 0.1 \\ 0 & 0.5 & 0.5 \end{pmatrix}^{52} = \left(\tfrac{1}{4} \quad \tfrac{1}{2} \quad \tfrac{1}{4} \right) \qquad \text{Using a calculator}$$

After one year there will be $\frac{1}{4} \cdot 180 = 45$ trucks in Springfield, $\frac{1}{2} \cdot 180 = 90$ trucks in Tulsa, and $\frac{1}{4} \cdot 180 = 45$ trucks in Little Rock.

b. The initial numbers of trucks in part (b) give

$$D_0 = \begin{pmatrix} \frac{90}{180} & \frac{30}{180} & \frac{60}{180} \end{pmatrix} = \begin{pmatrix} \frac{1}{2} & \frac{1}{6} & \frac{1}{3} \end{pmatrix}$$

so

$$D_{52} = \begin{pmatrix} \frac{1}{2} & \frac{1}{6} & \frac{1}{3} \end{pmatrix} \cdot \begin{pmatrix} 0.6 & 0.1 & 0.3 \\ 0.2 & 0.7 & 0.1 \\ 0 & 0.5 & 0.5 \end{pmatrix}^{52} = \begin{pmatrix} \frac{1}{4} & \frac{1}{2} & \frac{1}{4} \end{pmatrix} \qquad \text{Using a calculator}$$

This is the same distribution found in part (a), so again there will be 45 trucks in Springfield, 90 trucks in Tulsa, and 45 trucks in Little Rock.

c. For the initial numbers in part (c), we have

$$D_0 = \begin{pmatrix} 0 & \frac{180}{180} & 0 \end{pmatrix} = \begin{pmatrix} 0 & 1 & 0 \end{pmatrix}$$

so

$$D_{52} = \begin{pmatrix} 0 & 1 & 0 \end{pmatrix} \cdot \begin{pmatrix} 0.6 & 0.1 & 0.3 \\ 0.2 & 0.7 & 0.1 \\ 0 & 0.5 & 0.5 \end{pmatrix}^{52} = \begin{pmatrix} \frac{1}{4} & \frac{1}{2} & \frac{1}{4} \end{pmatrix} \qquad \text{Using a calculator}$$

Again we get the same distribution as in parts (a) and (b), so again there will be 45 trucks in Springfield, 90 trucks in Tulsa, and 45 trucks in Little Rock.

As we expected, the long-term distribution of the trucks did not depend on the initial distribution. Moreover, further mixing will not change the result:

$$\begin{pmatrix} \frac{1}{4} & \frac{1}{2} & \frac{1}{4} \end{pmatrix} \cdot \begin{pmatrix} 0.6 & 0.1 & 0.3 \\ 0.2 & 0.7 & 0.1 \\ 0 & 0.5 & 0.5 \end{pmatrix} = \begin{pmatrix} \frac{1}{4} & \frac{1}{2} & \frac{1}{4} \end{pmatrix} \qquad \text{This distribution remains unchanged by another transition}$$

A distribution such as this that remains unchanged by a transition is called a *steady-state distribution*.

Steady-State Distribution

A state distribution vector D is a *steady-state distribution* for the Markov chain with transition matrix T if $D \cdot T = D$.

Practice Problem 1 Is $D = \begin{pmatrix} \frac{1}{2} & \frac{1}{2} \end{pmatrix}$ a steady-state distribution for the Markov chain with transition matrix $T = \begin{pmatrix} \frac{1}{4} & \frac{3}{4} \\ \frac{3}{4} & \frac{1}{4} \end{pmatrix}$?

➤ Solution on page 499

The Fundamental Theorem of Regular Markov Chains

For the truck rental company in Example 2 we obtained the same long-run distribution regardless of the initial distribution. Can we obtain this long-run distribution directly from the transition matrix? The answer is "yes," and in two different ways.

For a regular Markov chain, calculating high powers T^k of the transition matrix T will result in rows that "settle down" to this steady-state distribution.

Graphing Calculator Exploration

To see whether the transition matrix for the truck rental company in Example 2 has the property that the rows of higher powers "settle down" to the steady-state distribution, we may calculate as follows.

```
MATRIX[A]  3 X3
[.6    .1    .3    ]
[.2    .7    .1    ]
[0     .5    .5    ]
```

a. Enter the matrix $T = \begin{pmatrix} 0.6 & 0.1 & 0.3 \\ 0.2 & 0.7 & 0.1 \\ 0 & 0.5 & 0.5 \end{pmatrix}$ in matrix [A].

b. Then calculate T^k for $k = 2, 4, 26,$ and 52 (a year of weekly transitions):

```
[A]²
[[.38  .28  .34]
 [.26  .56  .18]
 [.1   .6   .3 ]]
```

```
[A]^4
[[.2512  .4672 ....
 [.2624  .4944 ....
 [.224   .544  ....
```

```
[A]^26
[[.2500000004 ....
 [.2499999997 ....
 [.2500000001 ....
```

```
[A]^52
[[.25  .5  .25]
 [.25  .5  .25]
 [.25  .5  .25]]
```

Notice that the rows in successive screens all seem to be approaching the same distribution $\begin{pmatrix} \frac{1}{4} & \frac{1}{2} & \frac{1}{4} \end{pmatrix}$, with the last screen (for T^{52}) showing exactly this steady-state distribution, the same one that we found in Example 2.

Why does each column "settle down" to a single value as the matrix is raised to higher and higher powers? For a regular transition matrix, there is a power (call it n) such that all entries of T^n are positive. Consider the matrix product $T^n \cdot T^k$, which involves multiplying rows of T^n by columns of T^k. Each row of T^n consists of positive numbers adding to 1, so multiplying them by a column of T^k amounts to taking a *weighted average* of all of the column entries, which will bring the extremes of the column closer together. As k increases, the resulting column entries will therefore converge to a single "averaged" value (see page 410). This means that as we raise T to higher and higher powers, each column will tend towards a single value. This column value, being the long-run probability of entering into a state, must be the long-run probability of being in that state. Therefore, the rows must tend to the steady-state distribution, which is therefore unique. Note that this reasoning would fail if any of the elements of T^n were zero, which is why regularity is needed. This result is so important that it is called the Fundamental Theorem of Regular Markov Chains.

Fundamental Theorem of Regular Markov Chains

A regular Markov chain with transition matrix T has exactly one steady-state distribution D solving $D \cdot T = D$. Higher powers of the transition matrix T approximate arbitrarily closely a matrix each of whose rows is equal to D.

How to Solve $D \cdot T = D$

How can we find the steady-state distribution D that solves $D \cdot T = D$ without having to calculate arbitrarily high powers of the transition matrix? We begin by writing the equation with zero on the right:

$D \cdot T - D = 0$	From $D \cdot T = D$
$D \cdot (T - I) = 0$	Factoring (since $D \cdot I = D$)
$(T - I)^t \cdot D^t = 0$	Taking transposes since $(A \cdot B)^t = B^t \cdot A^t$
$(T^t - I) \cdot D^t = 0$	Since $(A - B)^t = A^t - B^t$ and $I^t = I$

Since D is a state distribution vector, the sum of the entries of D must be 1, which we can write as the matrix equation

$$(1 \cdots 1) \cdot D^t = 1 \qquad \text{Entries of } D \text{ add up to 1}$$

Combining these last two matrix equations into one large matrix equation, we need to solve

$$\begin{pmatrix} T^t - I \\ 1 \cdots 1 \end{pmatrix} \cdot D^t = \begin{pmatrix} 0 \\ 1 \end{pmatrix} \qquad \begin{pmatrix} 0 \\ 1 \end{pmatrix} \text{ means } \begin{pmatrix} 0 \\ \vdots \\ 0 \\ 1 \end{pmatrix}$$

We may now solve for D^t by row-reducing the augmented matrix $\left(\begin{array}{c|c} T^t - I & 0 \\ \hline 1 \cdots 1 & 1 \end{array}\right)$ to obtain $\left(\begin{array}{c|c} I & D^t \\ \hline 0 & 0 \end{array}\right)$, just as we did in Section 3 of Chapter 3 (see page 191).

Steady-State Distribution for a Regular Markov Chain

The steady-state distribution D for a *regular* Markov chain with transition matrix T may be found by row-reducing the augmented matrix

$$\left(\begin{array}{c|c} T^t - I & 0 \\ \hline 1 \cdots 1 & 1 \end{array}\right) \quad \text{to obtain} \quad \left(\begin{array}{c|c} I & D^t \\ \hline 0 & 0 \end{array}\right)$$

and then transposing all but the bottom entry in the last column.

EXAMPLE 3

$$T = \begin{pmatrix} 0.6 & 0.1 & 0.3 \\ 0.2 & 0.7 & 0.1 \\ 0 & 0.5 & 0.5 \end{pmatrix}$$

FINDING A STEADY-STATE DISTRIBUTION

Find the steady-state distribution for the transition matrix from the truck rental situation described in Example 2.

Solution

Following the procedure described above, we transpose T by changing rows into columns and then calculate $T^t - I$:

$$T^t - I = \begin{pmatrix} 0.6 & 0.2 & 0 \\ 0.1 & 0.7 & 0.5 \\ 0.3 & 0.1 & 0.5 \end{pmatrix} - \begin{pmatrix} 1 & 0 & 0 \\ 0 & 1 & 0 \\ 0 & 0 & 1 \end{pmatrix} = \begin{pmatrix} -0.4 & 0.2 & 0 \\ 0.1 & -0.3 & 0.5 \\ 0.3 & 0.1 & -0.5 \end{pmatrix}$$

We need to solve the matrix equation

$$\begin{pmatrix} -0.4 & 0.2 & 0 \\ 0.1 & -0.3 & 0.5 \\ 0.3 & 0.1 & -0.5 \\ 1 & 1 & 1 \end{pmatrix} \begin{pmatrix} d_1 \\ d_2 \\ d_3 \end{pmatrix} = \begin{pmatrix} 0 \\ 0 \\ 0 \\ 1 \end{pmatrix} \qquad \begin{pmatrix} T^t - I \\ 1 \cdots 1 \end{pmatrix} \cdot D^t = \begin{pmatrix} 0 \\ 1 \end{pmatrix}$$

We row-reduce the augmented matrix

$$\left(\begin{array}{ccc|c} -0.4 & 0.2 & 0 & 0 \\ 0.1 & -0.3 & 0.5 & 0 \\ 0.3 & 0.1 & -0.5 & 0 \\ 1 & 1 & 1 & 1 \end{array}\right) \quad \text{to obtain} \quad \left(\begin{array}{ccc|c} 1 & 0 & 0 & \frac{1}{4} \\ 0 & 1 & 0 & \frac{1}{2} \\ 0 & 0 & 1 & \frac{1}{4} \\ \hline 0 & 0 & 0 & 0 \end{array}\right)$$

$$\begin{array}{c} \left(\begin{array}{c|c} T^t - I & 0 \\ \hline 1 \cdots 1 & 1 \end{array}\right) \\ \text{reduces to} \\ \left(\begin{array}{c|c} I & D^t \\ \hline 0 & 0 \end{array}\right) \end{array}$$

(omitting the details). The last column (omitting the bottom zero) is the transpose D^t of the steady-state distribution

$$D = \begin{pmatrix} \frac{1}{4} & \frac{1}{2} & \frac{1}{4} \end{pmatrix}$$

This answer agrees with the fractions that we found in Example 2 (on pages 493–494) and also with the high powers of the transition matrix calculated in the Graphing Calculator Exploration on page 495.

Practice Problem 2

Use row-reduction to find the steady-state distribution of the Markov chain with transition matrix $T = \begin{pmatrix} 0 & 1 \\ \frac{1}{2} & \frac{1}{2} \end{pmatrix}$. ➤ **Solution on next page**

Graphing Calculator Exploration

We may solve Example 3 using a graphing calculator. The program* MARKOV displays powers of a transition matrix T, the augmented matrix $\left(\begin{array}{c|c} T^t - I & 0 \\ \hline 1 \cdots 1 & 1 \end{array} \right)$, and the row-reduced form showing the steady-state distribution. To find the steady-state distribution in Example 3, we proceed as follows:

a. Enter the T in matrix [A].

b. Run the program MARKOV and press ENTER several times to display T^k for $k = 1, 2, 3, 32$, and 64, and then $\left(\begin{array}{c|c} T^t - I & 0 \\ \hline 1 \cdots 1 & 1 \end{array} \right)$ followed by its row-reduced form.

```
MATRIX[A] 3 X3
[.6    .1    .3   ]
[.2    .7    .1   ]
[0     .5    .5   ]
```

```
[A]
[[3/5 1/10 3/10…
 [1/5 7/10 1/10…
 [0   1/2  1/2 …
```

```
[A]^64
[[1/4 1/2  1/4]
 [1/4 1/2  1/4]
 [1/4 1/2  1/4]]
```

```
    (A^T-I)X=0
AND (1..1)X=1 IS
[[-2/5  1/5   0 …
 [1/10 -3/10 1/…
 [3/10  1/10  -1…
 [1     1     1 …
```

```
SOLUTION
[[1 0 0 1/4]
 [0 1 0 1/2]
 [0 0 1 1/4]
 [0 0 0 0  ]]
```

---

* See the Preface for information on how to obtain this and other graphing calculator programs.

The steady state distribution $\left(\frac{1}{4} \quad \frac{1}{2} \quad \frac{1}{4}\right)$ shown in the right-hand column agrees with the answer found earlier.

7.2 Section Summary

A transition matrix is *regular* if some power of it has all positive entries. The long-term behavior of a regular Markov chain with transition matrix T is the solution D of the *steady-state distribution* equation

$$D \cdot T = D$$

D may be found by row-reducing the augmented matrix

$$\left(\begin{array}{c|c} T^t - I & 0 \\ \hline 1 \cdots 1 & 1 \end{array}\right) \qquad \text{to obtain} \qquad \left(\begin{array}{c|c} I & D^t \\ \hline 0 & 0 \end{array}\right)$$

and then transposing the last column (omitting the bottom zero).

➤ **Solutions to Practice Problems**

1. $\left(\frac{1}{2} \quad \frac{1}{2}\right) \cdot \begin{pmatrix} \frac{1}{4} & \frac{3}{4} \\ \frac{3}{4} & \frac{1}{4} \end{pmatrix} = \left(\frac{1}{2} \cdot \frac{1}{4} + \frac{1}{2} \cdot \frac{3}{4} \quad \frac{1}{2} \cdot \frac{3}{4} + \frac{1}{2} \cdot \frac{1}{4}\right) = \left(\frac{4}{8} \quad \frac{4}{8}\right) = \left(\frac{1}{2} \quad \frac{1}{2}\right)$

Therefore, $\left(\frac{1}{2} \quad \frac{1}{2}\right)$ *is* a steady-state distribution for the transition matrix.

2. $T^t - I = \begin{pmatrix} 0 & 1 \\ \frac{1}{2} & \frac{1}{2} \end{pmatrix}^t - \begin{pmatrix} 1 & 0 \\ 0 & 1 \end{pmatrix} = \begin{pmatrix} 0 & \frac{1}{2} \\ 1 & \frac{1}{2} \end{pmatrix} - \begin{pmatrix} 1 & 0 \\ 0 & 1 \end{pmatrix} = \begin{pmatrix} -1 & \frac{1}{2} \\ 1 & -\frac{1}{2} \end{pmatrix}$

so $\left(\begin{array}{c|c} T^t - I & 0 \\ \hline 1 \cdots 1 & 1 \end{array}\right) = \left(\begin{array}{cc|c} -1 & \frac{1}{2} & 0 \\ 1 & -\frac{1}{2} & 0 \\ \hline 1 & 1 & 1 \end{array}\right)$.

One of the many ways to row-reduce this matrix (all leading to the same result) is:

$$\begin{array}{c} \\ R_2 + R_1 \rightarrow \\ R_3 + R_1 \rightarrow \end{array} \left(\begin{array}{cc|c} -1 & \frac{1}{2} & 0 \\ 0 & 0 & 0 \\ 0 & \frac{3}{2} & 1 \end{array}\right) \Rightarrow \begin{array}{c} -R_1 \rightarrow \\ R_3 \rightarrow \\ R_2 \rightarrow \end{array} \left(\begin{array}{cc|c} 1 & -\frac{1}{2} & 0 \\ 0 & \frac{3}{2} & 1 \\ 0 & 0 & 0 \end{array}\right) \Rightarrow \frac{2}{3} R_2 \rightarrow \left(\begin{array}{cc|c} 1 & -\frac{1}{2} & 0 \\ 0 & 1 & \frac{2}{3} \\ 0 & 0 & 0 \end{array}\right) \Rightarrow \begin{array}{c} R_1 + \frac{1}{2}R_2 \rightarrow \\ \\ \end{array} \left(\begin{array}{cc|c} 1 & 0 & \frac{1}{3} \\ 0 & 1 & \frac{2}{3} \\ 0 & 0 & 0 \end{array}\right)$$

The steady-distribution is $D = \left(\frac{1}{3} \quad \frac{2}{3}\right)$.

7.2 Exercises

Identify each transition matrix as "regular" or "not regular."

1. $\begin{pmatrix} 0.15 & 0.85 \\ 0.40 & 0.60 \end{pmatrix}$

2. $\begin{pmatrix} 0 & 0 & 1 \\ 0 & 0 & 1 \\ 0.5 & 0.5 & 0 \end{pmatrix}$

3. $\begin{pmatrix} 0.75 & 0.25 \\ 1 & 0 \end{pmatrix}$

4. $\begin{pmatrix} 0 & 0.4 & 0.6 \\ 0 & 0.8 & 0.2 \\ 1 & 0 & 0 \end{pmatrix}$

5. $\begin{pmatrix} 0.75 & 0.25 \\ 0 & 1 \end{pmatrix}$

6. $\begin{pmatrix} 1 & 0 & 0 \\ 0 & 0.8 & 0.2 \\ 0 & 0.4 & 0.6 \end{pmatrix}$

7. $\begin{pmatrix} 0.7 & 0.3 & 0 \\ 0 & 0.3 & 0.7 \\ 0.7 & 0.3 & 0 \end{pmatrix}$

8. $\begin{pmatrix} 1 & 0 & 0 \\ 0 & 0 & 1 \\ 0 & 1 & 0 \end{pmatrix}$

For each transition matrix T:
a. Find the steady-state distribution.
b. Calculate T^k for $k = 5, 10, 20,$ and 50 to verify that the rows of T^k approach the steady-state distribution.

9. $\begin{pmatrix} 0.4 & 0.6 \\ 0.6 & 0.4 \end{pmatrix}$

10. $\begin{pmatrix} 0.6 & 0.4 \\ 0.1 & 0.9 \end{pmatrix}$

11. $\begin{pmatrix} 0.55 & 0.45 \\ 0.30 & 0.70 \end{pmatrix}$

12. $\begin{pmatrix} 0.45 & 0.55 \\ 0.70 & 0.30 \end{pmatrix}$

13. $\begin{pmatrix} 0.1 & 0.8 & 0.1 \\ 0.2 & 0.7 & 0.1 \\ 0.4 & 0.5 & 0.1 \end{pmatrix}$

14. $\begin{pmatrix} 0.7 & 0.2 & 0.1 \\ 0.1 & 0.3 & 0.6 \\ 0.5 & 0.3 & 0.2 \end{pmatrix}$

15. $\begin{pmatrix} 0 & 0.6 & 0.4 \\ 0.1 & 0.9 & 0 \\ 0.3 & 0.5 & 0.2 \end{pmatrix}$

16. $\begin{pmatrix} 0 & 0.6 & 0.4 \\ 0.2 & 0.8 & 0 \\ 0 & 0.6 & 0.4 \end{pmatrix}$

For each regular transition matrix T:
a. Construct a state-transition diagram and find the smallest number k of transitions needed to move from the given state S_i to state S_j and the probability of that transition.
b. Use this k to verify that T^k has a positive transition probability in row i and column j equal to the probability found in (a).

17. $\begin{pmatrix} 0 & 0 & 1 \\ 0.5 & 0.5 & 0 \\ 0 & 1 & 0 \end{pmatrix}$, $S_2 \to S_3$

18. $\begin{pmatrix} 0 & 0.5 & 0.5 \\ 1 & 0 & 0 \\ 0.5 & 0 & 0.5 \end{pmatrix}$, $S_2 \to S_2$

19. $\begin{pmatrix} 0 & 0 & 1 \\ 0.5 & 0 & 0.5 \\ 0 & 1 & 0 \end{pmatrix}$, $S_1 \to S_1$

20. $\begin{pmatrix} 0 & 0 & 0 & 1 \\ 0 & 0 & 1 & 0 \\ 0.5 & 0 & 0 & 0.5 \\ 0 & 1 & 0 & 0 \end{pmatrix}$, $S_4 \to S_1$

Represent each situation as a Markov chain by constructing a state-transition diagram and the corresponding transition matrix. Find the steady-state distribution and interpret it in terms of the original situation. Be sure to state your final answer in terms of the original question.

21. SOCIAL SCIENCE: Voting Patterns The students in the political science summer program at Edson State College are studying voting patterns in Marston County. Half of the students reviewed voter records at the County Clerk's Office and found that a person who voted in an election has an 80% chance of voting in the next election, while someone who did not vote in an election has a 30% chance of voting in the next election. The other half of the students conducted surveys door to door and at shopping centers and found that voter perceptions were somewhat different: 90% of those who claimed to have voted in the last election said they would vote in the next, while 40% of those who said they hadn't voted in the last election said they would vote in the next. If these findings are valid for predicting long-term trends, which survey is consistent with the national average of about 61% of eligible voters actually voting? (*Source:* Committee for the Study of the American Electorate)

22. SOCIAL SCIENCE: Population Dynamics Life is so good in Lucas, Marion, and Warren counties in upstate Maine that no one ever moves away from the tri-county region. However, each year 4% of the Lucas residents move to Marion and 2% move to Warren; 2% of the Marion residents move to Lucas and 2% move to Warren; and 2% of the Warren residents move to Lucas and 1% move to Marion. After many years, how many reside in each county if the combined population of these three counties is 11,200?

23. **SOCIAL SCIENCE: Mass Transit** Commuters in the Pittsburgh metropolitan area either drive alone, join a car pool, or take the bus. Each month, of those who drive by themselves, 20% join a car pool, 30% switch to the bus, and the rest continue driving alone; of those who are in a car pool, 30% switch to driving alone, 20% switch to the bus, and the rest stay in a car pool; and of those who take the bus, 20% switch to driving alone, 30% join a car pool, and the rest continue taking the bus. After many months, how many of the three million commuters will be driving alone?

24. **GENERAL: Art Gallery Shows** The Harmon Gallery's showing of paintings by Clyberg, Stevensen, and Georgan attracted 1100 art dealers and collectors on its opening night. The security staff reported the following pattern of crowd movement every five minutes throughout the evening: Of the crowd in the Clyberg room, 10% stayed, 30% moved on to the Stevensen exhibit, and 60% went for more hors d'oeuvres. Of those in the Stevensen room, 30% stayed, 30% moved to the Clyberg exhibit, 20% moved to the Georgan exhibit, and 20% went for more hors d'oeuvres. Of those in the Georgan room, 20% stayed, 70% moved on to the Stevensen exhibit, and 10% went for more hors d'oeuvres. And of those in the central refreshments room, 40% stayed for more hors d'oeuvres, 10% went to the Clyberg exhibit, 20% went to the Stevensen exhibit, and 30% went to the Georgan exhibit. After several hours of milling about in this fashion, how many people were in each room at the gallery?

25. **BUSINESS: Market Share** The Peerless Products Corporation has decided to market a new toothpaste, DentiMint, designed to compete successfully with the market leaders: ConfiDent, SuperSmile, and MaxiWhite. Test marketing results from several cities indicate that each week, of those who used ConfiDent the previous week, 40% will buy it again, 20% will switch to SuperSmile, 30% will switch to MaxiWhite, and 10% will switch to DentiMint. Of those who used SuperSmile the previous week, 40% will buy it again, 20% will switch to ConfiDent, 10% will switch to MaxiWhite, and 30% will switch to DentiMint. Of those who used MaxiWhite the previous week, 40% will buy it again, 30% will switch to ConfiDent, 10% will switch to SuperSmile, and 20% will switch to DentiMint. And of those who used DentiMint the previous week, 40% will buy it again, 10% will switch to ConfiDent, 10% will switch to SuperSmile, and 40% will switch to MaxiWhite. If these buying patterns continue, what will the long-term market share be for DentiMint toothpaste?

26. **BUSINESS: Apple Harvests** The apple harvest in the Shenandoah Valley is rated as excellent, average, or poor. Following an excellent harvest, the chances of having an excellent, average, or poor harvest are 0.5, 0.3, and 0.2, respectively. Following an average harvest, the chances of having an excellent, average, or poor harvest are 0.3, 0.5, and 0.2, respectively. And following a poor harvest, the chances of having an excellent, average, or poor harvest are 0.6, 0.2, and 0.2, respectively. Assuming these trends continue, find the long-term chance for an excellent harvest.

27. **BUSINESS: Car Insurance** The records of an insurance broker in Boston classify the auto policy holders as preferred, satisfactory, poor, or in the assigned-risk pool. Each year 20% of the preferred policies are downgraded to satisfactory; 30% of the satisfactory policies are upgraded to preferred and another 20% are downgraded to poor; 60% of the poor policies are upgraded to satisfactory but 30% are placed in the assigned-risk pool; and just 20% of those in the assigned-risk pool are moved up to the poor classification. Assuming that these trends continue for many years, what percentage of the broker's auto policies are rated satisfactory or better?

28. GENERAL: Rumor Accuracy When repeating a rumor, gossips sometimes make mistakes and spread misinformation instead. Suppose the residents of a small town repeat a rumor that the mayor has been convicted of tax fraud at a trial held in the state capital, but that when they pass it on, 1% of the gossips reverse the result of whatever they were told. What is the long-term accuracy of the gossips' information?

29. BUSINESS: Renters Who Move The rental records at a Denver apartment building show that each year 78% of those who lived in the building the entire previous year will remain for the next year, while only 44% of those who moved in during the year will remain for the

next year. If these observations represent a long-term trend, are they consistent with the U.S. Census statistic that about 34.3% of renters nationwide move each year?

30. BUSINESS: Homeowners Who Move The property transfer records at a real estate agency in the Salt Lake City suburbs show that each year 96% of those who owned their own house the entire previous year will remain for the next year, while only 46% of those who moved into their own house during the year will remain for the next year. If these observations represent a long-term trend, are they consistent with the U.S. Census statistic that about 8.9% of homeowners nationwide move each year?

7.3 ABSORBING MARKOV CHAINS

Introduction

A state is *absorbing* if, once you enter it, you cannot leave it. If a Markov chain has several absorbing states, will it always end up in one of them? If so, which one and how soon? Unlike regular Markov chains, where the initial distribution has no influence on the long-term outcome, with absorbing states the initial distribution *does* affect when and where the eventual absorption occurs. As in the regular case, we will see that much useful information is given by high powers of the transition matrix.

Absorbing Markov Chains

For an *absorbing state,* the transition probabilities are 1 from that state to itself and 0 to every other state. A Markov chain is *absorbing* if it has at least one absorbing state *and* if from every nonabsorbing state it is possible to reach an absorbing state (in some number of steps).

EXAMPLE 1

FINDING WHETHER A MARKOV CHAIN IS ABSORBING

Determine whether each Markov chain is absorbing.

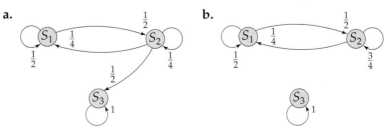

a.

b.

Solution

a. This *is* an absorbing Markov chain since it has an absorbing state, S_3, which can be reached from S_1 in 2 steps and from S_2 in 1 step.

b. This is *not* an absorbing Markov chain. Although S_3 is an absorbing state, it cannot be reached from either S_1 or S_2 (the nonabsorbing states).

EXAMPLE 2 **WETLANDS POLLUTION**

The Southern Electric Company has filed an application with the Environmental Protection Agency to store heavy-metal waste (mostly cadmium and mercury) in a clay-lined storage pool at its manufacturing plant in Mayfield. An independent engineering consultant has certified that only 1% of the waste will leach out of the pool into the ground water each year. But a hydrologist working with the local chapter of People for a Cleaner Planet has pointed out that each year 2% of the contaminants in the ground water reach the Marlin Memorial Wetlands Conservation Area, where they remain indefinitely. Represent this situation by a transition diagram and matrix. Show that the wetlands are an absorbing state for the pollution.

Solution

The waste can be in one of three locations: at the plant (state P), in the ground water (state G), or in the wetlands (state W). The given information is shown in the following state-transition diagram:

$$T = \begin{array}{c} \\ \text{Current state} \end{array} \begin{array}{c} \\ P \\ G \\ W \end{array} \overset{\begin{array}{ccc} \text{Next state} \\ P \quad G \quad W \end{array}}{\begin{pmatrix} 0.99 & 0.01 & 0 \\ 0 & 0.98 & 0.02 \\ 0 & 0 & 1 \end{pmatrix}}$$

Since these changes occur each year, we have a Markov chain with transition matrix T. W is an absorbing state because once the contamination reaches the Wetlands, it does not leave, and the other states P and G are nonabsorbing. The state-transition diagram shows that it is possible for some of the waste to move from the plant storage pool to the groundwater and on to the wetlands, so this *is* an absorbing Markov chain.

$$T^2 = \begin{pmatrix} 0.9801 & 0.0197 & 0.0002 \\ 0 & 0.9604 & 0.0396 \\ 0 & 0 & 1 \end{pmatrix}$$

We could also use the transition matrix to show that from each nonabsorbing state it is possible to reach the absorbing state. Squaring the transition matrix T gives the matrix on the left, and since the last column (corresponding to the wetlands) is all positive, it is possible to reach the wetlands from any other state in two steps.

How much of the heavy-metal waste will reach the Wetlands, and how soon is it expected to get there? Will *all* the contamination ultimately reach the wetlands? In Example 4 on pages 506–507 we will answer these important questions.

Standard Form

The transition matrix of an absorbing Markov chain is in *standard form* if it is written so that the absorbing states are listed *before* the nonabsorbing states. Since the absorbing states themselves may be in any order (as may the nonabsorbing states), the standard form is not necessarily unique.

Standard Form of an Absorbing Transition Matrix

An absorbing transition matrix is in *standard form* if the absorbing states appear before the nonabsorbing states, in which case it takes the form

$$
\begin{array}{c}
\text{Absorbing} \{ \\
\text{Nonabsorbing} \{
\end{array}
\overbrace{}^{\text{Absorbing}} \quad \overbrace{}^{\text{Nonabsorbing}}
$$

$$
\begin{array}{c}
\text{Absorbing} \{ \\
\text{Nonabsorbing} \{
\end{array}
\left(
\begin{array}{c|c}
I & 0 \\
\hline
R & Q
\end{array}
\right)
\qquad
\begin{array}{l}
I = \text{identity matrix} \\
0 = \text{zero matrix}
\end{array}
$$

R is the matrix of transition probabilities from the nonabsorbing states to the absorbing states and Q is the matrix of transition probabilities from nonabsorbing states to nonabsorbing states.

Any absorbing transition matrix can be rewritten in standard form by reordering the rows so the absorbing states are above the nonabsorbing states and then reordering the columns to have the same order as the rows.

EXAMPLE 3

REWRITING A TRANSITION MATRIX IN STANDARD FORM

Rewrite the following absorbing transition matrix in standard form and identify the matrices R and Q.

$$
\begin{pmatrix}
0.80 & 0.03 & 0.10 & 0.02 & 0.05 \\
0 & 1 & 0 & 0 & 0 \\
0.10 & 0.07 & 0.20 & 0.03 & 0.60 \\
0 & 0 & 0 & 1 & 0 \\
0.05 & 0.01 & 0.10 & 0.04 & 0.80
\end{pmatrix}
$$

Solution

We first identify the absorbing states. Since the row of an absorbing state consists of all zeros except one 1 in the *same column as the row*, the second and fourth rows represent absorbing states, which we will call A_1 and A_2. The other rows represent the nonabsorbing states, which we will call N_1, N_2 and N_3:

$$
\begin{array}{c}
N_1 \\
\text{Absorbing} \rightarrow A_1 \\
\text{rows (1's on} \\
\text{the main} \quad N_2 \\
\text{diagonal)} \rightarrow A_2 \\
N_3
\end{array}
\begin{pmatrix}
0.80 & 0.03 & 0.10 & 0.02 & 0.05 \\
0 & 1 & 0 & 0 & 0 \\
0.10 & 0.07 & 0.20 & 0.03 & 0.60 \\
0 & 0 & 0 & 1 & 0 \\
0.05 & 0.01 & 0.10 & 0.04 & 0.80
\end{pmatrix}
$$

Putting the rows in the order A_1, A_2, N_1, N_2, N_3, we have

$$
\begin{array}{cccccc}
 & N_1 & A_1 & N_2 & A_2 & N_3 \\
A_1 & 0 & 1 & 0 & 0 & 0 \\
A_2 & 0 & 0 & 0 & 1 & 0 \\
N_1 & 0.80 & 0.03 & 0.10 & 0.02 & 0.05 \\
N_2 & 0.10 & 0.07 & 0.20 & 0.03 & 0.60 \\
N_3 & 0.05 & 0.01 & 0.10 & 0.04 & 0.80
\end{array}
$$

First fix the rows

and then putting the columns in the same order gives

$$
\begin{array}{cccccc}
 & A_1 & A_2 & N_1 & N_2 & N_3 \\
A_1 & 1 & 0 & 0 & 0 & 0 \\
A_2 & 0 & 1 & 0 & 0 & 0 \\
N_1 & 0.03 & 0.02 & 0.80 & 0.10 & 0.05 \\
N_2 & 0.07 & 0.03 & 0.10 & 0.20 & 0.60 \\
N_3 & 0.01 & 0.04 & 0.05 & 0.10 & 0.80
\end{array}
$$

Then fix the columns

$\underbrace{\hspace{2cm}}_{R}$ $\underbrace{\hspace{3cm}}_{Q}$

R is the matrix of transition probabilities from the nonabsorbing states to the absorbing states, while Q is the matrix of transition probabilities from the nonabsorbing states to the nonabsorbing states:

$$
R = \begin{array}{c} N_1 \\ N_2 \\ N_3 \end{array}
\begin{array}{cc}
A_1 & A_2 \\
\begin{pmatrix}
0.03 & 0.02 \\
0.07 & 0.03 \\
0.01 & 0.04
\end{pmatrix}
\end{array}
\qquad
Q = \begin{array}{c} N_1 \\ N_2 \\ N_3 \end{array}
\begin{array}{ccc}
N_1 & N_2 & N_3 \\
\begin{pmatrix}
0.80 & 0.10 & 0.05 \\
0.10 & 0.20 & 0.60 \\
0.05 & 0.10 & 0.80
\end{pmatrix}
\end{array}
$$

Practice Problem 1

Rewrite the transition matrix

$$\begin{array}{c} \\ P \\ G \\ W \end{array}\begin{array}{c} P \quad\quad G \quad\quad W \\ \begin{pmatrix} 0.99 & 0.01 & 0 \\ 0 & 0.98 & 0.02 \\ 0 & 0 & 1 \end{pmatrix} \end{array}$$

from Example 2 in standard form.

➤ Solution on page 511

Transition Times and Absorption Probabilities

Just as with regular Markov chains, the long-term behavior of an absorbing chain can be seen from high powers of the transition matrix. However, even more information about the ultimate behavior of the chain is provided by a theorem about the limiting form of these powers. The following results are proved in Exercises 41–50.

Expected Transition Times and Long-Term Absorption Probabilities

For an absorbing transition matrix in standard form

$$T = \left(\begin{array}{c|c} I & 0 \\ \hline R & Q \end{array}\right)$$

Absorbing states before nonabsorbing states

high powers of T approach a matrix of the form

$$T^* = \left(\begin{array}{c|c} I & 0 \\ \hline F \cdot R & 0 \end{array}\right)$$

T^* gives the long-term transition probabilities

where the matrix F is given by $F = (I - Q)^{-1}$ and is called the *fundamental matrix* of the chain. It provides the following information:

1. The entry in row i and column j of F gives the expected number of times that the chain, if it begins in state N_i, will be in state N_j before being absorbed.

2. The sum of the entries in row i of F gives the expected number of times that the chain, if it begins in state N_i, will be in the nonabsorbing states before being absorbed.

3. The entry in row i and column j of the matrix $F \cdot R$ gives the probability that the chain, if it begins in state N_i, will be absorbed into the state A_j.

EXAMPLE 4

FINDING EXPECTED TIME UNTIL ABSORPTION

Find the expected number of years until the heavy-metal waste in the plant (state P) will reach the Wetlands Conservation Area (state W) for the Markov chain described in Example 2.

Solution

Writing the transition matrix from Example 2 in standard form as

$$T = \begin{matrix} & \begin{matrix} W & & G & & P \end{matrix} \\ \begin{matrix} W \\ G \\ P \end{matrix} & \left(\begin{array}{c|cc} 1 & 0 & 0 \\ \hline 0.02 & 0.98 & 0 \\ 0 & 0.01 & 0.99 \end{array} \right) \end{matrix}$$

Listing first the absorbing state W (for "wetlands")

we have

$$Q = \begin{matrix} & \begin{matrix} G & & P \end{matrix} \\ \begin{matrix} G \\ P \end{matrix} & \left(\begin{array}{cc} 0.98 & 0 \\ 0.01 & 0.99 \end{array} \right) \end{matrix}$$

From $T = \left(\begin{array}{c|c} I & 0 \\ \hline R & Q \end{array} \right)$

and

$$I - Q = \begin{pmatrix} 1 & 0 \\ 0 & 1 \end{pmatrix} - \begin{pmatrix} 0.98 & 0 \\ 0.01 & 0.99 \end{pmatrix} = \begin{pmatrix} 0.02 & 0 \\ -0.01 & 0.01 \end{pmatrix}$$

We calculate the fundamental matrix $F = (I - Q)^{-1}$ by row-reducing "$(A|I)$ to obtain $(I|A^{-1})$" (see page 212).

$$\left(\begin{array}{cc|cc} 0.02 & 0 & 1 & 0 \\ -0.01 & 0.01 & 0 & 1 \end{array} \right) \qquad (A|I)$$

$$R_2 + \tfrac{1}{2}R_1 \rightarrow \left(\begin{array}{cc|cc} 0.02 & 0 & 1 & 0 \\ 0 & 0.01 & \tfrac{1}{2} & 1 \end{array} \right)$$

$$\begin{matrix} 50R_1 \rightarrow \\ 100R_2 \rightarrow \end{matrix} \left(\begin{array}{cc|cc} 1 & 0 & 50 & 0 \\ 0 & 1 & \underbrace{50 \quad 100}_{F} \end{array} \right) \qquad (I|A^{-1})$$

The fundamental matrix F is the matrix on the right of the bar in the row-reduced matrix. From part (1) of the box on the previous page, its entries have the following interpretations.

$$F = \begin{matrix} & \begin{matrix} G & P \end{matrix} \\ \begin{matrix} G \\ P \end{matrix} & \begin{pmatrix} 50 & 0 \\ 50 & 100 \end{pmatrix} \end{matrix}$$

Beginning in G, expected time in G
Beginning in G, expected time in P
Beginning in P, expected time in G
Beginning in P, expected time in P

Note the pattern:
Beginning in (row), expected time in (column)

From part (2) of the box on the previous page, the expected number of years until the heavy-metal waste in the plant (state P) is absorbed into the wetlands (the absorbing state W) is the sum of the numbers in row P: $50 + 100 = 150$ years.

Practice Problem 2

How long is it expected that the heavy-metal waste will spend in the plant storage pool (state P) before absorption? How long in the ground water (state G)? ➤ Solution on page 511

Graphing Calculator Exploration

(*Continuation of Example 3*) To find the probability that the absorbing Markov chain in Example 3 (pages 504–505) will be absorbed into state A_2 given that it started in state N_3, proceed as follows.

Enter the matrices Q and R from the solution of Example 3 into the calculator as [A] and [B], along with a 3×3 identity matrix in [I], as shown in the first three screens below.

```
MATRIXCAJ 3 X3     MATRIXCBJ 3 X2     MATRIXCIJ 3 X3     (CIJ-CAJ)⁻¹*CBJ
C.8  .1  .05  J    C.03 .02    J      C1   0   0   J      CC.46 .54J
C.1  .2  .6   J    C.07 .03    J      C0   1   0   J       C.43 .57J
C.05 .1  .8   J    C.01 .04    J      C0   0   1   J       C.38 .62JJ

3,3=.8             3,2=.04            3,3=1
```

Then calculate $F \cdot R$ (with $F = (I - Q)^{-1}$) as $([I] - [A])^{-1} * [B]$, as shown in the last screen above.

By part (3) of the result on page 506, the probability that the chain, beginning in state N_3, will be absorbed into state A_2 is the entry in row 3 and column 2 of $F \cdot R$, which is 0.62.

$$F \cdot R = \begin{matrix} & A_1 & A_2 \\ N_1 & 0.46 & 0.54 \\ N_2 & 0.43 & 0.57 \\ N_3 & 0.38 & 0.62 \end{matrix}$$

Practice Problem 3

From the matrix $F \cdot R$ in the preceding Graphing Calculator Exploration, what is the probability that the chain, if it begins in state N_1, will be absorbed into state A_2? ➤ Solution on page 511

Both the preceding Graphing Calculator Exploration and Practice Problem 3 found probabilities of being absorbed into state A_2, but beginning in different states. The fact that we obtained two different probabilities (0.62 and 0.54) shows that for an absorbing Markov chain, the initial state *does* affect the ultimate behavior. This is quite different from a *regular* Markov chain, where the initial state had no effect on the ultimate distribution.

EXAMPLE 5   **FARM SIZE DISTRIBUTION***

An analysis of Census data for farms in the Grand Forks metropolitan area from 1950 to 2000 found the following probabilities for transitions between small (state S) family-owned farms of 50 to 200 acres, medium (state M) family-owned farms of 201 to 600 acres, large (state L) family-owned farms of 601 to 1000 acres, housing developments (state H), and corporate-owned agribusiness farms (state C). Notice that once a farm becomes a housing development or corporate-owned, it never returns to family-owned farm status again.

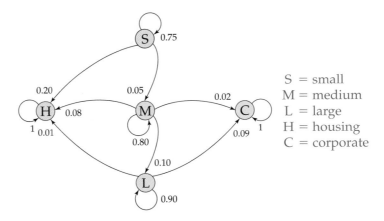

S = small
M = medium
L = large
H = housing
C = corporate

Assuming that these yearly transition patterns continue, what is the expected remaining lifetime of a medium family-owned farm (that is, the time until it becomes a housing development or owned by a corporation)? What is the probability that it will become a housing development?

Solution

We represent the situation as a Markov chain. Since H and C are absorbing states, one standard form for the transition matrix is

$$
\begin{array}{c c}
 & \begin{matrix} H & C & S & M & L \end{matrix} \\
\begin{matrix} H \\ C \\ S \\ M \\ L \end{matrix} &
\left(\begin{array}{cc|ccc}
1 & 0 & 0 & 0 & 0 \\
0 & 1 & 0 & 0 & 0 \\
\hline
0.20 & 0 & 0.75 & 0.05 & 0 \\
0.08 & 0.02 & 0 & 0.80 & 0.10 \\
0.01 & 0.09 & 0 & 0 & 0.90
\end{array}\right)
\end{array}
$$

---

* This example is based on ideas in "Projection of Farm Numbers for North Dakota with Markov Chains" by Ronald D. Krenz, *Agricultural Economics Research* **XVI**(3): 77–83.

So

$$
\begin{array}{c}
\begin{array}{cc} H & C \end{array} \\
R = \begin{array}{c} S \\ M \\ L \end{array}\!\!\begin{pmatrix} 0.20 & 0 \\ 0.08 & 0.02 \\ 0.01 & 0.09 \end{pmatrix}
\end{array}
\quad \text{and} \quad
\begin{array}{c}
\begin{array}{ccc} S & M & L \end{array} \\
Q = \begin{array}{c} S \\ M \\ L \end{array}\!\!\begin{pmatrix} 0.75 & 0.05 & 0 \\ 0 & 0.80 & 0.10 \\ 0 & 0 & 0.90 \end{pmatrix}
\end{array}
$$

Using a calculator or pencil and paper, we find that the fundamental matrix $F = (I - Q)^{-1}$ is

$$
\left(\begin{pmatrix} 1 & 0 & 0 \\ 0 & 1 & 0 \\ 0 & 0 & 1 \end{pmatrix} - \begin{pmatrix} 0.75 & 0.05 & 0 \\ 0 & 0.80 & 0.10 \\ 0 & 0 & 0.90 \end{pmatrix} \right)^{-1} = \begin{array}{c}
\begin{array}{ccc} S & M & L \end{array} \\
\begin{array}{c} S \\ M \\ L \end{array}\!\!\underbrace{\begin{pmatrix} 4 & 1 & 1 \\ 0 & 5 & 5 \\ 0 & 0 & 10 \end{pmatrix}}_{F}
\end{array}
$$

F gives expected times

and $F \cdot R$ is

$$
\begin{array}{c} S \\ M \\ L \end{array}\!\!\begin{pmatrix} 4 & 1 & 1 \\ 0 & 5 & 5 \\ 0 & 0 & 10 \end{pmatrix} \cdot \begin{array}{c}
\begin{array}{cc} H & C \end{array} \\
\begin{pmatrix} 0.20 & 0 \\ 0.08 & 0.02 \\ 0.01 & 0.09 \end{pmatrix}
\end{array} = \begin{array}{c}
\begin{array}{cc} H & C \end{array} \\
\begin{array}{c} S \\ M \\ L \end{array}\!\!\underbrace{\begin{pmatrix} 0.89 & 0.11 \\ 0.45 & 0.55 \\ 0.10 & 0.90 \end{pmatrix}}_{F \cdot R}
\end{array}
$$

$F \cdot R$ gives absorption probabilities from *row* to *column*

The expected remaining lifetime of a medium family-owned farm (until it becomes housing or corporate-owned) is $0 + 5 + 5 = 10$ years, and the probability that it will become a housing development is 0.45, or 45%.

Practice Problem 4

From these matrices, what is the expected remaining lifetime of a *small* family-owned farm (until it becomes housing or corporate-owned), and what is the probability that it will become corporate-owned?

➤ Solution on next page

7.3 Section Summary

For an *absorbing* state, the transition probabilities are 1 from itself to itself and 0 to every other state. An *absorbing Markov chain* has at least one absorbing state and from each nonabsorbing state it is possible to reach some absorbing state (in some number of moves). The transition matrix is in *standard form* if the absorbing states are listed before the nonabsorbing states.

$$
\begin{array}{c}
 \\
\text{Absorbing \{} \\
\text{Nonabsorbing \{}
\end{array}
\overbrace{\begin{array}{c|c}
\text{Absorbing} & \text{Nonabsorbing} \\
\hline
I & 0 \\
\hline
R & Q
\end{array}}
$$

R gives the transition probabilities from the nonabsorbing states to the absorbing states, and Q gives the transition probabilities from the non-absorbing states to the nonabsorbing states. Much information is given by the fundamental matrix $F = (Q - I)^{-1}$ and the product $F \cdot R$:

$$
F = (Q - I)^{-1} = \begin{array}{c} \\ N_1 \\ \vdots \end{array}\overset{N_1 \dots}{\left(\right)} \qquad F \cdot R = \begin{array}{c} \\ N_1 \\ \vdots \end{array}\overset{A_1 \dots}{\left(\right)}
$$

1. The entry in row i and column j of F gives the expected number of times that the chain, if it begins in state N_i, will be in state N_j before being absorbed.

2. The sum of the entries in row i of F gives the expected number of times that the chain, if it begins in state N_i, will be in the non-absorbing states before being absorbed.

3. The entry in row i and column j of $F \cdot R$ gives the probability that the chain, if it begins in state N_i, will be absorbed into the state A_j.

➤ **Solutions to Practice Problems**

1. Both
$$
\begin{array}{c} \\ W \\ P \\ G \end{array}
\overset{\begin{array}{ccc} W & P & G \end{array}}{\left(\begin{array}{ccc} 1 & 0 & 0 \\ 0 & 0.99 & 0.01 \\ 0.02 & 0 & 0.98 \end{array} \right)}
\quad \text{and} \quad
\begin{array}{c} \\ W \\ G \\ P \end{array}
\overset{\begin{array}{ccc} W & G & P \end{array}}{\left(\begin{array}{ccc} 1 & 0 & 0 \\ 0.02 & 0.98 & 0 \\ 0 & 0.01 & 0.99 \end{array} \right)}
\quad \text{are}
$$
standard forms of the transition matrix.

2. The expected times are 100 years in the plant storage pool and then 50 years in the ground water.

3. 0.54 (from the entry in row 1 and column 2 of $F \cdot R$).

4. The expected lifetime is $4 + 1 + 1 = 6$ years, and the probability that it will become corporate-owned is 0.11, or 11%.

7.3 Exercises

For each state-transition diagram, identify the absorbing and the nonabsorbing states. For the given starting state, find the smallest number of steps required to reach an absorbing state, and the probability of that sequence of steps.

1. Starting in state S_1

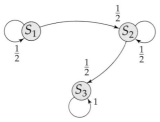

2. Starting in state S_2

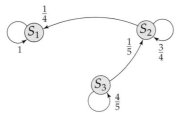

3. Starting in state S_4

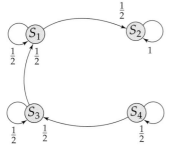

4. Starting in state S_4

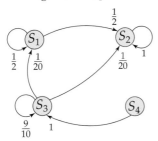

For each absorbing transition matrix T:

a. Draw a state-transition diagram, identify the absorbing and the nonabsorbing states, and find the smallest integer k such that after k transitions every nonabsorbing state can reach an absorbing state.

b. Verify your value for k by finding T^k and checking that every nonabsorbing state row has a positive probability in at least one absorbing state column.

5. $\begin{pmatrix} 1 & 0 & 0 \\ 0.20 & 0.55 & 0.25 \\ 0 & 0.25 & 0.75 \end{pmatrix}$

6. $\begin{pmatrix} 0.75 & 0 & 0 & 0.25 \\ 0 & 0.50 & 0.50 & 0 \\ 0 & 0 & 1 & 0 \\ 0 & 0.25 & 0.25 & 0.50 \end{pmatrix}$

7. $\begin{pmatrix} 0 & 1 & 0 & 0 \\ 0 & 1 & 0 & 0 \\ 0.2 & 0 & 0.6 & 0.2 \\ 0 & 0 & 0.2 & 0.8 \end{pmatrix}$

8. $\begin{pmatrix} 0.50 & 0.05 & 0.30 & 0.15 \\ 0 & 1 & 0 & 0 \\ 0.50 & 0 & 0.50 & 0 \\ 0 & 0 & 0 & 1 \end{pmatrix}$

Rewrite each absorbing transition matrix in standard form and identify the matrices R and Q.

9. $\begin{pmatrix} 0.88 & 0.02 & 0.10 \\ 0 & 1 & 0 \\ 0.04 & 0.16 & 0.80 \end{pmatrix}$

10. $\begin{pmatrix} 1 & 0 & 0 & 0 \\ 0.10 & 0.85 & 0.05 & 0 \\ 0.05 & 0.10 & 0.80 & 0.05 \\ 0 & 0 & 0 & 1 \end{pmatrix}$

11. $\begin{pmatrix} 0.80 & 0 & 0.20 & 0 & 0 \\ 0 & 1 & 0 & 0 & 0 \\ 0 & 0.20 & 0.50 & 0.30 & 0 \\ 0 & 0 & 0 & 1 & 0 \\ 0.20 & 0 & 0.05 & 0 & 0.75 \end{pmatrix}$

12.
$$\begin{pmatrix} 1 & 0 & 0 & 0 & 0 \\ 0 & 0.65 & 0 & 0.30 & 0.05 \\ 0 & 0 & 1 & 0 & 0 \\ 0.05 & 0.30 & 0.05 & 0.60 & 0 \\ 0 & 0 & 0 & 0 & 1 \end{pmatrix}$$

Find the fundamental matrix for each absorbing transition matrix in standard form. (You may use 🖩 if permitted by your instructor.)

13.
$$\left(\begin{array}{c|cc} 1 & 0 & 0 \\ \hline 0.4 & 0.4 & 0.2 \\ 0.2 & 0.2 & 0.6 \end{array} \right)$$

14.
$$\left(\begin{array}{c|cc} 1 & 0 & 0 \\ \hline 0.2 & 0.2 & 0.6 \\ 0.1 & 0.1 & 0.8 \end{array} \right)$$

15.
$$\left(\begin{array}{cc|cc} 1 & 0 & 0 & 0 \\ 0 & 1 & 0 & 0 \\ \hline 0.10 & 0.10 & 0.20 & 0.60 \\ 0.05 & 0.15 & 0.20 & 0.60 \end{array} \right)$$

16.
$$\left(\begin{array}{cc|cc} 1 & 0 & 0 & 0 \\ 0 & 1 & 0 & 0 \\ \hline 0.05 & 0 & 0.85 & 0.10 \\ 0.10 & 0.05 & 0.05 & 0.80 \end{array} \right)$$

17.
$$\left(\begin{array}{c|ccc} 1 & 0 & 0 & 0 \\ \hline 0.1 & 0.2 & 0.1 & 0.6 \\ 0.1 & 0 & 0.9 & 0 \\ 0.2 & 0.2 & 0 & 0.6 \end{array} \right)$$

18.
$$\left(\begin{array}{c|ccc} 1 & 0 & 0 & 0 \\ \hline 0.2 & 0.5 & 0.2 & 0.1 \\ 0.4 & 0 & 0.4 & 0.2 \\ 0.1 & 0 & 0.1 & 0.8 \end{array} \right)$$

19.
$$\left(\begin{array}{cc|ccc} 1 & 0 & 0 & 0 & 0 \\ 0 & 1 & 0 & 0 & 0 \\ \hline 0.10 & 0.10 & 0.20 & 0.10 & 0.50 \\ 0.05 & 0.05 & 0.10 & 0.80 & 0 \\ 0.20 & 0 & 0.20 & 0.10 & 0.50 \end{array} \right)$$

20.
$$\left(\begin{array}{cc|ccc} 1 & 0 & 0 & 0 & 0 \\ 0 & 1 & 0 & 0 & 0 \\ \hline 0.04 & 0.16 & 0.60 & 0.10 & 0.10 \\ 0.06 & 0.04 & 0 & 0.90 & 0 \\ 0.08 & 0.02 & 0.20 & 0 & 0.70 \end{array} \right)$$

For Exercises 21–30, use the results of Exercises 13–20 as follows: For each given initial state, find the expected number of times in the given state before absorption, and find the probability of absorption in the given final state using the indicated transition matrix.

Initial State	Expected Number of Times in State	Probability of Absorption into State	Use the Transition Matrix from Exercise
21. N_1	N_1	A_1	13
22. N_1	N_2	A_1	14
23. N_2	N_1	A_2	15
24. N_2	N_2	A_2	16
25. N_2	N_2	A_1	17
26. N_3	N_2	A_1	18
27. N_1	N_3	A_1	19
28. N_1	N_3	A_1	20
29. N_2	any N_i	A_2	19
30. N_2	any N_i	A_2	20

Represent each situation as an absorbing Markov chain by constructing a state-transition diagram and a transition matrix in standard form. Use the fundamental matrix to solve each problem. Be sure to state your final answers in terms of the original questions.

31. BIOMEDICAL: Smallpox To populations in the New World never before exposed to smallpox, its arrival with European explorers was devastating. Suppose the disease spread through a newly infected town in such a way that each month 20% of the uninfected population became sick, 15% of the sick recovered (and were therefore immune to further infections), and 35% of the sick perished.

a. How much of the town's original population survived?

b. For an infected person, what was the expected number of months for the disease to run its course?

32. GENERAL: Random Walks and Gambler's Ruin How long can two players gamble against each other before one wins all and the other is ruined? Suppose each player has $2 and they flip a coin with the loser paying $1 to the winner. Find the expected number of times the game will be played before it ends. [*Hint:* Assume the coin is fair, so the probability of winning or losing on each toss is $\frac{1}{2}$, and define five states for the game, each corresponding to one of the player's money: $0, $1, $2, $3, and $4. Which states are absorbing?]

33. BIOMEDICAL: Geriatric Care Each year at the Shady Oaks Assisted Living Facility, 10% of the independent residents are reclassified as requiring assistance, 2% die, and 8% are transferred to a nursing home, while 10% of those needing assistance die and 15% are transferred to a nursing home. What is the probability that a resident requiring assistance will ultimately be transferred to a nursing home?

34. GENERAL: Inner-City Teachers A study of teachers in the Chicago area found that every five years 20% of the inner-city teachers continued at their schools, 40% relocated to suburban schools, 20% accepted administrative jobs, and the remaining 20% either retired or quit; 10% of the suburban school teachers relocated to inner-city schools, 70% continued at their schools, 10% accepted administrative jobs, and the remaining 10% either retired or quit; and 10% of the teachers working in administrative positions returned to inner-city teaching, 20% relocated to suburban schools, 60% continued as administrators, and the remaining 10% either retired or quit. Assuming that these trends continue, find the expected number of years an inner-city school teacher will teach at an inner-city school before retiring or quitting.

35. BUSINESS: Term Insurance The term life insurance records of the All-County Insurance Company show that every five years those policies that were renewed are renewed again with probability 0.60, renewed with increased coverage with probability 0.20, closed with a death benefit payment with probability 0.07, and discontinued with probability 0.13, while those that had been renewed with increased coverage are renewed again with probability 0.30, renewed with increased coverage with probability 0.60, closed with a death benefit payment with probability 0.08, and discontinued with probability 0.02.

a. What is the expected number of years that a policy that has just been renewed will be in force?

b. What is the probability that a policy that is renewed with increased coverage will be closed with a death benefit payment?

36. BUSINESS: Management Training The Big-Beige-Box Computer Company has an in-house program of classroom work and apprenticeships that gives technicians entry to management careers. Each year 25% of those in the classroom move on to the apprenticeship program, 25% drop out, and the remainder continue in the classroom, while 40% of those in the apprenticeship program are promoted to supervisor positions, 10% drop out, and the remainder continue as apprentices. What percentage of those in the classroom will become supervisors?

37. BIOMEDICAL: Dental Work A dentist's records indicate that prior to extraction, a tooth needing no work remains that way for another year with probability 0.95, needs a filling with probability 0.03, needs a root canal with probability 0.01, and needs to be extracted with probability 0.01; a tooth with a filling one year needs no work the next with probability 0.90, needs a filling with probability 0.06, needs a root canal with probability 0.02, or needs to be extracted with probability 0.02; and (interestingly enough) a tooth with a root canal one year needs no work the next. Find the expected number of times a tooth needing a filling will need fillings before being extracted.

38. BIOMEDICAL: Genetics A rapidly mutating class of SMX viruses can be separated into distinct strains that are "successful," together with other "mutant" variations. After each year of replications, 90% of the successful strains are still successful, 5% have developed into mutants, and the remaining 5% have become extinct, while 20% of the previous mutants have further evolved to become successful strains, 40% are still mutants, and the remaining 40% have become extinct. What is the expected number of years until a mutant becomes extinct?

39. SOCIAL SCIENCE: Third-World Economic Development The Henderholf Charitable Trust provides either business advice or low-interest loans to small companies in developing economies around the world. Each year 9% of the companies they advise become successful and leave the program, 65% continue receiving advice, 25% enter the loan program, and 1% go bankrupt, while 7% of the companies receiving loans become successful and leave the program, 15% need no further loans and return to receiving advice, 75% continue receiving loans, and 3% go bankrupt. Is the Trust's claim that "more than three-quarters of the companies they help become successful" justified?

40. GENERAL: College Graduation Rates As the students at Edson State College earn more credits, they progress from freshmen to sophomores to juniors to seniors and then graduate, unless they withdraw from college sometime along the way. If each year 64% of each level moves on to the next (including graduation), 20% remain at their current level, and 16% withdraw, what percentage of incoming freshmen will graduate?

Explorations and Excursions

The following problems extend and augment the material presented in the text.

Proof of the Long-Term Behavior of an Absorbing Markov Chain

The following exercises establish the results on page 506 concerning the long-term behavior of an

absorbing Markov chain with transition matrix
$T = \left(\begin{array}{c|c} I & 0 \\ \hline R & Q \end{array} \right)$ in standard form.

41. Show that

a. $T^2 = \left(\begin{array}{c|c} I & 0 \\ \hline R & Q \end{array} \right) \cdot \left(\begin{array}{c|c} I & 0 \\ \hline R & Q \end{array} \right)$

$= \left(\begin{array}{c|c} I & 0 \\ \hline R + Q \cdot R & Q^2 \end{array} \right)$

$= \left(\begin{array}{c|c} I & 0 \\ \hline (I + Q) \cdot R & Q^2 \end{array} \right)$

b. $T^3 = T^2 \cdot T = \left(\begin{array}{c|c} I & 0 \\ \hline R + Q \cdot R + Q^2 \cdot R & Q^3 \end{array} \right)$

$= \left(\begin{array}{c|c} I & 0 \\ \hline (I + Q + Q^2) \cdot R & Q^3 \end{array} \right)$

42. Verify Exercise 41 for

$$T = \left(\begin{array}{cc|cc} 1 & 0 & 0 & 0 \\ 0 & 1 & 0 & 0 \\ 0.15 & 0.05 & 0.60 & 0.20 \\ 0.03 & 0.07 & 0.20 & 0.70 \end{array} \right).$$

43. Extend Exercise 41 to show that

$$T^m = \left(\begin{array}{c|c} I & 0 \\ \hline (I + Q + \cdots + Q^{m-1}) \cdot R & Q^m \end{array} \right).$$

44. Verify Exercise 43 for

$$T = \left(\begin{array}{cc|ccc} 1 & 0 & 0 & 0 & 0 \\ 0 & 1 & 0 & 0 & 0 \\ 0.02 & 0.08 & 0.60 & 0.20 & 0.10 \\ 0.07 & 0.03 & 0.10 & 0.70 & 0.10 \\ 0.15 & 0.05 & 0.20 & 0.40 & 0.20 \end{array} \right)$$ and $m = 4$.

45. Use the fact that T is absorbing to explain why every row of

$$(I + Q + \cdots + Q^{m-1}) \cdot R$$

has at least one positive entry for sufficiently large values of m. Then show that

$$I + Q + \cdots + Q^{m-1} \to (I - Q)^{-1}$$

as m increases. [*Hint*: Use Exercise 60 on page 224.]

46. Verify Exercise 45 for $Q = \begin{pmatrix} 0.11 & 0.02 \\ 0.05 & 0.10 \end{pmatrix}$ by calculating the *finite* sum $I + Q + Q^2 + Q^3 + Q^4$ and checking that rounding this value to four decimal places matches the value of

$$(I - Q)^{-1} = \begin{pmatrix} 1.1250 & 0.0250 \\ 0.0625 & 1.1125 \end{pmatrix}$$

47. Verify Exercise 45 for

$$Q = \begin{pmatrix} 0.12 & 0.02 & 0.02 \\ 0.03 & 0.08 & 0.08 \\ 0.01 & 0.01 & 0.01 \end{pmatrix}$$ by calculating the

finite sum $I + Q + Q^2 + Q^3 + Q^4 + Q^5 + Q^6 + Q^7$ and checking that rounding this value to six decimal places matches the value of

$$(I - Q)^{-1} = \begin{pmatrix} 1.137500 & 0.025000 & 0.025000 \\ 0.038125 & 1.088750 & 0.088750 \\ 0.011875 & 0.011250 & 1.011250 \end{pmatrix}$$

48. Adapt Exercises 31–34 on pages 490–491 to show that the entry in row i and column j of the fundamental matrix $F = (I - Q)^{-1}$ is the expected number of times that the chain, if it begins in state N_i, will be in state N_j before being absorbed. This established the first result in the box on page 506.

49. Explain how the second result in the box on page 506 follows from Exercise 48.

50. Explain how the third result in the box on page 506 follows from Exercise 48 and the fact that the entry in row i and column j of the matrix R is the probability that state N_i moves to state A_j in one transition.

Chapter Summary with Hints and Suggestions

Reading the text and doing the exercises in this chapter have helped you to master the following skills, which are listed by section (in case you need to review them) and are keyed to particular Review Exercises. Answers for all Review Exercises are given at the back of the book, and full solutions can be found in the Student Solutions Manual.

7.1 States and Transitions

- Construct a transition matrix from a state-transition diagram and identify the transition as "oscillating," "mixing," or "absorbing." (*Review Exercises 1–2.*)

- Construct a state-transition diagram from a transition matrix and identify the transition as "oscillating," "mixing," or "absorbing." (*Review Exercises 3–4.*)

- Calculate the kth state distribution vector for a Markov chain with transition matrix T and initial distribution D_0. (*Review Exercises 5–6.*)

$$D_k = D_0 \cdot T^k$$

- Find the expected number of times a Markov chain will be in a given state before moving to some other state. (*Review Exercises 7–8.*)

$$E = \frac{1}{1 - p}$$

- Use a Markov chain to solve an applied problem. (*Review Exercises 9–10.*)

7.2 Regular Markov Chains

- Identify a transition matrix as "regular" or "not regular." (*Review Exercises 11–14.*)

- Find the steady-state distribution of a regular Markov chain by row-reduction, and by calculating powers of the transition matrix (using ▦). *(Review Exercises 15–16.)*

$$D \cdot T = D$$

$$\left(\begin{array}{c|c} T^t - I & 0 \\ \hline 1 \cdots 1 & 1 \end{array}\right) \longrightarrow \left(\begin{array}{c|c} I & D^t \\ \hline 0 & 0 \end{array}\right)$$

- Find the smallest positive number k of transitions needed to move from state S_i to state S_j in a regular Markov chain, and then verify this value (using ▦) by checking that T^k has a positive transition probability in row i and column j. *(Review Exercises 17–18.)*

- Use a regular Markov chain to solve an applied problem. *(Review Exercises 19–20.)*

7.3 Absorbing Markov Chains

- Identify the absorbing and nonabsorbing states in a state-transition diagram, and find the smallest number of transitions needed to move from a given nonabsorbing state to an absorbing state. *(Review Exercises 21–22.)*

- Identify absorbing and nonabsorbing states from a transition matrix, and find the smallest number of transitions needed to ensure that every nonabsorbing state can reach an absorbing state. *(Review Exercises 23–24.)*

- Rewrite an absorbing Markov chain transition matrix in standard form. *(Review Exercises 25–26.)*

$$T = \left(\begin{array}{c|c} I & 0 \\ \hline R & Q \end{array}\right)$$

- Find the fundamental matrix of an absorbing Markov chain. *(Review Exercises 27–30.)*

$$F = (I - Q)^{-1}$$

- Use the fundamental matrix to find the expected number of times in nonabsorbing states before absorption and the probability of absorption into a particular absorbing state. *(Review Exercises 31–38.)*

- Use an absorbing Markov chain to solve an applied problem. *(Review Exercises 39–40.)*

Hints and Suggestions

- *(Overview)* We discussed three general types of Markov chains: *oscillating, mixing,* and *absorbing.* Oscillating chains always move back and forth. Mixing (that is, regular) chains settle down to a steady-state that is independent of how they began. Absorbing chains end in absorption, with time until absorption and the final absorbing state depending on where they began.

- An entry in a particular row and column of a transition matrix gives the probability of moving *from* the *row* state *to* the *column* state.

- For a regular Markov chain, the steady-state distribution D can be found in two ways: by raising the transition matrix to a high power so the rows become D, or by row-reducing $\left(\begin{array}{c|c} T^t - I & 0 \\ \hline 1 \cdots 1 & 1 \end{array}\right)$ to obtain $\left(\begin{array}{c|c} I & D^t \\ \hline 0 & 0 \end{array}\right)$, where the superscript t means *transpose* (not a power).

- An *absorbing* Markov chain is not just a chain with an absorbing state. To be an absorbing Markov chain, it must be possible to move from each nonabsorbing state to some absorbing state (possibly in several moves).

- When writing an absorbing transition matrix in standard form, be sure to keep the state names with the correct rows and columns as you switch the absorbing rows to the top positions and then as you put the columns in the same order.

- For an absorbing Markov chain we could find the long-term behavior by raising the transition matrix to a high power, but we obtain much more information from the results in the box on page 506. These results depend on the *fundamental matrix* $F = (I - Q)^{-1}$. F gives expected times and $F \cdot R$ gives probabilities, in each case from the *row* state to the *column* state.

- When doing an applied problem, always draw the state-transition diagram first (verifying that the probabilities of *leaving* each state add to 1), and then write the transition matrix (being sure that each *row* adds to 1).

- Watch out for careless errors like changing $\frac{1}{2}$ to 0.05 or 5% to 0.50.

- **Practice for Test:** Review Exercises 1, 3, 5, 7, 9, 13, 15, 19, 21, 25, 29, 35, 36, 39.

Review Exercises for Chapter 7

Practice test exercise numbers are in **green**.

7.1 States and Transitions

Construct a transition matrix T for the Markov chain represented by the state-transition diagram and identify the transition as "oscillating," "mixing," or "absorbing."

1.

2.
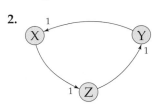

Construct a state-transition diagram for the Markov chain represented by the transition matrix and identify the transition as "oscillating," "mixing," or "absorbing."

3. $\begin{pmatrix} \frac{1}{3} & \frac{2}{3} \\ 0 & 1 \end{pmatrix}$

4. $\begin{pmatrix} 0 & \frac{2}{5} & \frac{3}{5} \\ 0 & \frac{4}{5} & \frac{1}{5} \\ \frac{1}{2} & \frac{1}{2} & 0 \end{pmatrix}$

For each Markov transition matrix T and initial state distribution vector D_0, calculate the next two state distribution vectors D_1 and D_2.

5. $T = \begin{pmatrix} 0.65 & 0.35 \\ 0.15 & 0.85 \end{pmatrix}$, $D_0 = (0.40 \quad 0.60)$

6. $T = \begin{pmatrix} 0 & 0.5 & 0.5 \\ 0.6 & 0.3 & 0.1 \\ 0.1 & 0.3 & 0.6 \end{pmatrix}$, $D_0 = (0 \quad 1 \quad 0)$

For each Markov transition matrix T and present state, find the expected number of times the chain will be in that state before moving to some other state.

7. $T = \begin{pmatrix} 0.95 & 0.05 \\ 0.45 & 0.55 \end{pmatrix}$, presently in state S_1

8. $T = \begin{pmatrix} 1 & 0 & 0 & 0 \\ 0.1 & 0.3 & 0 & 0.6 \\ 0.2 & 0 & 0.8 & 0 \\ 0.1 & 0.3 & 0 & 0.6 \end{pmatrix}$, presently in state S_3

Represent each situation as a Markov chain by constructing a state-transition diagram and the corresponding transition matrix. Be sure to state your final answer in terms of the original question.

9. **BUSINESS: Taxi Repairs** The Yellow-Top Cab Company has a large fleet of aging taxi cabs. Each day 11% break down and are removed from service, while 99% of those in the repair shop from yesterday are returned to service. What is the expected number of days a working cab will remain in service before breaking down?

10. **GENERAL: Two-Year Colleges** Each year 48% of the first-year students at the Marsten

Community College earn enough credits to become second-year students, 20% remain at the first-year level, and the remaining 32% withdraw. Also every year, 64% of the second-year students graduate, 20% remain at the second-year level, and the other 16% withdraw. If there are now 625 first-year and 250 second-year students, how many of these will be second-year students in two years?

7.2 Regular Markov Chains

Identify each transition matrix as "regular" or "not regular."

11. $\begin{pmatrix} 0.20 & 0.80 \\ 0.45 & 0.55 \end{pmatrix}$ **12.** $\begin{pmatrix} 0 & 1 \\ 1 & 0 \end{pmatrix}$

13. $\begin{pmatrix} 0 & 0.4 & 0.6 \\ 0 & 0.8 & 0.2 \\ 0.5 & 0 & 0.5 \end{pmatrix}$ **14.** $\begin{pmatrix} 0 & 0.2 & 0.8 \\ 0.1 & 0.2 & 0.7 \\ 0.3 & 0.2 & 0.5 \end{pmatrix}$

For each Markov transition matrix T:
a. Find the steady-state distribution.
b. Calculate T^k for $k = 5, 10, 20,$ and 50 to verify that the rows of T^k approach the steady-state distribution.

15. $\begin{pmatrix} 0.55 & 0.45 \\ 0.15 & 0.85 \end{pmatrix}$ **16.** $\begin{pmatrix} 0.3 & 0.6 & 0.1 \\ 0 & 0.4 & 0.6 \\ 0.3 & 0.6 & 0.1 \end{pmatrix}$

For each regular Markov transition matrix T:
a. Construct a state-transition diagram and find the smallest number k of steps needed to move from the given state S_i to state S_j and the probability of that sequence of steps.
b. Use this k to verify that T^k has a positive transition probability in row i and column j equal to the probability found in part (a).

17. $\begin{pmatrix} 0.4 & 0 & 0.6 \\ 0.3 & 0.7 & 0 \\ 0 & 0.3 & 0.7 \end{pmatrix}$, $S_1 \to S_2$

18. $\begin{pmatrix} 0.5 & 0.5 & 0 \\ 0.9 & 0 & 0.1 \\ 0.1 & 0 & 0.9 \end{pmatrix}$, $S_3 \to S_2$

Represent each situation as a Markov chain by constructing a state-transition diagram and a transition matrix. Find the steady-state distribution and interpret it in terms of the original situation. Be sure to state your final answer in terms of the original question.

19. BUSINESS: Customer Satisfaction Surveys of customer satisfaction with the repair service departments of dealers for a major car manufacturer show that of those rated "below average" one year, the next year 10% will remain that way, 70% will improve to "satisfactory," and the remaining 20% will improve to "excellent"; of those rated "satisfactory" one year, the next year 50% will remain that way, 20% will improve to "excellent," and the remaining 30% will slip to "below average"; and of those rated "excellent" one year, the next year 70% will remain that way and the remaining 30% will slip to "satisfactory." Assuming that this pattern has repeated for many years, how many of the manufacturer's 2660 dealers nationwide have service departments rated "excellent" by their customers?

20. GENERAL: Weather The old timers at Edna's Cafe in downtown Nora Springs will tell you the weather just keeps getting better and better for growing corn. In fact, if it was bad last year, there is a 20% chance it will now be terrific and a 60% chance it will be great this year; if last year was great, there is a 50% chance it will now be terrific and a 40% chance it will be great again this year; and if it was terrific last year, there is a 34% chance it will be that way again this year and a 54% chance it will be great this year. (Of course, they don't bother to mention that the weather might be bad this year because that could change their luck!) If they are right, how many of the next 25 years will have terrific weather for growing corn?

7.3 Absorbing Markov Chains

For each state-transition diagram, identify the absorbing and the nonabsorbing states. For the given starting state, find the smallest number of steps required to reach an absorbing state and the probability of that sequence of steps.

21. Starting in state S_1

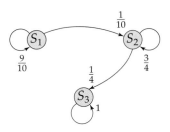

22. Starting in state S_4

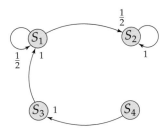

For each absorbing Markov transition matrix T:
a. Draw a state-transition diagram, identify the absorbing and the nonabsorbing states, and find the smallest integer k such that after k transitions, every nonabsorbing state can reach an absorbing state.
b. Verify your value for k by finding T^k and checking that every nonabsorbing state row has a positive probability in at least one absorbing state column.

23. $\begin{pmatrix} 0.6 & 0 & 0.2 & 0.2 \\ 0 & 1 & 0 & 0 \\ 0 & 0.5 & 0.5 & 0 \\ 0.5 & 0 & 0 & 0.5 \end{pmatrix}$

24. $\begin{pmatrix} 0.80 & 0 & 0 & 0.20 & 0 \\ 0 & 0.80 & 0.15 & 0 & 0.05 \\ 0 & 0 & 1 & 0 & 0 \\ 0.90 & 0.10 & 0 & 0 & 0 \\ 0 & 0 & 0 & 0 & 1 \end{pmatrix}$

Rewrite each absorbing Markov transition matrix in standard form and identify the matrices R and Q.

25. $\begin{pmatrix} 0.5 & 0.5 & 0 & 0 \\ 0 & 1 & 0 & 0 \\ 0.2 & 0 & 0.8 & 0 \\ 0 & 0 & 1 & 0 \end{pmatrix}$

26. $\begin{pmatrix} 0.6 & 0 & 0.3 & 0.1 \\ 0 & 1 & 0 & 0 \\ 0.1 & 0.1 & 0.8 & 0 \\ 0 & 0 & 0 & 1 \end{pmatrix}$

Find the fundamental matrix for each absorbing transition matrix in standard form. (You may use 🖩 if permitted by your instructor.)

27. $\left(\begin{array}{c|cc} 1 & 0 & 0 \\ \hline 0.3 & 0.3 & 0.4 \\ 0.1 & 0.1 & 0.8 \end{array} \right)$

28. $\left(\begin{array}{cc|cc} 1 & 0 & 0 & 0 \\ 0 & 1 & 0 & 0 \\ \hline 0.10 & 0 & 0.60 & 0.30 \\ 0.05 & 0.05 & 0.10 & 0.80 \end{array} \right)$

29. $\left(\begin{array}{c|ccc} 1 & 0 & 0 & 0 \\ \hline 0.1 & 0.7 & 0 & 0.2 \\ 0.1 & 0.1 & 0.8 & 0 \\ 0.2 & 0.4 & 0 & 0.4 \end{array} \right)$

30. $\left(\begin{array}{cc|ccc} 1 & 0 & 0 & 0 & 0 \\ 0 & 1 & 0 & 0 & 0 \\ \hline 0.05 & 0.15 & 0.20 & 0 & 0.60 \\ 0.10 & 0.10 & 0.20 & 0.50 & 0.10 \\ 0.10 & 0 & 0.10 & 0 & 0.80 \end{array} \right)$

For Exercises 31–38, use the results of Exercises 27–30 as follows: For each given initial state, find the expected number of times in the given state before absorption, and find the probability of absorption in the given final state using the indicated transition matrix.

	Initial State	Expected Number of Times in State	Probability of Absorption into State	Use the Transition Matrix from Exercise
31.	N_1	N_1	A_1	27
32.	N_2	N_1	A_1	27
33.	N_1	N_2	A_2	28
34.	N_2	N_2	A_1	28
35.	N_3	N_1	A_1	29
36.	N_2	any N_i	A_1	29
37.	N_1	N_2	A_2	30
38.	N_3	any N_i	A_1	30

Represent each situation as an absorbing Markov chain by constructing a state-transition diagram and a transition matrix in standard form. Use the fundamental matrix to solve each problem. Be sure to state your final answers in terms of the original questions.

39. BUSINESS: Computer Obsolescence A personal computer (PC) market analyst has estimated that each year 60% of the PCs from the previous year are still manufactured, 20% have been upgraded to newer models, and the remainder have been discontinued, while 30% of the new-model PCs from the previous year are still manufactured, 60% undergo further upgrading, and the remainder are dis-continued. How many years is a new model expected to be manufactured before it is discontinued?

40. BUSINESS: Car Loans The Final Federal Credit Corporation classifies its car loans as paid in full (and thus closed out), current, late, overdue, or bad (in which case they are sold to a collection agency). Each month 10% of the current accounts are paid in full, 10% are late, and the rest remain current; 3% of the late accounts are paid in full, 30% improve to be current, 50% become overdue, 7% become bad, and the rest remain late; 2% of the overdue accounts are paid in full, 10% become current, 20% improve to be late, 18% become bad, and the rest remain overdue. What is the probability that an overdue account will (ultimately) be paid in full?

8 Game Theory

Application Preview

Sherlock Holmes and James Moriarty

Near the end of *The Final Problem* by Sir Arthur Conan Doyle, the legendary detective Sherlock Holmes and his assistant Dr. Watson have finally collected enough evidence to prove the guilt of the criminal mastermind James Moriarty (who, incidentally, is a former professor of mathematics and author of a treatise on the binomial theorem). Knowing that Moriarty will now try to kill them, Holmes and Watson decide to flee by train to Dover where they can then escape by ship to France. As the train pulls out of the station they see Moriarty and conclude that he must have seen them as well. The train will make one intermediate stop in Canterbury, and Holmes and Watson must decide whether to get off at Canterbury or continue on to Dover. If Moriarty can guess where they will get off, he will go there and kill them. What should Holmes and Watson do to maximize their chance of avoiding Moriarty, and what should Moriarty do to bring about the opposite?

The material developed in this chapter will enable us to analyze Holmes's predicament, and on page 538 we will see that his decision to get off at Canterbury is, in fact, the most probable course of events. We will also be able to verify the famous remark of John von Neumann and Oskar Morgenstern, the inventors of game theory, that "Sherlock Holmes is as good as 48% dead when his train pulls out of Victoria Station."*

* *Theory of Games and Economic Behavior*, 3rd edition, by John von Neumann and Oskar Morgenstern, Princeton University Press, page 178.

8.1 TWO-PERSON GAMES AND SADDLE POINTS

Introduction

When faced with choices, we naturally wish for a clear way to decide what to do. Game theory provides a framework for analyzing situations in which opposing players make choices that affect the benefits received by each. The "games" that we consider include many of practical importance in business, economics, and political affairs. We restrict our attention to competitions involving just two players where one's gain is the other's loss. Such situations are called *two-person, zero-sum games* because whatever one player gains the other loses. Throughout this chapter, whenever we speak of a "game" we shall mean such a situation.*

Payoff Matrix

Column player's choices

Row player's choices { Payments to the row player }

A *game* consists of a list of choices for the first player and a list of choices for the second player, along with specified "payoffs" for the players that are determined by the choices they make. For a zero-sum game, one player's gain is the other player's loss, so we need only list the payoff to one player, the other *losing* that amount. We list the payoffs in a grid called the *payoff matrix*, with one player's possible choices corresponding to the *rows* of the matrix, the other player's choices corresponding to the *columns* of the matrix, and the entries in the matrix are *amounts paid to the row player by the column player*. A negative number means a loss for the row player and so a gain for the column player.

EXAMPLE 1 BUSINESS COMPETITION

Two competing coffee houses, Rachel's House of Java and Carla's Colombian Brew, are planning to enter a tri-city market by each locating a single new store in one of the towns of Lawrence (population 50,000), Moltonsville (population 90,000), and Northridge (population 70,000). If both stores are built in the same town, they will split the entire market equally. If not, the only store in town will get all of that town's market as well as half of that in any town without a store. What

---

* Fifty years after the original publication of the book by von Neumann and Morgenstern, the 1994 Nobel Memorial Prize in Economic Sciences was awarded to John Nash, John Harsanyi, and Reinhard Selten for their contributions to game theory. The contributions of John Nash were portrayed in the book *A Beautiful Mind* by Sylvia Nasar and in the movie of the same title.

are the possible choices for Rachel and for Carla? If the payoff is the difference between their markets (in thousands), find the payoff matrix.

Solution

Rachel's choices are L (for choosing Lawrence), M (for Moltonsville), and N (for Northridge), and similarly for Carla, so there are $3 \cdot 3 = 9$ different situations to consider. We list Rachel's choices as rows and Carla's as columns.

Should they both choose to locate in the same town, neither will have an advantage, giving zero payoffs for those choices:

		Carla		
		L	M	N
Rachel	L	0		
	M		0	
	N			0

Should Rachel choose L and Carla choose M, Rachel gets

$$\underbrace{50}_{\substack{\text{All of} \\ \text{Lawrence}}} + \underbrace{0}_{\substack{\text{None of} \\ \text{Moltonsville}}} + \underbrace{\tfrac{1}{2} \cdot 70}_{\substack{\text{Half of} \\ \text{Northridge}}} = 85 \qquad \text{In thousands of potential customers}$$

while Carla gets

$$\underbrace{0}_{\substack{\text{None of} \\ \text{Lawrence}}} + \underbrace{90}_{\substack{\text{All of} \\ \text{Moltonsville}}} + \underbrace{\tfrac{1}{2} \cdot 70}_{\substack{\text{Half of} \\ \text{Northridge}}} = 125$$

This is an advantage of $125 - 85 = 40$ thousand for Carla over Rachel, and we indicate this as -40 in the table because it is a *loss* for Rachel:

		Carla		
		L	M	N
Rachel	L	0	−40	
	M		0	
	N			0

Making similar calculations for the remaining five possibilities gives the payoff matrix for this game:

$$\begin{array}{c} & \begin{array}{ccc} \text{L} & \text{M} & \text{N} \end{array} \\ \begin{array}{c} \text{L} \\ \text{M} \\ \text{N} \end{array} & \begin{pmatrix} 0 & -40 & -20 \\ 40 & 0 & 20 \\ 20 & -20 & 0 \end{pmatrix} \end{array}$$

A different ordering of the choices would give a different ordering of the rows and columns

Practice Problem 1

Verify the bottom left-hand value in the above payoff matrix by calculating Rachel's advantage over Carla when Rachel chooses N and Carla chooses L.

➤ Solution on pages 529–530

Any game can be expressed as a matrix, and any matrix defines a game. Mathematically, a game is completely defined by its payoff matrix, calling the players R (choosing the row) and C (choosing the column), with no other explanation necessary. When setting up the matrix, it makes no difference which player chooses rows and which chooses columns as long as the entries are the payoffs to R, so that a positive entry is a gain for R (and a loss for C) while a negative entry is a loss for R (with the corresponding gain for C).

Optimal Strategy

The goal of game theory is to find the *optimal strategy* for each player in a game. Remember that R wants the *highest* number in the matrix, since R receives that amount, while C wants the *smallest* number, since C loses that amount. *Solving* a game means finding an optimal strategy for each player.

Remember: The smallest number in $\{2, -3, -5\}$ is -5.

Returning to the situation of Example 1, from Rachel's viewpoint row 2 is a good choice because the worst she could get would be zero. (There are better numbers, 40 and 20, but if Rachel repeatedly chose row 2 in hopes of getting them, Carla would choose column 2 to give Rachel only the zero.) From Carla's viewpoint, column 2 is appealing because the worst she could do is zero. (Carla would rather have the −40 and −20, since these mean positive payments to her, but if she chose column 2 repeatedly, Rachel would choose row 2 to give her only the zero.) Rachel's best strategy is row 2, and Carla's is column 2, both settling on the zero, corresponding to both locating in Moltonsville. These choices put them in equilibrium with each other: neither can gain by changing if the other stays with the same choice.

Carla chooses →

$$\begin{array}{c} & \begin{array}{ccc} \text{L} & \text{M} & \text{N} \end{array} \\ \begin{array}{c} \text{L} \\ \rightarrow\text{M} \\ \text{N} \end{array} & \begin{pmatrix} 0 & -40 & -20 \\ 40 & 0 & 20 \\ 20 & -20 & 0 \end{pmatrix} \end{array}$$

└─ Rachel chooses

Notice that the number at the intersection of the chosen row and column (here zero but it won't always be zero) is both the *minimum number in its row* and also the *maximum number in its column*. Such a number is called a *saddle point* of the game.

A saddle point is the lowest point along one curve and the highest along another

> ### Saddle Point and Value of a Game
>
> A *saddle point* of a game is an entry in the payoff matrix that is both the *minimum value in its row* and the *maximum value in its column*. If a game has a saddle point, it represents the *optimal strategy* for both players. The saddle point gives the *value* of the game.

A game with a saddle point is said to be *strictly determined*, and each player's selection of its corresponding row or column is called a *pure* optimal strategy. The *value* of the game (the saddle point) is a payoff to the row player, so a *positive* value means the game is favorable to the row player, a *negative* value means the game is favorable to the column player, and a *zero* value means that it is a "fair" game, favorable to neither. For the game in Example 1, the saddle point is the 0 in the Moltonsville row and column, so the pure optimal strategies for both Rachel and Carla are to locate their new stores in Moltonsville and split the market equally, for a game value of 0.

Finding Saddle Points

A game may have any number of saddle points: none, one, or many, but if it has many, they will all have the same value. There is a simple way to find the saddle points of a matrix: a saddle point is the minimum value in its row, so we first circle ○ the smallest entry in each row; a saddle point is also the maximum in its column, so we box □ the largest entry in each column; if an entry is *both circled and boxed*, it is a saddle point. The circled (or boxed) entries of G are called *the security values* of the rows (or the columns) since they represent the guaranteed return from choosing that row (or that column).

EXAMPLE **2**

FINDING SADDLE POINTS

Find the saddle points of the game $\begin{pmatrix} 2 & 2 & 2 & 5 & 4 \\ 0 & 2 & 1 & 3 & 3 \\ 1 & 2 & -1 & 4 & 2 \\ 2 & 4 & 2 & 3 & 5 \end{pmatrix}$.

Solution

First we circle the smallest value in each row.

$$\begin{pmatrix} ② & ② & ② & 5 & 4 \\ ① & 2 & 1 & 3 & 3 \\ 1 & 2 & ㊀① & 4 & 2 \\ ② & 4 & ② & 3 & 5 \end{pmatrix}$$

Then we box the largest value in each column.

$$\begin{pmatrix} ☐② & ② & ☐② & ☐5 & 4 \\ ① & 2 & 1 & 3 & 3 \\ 1 & 2 & ㊀① & 4 & 2 \\ ☐② & ☐4 & ☐② & 3 & ☐5 \end{pmatrix}$$

The saddle points are the circled *and* boxed numbers.

$$\begin{pmatrix} ☒② & ② & ☒② & ☐5 & 4 \\ ① & 2 & 1 & 3 & 3 \\ 1 & 2 & ㊀① & 4 & 2 \\ ☒② & ☐4 & ☒② & 3 & ☐5 \end{pmatrix}$$

This game has four saddle points. The value of the game is 2 and each player has two optimal strategies corresponding to choosing the row or column of a saddle point: R chooses row 1 or 4 and C chooses column 1 or 3.

Practice Problem 2

Find the saddle points of the game $\begin{pmatrix} 8 & 2 & 5 \\ 5 & 4 & 7 \\ 2 & 3 & 6 \\ 6 & 4 & 5 \end{pmatrix}$.

What are the pure optimal strategies?

➤ Solution on page 530

EXAMPLE 3

HOLMES AND MORIARTY

Recall from the Application Preview on page 523 that Sherlock Holmes and Dr. Watson can either leave the train at Canterbury (C) or continue on to Dover (D) for a ship to France. Moriarty has the same choices: Go to Canterbury or Dover. Construct a payoff matrix and find whether there is a pure optimal strategy for this situation.

Solution

Many payoffs matrices are possible, and we will try to construct a reasonable one. On a scale of 0 to 100, surely we should award all 100 points to Moriarty should he kill Holmes:

		Moriarty	
		C	D
Holmes	C	−100	
	D		−100

If Holmes goes to Dover but Moriarty stops at Canterbury, he and Dr. Watson will succeed in their escape and, since this is a success but not as permanent as a kill, we award 50 points to Holmes. If Holmes stops at Canterbury but Moriarty goes on to Dover, then Holmes has not yet escaped and Moriarty may still pursue him in England, so we call this a tie worth 0 points. Thus, we can represent the problem as a game with payoff matrix

$$\begin{pmatrix} -100 & 0 \\ 50 & -100 \end{pmatrix}$$

If we circle the smallest entry in each row and box the largest entry in each column, we see that there are no circled and boxed numbers. Therefore, this game does not have a saddle point (it is not strictly determined) and there are no pure optimal strategies.

$$\begin{pmatrix} \fbox{-100} & \boxed{0} \\ \boxed{50} & \fbox{-100} \end{pmatrix}$$

In the next section we will find a different kind of solution to this game.

8.1 Section Summary

A *zero-sum, two-person game* is a matrix with players R (choosing the row) and C (choosing the column). A positive entry represents a gain for R (and a loss for C) and a negative entry indicates a loss for R (and a gain for C).

A *saddle point* of a game is an entry that is both the smallest in its row and the largest in its column. A saddle point *solves the game* with an *optimal strategy* for each player consisting of the choice of the corresponding row or column. The *value* of the game is the saddle point entry.

A game may have no, one, or many saddle points.

➤ Solutions to Practice Problems

1. If Rachel (R) chooses Northridge (population 70,000) and Carla (C) chooses Lawrence (population 50,000), then

R gets: 0   + $\frac{1}{2} \cdot 90$ +   70   =   115   In thousands

            None of   Half of   All of
           Lawrence   Moltonsville Northridge

C gets: $\quad 50 \quad + \quad \frac{1}{2}\cdot 90 \quad + \quad 0 \quad = \quad 95$

All of Lawrence   Half of Moltonsville   None of Northridge

$$
\begin{array}{c}
 \\
L \\
M \\
N
\end{array}
\begin{array}{ccc}
L & M & N \\
\end{array}
\left(\begin{array}{ccc}
0 & -40 & -20 \\
40 & 0 & 20 \\
20 & -20 & 0
\end{array}\right)
$$

Therefore, R gets $115 - 95 = 20$ thousand more than C, verifying the positive 20 in row N, column L of the payoff matrix.

2. We circle the smallest entry in each row.   Then we box the largest entry in each column.

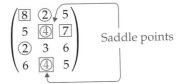

Saddle points

There are two saddle points. R has two optimal strategies: Choose row 2 or row 4. C has one optimal strategy: Choose column 2.

8.1 Exercises

For each situation, identify the two players, their possible choices, and construct a payoff matrix for their conflict.

1. **MANAGEMENT SCIENCE: TV Scheduling** In an attempt to gain more viewers, Channel 86 and Channel 7 are each trying to decide whether to schedule a quiz show or a reality series in their 8:00 prime time slot. Market research indicates that if Channel 86 chooses a quiz show it will gain 5% of the market if Channel 7 runs a quiz show and lose 8% if Channel 7 runs a reality series, while if Channel 86 chooses a reality series it will gain 10% if Channel 7 runs a quiz show and lose 10% if Channel 7 runs a reality series. [*Hint:* Use Q and R for quiz show and reality series.]

2. **GENERAL: Matching Pennies** Roy and Chuck each has a penny to place on the table showing either "heads" or "tails." If they are the same, Roy wins $1 from Chuck, and if they are different, Chuck wins $1 from Roy.

3. **MANAGEMENT SCIENCE: Market Share** Andersonville has two gas stations, Ralph's Qwik-Serv and Charlie's Gas-n-Go. Both

Ralph and Charlie are considering raising prices by 1¢, staying with their current prices, or lowering prices by 1¢. If they both make the same choice there will be no change in their market shares, but if they make different choices, the one with the lower price will gain 5% of the market for each penny difference in their prices.

4. **GENERAL: Scissors-Paper-Stone** In the children's game "scissors-paper-stone," two players show each other either 2 fingers (scissors), an open hand (paper), or a closed fist (stone), and the winner is determined by the rule "scissors cut paper, paper wraps stone, but stone breaks scissors." If both hands show the same, it is a tie.

For each game, identify the saddle point and determine the corresponding optimal strategy for each player.

5. $\begin{pmatrix} 3 & 7 \\ 2 & 5 \end{pmatrix}$ 6. $\begin{pmatrix} 1 & -2 \\ 3 & -1 \end{pmatrix}$

7. $\begin{pmatrix} 3 & 1 & 2 \\ 5 & -1 & -2 \end{pmatrix}$ 8. $\begin{pmatrix} 2 & 1 & 9 \\ 6 & 10 & 7 \end{pmatrix}$

9. $\begin{pmatrix} 3 & 2 & 6 \\ -3 & -2 & 7 \\ 4 & 1 & -1 \end{pmatrix}$ **10.** $\begin{pmatrix} 5 & -5 & -3 \\ -5 & 2 & -2 \\ 4 & 1 & -1 \end{pmatrix}$

11. $\begin{pmatrix} 24 & 8 & 7 & 1 \\ 21 & 15 & 16 & 26 \\ 11 & 13 & 18 & 10 \end{pmatrix}$

12. $\begin{pmatrix} 19 & 15 & -2 & -3 \\ -4 & 1 & 2 & 14 \\ 16 & 8 & 7 & 12 \\ 11 & 9 & 6 & 4 \end{pmatrix}$

Determine the optimal strategy for each situation by representing it as a game and finding the saddle point. State your final answer in the terms of the original question.

13. MANAGEMENT SCIENCE: Market Share
In an ongoing price war between Burger Haven (locally owned) and MacArches (a chain), both restaurant managers plan to change the price of a hamburger by 10¢. If they both raise their prices, there will be no change in their market shares, but if they both lower their prices, the chain's national advertising will ensure that MacArches gains 6% of the market. Again because of advertising, if Burger Haven lowers their price and MacArches raises theirs, Burger Haven will gain only 4% of the market, but if Burger Haven raises their price and MacArches lowers theirs, MacArches will gain 8% of the market. What should the managers do?

14. GENERAL: "Fingers-Only" Morra Two players show each other 1, 2, or 3 fingers. Whoever shows the most wins the sum of the fingers shown, but if they are the same, it is a tie. Is this a fair game? How should you play it?

15. SOCIAL SCIENCE: Political Campaigns
A Republican and a Democratic candidate are running for office in a heavily Republican county. A recent newspaper poll comparing their views on taxes and welfare reform showed that when compared on taxes or on welfare reform, the Republican leads by 15%. When comparing the Democrat on welfare reform to the Republican on taxes, the Republican leads by 20%. But when comparing the Democrat on taxes to the Republican on wel-

fare reform, the Republican leads by only 10%. What should each discuss at their next debate?

16. MANAGEMENT SCIENCE: Best Business Location Two auto-parts distributors plan to enter a tri-town market by each locating a store in one of the three towns of Amboy (population 50,000), Bradford (population 60,000), and Clayton (population 70,000). If both stores are built in the same town, they will split the entire market equally. But if the stores are built in different towns, each will get all of the market in its town and the store closer to the town without a store will get all of that town's business. The following map shows the distances between the towns. Where should the stores be built? [*Hint:* In your matrix, list numbers *in thousands*.]

Explorations and Excursions

The following problems extend and augment the material presented in the text.

More About Holmes and Moriarty

17. Perhaps Example 3 (see pages 528–529) did not have a saddle point because we awarded too many points for a kill or not enough for a tie. Since a kill is more impressive than an escape, pick any two numbers $k > l > 0$ and form the game $\begin{pmatrix} -k & 0 \\ l & -k \end{pmatrix}$. Show that no such game has a saddle point.

2 × 2 Games and Saddle Points

Consider the 2×2 game $G = \begin{pmatrix} a & b \\ c & d \end{pmatrix}$.

18. Show that if $(a + d) - (b + c) = 0$, then G has at least one saddle point.
[*Hint:* Let $b = a + i$ and $c = a + j$. Solve $(a + d) - (b + c) = 0$ for d to show that $d = a + (i + j)$. Thus
$$G = \begin{pmatrix} a & a + i \\ a + j & a + (i + j) \end{pmatrix}.$$
(continues)

Consider the following cases:
a. $i = 0$ and $j = 0$
b. $i = 0$ and $j \neq 0$
c. $i \neq 0$ and $j = 0$
d. $i \neq 0$ and $j \neq 0$ with $i \cdot j > 0$
e. $i \neq 0$ and $j \neq 0$ with $i \cdot j < 0$.]

19. Deduce from Exercise 18 that if G does not have a saddle point, then $(a + d) - (b + c) \neq 0$.

20. Let $G = \begin{pmatrix} 3 & 4 \\ 0 & 2 \end{pmatrix}$. Show that 3 is the saddle point. Since $(3 + 2) - (4 + 0) = 1$, is this a contradiction to Exercise 19?

8.2 MIXED STRATEGIES

Introduction

We know that if a game has a saddle point, then the optimal strategies for the row and column players are to choose the row and column corresponding to that saddle point. But what if there is no saddle point? If the game is played several times and one player repeatedly makes a particular choice, the other will quickly notice and shift to the best response to the first player's choice. But then the first player will notice the consistency of the other player's choices and will react accordingly. Such continuing reacting and shifting cannot be the optimal solution because it does not represent an equilibrium between the desires of the opposing players. Thus, we must consider *random* changes between the possible choices, since only random actions are not susceptible to pattern detection.

Mixed Strategies and Expected Values

How do we change randomly among choices? For each possible choice we select a probability between 0 and 1 (inclusive) such that the sum of the probabilities is 1. (More later on how to choose the probabilities.) Such a collection of probabilities for the possible choices is called a *mixed strategy*. But if the row and column players each use mixed strategies, how can we find the value of the outcome? As we saw on page 409, the *long-term average winnings* from repeated plays of a game may be found by multiplying each possible outcome by its probability and adding the products. The result is called the *expected value* of the game.

EXAMPLE 1

FINDING THE EXPECTED VALUE OF A GAME

For the game $\begin{pmatrix} -1 & 2 \\ 2 & -2 \end{pmatrix}$, suppose that the row player R uses a mixed strategy with probabilities $\frac{2}{5}$ and $\frac{3}{5}$ (meaning choosing the first row with probability $\frac{2}{5}$ and the second row with probability $\frac{3}{5}$) while the column player C uses a mixed strategy with probabilities $\frac{1}{2}$ and $\frac{1}{2}$. What is the expected value of one play of the game?

Solution

If C chooses the first column, then R can obtain the payoffs in the first column, $\begin{pmatrix} -1 \\ 2 \end{pmatrix}$, with probabilities $\frac{2}{5}$ and $\frac{3}{5}$, for an expected value of

$$\tfrac{2}{5} \cdot (-1) + \tfrac{3}{5} \cdot 2 = \tfrac{-2+6}{5} = \tfrac{4}{5}$$

See page 409 to review how to calculate expected value

In matrix notation, this is just the product of a row of probabilities with a column of payoffs:

$$\begin{pmatrix} \tfrac{2}{5} & \tfrac{3}{5} \end{pmatrix}\begin{pmatrix} -1 \\ 2 \end{pmatrix} = \tfrac{2}{5} \cdot (-1) + \tfrac{3}{5} \cdot 2 = \tfrac{4}{5}$$

See page 201 to review row times column multiplication

Carrying this out for *each* column:

$$\begin{pmatrix} \tfrac{2}{5} & \tfrac{3}{5} \end{pmatrix}\begin{pmatrix} -1 & 2 \\ 2 & -2 \end{pmatrix} = \begin{pmatrix} \tfrac{4}{5} & -\tfrac{2}{5} \end{pmatrix}$$

See page 202 to review matrix multiplication

The two resulting numbers are the expected values for the columns, which we then multiply by the probabilities of choosing those columns, $\frac{1}{2}$ each, and add the results. This calculation, however, amounts to multiplying the above row $\begin{pmatrix} \tfrac{4}{5} & -\tfrac{2}{5} \end{pmatrix}$ by the column of probabilities $\begin{pmatrix} 1/2 \\ 1/2 \end{pmatrix}$, which is equivalent to multiplying *all three matrices together:*

$$\begin{pmatrix} \text{Expected} \\ \text{value} \end{pmatrix} = \begin{pmatrix} \tfrac{2}{5} & \tfrac{3}{5} \end{pmatrix}\begin{pmatrix} -1 & 2 \\ 2 & -2 \end{pmatrix}\begin{pmatrix} \tfrac{1}{2} \\ \tfrac{1}{2} \end{pmatrix} = \tfrac{1}{5}$$

Expected value of one play of the game

Graphing Calculator Exploration

The expected value found in Example 1 is easily calculated as a matrix product by entering the row probabilities in matrix [A], the payoff matrix in matrix [B], the column probabilities in matrix [C], and then finding the product [A][B][C].

```
MATRIX[A] 1 x2
[.4    .6    ]
```

```
MATRIX[B] 2 x2
[-1    2    ]
[2    -2    ]
```

```
MATRIX[C] 2 x1
[.5    ]
[.5    ]
```

```
[A] [B] [C]
          [[.2]]
```

The answer, 0.2, agrees with the answer $\frac{1}{5}$ we found in Example 1.

In general, the row player's mixed strategy consists of a probability for each row and the column player's mixed strategy consists of a probability for each column, with the results most conveniently represented as matrices.

Mixed Strategies and Expected Value

For a game G with m rows and n columns:

A mixed strategy for the row player consists of a row matrix:

$$r = (r_1 \ldots r_m)$$

Each number between 0 and 1 and adding to 1

A mixed strategy for the column player consists of a column matrix

$$c = \begin{pmatrix} c_1 \\ \vdots \\ c_n \end{pmatrix}$$

Each number between 0 and 1 and adding to 1

The expected value E for this mixed strategy is the matrix product

$$E = r \cdot G \cdot c$$

G is the payoff matrix

Practice Problem 1

$$\begin{pmatrix} -1 & 2 \\ 2 & -2 \end{pmatrix}$$

For the game in Example 1, suppose the row player chooses mixed strategy $r = \begin{pmatrix} \frac{1}{2} & \frac{1}{2} \end{pmatrix}$ and the column player chooses mixed strategy $c = \begin{pmatrix} \frac{4}{5} \\ \frac{1}{5} \end{pmatrix}$. What is the expected value of each play of the game?

➤ Solution on page 542

How do the *pure strategies* of the preceding section relate to the *mixed strategies* discussed here? Pure strategies are simply mixed strategies where one of the choices has probability 1 and the others have probability 0, so that the same row and column (corresponding to the saddle point) are always chosen.

Given a mixed strategy, how in practice does a player choose moves according to the probabilities? If there are only two possible choices and each has probability $\frac{1}{2}$, then it is simple enough to just flip a coin and make the first choice on "heads" and the other choice on "tails." For other probabilities it is usually easiest to base the decision on a *random number* from a table, calculator, or computer.

Graphing Calculator Exploration

A graphing calculator can help to make random choices for any strategy. For example, the strategy $r = \begin{pmatrix} \frac{1}{4} & \frac{1}{2} & \frac{1}{4} \end{pmatrix}$ separates the interval from 0 to 1 into three subintervals of lengths $\frac{1}{4}, \frac{1}{2}$, and $\frac{1}{4}$:

The random number command gives random numbers between 0 and 1.

then

and

The first random number, 0.3897, lies in the second of the above intervals, so you would play the *second* choice. The next random number, 0.8087, lies in third interval, so you would next play the *third* choice, and so on, using as many random numbers as you need. Of course, the row and the column player should each pick their own random number to decide their next moves.

Optimal Mixed Strategies for 2 × 2 Games

The row player wants to find the mixed strategy that *maximizes* the expected value per play (since the value goes to the row player), while the column player wants the mixed strategy that *minimizes* the expected value per play. How can we find such *optimal* mixed strategies for each player? In the next section, we will answer this question for *any* game, but in this section we will solve this problem for 2 × 2 games.

Let $\begin{pmatrix} a & b \\ c & d \end{pmatrix}$ be a 2 × 2 game without a saddle point. A mixed strategy $r = (r_1 \quad r_2)$ for R is really just $(r_1 \quad 1 - r_1)$ since $r_1 + r_2 = 1$.

(For example, if one probability is $\frac{1}{3}$, the other must be $1 - \frac{1}{3} = \frac{2}{3}$.) If C chooses the first column, $\begin{pmatrix} a \\ c \end{pmatrix}$, then the expected value is

$$E_1 = (r_1 \quad 1 - r_1)\begin{pmatrix} a \\ c \end{pmatrix} = ar_1 + c(1 - r_1) \qquad \text{For } 0 \le r_1 \le 1$$

Graphing values of $E_1 = ar_1 + c(1 - r_1)$ for values of r_1 between 0 and 1 gives a line between heights a and c as shown below.

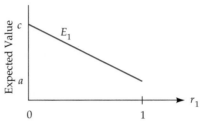

Expected value if C chooses column 1
and R uses mixed strategy $(r_1 \quad 1-r_1)$

Similarly, if C chooses the second column, the expected value is

$$E_2 = (r_1 \quad 1 - r_1)\begin{pmatrix} b \\ d \end{pmatrix} = br_1 + d(1 - r_1)$$

and the graph becomes

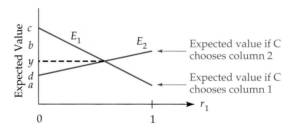

Expected value if C
chooses column 2

Expected value if C
chooses column 1

R uses mixed strategy $(r_1 \quad 1-r_1)$

In this graph, C chooses the line, but R chooses the *point* on the line (by choosing r_1), so if R chooses any point *other* than the intersection point, C will choose the lower line to decrease R's expected value. It is clear that R should choose the intersection point, in which case it does not matter which line C chooses. Solving for the intersection point of the lines:

$$ar_1 + c(1 - r_1) = br_1 + d(1 - r_1) \qquad \text{Setting } E_1 = E_2$$
$$(a + d - b - c)r_1 = d - c \qquad \text{Collecting like terms}$$

$$r_1 = \frac{d - c}{(a + d) - (b + c)} \qquad \text{Dividing to find } r_1$$

(Dividing by $(a + d) - (b + c)$ in the last step is permissible because if G does not have a saddle point, then $(a + d) - (b + c) \neq 0$ as shown in Exercise 19 on page 532.) The other probability for R's optimal mixed strategy is found by subtracting r_1 from 1:

$$r_2 = 1 - r_1 = 1 - \frac{d - c}{(a + d) - (b + c)} = \frac{a - b}{(a + d) - (b + c)}$$

The value of the game when R uses the mixed strategy (r_1, r_2) is then the height of the intersection point in the preceding graph, which is

$$v = ar_1 + cr_2 = \frac{a(d - c) + c(a - b)}{(a + d) - (b + c)} = \frac{ad - bc}{(a + d) - (b + c)}$$

A similar calculation gives the probabilities c_1 and c_2 for C's optimal strategy, which are included in the summary in the box below.

> **Optimal Mixed Strategy for a 2 × 2 Game**
>
> A game $\begin{pmatrix} a & b \\ c & d \end{pmatrix}$ without a saddle point has optimal mixed strategies:
>
> For row player R:
> $$\begin{cases} r_1 = \dfrac{d - c}{(a + d) - (b + c)} \\ r_2 = 1 - r_1 \end{cases}$$
>
> For column player C:
> $$\begin{cases} c_1 = \dfrac{d - b}{(a + d) - (b + c)} \\ c_2 = 1 - c_1 \end{cases}$$
>
> The value of the game is $v = \dfrac{ad - bc}{(a + d) - (b + c)}$.

EXAMPLE 2

CALCULATING AN OPTIMAL MIXED STRATEGY

Find the optimal mixed strategy and the value of the game $\begin{pmatrix} 7 & -2 \\ 3 & 6 \end{pmatrix}$.

Solution

For the row player we find probabilities

$$\begin{cases} r_1 = \dfrac{6 - 3}{(7 + 6) - (-2 + 3)} = \dfrac{3}{12} = \dfrac{1}{4} \\ r_2 = 1 - \dfrac{1}{4} = \dfrac{3}{4} \end{cases}$$

$\begin{cases} r_1 = \dfrac{d - c}{(a + d) - (b + c)} \\ r_2 = 1 - r_1 \end{cases}$

with $a = 7$, $b = -2$, $c = 3$, and $d = 6$

For the column player we find

$$\begin{cases} c_1 = \dfrac{6 - (-2)}{(7 + 6) - (-2 + 3)} = \dfrac{8}{12} = \dfrac{2}{3} \\ c_2 = 1 - \dfrac{2}{3} = \dfrac{1}{3} \end{cases} \qquad \begin{cases} c_1 = \dfrac{d - b}{(a + d) - (b + c)} \\ c_2 = 1 - c_1 \end{cases}$$
with $a = 7$, $b = -2$, $c = 3$, and $d = 6$

The value of the game is

$$v = \frac{7 \cdot 6 - (-2) \cdot 3}{(7 + 6) - (-2 + 3)} = \frac{48}{12} = 4 \qquad v = \frac{ad - bc}{(a + d) - (b + c)}$$
with $a = 7$, $b = -2$, $c = 3$, and $d = 6$

Therefore, the row player should play rows 1 and 2 with probabilities $\frac{1}{4}$ and $\frac{3}{4}$ and the column player should play columns 1 and 2 with probabilities $\frac{2}{3}$ and $\frac{1}{3}$. The value of the game is 4. (That is, the game is favorable to the row player, who, in the long run, should expect a gain of about 4 per play.)

Practice Problem 2   Find the optimal mixed strategy and value of the game $\begin{pmatrix} 3 & 0 \\ -2 & 1 \end{pmatrix}$.

➤ Solution on page 542

EXAMPLE 3

OPTIMAL MIXED STRATEGIES FOR HOLMES AND MORIARTY

Find the optimal mixed strategy and the value of the game for the Holmes and Moriarty conflict described in Example 3 on page 528.

Solution

$$\begin{pmatrix} -100 & 0 \\ 50 & -100 \end{pmatrix}$$
Payoff matrix
from page 529

Using the formulas in the preceding box with the numbers from the payoff matrix (omitting the details), the optimal strategy for Holmes is to get off at Canterbury with probability $r_1 = \frac{3}{5}$ and to continue on to Dover with probability $r_2 = \frac{2}{5}$. Moriarty's optimal strategy is to go to Canterbury with probability $c_1 = \frac{2}{5}$ and to Dover with probability $c_2 = \frac{3}{5}$. (In *The Final Problem*, Holmes and Moriarty each choose their more likely option, with Holmes getting off at Canterbury and Moriarty going to Dover.) The value of the game (found again from the formula in the box) is -40, meaning that even with this optimal strategy the situation is unfavorable to Holmes.

The Spreadsheet Application shows the solution to this problem. Notice that the $r_1 = \frac{3}{5}$ and $c_1 = \frac{2}{5}$ found above correspond to the *saddle point* of the graph in the spreadsheet, and that the value in the corresponding cell is -40, which agrees with the value that we found for the game.

Spreadsheet Exploration

The following spreadsheet* calculates and graphs the expected values for 100 of the possible mixed strategies for the Holmes and Moriarty payoff matrix on the previous page. The 2×2 game is located in cells **A1:B2**. Probabilities for Holmes getting off the train in Canterbury are listed in cells **A5:A15**, and probabilities for Moriarty stopping at Canterbury are in cells **B4:L4**. Since there are only two choices for each player, the probabilities of their going to Dover (their other choice) are 1 minus the listed values. The 100 values in cells **B5:L15** are calculated by formulas similar to the one shown at the top for cell **F11**.

| F11 | | = | =$A$1*($A11)*(F$4)+$B$1*($A11)*(1-F$4)+$A$2*(1-$A11)*(F$4)+$B$2*(1-$A11)*(1-F$4) |

	A	B	C	D	E	F	G	H	I	J	K	L
1	-100	0										
2	50	-100										
3												
4		0	0.1	0.2	0.3	0.4	0.5	0.6	0.7	0.8	0.9	1
5	0	-100	-85	-70	-55	-40	-25	-10	5	20	35	50
6	0.1	-90	-78	-65	-53	-40	-28	-15	-2.5	10	22.5	35
7	0.2	-80	-70	-60	-50	-40	-30	-20	-10	0	10	20
8	0.3	-70	-63	-55	-48	-40	-33	-25	-18	-10	-2.5	5
9	0.4	-60	-55	-50	-45	-40	-35	-30	-25	-20	-15	-10
10	0.5	-50	-48	-45	-43	-40	-38	-35	-33	-30	-28	-25
11	0.6	-40	-40	-40	-40	-40	-40	-40	-40	-40	-40	-40
12	0.7	-30	-33	-35	-38	-40	-43	-45	-48	-50	-53	-55
13	0.8	-20	-25	-30	-35	-40	-45	-50	-55	-60	-65	-70
14	0.9	-10	-18	-25	-33	-40	-48	-55	-63	-70	-78	-85
15	1	0	-10	-20	-30	-40	-50	-60	-70	-80	-90	-100

Other Interpretations of Mixed Strategies

While we developed mixed strategies to give the best probabilities for choices that must be made repeatedly in competitive situations, they have other interpretations as well. Even if you are in the situation only once, there may be many pairs of people in similar situations, and the mixed strategies then give the *proportions* of people who should make each choice. In fact, it is not even necessary to have an actual opposing player. In many situations it is useful to imagine a *fictitious opponent*. For example, an investor in stocks and bonds might imagine the *economy* as the opponent, bringing about changes in economic conditions, and a strategy of $\left(\frac{2}{3} \quad \frac{1}{3}\right)$ might suggest investing $\frac{2}{3}$ of one's capital in bonds and $\frac{1}{3}$ in stocks. Other examples are farmers with unpredictable weather and doctors with uncertain diagnoses. While these situations may not involve an actual "opponent" who is actively trying to minimize your gain, using the optimal strategy will guarantee a certain average value no matter what happens.

EXAMPLE 4

OPTIMAL FARM POLICY

After carefully studying the economy and the likelihood of price supports, a midwestern farmer has estimated his profits per acre for both wheat and soybeans under recession and expansion conditions, as shown in the following table.

		Economy	
		Recession	Expansion
Farmer's Profits	Wheat	$10	$40
	Soybeans	$30	$20

How should the farmer manage his 10,000 acres in the face of uncertain economic conditions?

Solution

Let us view the economy as a fictitious opponent. The game $\begin{pmatrix} 10 & 40 \\ 30 & 20 \end{pmatrix}$ does not have a saddle point, and the farmer's optimal mixed strategy is

$$\begin{cases} r_1 = \dfrac{20 - 30}{(10 + 20) - (40 + 30)} = \dfrac{-10}{30 - 70} = \dfrac{1}{4} \\ r_2 = 1 - \dfrac{1}{4} = \dfrac{3}{4} \end{cases}$$

From the box on page 537

The farmer should devote $\frac{1}{4}$ of the farm to wheat (that is, 2500 acres) and $\frac{3}{4}$ to soybeans (that is, 7500 acres). With this mixed strategy, should the economy be in recession, the farmer's profits will be

$$\$10 \cdot 2500 + \$30 \cdot 7500 = \$250,000$$

while if it is in expansion, the farmer's profit will still be

$$\$40 \cdot 2500 + \$20 \cdot 7500 = \$250,000$$

The profit of $250,000 represents a security level for the farmer because, while he could do better if he knew the future state of the economy, there is no way he can receive less.

8.2 Section Summary

Games without saddle points can be solved by *optimal mixed strategies* that specify the probabilities for the various possible courses of action. For a game G and mixed strategies

$$r = (r_1 \ldots r_m) \quad \text{for the row player and}$$

$$c = \begin{pmatrix} c_1 \\ \vdots \\ c_n \end{pmatrix} \quad \text{for the column player,}$$

the expected value is the matrix product $E = r \cdot G \cdot c$.

For a 2×2 game $G = \begin{pmatrix} a & b \\ c & d \end{pmatrix}$ without a saddle point, the optimal mixed strategies are

For row player R: $\qquad r_1 = \dfrac{d - c}{(a + d) - (b + c)}$ and $r_2 = 1 - r_1$

For column player C: $\qquad c_1 = \dfrac{d - b}{(a + d) - (b + c)}$ and $c_2 = 1 - c_1$

The value of the game is $\quad v = \dfrac{ad - bc}{(a + d) - (b + c)}.$

A mixed strategy may be interpreted as the *proportions of a resource to assign to the various choices.* Modeling a conflicting-choice situation as a game against a fictitious opponent finds the security level that makes the best of a bad situation.

➤ **Solutions to Practice Problems**

1. $\underbrace{\left(\begin{smallmatrix}\frac{1}{2} & \frac{1}{2}\end{smallmatrix}\right)\begin{pmatrix}-1 & 2 \\ 2 & -2\end{pmatrix}}_{\left(\begin{smallmatrix}\frac{1}{2} & 0\end{smallmatrix}\right)}\begin{pmatrix}\frac{4}{5} \\ \frac{1}{5}\end{pmatrix} = \left(\begin{smallmatrix}\frac{1}{2} & 0\end{smallmatrix}\right)\begin{pmatrix}\frac{4}{5} \\ \frac{1}{5}\end{pmatrix} = \left(\frac{2}{5}\right)$

The expected value for each play is $\frac{2}{5}$.

2. Since $G = \begin{pmatrix} 3 & 0 \\ -2 & 1 \end{pmatrix}$ does not have a saddle point, we substitute $a = 3$, $b = 0$, $c = -2$, and $d = 1$ into the formulas from the box on page 537.

$$(a + d) - (b + c) = (3 + 1) - (0 - 2) = 4 + 2 = 6$$

For the row player: $\quad r_1 = \dfrac{1 - (-2)}{6} = \dfrac{3}{6} = \dfrac{1}{2}$ and $r_2 = 1 - \dfrac{1}{2} = \dfrac{1}{2}$

For the column player: $\quad c_1 = \dfrac{1 - 0}{6} = \dfrac{1}{6}$ and $c_2 = 1 - \dfrac{1}{6} = \dfrac{5}{6}$

The value of the game is $\quad \dfrac{3 \cdot 1 - 0 \cdot (-2)}{6} = \dfrac{1}{2}$

8.2 Exercises

For each game and mixed strategies, find the expected value.

Let $G = \begin{pmatrix} 2 & -2 \\ 1 & 3 \end{pmatrix}$.

1. $r = \left(\begin{smallmatrix}\frac{1}{2} & \frac{1}{2}\end{smallmatrix}\right)$ and $c = \begin{pmatrix}\frac{1}{2} \\ \frac{1}{2}\end{pmatrix}$

2. $r = \left(\begin{smallmatrix}\frac{3}{4} & \frac{1}{4}\end{smallmatrix}\right)$ and $c = \begin{pmatrix}\frac{2}{5} \\ \frac{3}{5}\end{pmatrix}$

Let $G = \begin{pmatrix} 2 & -4 \\ -1 & 3 \end{pmatrix}$.

3. $r = \left(\begin{smallmatrix}\frac{4}{5} & \frac{1}{5}\end{smallmatrix}\right)$ and $c = \begin{pmatrix}\frac{2}{5} \\ \frac{3}{5}\end{pmatrix}$

4. $r = \left(\begin{smallmatrix}\frac{1}{3} & \frac{2}{3}\end{smallmatrix}\right)$ and $c = \begin{pmatrix}\frac{7}{10} \\ \frac{3}{10}\end{pmatrix}$

Let $G = \begin{pmatrix} 3 & 1 & 2 \\ -6 & 4 & -1 \end{pmatrix}$.

5. $r = \left(\begin{smallmatrix}\frac{1}{2} & \frac{1}{2}\end{smallmatrix}\right)$ and $c = \begin{pmatrix}\frac{1}{3} \\ \frac{1}{3} \\ \frac{1}{3}\end{pmatrix}$

6. $r = \left(\begin{smallmatrix}\frac{1}{6} & \frac{5}{6}\end{smallmatrix}\right)$ and $c = \begin{pmatrix}\frac{1}{2} \\ \frac{1}{4} \\ \frac{1}{4}\end{pmatrix}$

Let $G = \begin{pmatrix} 9 & 1 \\ -9 & 3 \\ -1 & 2 \end{pmatrix}$.

7. $r = \left(\begin{smallmatrix}\frac{1}{2} & 0 & \frac{1}{2}\end{smallmatrix}\right)$ and $c = \begin{pmatrix}\frac{3}{5} \\ \frac{2}{5}\end{pmatrix}$

8. $r = \left(\begin{smallmatrix}\frac{1}{8} & \frac{3}{8} & \frac{1}{2}\end{smallmatrix}\right)$ and $c = \begin{pmatrix}\frac{3}{4} \\ \frac{1}{4}\end{pmatrix}$

For each 2×2 game, find the optimal strategy for each player and the value of the game. Be sure to check for saddle points before using the formulas.

9. $\begin{pmatrix} 2 & -1 \\ 1 & 2 \end{pmatrix}$ **10.** $\begin{pmatrix} 1 & -1 \\ -2 & 2 \end{pmatrix}$

11. $\begin{pmatrix} 3 & 1 \\ -4 & 2 \end{pmatrix}$ **12.** $\begin{pmatrix} 6 & -3 \\ 2 & 5 \end{pmatrix}$

13. $\begin{pmatrix} 3 & -3 \\ -1 & 3 \end{pmatrix}$ **14.** $\begin{pmatrix} 5 & -9 \\ -1 & 5 \end{pmatrix}$

Represent each situation as a game and find the optimal strategy for each player. State your final answer in the terms of the original question.

15. SOCIAL SCIENCE: Political Campaigns
Political scientists distinguish between two kinds of issues in elections campaigns: "positional" issues (voters have sharply divided and incompatible views) and "valence" issues (voters agree on goals but are divided on the best ways to achieve them). In a race between the incumbent and the challenger for state governor, a marketing survey found that if the incumbent focused on positional issues, she trailed the challenger by 5% if he focused on the same kind of issues, but she led him by 5% if he concentrated on valence issues. If she focused on valence issues, she trailed him by 10% if he focused on the same kind of issues, but she led him by 20% if he concentrated on positional issues. In designing the incumbent's TV ads for the campaign, what proportion should focus on positional issues?

16. GENERAL: "Odd and Even" Morra Josh and Justin show each other 1 or 2 fingers. If the sum of the fingers shown is odd, Josh wins the sum, but if it is even, Justin wins it. Who has the advantage? What is the probability that Josh will show 2 fingers the next time they play?

17. BUSINESS: Farm Management A farmer grows apples on her 600 acre farm and must cope with occasional infestations of worms. If she refrains from using pesticides, she can get a premium for "organically grown" produce and her profits per acre increase by $600 if there is no infestation, but decrease by $400 if there is. If she does use pesticides and there is

an infestation, her crop is saved and the resulting apple shortage (since other farms are decimated) raises her profits by $500 per acre. Otherwise, her profits remain at their usual levels. How should she divide her farm into a "pesticide-free" zone and a "pesticide-use" zone? What will be her expected increase in profits per acre with this strategy?

18. GENERAL: *Modified* Matching Pennies
Rose and Carmen each have a penny to place on the table showing "heads" or "tails." If the coins are the same, Rose wins $1 from Carmen, while if the coins are different, Carmen wins $1 from Rose; they have also decided that if the winner has played "heads," then the winner gets $2. What is the probability that Carmen will play "tails" on the next round?

19. GENERAL: Quizzes in English Class In a section of Freshman Composition at your college, the English professor gives a daily quiz on either vocabulary or writing, and allows his students to bring either a dictionary or a grammar textbook (but not both) to use during the quiz. Joe estimates that if he brings a dictionary, he can get a 100 on a vocabulary quiz but only an 80 on a paragraph revision, while with his grammar textbook, he can get a 90 on the revision but only a 70 on the vocabulary quiz. How should he decide each day what to bring and what grade can he be expected to earn?

20. GENERAL: Guess Which Coin Cory has a penny and a nickel in his pocket. He picks one coin and hides it in his fist. If Ray can guess which coin it is, Cory gives it to him, but if Ray is wrong, he pays Cory the value of his guess. Is this game fair? How often should Ray guess it's a penny?

Explorations and Excursions

The following problems extend and augment the material presented in the text.

Dominant Rows and Columns

A row of a game *dominates* another if every entry in it is greater than or equal to the corresponding entry in the other row. Since the dominated row is "smaller," the row player will never choose it and removing it from the game will not alter the rows that R will choose.

Solve each game by first removing every dominated row and then solving the resulting 2×2 game. State the optimal strategy in terms of the original game.

21. $\begin{pmatrix} 3 & -1 \\ 1 & 3 \\ 2 & 4 \end{pmatrix}$
22. $\begin{pmatrix} 4 & -2 \\ 3 & -3 \\ 0 & 4 \\ 1 & 5 \end{pmatrix}$

A column of a game *dominates* another if every entry in it is less than or equal to the corresponding entry in the other column. Since the dominated column is "bigger," the column player will never choose it and removing it from the game will not alter the columns that C will choose.

Solve each game by first removing every dominated column and then solving the resulting 2×2 game. State the optimal strategy in terms of the original game.

23. $\begin{pmatrix} 1 & 2 & -5 \\ -2 & 1 & 2 \end{pmatrix}$
24. $\begin{pmatrix} 2 & 3 & -2 & -1 \\ 1 & 2 & 3 & 4 \end{pmatrix}$

Solve each game by first removing dominated rows and columns. State the optimal strategy in terms of the original game.

25. $\begin{pmatrix} 4 & 5 & 3 \\ 3 & 4 & 2 \\ 2 & 3 & 1 \end{pmatrix}$
26. $\begin{pmatrix} 1 & -2 & 1 \\ 3 & -1 & 2 \\ 2 & -2 & 3 \end{pmatrix}$

27. $\begin{pmatrix} 4 & -2 & 2 \\ -3 & 3 & 4 \\ 3 & -3 & 1 \end{pmatrix}$
28. $\begin{pmatrix} 5 & 4 & -4 \\ -1 & 1 & 3 \\ 2 & 1 & 4 \end{pmatrix}$

Changing the Value but Not the Strategy

29. If the same number is added to (or subtracted from) every entry of a game, the players will still make the same choices because the positions of the smaller and larger payoffs will not be changed. Check that each of the following games has the same optimal strategy as Exercise 9 but that the value has been changed. Can you predict the change in the value from Exercise 9 without calculating the values?

   a. $\begin{pmatrix} 3 & 0 \\ 2 & 3 \end{pmatrix}$
   b. $\begin{pmatrix} 4 & 1 \\ 3 & 4 \end{pmatrix}$

   c. $\begin{pmatrix} 1 & -2 \\ 0 & 1 \end{pmatrix}$
   d. $\begin{pmatrix} 0 & -3 \\ -1 & 0 \end{pmatrix}$

Holmes is 48% Dead

30. Use the optimal strategy $r = \begin{pmatrix} \frac{3}{5} & \frac{2}{5} \end{pmatrix}$ and $c = \begin{pmatrix} \frac{2}{5} \\ \frac{3}{5} \end{pmatrix}$ from Example 3 on page 538 to show that the remark by von Neumann and Morgenstern quoted in the Application Preview on page 523 is, in fact, correct. [*Hint:* Holmes and Moriarty are deciding *independently* what to do.]

8.3 GAMES AND LINEAR PROGRAMMING

Introduction

In the previous section we found the optimal strategy for a 2×2 game. In this section we will see how to find the optimal strategy for *any* game. Rather than develop a collection of complicated formulas analogous to those on page 537, we will show that a game can be expressed as a pair of dual linear programming problems (see pages 296–297), which can then be solved by the simplex method that we developed in Sections 3 and 4 of Chapter 4.

The idea of representing a game as a pair of dual linear programming problems follows from the observation that the profits P to be maximized and the costs C to be minimized always satisfy $P \leq C$ (see Exercise 35 on page 312). These dual problems are a game in which one player tries to increase the value of P while the other tries to decrease the value of C. If the first succeeds in making P as large as possible while the second succeeds in making C as small as possible, they will have found the solution to both problems at feasible points giving $P = C$. The Duality Theorem (see page 298) states that the solutions of both dual problems appear in the final simplex tableau that solves the maximum problem.

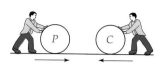

Games as Linear Programming Problems

$$\begin{pmatrix} 3 & 7 \\ 4 & 1 \end{pmatrix}$$

To see how to express a game as a linear programming problem, consider the game shown on the left. Given any strategy $(r_1 \quad r_2)$ for R, we may calculate the expected value depending on whether C chooses the first or second column:

<table>
<tr><td>If C chooses the first column</td><td>If C chooses the second column</td></tr>
</table>

$$E_1 = (r_1 \quad r_2)\begin{pmatrix} 3 \\ 4 \end{pmatrix} = 3r_1 + 4r_2 \qquad E_2 = (r_1 \quad r_2)\begin{pmatrix} 7 \\ 1 \end{pmatrix} = 7r_1 + 1r_2$$

Since C will choose the column to give R the *lower* of these expected gains, the value will be at or below *both* of these expected values, so that R wants to solve the problem

Maximize v R tries to maximize v

Subject to
$$\begin{cases} 3r_1 + 4r_2 \geq v \\ 7r_1 + 1r_2 \geq v \\ r_1 + r_2 = 1 \\ r_1 \geq 0 \quad \text{and} \quad r_2 \geq 0 \end{cases}$$

C keeps v at or below both expected values

r_1 and r_2 are probabilities

We now must make this into a *standard* linear programming problem (see page 273). Since every payoff for this game is positive, the value v is also positive, so we may divide the inequalities by v without changing their sense.

$$\begin{cases} 3r_1 + 4r_2 \geq v \\ 7r_1 + 1r_2 \geq v \end{cases} \quad \text{becomes} \quad \begin{cases} 3\left(\dfrac{r_1}{v}\right) + 4\left(\dfrac{r_2}{v}\right) \geq 1 \\ 7\left(\dfrac{r_1}{v}\right) + 1\left(\dfrac{r_2}{v}\right) \geq 1 \end{cases} \quad \text{Dividing by } v$$

Using new variables $y_1 = r_1/v$ and $y_2 = r_2/v$ (which are nonnegative), these inequalities take the simpler form $3y_1 + 4y_2 \geq 1$ and $7y_1 + 1y_2 \geq 1$. Furthermore, the equality $r_1 + r_2 = 1$ divided by

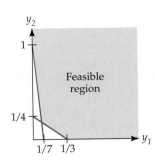

Feasible region

v becomes $y_1 + y_2 = \frac{1}{v}$. Since maximizing the positive quantity v is the same as minimizing $\frac{1}{v}$ (a bigger v means a smaller $\frac{1}{v}$), R wants to solve the standard linear programming problem

$$\text{Minimize} \quad y_1 + y_2$$

$$\text{Subject to} \quad \begin{cases} 3y_1 + 4y_2 \geq 1 \\ 7y_1 + 1y_2 \geq 1 \\ y_1 \geq 0 \quad \text{and} \quad y_2 \geq 0 \end{cases}$$

Now consider the game from the point of view of the column player. For any strategy $\begin{pmatrix} c_1 \\ c_2 \end{pmatrix}$ for C, we can calculate the expected value depending on the row that R chooses:

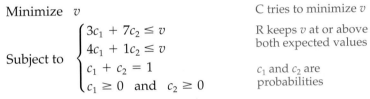

If R chooses the first row

$$\begin{pmatrix} 3 & 7 \end{pmatrix}\begin{pmatrix} c_1 \\ c_2 \end{pmatrix} = 3c_1 + 7c_2$$

If R chooses the second row

$$\begin{pmatrix} 4 & 1 \end{pmatrix}\begin{pmatrix} c_1 \\ c_2 \end{pmatrix} = 4c_1 + 1c_2$$

$$\begin{pmatrix} 3 & 7 \\ 4 & 1 \end{pmatrix}$$

C seeks the *smallest* possible value, while R chooses the row to keep it at or above these expectations, so that C wants to solve the problem

$$\text{Minimize} \quad v \qquad \text{C tries to minimize } v$$

$$\text{Subject to} \quad \begin{cases} 3c_1 + 7c_2 \leq v \\ 4c_1 + 1c_2 \leq v \\ c_1 + c_2 = 1 \\ c_1 \geq 0 \quad \text{and} \quad c_2 \geq 0 \end{cases}$$

R keeps v at or above both expected values

c_1 and c_2 are probabilities

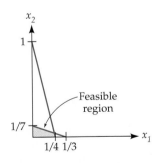

Feasible region

Using new variables $x_1 = c_1/v$ and $x_2 = c_2/v$, we can write this problem as a standard linear programming problem

$$\text{Maximize} \quad x_1 + x_2$$

$$\text{Subject to} \quad \begin{cases} 3x_1 + 7x_2 \leq 1 \\ 4x_1 + 1x_2 \leq 1 \\ x_1 \geq 0 \quad \text{and} \quad x_2 \geq 0 \end{cases}$$

$$x_1 + x_2 = \frac{1}{v}$$

$$3\left(\frac{c_1}{v}\right) + 7\left(\frac{c_2}{v}\right) \leq \frac{v}{v}$$

$$4\left(\frac{c_1}{v}\right) + 1\left(\frac{c_2}{v}\right) \leq \frac{v}{v}$$

Notice that the problems for R and C are *dual* linear programming problems:

For R:

$$\text{Maximize} \quad x_1 + x_2$$

$$\text{Subject to} \quad \begin{cases} 3x_1 + 7x_2 \leq 1 \\ 4x_1 + 1x_2 \leq 1 \\ x_1 \geq 0 \quad \text{and} \quad x_2 \geq 0 \end{cases}$$

For C:

$$\text{Minimize} \quad y_1 + y_2$$

$$\text{Subject to} \quad \begin{cases} 3y_1 + 4y_2 \geq 1 \\ 7y_1 + 1y_2 \geq 1 \\ y_1 \geq 0 \quad \text{and} \quad y_2 \geq 0 \end{cases}$$

See pages 296–297 to review dual linear programming problems

The maximum problem $\left(x_1 + x_2 = \frac{1}{v}\right)$ comes from the column player's attempt to minimize v, and the minimum problem

$\left(y_1 + y_2 = \frac{1}{v}\right)$ results from the row player's desire to maximize the same v. Since the simplex method solves both problems at the same time, we may solve the original game by constructing the initial simplex tableau and pivoting to find the optimal values for the variables $x_1, x_2, y_1,$ and y_2 and then recovering the probabilities $r_1, r_2, c_1,$ and c_2 for the optimal strategies. The initial simplex tableau (see page 276) is:

	x_1	x_2	s_1	s_2	
s_1	3	7	1	0	1
s_2	4	1	0	1	1
	-1	-1	0	0	0

	X	S	
S	A	I	b
	$-c^t$	0	0

with

$$A = \begin{pmatrix} 3 & 7 \\ 4 & 1 \end{pmatrix}, \quad b = \begin{pmatrix} 1 \\ 1 \end{pmatrix}, \quad c^t = (1 \quad 1)$$

Pivoting twice (first at column 1 and row 2, and then at column 2 and row 1), we arrive at the final tableau, with the resulting x- and y-values as shown.

	x_1	x_2	s_1	s_2	
x_2	0	1	$4/25$	$-3/25$	$1/25$
x_1	1	0	$-1/25$	$7/25$	$6/25$
	0	0	$3/25$	$4/25$	$7/25$

$x_2 = 1/25$
$x_1 = 6/25$

$y_1 = 3/25, \quad y_2 = 4/25$ ⟵ Maximum is $7/25$

Since the maximum value, $7/25$, is the reciprocal $\frac{1}{v}$ of the value v of the game, we have $v = \frac{25}{7}$. We can recover the strategy from the x's and the y's by multiplying back by the value we have found for v:

$$r_1 = v \cdot y_1 = \frac{25}{7} \cdot \frac{3}{25} = \frac{3}{7}$$

$$r_2 = v \cdot y_2 = \frac{25}{7} \cdot \frac{4}{25} = \frac{4}{7}$$

and

$$c_1 = v \cdot x_1 = \frac{25}{7} \cdot \frac{6}{25} = \frac{6}{7}$$

$$c_2 = v \cdot x_2 = \frac{25}{7} \cdot \frac{1}{25} = \frac{1}{7}$$

The value $v = \frac{25}{7}$ and the optimal strategies $\frac{3}{7}, \frac{4}{7}$ for R and $\frac{6}{7}, \frac{1}{7}$ for C solve the game.

For our first example we chose a 2×2 game, which also can be solved by the formulas on page 537 (giving the same answer, as you may easily check). However, the linear programming method we developed here is not restricted to 2×2 games and may be used to solve a game of *any* size.

The general procedure for solving any game by expressing it as a linear programming problem is described in the following box. The first step ensures that the value of the game is positive so that division by v will not alter the sense of the inequalities.

The Simplex Method Solution of Any Game

To solve a game of any size:

1. If any entries in the payoff matrix G are less than 1, add a constant k to every entry to make every entry at least 1. We write $G + k$ for this new matrix.

2. Solve the linear programming problem on the left below by applying the simplex method (see page 285) to the initial tableau on the right:

$$\text{Maximize } (\overbrace{1 \ \ldots \ 1}^{n \text{ 1's}})X$$

$$\text{Subject to } \left\{ \begin{array}{l} (G + k)X \leq \left. \begin{pmatrix} 1 \\ \vdots \\ 1 \end{pmatrix} \right\} m \text{ 1's} \\ X \geq 0 \end{array} \right.$$

$$\left. \begin{array}{c|cc|c} & X & S & \\ \hline S & G + k & I & \begin{matrix} 1 \\ \vdots \\ 1 \end{matrix} \\ \hline & \underbrace{-1 \ldots -1}_{n-1\text{'s}} & \underbrace{0 \ldots 0}_{m \ 0\text{'s}} & 0 \end{array} \right\} m \text{ 1's}$$

3. In the final tableau (from Step 2):

   a. The maximum value V of the objective function is in the bottom-right corner of the final tableau.

   b. The x-variables for the maximum problem take the following values: The basic variables take the values on the far right of the tableau, and the nonbasic variables take the value 0.

   c. The y-variables for the (dual) minimum problem take the values at the bottom of the tableau in the slack variable columns.

4. The value of the game is $v = \dfrac{1}{V} - k$. The optimal strategies are:

   For the row player R:
   $$r_1 = \frac{y_1}{V}, \ r_2 = \frac{y_2}{V}, \ \ldots, \ r_m = \frac{y_m}{V}$$

   For the column player C:
   $$c_1 = \frac{x_1}{V}, \ c_2 = \frac{x_2}{V}, \ \ldots, \ c_n = \frac{x_n}{V}$$

   Using x's and y's for the final x- and y-values

The optimal strategy found by the simplex method may be either pure or mixed.

EXAMPLE **1**

SOLVING A GAME BY THE SIMPLEX METHOD

Use the simplex method to find the optimal strategies for the game

$$G = \begin{pmatrix} 1 & -2 & 1 \\ 2 & -1 & -2 \\ -1 & 2 & 1 \end{pmatrix}.$$

Solution

Step 1. Since the smallest entry in the matrix is -2, adding 3 to every element of the matrix will make each at least 1:

$$\begin{pmatrix} 4 & 1 & 4 \\ 5 & 2 & 1 \\ 2 & 5 & 4 \end{pmatrix}$$

$G + k$
with

Step 2. The initial simplex tableau is

	x_1	x_2	x_3	s_1	s_2	s_3	
s_1	4	1	4	1	0	0	1
s_2	⑤	2	1	0	1	0	1
s_3	2	5	4	0	0	1	1
	-1	-1	-1	0	0	0	0

See page 276 for more about the initial simplex tableau

The first pivot (on the 5 in column 1 and row 2) yields

	x_1	x_2	x_3	s_1	s_2	s_3	
s_1	0	$-3/5$	⑯/5	1	$-4/5$	0	1/5
x_1	1	2/5	1/5	0	1/5	0	1/5
s_3	0	21/5	18/5	0	$-2/5$	1	3/5
	0	$-3/5$	$-4/5$	0	1/5	0	1/5

See page 280 for more about the pivot operation

The next pivot (on the 16/5 in column 3 and row 1) yields

	x_1	x_2	x_3	s_1	s_2	s_3	
x_3	0	$-3/16$	1	5/16	$-1/4$	0	1/16
x_1	1	7/16	0	$-1/16$	1/4	0	3/16
s_3	0	㊴/8	0	$-9/8$	1/2	1	3/8
	0	$-3/4$	0	1/4	0	0	1/4

The next pivot (on the 39/8 in column 2 and row 3) gives the final tableau:

	x_1	x_2	x_3	s_1	s_2	s_3	
x_3	0	0	1	7/26	−3/13	1/26	1/13
x_1	1	0	0	1/26	8/39	−7/78	2/13
x_2	0	1	0	−3/13	4/39	8/39	1/13
	0	0	0	1/13	1/13	2/13	4/13

$$x_3 = \tfrac{1}{13}$$
$$x_1 = \tfrac{2}{13}$$
$$x_2 = \tfrac{1}{13}$$

$$y_1 = \tfrac{1}{13}, \quad y_2 = \tfrac{1}{13}, \quad y_3 = \tfrac{2}{13}$$

$$V = \tfrac{4}{13}$$

Step 3. $V = \tfrac{4}{13}$ with $x_1 = \tfrac{2}{13},\ x_2 = \tfrac{1}{13},\ x_3 = \tfrac{1}{13}$
and $y_1 = \tfrac{1}{13},\ y_2 = \tfrac{1}{13},\ y_3 = \tfrac{2}{13}.$

Step 4. The value of the game is:

$$v = \frac{1}{4/13} - 3 = \frac{13}{4} - 3 = \frac{1}{4} \qquad\qquad v = \frac{1}{V} - k$$

The optimal strategy is

$$r_1 = \frac{1/13}{4/13} = \frac{1}{4}, \quad r_2 = \frac{1/13}{4/13} = \frac{1}{4}, \quad \text{and} \quad r_3 = \frac{2/13}{4/13} = \frac{1}{2} \qquad r_i = \frac{y_i}{V}$$

$$c_1 = \frac{2/13}{4/13} = \frac{1}{2}, \quad c_2 = \frac{1/13}{4/13} = \frac{1}{4}, \quad \text{and} \quad c_3 = \frac{1/13}{4/13} = \frac{1}{4} \qquad c_i = \frac{x_i}{V}$$

The row player should play the rows with probabilities $\tfrac{1}{4}, \tfrac{1}{4}, \tfrac{1}{2}$ and the column player should play the columns with probabilities $\tfrac{1}{2}, \tfrac{1}{4}, \tfrac{1}{4}$, for a game value of $\tfrac{1}{4}$ (favorable to the row player).

Practice Problem

Use the simplex method to find the optimal strategy for the game
$\begin{pmatrix} 3 & 2 & 1 \\ 4 & 5 & 6 \end{pmatrix}.$

➤ **Solution on page 554**

Graphing Calculator Exploration

The program* GameLP constructs and solves the simplex tableau for a game with all entries at least 1. To see that the particular constant k added to G in Step 1 does not change the solution of the game, we use GameLP to solve $G + 4 = \begin{pmatrix} 5 & 2 & 5 \\ 6 & 3 & 2 \\ 3 & 6 & 5 \end{pmatrix}$ for the game solved in Example 1 as follows.

a. Enter the $G + k$ game as matrix [A] and then run the GameLP program:

```
MATRIX[A] 3 ×3
[5      2      5      ]
[6      3      2      ]
[3      6      5      ]
```
and then
```
prgmGAMELP
```

b. Press ENTER twice and select YES or NO depending on whether or not you wish to see all the intermediate tableaux calculations. Choosing NO, we arrive at the final tableau, and then ENTER for the final answer.

Notice that in spite of the different value for k, the optimal strategy is still the same. Why is the value of the game now $4\frac{1}{4}$ instead of the $\frac{1}{4}$ found in Example 1?

Every Game Has a Solution

For any payoff matrix we may add a constant to make every entry at least 1. The linear programming problem (in Step 2 of the box on page 548) has a bounded region with the origin as a feasible point, so the

---

* See the Preface for information on how to obtain this and other graphing calculator programs.

simplex method *will* find an optimal strategy for the game. We therefore have the following important result.

Fundamental Theorem of Game Theory

Every two-person zero-sum game has an optimal strategy.

EXAMPLE **2**

REAL ESTATE SPECULATION

An investor's option to purchase 5000 acres of undeveloped land expires next Friday. The investor may have the land zoned for agricultural, residential, commercial, or industrial use. Rumors indicate that either or both an interstate highway and a nuclear power plant may be built near this property within the next few years. The investor estimates the returns (in percentages gained) for the various zoning possibilities as shown in the following table. Should the investor exercise the option to buy? If so, how would he prefer the land to be zoned?

		Nearby Construction			
		None	Highway	Power Plant	Both
Zoning	Agricultural	4%	8%	−8%	−4%
	Residential	−5%	12%	−6%	−9%
	Commercial	−3%	6%	−4%	−7%
	Industrial	−2%	5%	6%	8%

Solution

By imagining a fictitious opponent, the investor can find the best that can be achieved no matter what happens. Adding 10 to every entry in the table above, the initial simplex tableau for the corresponding game is:

	x_1	x_2	x_3	x_4	s_1	s_2	s_3	s_4	
s_1	14	18	2	6	1	0	0	0	1
s_2	5	22	4	1	0	1	0	0	1
s_3	7	16	6	3	0	0	1	0	1
s_4	8	15	16	18	0	0	0	1	1
	−1	−1	−1	−1	0	0	0	0	0

Pivoting at column 1, row 1 and then at column 3, row 4, we reach the final tableau:

	x_1	x_2	x_3	x_4	s_1	s_2	s_3	s_4	
x_1	1	$129/104$	0	$15/52$	$1/13$	0	0	$-1/104$	$7/104$
s_2	0	$1511/104$	0	$-227/52$	$-3/13$	1	0	$-23/104$	$57/104$
s_3	0	$563/104$	0	$-225/52$	$-4/13$	0	1	$-35/104$	$37/104$
x_3	0	$33/104$	1	$51/52$	$-1/26$	0	0	$7/104$	$3/104$
	0	$29/52$	0	$7/26$	$1/26$	0	0	$3/52$	$5/52$

The maximum value is

$$V = \frac{5}{25} \quad \text{with} \quad \begin{cases} x_1 = \frac{7}{104} \\ x_2 = 0 \\ x_3 = \frac{3}{104} \\ x_4 = 0 \end{cases} \quad \begin{array}{l} y_1 = \frac{1}{26} \\ y_2 = 0 \\ y_1 = 0 \\ y_4 = \frac{3}{52} \end{array} \quad \begin{array}{l} x_2 \text{ and } x_4 \text{ are} \\ \text{not in the basis} \end{array}$$

The value of the game is $v = \frac{1}{5/52} - 10 = \frac{2}{5}$. The optimal strategy for the investor is

$$r_1 = \tfrac{2}{5}, \quad r_2 = 0, \quad r_3 = 0, \quad r_4 = \tfrac{3}{5} \qquad \begin{array}{l} \text{Dividing each } y\text{-value by } 5/52 \\ \text{(or multiplying each by } 52/5) \end{array}$$

Since the value of this game is positive, the investor should exercise the option and buy the land. Of the 5000 acres, 2000 $\left(\text{that is, } \tfrac{2}{5}\right)$ should be zoned agricultural and the other 3000 should be zoned industrial.

8.3 Section Summary

Any game G has an optimal strategy that can be calculated using the simplex method. If any of the entries of G are less than 1, we add a positive number k to every entry to form a new game $G + k$ that has the same optimal strategy but with value increased by the amount k. The simplex method solution of the tableau finds the maximum value V of the objective function, the values of the maximum problem variables $x_1, x_2, \ldots, x_n$, and the values of the dual minimum problem variables $y_1, y_2, \ldots, y_m$. The value of the game G is $v = \dfrac{1}{V} - k$ and the optimal strategy is

	X	S	
			1
S	$G + k$	I	$\vdots$
			1
	$-1 \ldots -1$	$0 \ldots 0$	0

$$r_1 = \frac{y_1}{V}, \ldots, r_m = \frac{y_m}{V} \quad \text{and} \quad c_1 = \frac{x_1}{V}, \ldots, c_n = \frac{x_n}{V}$$

➤ **Solution to Practice Problem**

Since every entry of this 2×3 game is at least one, we do not need to add anything to the matrix. The initial simplex tableau is

	x_1	x_2	x_3	s_1	s_2	
s_1	3	2	1	1	0	1
s_2	④	5	6	0	1	1
	-1	-1	-1	0	0	0

Pivoting on the 4 in column 1, row 2, gives the final tableau:

	x_1	x_2	x_3	s_1	s_2	
s_1	0	$-7/4$	$-7/2$	1	$-3/4$	$1/4$
x_1	1	$5/4$	$3/2$	0	$1/4$	$1/4$
	0	$1/4$	$1/2$	0	$1/4$	$1/4$

The maximum value of the objective function is

$$V = \frac{1}{4} \quad \text{with} \quad \begin{cases} x_1 = \frac{1}{4} \\ x_2 = 0 \\ x_3 = 0 \end{cases} \quad \text{and} \quad \begin{cases} y_1 = 0 \\ y_2 = \frac{1}{4} \end{cases} \quad \begin{array}{l} x_2 \text{ and } x_3 \text{ are} \\ \text{not in the basis} \end{array}$$

The value of the game is $v = \frac{1}{1/4} - 0 = 4$ and the optimal strategy is

$$\begin{cases} r_1 = \frac{0}{1/4} = 0 \\ r_2 = \frac{1/4}{1/4} = 1 \end{cases} \quad \text{and} \quad \begin{cases} c_1 = \frac{1/4}{1/4} = 1 \\ c_2 = \frac{0}{1/4} = 0 \\ c_3 = \frac{0}{1/4} = 0 \end{cases}$$

Notice that this optimal strategy is the pure strategy that selects the saddle point in row 2 and column 1:

$$\begin{pmatrix} 3 & 2 & ① \\ ④ & ⑤ & ⑥ \end{pmatrix}$$

8.3 Exercises

Use the simplex method to find the solution of each game.

 You may use the GameLP or the PIVOT program if permitted by your instructor.

1. $\begin{pmatrix} 3 & 1 \\ 2 & 4 \end{pmatrix}$

2. $\begin{pmatrix} 2 & -1 \\ -4 & 2 \end{pmatrix}$

3. $\begin{pmatrix} -3 & 6 \\ 2 & -4 \end{pmatrix}$

4. $\begin{pmatrix} 6 & -3 \\ 2 & 3 \end{pmatrix}$

5. $\begin{pmatrix} 1 & -4 & -3 \\ -1 & 5 & 3 \end{pmatrix}$

6. $\begin{pmatrix} 1 & 3 & 2 \\ 4 & -1 & 0 \end{pmatrix}$

7. $\begin{pmatrix} -1 & 1 & 2 \\ 0 & 1 & -1 \\ 2 & 0 & 3 \end{pmatrix}$

8. $\begin{pmatrix} 3 & -1 & 1 \\ 0 & 1 & 0 \\ 2 & -1 & 4 \end{pmatrix}$

9. $\begin{pmatrix} -3 & 3 & 1 & 3 \\ 2 & 1 & 2 & -2 \\ -1 & 2 & 2 & 2 \end{pmatrix}$

10. $\begin{pmatrix} 2 & -3 & 4 & -1 \\ -3 & 3 & 2 & -2 \\ 1 & -2 & 2 & 1 \end{pmatrix}$

Determine an optimal strategy for each situation by representing it as a game and using the simplex method. State your final answer in the terms of the original question.

11. **PERSONAL FINANCE: Portfolio Management** An investment counselor recommends three mutual funds (in precious metals, municipal bonds, and technology stocks) and estimates their potential returns under bull, stagnant, and bear market conditions, as shown in the table. How much of a $50,000 retirement portfolio should be invested in each fund and what is the expected minimum return?

Market Conditions

		Bull	Stagnant	Bear
Investment Returns	Precious Metals	1%	1%	4%
	Municipal Bonds	2%	4%	2%
	Technology Stocks	6%	2%	3%

12. **GENERAL: Modified Scissors-Paper-Stone** Eduardo likes to play "scissors, paper, stone" (see Exercise 4 on page 530), but he thinks "scissors" are "so cool" that they should win more. Suppose that each "win" is worth $1 but a win using scissors is worth $2. Is this game fair? What is the probability of Eduardo playing "scissors" on any particular play of this game?

13. **SOCIAL SCIENCE: Political Campaigns** Midway through the fall campaign, the mayor's election manager conducted focus groups to estimate the challenger's perceived lead on a variety of paired issues. The results

in percentage lead are shown in the table. On what issues may the challenger be expected to concentrate during the rest of the campaign?

		Challenger			
		Lower Taxes	Downtown Mall	More Police	Better Schools
Mayor	Recycling	−2	3	2	−1
	New Stadium	−1	−2	0	0
	Mass Transit	−1	1	−2	0

14. **GENERAL: Card Game** The dealer and the gambler both have two red cards and one black card. Each chooses and plays a card, with the gambler winning the round if the cards played have the same color and the dealer winning if they are different. After all three rounds are played, the player who won the most rounds earns from the other in dollars the *difference* between the numbers of rounds they each won. Is this a fair game? How frequently should the gambler play the black card first?

15. **GENERAL: Guess which Coin** Adana has a dime, a quarter, and a half-dollar in her pocket. She picks one coin and hides it in her fist. If Daryl can guess which coin it is, Adana gives it to him, but if Daryl is wrong, he pays Adana the value of his guess. Is this game fair? How often should Adana hide the quarter in her hand? [*Hint:* After setting up the payoff matrix G, solve $G + 100$ by the simplex method.]

16. **GENERAL: "2 Finger" Morra** Two players show each other either 1 or 2 fingers and say a number at the same time. If both say the correct sum or if both are wrong, no money is exchanged. If only one is correct, that player wins that amount. Is this a fair game? How should you play it? [*Hint:* One possible play is "show 1 finger and say 2."]

Explorations and Excursions

The following problems extend and augment the material presented in the text.

"Optimal" May Not Be "Unique"

The optimal strategy for a game is not necessarily unique although the value is.

17. Let $G = \begin{pmatrix} 4 & 1 & 3 \\ 2 & 2 & 4 \\ 2 & 1 & 5 \end{pmatrix}$.

    a. Use the simplex method to show that the value is 2 with an optimal strategy

$$r = (0 \quad 1 \quad 0) \quad \text{and} \quad c = \begin{pmatrix} \frac{1}{3} \\ \frac{2}{3} \\ 0 \end{pmatrix}.$$

    b. Show that the 2 in row 2 and column 2 is the only saddle point of G and so another optimal strategy is

$$r = (0 \quad 1 \quad 0) \quad \text{and} \quad c = \begin{pmatrix} 0 \\ 1 \\ 0 \end{pmatrix}.$$

18. For the game $G = \begin{pmatrix} 2 & 2 & 2 & 5 & 4 \\ 0 & 2 & 1 & 3 & 3 \\ 1 & 2 & -1 & 4 & 2 \\ 2 & 4 & 2 & 3 & 5 \end{pmatrix}$

from Example 2 in Section 1 (see pages 527–528), show that

$$r = \begin{pmatrix} \frac{1}{2} & 0 & 0 & \frac{1}{2} \end{pmatrix} \quad \text{and} \quad c = \begin{pmatrix} \frac{1}{3} \\ 0 \\ \frac{2}{3} \\ 0 \\ 0 \end{pmatrix}$$

is an optimal strategy.

Chapter Summary with Hints and Suggestions

Reading the text and doing the exercises in this chapter have helped you to master the following skills, which are listed by section (in case you need to review them) and are keyed to particular Review Exercises. Answers for all Review Exercises are given at the back of the book, and full solutions can be found in the Student Solutions Manual.

8.1 Two-Person Games and Saddle Points

- Construct a payoff matrix for a game by identifying the players, their possible choices, and the resulting gains and losses.
  (Review Exercises 1–2)

- Find the saddle point and optimal strategy of a game. *(Review Exercises 3–6)*

  ○ smallest in row   □ largest in column

- Determine an optimal strategy for a conflict situation by representing it as a game and then solving the game. *(Review Exercises 7–8)*

8.2 Mixed Strategies

- Find the expected value of a mixed strategy for a game. *(Review Exercises 9–10)*
$$E = r \cdot G \cdot c$$

- Find an optimal strategy for a 2×2 game by identifying the saddle points or using the formulas for the optimal mixed strategy. *(Review Exercises 11–12)*

$$G = \begin{pmatrix} a & b \\ c & d \end{pmatrix}$$

$$r_1 = \frac{d - c}{(a + d) - (b + c)}, \quad r_2 = 1 - r_1$$

$$c_1 = \frac{d - b}{(a + d) - (b + c)}, \quad c_2 = 1 - c_2$$

- Determine an optimal strategy for a conflict situation by representing it as a game and then solving the game. *(Review Exercises 13–14)*

8.3 Games and Linear Programming

- Find an optimal strategy for a game using the simplex method. *(Review Exercises 15–18)*

$$
\begin{array}{c|cc|c}
 & X & S & \\
\hline
S & G+k & I & 1 \\
\hline
 & -1 & 0 & 0
\end{array}
\Rightarrow
\left.
\begin{array}{c|cc|c}
 & X & S & \\
\hline
 & \vdots & \vdots & \vdots \\
\hline
 & \cdots & Y^* & V
\end{array}
\right\}
\begin{array}{l}
c_i = \dfrac{x_i}{V} \\[6pt]
r_i = \dfrac{y_i}{V}
\end{array}
$$

- Determine an optimal strategy for a conflict situation by representing it as a game and then using the simplex method. *(Review Exercises 19–20)*

Hints and Suggestions

- In a *two-person zero-sum* game, whatever one player gains the other loses. In the payoff matrix the entries represent gains for the row player (and so losses for the column player). Therefore, negative numbers represent losses for the row player (and gains for the column player).

- Always first check a game for saddle points. A saddle point is a row minimum and a column maximum, and represents a *pure optimal solution* for a game. For a game with no saddle points, use the formulas on page 537 for 2×2 games and use the linear programming method on page 548 for larger games.

- For a game you do not need to have an actual opponent actively trying to minimize your gain. You may have a game against *uncertainty* (such as weather or economic conditions) as long as you can estimate probabilities for the various random occurrences.

- **Practice for Test:** Review Exercises 2, 4, 8, 11, 13, 15, 20.

Review Exercises for Chapter 8

Practice test exercise numbers are in green.

8.1 Two-Person Games and Saddle Points

For each situation, identify the two players, their possible choices, and construct a payoff matrix for their conflict.

1. **BUSINESS: Retail Competition** Ever since the Sports-R-Us discount store opened, Bill's Pro Shop has been struggling to retain customers. Both stores may or may not hold Memorial Day sales this year. If Bill's Pro Shop does and the discount store doesn't, he'll gain back 5% of the local market, but if they also have a sale, he'll lose 10% of the market. If he doesn't have a sale and they don't either, he'll still lose 1% of the market, but if they have a sale, he'll lose 20%.

2. **GENERAL: "Difference" Morra** Two players show each other 1, 2 or 3 fingers. If the dif-

ference of the numbers is 0, no money is exchanged; if the difference is 1, the player of the smaller number gets $1; while if the difference is 2, the player of the larger gets $2.

For each game, identify the saddle point and determine the corresponding optimal strategy.

3. $\begin{pmatrix} -2 & 1 & 5 \\ 7 & 2 & 4 \end{pmatrix}$

4. $\begin{pmatrix} 4 & 7 & 3 \\ 5 & 0 & 2 \\ -2 & 8 & -1 \end{pmatrix}$

5. $\begin{pmatrix} 1 & 2 & 9 & -5 \\ 4 & 3 & 8 & 12 \\ -1 & -6 & 14 & -9 \end{pmatrix}$

6. $\begin{pmatrix} -1 & 7 & 2 & 0 \\ 15 & 5 & 4 & 10 \\ -10 & -4 & -3 & 13 \end{pmatrix}$

Determine the optimal strategy for each situation by representing it as a game and finding the saddle point. State your final answer in the terms of the original question.

7. **SOCIAL SCIENCE: Political Campaigns**
Mr. Brown is challenging Mr. Clark for a seat on the school board. As Mr. Brown sees it, the central issues are the budget and achieving higher standards, but Mr. Clark prefers to talk about the success of the football team during his twenty years on the board. Every time they debate and Mr. Brown can keep the discussion on the budget or on standards, he leads Mr. Clark in voter approval polls by 15%. If Mr. Clark continues to hold forth on budgets or standards but Mr. Brown starts expressing his views on football, Mr. Brown still leads by 10%. But if Mr. Clark talks about football, he leads by 20% no matter what Mr. Brown tries to discuss. Should Mr. Brown agree to any more debates before the election?

8. **BUSINESS: Store Location** Two national video-game chains plan to enter a four-city market by each granting a franchise for a new store in one of the cities of Fulton (population 50 thousand), Gardner (population 70 thousand), Hillcrest (population 40 thousand), and Irving (population 60 thousand). If both stores are built in the same city, they will split the entire market equally. But if the stores are built in different cities, each will get all of the market in its city while cities without a store will send all of their business to the store in the closest city. The distances between the cities are shown in the following map. Where should the stores be built?

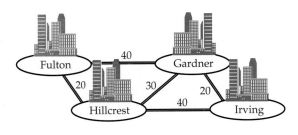

8.2 Mixed Strategies

For the game $G = \begin{pmatrix} 3 & 1 \\ -6 & 4 \end{pmatrix}$, find the expected value of each mixed strategy.

9. $r = \begin{pmatrix} \frac{1}{3} & \frac{2}{3} \end{pmatrix}$ and $c = \begin{pmatrix} \frac{1}{2} \\ \frac{1}{2} \end{pmatrix}$

10. $r = \begin{pmatrix} \frac{5}{6} & \frac{1}{6} \end{pmatrix}$ and $c = \begin{pmatrix} \frac{1}{4} \\ \frac{3}{4} \end{pmatrix}$

Find the optimal strategy for each 2×2 game. Be sure to check for saddle points before attempting to use the mixed strategy formulas.

11. $\begin{pmatrix} 8 & -2 \\ -4 & 2 \end{pmatrix}$ 12. $\begin{pmatrix} 5 & -5 \\ -7 & 13 \end{pmatrix}$

Determine an optimal strategy for each situation after representing it as a game. State your final answer in the terms of the original question.

13. **BUSINESS: Ranch Management** A rancher tends both cattle and sheep on his 30 square mile ranch but worries that bad weather could ruin his profits. He estimates that if the summer is too dry, the profit per head will be decreased by $8 for cattle and by $13 for sheep, and if the weather is too wet, the profit per head will be decreased by $17 for cattle and by $7 for sheep. How should he manage his herd sizes and what is the corresponding worst-case loss per head?

14. **PERSONAL FINANCE: Speculative Investments** A stock broker estimates the potential returns on two energy stocks, Bright-Tech Solar Power and Pacific West Electric, in view of possible federal price controls on energy, as shown in the table. How much of a $36,000 speculation account should be invested in each and what is the worst-case expected return?

		Federal Price Controls?	
		Yes	No
Investment Returns	Solar Power	6%	1%
	Electric	−2%	5%

8.3 Games and Linear Programming

Use the simplex method to find the solution of each game. You may use the GameLP or PIVOT program if permitted by your instructor.

15. $\begin{pmatrix} 4 & -1 \\ -2 & 3 \end{pmatrix}$ **16.** $\begin{pmatrix} 2 & -3 & -8 \\ -1 & 2 & 4 \end{pmatrix}$

17. $\begin{pmatrix} 4 & -2 & 4 \\ -1 & 4 & 1 \\ 7 & -2 & 1 \end{pmatrix}$ **18.** $\begin{pmatrix} 3 & -1 & 2 & 1 \\ 2 & 0 & -1 & 0 \\ 0 & 1 & 0 & -1 \end{pmatrix}$

Determine an optimal strategy for each situation by representing it as a game and using the simplex method. State your final answer in the terms of the original question.

19. GENERAL: Math Quizzes In one section of this course the instructor gives a daily quiz on just one of recent numerical calculations, concepts, or definitions, and allows her students the choice of bringing either their calculator or textbook (but not both) to use during the quiz. Jack estimates that if he brings his calculator, he can get a 100 on a numerical calculation, a 70 on concepts, or a 60 on definitions, but that with his textbook he could get a 60 on a numerical calculation, an 80 on concepts, or a 100 on definitions. How often can Jack expect a quiz on concepts, and what grade can he expect to earn?

20. GENERAL: Card Game The dealer has three cards, one king and two queens, and can lay face down any two of them. The gambler attempts to guess one at a time what each card is before it is turned over. If both guesses are correct, the gambler wins $2. But if both are wrong, the gambler pays the dealer $5. And if only one guess is wrong, the gambler pays the dealer $1. How should the gambler guess the cards?

Appendix

NORMAL PROBABILITIES USING TABLES

This appendix shows how to find normal probabilities from tables. It replaces pages 463–469 (through Practice Problem 3) of Chapter 6 (Statistics) for readers who do not have graphing calculators that calculate normal probabilities.

z-Scores

The normal distribution with *mean* 0 and *standard deviation* 1 has special significance and is called the *standard* normal distribution (the word *standard* indicating mean 0 and standard deviation 1). The letter z is traditionally used for the variable in standard normal calculations. Since the mean is zero, the highest point on the curve is at $z = 0$, and the curve is symmetric about this value.

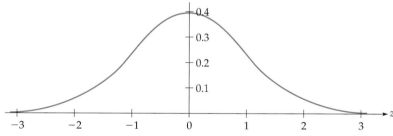

The standard normal distribution (mean 0 and standard deviation 1)

From the diagram on page 462 with $\mu = 0$ and $\sigma = 1$ we have the following facts about the standard normal distribution: More than

68% of its probability is between −1 and +1

95% of its probability is between −2 and +2

99% of its probability is between −3 and +3

A-1

To change any x-value for a normal distribution with mean μ and standard deviation σ into the corresponding z-value for the *standard normal distribution*, we subtract the mean and then divide by the standard deviation. The resulting number is called the *z-score* and this process is called *standardizing*.

---

z-Score

$$z = \frac{x - \mu}{\sigma}$$

Subtracting the mean μ and dividing by the standard deviation σ

The z-score gives the number of standard deviations the value is from the mean.

EXAMPLE **1** **FINDING z-SCORES FOR HEIGHTS OF MEN**

The heights of American men are approximately normally distributed with mean $\mu = 68.1$ inches and standard deviation $\sigma = 2.7$ inches.* Find the z-scores corresponding to the following heights (in inches) and interpret the results:

a. 73.5   **b.** 62.7

Solution

a. $z = \dfrac{73.5 - 68.1}{2.7} = \dfrac{5.4}{2.7} = 2$ Using $z = (x - \mu)/\sigma$ with $\mu = 68.1$ and $\sigma = 2.7$

Therefore, a height of 73.5 inches is two standard deviations *above* the mean.

b. $z = \dfrac{62.7 - 68.1}{2.7} = \dfrac{-5.4}{2.7} = -2$

Therefore, a height of 62.7 inches is two standard deviations *below* the mean.

Since the standard normal distribution has more than 95% of its probability between −2 and +2, these results mean that more than 95% of men have heights between 62.7 inches and 73.5 inches (5 feet 2.7 inches and 6 feet 1.5 inches).

---

* These and other data in the appendix are from *Handbook of Human Factors and Ergonomics* by Gavriel Salvendy (Wiley).

Practice Problem 1

The weights of American women are approximately normally distributed with mean 134.7 pounds and standard deviation 30.4 pounds. Find the z-scores for the following weights, and interpret your answers:

a. 165.1 lbs.   **b.** 104.3 lbs.

➤ **Solutions on page A-8**

On page A-10 of this appendix is a brief table for the normal distribution, giving the probability that a standard normal random variable takes a value between 0 and any given positive number. For example, to find $P(0 \le Z \le 1.24)$ (in words, the probability that a standard normal random variable has a value between 0 and 1.24), we locate in the table the row headed **1.2** and the column headed **0.04** (the second decimal place), and $P(0 \le Z \le 1.24)$ is the number where this row and column intersect, **0.3925**.

In the table on page A-10, the row headed **1.2** and the column headed **0.04** intersect at the table value **0.3925**

Therefore,

$$P(0 \le Z \le 1.24) \approx 0.3925$$

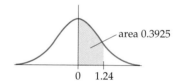

area 0.3925

Probabilities for other intervals can be found by adding and subtracting such areas and using the symmetry of the normal curve, as the following examples illustrate.

EXAMPLE 2

FINDING A PROBABILITY FOR A NORMAL RANDOM VARIABLE

Using the mean and standard deviation for the heights of men given in Example 1, find the proportion of men who are between 5 feet 9 inches and 6 feet tall.

Solution

Converting to inches, we want the probability of a man's height being between 69 inches and 72 inches. We must convert these numbers to the corresponding z-scores for a *standard* normal random variable by subtracting the mean and dividing by the standard deviation:

$x = 69$ corresponds to $\quad z = \dfrac{69 - 68.1}{2.7} \approx 0.33 \quad$ Using $z = \dfrac{x - \mu}{\sigma}$

$x = 72$ corresponds to $\quad z = \dfrac{72 - 68.1}{2.7} \approx 1.44 \quad$ with $\mu = 68.1$ and $\sigma = 2.7$

Using these values, we then want $P(0.33 \le Z \le 1.44)$, which is equivalent to the shaded area in the first of the following graphs, which is equal to the difference between the next two areas on the right. These two areas (or probabilities) are found from the table on page A-10, with the calculations shown below.

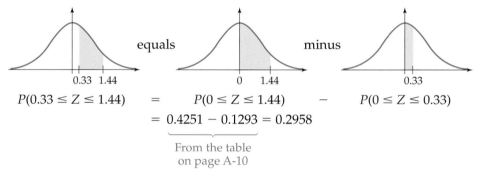

equals   minus

$$P(0.33 \le Z \le 1.44) \quad = \quad P(0 \le Z \le 1.44) \quad - \quad P(0 \le Z \le 0.33)$$

$$= 0.4251 - 0.1293 = 0.2958$$

From the table
on page A-10

Therefore, about 30% of American men are between 5 feet 9 inches and 6 feet tall.

Practice Problem 2

Using the mean and standard deviation for weights of women given in Practice Problem 1, find the proportion of women who weigh between 130 and 150 pounds. ➤ **Solution on page A-8**

The Normal and Binomial Distributions

As we saw on page 412, the binomial distribution has a kind of "bell" shape, with a peak at the expected value $\mu = np$ and falling on both sides to very small probabilities further away from the mean.

$n = 20, p = 0.4$ $\qquad\qquad$ $n = 15, p = 0.5$ $\qquad\qquad$ $n = 25, p = 0.7$

Several binomial distributions with different values for n and p

Since both the binomial and the normal distributions have similar "mound" shapes, could they be related? In the eighteenth century, Abraham de Moivre (1667–1754) and Pierre-Simon Laplace (1749–1827) discovered and proved that for any choice of p between 0 and 1, the binomial distribution *approaches* the normal distribution as n becomes large. This fundamental fact is known as the *de Moivre–Laplace theorem*.

   We may use the de Moivre–Laplace theorem to approximate binomial distributions by the normal distribution.

Normal Approximation to the Binomial

Let X be a binomial random variable with parameters n and p. If $np > 5$ and $n(1 - p) > 5$, then the distribution of X is approximately normal with mean $\mu = np$ and standard deviation $\sigma = \sqrt{np(1 - p)}$.

Approximating the *discrete* binomial distribution by the *continuous* normal distribution requires an adjustment: To have the same width for a "slice" of the normal distribution as for the binomial, we adopt the convention that the binomial probability $P(X = x)$ corresponds to the area under the normal distribution curve from $x - \frac{1}{2}$ to $x + \frac{1}{2}$. This is called the *continuous correction*.

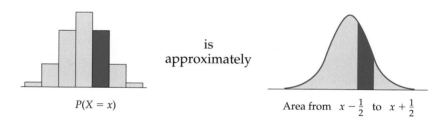

| $P(X = x)$ | is approximately | Area from $x - \frac{1}{2}$ to $x + \frac{1}{2}$ |

EXAMPLE 3

NORMAL APPROXIMATION OF A BINOMIAL PROBABILITY

Estimate $P(X = 12)$ for a binomial random variable X with $n = 25$ and $p = 0.6$ using the corresponding normal distribution.

Solution

Since

$$np = 25 \cdot 0.6 = 15 \quad \text{and} \quad n(1 - p) = 25 \cdot 0.4 = 10$$

are both greater than 5, we may use the normal distribution with

$$\mu = np = 25 \cdot 0.6 = 15$$

and

$$\sigma = \sqrt{np(1-p)} = \sqrt{25 \cdot 0.6 \cdot 0.4} = \sqrt{6} \approx 2.45$$

By the continuous correction we interpret the event $X = 12$ as $11.5 \leq X < 12.5$ to include numbers that would round to 12. Converting these x-values into z-scores, we find that

$x = 11.5$ corresponds to $z = \dfrac{11.5 - 15}{2.45} \approx -1.43$ Using $z = \dfrac{x - \mu}{\sigma}$

$x = 12.5$ corresponds to $z = \dfrac{12.5 - 15}{2.45} \approx -1.02$ with $\mu = 15$ and $\sigma = 2.45$

    The probability $P(-1.43 \leq Z \leq -1.02)$ is represented below by the shaded area in the first graph, which, by symmetry, is equivalent to the second graph, which in turn is equivalent to the difference between the third and fourth graphs. The calculation with the probabilities found from the normal table is shown below.

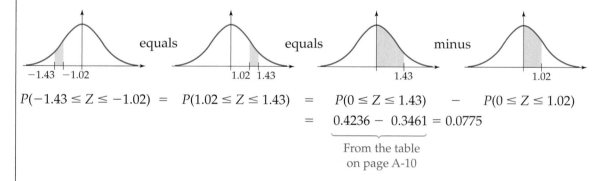

equals   equals   minus

-1.43 -1.02 1.02 1.43 1.43 1.02

$P(-1.43 \leq Z \leq -1.02) = P(1.02 \leq Z \leq 1.43) = P(0 \leq Z \leq 1.43) - P(0 \leq Z \leq 1.02)$

$$= 0.4236 - 0.3461 = 0.0775$$

From the table
on page A-10

The required probability is approximately 0.078.

The normal approximation of 0.0775 is indeed very close to the actual value calculated from the binomial probability formula (see page 412): $_{25}C_{12}(0.6)^{12}(1 - 0.6)^{13} \approx 0.076$.

EXAMPLE 4

MANAGEMENT MBAs

At a major Los Angeles accounting firm, 73% of the managers have MBA degrees. In a random sample of 40 of these managers, what is the probability that between 27 and 32 will have MBAs?

Solution

Because each manager either has or does not have an MBA, presumably independently of each other, the question asks for the probability that a binomial random variable X with $n = 40$ and $p = 0.73$ satisfies $27 \le X \le 32$. Since both $np = 40 \cdot 0.73 = 29.2$ and $n(1 - p) = 40 \cdot 0.27 = 10.8$ are greater than 5, this probability can be approximated as the area under the normal distribution curve with mean $\mu = np = 40 \cdot 0.73 = 29.2$ and standard deviation $\sigma = \sqrt{np(1 - p)} = \sqrt{40 \cdot 0.73 \cdot 0.27} \approx 2.81$ from $26\frac{1}{2}$ to $32\frac{1}{2}$ (again using the continuous correction to include values that would round to between 27 and 32). We convert the x-values to z-scores:

$$x = 26.5 \text{ corresponds to } z = \frac{26.5 - 29.2}{2.81} \approx -0.96$$

$$x = 32.5 \text{ corresponds to } z = \frac{32.5 - 29.2}{2.81} \approx 1.17$$

Using $z = \dfrac{x - \mu}{\sigma}$ with $\mu = 29.2$ and $\sigma = 2.81$

The probability $P(-0.96 \le Z \le 1.17)$ is represented by the shaded area in the first graph, which is equivalent to the sum of the next two areas, with the calculation shown below.

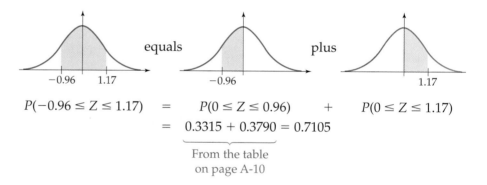

$$P(-0.96 \le Z \le 1.17) \quad = \qquad P(0 \le Z \le 0.96) \quad + \quad P(0 \le Z \le 1.17)$$
$$= \quad 0.3315 + 0.3790 = 0.7105$$

From the table on page A-10

The probability that the number of MBAs will be between 27 and 32 is 0.711, or about 71%.

The exact answer to the preceding problem, from summing the binomial distribution from 27 to 32, is 71.5%, so the normal approximation is very accurate.

Practice Problem **3**

A brand of imported VCR is known to have defective tape rewind mechanisms in 8% of the units imported last April. If Jerry's Discount Electronics received a shipment of 80 of these VCR's, what is the probability that 10 or more are defective? ➤ Solution on page A-9

After completing Practice Problem 3, you should return to page 469 to read the Section Summary and do the Exercises in Section 6.4.

► **Solutions to Practice Problems**

1. Using $\mu = 134.7$ and $\sigma = 30.4$:

   a. $z = \dfrac{165.1 - 134.7}{30.4} = \dfrac{30.4}{30.4} = 1$

   b. $z = \dfrac{104.3 - 134.7}{30.4} = \dfrac{-30.4}{30.4} = -1$

   Interpretation: More than 68% of women have weights between about 104 and 165 pounds.

2. We first change the weights into z-scores using $z = \dfrac{x - \mu}{\sigma}$ with $\mu = 134.7$ and $\sigma = 30.4$:

   $$x = 130 \quad \text{corresponds to} \quad z = \dfrac{130 - 134.7}{30.4} \approx -0.15$$

   $$x = 150 \quad \text{corresponds to} \quad z = \dfrac{150 - 134.7}{30.4} \approx 0.50$$

   Using these values, we then want $P(-0.15 \le Z \le 0.50)$, which is equivalent to the shaded area shown in the graph on the left below, which is equal to the *sum* of the following two shaded areas. The two areas are found from the table on page A-10, with the calculation shown below.

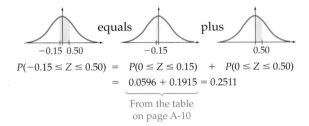

$$
\begin{aligned}
P(-0.15 \le Z \le 0.50) &= P(0 \le Z \le 0.15) \;+\; P(0 \le Z \le 0.50) \\
&= \underbrace{0.0596 + 0.1915}_{\substack{\text{From the table} \\ \text{on page A-10}}} = 0.2511
\end{aligned}
$$

Therefore, about 25% of American women weigh between 130 and 150 pounds.

3. Since $np = 80 \cdot 0.08 = 6.4$ and $n(1 - p) = 80 \cdot 0.92 = 73.6$ are each greater than 5, we may use the normal distribution with $\mu = np = 80 \cdot 0.08 \approx 6.4$ and $\sigma = \sqrt{np(1 - p)} = \sqrt{80 \cdot 0.08 \cdot 0.92} \approx 2.43$ to approximate $P(X \geq 10)$ as the area under the normal curve from 9.5 to 80.5 (corresponding to all 80 being defective, and again including rounding). Converting the x-values into z-scores using $z = \dfrac{x - \mu}{\sigma}$ with $\mu = 6.4$ and $\sigma = 2.43$:

$$x = 9.5 \quad \text{corresponds to} \quad z = \frac{9.5 - 6.4}{2.43} \approx 1.28$$

$$x = 80.5 \quad \text{corresponds to} \quad z = \frac{80.5 - 6.4}{2.43} \approx 30.49$$

The probability $P(1.28 \leq Z \leq 30.49)$ is represented by the shaded area on the left, which is equivalent to the difference between the two areas on the right with the calculation shown below.

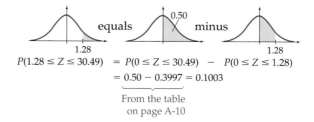

$$P(1.28 \leq Z \leq 30.49) \quad = P(0 \leq Z \leq 30.49) \quad - \quad P(0 \leq Z \leq 1.28)$$
$$= 0.50 - 0.3997 = 0.1003$$

From the table
on page A-10

The probability of at least ten defective VCR's is (about) 0.10, or 10%.

Area under the standard
normal distribution from 0 to z

x	0.00	0.01	0.02	0.03	0.04	0.05	0.06	0.07	0.08	0.09
0.0	0.0000	0.0040	0.0080	0.0120	0.0160	0.0199	0.0239	0.0279	0.0319	0.0359
0.1	0.0398	0.0438	0.0478	0.0517	0.0557	0.0596	0.0636	0.0675	0.0714	0.0754
0.2	0.0793	0.0832	0.0871	0.0910	0.0948	0.0987	0.1026	0.1064	0.1103	0.1141
0.3	0.1179	0.1217	0.1255	0.1293	0.1331	0.1368	0.1406	0.1443	0.1480	0.1517
0.4	0.1554	0.1591	0.1628	0.1664	0.1700	0.1736	0.1772	0.1808	0.1844	0.1879
0.5	0.1915	0.1950	0.1985	0.2019	0.2054	0.2088	0.2123	0.2157	0.2190	0.2224
0.6	0.2258	0.2291	0.2324	0.2357	0.2389	0.2422	0.2454	0.2486	0.2518	0.2549
0.7	0.2580	0.2612	0.2642	0.2673	0.2704	0.2734	0.2764	0.2794	0.2823	0.2852
0.8	0.2881	0.2910	0.2939	0.2967	0.2996	0.3023	0.3051	0.3078	0.3106	0.3133
0.9	0.3159	0.3186	0.3212	0.3238	0.3264	0.3289	0.3315	0.3340	0.3365	0.3389
1.0	0.3413	0.3438	0.3461	0.3485	0.3508	0.3531	0.3554	0.3577	0.3599	0.3621
1.1	0.3643	0.3665	0.3686	0.3708	0.3729	0.3749	0.3770	0.3790	0.3810	0.3820
1.2	0.3849	0.3869	0.3888	0.3907	0.3925	0.3944	0.3962	0.3980	0.3997	0.4015
1.3	0.4032	0.4049	0.4066	0.4082	0.4099	0.4115	0.4131	0.4147	0.4162	0.4177
1.4	0.4192	0.4207	0.4222	0.4236	0.4251	0.4265	0.4279	0.4292	0.4306	0.4319
1.5	0.4332	0.4345	0.4357	0.4370	0.4382	0.4394	0.4406	0.4418	0.4429	0.4441
1.6	0.4452	0.4463	0.4474	0.4484	0.4495	0.4505	0.4515	0.4525	0.4535	0.4545
1.7	0.4554	0.4564	0.4573	0.4582	0.4591	0.4599	0.4608	0.4616	0.4625	0.4633
1.8	0.4641	0.4649	0.4656	0.4664	0.4671	0.4678	0.4686	0.4693	0.4699	0.4706
1.9	0.4713	0.4719	0.4726	0.4732	0.4738	0.4744	0.4750	0.4756	0.4761	0.4767
2.0	0.4772	0.4778	0.4783	0.4788	0.4793	0.4798	0.4803	0.4808	0.4812	0.4817
2.1	0.4821	0.4826	0.4830	0.4834	0.4838	0.4842	0.4846	0.4850	0.4854	0.4857
2.2	0.4861	0.4864	0.4868	0.4871	0.4875	0.4878	0.4881	0.4884	0.4887	0.4890
2.3	0.4893	0.4896	0.4898	0.4901	0.4904	0.4906	0.4909	0.4911	0.4913	0.4916
2.4	0.4918	0.4920	0.4922	0.4925	0.4927	0.4929	0.4931	0.4932	0.4934	0.4936
2.5	0.4938	0.4940	0.4941	0.4943	0.4945	0.4946	0.4948	0.4949	0.4951	0.4952
2.6	0.4953	0.4955	0.4956	0.4957	0.4959	0.4960	0.4961	0.4962	0.4963	0.4964
2.7	0.4965	0.4966	0.4967	0.4968	0.4969	0.4970	0.4971	0.4972	0.4973	0.4974
2.8	0.4974	0.4975	0.4976	0.4977	0.4977	0.4978	0.4979	0.4979	0.4980	0.4981
2.9	0.4981	0.4982	0.4982	0.4983	0.4984	0.4984	0.4985	0.4985	0.4986	0.4986
3.0	0.4987	0.4987	0.4987	0.4988	0.4988	0.4989	0.4989	0.4989	0.4990	0.4990
3.1	0.4990	0.4991	0.4991	0.4991	0.4992	0.4992	0.4992	0.4992	0.4993	0.4993
3.2	0.4993	0.4993	0.4994	0.4994	0.4994	0.4994	0.4994	0.4995	0.4995	0.4995
3.3	0.4995	0.4995	0.4995	0.4996	0.4996	0.4996	0.4996	0.4996	0.4996	0.4997
3.4	0.4997	0.4997	0.4997	0.4997	0.4997	0.4997	0.4997	0.4997	0.4997	0.4998
3.5	0.4998	0.4998	0.4998	0.4998	0.4998	0.4998	0.4998	0.4998	0.4998	0.4998

Answers to Selected Exercises

Exercises 1.1 page 14

1. $\{x \mid 0 \le x < 6\}$ **3.** $\{x \mid x \le 2\}$

5. a. Increase by 15 units   **b.** Decrease by 10 units   **7.** $m = -2$ **9.** $m = \frac{1}{3}$ **11.** $m = 0$

13. Slope is undefined

15. $m = 3$, $(0, -4)$ **17.** $m = -\frac{1}{2}$, $(0, 0)$ **19.** $m = 0$, $(0, 4)$

      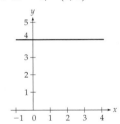

21. Slope and y-intercept do not exist.   **23.** $m = \frac{2}{3}$, $(0, -4)$ **25.** $m = -1$, $(0, 0)$

      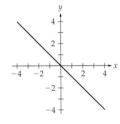

27. $m = 1$, $(0, 0)$ **29.** $m = \frac{1}{3}$, $(0, \frac{2}{3})$ **31.** $m = \frac{2}{3}$, $(0, -1)$

33.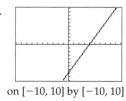

on $[-10, 10]$ by $[-10, 10]$

35.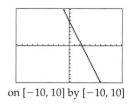

on $[-10, 10]$ by $[-10, 10]$

37.

on $[-160, 160]$ by $[-160, 160]$

39. $y = -2.25x + 3$ **41.** $y = 5x + 3$ **43.** $y = -4$ **45.** $x = 1.5$ **47.** $y = -2x + 13$
49. $y = -1$ **51.** $y = -2x + 1$ **53.** $y = \frac{3}{2}x - 2$
55. $y = -x + 5$, $y = -x - 5$, $y = x + 5$, $y = x - 5$
57. Substituting $(0, b)$ into $y - y_1 = m(x - x_1)$ gives $y - b = m(x - 0)$, or $y = mx + b$.
59. $(-b/m, 0)$, $m \neq 0$ **61. a.**   **b.**

on $[-5, 5]$ by $[-5, 5]$ on $[-5, 5]$ by $[-5, 5]$

63. Low: $[0, 8)$; average: $[8, 20)$; high: $[20, 40)$; critical $[40, \infty)$
65. a. 3 minutes 38.28 seconds   **b.** In about 2033
67. a. $y = 4x + 2$ **b.** \$10 million   **c.** \$22 million   **69. a.** $y = \frac{9}{5}x + 32$ **b.** 68°
71. a. $V = 50{,}000 - 2200t$ **b.** \$39,000   **c.**

on $[0, 20]$ by $[0, 50{,}000]$

73. Smaller populations increase towards the carrying capacity
75. a. 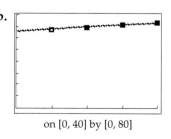   **b.** Men: 28.7 years;   **77. a.**   **b.** About 40%
women: 27.2 years   (from 39.8)
c. Men: 30 years;   **c.** About 60%
women: 28.9 years   (from 60.4)

79. b. $y_1 = 0.203x + 65.92$ **c.** 79.1 years

on $[0, 40]$ by $[0, 80]$

Exercises 1.2 page 28

1. 64 **3.** $\frac{1}{16}$ **5.** 8 **7.** $\frac{8}{5}$ **9.** $\frac{1}{32}$ **11.** $\frac{8}{27}$ **13.** 1 **15.** $\frac{4}{9}$ **17.** 5 **19.** 125 **21.** 8
23. 4 **25.** -32 **27.** $\frac{125}{216}$ **29.** $\frac{9}{25}$ **31.** $\frac{1}{4}$ **33.** $\frac{1}{2}$ **35.** $\frac{1}{8}$ **37.** $\frac{1}{4}$ **39.** $-\frac{1}{2}$ **41.** $\frac{1}{4}$
43. $\frac{4}{5}$ **45.** $\frac{64}{125}$ **47.** -243 **49.** 2.14 **51.** 274.37 **53.** -128 **55.** 6.25 **57.** 0.5
59. 0.4 **61.** 0.977 (rounded) **63.** 2.720 (rounded) **65.** x^{10} **67.** z^{27} **69.** x^8 **71.** w^5
73. y^5/x **75.** $27y^4$ **77.** $u^2v^2w^2$ **79.** 25.6 feet **81.** Costs will be multiplied by about 2.3.
83.

on $[0, 5]$ by $[0, 3]$

Capacity can be multiplied by about 3.2.

85. 125 beats per minute **87.**

on $[0, 200]$ by $[0, 150]$

Heart rate decreases more slowly as body weight increases.

89. About 42.6 thousand work-hours, or 42,600 work-hours, rounded to the nearest hundred hours.
91. a. About 32 times more ground motion **b.** About 8 times more ground motion
93. About 312 mph
95.

on $[0, 100]$ by $[0, 4]$

$x \approx 18.2$. Therefore, the land area must be increased by a factor of more than 18 to double the number of species.

97. b.

on $[-2, 32]$ by $[1{,}000, 3{,}500]$

$y_1 = 3261x^{-0.267}$ **c.** 1147 work-hours

Exercises 1.3 page 45

1. Yes **3.** No **5.** No **7.** No **9.** Domain: $\{x \mid x \le 0 \text{ or } x \ge 1\}$; Range: $\{y \mid y \ge -1\}$
11. a. $f(10) = 3$ **b.** $\{x \mid x \ge 1\}$ **c.** $\{y \mid y \ge 0\}$
13. a. $h(-5) = -1$ **b.** $\{z \mid z \ne -4\}$ **c.** $\{y \mid y \ne 0\}$
15. a. $h(81) = 3$ **b.** $\{x \mid x \ge 0\}$ **c.** $\{y \mid y \ge 0\}$ **17. a.** $f(-8) = 4$ **b.** $\mathbb{R}$ **c.** $\{y \mid y \ge 0\}$
19. a. $f(0) = 2$ **b.** $\{x \mid -2 \le x \le 2\}$ **c.** $\{y \mid 0 \le y \le 2\}$
21. a. $f(-25) = 5$ **b.** $\{x \mid x \le 0\}$ **c.** $\{y \mid y \ge 0\}$

23. **25.** **27.** **29.**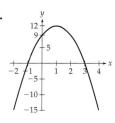

31. a. (20, 100) **b.**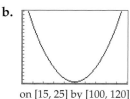

on [15, 25] by [100, 120]

33. a. (−40, −200) **b.**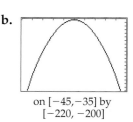

on [−45,−35] by [−220, −200]

35. $x = 7$, $x = -1$ **37.** $x = 3$, $x = -5$ **39.** $x = 4$, $x = 5$ **41.** $x = 0$, $x = 10$

43. $x = 5$, $x = -5$ **45.** $x = -3$ **47.** $x = 1$, $x = 2$ **49.** No solutions   **51.** No solutions

53. $x = -4$, $x = 5$ **55.** $x = 4$, $x = 5$ **57.** $x = -3$ **59.** No (real) solutions

61. $x = 1.14$, $x = -2.64$ **63. a.** The slopes are all 2, but the y-intercepts differ   **b.** $y = 2x - 8$

65. $C(x) = 4x + 20$ **67.** $P(x) = 15x + 500$ **69. a.** 17.7 lbs/in.^{2} **b.** 15,765 lbs/in.^{2}

71. 132 feet   **73. a.** 400 cells   **b.** 5200 cells   **75.** About 208 mph   **77.** 2.92 seconds

79. a. Break even at 40 units and at 200 units.

  b. Profit maximized at 120 units. Max profit is $12,800 per week.

81. a. Break even at 20 units and at 80 units.

  b. Profit maximized at 50 units. Max profit is $1800 per day.

83. $v = \dfrac{c}{w + a} - b$ **85. b.** 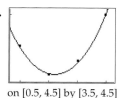   **c.** $y_1(5) = 4.575$, so about 4.6 million units

on [0.5, 4.5] by [3.5, 4.5]

87. a.

Since $(x - 1)^2$ is always positive or zero, the smallest value of $(x - 1)^2 + 2$ is 2 when $x = 1$.

  b.

Since $(x + 2)^2$ is always positive or zero, the smallest value of $(x + 2)^2 - 1$ is −1 when $x = -2$.

89. a. Since $a > 0$ and $\left(x + \dfrac{b}{2a}\right)^2$ is always positive or zero, the smallest value of

$f(x) = a\left(x + \dfrac{b}{2a}\right)^2 + c - \dfrac{b^2}{4a}$ is obtained when $x + \dfrac{b}{2a} = 0,$ that is, when $x = \dfrac{-b}{2a}.$

b. Since $a < 0$ and $\left(x + \dfrac{b}{2a}\right)^2$ is always positive or zero, the largest value of

$f(x) = a\left(x + \dfrac{b}{2a}\right)^2 + c - \dfrac{b^2}{4a}$ is obtained when $x + \dfrac{b}{2a} = 0,$ that is, when $x = \dfrac{-b}{2a}.$

Exercises 1.4 page 59

1. Domain: $\{x \mid x > 0 \text{ or } x < -4\}$; Range: $\{y \mid y > 0 \text{ or } y < -2\}$;
3. a. $f(-3) = 1$ **b.** $\{x \mid x \neq -4\}$ **c.** $\{y \mid y \neq 0\}$
5. a. $f(-1) = -\frac{1}{2}$ **b.** $\{x \mid x \neq 1\}$ **c.** $\{y \mid y \leq 0 \text{ or } y \geq 4\}$
7. a. $f(2) = 1$ **b.** $\{x \mid x \neq 0, x \neq -4\}$ **c.** $\{y \mid y > 0 \text{ or } y \leq -3\}$
9. a. $g(-5) = 3$ **b.** $\mathbb{R}$ **c.** $\{y \mid y \geq 0\}$ **11.** $x = 0,\ x = -3,\ x = 1$
13. $x = 0,\ x = 2,\ x = -2$ **15.** $x = 0,\ x = 3$ **17.** $x = 0,\ x = 5$ **19.** $x = 0,\ x = 3$
21. $x = -2,\ x = 0,\ x = 4$ **23.** $x = -1,\ x = 0,\ x = 3$ **25.** $x = 0,\ x = 3$ **27.** $x = -5,\ x = 0$
29. $x = 0,\ x = 1$ **31.** $x \approx -1.79,\ x = 0,\ x \approx 2.79$
33.　　　　　　　　　　　**35.**　　　　　　　　　　**37.**

39. Polynomial   **41.** Piecewise linear   **43.** Polynomial   **45.** Rational   **47.** Piecewise linear
49. Polynomial   **51.** None (not a polynomial because of the fractional exponent)
53. a. y_3 **b.** y_1 **c.**   **d.** $(10, 1000).$

55. a. $y = \text{INT}(x)$ on $[-5, 5]$ by $[-5, 5]$ Note that each line segment in this graph includes its left endpoint but excludes its right endpoint, so it should be drawn like ●──○.

b. Domain: $\mathbb{R}$; Range: $\{..., -3, -2, -1, 0, 1, 2, 3,...\}$, that is, the set of all integers.

57. a. $(7x - 1)^5$ **b.** $7x^5 - 1$ **59. a.** $\dfrac{1}{x^2 + 1}$ **b.** $\left(\dfrac{1}{x}\right)^2 + 1$

61. a. $(\sqrt{x} - 1)^3 - (\sqrt{x} - 1)^2$ **b.** $\sqrt{x^3 - x^2} - 1$ **63. a.** $\dfrac{(x^2 - x)^3 - 1}{(x^2 - x)^3 + 1}$ **b.** $\left(\dfrac{x^3 - 1}{x^3 + 1}\right)^2 - \dfrac{x^3 - 1}{x^3 + 1}$

65. a. $f(g(x)) = acx + ad + b$ **b.** Yes   **67. a.** 2.70481   **b.** 2.71815   **c.** 2.71828
d. Yes, 2.71828
69. Shifted left 3 units and up 6 units

71. a. $300   **b.** $500   **c.** $2000   **d.**

73. a. $f\left(\frac{2}{3}\right) = 7$, $f\left(1\frac{1}{3}\right) = 14$, $f(4) = 29$,   and $f(10) = 53$ **b.**

75. $R(v(t)) = 2(60 + 3t)^{0.3}$, $R(v(10)) \approx 7.714$ million dollars
77. a. 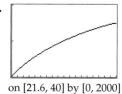   **b.** About $x = 27.9$ mpg

on [21.6, 40] by [0, 2000]

Exercises 1.5 page 73

1. a. 7.389   **b.** 0.135   **c.** 1.649   **3. a.** e^3 **b.** e^2 **c.** e^5
5.   **7.**   **9.** 5.697 (rounded)

11. a. e^x **b.** e^x **c.** e^x **d.** e^x **e.** e^x will exceed any power of x for large enough values of x.
13. a. $2143.59   **b.** $2203.76   **c.** $2225.54   **15. a.** $2196   **b.** $2096
17. Yes, the trust fund would be worth $1,000,508.52.
19. 9.9% compounded continuously (yielding about $1.1041) is better than 10% compounded quarterly (yielding about $1.1038). While the difference is small for $1 for one year, it would be more significant for larger sums and longer times.
21. Yes, it will be worth $1000.18   **23.** $3105.41   **25. a.** $3570 (rounded)   **b.** $16,125 (rounded)

27. 7.0 billion   **29. a.** 0.53 (the chances are better than 50–50)   **b.** 0.70 (quite likely)
31. a. 0.267 or 26.7%   **b.** 0.012 or 1.2%   **33. a.** 1.3 milligrams   **b.** 0.84 milligram   **35.** 208
37. a. About 153 degrees   **b.** About 123 degrees   **39.** 38   **41.** By about 25%
43. a.   **b.** In about 2060 (from $x \approx 60.2$)

Exercises 1.6 page 88

1. a. 5   **b.** -2 **c.** $\frac{1}{2}$ **3. a.** 5   **b.** -1 **c.** $\frac{1}{3}$ **5. a.** 0   **b.** 1   **c.** $\frac{2}{3}$
7. a. 2   **b.** -1 **c.** $\frac{1}{2}$ **9. a.** 1.348   **b.** 3.105   **11.** $\ln x$ **13.** $2 \ln x$ or $\ln x^2$ **15.** $\ln x$
17. $3x$ **19.** $7x$ **21.** 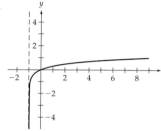   **23.** Domain: $\{x \mid x > 1 \text{ or } x < -1\}$

Range: $\mathbb{R}$

25. a. In about 1.9 years   **b.** In about 3.9 years   **27.** In about 2.5 years
29. About 31,400 years   **31.** About 1.7 million years   **33.** In 8 years
35. 228 million years
37. a. $\log_b b^x = x$, property 8 of logarithms   **b.** Follows directly from (a)
   c. Using the change-of-base formula, cancellation

Chapter 1 Review Exercises page 92

1. $\{x \mid 2 < x \le 5\}$ ⊶———● 2   5   **2.** $\{x \mid -2 \le x < 0\}$ ●———⊶ -2 0
3. $\{x \mid x \ge 100\}$ ⊶——— 100   **4.** $\{x \mid x \le 6\}$ ———⊶ 6
5. Hurricane: $[74, \infty)$; storm: $[55, 74)$; gale: $[38, 55)$; small craft warning: $[21, 38)$
6. a. $(0, \infty)$ **b.** $(-\infty, 0)$ **c.** $[0, \infty)$ **d.** $(-\infty, 0]$ **7.** $y = 2x - 5$ **8.** $y = -3x + 3$
9. $x = 2$ **10.** $y = 3$ **11.** $y = -2x + 1$ **12.** $y = 3x - 5$ **13.** $y = 2x - 1$ **14.** $y = -\frac{1}{2}x + 1$
15. a. $V = 25{,}000 - 3000t$ **b.** $13{,}000   **16. a.** $V = 78{,}000 - 5000t$ **b.** $38{,}000
17. b.   The regression line fits the data well.

   c. 13.7 million tons in the year 2010 [from $y_1(35)$]

on [0, 25] by [0, 30]

18. b.

on [0, 40] by [0, 80]

The regression line fits the data reasonably well.

c. 81 to 1 in the year 2010 [from $y_1(50) = 81.4$]

19. 36 **20.** $\frac{3}{4}$ **21.** 8 **22.** 10 **23.** $\frac{1}{27}$ **24.** $\frac{1}{1000}$ **25.** $\frac{9}{4}$ **26.** $\frac{64}{27}$ **27.** 13.97 **28.** 112.32
29. a. $f(11) = 2$ **b.** $\{x \mid x \geq 7\}$ **c.** $\{y \mid y \geq 0\}$ **30. a.** $g(-1) = \frac{1}{2}$ **b.** $\{t \mid t \neq -3\}$ **c.** $\{y \mid y \neq 0\}$
31. a. $h(16) = \frac{1}{8}$ **b.** $\{w \mid w > 0\}$ **c.** $\{y \mid y > 0\}$ **32. a.** $w(8) = \frac{1}{16}$ **b.** $\{z \mid z \neq 0\}$ **c.** $\{y \mid y > 0\}$
33. Yes **34.** No

35.   **36.**   **37.**   **38.**

39. $x = 0, \quad x = -3$ **40.** $x = 5, \quad x = -1$ **41.** $x = -2, \quad x = 1$ **42.** $x = 1, \quad x = -1$
43. a. Vertex: $(5, -50)$ **b.**   **44. a.** Vertex: $(-7, -64)$ **b.**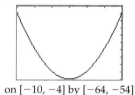

on [2, 8] by [−50, −40]

on [−10, −4] by [−64, −54]

45. $C(x) = 45 + 0.12x$ **46.** $I(t) = 800t$ **47.** $T(x) = 70 - \dfrac{x}{300}$
48. $C(t) = 27 + 0.58t$; In about 5 years from 2000, so in 2005.
49. a. Break even at 15 and 65 units. **b.** Profit maximized at 40 units. Max profit is $1250 per week.
50. a. Break even at 150 and 450 units **b.** Profit maximized at 300 units. Max profit is $67,500 per month.
51. a. $f(-1) = 1$ **b.** $\{x \mid x \neq 0 \text{ and } x \neq 2\}$ **c.** $\{y \mid y > 0 \text{ or } y \leq -3\}$
52. a. $f(-8) = \frac{1}{2}$ **b.** $\{x \mid x \neq 0 \text{ and } x \neq -4\}$ **c.** $\{y \mid y > 0 \text{ or } y \leq -4\}$
53. a. $g(-4) = 0$ **b.** $\mathbb{R}$ **c.** $\{y \mid y \geq -2\}$ **54. a.** $g(-5) = -10$ **b.** $\mathbb{R}$ **c.** $\{y \mid y \leq 0\}$
55. $x = 0, \quad x = 1, \quad x = -3$ **56.** $x = 0, \quad x = 2, \quad x = -4$ **57.** $x = 0, \quad x = 5$ **58.** $x = 0, \quad x = 2$

59.
60.
61.
62.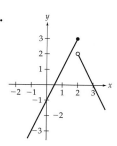

63. a. $f(g(x)) = \left(\dfrac{1}{x}\right)^2 + 1 = \dfrac{1}{x^2} + 1$ **b.** $g(f(x)) = \dfrac{1}{x^2 + 1}$

64. a. $f(g(x)) = \sqrt{5x - 4}$ **b.** $g(f(x)) = 5\sqrt{x} - 4$

65. a. $f(g(x)) = \dfrac{x^3 + 1}{x^3 - 1}$ **b.** $g(f(x)) = \left(\dfrac{x + 1}{x - 1}\right)^3$ **66. a.** $f(g(x)) = |x + 2|$ **b.** $g(f(x)) = |x| + 2$

67. a. $f(g(x)) = x$ **b.** $g(f(x)) = x$ **68. a.** $f(g(x)) = 5$ **b.** $g(f(x)) = 12$

69. $A(p(t)) = 2(18 + 2t)^{0.15},\quad A(p(4)) \approx \$3.26$ million

70. a. $x = -1,\quad x = 0,\quad x = 3$ **71. a.** $x = -3,\quad x = 0,\quad x = 1$ **72. a.** The points suggest a parabolic (quadratic) curve.

b.

on $[-5, 5]$ by $[-5, 5]$

b.

on $[-5, 5]$ by $[-5, 5]$

b.

on $[0.5, 5.5]$ by $[1.5, 3]$

c. $3.6 million, $4.8 million

73.
74.

75. $13,468.55   **76.** $2473   **77. a.** $8602 (rounded)   **b.** $18,783 (rounded)

78. a. $11,765 (rounded)   **b.** $58,566 (rounded)   **79.** $1616.07   **80.** $1,290,461.62

81. a. 3   **b.** −3   **c.** 3   **d.** $\frac{1}{4}$ **82. a.** $\frac{1}{2}$ **b.** 8   **c.** −1   **d.** $\frac{3}{2}$ **83.** $f(x) = \ln x$

84. $f(x) = 2x - 1$ **85. a.** In about 2.8 years   **b.** In about 5.6 years

86. a. In about 1.6 years   **b.** In about 3.2 years   **87.** About 50.7 million years

88. About 1.85 million years

Exercises 2.1 page 107

1. $1050 **3.** $3120 **5.** $298.57 **7.** $15.36 **9.** $2550 **11.** $7751.58 **13.** $3250.26
15. 6.4% **17.** $2500 **19.** 4 years **21.** $7500 **23.** $9450.90 **25.** $4669.41 **27.** 6 years
29. 8 years 9 months **31.** 20 years **33.** 5.48%
35. The $1000 is the same as the interest, so you would be agreeing to pay $1000 of interest on a loan of $0.
37. $950.40 **39.** 58.25% **41.** 17.89%

Exercises 2.2 page 117

1. a. $32,383.87 **b.** $33,120.59 **3. a.** $30,352.10 **b.** $31,046.39
5. a. $16,713.76 **b.** $16,799.62 **7. a.** $11,308.73 **b.** $10,926.92
9. a. $12,119.55 **b.** $11,917.53 **11. a.** $8106.29 **b.** $8057.40
13. 17 years and 9 months **15.** 8 years and 9 months **17.** 5 years and 11 months
19. About 8 years; 8.04 rounds up to 9 years **21.** About 9.11 years; 8 years and 41 weeks
23. About 11.80 years; 11 years 5 months **25.** 19.56% **27.** 8.75% **29.** 9.90%
31. $6827.69 **33.** $14,467.34 **35.** 30 years and 6 months **37.** 17.45%
39. The bond gives the greater return (6.80%; the CD returns only 6.62%). **41.** 32 years and 48 weeks
43. People's State Bank offers the higher effective yield (4.27%; Statewide Federal's effective rate is 4.18%).
45. $P(1 + r)^2 - P(1 + 2r) = P((1 + r)^2 - (1 + 2r)) = P(1 + 2r + r^2 - 1 - 2r) = P(r^2)$
47. $P(1 + r_e)^t = P(1 + (1 + r/m)^m - 1)^t = P((1 + r/m)^m)^t = P(1 + r/m)^{mt}$
49. 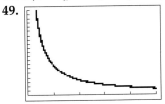 The two curves match each other very closely.

Exercises 2.3 page 128

1. $20,724.67 **3.** $68,852.94 **5.** $514,647.37 **7.** $32.20 **9.** $172.81 **11.** $345.85
13. 24 years **15.** 22 years and 6 months **17.** 7 years and 2 months
19. Joe will have $475,172.10, and Jill will have $546,822.42. **21.** $4001.27 **23.** $6.39
25. 1 year and 9 months **27.** 13.2% (from $x \approx 0.1319$)

Exercises 2.4 page 139

1. $105,353.72 **3.** $95,896.48 **5.** $212,572.08 **7.** $584.59 **9.** $116 **11.** $872.41
13. $85,934.02 **15.** $2719.69 **17.** $10,284.16 **19.** $4,306,638 **21.** $14,204,378 **23.** $70,777.95
25. $468,407.35; John should buy $500,000 of life insurance.
27. a. $3334.58 **b.** It saves $334.24 each month. **c.** The longer mortgage costs an extra $2,640,705.84.
29. $31.53 **31.** $115.66 **33.** $41,640.29 **35.** 11.9% **37.** $2531.31
45. The amortization table uses an annual payment of $7518.83 and the unpaid balance after 3 years is
$38,879.49, a difference of 2¢ compared to Exercise 16. The final payment is $7518.87, a correction of
4¢ from the others in the table.

Chapter 2 Review Exercises page 144

1. $217.50 **2.** $52.90 **3.** $2250 **4.** $336.88 **5.** $10,212.75 **6.** $1491.53 **7.** $2138.56
8. $4288.61 **9.** 6.8% **10.** $1800 **11.** 1 year and 6 months **12.** 4 years **13.** 5 years
14. $4500 **15.** $9394.08 **16.** $47,267.91 **17.** 42.86% **18.** 21.43% **19.** 6.6% **20.** $500,000
21. 11.1%; $1110 **22.** $31,535.24 **23.** $264,205.92 **24.** $370,893.09 **25.** $10,749.24
26. No; it would only be worth $110 billion **27.** $10,198.43 **28.** $18,564.23 **29.** $71,056.01
30. $12,702.50 **31.** $8466.50 **32.** 18 years and 6 months **33.** 4 years and 2 months **34.** 8 years
35. 3 years **36.** 18 years **37.** 9 years; 9 years (from 8.84) **38.** 12 years; 11 years and 7 months
39. 6 years; 5 years and 44 weeks **40.** 6 years; 7 years (from 6.12) **41.** 7.2 years; 8 years (from 7.27)
42. 13.92% **43.** 28.19% **44.** 106.64% **45.** 13.51% **46.** 2.94% **47.** $96,304.25
48. $19,541.30 **49.** $33,322.04 **50.** $935,211.12 **51.** $715,609.58 **52.** $1559.49 **53.** $205.98
54. $106.28 **55.** $168.76 **56.** $19.51 **57.** 26 years **58.** 24 years and 6 months
59. 5 years and 1 month **60.** 15 years and 7 months **61.** 13 years and 6 months **62.** 9.80%
63. 2.97% **64.** 9.42% **65.** 6.58% **66.** 3.00% (with 365 days per year) **67.** $319,583.90
68. $20,923,409 **69.** $30,147,922 **70.** $6,773,662.45 **71.** $158,891 **72.** $106.12 **73.** $46.14
74. $423.17 **75.** $1180.05
76. $548.25; do not confuse paying off a current debt with accumulating money in the future.
77. $75,148.14 **78.** $142,742.00 **79.** $5449.20 **80.** $523,692.95 **81.** $143,927.90

Exercises 3.1 page 162

1. $\begin{cases} x + y = 18 \\ x - y = 2 \end{cases}$ **3.** $\begin{cases} x - y = 6 \\ x + y = 40 \end{cases}$ **5.** $\begin{cases} x + y = 30 \\ x + 5y = 70 \end{cases}$ **7.** $\begin{cases} x + y = 100 \\ 10x + 5y = 650 \end{cases}$ **9.** $\begin{cases} x + y = 225 \\ x - 2y = 0 \end{cases}$

11. The solution is $x = 4$, $y = 2$. **13.** The solution is $x = 3$, $y = 2$.

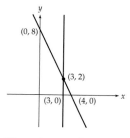

15. The solution is $x = 2$, $y = -2$. **17.** There is no solution. The equations are inconsistent.

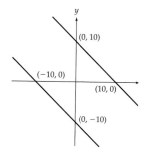

19. There are infinitely many solutions that may be parameterized as $x = 10 - t$, $y = t$. The system is dependent.

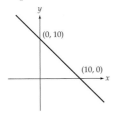

21. $x = 4$, $y = 3$.   **23.** $x = 5$, $y = 10$.   **25.** $x = 10$, $y = -10$.   **27.** $x = 4$, $y = 9$.

29. There are infinitely many solutions that may be parameterized as $x = 10 + t$, $y = t$. The system is dependent.

31. $x = 3$, $y = 8$.   **33.** $x = 4$, $y = 3$.   **35.** $x = 8$, $y = 3$.   **37.** $x = 15$, $y = 6$.

39. There is no solution. The system is inconsistent.

41. 34 investors contributed $5000 and 26 contributed $10,000.

43. There are 34 nickels and 26 dimes in the jar.

45. The retired couple should invest $2000 in the money market account and $8000 in the stock mutual fund.

47. The concession stand sold 1800 sodas and 1200 hot dogs.

49. The federal tax is $4900 and the state tax is $900.

51. The required calcium and phosphorus can be provided by just 5 tablets of supplement B (with none of supplement A) each day.

Exercises 3.2 page 176

1. 3×2; 1; 6 **3.** 3×4; 3; -6 **5.** 4×4; 1; 1; 1; 1; 0 **7.** 1×4; 6 **9.** 5×1; 8; 6

11. $\begin{pmatrix} 1 & 2 & | & 2 \\ 3 & 4 & | & 12 \end{pmatrix}$ **13.** $\begin{pmatrix} -4 & 3 & | & 84 \\ 5 & -2 & | & 70 \end{pmatrix}$ **15.** $\begin{pmatrix} 3 & -2 & | & 24 \\ 1 & 0 & | & 6 \end{pmatrix}$ **17.** $\begin{pmatrix} 5 & -15 & | & 30 \\ -4 & 12 & | & 24 \end{pmatrix}$

19. $\begin{pmatrix} 1 & 0 & | & 20 \\ 0 & 1 & | & 30 \end{pmatrix}$ **21.** $\begin{cases} x + y = 9 \\ y = 4 \end{cases}$ **23.** $\begin{cases} -4x + 3y = -60 \\ x - 2y = 20 \end{cases}$ **25.** $\begin{cases} x - 3y = -70 \\ x + y = 10 \end{cases}$

27. $\begin{cases} 2x + y = 6 \\ x + 2y = -6 \end{cases}$ **29.** $\begin{cases} 20x - 15y = 60 \\ -16x + 12y = -48 \end{cases}$ **31.** $\begin{matrix} R_2 \to \\ R_1 \to \end{matrix} \begin{pmatrix} 5 & 6 & | & 30 \\ 3 & 4 & | & 24 \end{pmatrix}$

33. $R_1 - R_2 \to \begin{pmatrix} 2 & 2 & | & -4 \\ 6 & 5 & | & 60 \end{pmatrix}$ **35.** $R_1 - 3R_2 \to \begin{pmatrix} 2 & 0 & | & -24 \\ 1 & 2 & | & 18 \end{pmatrix}$

37. $R_1 - R_2 \to \begin{pmatrix} 6 & 6 & | & -12 \\ 0 & 1 & | & -72 \end{pmatrix}$ **39.** $\frac{1}{8}R_2 \to \begin{pmatrix} 1 & -2 & | & -42 \\ 0 & 1 & | & 15 \end{pmatrix}$

41. $x = 7$, $y = -3$.

43. No solution. The system is inconsistent.

45. No solution. The system is inconsistent.

47. $x = 3 - 2t$, $y = t$ (infinitely many solutions). The system is dependent.

49. $x = t$, $y = -3$ (infinitely many solutions). The system is dependent.

51. $x = 3$, $y = 2$.   **53.** $x = 3$, $y = 1$.   **55.** $x = 1$, $y = 2$.   **57.** $x = 3$, $y = 2$.

59. $x = 9$, $y = 2$.   **61.** No solution. The system is inconsistent.   **63.** $x = 8$, $y = -15$.

65. $x = 4$, $y = 15$. **67.** $x = 9 + 3t$, $y = t$ (infinitely many solutions). The system is dependent.

69. $x = 21$, $y = -4$.

71. The commodities speculator invested $10,000 in soybean futures and $5000 in corn futures.

73. The older brother receives $4.8 million and the younger brother receives $2.4 million, leaving $4.8 million for their sister.

75. 175 members invested $5000 and 112 members invested $25,000.

77. The dietician probably wants positive whole numbers for the solution, so the possibilities are

Cans of NutraDrink:	12	8	4	0
Tablets of VitaPills:	0	5	10	15

79. The campaign manager should use 7 TV ads and 30 radio ads.

81. $x = 18$, $y = 16$.

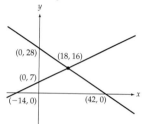

83. No solution. The system is inconsistent.

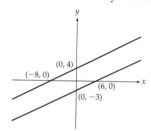

85. No solution. The system is inconsistent.

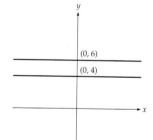

87. $x = 14 - \frac{7}{2}t$, $y = t$ (infinitely many solutions). The system is dependent.

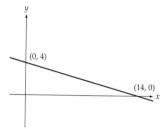

89. $x = t$, $y = 6$ (infinitely many solutions). The system is dependent.

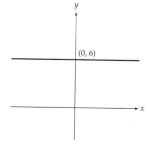

91. $\begin{pmatrix} 3 & -10 & | & -65 \\ -4 & 13 & | & 84 \end{pmatrix} \rightarrow \begin{matrix} R_2 \rightarrow \\ R_1 \rightarrow \end{matrix} \begin{pmatrix} -4 & 13 & | & 84 \\ 3 & -10 & | & -65 \end{pmatrix} \rightarrow \begin{matrix} R_2 \rightarrow \\ R_1 \rightarrow \end{matrix} \begin{pmatrix} 3 & -10 & | & -65 \\ -4 & 13 & | & 84 \end{pmatrix}$

93. $\begin{pmatrix} 3 & -10 & | & -65 \\ -4 & 13 & | & 84 \end{pmatrix} \rightarrow \begin{matrix} 5R_1 \rightarrow \end{matrix} \begin{pmatrix} 15 & -50 & | & -325 \\ -4 & 13 & | & 84 \end{pmatrix} \rightarrow \begin{matrix} \frac{1}{5}R_1 \rightarrow \end{matrix} \begin{pmatrix} 3 & -10 & | & -65 \\ -4 & 13 & | & 84 \end{pmatrix}$

95. $\begin{pmatrix} 3 & -10 & | & -65 \\ -4 & 13 & | & 84 \end{pmatrix} \rightarrow \begin{matrix} \frac{1}{4}R_2 \rightarrow \end{matrix} \begin{pmatrix} 3 & -10 & | & -65 \\ -1 & 13/4 & | & 21 \end{pmatrix} \rightarrow \begin{matrix} 4R_2 \rightarrow \end{matrix} \begin{pmatrix} 3 & -10 & | & -65 \\ -4 & 13 & | & 84 \end{pmatrix}$

97. $\begin{pmatrix} 3 & -10 & | & -65 \\ -4 & 13 & | & 84 \end{pmatrix} \rightarrow \begin{matrix} R_1 + R_2 \rightarrow \end{matrix} \begin{pmatrix} -1 & 3 & | & 19 \\ -4 & 13 & | & 84 \end{pmatrix} \rightarrow \begin{matrix} R_1 - R_2 \rightarrow \end{matrix} \begin{pmatrix} 3 & -10 & | & -65 \\ -4 & 13 & | & 84 \end{pmatrix}$

99. $\begin{pmatrix} 3 & -10 & | & -65 \\ -4 & 13 & | & 84 \end{pmatrix} \rightarrow \begin{matrix} R_1 - R_2 \rightarrow \end{matrix} \begin{pmatrix} 7 & -23 & | & -149 \\ -4 & 13 & | & 84 \end{pmatrix} \rightarrow \begin{matrix} R_1 + R_2 \rightarrow \end{matrix} \begin{pmatrix} 3 & -10 & | & -65 \\ -4 & 13 & | & 84 \end{pmatrix}$

Exercises 3.3 page 192

1. $\begin{pmatrix} 1 & 1 & 1 & | & 4 \\ 1 & 2 & 1 & | & 3 \\ 1 & 2 & 2 & | & 5 \end{pmatrix}$ **3.** $\begin{pmatrix} 2 & -1 & 2 & | & 11 \\ -1 & 1 & -3 & | & -12 \\ 2 & -2 & 7 & | & 27 \end{pmatrix}$ **5.** $\begin{pmatrix} 2 & 1 & 5 & 4 & 5 & | & 2 \\ 1 & 1 & 3 & 3 & 3 & | & -1 \end{pmatrix}$

7. $\begin{cases} 4x_1 + 3x_2 + 2x_3 = 11 \\ 3x_1 + 3x_2 + x_3 = 6 \\ x_1 - 2x_2 + 3x_3 = 13 \end{cases}$ **9.** $\begin{cases} 2x_1 + x_2 + x_3 = 7 \\ 2x_1 + 2x_2 + x_3 = 6 \\ 3x_1 + 3x_2 + 2x_3 = 10 \end{cases}$ **11.** $\begin{cases} 8x_1 + 3x_2 - 2x_3 + 19x_4 = 15 \\ 3x_1 + x_2 - x_3 + 7x_4 = 6 \end{cases}$

13. $x_1 = 4, \ x_2 = 5, \ x_3 = -4.$

15. The system is inconsistent, so no solution.

17. The system is dependent, with solution $x_1 = -5 + t, \ x_2 = 5 - t, \ x_3 = t.$

19. $\begin{pmatrix} 1 & 0 & 0 & | & 1 \\ 0 & 1 & 0 & | & 2 \\ 0 & 0 & 1 & | & 3 \end{pmatrix}$ **21.** $\begin{pmatrix} 1 & 0 & 0 & | & 1 \\ 0 & 1 & 0 & | & 2 \\ 0 & 0 & 1 & | & 3 \end{pmatrix}$ **23.** $\begin{pmatrix} 1 & 2 & 0 & | & 3 \\ 0 & 0 & 1 & | & -3 \\ 0 & 0 & 0 & | & 0 \end{pmatrix}$

25. $x_1 = 1, \ x_2 = -2, \ x_3 = 3.$

27. $x_1 = 1, \ x_2 = -2, \ x_3 = 3.$

29. $x_1 = 1, \ x_2 = 1, \ x_3 = 1, \ x_4 = 1.$

31. The system is inconsistent, with no solution.

33. The system is dependent, with solution $x_1 = 3 + 7t, \ x_2 = 4 - 10t, \ x_3 = t.$

35. $x_1 = 2, \ x_2 = 1, \ x_3 = -2, \ x_4 = 3.$

37. The system is dependent and inconsistent, so *no* solution.

39. The system is dependent, with solution $x_1 = 2 + t_1 - t_2, \ x_2 = 1 - 2t_1 - t_2, \ x_3 = t_1, \ x_4 = t_2.$

41. The gardener should use 2 bags of GrowRite, 1 bag of MiracleMix, and 3 bags of GreatGreen.

43. Let m be the number of $25,000 members in the partnership. Then there are $200 + 3m$ members at $5000 and $500 - 4m$ members at $10,000. Because the number of each can never be negative, there can be no more than 125 of the $25,000 members in the partnership.

45. a. The federal tax is $80,510, the state tax is $16,490, and the city tax is $2490.
  b. The effective combined tax rate is (about) 55.3%.

47. a. Letting B be the number of blouses and S be the number of skirts, the shop can make $60 + \frac{6}{5}B - \frac{4}{5}S$ scarves and $60 - \frac{9}{10}B - \frac{2}{5}S$ dresses (where none of these quantities are negative).
  b. 76 scarves and 38 dresses.

49. Letting R be the number of radio ads, the number of TV ads is $20 - \frac{1}{5}R$, and the number of newspaper ads is 60. The promotional director may choose R to be 0, 5, 10, 15, 20, . . . , 85, 90, 95, or 100 because these are the only values for R that will result in a whole number of TV ads, while ensuring that neither value is negative.

51. $\begin{cases} x_1 + 2x_2 = 3 \\ x_2 = 1 \end{cases}$ **53.** All are equivalent to $\left(\begin{array}{cc|c} 1 & 0 & 1 \\ 0 & 1 & 1 \end{array}\right)$ and so to each other.

55. $\left(\begin{array}{cc|c} 1 & 0 & 1 \\ 0 & 1 & 1 \end{array}\right)$ **57.** $x_1 = 1, \ x_2 = 1$. **59.** $x_1 = 1, \ x_2 = 1, \ x_3 = 2$.

61. $x_1 = 3, \ x_2 = 2, \ x_3 = 4, \ x_4 = 1$. **63.** $x_1 = 3, \ x_2 = -1, \ x_3 = 2$.

Exercises 3.4 page 207

1. $\begin{pmatrix} 1 & 4 & 7 \\ 2 & 5 & 8 \\ 3 & 6 & 9 \end{pmatrix}$ **3.** $\begin{pmatrix} 3 & 18 & 24 \\ 12 & 6 & 21 \\ 27 & 15 & 9 \end{pmatrix}$ **5.** $\begin{pmatrix} -9 & -8 & -7 \\ -6 & -5 & -4 \\ -3 & -2 & -1 \end{pmatrix}$ **7.** $\begin{pmatrix} 2 & 8 & 11 \\ 8 & 7 & 13 \\ 16 & 13 & 12 \end{pmatrix}$

9. $\begin{pmatrix} -1 & 4 & 5 \\ 0 & -4 & 1 \\ 2 & -3 & -7 \end{pmatrix}$ **11.** (6) **13.** $\begin{pmatrix} 3 & 4 \\ 6 & 8 \end{pmatrix}$ **15.** $\begin{pmatrix} -2 \\ 5 \end{pmatrix}$ **17.** $\begin{pmatrix} 6 & 6 \\ 7 & 7 \end{pmatrix}$ **19.** $\begin{pmatrix} 11 & 7 & 11 \\ 6 & 7 & 6 \\ -1 & 3 & -1 \end{pmatrix}$

21. $\begin{pmatrix} 3 & -1 & 4 \\ 2 & 1 & 3 \end{pmatrix}$ **23.** $\begin{pmatrix} 2 & 8 \\ 4 & 1 \\ 5 & -4 \end{pmatrix}$ **25.** $\begin{pmatrix} -1 & -4 & -1 \\ -2 & 6 & -1 \end{pmatrix}$ **27.** $\begin{pmatrix} 7 & 1 & 4 \\ 0 & -1 & 0 \\ 4 & 2 & 2 \end{pmatrix}$

29. $\begin{pmatrix} 5 & 5 & 6 \\ 7 & -7 & 4 \end{pmatrix}$ **31.** $\begin{pmatrix} 1 & 5 & 4 \\ 1 & 1 & 1 \\ 2 & 3 & 3 \end{pmatrix}\begin{pmatrix} x_1 \\ x_2 \\ x_3 \end{pmatrix} = \begin{pmatrix} 6 \\ 4 \\ 9 \end{pmatrix}$ **33.** $\begin{pmatrix} 4 & 3 & -1 \\ 3 & 3 & 2 \\ 2 & 1 & -3 \end{pmatrix}\begin{pmatrix} x_1 \\ x_2 \\ x_3 \end{pmatrix} = \begin{pmatrix} 2 \\ 9 \\ -6 \end{pmatrix}$

35. $\begin{pmatrix} 5 & 2 & -4 & 1 & 5 \\ 3 & 1 & -3 & 1 & 3 \end{pmatrix}\begin{pmatrix} x_1 \\ x_2 \\ x_3 \\ x_4 \\ x_5 \end{pmatrix} = \begin{pmatrix} 7 \\ 5 \end{pmatrix}$ **37.** $\begin{cases} 5x_1 + 9x_2 + 9x_3 = 11 \\ 4x_1 + 7x_2 + 6x_3 = 9 \\ 3x_1 + 5x_2 + 3x_3 = 8 \\ 4x_1 + 7x_2 + 5x_3 = 10 \end{cases}$ **39.** $\begin{cases} 5x_1 + 4x_2 + 7x_3 + 6x_4 = 18 \\ 2x_1 + 2x_2 + 3x_3 + 3x_4 = 9 \\ 4x_1 + 3x_2 + 5x_3 + 5x_4 = 16 \\ 3x_1 + 2x_2 + 3x_3 + 3x_4 = 11 \end{cases}$

41. Let P be the "price" matrix of selling prices and C be the "commission" matrix so that $C = 0.15P$. Choosing P to be a 3×3 matrix with the rows representing the manufacturers SlumberKing, DreamOn, and RestEasy in that order and the columns representing the models "economy," "best," and "deluxe" in that order, $P = \begin{pmatrix} 300 & 350 & 500 \\ 350 & 400 & 550 \\ 400 & 500 & 700 \end{pmatrix}$ and $C = \begin{pmatrix} 45.00 & 52.50 & 75.00 \\ 52.50 & 60.00 & 82.50 \\ 60.00 & 75.00 & 105.00 \end{pmatrix}$.

43. Let D be the "dealer invoice" matrix and S be the "sticker price" matrix so that the "markup" matrix M is $M = S - D$. Choosing D to be a 2×4 matrix with the rows representing the sales lots in Oakdale and Roanoke in that order and the columns representing the vehicle models "sedan," "station wagon," "van," and "pickup truck" in that order, $D = \begin{pmatrix} 15,000 & 19,000 & 23,000 & 25,000 \\ 15,000 & 19,000 & 23,000 & 25,000 \end{pmatrix}$ and $S = \begin{pmatrix} 18,900 & 22,900 & 26,900 & 29,900 \\ 19,900 & 21,900 & 27,900 & 28,900 \end{pmatrix}$ so then $M = \begin{pmatrix} 3900 & 3900 & 3900 & 4900 \\ 4900 & 2900 & 4900 & 3900 \end{pmatrix}$.

45. Let L be the "labor costs" matrix and M be the "materials cost" matrix so that the "total cost" matrix T is $T = L + M$. Choosing L to be a 2×3 matrix of values in pennies with the rows representing the countries Costa Rica and Honduras in that order and the columns representing the apparel items "shorts," "tee-shirts," and "caps" in that order,

$$L = \begin{pmatrix} 75 & 25 & 45 \\ 80 & 20 & 55 \end{pmatrix} \text{ and } M = \begin{pmatrix} 160 & 95 & 115 \\ 150 & 80 & 110 \end{pmatrix} \text{ so then } T = \begin{pmatrix} 235 & 120 & 160 \\ 230 & 100 & 165 \end{pmatrix}.$$

47. Let P be the "sale price" row matrix with values in dollars and N be the "number of items" column matrix so that the "total cost" matrix C is $C = P \cdot N$. Choosing the columns of P and the rows of N to represent the items "bottles of soda," "bottles of pickles," "packages of hot dogs," and "bags of chips" in

that order, $P = (0.89 \quad 1.29 \quad 2.39 \quad 1.69)$ and $N = \begin{pmatrix} 12 \\ 2 \\ 3 \\ 4 \end{pmatrix}$ so then $C = (27.19)$. The total cost of

these items at these prices is $27.19.

49. Let T be the "time" matrix and L be the "labor hourly cost" matrix so that the "production cost" matrix C is $C = T \cdot L$. Choosing T to be a 3×3 matrix with the rows representing the furniture items "table," "chair," and "desk" in that order and the columns representing the manufacturing steps

"cutting and milling," "assembly," and "finishing" in that order, $T = \begin{pmatrix} 2 & 1 & 2 \\ 1.5 & 1 & 0.5 \\ 3 & 2 & 3 \end{pmatrix}$, and choosing

L to be a 3×2 matrix with the rows representing the manufacturing steps "cutting and milling," "assembly," and "finishing" in that order and the columns representing the factory locations

"Wytheville" and "Andersen" in that order, $L = \begin{pmatrix} 9 & 10 \\ 14 & 13 \\ 13 & 12 \end{pmatrix}$, so then $C = \begin{pmatrix} 58 & 57 \\ 34 & 34 \\ 94 & 92 \end{pmatrix}$. For these

choices of T and L, the rows of C represent the furniture items "table," "chair," and "desk" in that order, and the columns represent the factory locations "Wytheville" and "Andersen" in that order.

51. Begin with the augmented matrix $\begin{pmatrix} a & b & | & h \\ c & d & | & k \end{pmatrix}$ and apply the "symbolic" row operations

$dR_1 - bR_2 \to R_1$, $(ad - bc)R_2 \to R_2$, $R_2 - cR_1 \to R_2$, and $\frac{1}{d}R_2 \to R_2$ to obtain

$\begin{pmatrix} ad - bc & 0 & | & hd - bk \\ 0 & ad - bc & | & ak - hc \end{pmatrix}$. Dividing through by $ad - bc$ gives the identity matrix on the left of

the bar with the desired solution on the right.

Exercises 3.5 page 220

1. $\begin{pmatrix} 1 & 2 \\ -1 & -1 \end{pmatrix}\begin{pmatrix} -1 & -2 \\ 1 & 1 \end{pmatrix} = \begin{pmatrix} 1 & 0 \\ 0 & 1 \end{pmatrix}$ so this *is* a matrix and its inverse.

3. $\begin{pmatrix} 1 & 1 & 0 \\ 2 & 1 & 1 \\ 1 & 0 & 0 \end{pmatrix}\begin{pmatrix} 0 & 0 & 1 \\ 1 & 0 & -1 \\ -1 & 1 & -1 \end{pmatrix} = \begin{pmatrix} 1 & 0 & 0 \\ 0 & 1 & 0 \\ 0 & 0 & 1 \end{pmatrix}$ so this *is* a matrix and its inverse.

5. $\begin{pmatrix} 4 & 6 & 3 \\ 3 & 4 & 1 \\ 5 & 7 & 3 \end{pmatrix}\begin{pmatrix} -5 & -3 & 6 \\ 4 & 3 & -5 \\ -1 & -2 & 2 \end{pmatrix} = \begin{pmatrix} 1 & 0 & 0 \\ 0 & 1 & 0 \\ 0 & 0 & 1 \end{pmatrix}$ so this *is* a matrix and its inverse.

7. $\begin{pmatrix} 10 & -4 & -7 \\ -7 & 3 & 5 \\ 4 & -1 & -3 \end{pmatrix}\begin{pmatrix} 4 & 5 & -1 \\ 1 & 2 & 1 \\ 5 & 6 & 2 \end{pmatrix} = \begin{pmatrix} 1 & 0 & -28 \\ 0 & 1 & 20 \\ 0 & 0 & -11 \end{pmatrix}$ so this is *not* a matrix and its inverse.

9. $\begin{pmatrix} 1 & -3 \\ 0 & 1 \end{pmatrix}$ **11.** $\begin{pmatrix} -1 & 2 \\ 6 & -11 \end{pmatrix}$ **13.** $\begin{pmatrix} 0 & 1 & -2 \\ 1 & -1 & 2 \\ 0 & -1 & 3 \end{pmatrix}$ **15.** Singular matrix

17. $\begin{pmatrix} 1 & 0 & 0 & -1 \\ 0 & 1 & 0 & 0 \\ -1 & 0 & 1 & 1 \\ 0 & -1 & 0 & 1 \end{pmatrix}$ **19.** $\begin{pmatrix} 1 & -3 & 1 \\ 0 & 2 & -1 \\ -1 & 0 & 1 \end{pmatrix}$ **21.** $\begin{pmatrix} 1 & 2 & 0 \\ -2 & -2 & 1 \\ -5 & -7 & 2 \end{pmatrix}$ **23.** Singular matrix

25. $\begin{pmatrix} x_1 \\ x_2 \end{pmatrix} = \begin{pmatrix} -1 & 2 \\ 6 & -11 \end{pmatrix}\begin{pmatrix} 9 \\ 5 \end{pmatrix} = \begin{pmatrix} 1 \\ -1 \end{pmatrix}$ so $\begin{cases} x_1 = 1 \\ x_2 = -1 \end{cases}$

27. $\begin{pmatrix} x_1 \\ x_2 \\ x_3 \end{pmatrix} = \begin{pmatrix} 0 & 1 & -2 \\ 1 & -1 & 2 \\ 0 & -1 & 3 \end{pmatrix}\begin{pmatrix} 2 \\ 5 \\ 2 \end{pmatrix} = \begin{pmatrix} 1 \\ 1 \\ 1 \end{pmatrix}$ so $\begin{cases} x_1 = 1 \\ x_2 = 1 \\ x_3 = 1 \end{cases}$

29. $x_1 = 1,\ x_2 = 1,\ x_3 = 1.$ **31.** $x_1 = 4,\ x_2 = -1,\ x_3 = -2.$

33. $x_1 = 2,\ x_2 = -1,\ x_3 = 2,\ x_4 = 1.$ **35.** $x_1 = 1,\ x_2 = -2,\ x_3 = 1,\ x_4 = 2.$

37. $x_1 = 2,\ x_2 = 4,\ x_3 = 1.$ **39.** $x_1 = -5,\ x_2 = 10,\ x_3 = -20,\ x_4 = 15.$

41. The multiplex sold 150 adult tickets and 350 child tickets for Film No. 1; 200 adult tickets and 200 child tickets for Film No. 2; 250 adult tickets and 200 child tickets for Film No. 3; 400 adult tickets and 100 child tickets for Film No. 4; and 600 adult tickets and no child tickets for Film No. 5.

43. The Gold Mines partnership consists of 250 members at $1000, 150 members at $5000, and 100 members at $10,000; the Oil Wells partnership consists of 150 members at $1000, 350 members at $5000, and 200 members at $10,000; and the Modern Art partnership consists of 325 members at $1000, 115 members at $5000, and 160 members at $10,000.

45. Billy needs 4 drops of Supplement 1, no drops of Supplement 2, 5 drops of Supplement 3, and 2 drops of Supplement 4; Susie needs 2 drops of Supplement 1, 2 drops of Supplement 2, 3 drops of Supplement 3, and 6 drops of Supplement 4; and Jimmy needs 3 drops of Supplement 1, 1 drop of Supplement 2, no drops of Supplement 3, and 8 drops of Supplement 4.

47. Mr. and Mrs. Jordan should invest $100,000 in the stock fund, $150,000 in the money market fund, and $50,000 in the bond fund; Mr. and Mrs. French should invest $78,300 in the stock fund, $51,100 in the money market fund, and $105,500 in the bond fund; and Mrs. Daimen should invest $90,000 in the stock fund, $105,000 in the money market fund, and $75,000 in the bond fund.

49. The mass transit manager should assign 120 subway cars, 50 buses, and 10 jitneys to Brighton; 100 subway cars, 30 buses, and 20 jitneys to Conway; 100 subway cars, 40 buses, and 10 jitneys to Longwood; and 110 subway cars, 40 buses, and 15 jitneys to Oakley.

51. $\dfrac{1}{ad-bc}\begin{pmatrix} d & -b \\ -c & a \end{pmatrix}\begin{pmatrix} a & b \\ c & d \end{pmatrix} = \dfrac{1}{ad-bc}\begin{pmatrix} da-bc & db-bd \\ -ca+ac & -cb+ad \end{pmatrix} = \dfrac{1}{ad-bc}\begin{pmatrix} ad-bc & 0 \\ 0 & ad-bc \end{pmatrix} = \begin{pmatrix} 1 & 0 \\ 0 & 1 \end{pmatrix}$

53. $(A \cdot B)(B^{-1} \cdot A^{-1}) = A\underbrace{BB^{-1}}_{I}A^{-1} = AIA^{-1} = AA^{-1} = I$ and similarly for $(B^{-1} \cdot A^{-1})(A \cdot B)$

55. If $B \cdot A = I$, then multiplying on the right by A^{-1} gives $B \cdot A \cdot A^{-1} = I \cdot A^{-1}$ so that $B \cdot I = A^{-1}$ and $B = A^{-1}$. Then $A \cdot B = A \cdot A^{-1} = I$.

57. The formula in the box with $a = 1$, reversing the order on the left, and multiplying numerator and denominator on the right by -1 gives the required equation, which can be written $1 + x + x^2 + \cdots + x^{n-1} = (1 - x^n)(1 - x)^{-1}$. The corresponding matrix equation is then $1 + A + A^2 + \cdots + A^{n-1} = (1 - A^n)(1 - A)^{-1}$.

59. Using the result of Exercise 58:
$A^3 = A \cdot A^2$ is substochastic of order $w \cdot w^2 = w^3$
$A^4 = A \cdot A^3$ is substochastic of order $w \cdot w^3 = w^4$
$\vdots$
$A^n = A \cdot A^{n-1}$ is substochastic of order $w \cdot w^{n-1} = w^n$

Exercises 3.6 page 233

1. $\begin{pmatrix} 0.30 & 0.50 \\ 0.40 & 0.10 \end{pmatrix}$ for sectors A and L

3. $\begin{pmatrix} 0.20 & 0.10 & 0.15 \\ 0.10 & 0.30 & 0.20 \\ 0.40 & 0.15 & 0.10 \end{pmatrix}$ for sectors C, E, and L

5.

7.

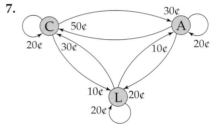

9. $Y = \begin{pmatrix} 71 \\ 37 \end{pmatrix}$ **11.** $Y = \begin{pmatrix} 89 \\ 118 \\ 101 \end{pmatrix}$ **13.** $X = \begin{pmatrix} 150 \\ 120 \end{pmatrix}$ **15.** $X = \begin{pmatrix} 80 \\ 60 \\ 100 \end{pmatrix}$

17. Heavy industry production must be $300 million and light industry production must be $280 million.

19. The current excess production from these sectors is $115 million (made up of $20 million from heavy industry, $50 million from light industry, and $45 million from the railroads). Each $10 million increase in heavy industry production raises the excess production by $1 million (but because the heavy industry production consumes both light industry and railroad production, this increase is composed of an $8 million increase in heavy industry production together with decreases of $5 million from light industry and $2 million from the railroads). When the heavy industry production level reaches $200 million, all light industry excess production will be consumed by the heavy industry sector, and no further expansion will be possible without expanding the light industry production.

21. $y = 9x + 34$ **23.** $y = -29x + 92$ **25.** $y = 9x + 165$

27. His sales in the fifth month may be expected to be $910,000.

29. At 79¢ per 8-ounce bag, the manufacturer can expect 1711 sales per 20,000 customers.

Chapter 3 Review Exercises page 239

1. Let x be the number of 30-day advance sale tickets and let y be the number of full-fare tickets:
$\begin{cases} x + y = 30 \\ 79x + 159y = 3970 \end{cases}$

2. Let x be the number of cows and let y be the number of horses: $\begin{cases} x + y = 420 \\ x - 2y = 0 \end{cases}$

3. $x = 13$, $y = 5$.

4. There are infinitely many solutions that may be parameterized as $x = 18 + 2t$, $y = t$. The system is dependent.

5. $x = 14$, $y = 30$.   **6.** No solution. The system is inconsistent.

7. $x = 9$, $y = 3$.   **8.** $x = -15$, $y = 24$.

9. He has 10 rosebushes and 15 tomato plants in his garden.

10. The fraternity should use 2 cars and 2 vans for the trip to Orlando.   **11.** 3×3; 5; 6; 4

12. 4×4; 7; 10; 13   **13.** $3R_2 \to \begin{pmatrix} 3 & 4 & | & 12 \\ 3 & 6 & | & 6 \end{pmatrix}$ **14.** $R_1 + R_2 \to \begin{pmatrix} 1 & -1 & | & 8 \\ 2 & -3 & | & 6 \end{pmatrix}$

15. $x = 15$, $y = 16$.   **16.** $x = 20$, $y = 30$.   **17.** No solution. The system is inconsistent.

18. $x = 3 + \frac{3}{4}t$, $y = t$ (infinitely many solutions). The system is dependent.

19. The pharmacist filled 63 prescriptions for antibiotics and 29 prescriptions for cough suppressants.

20. Each old *Life* magazine is $6 and each old *The New Yorker* is $5.

21. $x_1 = 3$, $x_2 = -3$, $x_3 = 6$.

22. The system is dependent, with solution $x_1 = 4 - t$, $x_2 = t$, $x_3 = 2$.

23. $\begin{pmatrix} 1 & 0 & 0 & | & 4 \\ 0 & 1 & 0 & | & -3 \\ 0 & 0 & 1 & | & 2 \end{pmatrix}$ **24.** $\begin{pmatrix} 1 & 0 & 0 & 1 & | & 0 \\ 0 & 1 & 0 & -1 & | & 0 \\ 0 & 0 & 1 & 0 & | & 0 \\ 0 & 0 & 0 & 0 & | & 1 \end{pmatrix}$

25. $x_1 = 4$, $x_2 = 3$, $x_3 = 1$, $x_4 = 2$.

26. The system is dependent, with solution $x_1 = 2 - t$, $x_2 = 3 - t$, $x_3 = t$, $x_4 = 4$.

27. The system is inconsistent, so there is no solution.

28. The system is inconsistent, so there is no solution.

29. The Nyack Nursery should raise 48 dahlias, 328 chrysanthemums, and 48 daisies.

30. Each hot dog costs $1. If the mirror house is $1.50 and a soda is $1, then a go-kart ride costs $4.50.

31. $\begin{pmatrix} 4 & -6 \\ 6 & -4 \end{pmatrix}$ **32.** $\begin{pmatrix} 1 & 2 & 3 & 4 \\ -4 & -3 & -2 & -1 \\ 1 & 2 & 3 & 4 \\ -4 & -3 & -2 & -1 \end{pmatrix}$ **33.** $\begin{pmatrix} 15 & 25 \\ 5 & 2 \end{pmatrix}$ **34.** $\begin{pmatrix} 12 & 3 & -2 \\ -2 & 12 & 11 \\ 9 & 3 & 1 \end{pmatrix}$

35. $\begin{pmatrix} 1 & 4 & 1 \\ 2 & 8 & 3 \\ 1 & 5 & 2 \end{pmatrix} \begin{pmatrix} x_1 \\ x_2 \\ x_3 \end{pmatrix} = \begin{pmatrix} 15 \\ 26 \\ 17 \end{pmatrix}$ **36.** $\begin{pmatrix} 2 & 3 & -1 & 1 \\ 5 & 4 & 1 & 2 \\ 2 & 1 & 1 & 1 \end{pmatrix} \begin{pmatrix} x_1 \\ x_2 \\ x_3 \\ x_4 \end{pmatrix} = \begin{pmatrix} 20 \\ 35 \\ 12 \end{pmatrix}$

37. Let T be the "this year" matrix and L be the "last year" matrix so that the "growth" matrix G is $G = T - L$. Choosing T to be a 3×2 matrix with the rows representing the grandchildren Thomas, Richard, and Harriet in that order and the columns representing their heights and weights in that order,

$$T = \begin{pmatrix} 61 & 90 \\ 54 & 75 \\ 47 & 60 \end{pmatrix} \text{ and } L = \begin{pmatrix} 58 & 80 \\ 52 & 70 \\ 46 & 55 \end{pmatrix} \text{ so then } G = \begin{pmatrix} 3 & 10 \\ 2 & 5 \\ 1 & 5 \end{pmatrix}.$$

38. Let N be the "number of items needed" column matrix and P be the "price" matrix so that the "cost of her order" matrix C is $C = P \cdot N$. Choosing the rows of N to represent the items "jacket," "blouse," "skirt," and "slacks" in that order, N is the 4×1 matrix $N = \begin{pmatrix} 200 \\ 300 \\ 250 \\ 175 \end{pmatrix}$. Then the columns of P must

also represent the items "jacket," "blouse," "skirt," and "slacks" in that order, so P is the 2×4 matrix $P = \begin{pmatrix} 195 & 85 & 145 & 130 \\ 190 & 90 & 150 & 125 \end{pmatrix}$ and the rows represent the prices of the East Coast designer and the Italian

team in the order. Then $C = \begin{pmatrix} 123{,}500 \\ 124{,}375 \end{pmatrix}$. The cost of her order from the East Coast designer is \$123,500, and from the Italian team it is \$124,375.

39. $\begin{pmatrix} 1 & 2 & 3 \\ 1 & 1 & 1 \\ 0 & 1 & 3 \end{pmatrix} \begin{pmatrix} -2 & 3 & 1 \\ 3 & -3 & -2 \\ -1 & 1 & 1 \end{pmatrix} = \begin{pmatrix} 1 & 0 & 0 \\ 0 & 1 & 0 \\ 0 & 0 & 1 \end{pmatrix}$ so this *is* a matrix and its inverse.

40. $\begin{pmatrix} -3 & 0 & 1 \\ 1 & 3 & 1 \\ -3 & 2 & 2 \end{pmatrix} \begin{pmatrix} -4 & -2 & 3 \\ 5 & 3 & -4 \\ -11 & -6 & 8 \end{pmatrix} = \begin{pmatrix} 1 & 0 & -1 \\ 0 & 1 & -1 \\ 0 & 0 & -1 \end{pmatrix}$ so this is *not* a matrix and its inverse.

41. $\begin{pmatrix} 3 & 0 & -1 \\ -2 & 0 & 1 \\ -6 & 1 & 3 \end{pmatrix}$　　**42.** Singular matrix　　**43.** $\begin{pmatrix} x_1 \\ x_2 \end{pmatrix} = \begin{pmatrix} 1 & -4 \\ -3 & 13 \end{pmatrix} \begin{pmatrix} 33 \\ 8 \end{pmatrix} = \begin{pmatrix} 1 \\ 5 \end{pmatrix}$ so $\begin{cases} x_1 = 1 \\ x_2 = 5 \end{cases}$

44. $\begin{pmatrix} x_1 \\ x_2 \end{pmatrix} = \begin{pmatrix} -5 & 8 \\ 2 & -3 \end{pmatrix} \begin{pmatrix} 25 \\ 16 \end{pmatrix} = \begin{pmatrix} 3 \\ 2 \end{pmatrix}$ so $\begin{cases} x_1 = 3 \\ x_2 = 2 \end{cases}$

45. $\begin{pmatrix} x_1 \\ x_2 \\ x_3 \end{pmatrix} = \begin{pmatrix} 1 & 1 & -3 \\ 0 & 1 & -1 \\ -2 & -3 & 8 \end{pmatrix} \begin{pmatrix} 11 \\ 7 \\ 5 \end{pmatrix} = \begin{pmatrix} 3 \\ 2 \\ -3 \end{pmatrix}$ so $\begin{cases} x_1 = 3 \\ x_2 = 2 \\ x_3 = -3 \end{cases}$

46. $\begin{pmatrix} x_1 \\ x_2 \\ x_3 \\ x_4 \end{pmatrix} = \begin{pmatrix} 1 & 0 & 0 & -1 \\ 0 & 1 & -2 & 0 \\ 0 & -2 & 6 & -1 \\ -1 & 0 & -1 & 2 \end{pmatrix} \begin{pmatrix} 7 \\ 10 \\ 4 \\ 6 \end{pmatrix} = \begin{pmatrix} 1 \\ 2 \\ -2 \\ 1 \end{pmatrix}$ so $\begin{cases} x_1 = 1 \\ x_2 = 2 \\ x_3 = -2 \\ x_4 = 1 \end{cases}$

47. The Kingman store can display 7 living room suites (2 in the window and 5 more only on the showroom floor) and 8 bedroom suites (3 in the window and 5 more only on the showroom floor); the Prescott store can display 11 living room suites (3 in the window and 8 more only on the showroom floor) and 10 bedroom suites (3 in the window and 7 more only on the showroom floor); and the Holbrook store can display 9 living room suites (3 in the window and 6 more only on the showroom floor) and 12 bedroom suites (2 in the window and 10 more only on the showroom floor).

48. Mr. Dahlman's taxes are $17,100 (federal), $7600 (state), and $3600 (city); Mrs. Farrell's taxes are $8550 (federal), $3800 (state), and $1800 (city); Ms. Mazlin's taxes are $13,680 (federal), $6080 (state), and $2880 (city); and Mr. Seidner's taxes are $25,650 (federal), $11,400 (state), and $5400 (city).

49. $\begin{pmatrix} 0.15 & 0.25 \\ 0.30 & 0.20 \end{pmatrix}$ for sectors A and B   **50.** $\begin{pmatrix} 0.10 & 0.20 & 0.30 & 0 \\ 0.20 & 0.10 & 0 & 0.20 \\ 0.30 & 0 & 0.10 & 0.30 \\ 0 & 0.20 & 0.30 & 0.10 \end{pmatrix}$ for sectors A, B, C, and D

51.

52.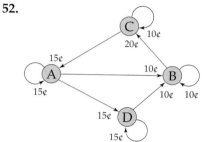

53. $Y = \begin{pmatrix} 371 \\ 449 \\ 419 \end{pmatrix}$ **54.** $Y = \begin{pmatrix} 170 \\ 115 \\ 160 \\ 143 \end{pmatrix}$ **55.** $X = \begin{pmatrix} 364 \\ 436 \end{pmatrix}$ **56.** $X = \begin{pmatrix} 457 \\ 605 \\ 529 \end{pmatrix}$

57. Each division must produce $4 million.
58. The other sectors of the economy can use $841 million of domestic oil and $1948 million of foreign oil, but all the military protection budget is consumed in the production of this output.
59. $y = 7x + 9$ **60.** $y = 73x + 14$ **61.** $y = 13x + 13$ **62.** $y = 19x - 52$
63. They can expect sales of $4405.   **64.** She can expect 12 minor accidents.

Exercises 4.1 page 256

1. b   **3.** c   **5.** a   **7.** b
9. The vertices are (0, 0), (40, 0), and (0, 20). The region is bounded.   **11.** The vertices are (0, 0), (10, 0), (10, 30), and (0, 10). The region is bounded.

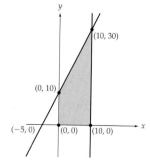

13. The vertices are $(0, 0)$, $(6, 0)$, $(4, 2)$, and $(0, 4)$. The region is bounded.

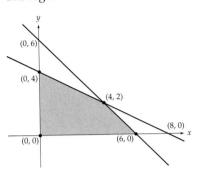

15. The vertices are $(4, 0)$ and $(0, 10)$. The region is unbounded.

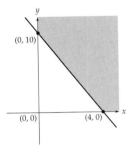

17. The vertices are $(3, 4)$ and $(0, 8)$. The region is unbounded.

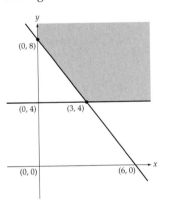

19. The vertices are $(8, 0)$, $(2, 6)$, and $(0, 12)$. The region is unbounded.

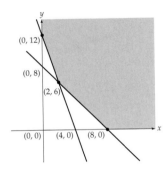

21. The vertices are $(30, 0)$, $(40, 0)$, $(0, 80)$, and $(0, 10)$. The region is bounded.

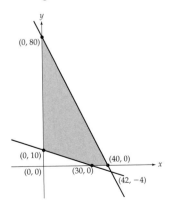

23. The vertices are $(15, 0)$ and $(0, 30)$. The region is unbounded.

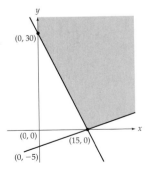

25. Let x be the number of goats and let y be the number of llamas.

$$\begin{cases} 2x + 5y \le 400 & \text{(Land)} \\ 100x + 80y \le 13{,}200 & \text{(Money)} \\ x \ge 0, \ y \ge 0 & \text{(Nonnegativity)} \end{cases}$$

The vertices are $(0, 0)$, $(132, 0)$, $(100, 40)$, and $(0, 80)$.

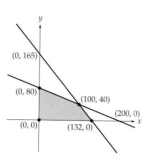

27. Let x be the number of dinghies and let y be the number of rowboats.

$$\begin{cases} 2x + 3y \le 120 & \text{(Metal work)} \\ 2x + 2y \le 100 & \text{(Painting)} \\ x \ge 0, \ y \ge 0 & \text{(Nonnegativity)} \end{cases}$$

The vertices are $(0, 0)$, $(50, 0)$, $(30, 20)$, and $(0, 40)$.

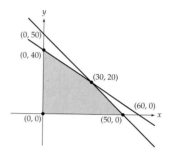

29. Let x be the number of servings of SugarSnaks and let y be the number of bags of Gobbl'Ems.

$$\begin{cases} 5x + 8y \le 80 & \text{(Fat)} \\ 125x + 250y \le 2250 & \text{(Calories)} \\ x \ge 0, \ y \ge 0 & \text{(Nonnegativity)} \end{cases}$$

The vertices are $(0, 0)$, $(16, 0)$, $(8, 5)$, and $(0, 9)$.

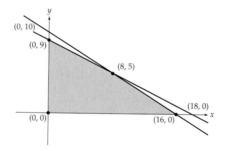

31. Let x be the number of hours the Ohio factory operates and let y be the number of hours the Pennsylvania factory operates.

$$\begin{cases} 4x + 4y \le 64 & \text{(Sulfur dioxide)} \\ 5x + 3y \le 60 & \text{(Particulates)} \\ x \ge 0, \ y \ge 0 & \text{(Nonnegativity)} \end{cases}$$

The vertices are $(0, 0)$, $(12, 0)$, $(6, 10)$, and $(0, 16)$.

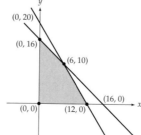

33. Let x be the amount of money (in millions of dollars) invested in stock funds and let y be the amount of money (in millions of dollars) invested in bond funds.

$$\begin{cases} x + y \leq 8 & \text{(Money)} \\ x \leq y & \text{(Limit risk)} \\ x \geq 0, \ y \geq 0 & \text{(Nonnegativity)} \end{cases}$$

The vertices are $(0, 0)$, $(4, 4)$, and $(0, 8)$.

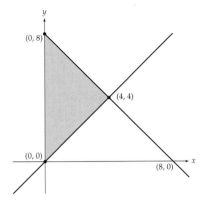

Exercises 4.2 page 269

1. The maximum is 45 (when $x = 15$, $y = 15$). **3.** The minimum is 30 (when $x = 10$, $y = 0$).

5. The maximum is 15 (when $x = 15$, $y = 0$). **7.** The minimum is 60 (when $x = 0$, $y = 60$).

9. There is no maximum because the region is unbounded in the positive y direction and P increases for increasing y values.

11. The maximum is 400 (when $x = 0$, $y = 10$). **13.** The minimum is 150 (when $x = 10$, $y = 0$).

15. The maximum is 160 (when $x = 15$, $y = 20$). **17.** The minimum is 232 (when $x = 6$, $y = 16$).

19. The maximum is 230 (when $x = 40$, $y = 10$). **21.** The maximum is 1080 (when $x = 0$, $y = 90$).

23. The minimum is 885 (when $x = 33$, $y = 9$).

25. The rancher should raise 100 goats and 40 llamas to obtain the greatest possible profit of $9600.

27. The company should manufacture 32 prams and 9 yawls to obtain the greatest possible profit of $6420.

29. The county should operate the Norton incinerator 4 hours each day and the Wiseburg incinerator 6 hours each day to obtain the least cost of $620.

31. Each bunny should receive 30 handfuls of greens and 16 drops of supplement each week to minimize the owner's costs at 76¢.

33. No grassland and 2400 acres of forest should be reclaimed this year to raise the greatest possible amount of $360,000 in long-term leases for use in next year's reclamation efforts.

Exercises 4.3 page 289

1.

	x_1	x_2	s_1	s_2	
s_1	3	2	1	0	12
s_2	6	1	0	1	15
P	−8	−9	0	0	0

3.

	x_1	x_2	s_1	s_2	s_3	
s_1	4	3	1	0	0	12
s_2	5	2	0	1	0	20
s_3	1	6	0	0	1	12
P	−13	−7	0	0	0	0

5.

	x_1	x_2	x_3	x_4	s_1	s_2	
s_1	2	1	1	3	1	0	6
s_2	1	4	-2	1	0	1	8
P	-5	2	-10	5	0	0	0

7.

	x_1	x_2	x_3	s_1	s_2	s_3	
s_1	8	1	4	1	0	0	32
s_2	3	5	7	0	1	0	30
s_3	6	2	9	0	0	1	28
P	-10	-20	-15	0	0	0	0

9. The smallest negative entry in the bottom row is -8, so the pivot column is column 2. The first row may not be considered for the pivot row since the pivot column entry is 0. The pivot row is row 2 and the pivot element is the 1 in column 2 and row 2. Since the pivot element is already 1, the pivot operation consists of one row operation:

		x_1	x_2	s_1	s_2	
	s_1	2	0	1	0	4
	x_2	1	1	0	1	5
$R_3 + 8R_2 \rightarrow$	P	1	0	0	8	40

11. The smallest negative entry in the bottom row is -5, so the pivot column is column 2. The ratios are $\frac{12}{2} = 6$ and $\frac{8}{1} = 8$, so the pivot row is row 1. The pivot element is the 2 in column 2 and row 1 of the simplex tableau. The first step of the pivot operation is the row operation $\frac{1}{2}R_1 \rightarrow R_1$ to replace the pivot element with a 1. Then:

		x_1	x_2	x_3	x_4	s_1	s_2	
	x_2	2	1	3	1	1/2	0	6
$R_2 - R_1 \rightarrow$	s_2	1	0	-1	0	-1/2	1	2
$R_3 + 5R_1 \rightarrow$	P	6	0	21	2	5/2	0	30

13. No pivot column because there are no negative entries in the bottom row. The solution has been found: The maximum is 90 when $x_1 = 10$, $x_2 = 15$, $x_3 = 0$, and $x_4 = 0$ (and $s_1 = 10$, $s_2 = 0$, $s_3 = 0$).

15. The smallest negative entry in the bottom row is -10 so the pivot column is column 5. Since all of the entries in the pivot column are zero or negative, there is no pivot row. There is no maximum value.

17. The maximum is 28 when $x_1 = 0$, $x_2 = 14$.

19. The maximum is 2000 when $x_1 = 0$, $x_2 = 50$, $x_3 = 0$.

21. There is no maximum (the second pivot column is column 2 but there is no pivot row).

23. The maximum is 300 when $x_1 = 0$, $x_2 = 20$, $x_3 = 0$, $x_4 = 40$.

25. The maximum is 16 when $x_1 = 3$, $x_2 = 2$.

27. The maximum is 12,600 when $x_1 = 0$, $x_2 = 160$, $x_3 = 60$.

29. The maximum is 30 when $x_1 = 5$, $x_2 = 5$, $x_3 = 5$.

31. The shop should rebuild no carburetors, 35 fuel pumps, and 20 alternators for a greatest possible profit of $690.

33. The recycling center should accept 300 crates of paper products and 500 crates of glass bottles to raise the greatest possible amount of $59 each week.

35. The company should process no agates, 49 trays of onyxes, and 7 trays of garnets each day to obtain the greatest possible profit of $581.

37. The politician should run 3 daytime ads, 7 prime time ads, and no late night ads to reach 47,000 voters, the most possible.

39. The farmer should plant 30 acres of corn, 90 acres of peanuts, and 120 acres of soybeans for the greatest possible profit of $43,500.

43. The final tableau is

	x_1	x_2	x_3	x_4	s_1	s_2	s_3	
x_3	0	0	1	0	0	0	1	1
s_1	0	-8	0	30	1	-2	3	3
x_1	1	-24	0	6	0	2	1	1
P	0	2	0	21/2	0	3/2	5/4	5/4

The maximum is $\frac{5}{4}$ when $x_1 = 1$, $x_2 = 0$, $x_3 = 1$, $x_4 = 0$.

45. No.   **47.** The maximum is 100 when $x_1 = 100$, $x_2 = 0$.

49. Maximize $P = (1 \quad 10 \quad 100 \quad 1000) \begin{pmatrix} x_1 \\ x_2 \\ x_3 \\ x_4 \end{pmatrix}$

Subject to $\begin{pmatrix} 0 & 0 & 0 & 1 \\ 0 & 0 & 1 & 20 \\ 0 & 1 & 20 & 200 \\ 1 & 20 & 200 & 2000 \end{pmatrix} \begin{pmatrix} x_1 \\ x_2 \\ x_3 \\ x_4 \end{pmatrix} \leq \begin{pmatrix} 1 \\ 100 \\ 10,000 \\ 1,000,000 \end{pmatrix}$ and $\begin{pmatrix} x_1 \\ x_2 \\ x_3 \\ x_4 \end{pmatrix} \geq 0.$

The maximum is 1,000,000 when $x_1 = 1,000,000$, $x_2 = 0$, $x_3 = 0$, $x_4 = 0$.

Exercises 4.4 page 309

1. The dual problem is maximize $P = 21x_1$ subject to $\begin{cases} 3x_1 + 4x_2 \leq 84 \\ x_1 - x_2 \leq 21 \\ x_1 \geq 0, \quad x_2 \geq 0 \end{cases}$

with initial simplex tableau

	x_1	x_2	s_1	s_2	
s_1	3	4	1	0	84
s_2	1	-1	0	1	21
P	-21	0	0	0	0

3. The dual problem is maximize $P = 180x_1 + 120x_2$ subject to $\begin{cases} x_1 + 4x_2 \leq 60 \\ 2x_1 + 5x_2 \leq 100 \\ 3x_1 + 6x_2 \leq 300 \\ x_1 \geq 0, \quad x_2 \geq 0 \end{cases}$

with initial simplex tableau

	x_1	x_2	s_1	s_2	s_3	
s_1	1	4	1	0	0	60
s_2	2	5	0	1	0	100
s_3	3	6	0	0	1	300
P	-180	-120	0	0	0	0

5. The dual problem is maximize $P = 150x_1 + 100x_2 + 228x_3$ subject to $\begin{cases} 3x_1 + x_2 + 3x_3 \leq 3 \\ 2x_1 + 4x_2 + 4x_3 \leq 20 \\ x_1 \geq 0, \quad x_2 \geq 0, \quad x_3 \geq 0 \end{cases}$

with initial simplex tableau

	x_1	x_2	x_3	s_1	s_2	
s_1	3	1	3	1	0	3
s_2	2	4	4	0	1	20
P	-150	-100	-228	0	0	0

7. The dual problem is maximize $P = -30x_1 + 30x_2$ subject to $\begin{cases} -x_1 + x_2 \leq 15 \\ x_1 + 2x_2 \leq 20 \\ 2x_1 + x_2 \leq 5 \\ x_1 \geq 0, \ x_2 \geq 0 \end{cases}$

with initial simplex tableau

	x_1	x_2	s_1	s_2	s_3	
s_1	-1	1	1	0	0	15
s_2	1	2	0	1	0	20
s_3	2	1	0	0	1	5
P	30	-30	0	0	0	0

9. The minimum is 300 when $y_1 = 20$, $y_2 = 0$.   **11.** The minimum is 45 when $y_1 = 0$, $y_2 = 15$.
13. The minimum is 43 when $y_1 = 7$, $y_2 = 3$.   **15.** The minimum is 10 when $y_1 = 0$, $y_2 = 10$.
17. The minimum is 900 when $y_1 = 0$, $y_2 = 40$, $y_3 = 10$.
19. The minimum is 98 when $y_1 = 2$, $y_2 = 2$, $y_3 = 0$.   **21.** There is no minimum.
23. The minimum is 2376 when $y_1 = 6$, $y_2 = 12$, $y_3 = 6$.
25. The athlete should use 8 Bulk-Up Bars and 4 cans of Power Drink to receive the needed fat and protein at the least possible cost of $5.64.
27. The office manager should buy 45 packages from Jack's Office Supplies and 30 packages from John's Discount to restock the store room at the least possible cost of $1230.
29. The project engineer should use 350 heavy-duty dump truck loads of dirt, 300 heavy-duty dump truck loads of crushed rock, no regular dump truck loads of dirt, and 150 regular dump truck loads of crushed rock to get the project finished on time at the least possible cost of $66,000.
31. The farmer should buy 8000 pounds of Miracle Mix and 18,000 pounds of the store brand to meet the fertilizer needs of his field at the least possible cost of $2640.
33. The Kentucky warehouse should ship 200 cartons to Kansas, 200 cartons to Texas, and none to Oregon, and the Utah warehouse should ship none to Kansas, 100 cartons to Texas, and 100 cartons to Oregon to incur the smallest possible shipping cost of $1600.
39. The final tableau shows that the solutions of the minimum and maximum problems are both 3390 with the variables taking the following values.

Minimum problem	$y_1 = 7$	$y_2 = 3$	$y_3 = 1$	$y_4 = 0$	$t_1 = 0$	$t_2 = 0$	$t_3 = 0$
Maximum problem	$s_1 = 0$	$s_2 = 0$	$s_3 = 0$	$s_4 = 75$	$x_1 = 60$	$x_2 = 45$	$x_3 = 30$

Thus one of each pair x_1 and t_1, x_2 and t_2, x_3 and t_3, s_1 and y_1, s_2 and y_2, s_3 and y_3, and s_4 and y_4 is zero.

Exercises 4.5 page 322

1.

	x_1	x_2	x_3	s_1	s_2	
s_1	3	2	8	1	0	96
s_2	-5	-1	-6	0	1	-30
P	-15	-20	-18	0	0	0

The dual pivot element is the -1 in row 2 and column 2.

3.

	x_1	x_2	x_3	s_1	s_2	
s_1	-2	-1	-3	1	0	-30
s_2	-1	-1	-2	0	1	-20
P	-6	-4	-6	0	0	0

The dual pivot element is the -1 in row 1 and column 2.

5.

	x_1	x_2	x_3	s_1	s_2	s_3	
s_1	−1	−1	−1	1	0	0	−8
s_2	1	2	3	0	1	0	30
s_3	1	2	1	0	0	1	18
P	−20	−30	−10	0	0	0	0

The dual pivot element is the −1 in row 1 and column 2.

7.

	x_1	x_2	x_3	x_4	s_1	s_2	s_3	
s_1	−1	−1	−1	−1	1	0	0	−30
s_2	−2	−3	−2	−1	0	1	0	−20
s_3	4	2	1	2	0	0	1	80
P	−3	−2	−5	−4	0	0	0	0

The dual pivot element is the −1 in row 1 and column 3.

9. The maximum is 110 when $x_1 = 20$, $x_2 = 30$.

11. The maximum is 1200 when $x_1 = 0$, $x_2 = 60$, $x_3 = 0$.

13. There is no maximum.   **15.** The maximum is 270 when $x_1 = 2$, $x_2 = 20$, $x_3 = 0$.

17. The maximum is 13 when $x_1 = 8$, $x_2 = 5$.

19. The maximum is 720 when $x_1 = 0$, $x_2 = 0$, $x_3 = 60$, $x_4 = 0$.

21. The maximum is 808 when $x_1 = 0$, $x_2 = 14$, $x_3 = 0$, $x_4 = 24$.

23. The maximum is 30 when $x_1 = 10$, $x_2 = 40$, $x_3 = 0$.

25. The retired couple should invest $15,000 in certificates of deposit and $5000 in treasury bonds.

27. The manager should place 5 newspaper ads, 15 radio commercials, and 5 TV spots to reach 295,000 potential customers, the greatest possible number.

29. The furniture shop should manufacture 20 desks, 10 tables, and 80 chairs to obtain the greatest possible profit of $7620.

31. The farmer should grow 100 acres of wheat, 300 acres of barley, and 100 acres of oats to obtain the greatest possible profit of $21,500.

33. The electric power plant should purchase 40,000 tons of low-sulfur coal and 50,000 tons of high-sulfur coal to obtain the most energy of 2,300,000 million BTUs.

35. a. The minimum of C is 12 when $x = 4$, $y = 0$, while the maximum of P is −12, also when $x = 4$, $y = 0$. Yes.

b. The simplex tableaux for the dual of this problem are

	x_1	x_2	s_1	s_2	
s_1	−1	2	1	0	3
s_2	−1	1	0	1	4
P	10	−8	0	0	0

and then

	x_1	x_2	s_1	s_2	
x_2	−1/2	1	1/2	0	3/2
s_2	−1/2	0	−1/2	1	5/2
P	6	0	4	0	12

c. The simplex tableaux for this nonstandard problem are

	x_1	x_2	s_1	s_2	
s_1	1	1	1	0	10
s_2	−2	−1	0	1	−8
P	3	4	0	0	0

and then

	x_1	x_2	s_1	s_2	
s_1	0	1/2	1	1/2	6
x_1	1	1/2	0	−1/2	4
P	0	5/2	0	3/2	−12

d. Yes; yes and yes.

37. The maximum is 16 when $x_1 = 2$, $x_2 = 2$.   **39.** The maximum is 16 when $x_1 = 2$, $x_2 = 2$.

41. The maximum is 23 when $x_1 = 1$, $x_2 = 4$.

Exercises 4.6 page 337

1.

	x_1	x_2	x_3	s_1	s_2	a_1	
s_1	3	2	8	1	0	0	96
a_1	5	1	6	0	-1	1	30
P	$-5M - 15$	$-M - 20$	$-6M - 18$	0	M	0	$-30M$

3.

	x_1	x_2	x_3	s_1	s_2	a_1	a_2	
a_1	2	1	3	-1	0	1	0	30
a_2	1	1	2	0	-1	0	1	20
P	$-3M - 6$	$-2M - 4$	$-5M - 6$	M	M	0	0	$-50M$

5.

	x_1	x_2	x_3	s_1	s_2	s_3	a_1	
s_1	1	2	3	1	0	0	0	30
s_2	1	2	1	0	1	0	0	18
a_1	1	1	1	0	0	-1	1	8
P	$-M - 20$	$-M - 30$	$-M - 10$	0	0	M	0	$-8M$

7.

	x_1	x_2	x_3	x_4	s_1	s_2	a_1	a_2	a_3	
a_1	1	1	1	1	-1	0	1	0	0	30
a_2	2	3	2	1	0	-1	0	1	0	20
a_3	4	2	1	2	0	0	0	0	1	80
P	$-7M - 3$	$-6M - 2$	$-4M - 5$	$-4M - 4$	M	M	0	0	0	$-130M$

9. The maximum is 110 when $x_1 = 20, \ x_2 = 30$.

11. The maximum is 1200 when $x_1 = 0, \ x_2 = 60, \ x_3 = 0$.

13. There is no maximum.   **15.** The maximum is 270 when $x_1 = 2, \ x_2 = 20, \ x_3 = 0$.

17. The maximum is 13 when $x_1 = 8, \ x_2 = 5$.

19. The maximum is 480 when $x_1 = 0, \ x_2 = 40, \ x_3 = 20, \ x_4 = 0$.

21. The maximum is 808 when $x_1 = 0, \ x_2 = 14, \ x_3 = 0, \ x_4 = 24$.

23. The maximum is 30 when $x_1 = 10, \ x_2 = 40, \ x_3 = 0$.

25. The retired couple should invest \$15,000 in certificates of deposit and \$5000 in Treasury bonds.

27. The manager should place 5 newspaper ads, 15 radio commercials, and 5 TV spots to reach 295,000 potential customers, the greatest possible number.

29. The furniture shop should manufacture 20 desks, 10 tables, and 80 chairs to obtain the greatest possible profit of \$7620.

31. The farmer should grow 100 acres of wheat, 300 acres of barley, and 100 acres of oats to obtain the greatest possible profit of \$21,500.

33. The electric power plant should purchase 40,000 tons of low-sulfur coal and 50,000 tons of high-sulfur coal to obtain the most energy of 2,300,000 million BTUs.

35. a. The minimum of C is 12 when $x = 4, \ y = 0$, while the maximum of P is -12, also when $x = 4$, $y = 0$. Yes.

37. The maximum is 16 when $x_1 = 2, \ x_2 = 2$.   **39.** The maximum is 23 when $x_1 = 1, \ x_2 = 4$.

Chapter 4 Review Exercises page 344

1.

(8, 0)

(0, −12)

2.

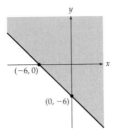

(−6, 0)

(0, −6)

3. The vertices are $(0, 0)$, $(20, 0)$, and $(0, 20)$. The region is bounded.

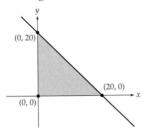

(0, 20)

(20, 0)

(0, 0)

4. The vertices are $(10, 0)$ and $(0, 5)$. The region is unbounded.

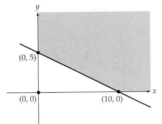

(0, 5)

(0, 0)   (10, 0)

5. The vertices are $(0, 0)$, $(10, 0)$, $(5, 10)$, and $(0, 15)$. The region is bounded.

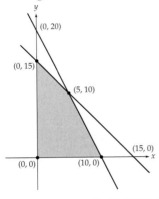

(0, 20)

(0, 15)

(5, 10)

(15, 0)

(0, 0)   (10, 0)

6. The vertices are $(8, 0)$, $(4, 4)$, and $(0, 12)$. The region is unbounded.

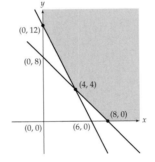

(0, 12)

(0, 8)

(4, 4)

(8, 0)

(0, 0)   (6, 0)

7. The vertices are $(2, 0)$, $(8, 0)$, $(4, 6)$, and $(2, 8)$. The region is bounded.

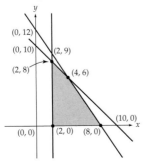

(0, 12)

(0, 10)   (2, 9)

(2, 8)   (4, 6)

(10, 0)

(0, 0)   (2, 0)   (8, 0)

8. The vertices are $(0, 0)$, $(9, 0)$, $(9, 1)$, $(3, 7)$, and $(0, 4)$. The region is bounded.

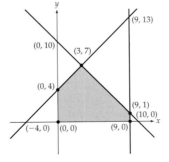

(9, 13)

(0, 10)   (3, 7)

(0, 4)

(9, 1)
(10, 0)

(−4, 0)   (0, 0)   (9, 0)

9. Let x be the number of Irish setters and let y be the number of Labrador retrievers.

$$\begin{cases} \frac{1}{2}x + \frac{3}{4}y \le 6 & \text{(Time in hours)} \\ 8x + 8y \le 80 & \text{(Dog treats)} \\ x \ge 0, \ y \ge 0 & \text{(Nonnegativity)} \end{cases}$$

The vertices are $(0, 0)$, $(10, 0)$, $(6, 4)$, and $(0, 8)$.

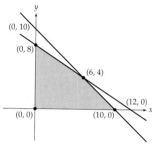

10. Let x be the number of pounds of SongBird brand bird seed and let y be the number of pounds of MeadowMix brand bird seed.

$$\begin{cases} 2x + 4y \ge 104 & \text{(Sunflower hearts)} \\ 3x + 2y \ge 84 & \text{(Crushed peanuts)} \\ x \ge 0, \ y \ge 0 & \text{(Nonnegativity)} \end{cases}$$

The vertices are $(52, 0)$, $(16, 18)$, and $(0, 42)$.

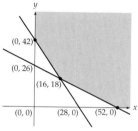

11. The maximum is 40 when $x = 0$, $y = 10$.   **12.** The minimum is 10 when $x = 0$, $y = 5$.

13. There is no maximum because the region is unbounded to the right and upward, and P increases in these directions.

14. The minimum is 60 when $x = 30$, $y = 0$.   **15.** The maximum is 240 when $x = 12$, $y = 0$.

16. The minimum is 2400 when $x = 48$, $y = 24$.   **17.** The maximum is 350 when $x = 4$, $y = 9$.

18. The minimum is 140 when $x = 20$, $y = 0$.

19. The shop should make 24 wall clock cases and 15 mantle clock cases to obtain the greatest possible profit of $9000.

20. The cat's nutritional needs can be met by 30 ounces of canned food and 4 ounces of dry food at a least cost of $1.82.

21.

	x_1	x_2	x_3	s_1	s_2	
s_1	5	2	3	1	0	30
s_2	2	3	4	0	1	24
P	-4	-10	-9	0	0	0

22.

	x_1	x_2	s_1	s_2	s_3	
s_1	1	1	1	0	0	10
s_2	2	1	0	1	0	14
s_3	1	-1	0	0	1	4
P	-5	-7	0	0	0	0

23. The smallest negative entry in the bottom row is -8, so the pivot column is column 3. The ratios are $\frac{6}{1} = 6$, $\frac{8}{2} = 4$, and (omit), so the pivot row is row 2. The pivot element is the 2 in column 3 and row 2 of the simplex tableau. The first step of the pivot operation is the row operation $\frac{1}{2}R_2 \to R_2$ to replace the pivot element with a 1. Then:

	x_1	x_2	x_3	s_1	s_2	s_3	
$R_1 - R_2 \to s_1$	$-1/2$	$1/2$	0	1	$-1/2$	0	2
x_3	$3/2$	$1/2$	1	0	$1/2$	0	4
$R_3 + 3R_2 \to s_3$	$11/2$	$5/2$	0	0	$3/2$	1	18
$R_4 + 8R_2 \to P$	6	8	0	0	4	0	32

24. No pivot column because there are no negative entries in the bottom row. The solution has been found: The maximum is 150 when $x_1 = 0$ and $x_2 = 10$ (and $s_1 = 8$, $s_2 = 42$, and $s_3 = 0$).

25. The maximum is 96 when $x_1 = 0$, $x_2 = 12$, $x_3 = 0$.

26. The maximum is 90 when $x_1 = 0$, $x_2 = 9$, $x_3 = 0$.

27. The maximum is 37 when $x_1 = 2$, $x_2 = 3$, $x_3 = 0$, $x_4 = 0$.

28. The maximum is 61 when $x_1 = 8$, $x_2 = 0$, $x_3 = 7$.

29. The publisher should print no trade edition copies, no book club edition copies, and 250,000 paperback edition copies for the greatest possible profit of $1,000,000.

30. The plastics factory should produce 20 batches of toy racing cars, no toy jet airplanes, and 10 batches of toy speed boats for the greatest possible profit of $3500.

31. It is a standard problem because $\begin{pmatrix} 10 \\ 15 \\ 20 \end{pmatrix} \geq 0$. The dual problem is

$$\text{Maximize } P = 40x_1 + 30x_2$$

$$\text{Subject to } \begin{cases} 2x_1 + x_2 \leq 10 \\ -x_1 + x_2 \leq 15 \\ x_1 + 3x_2 \leq 20 \\ x_1 \geq 0 \text{ and } x_2 \geq 0 \end{cases}$$

The initial simplex tableau is:

	x_1	x_2	s_1	s_2	s_3	
s_1	2	1	1	0	0	10
s_2	-1	1	0	1	0	15
s_3	1	3	0	0	1	20
P	-40	-30	0	0	0	0

32. It is a standard problem because $\begin{pmatrix} 30 \\ 20 \end{pmatrix} \geq 0$. The dual problem is

$$\text{Maximize } P = 210x_1 + 252x_2 - 380x_3$$

$$\text{Subject to } \begin{cases} 5x_1 + 3x_2 - 5x_3 \leq 30 \\ 2x_1 + 4x_2 - 4x_3 \leq 20 \\ x_1 \geq 0, \ x_2 \geq 0, \text{ and } x_3 \geq 0 \end{cases}$$

The initial simplex tableau is:

	x_1	x_2	x_3	s_1	s_2	
s_1	5	3	-5	1	0	30
s_2	2	4	-4	0	1	20
P	-210	-252	380	0	0	0

33. It is a standard problem because $\begin{pmatrix} 42 \\ 36 \end{pmatrix} \geq 0$. The dual problem is

$$\text{Maximize } P = 84x_1 + 18x_2$$

$$\text{Subject to } \begin{cases} 7x_1 + x_2 \leq 42 \\ 4x_1 + x_2 \leq 36 \\ x_1 \geq 0 \text{ and } x_2 \geq 0 \end{cases}$$

The initial simplex tableau is:

	x_1	x_2	s_1	s_2	
s_1	7	1	1	0	42
s_2	4	1	0	1	36
P	-84	-18	0	0	0

34. It is a standard problem because $\begin{pmatrix} 130 \\ 40 \\ 98 \end{pmatrix} \geq 0$. The dual problem is

$$\text{Maximize} \quad P = 51x_1 + 60x_2 + 57x_3$$

$$\text{Subject to} \quad \begin{cases} 3x_1 + 4x_2 + 3x_3 \leq 130 \\ x_1 + x_2 + x_3 \leq 40 \\ 2x_1 + 2x_2 + 3x_3 \leq 98 \\ x_1 \geq 0, \ x_2 \geq 0, \ \text{and} \ x_3 \geq 0 \end{cases}$$

The initial simplex tableau is:

	x_1	x_2	x_3	s_1	s_2	s_3	
s_1	3	4	3	1	0	0	130
s_2	1	1	1	0	1	0	40
s_3	2	2	3	0	0	1	98
P	-51	-60	-57	0	0	0	0

35. The minimum is 192 when $y_1 = 8, \ y_2 = 5, \ y_3 = 0$.

36. The minimum is 510 when $y_1 = 11, \ y_2 = 5$.   **37.** There is no solution.

38. The minimum is 60 when $y_1 = 0, \ y_2 = 12, \ y_3 = 0$.

39. The student should buy no single pens, 3 packages of ink cartridges, and 3 "writer's combo" packages to spend the least possible amount of $6.24.

40. The pasta company should purchase 18 small-capacity machines and 8 large-capacity machines to expand its linguini production at the least possible cost of $138,000.

41.

	x_1	x_2	s_1	s_2	s_3	
s_1	4	1	1	0	0	40
s_2	2	3	0	1	0	60
s_3	-1	-1	0	0	1	-10
P	-80	-30	0	0	0	0

The dual pivot element is the -1 in row 3 and column 1.

42.

	x_1	x_2	x_3	x_4	s_1	s_2	
s_1	2	3	-3	1	1	0	30
s_2	-7	5	-1	-5	0	1	-35
P	-4	-8	-6	-10	0	0	0

The dual pivot element is the -1 in row 2 and column 3.

43. The maximum is 40 when $x_1 = 0, \ x_2 = 10$.

44. The maximum is 280 when $x_1 = 35, \ x_2 = 0, \ x_3 = 0$.

45. The maximum is 48 when $x_1 = 8, \ x_2 = 0$.

46. The maximum is 300 when $x_1 = 25, \ x_2 = 0, \ x_3 = 0$.

47. The maximum is 59 when $x_1 = 4$, $x_2 = 5$.

48. The maximum is 300 when $x_1 = 0$, $x_2 = 0$, $x_3 = 30$.

49. The sawmill should produce 6 thousand board-feet of rough-cut lumber and 4 thousand board-feet of finished-grade boards each day to obtain the greatest possible profit of $1120.

50. The money manager should invest $100,000 in U.S. bonds and $50,000 in Canadian stocks (and nothing in U.S. stocks or in Canadian bonds) to obtain the greatest possible return of $10,500.

51.

	x_1	x_2	s_1	s_2	s_3	a_1	
s_1	4	1	1	0	0	0	40
s_2	2	3	0	1	0	0	60
a_1	1	1	0	0	-1	1	10
P	$-M - 80$	$-M - 30$	0	0	M	0	$-10M$

52.

	x_1	x_2	x_3	x_4	s_1	s_2	a_1	
s_1	2	3	-3	1	1	0	0	30
a_1	7	-5	1	5	0	-1	1	35
P	$-7M - 4$	$5M - 8$	$-M - 6$	$-5M - 10$	0	M	0	$-35M$

53. The maximum is 40 when $x_1 = 0$, $x_2 = 10$.

54. The maximum is 280 when $x_1 = 35$, $x_2 = 0$, $x_3 = 0$.

55. The maximum is 48 when $x_1 = 8$, $x_2 = 0$.

56. The maximum is 300 when $x_1 = 25$, $x_2 = 0$, $x_3 = 0$.

57. The maximum is 59 when $x_1 = 4$, $x_2 = 5$.

58. The maximum is 300 when $x_1 = 0$, $x_2 = 0$, $x_3 = 30$.

59. The sawmill should produce 6 thousand board-feet of rough-cut lumber and 4 thousand board-feet of finished-grade boards each day to obtain the greatest possible profit of $1120.

60. The money manager should invest $100,000 in U.S. bonds and $50,000 in Canadian stocks (and nothing in U.S. stocks and in Canadian bonds) to obtain the greatest possible return of $10,500.

Exercises 5.1 page 360

1. 27   **3.** 60   **5.** 44   **7.** 31   **9.** 17

11.

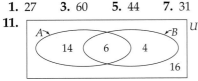

13. 24   **15.** 20   **17.** 30

19.

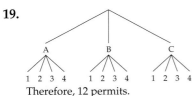

Therefore, 12 permits.

21.

Windbreakers — Red, Blue; Ski Jackets — Red, Blue; Overcoats — Red, Blue
Therefore, 6 kinds.

23. 17   **25.** 200   **27.** $25 \cdot 9 \cdot 9 = 2025$ **29.** $15^8 = 2{,}562{,}890{,}625$ **31.** $10^4 = 10{,}000$

33. $3 \cdot 4 \cdot 4 \cdot 5 = 240$

35. According to the Addition Principle, the number of cars that were either convertibles or had out-of-state plates is $12 + 15 - 5 = 22$, which is impossible since he searched only 20 cars.

37. $2^6 = 64$ **39.** $2^{10} - 1 = 1024 - 1 = 1023$

41. The members of $(A \cup B)^c$ are $8, 9,$ and $10,$ and these are also the members of $A^c \cap B^c$.

43.

$(A \cup B)^c = $ ▧ $A^c = $ ▨ $B^c = $ ▤
$A^c \cap B^c = $ ▧
Note that the ▧ shadings agree.

Exercises 5.2 page 372

1. a. 2   **b.** 720   **3.** 720   **5.** 39,800   **7.** 120   **9.** 8

11. a. 17,160   **b.** 154,440   **c.** 1,235,520   **13. a.** 10   **b.** 30,240   **c.** 3,628,800

15. 15   **17.** 35   **19. a.** 330   **b.** 462   **c.** 462   **d.** 330

21. a. 12   **b.** 66   **c.** 924   **d.** 12   **e.** 1

23. $8! = 40,320$ **25.** $_{12}P_4 = 11,880$ **27.** $8 \cdot {_9}P_6 = 483,840$

29. $36 \cdot 35 \cdot 34 \cdot 33 = 1,413,720; \ 34 \cdot 33 \cdot 32 \cdot 31 = 1,113,024$ **31.** $_6P_2 = 30$ **33.** $_{12}C_5 = 792$

35. $_{13}C_5 = 1287; \ 4 \cdot 1287 = 5148$ **37.** $(_{10}C_4)(_{12}C_4) = 103,950$ **39.** $_{12}C_{10} = {_{12}}C_2 = 66$

41. $_{100}C_5 = 75,287,520; \ \frac{100 \cdot 98 \cdot 96 \cdot 94 \cdot 92}{5!} = 67,800,320$ **43. a.** 362,880   **b.** Over 31 years

45. Choosing one factor from n factors can be done in $_nC_1$ ways, choosing two factors from n can be done in $_nC_2$ ways, and so on. The left-hand sides of the equations are equal, and since $_nC_0 = {_n}C_n = 1,$ so are the right-hand sides.

47. Using the fifth row of Pascal's triangle: $(x + y)^4 = x^4 + 4x^3y + 6x^2y^2 + 4xy^3 + y^4.$

49. a. The left-hand side of the binomial theorem becomes $(1 + 1)^n = 2^n$ and the right-hand side, after omitting the 1s, becomes $_nC_0 + {_n}C_1 + {_n}C_2 + \cdots + {_n}C_n.$

b. The left-hand side is the total number of committees that can be made from n people (including an "empty" committee and the committee of all n), and the right-hand side gives the number of committees of size 0, size 1, size 2, and so on up to size n, which must add up to the *total* number of committees.

Exercises 5.3 page 384

1. $\{(A, B), (A, C), (A, D), (B, C), (B, D), (C, D)\}; \ \{(A, B), (A, D), (B, C), (C, D)\}$

3. $\{(S_1, J_1), (S_1, J_2), (S_2, J_1), (S_2, J_2), (S_3, J_1), (S_3, J_2), (S_4, J_1), (S_4, J_2)\}$

5. $\{(R, R), (R, G), (R, B), (G, R), (G, G), (G, B), (B, R), (B, G), (B, B)\};$
$\{(R, G), (R, B), (G, R), (G, B), (B, R), (B, G)\}$

7. a. $\{3\}$ **b.** $\{2, 3, 4, 6\}$ **9. a.** $\{(6, 2), (6, 4), (6, 6), (2, 6), (4, 6)\}$ **b.** $\varnothing$ **11. a.** 0.80   **b.** 1

13. a. 0.20   **b.** 0.15   **15.** $1/(_{12}C_3) = \frac{1}{220}, \ 1/(_{12}P_3) = \frac{1}{1320}$ **17.** $P(R) = \frac{6}{10} = \frac{3}{5}, \ P(B) = \frac{4}{10} = \frac{2}{5}$

19. $P(6) = \frac{1}{2}, \ P(8) = \frac{1}{4}, \ P(12) = \frac{1}{4}$ **21. a.** $\frac{1}{20}$ **b.** $\frac{19}{20}$ **23.** 32%

25. $63\% + 48\% - 15\% + 10\% = 106\%$ when it should add to 100%.

27. a. $\frac{_4C_3}{_{12}C_3} = \frac{1}{55}$ **b.** $\frac{_4C_3}{_{12}C_3} + \frac{_8C_3}{_{12}C_3} = \frac{3}{11}$ **29.** $1 - \frac{_9C_3}{_{10}C_3} = \frac{3}{10}$ **31.** $\frac{_{48}C_1}{_{52}C_5} \approx 0.000018$

33. $\frac{7 \cdot 6 \cdot 5 \cdot 4 \cdot 3}{7^5} \approx 0.15$ **35.** $\frac{_{90}C_{10}}{_{100}C_{10}} \approx 0.33$ **37.** $1 - \frac{_{98}C_8}{_{100}C_8} \approx 0.15$ **39.** $\frac{_4C_1}{_6C_1} = \frac{2}{3}$

41. Yes   **43.** $P(H) = \frac{1}{2}$ and $P(T) = \frac{1}{2}$ give $\frac{1}{2} : \frac{1}{2}$ or $1 : 1.$

45. $P(E^c) = \frac{m - n}{m}$ so the odds are $\frac{n}{m} : \frac{m - n}{m},$ or $n : (m - n).$

47. Since the probabilities must add to 1, we divide the odds by $n + m,$ so $P(E) = \frac{n}{n + m}.$

49. $\frac{7}{11}$ **51.** $\frac{7}{9}$

Exercises 5.4 page 397

1. a. $\frac{0.2}{0.4} = \frac{1}{2}$ **b.** $\frac{0.2}{0.6} = \frac{1}{3}$ **3.** Both parts require first finding $P(A \cap B) = 0.3$. **a.** $\frac{3}{5}$ **b.** $\frac{3}{4}$

5. $\frac{2}{3}$ **7.** $\frac{1}{3}$ **9.** $\frac{1}{3}$ **11.** $\frac{0.38}{0.68} \approx 0.56$ **13.** $0.95 \cdot 0.20 + 0.70 \cdot 0.80 = 0.75$ **15.** 0.16

17. Independent $[P(A) \cdot P(B) = \frac{1}{2} \cdot \frac{1}{2}$ and $P(A \cap B) = \frac{1}{4}]$

19. Dependent $[P(A) \cdot P(B) = \frac{1}{2} \cdot \frac{3}{36} = \frac{1}{24}$ but $P(A \cap B) = \frac{2}{36} = \frac{1}{18}]$

21. $\left(\frac{1}{6}\right)^3 = \frac{1}{216}$ **23.** $\frac{0.9}{1 - 0.1 \cdot 0.3} \approx 0.928$, so about 93% **25. a.** $\left(\frac{1}{6}\right)^3 = \frac{1}{216}$ **b.** $\frac{6}{216} = \frac{1}{36}$ **c.** $\frac{35}{36}$

27. $\frac{1}{3}$ **29.** 0.636 **31. a.** pq **b.** $1 - (1 - p)(1 - q)$ **c.** 0.25 and 0.75, 0.81 and 0.99, parallel

33. b. 23 **35.** $P(A \cap B^c) = P(A) - P(A \cap B) = P(A) - P(A) \cdot P(B) = P(A)[1 - P(B)] = P(A) \cdot P(B^c)$

37. $P(A \cap B) = \frac{1}{4} = \frac{1}{2} \cdot \frac{1}{2} = P(A) \cdot P(B)$

$P(A \cap C) = \frac{1}{4} = \frac{1}{2} \cdot \frac{1}{2} = P(A) \cdot P(C)$

$P(B \cap C) = \frac{1}{4} = \frac{1}{2} \cdot \frac{1}{2} = P(B) \cdot P(C)$

$P(A \cap B \cap C) = \frac{1}{4}$ but $P(A) \cdot P(B) \cdot P(C) = \frac{1}{2} \cdot \frac{1}{2} \cdot \frac{1}{2} = \frac{1}{8}$

Exercises 5.5 page 404

1. $\frac{0.55 \cdot 0.60}{0.55 \cdot 0.60 + 0.65 \cdot 0.40} \approx 0.559$, so about 56% **3.** $\frac{0.95 \cdot 0.001}{0.95 \cdot 0.001 + 0.05 \cdot 0.999} \approx 0.019$, so about 2%

5. $\frac{0.02 \cdot 0.3}{0.03 \cdot 0.5 + 0.02 \cdot 0.3 + 0.01 \cdot 0.2} \approx 0.261$, so about 26%

7. $\frac{0.90 \cdot \frac{30}{300,000,000}}{0.90 \cdot \frac{30}{300,000,000} + 0.0005 \cdot \frac{299,999,970}{300,000,000}} \approx 0.00018$, so about 0.02%

9. $\frac{0.05 \cdot 0.99}{0.05 \cdot 0.99 + 0.95 \cdot 0.01} \approx 0.839$, so about 84% **11.** $\frac{0.70 \cdot 0.2}{0.90 \cdot 0.5 + 0.85 \cdot 0.3 + 0.70 \cdot 0.2} \approx 0.166$, so about 17%

13. $\frac{0.75 \cdot 0.02}{0.75 \cdot 0.02 + 0.3 \cdot 0.98} \approx 0.049$, so about 5% **15.** $\frac{0.006}{0.009 + 0.01 + 0.008 + 0.0075 + 0.006} \approx 0.148$, so about 15%

17. a. $P(Y) = p \cdot \frac{1}{2} + p \cdot \frac{1}{4} + (1 - p) \cdot \frac{1}{4}$ **b.** $p = 2P(Y) - \frac{1}{2}$ **c.** 74%

Exercises 5.6 page 416

1.

x	0	1	2	3
$P(X = x)$	$\frac{1}{8}$	$\frac{3}{8}$	$\frac{3}{8}$	$\frac{1}{8}$

3.

x	-1	11
$P(X = x)$	$\frac{3}{4}$	$\frac{1}{4}$

5.

x	-12	2	3
$P(X = x)$	$\frac{1}{6}$	$\frac{1}{2}$	$\frac{1}{3}$

7.

x	1	2	3	4	5	6
$P(X = x)$	$\frac{1}{36}$	$\frac{1}{12}$	$\frac{5}{36}$	$\frac{7}{36}$	$\frac{1}{4}$	$\frac{11}{36}$

9. $E(X) = 0 \cdot \frac{1}{8} + 1 \cdot \frac{3}{8} + 2 \cdot \frac{3}{8} + 3 \cdot \frac{1}{8} = \frac{3}{2}$ **11.** $E(X) = 2$
13. $E(X) = 0$ **15.** $E(X) = \frac{161}{36} \approx 4.5$ **17.** $E(X) = 3\frac{1}{4}$ **19.** No ($72,000 versus $74,000)
21. $E(X) = 20 \cdot \frac{1}{2} = 10$ **23.** $E(X) = 16$

25. 0.78125, or about 78%   **27.** $_8C_4\left(\frac{1}{2}\right)^8 = \frac{70}{256} \approx 0.273$,   or about 27%
29. Most likely: 2 heads, with probability 0.329 (approx.)
    Least likely: 6 heads, with probability 0.0014 (approx.)
31. Most likely: 1 six, with probability 0.372 (approx.), $E(X) = \frac{4}{3}$ **33.** $14,000   **35.** $E(X) = 0.2$
37. $P(7 \le X \le 10) \approx 0.172$ **39.** 0.972 (compared to 0.90 for transmitting a single digit)
41. $(_{48}C_{13}/_{52}C_{13})^3 \approx 0.028$,   or about 3%
43. a. $_3C_2\left(\frac{3}{5}\right)^2\left(\frac{2}{5}\right)^1 + _3C_3\left(\frac{3}{5}\right)^3\left(\frac{2}{5}\right)^0 \approx 0.648$,   or about 65%   **b.** 0.682, or about 68%   **c.** 0.710, or about 71%
45. 5   **47.** 0.999996, 0.9998
49. a. $\frac{5}{16}$ **b.** $\left(\frac{11}{16}\right)^4 \approx 0.223$ or about 22%   **c.** $\left(\frac{11}{16}\right)^8 \approx 0.05$ or about 5%
53. By the definition of expected value.
55. The expected value of a sum is the sum of the expected values (from Exercise 54) and the formula for
    the expected value of a binomial random variable with parameters 1 and p (from Exercise 53).

Chapter 5 Review Exercises page 422

1. a. 17   **b.** 32   **c.** 21   **d.** 15   **2.** 110   **3. a.** $26^3 = 17,576$ **b.** $26 \cdot 25 \cdot 24 = 15,600$
4. a. 120   **b.** 20   **5.** $_{20}C_4 = 4845$, $_{20}P_4 = 116,280$ **6.** $_{13}C_5 = 1287$ **7.** $_{20}C_5 = 15,504$
8. $\{(C_1, S_1, H_1), (C_1, S_1, H_2), (C_1, S_2, H_1), (C_1, S_2, H_2), (C_2, S_1, H_1), (C_2, S_1, H_2), (C_2, S_2, H_1), (C_2, S_2, H_2)\}$
9. a. $\{(B, B), (B, Y), (B, R), (Y, B), (Y, Y), (Y, R), (R, B), (R, Y), (R, R)\}$
    b. $\{(B, Y), (B, R), (Y, B), (Y, R), (R, B), (R, Y)\}$
10. a. $\{(H, H), (H, T), (T, H)\}$ **b.** $\{(H, T), (T, H), (T, T)\}$ **c.** $\{(H, T), (T, H)\}$
11. a. $\frac{1}{4}$ **b.** $\frac{3}{4}$ **c.** $\frac{1}{2}$ **12.** $1/_{15}C_2 = \frac{1}{105}$, $1/_{15}P_2 = \frac{1}{210}$ **13.** $P(1) = P(3) = P(5) = P(7) = \frac{1}{4}$
14. $P(1) = \frac{1}{2}$, $P(2) = P(4) = P(6) = \frac{1}{6}$ **15.** $P(R) = \frac{1}{6}$, $P(G) = \frac{1}{3}$, $P(B) = \frac{1}{2}$ **16.** 0.45
17. a. $\frac{1}{3}$ **b.** $\frac{1}{2}$ **18.** $_5C_4/_{40}C_5 \approx 0.00000760$
19. a. $_{12}C_5/_{52}C_5 \approx 0.000305$,   or about 0.03%   **b.** $_{40}C_5/_{52}C_5 \approx 0.253$,   or about 25%
20. a. $_{39}C_{13}/_{52}C_{13} \approx 0.0128$,   or about 1%   **b.** $_{40}C_{13}/_{52}C_{13} \approx 0.0189$,   or about 2%
21. $_{28}C_5/_{30}C_5 \approx 0.690$,   or about 69%
22. $_{48}C_2/_{50}C_4 \approx 0.0049$,   or about 0.5%; $_{48}C_4/_{50}C_4 \approx 0.8449$,   or about 84.5%
23. $_{28}C_1/_{30}C_3 = \frac{1}{145} \approx 0.00690$,   or about 0.7%
24. $_6C_2/_{10}C_2 = \frac{1}{3}$ **25. a.** $P(A$ given $B) = \frac{3}{4}$ **b.** $P(B$ given $A) = \frac{3}{5}$ **26.** $\frac{1}{3}$ **27.** $\frac{1}{7}$
28. 0.00235, or about 0.2%   **29.** $\frac{3}{16}$ **30.** 0.32   **31.** $0.8 \cdot 0.6 + 0.9 \cdot 0.4 = 0.84$ **32.** 0.983 (approx.)
33. a. No $\left(\frac{3}{4} \cdot \frac{3}{4} \ne \frac{1}{2}\right)$ **b.** Yes $\left(\frac{1}{2} \cdot \frac{1}{2} = \frac{1}{4}\right)$ **34.** 0.9999   **35. a.** $\frac{1}{32}$ **b.** $\frac{1}{16}$ **c.** $\frac{1}{16}$
36. $\frac{2}{3}$ **37.** $\frac{0.04 \cdot 0.25}{0.04 \cdot 0.25 + 0.03 \cdot 0.35 + 0.02 \cdot 0.40} \approx 0.351$,   or about 35%   **38.** 0.247,   or about 25%

39.

Value of X	Outcomes
$X = 5$	(H, H, H, H, H)
$X = 1$	$(H, T, T, T, T), (T, H, T, T, T), (T, T, H, T, T), (T, T, T, H, T), (T, T, T, T, H)$
$X = 0$	(T, T, T, T, T)

40.

x	34	-2
$P(X = x)$	$\frac{1}{4}$	$\frac{3}{4}$

41. $E(X) = 7$

42.

x	0	1	2
$P(X = x)$	$\frac{5}{18}$	$\frac{5}{9}$	$\frac{1}{6}$

43. $E(X) = \frac{8}{9}$

44. $E(X) = \$3.10$

45.

x	0	1	2	3	4
$P(X = x)$	$\frac{1}{625}$	$\frac{16}{625}$	$\frac{96}{625}$	$\frac{256}{625}$	$\frac{256}{625}$

$E(X) = \frac{16}{5} = 3\frac{1}{5}$

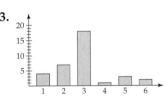

46. 0.711 (approx.), or about 71%   **47.** $360   **48.** 2   **49.** $E(X) = 12 \cdot 0.01 = 0.12$
50. 0.859, or about 86%   **51.** 0.633, or about 63%

Exercises 6.1 page 436

1. Nominal   **3.** Ordinal   **5.** Ordinal   **7.** Interval   **9.** Ratio

11.

1 = Never married
2 = Married
3 = Widowed
4 = Divorced

13.

1 = Business executives
2 = Developers
3 = Lawyers
4 = Doctors
5 = CPAs
6 = Retirees

15. Using 6 classes of width 8 beginning at 12.5:   **17.** Using 8 classes of width 10 beginning at 10.5:

The lighter
weight boxes
are in greatest
demand.

The highest
number of hits
were in classes
11–20 and 31–40.

19. Using 11 classes of width 32 beginning at 218.5:

More ticket prices are grouped at the upper end.

21. Using 5 classes of width 5 beginning at 35.5:

More are grouped in the middle.

23. Using 7 classes of width 3 beginning at 8.5:

They are grouped in the middle.

25. Using 10 classes of width 6 beginning at 15.5:

Not evenly distributed: teenagers and the elderly have more accidents.

Exercises 6.2 page 446

1. Mode = 9, median = 10, $\bar{x}$ = 11   **3.** Modes = 7 and 14 (bimodal), median = 9, $\bar{x}$ = 10
5. Mode = 19, median = 15, $\bar{x}$ = 15.2   **7.** Mode = 14, median = 13, $\bar{x}$ = 12
9. Mode = 19, median = 18, $\bar{x}$ = 16
11. Modes = 12 and 20 (bimodal), median = 15, $\bar{x}$ = 15.75, each in minutes
13. Mode = 10, median = 10, $\bar{x}$ = 11.1, each in hours   **15.** Mode = 9, median = 9, $\bar{x}$ = 10, each in weeks
17. The mode is 450, the median is 450, and the mean is 468. The unusually high 750 skews the mean higher.
19. The modes are 0.5, 0.6 and 0.8, the median is 1.1, and the mean is 3.8 (rounded, all in millions of dollars). The two very high salaries skew the mean.
21. There is no mode, the median is 41.05, and the mean is 55.35 (all in thousands of dollars). The relatively few extremely high salaries of the wealthiest people skew the mean higher.
25. $\bar{x} \approx \$44.95$

Exercises 6.3 page 457

1. 62   **3.** 20   **5.** 56   **7.** Minimum = 3
First quartile = 10
Median = 14
Third quartile = 21
Maximum = 23

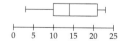

9. Minimum = 8
First quartile = 10
Median = 13
Third quartile = 17
Maximum = 26

11. Minimum = 8
First quartile = 12
Median = 17
Third quartile = 24
Maximum = 25

13. Minimum = 6
First quartile = 8
Median = 16
Third quartile = 18
Maximum = 24

15. Minimum = 5
First quartile = 8.5
Median = 15.5
Third quartile = 17.5
Maximum = 20

17. $s \approx 5.79$ (from $\bar{x} = 11$)   **19.** $s \approx 8.63$ (from $\bar{x} = 13$)

21. Range = 20
Minimum = 5
First quartile = 11
Median = 13   All in dollars
Third quartile = 17
Maximum = 25
$s \approx 4.64$

23. Range = 11
Minimum = 2
First quartile = 6   All in
Median = 8   numbers
Third quartile = 10   of people
Maximum = 13
$s \approx 2.55$

25. Range = 33
Minimum = 2
First quartile = 15.5
Median = 20.5   All in weeks
Third quartile = 24.5
Maximum = 35
$s \approx 8.19$

27.

Gender inequality increases with income

29. Family Sedans $\begin{cases} \text{Minimum} = 19 \\ \text{First quartile} = 21 \\ \text{Median} = 22 \\ \text{Second quartile} = 22.5 \\ \text{Maximum} = 24 \end{cases}$ SUVs $\begin{cases} \text{Minimum} = 10 \\ \text{First quartile} = 13.5 \\ \text{Median} = 16 \\ \text{Second quartile} = 18 \\ \text{Maximum} = 21 \end{cases}$

The worst family sedan is better
than three-quarters of the SUVs.

31. a. 5.02 and 1.62   **b.** Yes   **c.** No

Exercises 6.4 page 470

(*Note:* Answers may differ depending on rounding and on whether graphing calculators or tables were used.)
  1. 0.6827   **3.** 0.9973   **5.** 0.3413   **7.** 0.0674   **9.** 0.3361   **11.** $z = 1$ **13.** $z = 0$
15. $z = -1$ **17.** $z = -5$ **19.** $z = 2$
21. 0.0298 [from $\mu = 14$, $\sigma \approx 2.05$, after checking that $np > 5$ and $n(1 - p) > 5$]
23. 0.1883 [from $\mu = 14$, $\sigma \approx 2.05$, after checking that $np > 5$ and $n(1 - p) > 5$]
25. 0.0036 [from $\mu = 14$, $\sigma \approx 2.05$, after checking that $np > 5$ and $n(1 - p) > 5$]
27. 0.4029 [from $\mu = 14$, $\sigma \approx 2.05$, after checking that $np > 5$ and $n(1 - p) > 5$]
29. 0.9992 [from $\mu = 14$, $\sigma \approx 2.05$, after checking that $np > 5$ and $n(1 - p) > 5$]
31. About 0.89, or 89%
33. About 0.25, or 25% using $P(x \geq 101)$. [Alt. answers: 28% for $P(x > 100.5)$ or 31% for $P(x > 100)$]
35. About 0.31, or 31%
37. About 0.24, or 24% [from $\mu = 564$, $\sigma \approx 15$, after checking that $np > 5$ and $n(1 - p) > 5$]
39. About 0.26, or 26% [from $\mu = 80$, $\sigma \approx 6.93$, after checking that $np > 5$ and $n(1 - p) > 5$]
41. $\frac{8}{9} \approx 89\%$
43. The probability that the values of the random variable lie between 85 and 115 is at least $\frac{8}{9} \approx 89\%$.
45. 99.7%. This is a stronger result than the probability of $\frac{8}{9} \approx 89\%$ found in Exercise 43.

Chapter 6 Review Exercises page 474

  1. Nominal   **2.** Ordinal   **3.** Interval   **4.** Ratio

5.

1 = Army
2 = Navy
3 = Marines
4 = Air Force

6.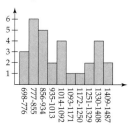

1 = Handguns
2 = Other firearms
3 = Knives
4 = Blunt instruments
5 = Hands

  7. Using 8 classes of width 4 beginning at 33.5:   **8.** Using 10 classes of width 79 beginning at 697.5:

  9. Mode = 11, median = 10, $\bar{x} = 9$ **10.** Modes = 6 and 13 (bimodal), median = 11, $\bar{x} = 10$
11. Mode = 13, median = 12.5, $\bar{x} = 11$ **12.** Mode = 8, median = 8, $\bar{x} = 8.50$ (all in dollars)
13. Mode = 4, median = 20, $\bar{x} = 17.4$ (all in numbers of calls)   **14.** 18

15. Minimum = 4
First quartile = 6
Median = 10
Third quartile = 14.5
Maximum = 20

16. $s = 4$

17. Range = 6
Minimum = 3
First quartile = 5
Median = 6
Third quartile = 8
Maximum = 9
$s \approx 1.96$

} All in minutes

18. Range = 40
Minimum = 114
First quartile = 124
Median = 131
Third quartile = 134
Maximum = 154
$s \approx 9.74$

} All in thousands of dollars

19. 0.8186
20. 0.3023
} Answers may differ slightly

21. $z = 2$ **22.** $z = -2$

23. 0.015 [from $\mu = 8.75$, $\sigma \approx 2.38$, after checking that $np > 5$ and $n(1 - p) > 5$]

24. 0.68 [from $\mu = 8.75$, $\sigma \approx 2.38$, after checking that $np > 5$ and $n(1 - p) > 5$]

25. About 0.66, or 66%   **26.** About 0.16, or 16%

27. About 0.63, or 63% [from $\mu = 6.9$, $\sigma \approx 2.61$, after checking that $np > 5$ and $n(1 - p) > 5$]

28. About 0.38, or 38% [from $\mu = 48$, $\sigma \approx 5.71$, after checking that $np > 5$ and $n(1 - p) > 5$]

Exercises 7.1 page 488

1. $\begin{array}{c}\quad A \;\; B \;\; C \\ A \\ B \\ C\end{array}\begin{pmatrix} 0 & 1 & 0 \\ 0 & 0 & 1 \\ 1 & 0 & 0 \end{pmatrix}$, oscillating

3. $\begin{array}{c}\quad D \;\; E \\ D \\ E\end{array}\begin{pmatrix} \frac{2}{5} & \frac{3}{5} \\ 1 & 0 \end{pmatrix}$, mixing

5. $\begin{array}{c}\quad F \;\; G \\ F \\ G\end{array}\begin{pmatrix} \frac{2}{5} & \frac{3}{5} \\ 0 & 1 \end{pmatrix}$, absorbing

7. Oscillating

9. Mixing

11. Absorbing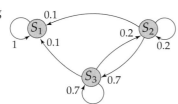

13. $D_1 = \left(\frac{7}{20} \quad \frac{13}{20}\right)$, $D_2 = \left(\frac{33}{80} \quad \frac{47}{80}\right)$

15. $D_1 = (0.35 \quad 0.25 \quad 0.40)$, $D_2 = (0.31 \quad 0.17 \quad 0.52)$ **17.** 5 **19.** $2\frac{1}{2}$

21. The expected number of minutes that a green light for traffic on Main Street will be green before changing to red is 4 minutes.

23. Next week, 71% will be eating cereal; the week after, 71.8% will be eating cereal.

25. Tomorrow, 25% will be fishing; the day after, 52.5% will be golfing.

27. In two years, approximately 75.4% of the portfolio will be "secure." Because "bankrupt" is an absorbing state for all the other states, ultimately every company in the portfolio will become bankrupt.

29. The expected number of weekends that the laundry room will be manageable, given that it is now manageable, is 2 weekends.

31. $\underbrace{p \cdot \cdots \cdot p}_{\substack{\text{Stays in } S_i \text{ and then} \\ \text{for } n-1 \\ \text{transitions}}} \cdot \underbrace{(1-p)}_{\substack{\text{leaves } S_i}} = p^{n-1}(1-p)$

33. $E = 1 \cdot (1-p) + 2 \cdot p(1-p) + 3 \cdot p^2(1-p) + \cdots = (1-p) + \underbrace{(2p - 2p^2)}_{+p} + \underbrace{(3p^2 - 3p^3)}_{+p^2} + \underbrace{\cdots}_{+p^3}$

$= 1 + p + p^2 + p^3 + \cdots$

Exercises 7.2 page 500

1. Regular **3.** Regular **5.** Not regular **7.** Regular

9. $D = \left(\frac{1}{2} \quad \frac{1}{2}\right)$, $T^{20} = \begin{pmatrix} 0.5 & 0.5 \\ 0.5 & 0.5 \end{pmatrix}$ **11.** $D = \left(\frac{2}{5} \quad \frac{3}{5}\right)$, $T^{20} = \begin{pmatrix} 0.4 & 0.6 \\ 0.4 & 0.6 \end{pmatrix}$

13. $D = \left(\frac{1}{5} \quad \frac{7}{10} \quad \frac{1}{10}\right)$, $T^{20} = \begin{pmatrix} 0.2 & 0.7 & 0.1 \\ 0.2 & 0.7 & 0.1 \\ 0.2 & 0.7 & 0.1 \end{pmatrix}$ **15.** $D = \left(\frac{1}{10} \quad \frac{17}{20} \quad \frac{1}{20}\right)$, $T^{50} = \begin{pmatrix} 0.10 & 0.85 & 0.05 \\ 0.10 & 0.85 & 0.05 \\ 0.10 & 0.85 & 0.05 \end{pmatrix}$

17. 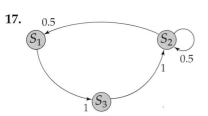 $k = 2$, with probability 0.5

19.

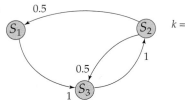

$k = 3$, with probability 0.5

21. The voter records lead to a steady-state distribution with 60% of the population voting, which is consistent with the national average (the surveys lead to a steady-state distribution with 80% of the population voting, which is not consistent with the national average).

23. One million of three million commuters will be driving alone.

25. The long-term market share for DentiMint toothpaste will be 24%.

27. 45% + 30% = 75% of the broker's auto policies are rated satisfactory or better.

29. Yes; a long-term moving probability of $\frac{1}{3}$ is consistent with the national average of about 34.3%.

Exercises 7.3 page 512

1. S_3 is absorbing; S_1 and S_2 are nonabsorbing. 2 steps, with probability $\frac{1}{2}\cdot\frac{1}{2} = \frac{1}{4}$

3. S_2 is absorbing; S_1, S_3, and S_4 are nonabsorbing. 3 steps, with probability $\frac{1}{2}\cdot\frac{1}{2}\cdot\frac{1}{2} = \frac{1}{8}$

5. a.

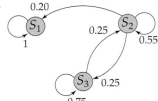

S_1 is absorbing;
S_2 and S_3 are nonabsorbing.
$k = 2$

b. $T^2 = \begin{pmatrix} 1 & 0 & 0 \\ 0.31 & 0.365 & 0.325 \\ 0.05 & 0.325 & 0.625 \end{pmatrix}$

7. a.

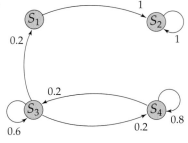

S_2 is absorbing;
S_1, S_3, and S_4 are nonabsorbing.
$k = 3$

b. $T^3 = \begin{pmatrix} 0 & 1 & 0 & 0 \\ 0 & 1 & 0 & 0 \\ 0.08 & 0.32 & 0.296 & 0.304 \\ 0.056 & 0.04 & 0.304 & 0.6 \end{pmatrix}$

9.
$$\begin{array}{c} & S_2 & S_1 & S_3 \\ S_2 & 1 & 0 & 0 \\ S_1 & 0.02 & 0.88 & 0.10 \\ S_3 & 0.16 & 0.04 & 0.80 \end{array}$$ Other answers are possible.

$$\underbrace{}_{R} \quad \underbrace{}_{Q}$$

11.
$$\begin{array}{c} & S_2 & S_4 & S_3 & S_1 & S_5 \\ S_2 & 1 & 0 & 0 & 0 & 0 \\ S_4 & 0 & 1 & 0 & 0 & 0 \\ S_3 & 0.20 & 0.30 & 0.50 & 0 & 0 \\ S_1 & 0 & 0 & 0.20 & 0.80 & 0 \\ S_5 & 0 & 0 & 0.05 & 0.20 & 0.75 \end{array}$$ Other answers are possible.

$$\underbrace{}_{R} \quad \underbrace{}_{Q}$$

13. $\begin{pmatrix} 2 & 1 \\ 1 & 3 \end{pmatrix}$ **15.** $\begin{pmatrix} 2 & 3 \\ 1 & 4 \end{pmatrix}$ **17.** $\begin{pmatrix} 2 & 2 & 3 \\ 0 & 10 & 0 \\ 1 & 1 & 4 \end{pmatrix}$ **19.** $\begin{pmatrix} 2 & 2 & 2 \\ 1 & 6 & 1 \\ 1 & 2 & 3 \end{pmatrix}$

21. 2; 1   **23.** 1; 0.70   **25.** 10; 1   **27.** 2; 0.70   **29.** $1 + 6 + 1 = 8$; 0.40

31. a. 30% of the town's original population survived.
  b. The expected time for the disease to run its course was 2 months.

33. With 0.60 probability, a resident requiring assistance will ultimately be transferred to a nursing home.

35. a. A policy that has just been renewed can be expected to continue for $4 + 2 = 6$ periods, so $6 \cdot 5 = 30$ years.
  b. A policy that is renewed with increased coverage will close with a death benefit payment with probability 0.53.

37. A tooth needing a filling can be expected to need 4 fillings before being extracted.

39. Since 80% of the companies receiving advice become successful and 76% of those receiving loans become successful, the trust's claim is justified.

Chapter 7 Review Exercises page 518

1. $\begin{array}{c} & A & B \\ A & \frac{2}{3} & \frac{1}{3} \\ B & 1 & 0 \end{array}$, mixing   **2.** $\begin{array}{c} & X & Y & Z \\ X & 0 & 0 & 1 \\ Y & 1 & 0 & 0 \\ Z & 0 & 1 & 0 \end{array}$, oscillating

3. Absorbing

4. Mixing

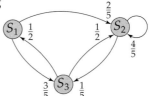

5. $D_1 = (0.35 \quad 0.65)$, $D_2 = (0.325 \quad 0.675)$ **6.** $D_1 = (0.60 \quad 0.30 \quad 0.10)$, $D_2 = (0.19 \quad 0.42 \quad 0.39)$

7. 20   **8.** 5

9. A working cab can be expected to remain in service for 100 days before breaking down.

10. Of these 875 students, 130 will be at the second-year level after 2 years.

11. Regular   **12.** Not regular   **13.** Regular   **14.** Regular

15. $D = \begin{pmatrix} \frac{1}{4} & \frac{3}{4} \end{pmatrix}$, $T^{50} = \begin{pmatrix} 0.25 & 0.75 \\ 0.25 & 0.75 \end{pmatrix}$

16. $D = \begin{pmatrix} \frac{3}{20} & \frac{1}{2} & \frac{7}{20} \end{pmatrix}$, $T^{20} = \begin{pmatrix} 0.15 & 0.50 & 0.35 \\ 0.15 & 0.50 & 0.35 \\ 0.15 & 0.50 & 0.35 \end{pmatrix}$

17.

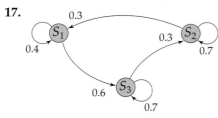

2 steps, with probability
$0.6 \cdot 0.3 = 0.18$

18.

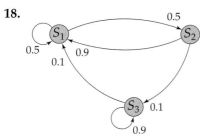

2 steps, with probability
$0.1 \cdot 0.5 = 0.05$

19. Of the manufacturer's 2660 dealers, $0.40 \cdot 2660 = 1064$ have service departments rated "excellent" by their customers.

20. Out of the next 25 years, 10 will have terrific weather for growing corn, if the old timers are right.

21. S_3 is absorbing; S_1 and S_2 are nonabsorbing. 2 steps, with probability $\frac{1}{10} \cdot \frac{1}{4} = \frac{1}{40}$

22. S_2 is absorbing; S_1, S_3, and S_4 are nonabsorbing. 3 steps, with probability $1 \cdot 1 \cdot \frac{1}{2} = \frac{1}{2}$

23. a.

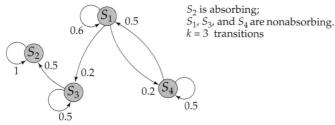

S_2 is absorbing;
S_1, S_3, and S_4 are nonabsorbing.
$k = 3$ transitions

b. $T^3 = \begin{pmatrix} 0.386 & 0.21 & 0.202 & 0.202 \\ 0 & 1 & 0 & 0 \\ 0 & 0.875 & 0.125 & 0 \\ 0.505 & 0.05 & 0.16 & 0.285 \end{pmatrix}$

24. a.

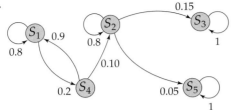

S_3 and S_5 are absorbing;
S_1, S_2, and S_4 are nonabsorbing.
$k = 3$ transitions

b. $T^3 = \begin{pmatrix} 0.8 & 0.032 & 0.003 & 0.164 & 0.001 \\ 0 & 0.512 & 0.366 & 0 & 0.122 \\ 0 & 0 & 1 & 0 & 0 \\ 0.738 & 0.082 & 0.027 & 0.144 & 0.009 \\ 0 & 0 & 0 & 0 & 1 \end{pmatrix}$

25.

$$
\begin{array}{c|cccc}
 & S_2 & S_1 & S_3 & S_4 \\
\hline
S_2 & 1 & 0 & 0 & 0 \\
S_1 & 0.5 & 0.5 & 0 & 0 \\
S_3 & 0 & 0.2 & 0.8 & 0 \\
S_4 & 0 & 0 & 1 & 0 \\
\end{array}
$$

$\underbrace{}_{R}\ \underbrace{}_{Q}$

Other answers are possible.

26.

$$
\begin{array}{c|cccc}
 & S_2 & S_4 & S_3 & S_1 \\
\hline
S_2 & 1 & 0 & 0 & 0 \\
S_4 & 0 & 1 & 0 & 0 \\
S_3 & 0.1 & 0 & 0.8 & 0.1 \\
S_1 & 0 & 0.1 & 0.3 & 0.6 \\
\end{array}
$$

$\underbrace{}_{R}\ \underbrace{}_{Q}$

Other answers are possible.

27. $\begin{pmatrix} 2 & 4 \\ 1 & 7 \end{pmatrix}$ **28.** $\begin{pmatrix} 4 & 6 \\ 2 & 8 \end{pmatrix}$ **29.** $\begin{pmatrix} 6 & 0 & 2 \\ 3 & 5 & 1 \\ 4 & 0 & 3 \end{pmatrix}$ **30.** $\begin{pmatrix} 2 & 0 & 6 \\ 1 & 2 & 4 \\ 1 & 0 & 8 \end{pmatrix}$ **31.** 2; 1   **32.** 1; 1

33. 6; 0.3   **34.** 8; 0.6   **35.** 4; 1   **36.** $3 + 5 + 1 = 9$; 1   **37.** 0; 0.3   **38.** $1 + 0 + 8 = 9$; 0.85

39. A new model PC can be expected to be manufactured for $3 + 4 = 7$ years before it is discontinued.

40. With probability 0.39 an overdue account will (ultimately) be paid in full.

Exercises 8.1 page 530

1. Channel 86 $\left\{\begin{array}{c} \\ \end{array}\right.$

$$
\begin{array}{cc}
 & \overbrace{}^{\text{Channel 7}} \\
 & \begin{array}{cc} Q & R \end{array} \\
\begin{array}{c} Q \\ R \end{array} & \begin{pmatrix} 5 & -8 \\ 10 & -10 \end{pmatrix}
\end{array}
$$

Other answers are possible.

3. Ralph $\left\{\begin{array}{c} \\ \\ \end{array}\right.$

$$
\begin{array}{cc}
 & \overbrace{}^{\text{Charlie}} \\
 & \begin{array}{ccc} R & S & L \end{array} \\
\begin{array}{c} R \\ S \\ L \end{array} & \begin{pmatrix} 0 & -5 & -10 \\ 5 & 0 & -5 \\ 10 & 5 & 0 \end{pmatrix}
\end{array}
$$

where R represents raising the price by 1¢,
S represents keeping it the same,
and L represents lowering the price by 1¢.
Other answers are possible.

5. The saddle point is the 3 in row 1 and column 1. The optimal strategy for R is to always choose 1 and for C to always choose 1.

7. The saddle point is the 1 in row 1 and column 2. The optimal strategy for R is to always choose 1 and for C to always choose 2.

9. The saddle point is the 2 in row 1 and column 2. The optimal strategy for R is to always choose 1 and for C to always choose 2.

11. The saddle point is the 15 in row 2 and column 2. The optimal strategy for R is to always choose 2 and for C to always choose 2.

13. Burger Haven $\begin{matrix} & \text{MacArches} \\ & \overbrace{\quad L \quad R \quad} \\ \begin{matrix} L \\ R \end{matrix} & \begin{pmatrix} -6 & 4 \\ -8 & 0 \end{pmatrix} \end{matrix}$ where L represents lowering the price by 10¢ and R represents raising the price by 10¢. The saddle point is the -6 in the first row and first column. Both restaurants should lower their prices.

15. Republican $\begin{matrix} & \text{Democrat} \\ & \overbrace{\quad T \quad W \quad} \\ \begin{matrix} T \\ W \end{matrix} & \begin{pmatrix} 15 & 20 \\ 10 & 15 \end{pmatrix} \end{matrix}$ where T represents "taxes" and W represents "welfare reform." The saddle point is the 15 in the first row and the first column. They should both discuss taxes at their next debate.

Exercises 8.2 page 542

1. 1   3. -1 5. $\frac{1}{2}$ 7. 3   9. $r = \begin{pmatrix} \frac{1}{4} & \frac{3}{4} \end{pmatrix}$ and $c = \begin{pmatrix} \frac{3}{4} \\ \frac{1}{4} \end{pmatrix}$ with a value of $\frac{5}{4}$

11. $r = \begin{pmatrix} \frac{3}{4} & \frac{1}{4} \end{pmatrix}$ and $c = \begin{pmatrix} \frac{1}{8} \\ \frac{7}{8} \end{pmatrix}$ with a value of $\frac{5}{4}$ 13. $r = \begin{pmatrix} \frac{2}{5} & \frac{3}{5} \end{pmatrix}$ and $c = \begin{pmatrix} \frac{3}{5} \\ \frac{2}{5} \end{pmatrix}$ with a value of $\frac{3}{5}$

15. Incumbent $\begin{matrix} & \text{Challenger} \\ & \overbrace{\quad P \quad V \quad} \\ \begin{matrix} P \\ V \end{matrix} & \begin{pmatrix} -5 & 5 \\ 20 & -10 \end{pmatrix} \end{matrix}$ so $r = \begin{pmatrix} \frac{3}{4} & \frac{1}{4} \end{pmatrix}$ and $c = \begin{pmatrix} \frac{3}{8} \\ \frac{5}{8} \end{pmatrix}$. 75% of the incumbent's TV ads should focus on positional issues.

17. $\begin{matrix} & \quad \text{No worms} \quad \text{Worms} \\ \begin{matrix} \text{No pesticides} \\ \text{Pesticides} \end{matrix} & \begin{pmatrix} 600 & -400 \\ 0 & 500 \end{pmatrix} \end{matrix}$ has value 200 and optimal mixed strategy $r = \begin{pmatrix} \frac{1}{3} & \frac{2}{3} \end{pmatrix}$ and $c = \begin{pmatrix} \frac{3}{5} \\ \frac{2}{5} \end{pmatrix}$. The farmer should set aside 200 acres for pesticide-free apples and use pesticides on the other 400 acres, and this strategy will increase her expected profits by $200 per acre.

19. Joe brings $\begin{matrix} & \text{Quiz on} \\ & \overbrace{\quad V \quad R \quad} \\ \begin{matrix} D \\ G \end{matrix} & \begin{pmatrix} 100 & 80 \\ 70 & 90 \end{pmatrix} \end{matrix}$ with $r = \begin{pmatrix} \frac{1}{2} & \frac{1}{2} \end{pmatrix}$ and $c = \begin{pmatrix} \frac{1}{4} \\ \frac{3}{4} \end{pmatrix}$ and the value is 85. Joe should flip a coin each day and take his dictionary if it is heads and the grammar textbook if it is tails. His expected grade is an 85.

21. Row 3 dominates row 2, so solve just $\begin{pmatrix} 3 & -1 \\ 2 & 4 \end{pmatrix}$; the optimal strategy is $r = \begin{pmatrix} \frac{1}{3} & 0 & \frac{2}{3} \end{pmatrix}$ and $c = \begin{pmatrix} \frac{5}{6} \\ \frac{1}{6} \end{pmatrix}$ with a value of $2\frac{1}{3}$.

23. Column 1 dominates column 2, so solve just $\begin{pmatrix} 1 & -5 \\ -2 & 2 \end{pmatrix}$; the optimal strategy is $r = \begin{pmatrix} \frac{2}{5} & \frac{3}{5} \end{pmatrix}$ and $c = \begin{pmatrix} \frac{7}{10} \\ 0 \\ \frac{3}{10} \end{pmatrix}$ with a value of $-\frac{4}{5}$.

25. $r = \begin{pmatrix} 1 & 0 & 0 \end{pmatrix}$ and $c = \begin{pmatrix} 0 \\ 0 \\ 1 \end{pmatrix}$ with a value of 3

27. $r = \begin{pmatrix} \frac{1}{2} & \frac{1}{2} & 0 \end{pmatrix}$ and $c = \begin{pmatrix} \frac{5}{12} \\ \frac{7}{12} \\ 0 \end{pmatrix}$ with a value of $\frac{1}{2}$

29. They all have the same optimal strategy, $r = \begin{pmatrix} \frac{1}{4} & \frac{3}{4} \end{pmatrix}$ and $c = \begin{pmatrix} \frac{3}{4} \\ \frac{1}{4} \end{pmatrix}$, as Exercise 9. The values are

a. $2\frac{1}{4}$ **b.** $3\frac{1}{4}$ **c.** $\frac{1}{4}$ and **d.** $-\frac{3}{4}$. Yes; just add the difference to $1\frac{1}{4}$.

Exercises 8.3 page 554

1. $r = \begin{pmatrix} \frac{1}{2} & \frac{1}{2} \end{pmatrix}$ and $c = \begin{pmatrix} \frac{3}{4} \\ \frac{1}{4} \end{pmatrix}$ with a value of $\frac{5}{2}$

3. $r = \begin{pmatrix} \frac{2}{5} & \frac{3}{5} \end{pmatrix}$ and $c = \begin{pmatrix} \frac{2}{3} \\ \frac{1}{3} \end{pmatrix}$ with a value of 0

5. $r = \begin{pmatrix} \frac{1}{2} & \frac{1}{2} \end{pmatrix}$ and $c = \begin{pmatrix} \frac{3}{4} \\ 0 \\ \frac{1}{4} \end{pmatrix}$ with a value of 0

7. $r = \begin{pmatrix} \frac{1}{14} & \frac{4}{7} & \frac{5}{14} \end{pmatrix}$ and $c = \begin{pmatrix} \frac{3}{14} \\ \frac{5}{7} \\ \frac{1}{14} \end{pmatrix}$ with a value of $\frac{9}{14}$

9. $r = \begin{pmatrix} 0 & \frac{3}{7} & \frac{4}{7} \end{pmatrix}$ and $c = \begin{pmatrix} \frac{4}{7} \\ 0 \\ 0 \\ \frac{3}{7} \end{pmatrix}$ with a value of $\frac{2}{7}$

11. $\begin{array}{c} \\ M \\ B \\ S \end{array}\begin{array}{c} \text{Bu} \quad \text{S} \quad \text{Br} \\ \begin{pmatrix} 1 & 1 & 4 \\ 2 & 4 & 2 \\ 6 & 2 & 3 \end{pmatrix} \end{array}$ has value $2\frac{3}{4}$ with optimal strategy $r = \begin{pmatrix} \frac{1}{4} & \frac{1}{2} & \frac{1}{4} \end{pmatrix}$ and $c = \begin{pmatrix} \frac{1}{24} \\ \frac{3}{8} \\ \frac{7}{12} \end{pmatrix}$. $12,500 should be invested in the precious metals fund, $25,000 in the municipal bonds fund, and $12,500 in the technology stocks fund, for an expected minimum return of 2.75%.

13. $\begin{pmatrix} -2 & 3 & 2 & -1 \\ -1 & -2 & 0 & 0 \\ -1 & 1 & -2 & 0 \end{pmatrix}$ has value -1 with $r = \begin{pmatrix} 0 & \frac{2}{3} & \frac{1}{3} \end{pmatrix}$ and $c = \begin{pmatrix} 1 \\ 0 \\ 0 \\ 0 \end{pmatrix}$. The challenger's strategy

will be to concentrate only on "lower taxes" for the rest of the campaign.

15. Daryl $\begin{array}{c} \\ \\ \end{array}$ $\begin{array}{c} \overbrace{}^{\text{Adana}} \\ \begin{array}{ccc} D & Q & H \end{array} \\ \begin{array}{c} D \\ Q \\ H \end{array} \begin{pmatrix} 10 & -10 & -10 \\ -25 & 25 & -25 \\ -50 & -50 & 50 \end{pmatrix} \end{array}$ has value $-6\frac{1}{4}$ with optimal strategy $r = \begin{pmatrix} \frac{5}{8} & \frac{1}{4} & \frac{1}{8} \end{pmatrix}$ and

$c = \begin{pmatrix} \frac{3}{16} \\ \frac{3}{8} \\ \frac{7}{16} \end{pmatrix}$. This is not a fair game since it has an advantage of 6.25¢ to Adana, who should hide the

quarter in her hand 37.5% of the time.

Chapter 8 Review Exercises page 557

1. Bill's Pro Shop $\begin{array}{c} \overbrace{}^{\text{Sports-R-Us}} \\ \begin{array}{cc} S & N \end{array} \\ \begin{array}{c} S \\ N \end{array} \begin{pmatrix} -10 & 5 \\ -20 & -1 \end{pmatrix} \end{array}$ where S represents a sale and N represents no sale; other answers

are possible.

2. $\begin{array}{c} \begin{array}{ccc} 1 & 2 & 3 \end{array} \\ \begin{array}{c} 1 \\ 2 \\ 3 \end{array} \begin{pmatrix} 0 & 1 & -2 \\ -1 & 0 & 1 \\ 2 & -1 & 0 \end{pmatrix} \end{array}$ where "1," and "2," and "3" represent the number of fingers shown; other answers

are possible.

3. The saddle point is the 2 in row 2 and column 2. The optimal strategy is for R to always choose 2 and for C to always choose 2.

4. The saddle point is the 3 in row 1 and column 3. The optimal strategy is for R to always choose 1 and for C to always choose 3.

5. The saddle point is the 3 in row 2 and column 2. The optimal strategy is for R to always choose 2 and for C to always choose 2.

6. The saddle point is the 4 in row 2 and column 3. The optimal strategy is for R to always choose 2 and for C to always choose 3.

7. Mr. Brown $\begin{array}{c} \overbrace{}^{\text{Mr. Clark}} \\ \begin{array}{ccc} B & S & F \end{array} \\ \begin{array}{c} B \\ S \\ F \end{array} \begin{pmatrix} 15 & 15 & -20 \\ 15 & 15 & -20 \\ 10 & 10 & -20 \end{pmatrix} \end{array}$ where B represents "budgets," S represents "standards," and F

represents "football." The three -20's are saddle points, so Mr. Brown may expect Mr. Clark to talk only about football for the rest of the campaign. He shouldn't bother having any more debates with Mr. Clark.

8. The relative market advantage is
$$\begin{array}{c}\\F\\G\\H\\I\end{array}\begin{array}{cccc}F & G & H & I\end{array}\\\begin{pmatrix} 0 & -40 & -120 & -40 \\ 40 & 0 & 40 & 100 \\ 120 & -40 & 0 & -40 \\ 40 & -100 & 40 & 0 \end{pmatrix}$$
with a saddle point in the second row and second column. Both video-game franchises should build their new stores in Gardner and split the quad-city market equally.

9. 0   **10.** $\frac{3}{2}$

11. $r = \begin{pmatrix} \frac{3}{8} & \frac{5}{8} \end{pmatrix}$ and $c = \begin{pmatrix} \frac{1}{4} \\ \frac{3}{4} \end{pmatrix}$ with a value of $\frac{1}{2}$

12. $r = \begin{pmatrix} \frac{2}{3} & \frac{1}{3} \end{pmatrix}$ and $c = \begin{pmatrix} \frac{3}{5} \\ \frac{2}{5} \end{pmatrix}$ with a value of 1

13. $\begin{array}{c}\\C\\S\end{array}\begin{array}{cc}D & W\end{array}\begin{pmatrix} -8 & -17 \\ -13 & -7 \end{pmatrix}$ has value $-\$11$ and optimal mixed strategy $r = \begin{pmatrix} \frac{2}{5} & \frac{3}{5} \end{pmatrix}$ and $c = \begin{pmatrix} \frac{2}{3} \\ \frac{1}{3} \end{pmatrix}$. 40% of his animals should be cattle and 60% should be sheep, for a worst-case loss of $11 per head.

14. $\begin{array}{c}\\S\\E\end{array}\begin{array}{cc}Y & N\end{array}\begin{pmatrix} 6 & 1 \\ -2 & 5 \end{pmatrix}$ has value $2\frac{2}{3}$ with optimal strategy $r = \begin{pmatrix} \frac{7}{12} & \frac{5}{12} \end{pmatrix}$ and $c = \begin{pmatrix} \frac{1}{3} \\ \frac{2}{3} \end{pmatrix}$. $21,000 should be invested in the Bright-Tech Solar Power stock and $15,000 in the Pacific West Electric stock for a worst-case expected return of $2\frac{2}{3}\%$.

15. $r = \begin{pmatrix} \frac{1}{2} & \frac{1}{2} \end{pmatrix}$ and $c = \begin{pmatrix} \frac{2}{5} \\ \frac{3}{5} \end{pmatrix}$ with a value of 1

16. $r = \begin{pmatrix} \frac{1}{3} & \frac{2}{3} \end{pmatrix}$ and $c = \begin{pmatrix} \frac{4}{5} \\ 0 \\ \frac{1}{5} \end{pmatrix}$ with a value of 0

17. $r = \begin{pmatrix} \frac{1}{5} & \frac{3}{5} & \frac{1}{5} \end{pmatrix}$ and $c = \begin{pmatrix} \frac{3}{10} \\ \frac{2}{5} \\ \frac{3}{10} \end{pmatrix}$ with a value of $\frac{8}{5}$

18. $r = \begin{pmatrix} \frac{1}{4} & \frac{1}{2} & \frac{1}{4} \end{pmatrix}$ and $c = \begin{pmatrix} 0 \\ \frac{1}{2} \\ 0 \\ \frac{1}{2} \end{pmatrix}$ with a value of 0

19. Jack brings $\begin{cases} C \\ T \end{cases}\begin{array}{ccc}N & C & D\end{array}\begin{pmatrix} 100 & 70 & 60 \\ 60 & 80 & 100 \end{pmatrix}$ has value 76 with optimal strategy $r = \begin{pmatrix} \frac{2}{5} & \frac{3}{5} \end{pmatrix}$ and $c = \begin{pmatrix} \frac{1}{5} \\ \frac{4}{5} \\ 0 \end{pmatrix}$. The students can expect a quiz on concepts 80% of the time and Jack's expected grade is 76.

20. Dealer $\left\{ \begin{array}{c} \text{QQ} \\ \text{QK} \\ \text{KQ} \end{array} \right.$
$\begin{array}{c} \text{Gambler} \\ \overbrace{\begin{array}{ccc} \text{QQ} & \text{QK} & \text{KQ} \end{array}} \\ \begin{pmatrix} -2 & 1 & 1 \\ 1 & -2 & 5 \\ 1 & 5 & -2 \end{pmatrix} \end{array}$
has value 1 with optimal strategy $r = \begin{pmatrix} 0 & \frac{4}{7} & \frac{3}{7} \end{pmatrix}$ and $c = \begin{pmatrix} 1 \\ 0 \\ 0 \end{pmatrix}$.

The gambler should always guess "queen, queen."

Index

FUNCTIONS

$$m = \frac{y_2 - y_1}{x_2 - x_1} \qquad\qquad y = mx + b \qquad\qquad ax + by = c$$

$$x^0 = 1 \qquad x^{-n} = \frac{1}{x^n} \qquad x^{m/n} = \sqrt[n]{x^m} = \left(\sqrt[n]{x}\right)^m \qquad x^m \cdot x^n = x^{m+n} \qquad \frac{x^m}{x^n} = x^{m-n} \qquad (x^m)^n = x^{m \cdot n}$$

$$y = ax^2 + bx + c \quad \text{has vertex at} \quad x = \frac{-b}{2a} \quad \text{and } x\text{-intercepts at} \quad x = \frac{-b \pm \sqrt{b^2 - 4ac}}{2a}$$

$$y = \log x \quad \text{means} \quad x = 10^y \qquad\qquad y = \ln x \quad \text{means} \quad x = e^y$$

$$\log 1 = 0 \quad \log(M \cdot N) = \log M + \log N \quad \log(M/N) = \log M - \log N \quad \log 10^x = x \quad \log(M^P) = P \cdot \log M$$

$$\ln 1 = 0 \quad \ln(M \cdot N) = \ln M + \ln N \qquad \ln(M/N) = \ln M - \ln N \qquad \ln e^x = x \quad \ln(M^P) = P \cdot \ln M$$

FINANCE

Simple Interest: $\qquad I = Prt \qquad\qquad A = P(1 + rt) \qquad\qquad r_s = \dfrac{r}{1 - rt}$

Compound Interest: $\quad A = P(1 + r/m)^{mt} \qquad r_e = (1 + r/m)^{mt} - 1 \qquad \left(\begin{array}{c}\text{Doubling}\\\text{time}\end{array}\right) \approx \dfrac{72}{r \times 100}$

Annuities and Amortization:

$$A = P\frac{(1 + r/m)^{mt} - 1}{r/m} \qquad mt = \frac{\log\left(\dfrac{A}{P}\dfrac{r}{m} + 1\right)}{\log(1 + r/m)} \qquad PV = P\frac{1 - (1 + r/m)^{-mt}}{r/m} \qquad P = D\frac{r/m}{1 - (1 + r/m)^{-mt}}$$

MATRICES

$$\begin{cases} ax + by = h \\ cx + dy = k \end{cases} \Leftrightarrow \left(\begin{array}{cc|c} a & b & h \\ c & d & k \end{array}\right) \xrightarrow[\text{reduce to}]{\text{may row}} \left(\begin{array}{cc|c} 1 & 0 & p \\ 0 & 1 & q \end{array}\right) \Leftrightarrow \begin{array}{l} x = p \\ y = q \end{array}$$

$$A \cdot A^{-1} = A^{-1} \cdot A = I \qquad (A\,|\,I) \xrightarrow[\text{reduce to}]{\text{may row}} (I\,|\,A^{-1}) \qquad AX = B \quad \text{solved by} \quad X = A^{-1}B$$

LINEAR PROGRAMMING

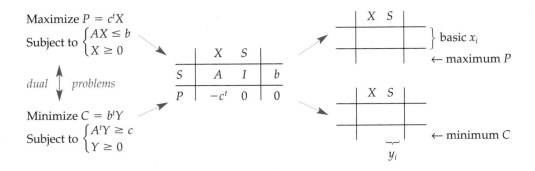